Principles and Practices of Ground Improvement

Principles and Practices of Ground Improvement

Jie Han

WILEY

This book is printed on acid-free paper.

Published by John Wiley & Sons, Inc., Hoboken, New Jersey
Published simultaneously in Canada

Library of Congress Cataloging-in-Publication Data:

Han, Jie, 1964-
Principles and practice of ground improvement / Jie Han.
pages cm
Includes bibliographical references and index.
ISBN 978-1-118-25991-7 (hardback); 978-1-118-41960-1 (ebk.); 978-1-118-42130-7 (ebk.)
1. Soil stabilization. 2. Foundations. I. Title.
TA749.H36 2014
624.1'51–dc23

2014018379

Cover images courtesy of Serge Varaksin and Jie Han
Cover design: Wiley

Printed in the United States of America

10 9 8 7 6 5 4 3 2 1

CONTENTS

PREFACE

Ground improvement is popular in many countries to solve difficult geotechnical problems, especially when construction necessarily occurs in problematic soils and under difficult geotechnical conditions. Many recent developments in equipment, materials, and design methods have made ground improvement technologies more effective, efficient, and economic. However, the state of practice for most ground improvement technologies is that the practice is ahead of theory. Some contractors have developed their proprietary technologies, design methods, and construction techniques for their competitive advantages. Most of the existing books on ground improvement are focused on the concept, application, and case study. However, few books have been devoted to the principles and design methods of ground improvement. This book covers both theoretical and practical aspects in the design and construction of a variety of ground improvement technologies commonly used in practice. This book includes detailed design procedures for most of the ground improvement methods, which enable their easy implementation in practice. The design examples and homework assignments in this book will help readers better understand the principles of each ground improvement technology and how to apply the principles to solve real problems. This book can be used as a textbook for upper-level undergraduate students and graduate students and a reference book for researchers and practicing engineers. It can also be used a guide for contractors and installers to implement ground improvement technologies in field.

Writing this book was part of my big dream when I left China in 1993 to pursue my Ph.D. degree at the Georgia Institute of Technology. At that time, I set three goals: (1) to obtain a Ph.D. degree in the United States, (2) to publish technical papers on internationally well-recognized journals, and (3) to write a book on ground improvement in English. The first two goals were fulfilled in the late 1990s. The last goal has taken much longer than I expected.

I started to become interested in ground improvement in 1985 when I took the class of ground improvement at Tongji University in China, taught by Prof. Shulin Ye, as part of my bachelor's degree. I became a master's student in 1986, studying under Prof. Ye on stone columns. After I received my master's degree in 1989, I became a faculty member at Tongji University, continuing my research in ground improvement, including stone columns, deep mixed columns, grouting, dynamic compaction, and micropiles. I was fortunate to get involved in writing the first Shanghai ground improvement design code. I also coauthored a textbook on ground improvement in Chinese with Prof. Shulin Ye and Prof. Guanbao Ye right before I left China. The selection of the Georgia Institute of Technology for my Ph.D. study was also related to ground improvement because I had read the research report *Design and Construction of Stone Columns*, written by then Prof. Richard D. Barksdale for the U.S. Federal Highway Administration. I arrived at Georgia Tech at a great time. I was exposed to many new and innovative subjects and ideas and learned a great deal about the scientific aspects of geotechnical engineering. My Ph.D. research on fiber-reinforced polymeric piles, under the supervision of Prof. J. David Frost, provided me the great opportunity of learning composite mechanics and geosynthetics. This led me to my first job in the United States at one of the leading geosynthetics manufacturers, the Tensar Corporation. I started with a design engineer and was promoted to a senior engineer and then manager of technology development. I was exposed to many practical problems related to geosynthetic-reinforced earth structures. After several attempts at obtaining faculty positions, I joined Widener University in 2001 and taught the first ground improvement class there. Since transfering to the University of Kansas, I have worked closely with my colleague, Prof. Robert L. Parsons, and been involved in many research projects, which were sponsored by the federal agencies,

Kansas Department of Transportation, and the geosynthetics and ground improvement industries.

Many people have made positive impacts on my career in geotechnical engineering, especially in ground improvement. In addition to my master's advisor, Prof. Ye, and my Ph.D. advisor, Prof. Frost, I have been fortunate to have opportunities to work with internationally recognized scholars Dr. J.P. Giroud, Prof. Dov Leshchinsky, Dr. James Collin, Prof. Mo Gabr, Prof. Hoe Ling and others. As part of the major U.S. Strategic Highway Research Program (SHRP) II R02 project team led by Prof. Vernon R. Schaefer and Mr. Ryan Berg, I had the opportunity to work with the top researchers and experts in ground improvement, including Prof. James Mitchell. I have also been fortunate enough to work with many talent and hard-working students and visiting scholars through the years.

I appreciate the review comments and suggestions from the experts in this field, including Dr. Dimiter Alexiew, Prof. Jinchun Chai, Dr. Xin Chen, Prof. Jian Chu, Prof. Masaki Kitazume, Prof. Dov Leshchinsky, Prof. Paul Mayne, Prof. Shuilong Shen, Prof. Burak Tanyu, Dr. Mark Wayne, Prof. Ming Xiao, and Mr. Kenny Yee. During the preparation of this book, I received great help from my current and former students, especially Yan Jiang, Shaymaa Kadhim, Deep Khatri, Xiaohui Sun, Fei Wang, and Zhen Zhang. I am also thankful for Mr. Serge Varaksin to provide the beautiful photo for the book cover. Many organizations and individuals have granted me permissions to republish their figures and tables in this book and their generous support is greatly appreciated.

Finally, I would like to acknowledge the editors at John Wiley & Sons, Inc. for their patience and great support. My heartfelt thanks go to my dear wife, Jing Ye, and two sons, Terry and Shawn, for their understanding and support for me to complete this book.

Please go to www.wiley.com/go/GroundImprovement for instructor materials.

CHAPTER 1

Introduction

1.1 INTRODUCTION

With civilization and urbanization, there have been increased demands for the use of land for better living and transportation. More and more houses, commercial buildings, high-rise office buildings, highways, railways, tunnels, levees, and earth dams have been constructed and will be continuously built in the future. As suitable constrsuction sites with favorable geotechnical conditions become less available, the need to utilize unsuitable or less suitable sites for construction increases. Engineers have faced increased geotechnical problems and challenges, such as bearing failure, large total and differential settlements, instability, liquefaction, erosion, and water seepage. The options to deal with problematic geomaterials and geotechnical conditions include: (1) avoiding the site, (2) designing superstructures accordingly, (3) removing and replacing problematic geomaterials with better and non-problematic geomaterials and (4) improving geomaterial properties and geotechnical conditions (Hausmann, 1990). It becomes increasingly necessary to improve geomaterials and geotechnical conditions for many projects.

Ground improvement has become an important part of geotechnical practice. Different terminologies have been used in the literature for ground improvement, such as soil improvement, soil stabilization, ground treatment, and ground modification. The term "ground improvement" has been most commonly used in the literature and practice and therefore adopted for this book.

1.2 PROBLEMATIC GEOMATERIALS AND CONDITIONS

1.2.1 Problematic Geomaterials

Geomaterials include all the materials used for geotechnical applications, which consist of natural geomaterials, processed or manufactured geomaterials, and improved geomaterials. Natural geomaterials are mainly soil and rock. O'Neill and Reese (1999) proposed a terminology of intermediate geomaterial, which has properties and behavior between soil and rock. Cohesive intermediate geomaterial has an unconfined compressive strength from 0.5 to 5.0 MPa, while a cohesionless intermediate geomaterial has the number of blow counts of a standard penetration test (SPT) between 50 and 100. Most rocks and intermediate geomaterials are strong and stiff and therefore suitable for geotechnical applications. However, natural soils, especially soft clay and silt, loose sand, expansive soil, collapsible soil, and frozen soil can be problematic to geotechnical applications.

Processed or manufactured geomaterials are produced from other materials. For example, crushed stone aggregates are produced from rock. Recycled asphalt pavement (RAP) aggregates are produced from aged asphalt pavements. Lightweight aggregates are produced by heating raw shale, clay or slate in a rotary kiln at high temperatures, causing the material to expand, and then cooling, crushing, and screening it for different applications. Processed or manufactured geomaterials are mainly used for fill materials, which have a wide variety, ranging from granular fill, lightweight fill, uncontrolled fill, recycled material, fly ash, solid waste, and bio-based byproducts to dredged material. Due to the large variations of fill materials, some of them can be used to improve soil properties (e.g., granular fill and fly ash), but others can be problematic to geotechnical applications (e.g., uncontrolled fill and sludge). Uncontrolled fill or uncompacted fill is mostly loose and underconsolidated; therefore, it settles under its own weight.

Improved geomaterials are the geomaterials treated hydraulically, mechanically, chemically, and biologically. For example, fibers can be mechanically mixed with sand or clay to form fiber-reinforced soil. Lime or cement can be added into soil to form lime or cement-stabilized soil. Denitrifying bacteria can be introduced into soil to generate tiny, inert nitrogen gas bubbles to reduce the degree of saturation of sand (He et al., 2013). As a result, the liquefaction potential of the sand is minimized. Improved geomaterials are often the end products of ground improvement; therefore, they are not problematic to geotechnical applications.

Table 1.1 lists problematic geomaterials and their potential problems. Some natural geomaterials and fill are the targets of ground improvement. When natural geomaterials are discussed in this book, soil and rock are often referred because these terms are commonly used in practice.

1.2.2 Problematic Conditions

In addition to problematic geomaterials, geotechnical problems may occur due to problematic conditions induced naturally and/or by human activities. Natural conditions include geologic, hydraulic, and climatic conditions, such

Table 1.1 Problematic Geomaterials and Potential Problems

Type of Geomaterial	Name	Potential Problems
Natural	Soft clay	Low strength, high compressibility, large creep deformation, low permeability
	Silt	Low strength, high compressibility, high liquefaction potential, low permeability, high erodibility
	Organic soil	High compressibility, large creep deformation
	Loose sand	Low strength, high compressibility, high liquefaction potential, high permeability, high erodibility
	Expansive soil	Large volume change
	Loess	Large volume change, high collapsible potential
Fill	Uncontrolled fill	Low strength, high compressibility, nonuniformity, high collapsible potential
	Dredged material	High water content, low strength, high compressibility
	Reclaimed fill	High water content, low strength, high compressibility
	Recycled material	Nonuniformity, high variability of properties
	Solid waste	Low strength, high compressibility, nonuniformity, and high degradation potential
	Bio-based by-product	Low strength, high compressibility, and high degradation potential

as earthquakes, cavities and sinkholes, floods, wind, and freeze–thaw cycles. Geotechnical conditions are part of geologic conditions, which exist close to the ground surface and are more related to construction and human activities. Examples of problematic geotechnical conditions are existence of problematic geomaterials, a high groundwater table, inclined bedrock, and steep natural slopes. Human activities, mainly the construction of superstructures, substructures, and earth structures, can change geotechnical conditions, which may cause problems for projects, for example, excavation, tunneling, pile driving, rapid drawdown of surface water, elevation of surface water by levees and dams, and groundwater withdrawal. Human activities can also change other conditions, such as the application of static, dynamic, and impact loads.

1.3 GEOTECHNICAL PROBLEMS AND FAILURES

Common geotechnical problems include bearing failure, large total and differential settlements, hydrocompression, ground heave, instability, liquefaction, erosion, and water seepage. The theoretical bases and reasons for these geotechnical problems are provided in Table 1.2.

Failures can happen if geotechnical problems are not properly addressed and become excessive, which typically results in significant financial loss, sometimes even cause loss of life.

1.4 GROUND IMPROVEMENT METHODS AND CLASSIFICATION

1.4.1 Historical Developments

Ground improvement methods have been used since ancient times. For example, about 6000 years ago (in the Neolithic Age), the Banpo people in China used rammed columns to support wooden posts in the ground (Chen et al., 1995). Soil compaction methods using rammers have also been employed since the Neolithic Age. Different types of rammers were used, from stone rammers (in the Neolithic Age) to iron rammers (about 1000 years ago). One type of rammer was operated by 8–12 people, each pulling a rope connected to the rammer to raise it and then letting it fall freely to pound the ground (Chen et al., 1995). About 3500 years ago, reeds in the form of bound cables (approximately 100 mm in diameter) were used in Iraq as horizontal drains for dissipation of pore water pressure in soil mass in high earth structures (Mittal, 2012). About 2000 years ago, The Romans used lime for roadway construction. More than 1000 years ago in the Han dynasty, Chinese people built earth retaining walls using local sand and weeds for border security and paths to the Western world. About 500 years ago (in the Ming dynasty of China), lime was mixed with clayey soil in proportion (typically 3:7 or 4:6 in volume) to form compacted lime–soil foundations for load support (Chen et al., 1995).

Modern ground improvement methods were developed since the 1920s. For example, the use of vertical sand drains to accelerate consolidation of soft soil was first proposed in 1925 and then patented in 1926 by Daniel D. Moran in the United States. Cotton fabric was used as reinforcement by South Carolina Highway Department in the United States for roadway construction in 1926. The vibro-flotation method was developed in Germany to densify loose cohesionless soil in 1937. The first type of prefabricated vertical drains was developed by Walter Kjellman in Sweden in 1947. Fernando Lizzi developed and patented the root pile method to underpin existing foundations in Italy in 1952. In the 1960s, there

Table 1.2 Geotechnical Problems and Possible Causes

Problem	Theoretical Basis	Possible Causes
Bearing failure	Applied pressure is higher than ultimate bearing capacity of soil	High applied pressure Inclined load Small loading area Low-strength soil
Large total and differential settlements	Hooke's law and particle re-arrangement	High applied pressure Large loading area Highly compressible soil Nonuniform soil Large creep deformation
Hydrocompression	High applied pressure is higher than threshold collapse stress	High applied pressure Collapsible soil Water
Ground heave	Swelling pressure is higher than applied pressure	Water Expansive soil Frozen soil Low temperature
Instability (sliding, overturning, and slope failure)	Shear stress is higher than shear strength; driving force is higher than resisting force; driving moment is higher than resisting moment	High earth structure Steep slope High water pressure Soft foundation soil High surcharge High loading rate
Liquefaction	Effective stress becomes zero due to increase of excess pore water pressure	Earthquake Loose silt and sand High groundwater table
Erosion	Shear stress induced by water is higher than maximum allowable shear strength of soil	Running water High speed of water flow Highly erodible soil (silt and sand)
Seepage	Dacy's law	High water head Permeable soil

were several developments of ground improvement methods, including the steel reinforcement for retaining walls by Henri Vidal in France, dynamic compaction by Louis Menard in France, deep mixing in Japan and Sweden, and jet grouting in Japan. In 1986, J. P. Giroud acclaimed the development from geotextiles to geosynthetics is a revolution in geotechnical engineering (Giroud, 1986).

1.4.2 Classification

Many ground improvement methods have been used in practice. The research team for the U.S. Strategic Highway Research Program (SHRP) II R02 project Geotechnical Solutions for Soil Improvement, Rapid Embankment Construction, and Stabilization of the Pavement Working Platform identified 46 ground improvement methods, as provided in Table 1.3 (Schaefer and Berg, 2012).

Different authors or organizations have classified ground improvement methods (Table 1.4) based on different criteria, including Mitchell (1981) in his state-of-the-art report for soil improvement, Hausmann (1990), Ye et al. (1994), the International Society of Soil Mechanics and Geotechnical Engineering (ISSMGE) TC17 committee (Chu et al., 2009), and the SHRP II R02 team led by Schaefer and Berg (2012). Clearly, each method of classification has its reasoning and advantages but also has its limitations. This situation results from the fact that several ground improvement methods can fit in one or more categories. For example, stone columns can serve the functions of densification, replacement, drainage, and reinforcement; however, the key function of stone columns for most applications is replacement. In this book, the method of classification proposed by Ye et al. (1994) is adopted with some minor modifications. In addition, the ground improvement methods can be grouped in terms of shallow and deep improvement in some categories or cut-and-fill improvement in other categories. In this book, shallow improvement is considered as having an improvement depth equal or less than 3 m, while deep improvement

Table 1.3 Ground Improvement Methods for Transportation Infrastructure

Aggregate columns	Fiber reinforcement in pavement systems	Micropiles
Beneficial reuse of waste materials	Geocell confinement in pavement systems	Onsite use of recycled pavement materials
Bio-treatment for subgrade	Geosynthetic reinforced construction platforms	Partial encapsulation
Blasting densification	Geosynthetic reinforced embankments	Prefabricated vertical drains (PVDs) and fill preloading
Bulk-infill grouting	Geosynthetic reinforcement in pavement systems	Rapid impact compaction
Chemical grouting/injection systems	Geosynthetic separation in pavement systems	Reinforced soil slopes
Chemical stabilization of subgrades and bases	Geosynthetics in pavement drainage	Sand compaction piles
Column-supported embankments	Geotextile encased columns	Screw-in soil nailing
Combined soil stabilization with vertical columns	High-energy impact rollers	Shoot-in soil nailing
Compaction grouting	Hydraulic fill + vacuum consolidation + geocomposite drains	Shored mechanically stabilized earth wall system
Continuous flight auger piles	Injected lightweight foam fill	Traditional compaction
Deep dynamic compaction	Intelligent compaction/roller integrated compaction monitoring	Vacuum preloading with and without PVDs
Deep mixing methods	Jet grouting	Vibro-compaction
Drilled/grouted and hollow bar soil nailing	Lightweight fill, expanded polystyrene (EPS) geofoam, low-density cementitious fill	Vibro-concrete columns
Electro-osmosis	Mechanical stabilization of subgrades and bases	
Excavation and replacement	Mechanically stabilized earth wall systems	

Source: Schaefer and Berg (2012).

Table 1.4 Classification of Ground Improvement Methods

Reference	Criterion	Categories
Mitchell (1981)	Construction/function	1. In situ deep compaction of cohesionless soils 2. Precompression 3. Injection and grouting 4. Admixtures 5. Thermal 6. Reinforcement
Hausmann (1990)	Process	1. Mechanical modification 2. Hydraulic modification 3. Physical and chemical modification 4. Modification by inclusions and confinement

(*continued*)

Table 1.4 ***(Continued)***

Reference	Criterion	Categories
Ye et al. (1994)	Function	1. Replacement 2. Deep densification 3. Drainage and consolidation 4. Reinforcement 5. Thermal treatment 6. Chemical stabilization
ISSMGE TC17 (Chu et al., 2009)	Soil type and inclusion	1. Ground improvement without admixtures in noncohesive soils or fill materials 2. Ground improvement without admixtures in cohesive soils 3. Ground improvement with admixtures or inclusions 4. Ground improvement with grouting type admixtures 5. Earth reinforcement
Schaefer and Berg (2012)	Application	1. Earthwork construction 2. Densification of cohesionless soils 3. Embankments over soft soils 4. Cutoff walls 5. Increased pavement performance 6. Sustainability 7. Soft ground drainage and consolidation 8. Construction of vertical support elements 9. Lateral earth support 10. Liquefaction mitigation 11. Void filling
This book	Function	1. Densification 2. Replacement 3. Drainage and consolidation 4. Chemical stabilization 5. Reinforcement 6. Thermal and biological treatment

has an improvement depth greater than 3 m. The fill reinforcement includes the methods using metallic or geosynthetic reinforcement for fill construction, while the in situ ground reinforcement includes the methods using ground anchors or soil nails for cut construction.

1.4.3 General Description, Function, and Application

Table 1.5 provides the general descriptions, benefits, and applications of most ground improvement methods to be discussed in this book.

1.5 SELECTION OF GROUND IMPROVEMENT METHOD

1.5.1 Necessity of Ground Improvement

When superstructures are to be built on ground, there are five foundation options (Figure 1.1): (a) bearing on natural ground, (b) bearing on replaced ground, (c) bearing on compacted/consolidated ground, and (d) bearing on composite ground, and (e) bearing on piles to deeper stratum. Options (b), (c), and (d) involve ground improvement methods. The final selection often depends on geotechnical condition, loading condition, performance requirement, and cost. Option (a) is preferred and also more economic when the load on the foundation is low and competent geomaterial exists near the ground surface. Option (e) is more suitable for high foundation loads on problematic geomaterials with high-performance requirements, which is often most expensive. Options (b), (c), and (d) are more suitable for intermediate conditions and requirements between option (a) and option (e).

There are also four options for earth retaining structures as shown in Figure 1.2: (a) unreinforced cut-and-fill slopes, (b) unreinforced cut-and-fill earth walls, (c) reinforced cut-and-fill slopes, and (d) reinforced cut-and-fill

Table 1.5 General Descriptions, Functions, and Applications of Ground Improvement Methods

Category	Subcategory	Method and Level of Establishment[a]	General Description	Benefit	Application
Densification	Shallow compaction	Traditional compaction Level = 5	Apply static or vibratory load on ground surface in a certain number of passes to densify problematic geomaterial	Increase density, strength, and stiffness; reduce deformation, permeability, collapsible potential, and ground heave	Suitable for a wide range of fills to a lift thickness of 0.3 m; used to compact fill
		High-energy impact roller compaction Level = 2	Apply a lifting and falling motion by a roller with high-energy impact on ground surface to densify or crush problematic geomaterial	Increase density, strength, and stiffness; reduce deformation, permeability, collapsible potential, and ground heave; crush rock and concrete into rubble	Suitable for a wide range of geomaterials to a depth of 2 m; used to improve subgrade and foundation soil and compact fill
		Rapid impact compaction Level = 2	Use an excavator to drop a weight repeatedly on ground surface to densify problematic geomaterial	Increase density, strength, and stiffness; reduce deformation, permeability, collapsible potential, and ground heave	Suitable for granular geomaterials up to 6 m deep; used to improve subgrade and foundation soil and compact fill
		Intelligent compaction Level = 2	Apply and adjust compaction energy based on on-board display from measurements in real time to densify problematic geomaterial	Increase density, strength and stiffness; reduce deformation, permeability, collapsible potential, and ground heave, identify areas of poor compaction, and maximize productivity	Suitable for granular geomaterials; used to improve subgrade and foundation soil and compact fill
	Deep compaction	Dynamic compaction Level = 5	Drop a heavy weight from a high distance to apply high energy on ground surface, causing liquefaction of saturated problematic geomaterial and densification of unsaturated problematic geomaterial	Increase density, strength and stiffness; reduce deformation, liquefaction, collapsible potential to a greater depth	Suitable for granular geomaterials, collapsible soil, and waste material with less than 15% fines to a depth of 10 m; used to improve foundations
		Vibro compaction Level = 5	Apply a vibratory force and/or water by a probe on surrounding problematic geomaterial, causing liquefaction and densification	Increase density, strength, and stiffness; reduce deformation, liquefaction, and collapsible potential to a greater depth	Suitable for clean sands with less than 15% silt or less than 2% clay to a typical depth of 5–15 m; used to improve foundations
Replacement	Shallow replacement	Overexcavation and replacement Level = 5	Remove problematic geomaterial and replace with good-quality geomaterial	Increase strength and stiffness; reduce deformation, liquefaction, collapsible, and ground heave potential	Suitable and economic for a wide range of geomaterials with limited area and limited depth (typically to 3 m deep and above groundwater table)

(*continued*)

Table 1.5 (***Continued***)

Category	Subcategory	Method and Level of Establishment[a]	General Description	Benefit	Application
Replacement	Deep replacement	Sand compaction columns Level = 5*	Displace problematic geomaterial by driving a casing into the ground and backfill the hole with sand (densified by vibration during casing withdrawal)	Increase bearing capacity and stability; reduce settlement and liquefaction potential; accelerate consolidation	Suitable for a wide range of geomaterials to a typical depth of 5–15 m; used to improve foundations
		Stone columns Level = 5*	Jet water or air to remove or displace problematic geomaterial by a probe and backfill the hole with stone to form a densified column by vibration	Increase bearing capacity and stability; reduce settlement and liquefaction potential; accelerate consolidation	Suitable for a wide range of geomaterials (undrained shear strength >15 kPa) to a typical depth of 5–10 m (up to 30 m); used to improve foundations
		Rammed aggregate columns Level = 4	Predrill a backfilled with aggregate, densified by ramming	Increase bearing capacity and stability; reduce settlement and liquefaction potential; accelerate consolidation	Suitable for a wide range of geomaterials to a typical depth of 5–10 m with a deep groundwater level; used to improve foundations
		Vibro-concrete columns Level = 3	Drive a vibrating probe to the ground to displace problematic geomaterial, replaced with concrete	Increase bearing capacity and stability; reduce settlement	Suitable and economic for very soft soil to a typical depth of 5–10 m; used to improve foundations
		Geosynthetic-encased columns Level = 2*	Drive a steel casing to the ground to displace problematic geomaterial, replaced with a geosynthetic casing and fill	Increase bearing capacity and stability; reduce settlement; accelerate consolidation	Suitable and economic for very soft soil (undrained shear strength <15 kPa) to a typical depth of 5–10 m; used to improve foundations
Drainage, dewatering, and consolidation	Drainage	Fill drains Level = 5*	Place a layer of permeable fill inside a roadway or earth structure	Reduce water pressure and collapsible and ground heave potential; accelerate consolidation; increase strength, stiffness, stability	Suitable for low permeability geomaterial; used for roads, retaining walls, slopes, and landfills
		Drainage geosynthetics Level = 4	Place a layer of nonwoven geotextile or geocomposite in ground or inside a roadway or earth structure	Reduce water pressure and collapsible and ground heave potential; accelerate consolidation; increase strength, stiffness, stability	Suitable for low permeability geomaterial; used for roads, retaining walls, slopes, and landfills
		Open pumping Level = 5	Use sumps, trenches, and pumps to remove a small amount of water inflow in open excavation	Remove water to ease construction	Suitable for a small area, relatively impermeable soil, and lowering of the groundwater table by a limited depth in open excavation

(*continued*)

Table 1.5 ***(Continued)***

Category	Subcategory	Method and Level of Establishment[a]	General Description	Benefit	Application
Drainage, dewatering, and consolidation	Dewatering	Well system Level = 4	Use well points and/or deep wells to remove a large amount of water inflow in open excavation	Remove water to ease construction and increase stability of excavation	Suitable for a large area, relatively permeable soil, and lowering of the groundwater table by a large depth for excavation
		Electro osmosis method Level = 2	Create electric gradients in soil by installing anode and cathode to induce water flow and collect and discharge the water by a cathode well point	Remove water to ease construction	Suitable for relatively impermeable silt or clayey soil
	Consolidation	Fill preloading Level = 5	Apply temporary surcharge on ground surface for a duration and then remove the surcharge for construction	Increase soil strength; reduce settlement	Suitable for saturated inorganic clay and silt; used to reduce settlement for foundation soil
		Vacuum preloading Level = 3	Apply vacuum pressure on ground surface and/or through drains into the ground for a desired duration and then remove the pressure for construction	Increase soil strength; reduce settlement	Suitable for saturated inorganic clay and silt; used to reduce settlement for foundation soil
Chemical stabilization	Shallow stabilization	Chemical stabilization of subgrade and base Level = 5	Mix lime, cement, and/or fly ash with subgrade and base course in field and then compact the mixture; have chemical reaction with soil particles to form a cementitious matrix	Increase strength and stiffness; reduce ground heave potential	Suitable for unsaturated clay and silt; mainly used for roadway construction with a typical lift thickness of 0.3 m or less
	Deep stabilization	Grouting Level = 3	Inject grout into ground to fill voids, densify soil, and have chemical reaction with soil particles to form a hardened mass	Increase strength and stiffness; reduce permeability, liquefaction, and ground heave potential	Different grout suitable for different geomaterial; mainly used for remedying measures or protective projects
		Jet grouting Level = 4	Inject high-pressure cement-based fluid into ground to cut and then mix with geomaterial to form a hardened column by chemical reaction with soil particles	Increase strength, stiffness, and stability; reduce permeability, liquefaction, and ground heave potential	Suitable for a wide range of geomaterials; mainly used for remedying measures and protective projects to a typical depth of 30 m or less
		Deep mixing Level = 4*	Mix cement or lime from surface to depth with geomaterial by mechanical blade to have chemical reaction with soil particles after mixed to form a cementitious matrix	Increase strength, stiffness, and stability; reduce permeability, liquefaction, and ground heave potential	Suitable for a wide range of geomaterials; mainly used for foundation support, earth retaining during excavation, containment, and liquefaction mitigation

(continued)

Table 1.5 **(*Continued*)**

Category	Subcategory	Method and Level of Establishment[a]	General Description	Benefit	Application
Reinforcement	Fill reinforcement	Geosynthetic-reinforced slopes Level = 5	Place geosynthetics in slope at different elevations during fill placement to provide tensile resistance	Increase stability	Suitable for low plasticity fill; mainly used for slope stability
		Geosynthetic-reinforced embankments Level = 5	Place high-strength geosynthetic at base of embankment to provide tensile resistance	Increase bearing capacity and stability	Suitable for embankments over soft foundation; mainly used for enhancing embankment stability
		Geosynthetic-reinforced column-supported embankments Level = 3	Place geosynthetic reinforcement over columns at base of embankment to support embankment load between columns	Reduce total and differential settlements; accelerate construction; increase stability	Suitable for embankments over soft foundation with strict settlement requirement and time constraint
		Mechanically stabilized earth walls Level = 5	Place geosynthetic or metallic reinforcements in wall at different elevations during fill placement to provide tensile resistance	Increase stability	Suitable for low plasticity free-draining fill
		Geosynthetic-reinforced foundations Level = 3*	Place geosynthetic reinforcements within fill under a footing to provide load support	Increase bearing capacity and reduce settlement	Suitable and economic for granular fill over soft soil with limited area and depth
		Geosynthetic-reinforced roads Level = 4	Place geosynthetic reinforcement on top of subgrade or within base course to provide lateral constraint	Increase bearing capacity and roadway life; reduce deformation and base thickness requirement	Suitable for granular bases over soft subgrade
	In-situ ground reinforcement	Ground anchors Level = 4*	Insert steel tendons with grout at end in existing ground to provide tensile resistance and prevent ground movement	Increase stability and resistance to uplift force	Suitable for granular soil or rock; used for temporary and permanent slopes and walls during excavation and substructures subjected to uplift force
		Soil nails Level = 4	Insert a steel bar with grout throughout the whole nail in existing ground to provide tensile resistance and prevent ground movement	Increase stability	Suitable for low plasticity stiff to hard clay, dense granular soil, and rock; used for temporary and permanent slopes and walls during excavation
		Micropiles Level = 4	Insert a steel reinforcing bar in a bored hole, grout in place to form a small diameter pile (<0.3 m) and provide vertical and lateral load capacities	Increase stability; protect existing, structures during ground movement	Suitable for a variety of geomaterials; used for slopes, walls, and unpinning of existing foundations

(*continued*)

Table 1.5 (*Continued*)

Category	Subcategory	Method and Level of Establishment[a]	General Description	Benefit	Application
Thermal and biological treatment		Ground freezing Level = 2	Remove heat from ground to reduce soil temperature below freezing point and turn geomaterial into solid	Increase strength; reduce water flow and ground movement	Suitable for saturated clay and sand; used for temporary protection during excavation
		Biological treating Level = 1	Utilize vegetation and roots to increase shear strength of soil or change soil properties by biomediated geochemical process, including mineral precipitation, gas generation, biofilm formation, and biopolymer generation	Increase strength and stiffness; reduce erodibility and liquefaction potential	Suitable for cohesive and cohesionless geomaterials; requires more research and field trial before it is adopted in practice

[a]Level of technology establishment: rating scale 1 = not established, 3 = averagely established, and 5 = well established (most of the ratings are based on the recommendations by the SHRP II R02 team; however, some ratings with an asterisk * are adjusted or added from the international perspective and the author's judgment).

earth walls. Among these options, cut and fill are two different situations. Option (a) is often adopted when there is open land. It is also least expensive and easy for vegetation. Options (b) and (d) are often adopted when there is limited space. Option (b) using slurry walls, gravity walls, or cantilever walls is often most expensive among all the options but gains large useful land. Reinforced earth walls are typically less expensive than slurry walls, gravity walls, and cantilever walls. Option (c) is between option (a) and options (b) and (d) in terms of land requirement/utilization and cost. Option (c) has flexibility of different slope angles and may still establish vegetation. Ground improvement methods can be used for options (b), (c), and (d).

For roadway construction, options for subgrade and base course (or ballast) can be: (1) natural subgrade and granular base, (2) lime/cement-stabilized subgrade and base, and (3) geosynthetic-reinforced subgrade and base.

For all applications, unimproved conditions should be evaluated against performance criteria first. If unimproved conditions satisfy the performance criteria, no ground improvement is needed; otherwise, ground improvement is required.

1.5.2 Factors for Selecting Ground Improvement Method

Selection of ground improvement method should consider the following conditions: (1) structural conditions,

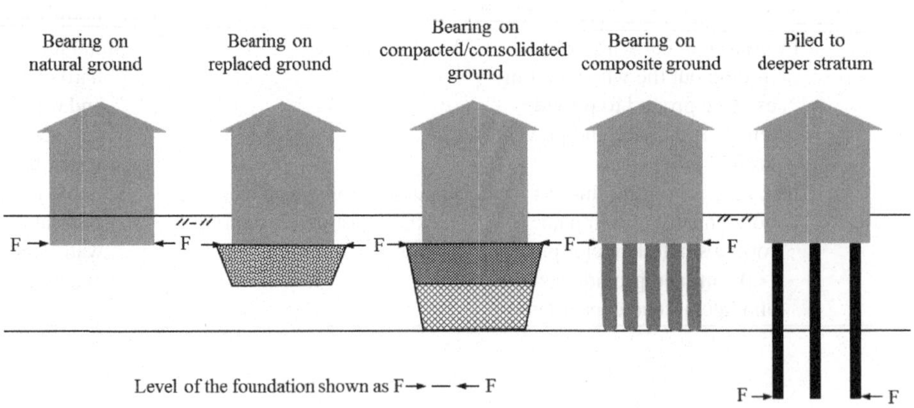

Figure 1.1 Options for foundations (modified from Mitchell and Jardine, 2002).

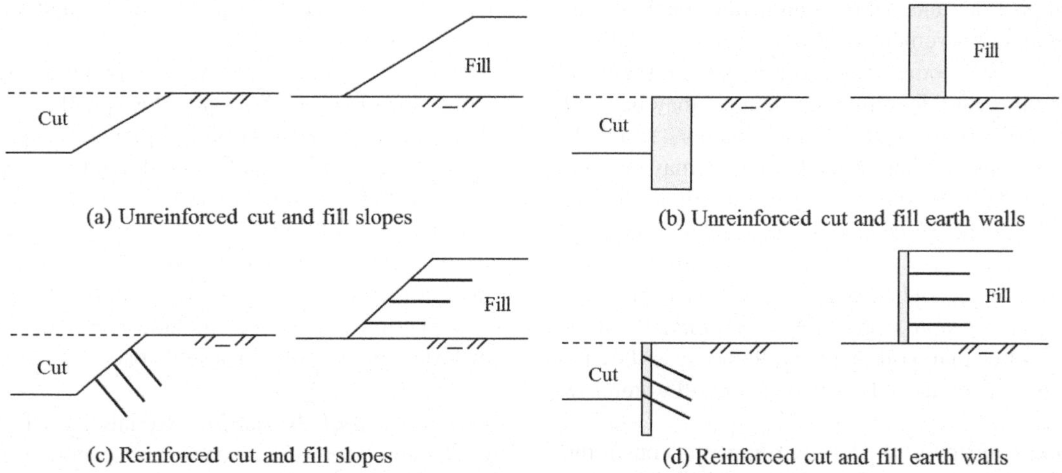

Figure 1.2 Options for earth retaining structures.

(2) geotechnical conditions, (3) environmental constraints, (4) construction conditions, and (5) reliability and durability.

Structural Conditions The structural conditions may include type, shape, and dimension of structure and footing, flexibility and ductility of structural and footing elements, type, magnitude, and distribution of loads, and performance requirements (e.g., total and differential settlements, lateral movement, and minimum factor of safety).

Geotechnical Conditions The geotechnical conditions may include geographic landscape, geologic formations, type, location, and thickness of problematic geomaterial, possible end-bearing stratum, age, composition, distribution of fill, and groundwater table. Soil type and particle size distribution are essential for preliminary selection of ground improvement methods as shown in Figure 1.3. This guideline is suitable for ground improvement methods for foundation support. The thickness and location of problematic geomaterial are also important for the selection of ground improvement methods. For example, when a thin problematic geomaterial layer exists at a shallow depth, the over excavation and replacement method is one of the most suitable and economic method. When a relatively thick loose cohesionless geomaterial layer exists near ground surface,

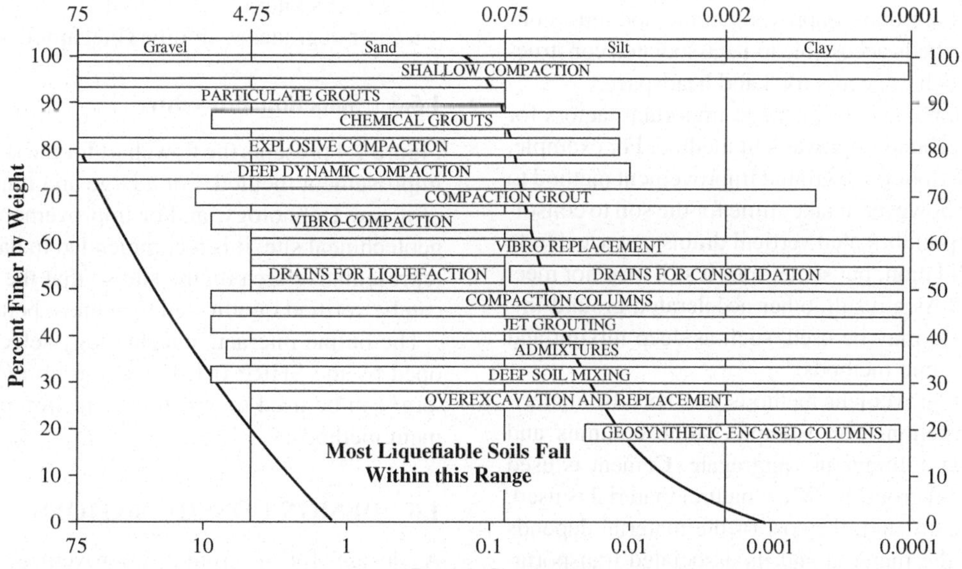

Figure 1.3 Available ground improvement methods for different soil types (modified from Schaefer et al., 2012).

dynamic compaction and vibro-compaction methods are suitable ground improvement methods. When a relatively thick soft cohesive geomaterial layer exists near ground surface, preloading and deep mixing methods may be used. When a site needs to be excavated, tieback anchors, soil nails, deep mixed columns, and jet-grouted columns may be used. When a site needs to be elevated, geosynthetic-reinforced slopes and walls can be good choices. The level of groundwater table often affects the selection of ground improvement methods. For example, when deep excavation happens in ground with a high groundwater table, deep mixed column walls may be better than soil nailed walls because they not only can retain the geomaterial but also can cut off water flow.

Environmental Constraints The environmental constraints may include limited vibration, noise, traffic, water pollution, deformation to existing structures, spoil, and headspace. For example, dynamic compaction induces vibration and noise, which may not be suitable in a residential area. The wet method to construct stone columns by water jetting produces spoil on site, which may be troublesome for a site with limited space. Under such a condition, the dry method may be used instead. Preloading induces settlements at nearby areas, which may be detrimental to existing structures.

Construction Conditions The selection of a ground improvement method should consider the following construction conditions: (1) site condition, (2) allowed construction time, (3) availability of construction material, (4) availability of construction equipment and qualified contractor, and (5) construction cost.

The selection of a ground improvement method must consider whether the site is accessible to its associated construction equipment, such as access road and headspace.

Construction time is one of the most important factors for the selection of a ground improvement method. For example, preloading is a cost-effective ground improvement method to improve soft soil; however, it takes time for the soil to consolidate. The use of prefabricated vertical drains can accelerate the rate of consolidation, but sometimes it still may not meet time requirement. As a result, other accelerated ground improvement methods may be used, such as deep mixing and vibro-concrete column methods.

Most ground improvement methods use specific materials during construction. For example, stone columns and rammed aggregate columns use aggregate. Cement is used for deep mixing and grouting. When natural material is used, such as aggregate or sand, the cost of the material depends on the source of the material and its associated transportation distance. For example, in a mountain area, aggregate is often less expensive; therefore, stone columns or aggregate columns are often a cost-effective solution. In general, the use of locally available material results in more cost-effective ground improvement.

To select a ground improvement method, engineers should gather information about possible qualified contractors and their available construction equipment. It is preferable to use a locally available qualified contractor because this will reduce the mobilization cost and the contractor is more familiar with local conditions.

Construction cost is always one of the key factors that dominate the selection of a ground improvement method. The construction cost should include mobilization, installation, material, and possible disposal costs.

Reliability and Durability Reliability of a ground improvement method depends on several factors, such as the level of establishment, variability of geotechnical and structural conditions, variability of construction material, quality of the contractor, quality of installation, and quality control and assurance. Several researchers have reported that samples from deep mixed columns have a high variability in terms of their unconfined compressive strengths. Automatic or computer-controlled installation processes can reduce the variability of improved geomaterials. The number of well-documented successful or failure case histories is also the evidence of the reliability of a specific ground improvement method.

Ground improvement methods are used for temporary and permanent structures. For permanent structures, the durability of the construction material should be evaluated or considered in the design. For example, geosynthetics have creep behavior. The corrosion of steel reinforcement with time reduces its thickness. The strength of cement-stabilized soil in seawater degrades with time (Ikegami et al., 2002).

1.5.3 Selection Procedure

Figure 1.4 presents the flowchart for the selection of a ground improvement method. For a large and important project, use of a new technology, and/or improvement of a complicated geotechnical site, it is recommended to have a field trial on a representative area on the site so that the design parameters can be verified or adjusted to achieve better performance.

The online interactive technology selection system developed by the SHRP II R02 team at http://www.geotechtools.org/ can be used to assist the selection of ground improvement methods.

1.6 DESIGN CONSIDERATIONS

A design for a ground improvement method typically requires the inputs on geometry of structures, geotechnical conditions, loading conditions, material characteristics, and performance criteria. Typical design parameters and

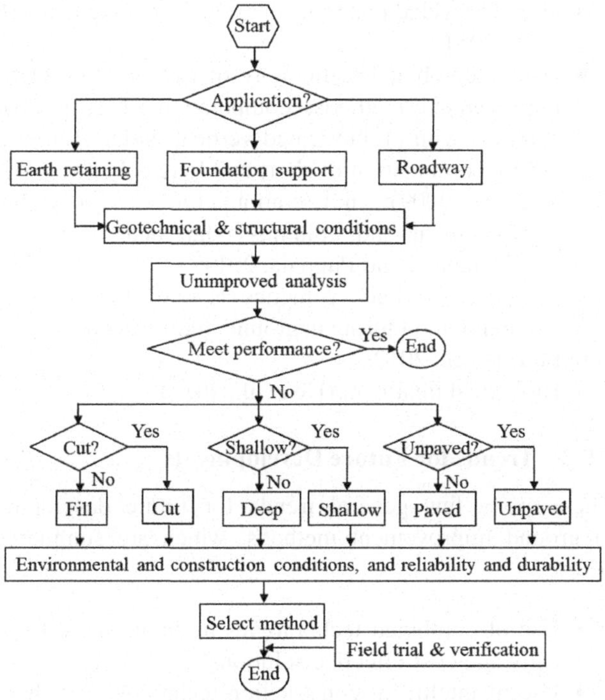

Figure 1.4 Flowchart for selection of ground improvement method.

maximum limits should be considered during design, such as diameter, spacing, and depth of columns, drop energy and improvement depth of compaction, and geosynthetic spacing. Trial-and-error methods may be used for some design procedures. The final design should meet the performance criteria. The outputs of a design typically include the size of improvement zone, plan layout, cross section, amount and properties of materials, and sometimes construction rate and sequence.

In this book, allowable strength design (ASD) is adopted instead of reliability-based design, considering the fact that design methods for most ground improvement methods have not been calibrated using the reliability-based approach due to limited test data. In other words, factors of safety instead of load and resistance factors or partial factors are used in this book.

1.7 CONSTRUCTION

Construction is one of the most important components of ground improvement. No matter how correct the concept is and how good the design is, the design of a ground improvement method must be implemented correctly in the field to reach its maximum performance extents. Chu et al. (2009) provide an overview of construction processes used in geotechnical engineering, including ground improvement methods. Construction should be delivered based on plans and specifications.

There are five types of specifications for transportation construction suggested by the American Association of State Highway and Transportation Officials (AASHTO) Highway Subcommittee on Construction Quality Construction Task Force in 2003 and are listed in Table 1.6. The method specifications have been mostly adopted for ground improvement methods. However, the method specifications with performance criteria have been increasingly used.

Table 1.6 Types of Construction Specifications

Type	Description
Method specifications	Specifications that require the contractor to produce and place a product using specified materials in definite proportions and specific types of equipment and methods under the direction of the agency.
End-result specifications	Specifications that require the contractor to take the entire responsibility for producing and placing a product. The agency's responsibility is to either accept or reject the final product or to apply a price adjustment commensurate with the degree of compliance with the specifications.
Quality assurance specifications	Specifications that require contractor quality control and agency acceptance activities throughout production and placement of a product. Final acceptance of the product is usually based on a statistical sampling of the measured quality level for key quality characteristics.
Performance-related specifications	Specifications that use quantified quality characteristics and life cycle cost (LCC) relationships that are correlated to product performance.
Performance-based specifications	Specifications that describe the desired levels of fundamental engineering properties that are predictors of performance and appear in primary prediction relationships.

Source: AASHTO (2003).

1.8 QUALITY CONTROL AND ASSURANCE

Both quality control and quality assurance ensure the quality of construction; however, they are done at different stages and by different entities. The SHRP II R02 project states: "Quality Control refers to procedures, measurements, and observations used by the contractor to monitor and control the construction quality such that all applicable requirements are satisfied. Quality Assurance refers to measurements and observations by the owner or the owner's engineer to provide assurance to the owner that the facility has been constructed in accordance with the plans and specifications" (Han et al., 2012). Quality control is done during construction while quality assurance is done during construction as well as at the end or after completion of construction. Automatic or computer-controlled installation processes and data collection systems can reduce the variability of improved geomaterials and avoid human errors so that the construction can be better controlled. Quality assurance often involves in situ testing and field monitoring.

1.9 RECENT ADVANCES AND TRENDS FOR FUTURE DEVELOPMENTS

1.9.1 Recent Advances

There have been many recent advances in ground improvement methods. Chu et al. (2009) pointed out that manufacturers have made significant contributions to these recent advances due to their constant innovations and improvements in the equipment. At the same time, researchers have helped improve design methods. Below are a few highlights of the recent advances:

- Different types of column technologies, such as geosynthetic-encased stone columns (Alexiew et al., 2003), controlled modulus (stiffness) columns, hollow concrete columns (Liu et al., 2003), multiple stepped columns (Borel, 2007; Liu, 2007a), X-shape (Liu, 2007b) or Y-shape (Chen et al., 2010) concrete columns, grouted stone columns (Liu, 2007a), T-shaped DM columns (Liu et al., 2012); and composite columns (Jamsawang et al., 2008; Zheng et al., 2009)
- Column-supported embankments (Han and Gabr, 2002; Filz et al., 2012)
- Online interactive technology selection system developed by the SHRP II R02 team (Schaefer and Berg, 2012)
- Cutter soil mixing method to construct trench walls (Mathieu et al., 2006)
- Horizontal twin-jet grouting method (Shen et al., 2013)
- Intelligent compaction on unbound geomaterial (White et al., 2007)
- Use of recycled materials in ground improvement (Han et al., 2011)
- Use of combined technology of two or more ground improvement methods: combination of deep mixed columns with prefabricated vertical drains (Liu et al., 2008); combination of short and long columns (Huang and Li, 2009); and combination of geosynthetic reinforcement and columns (Han and Gabr, 2002; Madhyannapu and Puppala, 2014)
- Sensor-enabled geosynthetics (Hatami et al., 2009)
- Computer monitoring in ground improvement construction (Bruce, 2012)
- Biological treatment (He et al., 2013)

1.9.2 Trends for Future Developments

There are a few general trends for future development in ground improvement methods, which are summarized below:

- Use of combined technologies to create more technically and cost-effective solutions
- Use of intelligent construction technologies with sensors and computer monitoring to improve efficiency and quality of ground improvement
- Use of recycled materials and other alternative materials to make ground improvement methods more sustainable
- Use of end-result or performance-based specifications
- Application of biological treatment in field

1.10 ORGANIZATION OF BOOK

This book has 10 chapters. This chapter is an introduction, which provides an overview of ground improvement methods. Chapter 2 reviews geotechnical materials, testing, and design, which are the bases for the following chapters. Chapters 3–10 are presented based on the classification of ground improvement methods in terms of their functions. Chapter 3 discusses shallow and deep compaction. Chapters 4 and 5 discuss overexcavation and replacement (i.e., shallow replacement) and deep replacement. Chapters 6 and 7 discuss drainage and dewatering and preloading and consolidation. Chapter 8 discusses deep chemical stabilization by grouting and deep mixing. Chapters 9 and 10 discuss in situ ground reinforcement (for cut situations) and fill reinforcement (for fill situations).

PROBLEMS

1.1. Give examples of three geotechnical problems that can be caused by problematic geomaterials.

1.2. List five possible geotechnical problems caused by human activities.

1.3. How can ground improvement methods be classified in terms of their functions?

1.4. What is the basic principle for soil liquefaction? List five possible ground improvement methods that can be used to mitigate soil liquefaction and explain why.

1.5. List five possible methods for shallow ground improvement.

1.6. List five possible methods for deep ground improvement.

1.7. A project site has a 5-m-thick loose gravel layer near ground surface that needs to be improved for foundation support. Which methods may be used for ground improvement? Why?

1.8. Explain why different ground improvement methods are needed for cut-and-fill walls.

1.9. What are the types of construction specifications possibly used in practice?

1.10. Explain why quality control and assurance are so important for ground improvement methods.

1.11. What are the future trends of ground improvement?

REFERENCES

AASHTO (American Association of State Highway and Transportation Officials) (2003). *Major Types of Transportation Construction Specifications: A Guideline to Understanding Their Evolution and Application.* A report of the AASHTO Highway Subcommittee on Construction Quality Construction Task Force, Washigton, DC.

Alexiew, D., Horgan, G.J., and Brokemper, D. (2003). "Geotextile encased columns (GEC): Load capacity and geotextile selection." In *Foundations: Innovations, Observations, Design and Practice.* Thomas Telford, London.

Borel, S. (2007). Soil mixing innovations: Geomix, SpringSol and Trenchmix. Presentation at Joint BGA/CFMS Meeting, London.

Bruce, D.A. (2012). "Computer monitoring in the grouting industry." *Geotechnical Engineering State of the Art and Practice, Keynote Lectures from GeoCongress 2012*, Geotechnical Special Publication No. 226, 549–565.

Chen, R.P., Xu, Z.Z, Chen, Y.M., Ling, D.S., and Zhu, B. (2010). "Field tests of pile-supported embankments on soft ground." *J. Geotech. Geoenviron. Eng.,* 136(6): 777–785.

Chen, Z.Y., Ye, S.L., and Wu, X.M. (1995). "A glimpse of some earthworks and foundations in ancient China." *Proceedings of the 10th Asian Regional Conference on Soil Mechanics and Foundation Engineering*, Beijing, China.

Chu, J., Varaksin, S., Klotz, U., and Menge, P. (2009). "Construction processes—State of the art report." *Proceedings of the 17th International Conference on Soil Mechanics and Geotechnical Engineering*, Alexandria, Egypt, 5–9 Oct., 3006–3135.

Filz, G., Sloan, J., McGuire, M.P., Collin, J., and Smith, M. (2012). "Column-supported embankments: Settlement and load transfer." *Geotechnical Engineering State of the Art and Practice, Keynote Lectures from GeoCongress 2012*, Geotechnical Special Publication No. 226, 54–77.

Giroud, J.P. (1986). "From geotextiles to geosynthetics: A revolution in geotechnical engineering." *Proceedings of the Third International Conference on Geotextiles*, 1, Vienna, Austria, April, 1–18.

Han, J. and Gabr, M.A. (2002). "Numerical analysis of geosynthetic-reinforced and pile-supported earth platforms over soft soil." *J. Geotech. Geoenviron. Eng.,* 128(1): 44–53.

Han, J., Pokharel, S.K., Yang, X., Manandhar, C., Leshchinsky, D., Halahmi, I., and Parsons, R.L. (2011). "Performance of geocell-reinforced RAP bases over weak subgrade under full-scale moving wheel loads." *J. Mat. Civil Eng.* 23(11): 1525–1534.

Han, J., Thakur, J.K., Corey, R., Christopher, B.R., Khatri, D., and Acharya, B. (2012). "Assessment of QC/QA technologies for evaluating properties and performance of geosynthetics in roadway systems." GeoCongress 2012, Oakland, California, USA, March 25–29, *State of the Art and Practice in Geotechnical Engineering*, Geotechnical Special Publication No. 225, R.D. Hryciw, A. Athanasopoulos-Zekkos, and N. Yesiller (eds.), 1350–1359.

Hatami, K., Grady, B., and Ulmer, M. (2009). "Sensor-enabled geosynthetics: Use of conducting carbon networks as geosynthetic sensors." *Journal of Geotechnical and Geoenvironmental Engineering,* 135(7): 863–874.

He, J, Chu, J., and Ivanov, V. (2013). "Mitigation of liquefaction of saturated sand using biogas." *Geotechnique*, 63(4): 267–275.

Huang, M. and Li, B. (2009). "Numerical analysis of pile-supported earth platforms with nonuniform piles." *Advances in Ground Improvement*, ASCE GSP No. 188: 46–54.

Huasmann, M.R. (1990). *Engineering Principles of Ground Modification.* McGraw-Hill, New York.

Ikegami, M., Masuda, K., Ichiba, T., Tsuruya, H., Satoh S., and Terashi, M. (2002). "Physical properties and strength of cement-treated marine clay after 20 years." *Proc. of the 55th Annual Conference of the Japan Society of Civil Engineers*, 3(61): 123–124 (in Japanese).

Jamsawang, P., Bergado, D.T., Bhandari, A., and Voottipruex, P. (2008). "Investigation and simulation of behavior of stiffened deep cement mixing (SDCM) piles." *Int. J. Geotech. Eng.*, 2(3): 229–246.

Liu, H.L. (2007a). "New piling techniques for soil improvement in China." *Proc 13 Asian Regional Conf on Soil Mech and Geot. Eng.*, Kolkata.

Liu, H.L. (2007b). Construction method of X shaped pile. Chinese Patent 2007100203063 (in Chinese).

Liu, H.L., Fei, K., Ma, X.H., and Gao, Y.F. (2003). "Large-diameter driven cast-in-place concrete thin-wall pipe pile (I) research and development." *J. Rock Soil Mech.*, 24: 164–198 (in Chinese).

Liu, S.Y., Han, J., Zhang, D.W., and Hong, Z.S. (2008). "A new DJM-PVD combined method for soft ground improvement." *Geosynth. Int. J.*, 15(1): 43–54.

Madhyannapu, R.S. and Puppala, A.J. (2014). "Design and construction guidelines for deep soil mixing to stabilize expansive soils."*Journal of Geotechnical and Geoenvironmental Engineering* 140(9):119–128.

Mathieu, F, Borel, S., and Lefebvre, L. (2006). "CSM: An innovative solution for mixed-in-situ retaining walls, cut-off walls and soil improvement." *10th International Conference on Piling and Deep Foundations*, DFI/EFFC.

Mitchell, J.K. (1981). "Soil improvement—State-of-the-art report." *Proceedings of the 10th International Conference on Soil Mechanics and Foundation Engineering*, Stockholm, 4, 509–565.

Mitchell, J.M. and Jardine, F.M. (2002). *A Guide to Ground Treatment*. CIRIA Publication C573, Construction Industry Research and Information Association, London.

Mittal, S. (2012). "Recent experiences of ground stabilization techniques." In *Emerging Trends in Geotechnical Engineering*, A.M. Krishna (ed.),Workshop on Emerging Trends in Geotechnical Engineering (ETGE 2012), 8 June, Guwahati, India, 45–60.

O'Neil, M.W. and Reese, L.C. (1999). *Drilled Shafts: Construction Procedures and Design Methods*, FHWA Publication No. FHWA-IF-99-025, Federal Highway Administration, Washington, DC.

Schaefer, V.R. and Berg, R.R. (2012). *Geotechnical Solutions for Soil Improvement, Rapid Embankment Construction, and Stabilization of the Pavement Working Platform.* Final Phase 2 Summary Report by SHRP II R02 team, submitted to the U.S. Strategic Highway Research Program (SHRP) II.

Schaefer, V.R., Mitchell, J.R., Berg, R.R., Filz, G.M., and Douglas, S.C. (2012). "Ground improvement in the 21st century: A comprehensive web-based information system." *Geotechnical Engineering State of the Art and Practice, Keynote Lectures from GeoCongress 2012*, Geotechnical Special Publication No. 226, 272–293.

Shen, S.L., Wang, Z.F., Horpibulsuk, S, and Kim, Y.H. (2013). "Jet-grouting with a newly developed technology: Twin-Jet Method." *Eng. Geol.*, 152(1): 87–95.

White, D.J., Vennapusa, P.K.R., and Thompson, M.J. (2007). *Field Validation of Intelligent Compaction Monitoring Technology for Unbound Materials.* Center for Transportation Research and Education at Iowa State University, Ames, IA.

Ye, S.L., Han, J., and Ye, G.B. (1994). *Ground Improvement and Underpinning Technologies*, 2nd ed. China Building Industry Press, Beijing.

Zheng, G., Liu, S.-Y., and Chen, R.-P. (2009). "State of advancement of column-type reinforcement element and its application in China." *Advances in Ground Improvement*, Geotechnical Special Publication No. 188, J. Han, G. Zheng, V.R. Schaefer, M.S. Huang (eds.), Proceedings of the US-China Workshop on Ground Improvement Technologies, March 14, Orlando, FL; 12–25.

CHAPTER 2

Geotechnical Materials, Testing, and Design

2.1 INTRODUCTION

In geotechnical engineering, there are three types of geomaterials: (1) natural geomaterial, (2) processed or manufactured geomaterial, and (3) improved geomaterial. In addition to geomaterials, other materials are used for geotechnical applications, such as geosynthetics, cement, lime, and concrete. Geosynthetics are placed within or at boundaries of geomaterials to improve their hydraulic and/or mechanical properties and performance. Geosynthetcis have become an important geotechnical material for ground improvement. Cement and lime can be mixed with soil to form a stabilized soil. Concrete has been used to construct foundations, retaining walls, soil nails, piles, and other elements in geotechnical engineering. Cement and lime will be discussed in Chapter 8 and concrete is commonly discussed in civil engineering textbooks; therefore, they will not be discussed in this chapter. The material properties most relevant to geotechnical applications include physical, mechanical, and hydraulic properties. These properties are often determined by laboratory tests and/or in situ tests. This chapter will briefly discuss geomaterials, geosynthetics, and laboratory and in situ tests.

In addition to geotechnical materials and tests, ground improvement involves geotechnical design, such as shallow foundations, slope stability, earth-retaining structures, and liquefaction. This chapter will briefly review these subjects.

2.2 GEOMATERIALS AND PROPERTIES

2.2.1 Classifications

Natural geomaterial, processed geomaterial, and improved geomaterial have similarities but also differences. There is no unique classification system available for all these geomaterials. When natural geomaterials are discussed, soil and rock are often referred to. Fill is the most commonly used processed geomaterial. To be consistent with the current practice in geotechnical engineering, the terms, soil, rock, and fill, are mostly used in this book. The term, geomaterial, is used when other geomaterial is included.

In geotechnical engineering, the Unified Soil Classification System (USCS) and the AASHTO soil classification system are commonly used to classify natural soils. Based on the USCS, a natural soil is often classified as coarse-grained soil or fine-grained soil. The coarse-grained soil, which has 50% or more particles retained on the U.S. No. 200 (i.e., opening size of 0.075 mm) sieve, includes gravel and sand. The fine-grained soil, which has more than 50% particles passing the U.S. No. 200 sieve, includes silt and clay. The particles retained on the US No. 4 sieve are called gravel while the particles passing the US No. 4 sieve but retained on the US No. 200 sieve are called sand. The particles passing the U.S. No. 200 sieve are called fines (silt or clay). Grain size distribution is used to describe the uniformity of coarse-grained soils, while the Atterberg limits are used to describe the consistency of fine-grained soils. A soil with a wide range of particle sizes is called well-graded, while a soil with a narrow range of particle sizes is called poorly-graded. When coarse-grained soils have less than 5% fines, they are often referred to as clean gravel or clean sand. When fine-grained soils have liquid limits greater than 50, they are classified as high plasticity soils. Common soil types involved in ground improvement include low-plasticity clay (symbol: CL), high-plasticity clay (CH), silty sand (SM), poorly-graded sand (SP), well-graded sand (SW), poorly-graded gravel (GP), and well-graded gravel (GW). AASHTO classifies soils into granular materials (35% or less passing the 0.075 mm sieve) with soil groups of A-1 (stone fragments, gravel, and sand), A-2 (silty or clayey gravel and sand), and A-3 (fine sand) and silt-clay materials (>35% passing the 0.075 mm sieve) with soil groups of A-4 and A-5 (silty soils), and A-6 and A-7 (clayey soils). Detailed soil classifications can be found in soil mechanics textbooks, ASTM, and AASHTO standards. However, these two soil classification systems are not necessarily suitable for other geomaterials. For example, recycled asphalt pavement (RAP) aggregates contain coarse aggregate, fine aggregate, maybe fines, but also asphalt. The inclusion of asphalt makes these two soil classification systems invalid. Solid waste material is another example, which makes the soil classification systems invalid, because it may consist of soil, glass, metal, paper, and biodegradable food waste.

For all geomaterials, they can be classified as cohesionless geomaterials or cohesive geomaterials based on their shear strengths. Cohesionless geomaterials have zero or nearly zero cohesion when they are dry or fully saturated, while cohesive geomaterials have cohesion at all states (dry, saturated, and unsaturated) according to the total stress concept. Based on permeability, they may be classified as pervious geomaterials, semipervious geomaterials, and impervious geomaterials, according to Lukas (1995). The pervious geomaterials have high permeability, while the impervious geomaterials have low permeability. Semipervious geomaterials have permeability between pervious and impervious geomaterials.

2.2.2 Physical Properties

Physical properties include particle size, specific gravity, void ratio, relative density, unit weight, moisture content, degree of saturation, and Atterberg limits. Figure 2.1 shows a few useful particle sizes from a grain size distribution curve. Using D_{60} as an example, it corresponds to the sieve size 60% particles pass. D_{50} is often called the mean particle size. D_{10} is referred to as the effective particle size because it affects the permeability of a soil. D_{10}, D_{30}, and D_{60} are used for defining the gradation of a soil. D_{15} and D_{85} are used for filter design.

Geomaterial is often represented by a phase diagram, as shown in Figure 2.2, which consists of solid, liquid (or water), and air. Air has volume, V_a, but does not have weight.

Specific gravity of a geomaterial, G_s, is defined as

$$G_s = \frac{\gamma_s}{\gamma_w} \tag{2.1}$$

where γ_s is the unit weight of solid (i.e., W_s/V_s) and γ_w is the unit weight of water (W_w/V_w). W_s and W_w are weights of solid and water respectively while V_s and V_w are volumes of solid and water respectively. For soil and rock, the specific gravity mostly ranges from 2.60 to 2.75. However, for processed geomaterials, the specific gravity can vary widely. For example, alumina red mud sand has a specific gravity from 3.16 to 3.27 (Oweis and Khera, 1998). Lightweight aggregates have specific gravities from 1.25 to 1.40 (Holm and Valsangkar, 1993). Specific gravity is an important parameter to determine the void ratio of a geomaterial and the zero air void line for the compaction curve to be discussed later.

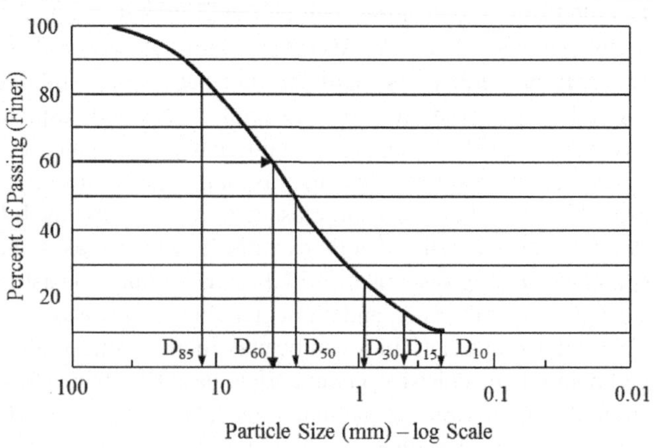

Figure 2.1 Useful particle sizes.

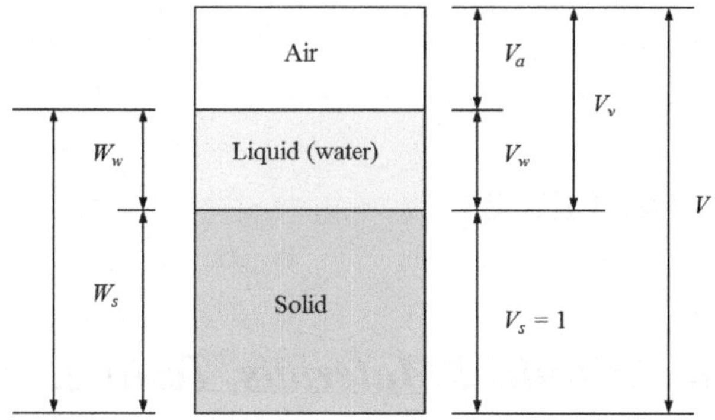

Figure 2.2 Phase diagram.

To evaluate the density of a geomaterial, void ratio, defined as the ratio of volume of void to volume of solid (i.e., V_v/V_s), is commonly used. The American Society for Testing and Materials (ASTM) ASTM D4253 and ASTM D4254 can be used to determine the maximum and minimum void ratios of a soil with up to 15% (by dry mass) of soil particles passing the No. 200 sieve size. With the maximum and minimum void ratios, the relative density of a geomaterial, D_r, can be calculated as follows:

$$D_r = \frac{e_{max} - e}{e_{max} - e_{min}} \tag{2.2}$$

where e_{max} = maximum void ratio
e_{min} = minimum void ratio
e = void ratio of soil in place

The qualitative description of degree of density of a coarse-grained soil is provided in Table 2.1. The soil is typically considered dense if its relative density, $D_r > 70\%$.

Unit weight, defined as the ratio of weight to volume of a geomaterial, can be easily measured in laboratory and field; therefore, it is a good quality assurance parameter for ground improvement, such as shallow compaction. Unit weight can be dry, moist, or saturated. The dry unit weight and the moist unit weight have a relationship with moisture content

Table 2.1 Qualitative Description of Degree of Density

D_r(%)	Description
0–25	Very loose
25–50	Loose
50–70	Medium dense
70–85	Dense
85–100	Very dense

Table 2.2 Degree of Saturation of Geomaterial

Condition	Degree of saturation (%)
Oven-dry (unsuitable for compaction)	0
Air-dry (very difficult for compaction)	1–30
Slightly moist (difficult for compaction)	30–60
Moist (easy for compaction)	60–90
Wet (difficult for compaction)	90–99
Saturated (unsuitable for compaction)	100

(i.e., percentage of mass of water to mass of solid) as follows:

$$\gamma_d = \frac{\gamma}{1 + w} \tag{2.3}$$

where γ_d = dry unit weight
γ = moist unit weight
w = moisture content

Degree of saturation is defined as the percentage of water volume to total void volume (i.e., V_w/V_v). This is an important parameter for compaction. Table 2.2 shows the condition of a geomaterial based on the degree of saturation. A high degree of saturation makes compaction of a geomaterial difficult because less void can be compressed. A low degree of saturation also makes compaction of a geomaterial difficult because high friction exists between particles.

There is a useful relationship among specific gravity, G_s; moisture content, w; degree of saturation, S_r; and void ratio, e, of a geomaterial as follows:

$$G_s w = S_r e \tag{2.4}$$

The Atterberg limits include plastic limit, liquid limit, and shrinkage limit, which are determined for soil particles passing the U.S. No. 40 sieve. Plastic limit, PL, is the moisture content below which the geomaterial cannot be shaped. Liquid limit, LL, is the moisture content above which the geomaterial becomes flowable. The shrinkage limit, *SL*, is the moisture content below which no volume change happens. The plasticity index, PI, is the difference between LL and PL as follows:

$$\text{PI} = \text{LL} - \text{PL} \tag{2.5}$$

LL and PL are used to classify soils using the plasticity chart based on the USCS as shown in Figure 2.3. The "A" line separates silts (symbol: M) from clays (symbol: C) while the "U" line is the upper limit of test data. LL = 50 is the boundary line to separate low (symbol: L) and high (symbol: H) plasticity soils. The symbol "O" stands for organic soils.

In addition to soil classification, PI is an important property that affects geomaterial behavior, such as swelling/shrinkage potential, permeability, strength, and compactability. This parameter is used to select quality backfill materials for slopes and walls.

Example 2.1

A soil sample has 35% particles retained on the U.S. No. 200 sieve. The plastic limit and liquid limit of this soil are 24 and 42, respectively. Classify this soil using the USCS.

Solution

Since this soil has 35% particles retained on the U.S. No. 200 sieve, the percent of particles passing the U.S. No. 200 sieve is 100% − 35% = 65%. It is a fine-grained soil. Use the Atterberg limits for soil classification.

Plasticity index PI = 42 − 24 = 18. Based on the plasticity chart as shown in Figure 2.3, this soil can be classified as CL (low-plasticity clay).

Example 2.2

A saturated soil sample has moisture content of 37% and its specific gravity is 2.69. What is the void ratio of this sample?

Solution

The following relationship can be used for this calculation:

$$G_s w = S_r e$$

Since it is a saturated sample, $S_r = 1$. Therefore,

$$e = G_s w = 2.69 \times 0.37 = 0.995$$

2.2.3 Mechanical Properties

Mechanical properties of geomaterials can be described in terms of stress–strain relationships. For practical applications, modulus (or compression index) and strength are two important parameters.

When a geomaterial is subjected to a stress, it follows the effective stress theory, i.e.:

$$\sigma' = \sigma - u \tag{2.6}$$

where σ' = effective stress
σ = total stress
u = pore water pressure

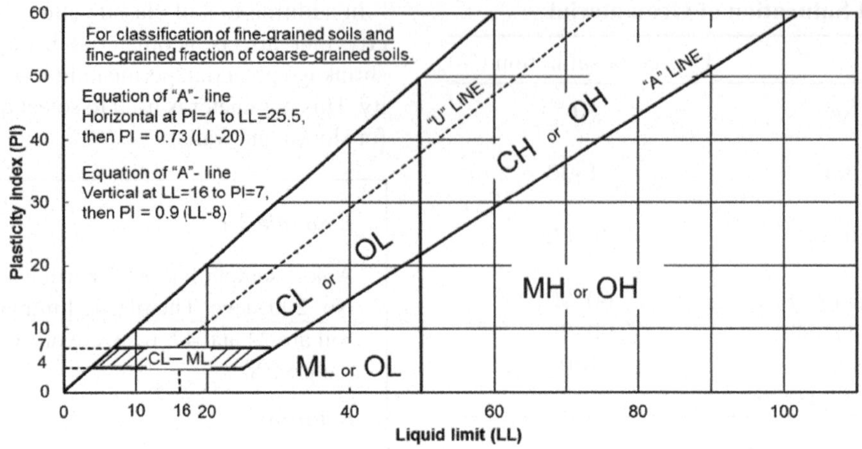

Figure 2.3 Plasticity chart (with permission from ASTM, copyright ASTM International.)

The effective stress is the average interparticle contact stress. The stress-induced excess pore water pressure dissipates with time. This process is called consolidation. Ground improvement by preloading utilizes this concept. Consolidation tests as shown in Figure 2.4 are used to determine the modulus or compression index and consolidation coefficient of a geomaterial.

Figure 2.5 shows an $e - \log \sigma'$ plot, and σ'_{z0} is the overburden stress of the geomaterial at a depth, z; p_c is the preconsolidation stress, which is the maximum effective stress the geomaterial has experienced. The ratio of p_c to σ'_{z0} is called overconsolidation ratio, OCR. The state of the soil with OCR = 1 is called normally consolidated (NC), that with OCR > 1 is overconsolidated (OC), and that with OCR < 1 is underconsolidated. Natural geomaterials are often NC, while processed geomaterials without compaction are mostly underconsolidated. The first portion with a slope, C_r, is referred to as the rebound or recompression line, while the second portion with a slope, C_c, is referred to as the virgin compression line.

The coefficient of consolidation is one of the important parameters to estimate the rate of consolidation and defined as follows:

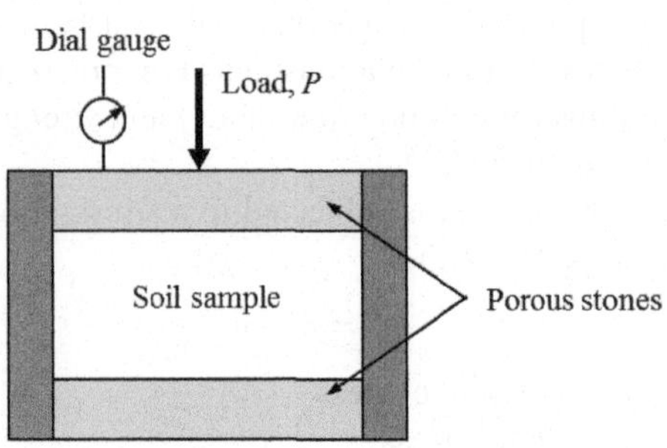

Figure 2.4 Consolidation test.

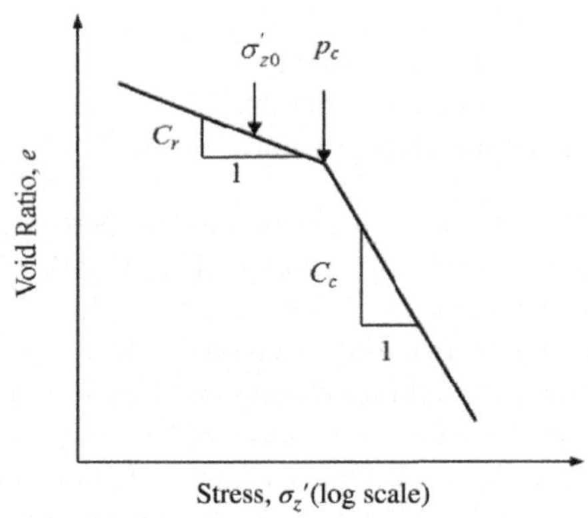

Figure 2.5 $e - \log \sigma'$ plot.

$$c_v = \frac{k(1+e_0)}{\gamma_w a_v} = \frac{k}{\gamma_w m_v} \tag{2.7}$$

$$m_v = \frac{\Delta e}{1+e_0} \cdot \frac{1}{\Delta \sigma'_z} \tag{2.8}$$

where
γ_w = unit weight of water
a_v = coefficient of compressibility
e_0 = initial void ratio
m_v = coefficient of volumetric compressibility
Δe = change of void ratio
$\Delta \sigma'_z$ = change of vertical stress
k = permeability of geomaterial

Figure 2.6 shows that a_v or m_v depends on the vertical stress; therefore, a_v is not constant and dependent on the stress level; c_v can be determined by consolidation tests. The stress level corresponding to a field condition should be chosen to determine the c_v value.

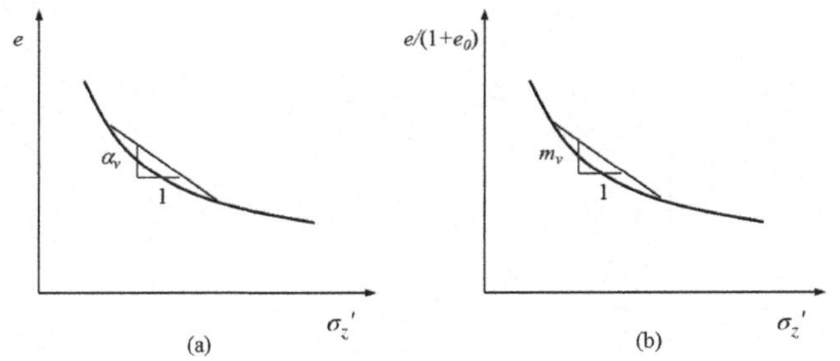

Figure 2.6 Coefficient of compressibility.

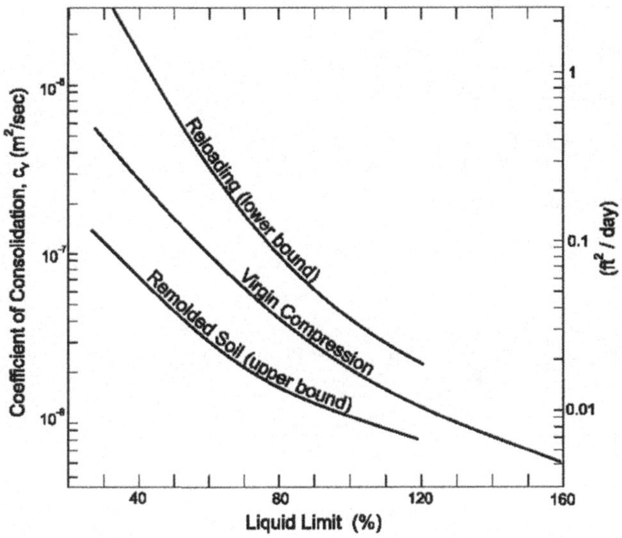

Figure 2.7 Coefficient of consolidation versus liquid limit (NAVFAC, 1982).

Figure 2.7 may be used for a preliminay design. It is shown that remolding of soil reduces the c_v value. Installation during ground improvement may cause disturbance or remolding of soil. As a result, the c_v value is reduced. This reduction should be considered in ground improvement design.

When the pore water pressure is equal to the total stress, the effective stress becomes zero. The state of a geomaterial with a zero effective stress is called liquefaction, which can be induced by dynamic or static loading. Ground improvement methods using vibratory forces in saturated cohesionless geomaterials can induce liquefaction.

Direct shear and triaxial shear tests as shown in Figure 2.8 are two commonly used laboratory tests to evaluate mechanical properties of geomaterials. Geomaterials can have three drainage conditions under loading: (1) undrained, (2) drained, and (3) partially drained. A partially drained condition is difficult to simulate and control; therefore, most tests are done under an undrained or drained condition. The undrained condition simulates geomaterial behavior during rapid loading, while the drained condition simulates long-term behavior of a geomaterial. Since triaxial shear tests have a better control on stress states and drainage conditions, they should be used to evaluate geomaterial behavior. Since direct shear tests can be easily performed, they are more commonly used in practice.

Figure 2.9 shows the typical stress–strain and volume relationships of the same geomaterial prepared at different states. For a dense or OC geomaterial under a drained condition, there is a peak shear strength and dilation at large displacement or strain. However, for a loose or NC geomaterial under a drained condition, there is no peak shear strength and compression. Clearly, the dense or OC geomaterial has a higher peak strength than the loose or NC geomaterial. However, they have the same steady-state or critical-state shear strength (also called constant volume strength or residual strength) at the constant volume if they are sheared at the same normal or confining stress.

At different confining stresses, the steady state of a geomaterial can be determined. A steady-state line (SSL) as shown in Figure 2.10 can be drawn. The soil behavior depends not only on the density but also on the stress level and the drained condition. Figure 2.10(a) shows that when

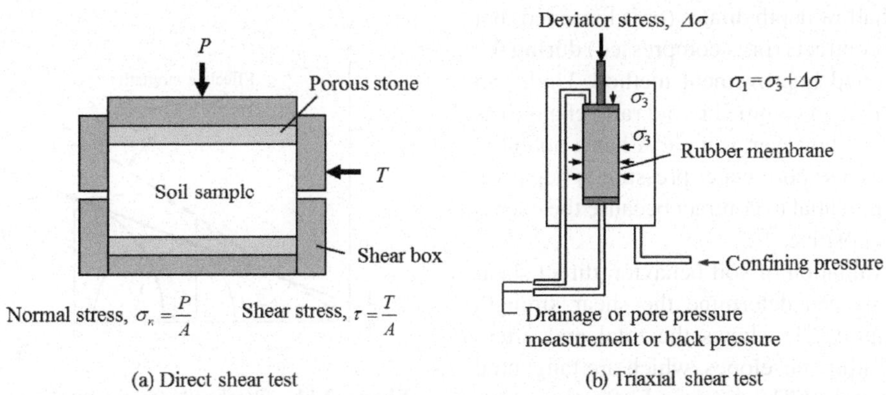

Figure 2.8 (a) Direct shear and (b) triaxial shear tests.

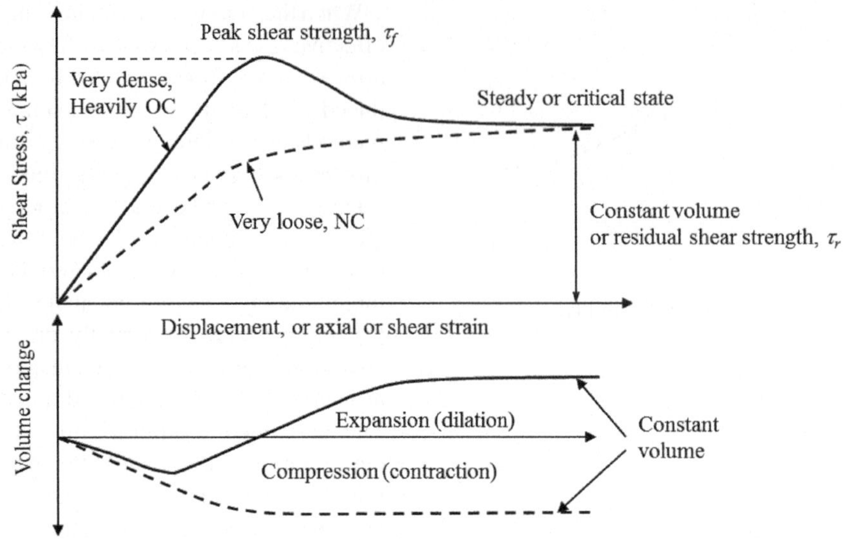

Figure 2.9 Stress–strain and volume relationships.

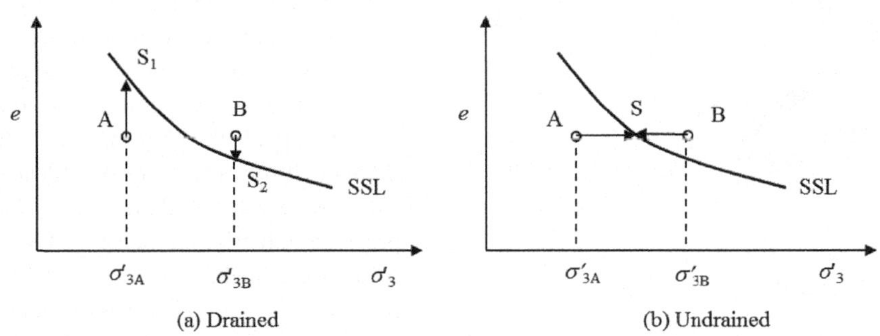

Figure 2.10 Soil behavior: (a) drained and (b) underained.

the same geomaterial sample (i.e., the same void ratio, e) is tested at two different confining stresses under a drained condition (i.e., σ'_3 does not change during the shear test), sample A at a lower confining stress dilates toward the SSL, while sample B at the higher confining stress contracts toward the SSL. This result can be used to explain the phenomenon why the geomaterial at a shallow depth dilates (or is loosened) but that at the deep depth contracts (or is compressed) during the densification by a ground improvement method. Under an undrained condition (i.e., no volume or void ratio change) as shown in Figure 2.10(b), however, sample A has a potential to dilate because the excess pore water pressure is negative, while sample B has a potential to contract because the excess pore water pressure is positive.

In addition to the evaluation of soil behavior, direct shear and triaxial shear tests can determine the shear strength of a geomaterial. Figure 2.11 shows the total and effective Mohr–Coulomb failure envelopes, which are tangential to the failure Mohr circles. The total and effective shear strengths are expressed as follows:

$$\tau_f = c + \sigma_n \tan\phi \tag{2.9}$$

$$\tau_f = c' + \sigma'_n \tan\phi' \tag{2.10}$$

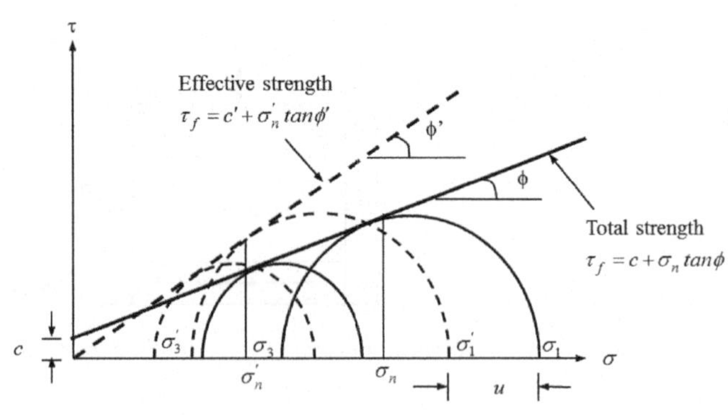

Figure 2.11 Total and effective Mohr–Coulomb failure envelopes.

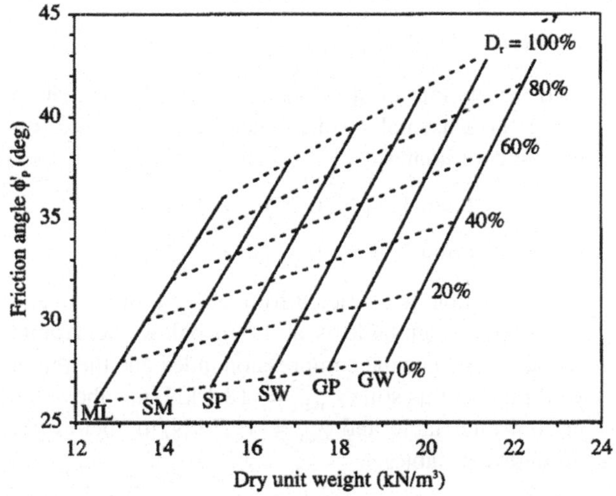

Figure 2.12 Effective peak friction angle of coarse-grained soil (NAVFAC, 1982).

The difference in the Mohr circles for total and effective stresses is the pore water pressure, u. For a normally consolidated geomaterial, the total Mohr–Coulomb failure envelope may have a cohesion, c, and a friction angle, ϕ, while the effective Mohr–Coulomb failure envelope has an effective cohesion equal to zero and an effective angle, ϕ'. It should be pointed out that an overconsolidated or dense geomaterial under low normal or confining stresses has a nonlinear initial portion. Direct shear and triaxial shear tests should be conducted at the stress levels corresponding to field conditions.

A design chart based on dry unit weight and relative density as shown in Figure 2.12 can be used to estimate the effective peak friction angle of coarse-grained soil.

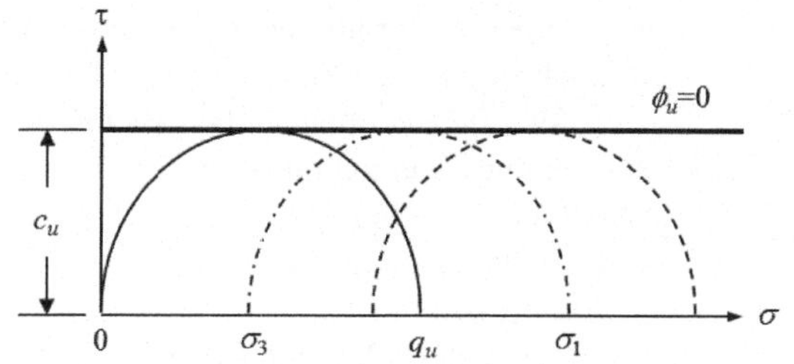

Figure 2.14 Unconfined compression and unconsolidated undrained tests.

A design chart based on plasticity index as shown in Figure 2.13 can be used to estimate the effective friction angle of normally consolidated fine-grained soil at constant volume.

Under an undrained condition, cohesive geomaterials can also be evaluated using unconfined compression tests or unconsolidated undrained (UU) tests. From these tests, undrain shear strength, c_u, and unconfined compressive strength, q_u, can be determined as shown in Figure 2.14. A typical relationship between c_u and q_u is as follows:

$$c_u = \frac{q_u}{2} \tag{2.11}$$

The undrained shear strength of a saturated cohesive soil can be estimated by (Ladd, 1991):

$$\frac{c_u}{\sigma'_{zc}} = \chi(\mathrm{OCR})^{\Lambda} \tag{2.12}$$

where c_u = undrained shear strength of saturated cohesive soil

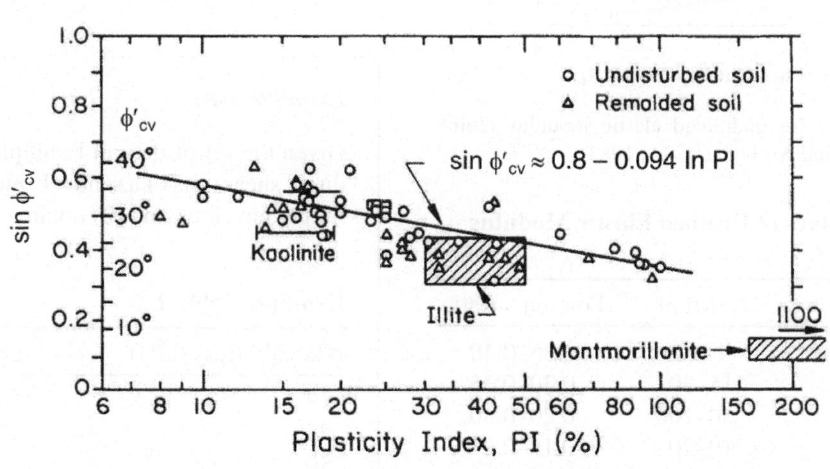

Figure 2.13 Effective friction angle of normally consolidated soil (Mitchell and Soga, 2005).

χ = 0.22 ± 0.03 for homogeneous sedimentary clays (above A line) or $\chi = 0.25 \pm 0.05$ for silts and organic clays (below A line)
OCR = overconsolidation ratio
$\Lambda = 0.88(1 - C_r/C_c)$
C_r = recompression index
C_c = compression index
σ'_{zc} = effective consolidation stress

Unconfined compression tests are commonly used to determine unconfined compressive strengths of chemically stabilized geomaterials.

The undrained elastic modulus of a cohesive soil can be estimated using the following relationship with Figure 2.15:

$$E_u = K_c c_u \tag{2.13}$$

The undrained Poisson ratio of clay is 0.5.

Typical drained elastic moduli and Poisson ratios of clay and sand are provided in Table 2.3.

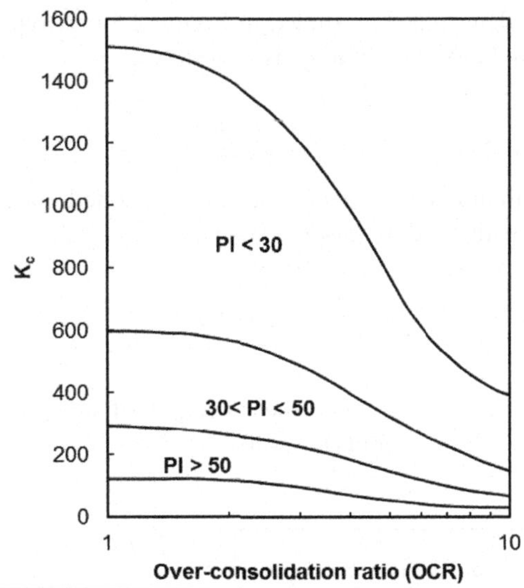

Figure 2.15 Constant K_c for undrained elastic modulus (Joint Department of the Army and Air Force, USA, 1983).

Table 2.3 Typical Values of Drained Elastic Modulus and Poisson's Ratio

Soil Type	Description	E (MPa)	Poisson's Ratio
	Soft	1–15	0.35–0.40
Clay	Medium	15–30	0.30–0.35
	Stiff	30–100	0.20–0.30
Sand	Loose	10–20	0.15–0.25
	Medium	20–40	0.25–0.30
	Dense	40–80	0.25–0.35

Source: Budhu (2000).

Example 2.3

Two data points on a one-dimensional consolidation curve for a normally consolidated clay have the following coordinates:

Point 1: $e_1 = 0.7,\ p_1 = 100$ kPa

Point 2: $e_2 = 0.6,\ p_2 = 300$ kPa

If the sample is extracted from a depth of 4.5 m and the groundwater table is at 1.5 m below the ground surface, calculate the compression index and the initial void ratio of this soil ($\gamma_{moist} = 18.1\ \text{kN/m}^3$ above the groundwater table and $\gamma_{sat} = 19.7\ \text{kN/m}^3$ below the groundwater table).

Solution

The compression index is

$$C_c = \frac{e_1 - e_2}{\log(p_2/p_1)} = \frac{0.7 - 0.6}{\log(300/100)} = 0.21$$

The effective overburden stress at the depth of 4.5 m is

$$\sigma'_{z0} = 1.5 \times 18.1 + 3.0 \times 19.7 - 3.0 \times 9.81 = 56.8\ \text{kPa}$$

We know

$$C_c = \frac{e_0 - e_2}{\log(p_2/p_0)}$$

The initial void ratio is

$$e_0 = e_2 + C_c \log\left(\frac{p_2}{p_0}\right) = e_2 + C_c \log\left(\frac{p_2}{\sigma'_{z0}}\right)$$

$$= 0.6 + 0.21 \times \log\left(\frac{300}{56.8}\right) = 0.752$$

Example 2.4

Given the set of data in Example Table 2.1 from four direct shear tests of a sand: (1) plot the Mohr–Coulomb failure envelope and (2) obtain the friction angle.

Example Table 2.1

Normal Stress (kPa)	Peak Shear Stress (kPa)
17	15
35	32
70	65
140	125

Solution

Based on the provided normal stresses and peak shear stresses, the Mohr–Coulomb failure envelope is plotted in Example Figure 2.1. Since it is a sand, the failure envelop passes the origin. The peak friction angle of the sand is

$$\phi_p = \tan^{-1}(0.9005) = 42.0°$$

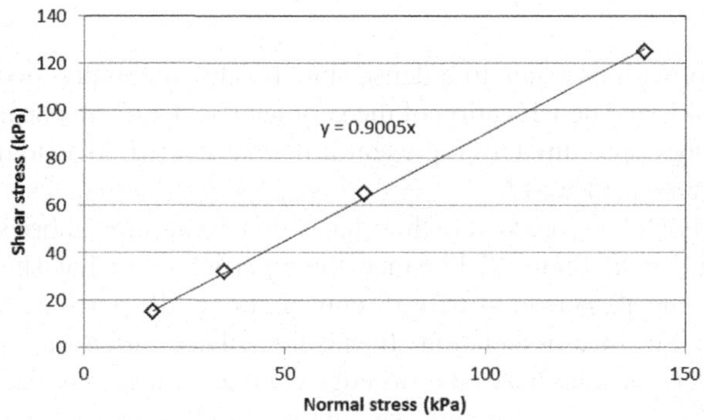

Example Figure 2.1

From Figure 2.13,

$$\sin \phi'_{cv} = 0.8 - 0.094 \ln \text{PI}$$
$$= 0.8 - 0.094 \times \ln(35) = 0.466$$

Therefore, $\phi'_{cv} = 27.8°$.

Example 2.5

A normally consolidated inorganic clay sample extracted from a depth of 10.0 m in ground has a saturated unit weight of 17.5 kN/m^3 and a plasticity index of 35. The groundwater table is at the ground surface. Estimate the undrained shear strength, undrained elastic modulus, and effective friction angle of this clay.

Solution

The effective overburden stress of this sample in the ground is

$$\sigma'_{z0} = \gamma' z = (17.5 - 9.81) \times 10.0 = 77.0 \text{ kPa}$$

Since it is an inorganic clay, $\chi = 0.22 \pm 0.03$. Select $\chi = 0.22$. Therefore, the undrained shear strength of the NC sample (i.e., OCR = 1.0) is

$$c_u = \chi(\text{OCR})^{\Lambda}\sigma'_{z0} = 0.22 \times 1 \times 77.0 = 16.9 \text{ kPa}$$

The constant K_c for undrained elastic modulus at OCR = 1 and PI = 35 is from 300–600. Select K_c = 500 because PI = 35 is close to the upper bound corresponding to PI = 30.

$$E_u = 500c_u = 500 \times 16.9 = 8450 \text{ kPa}$$

2.2.4 Hydraulic Properties

The rate of water flow through a geomaterial depends on the permeability (also called hydraulic conductivity) of the material. Darcy's law is used to estimate average velocity, v, of flow as follows:

$$v = ki \quad (2.14)$$

where k = permeability of geomaterial
i = hydraulic gradient

The quantity of water flow, Q_w, can be calculated as follows:

$$Q_w = kiA \quad (2.15)$$

where A is the cross-sectional area of water flow.

Permeability is often determined by constant head or falling head tests in the laboratory or pumping tests in the field. Typical permeability values for different geomaterials are provided in Table 2.4.

Louden (1952) verified the Hazen relationship between permeability and D_{10} for silty sand to coarse sand as follows:

$$k = 0.01 D_{10}^2 \quad (2.16)$$

where k = soil permeability (m/s)
D_{10} = sieve size of 10% particle passing (mm)

Table 2.4 Typical Values of Permeability of Geomaterials

Geomaterial	Range of Permeability (m/s)
Gravel	$10^{-2} - 10^{-1}$
Clean sand	$10^{-5} - 10^{-4}$
Clean sand and gravel mixture	$10^{-5} - 10^{-3}$
Medium to coarse sand	$10^{-4} - 10^{-3}$
Very fine to fine sand	$10^{-6} - 10^{-5}$
Silty sand	$10^{-7} - 10^{-4}$
Glacial till	$10^{-12} - 10^{-6}$
Homogenous clay	$10^{-11} - 10^{-9}$
Shale	$10^{-13} - 10^{-9}$
Sandstone	$10^{-10} - 10^{-6}$
Limestone	$10^{-9} - 10^{-6}$
Fractured rock	$10^{-8} - 10^{-4}$

Sources: Freeze and Cherry (1979) and Peck et al. (1974).

Example 2.6

A medium sand with $D_{10} = 0.1$ mm. Estimate the permeability of this sand.

Solution

The Hazen relationship can be used to estimate this permeability:

$$k = 0.01\,D_{10}^2 = 0.01 \times 0.1^2 = 1.0 \times 10^{-4}\ \text{m/s}.$$

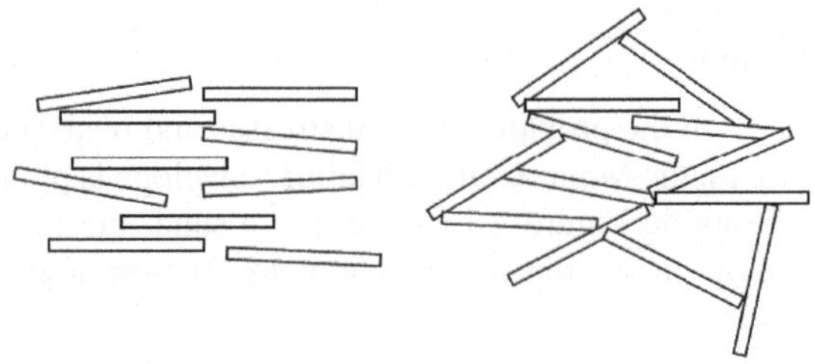

Figure 2.17 Clay fabric: (a) dispersion and (b) flocculation.

2.2.5 Compaction of Geomaterial

The phase diagram in Figure 2.2 shows that a geomaterial consists of solid, liquid (mostly water), and air. Particles are packed in different patterns, which result in different void ratios and densities. Mitchell and Soga (2005) show five possible packing patterns of uniform particles in Figure 2.16. The simple cubic packing pattern results in the largest void ratio (loosest state), while the pyramidal and tetrahedral packing patterns result in the smallest void ratio (i.e., densest state). The void is reduced from the loosest state to the densest state. When particles are deposited or placed, they are packed in a certain pattern. The packing pattern can be changed by applying a force or vibration. The change of the packing pattern from a loose state to a dense state results in compression of voids and densification of the geomaterial. Uniform particles may approximately represent a poorly graded cohesionless granular material.

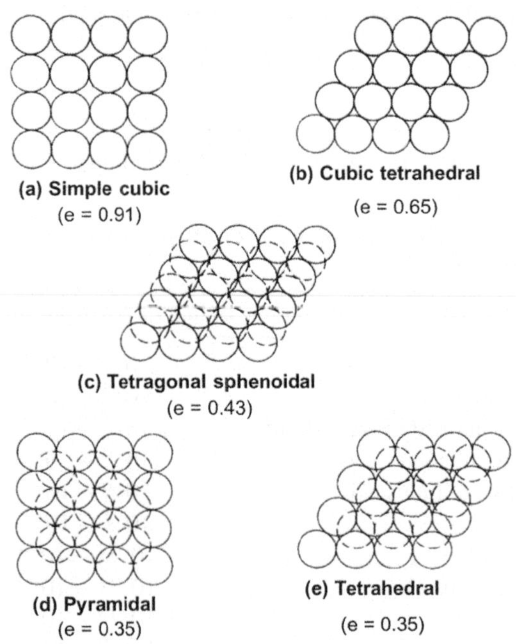

Figure 2.16 Packing pattern of uniform particles and void ratio: (a) Simple eubic, (b) cubic tetrahedral, (c) tetragonal sphenoidal, (d) pyramidal, and tetrahedral (modified from Mitchell and Soga, 2005).

For clays, however, they have two basic clay fabrics as shown in Figure 2.17 since their particles are flat sheets. In the dispersed structure, clay particles do not have any face-to-face association. In the flocculated structure, however, particles have edge-to-edge or edge-to-face association. The dispersed structure has a large number of small voids, while the flocculated structure has certain large voids with few small voids.

Due to the difference in soil structure, cohesionless and cohesive geomaterials can be densified by different means. Cohesionless geomaterials, consisting of large particles, can be effectively densified by vibration to rearrange particle packing patterns. The soil fabrics of cohesive geomaterials can be effectively changed by high pressure. Vibration is not effective to change soil fabrics of cohesive geomaterials. It is also true that high pressure within a small area is not effective to compress cohesionless geomaterials because they will fail under high pressure due to the low confining stress. Two different test methods have been developed to evaluate the maximum densities of the cohesionless and cohesive geomaterials.

For cohesionless geomaterials with 15% or less by dry mass of particles passing the U.S. No. 200 sieve, the maximum index density method (ASTM D4253) should be used to determine the maximum density. This method densifies the geomaterial by vibration. The maximum index density test is performed at either a dry condition or a saturated condition.

For cohesive geomaterials, two compaction tests can be performed in laboratory: standard Proctor test (ASTM D698) and modified Proctor test (ASTM D1557). These two methods use impact loads at different compaction energy levels as shown in Table 2.5 with a mold diameter of 100 mm. They are performed at different moisture contents. Figure 2.18 shows a typical compaction curve, which has a maximum dry unit weight and its corresponding optimum moisture content. Typically, five compaction tests are needed, of which at least two data points are on each side of the optimum.

Table 2.5 Parameters for Laboratory Compaction Tests

Type of Test	Mass of Hammer (kg)	Drop Distance (mm)	Layers	Blows per Layer
Standard Proctor test	2.5	305	3	25
Modified Proctor test	4.5	455	5	25

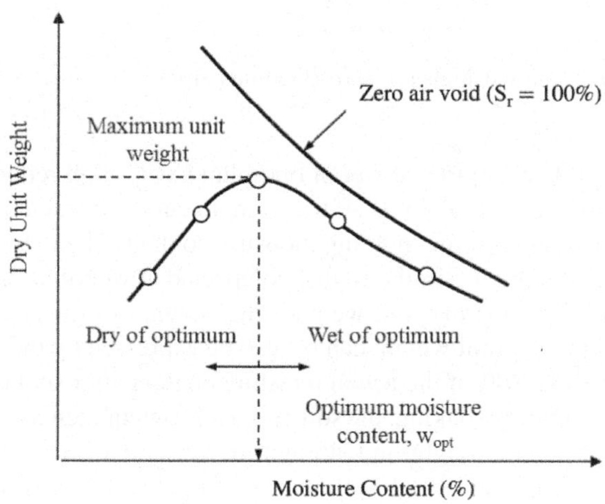

Figure 2.18 Typical compaction curve.

The shape of the typical compaction curve can be explained as follows. At low moisture content, water film around soil particles is thin so that there are large bonding forces between particles. Capillary force is high at low moisture content. In addition, friction between particles is high at low moisture content. All these factors make particle rearrangement difficult at low moisture content. With an increase of moisture content, the water film becomes thicker. As a result, the bonding force, the capillary force, and the friction decrease so that particles can be easily rearranged into a dense state. All these become the optimum condition for particle rearrangement at the optimum moisture content. With further increase of the moisture content, most of the air voids are occupied by water. The remaining air voids to be compressed become less. In addition, the soil becomes too soft to carry the impact force so that shear failure happens during compaction. All these result in a loose state. Since the zero air void line is a theoretical line with 100% saturation, no test data should be above this line. The dry unit weight can be expressed in terms of degree of saturation as follows:

$$\gamma_d = \frac{G_s\gamma_w}{1 + wG_s/S_r} \tag{2.17a}$$

When the degree of saturation, S_r = 100%, the zero air void line can be determined using the following equation:

$$\gamma_{dzav} = \frac{G_s\gamma_w}{1 + wG_s} \tag{2.17b}$$

where γ_{dzav} is the dry unit weight at the zero air void.

Research reveals that an increase of the moisture content and/or compaction effort changes the soil fabric from flocculation to dispersion as shown in Figure 2.19. This fabric change will affect the geomaterial behavior. Figure 2.19 also shows that an increase of the compaction effort increases the maximum dry unit weight and reduces the optimum moisture content.

Figure 2.20 shows that the geomaterial compacted at wet of optimum has a low permeability than that at dry of optimum.

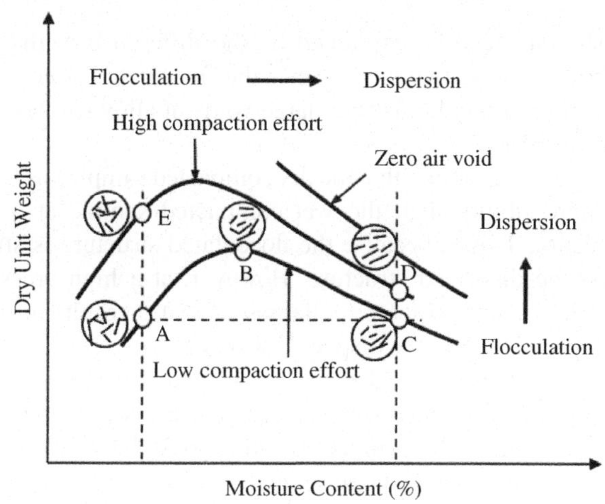

Figure 2.19 Soil fabrics at different moisture contents and compaction efforts (modified from Lambe, 1958).

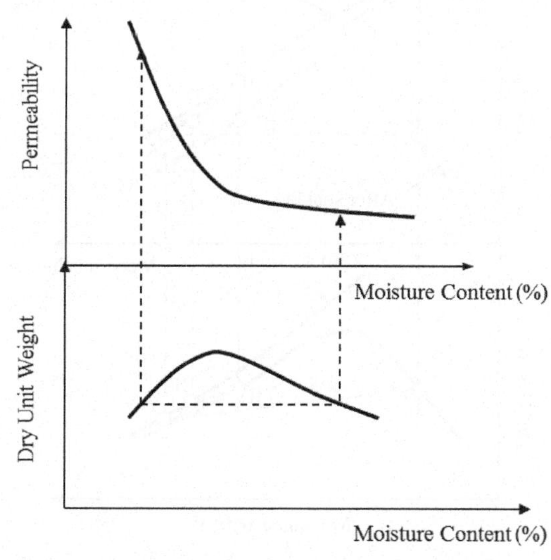

Figure 2.20 Effect on permeability (after Lambe, 1958).

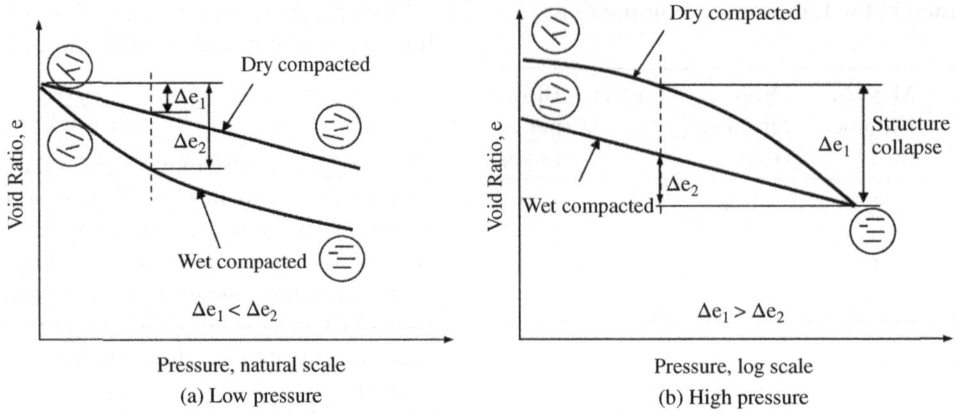

Figure 2.21 Effect on compressibility: (a) low pressure and (b) high pressure (modified from Lambe, 1958).

This result can be explained by the difference in the soil fabrics at these two sides. Since the flocculated structure at dry of optimum has certain large voids, it allows water flow more easily.

Figure 2.21 shows that the dry compacted sample has lower compressibility than the wet compacted sample at a low pressure. This is because the flocculated structure is stiffer than the dispersed structure. However, at a high pressure, the flocculated structure collapses so that it results in more compression than the dispersed structure.

Figure 2.22 shows that the compacted soil at dry of optimum has a higher strength than that at wet of optimum. However, after the compacted soil is soaked, the strength at dry of optimum has a large strength reduction from Point A to Point A' while that at wet of optimum has a small reduction from Point B to Point B', if the strength curve after soaking is plotted against the molding moisture content. Their strength after soaking is similar to that compacted at wet of optimum with the same dry unit weight. The changes of the strength and the dry unit weight can be also be expressed by the lines AA'' and BB'', if the actual moisture content after soaking is used. During soaking, the soil moisture content increases and the dry unit weight and strength decrease due to swelling. The sample at dry of optimum has larger reductions in the unit weight and strength and a larger increase in the moisture content than that at wet of optimum due to larger swelling at dry of optimum. The magnitude of the unit weight decrease also depends on the level of overburden stress.

Figure 2.23 shows that the soil compacted at dry of optimum has higher swell potential but lower shrinkage potential

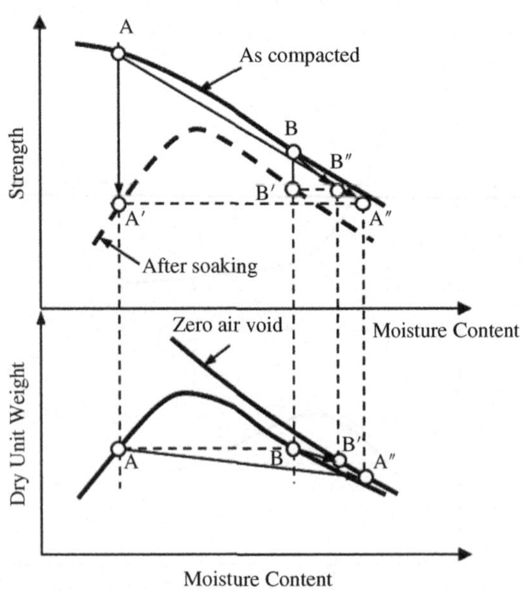

Figure 2.22 Effect of soaking on strength (modified from Lambe, 1958).

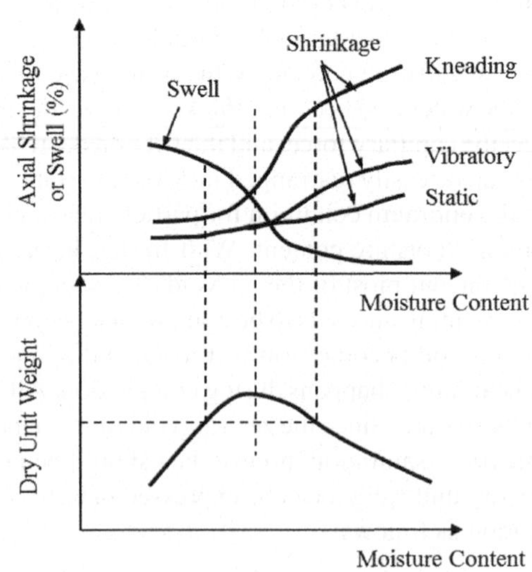

Figure 2.23 Effect on swell and shrinkage (after Seed and Chan, 1959).

than that at wet of optimum. These results are understandable because the soil at wet of optimum cannot absorb more water but can lose much water. Figure 2.23 also shows that the method of compaction affects the shrinkage potential. The kneading compaction results in the highest shrinkage potential while the static compaction results in the lowest shrinkage potential.

Example 2.7

A soil sample is compacted at a moisture content of 15%. The unit weight of the moist sample is 17.6 kN/m^3. The specific gravity of this soil is 2.69. Calculate the dry unit weight, the degree of saturation, and the corresponding zero air dry unit weight of this sample.

Solution

The dry unit weight of the sample is

$$\gamma_d = \frac{\gamma}{1+w} = \frac{17.6}{1+0.15} = 15.3\,\text{kNm}^3$$

The degree of saturation of the sample can be solved from the following equation:

$$\gamma_d = \frac{G_s\gamma_w}{1+wG_s/S_r} = \frac{2.69\times 9.81}{1+0.15\times 2.69/S_r} = \frac{26.39}{1+0.40/S_r} = 15.3$$

Hence, $S_r = 0.55 = 55\%$.

The zero air dry unit weight is

$$\gamma_{\text{dzav}} = \frac{G_s\gamma_w}{1+wG_s} = \frac{2.69\times 9.81}{1+0.15\times 2.69} = \frac{26.39}{1+0.40} = 18.9\ \text{kN/m}^3$$

2.3 GEOSYNTHETICS AND PROPERTIES

2.3.1 Type of Geosynthetic

Geosynthetics are factory-manufactured products (mainly polymers) used for geotechnical applications. They can be classified into the following types: geotextile, geogrid, geonet, geomembrane, geosynthetic clay liner, geocell, geofoam, and geocomposite. Except geocell manufactured in a three-dimensional honeycomb shape and geofoam in a cubic block, all other geosynthetics are two-dimensional or planar.

Geotextile can be nonwoven, woven, or knitted as shown in Figure 2.24. Nonwoven geotextile is formed by random arrangement of fibers that are bonded together by heat melting, needle punching, or resin. Woven geotextile is manufactured through a weaving process in which fibers are arranged essentially at right angles to each other in varying configurations. Knitted geotextile is manufactured by interlooping fibers. Geotextile is mostly made of polypropylene (PP) polymers. Other polymers are polyester [e.g., polyethylene terephthalate (PET)], polyethylene (PE), and polyamide (nylon).

(a) Nonwoven (b) Woven

Figure 2.24 Geotextile products.

Geogrid can be uniaxial, biaxial, or triaxial and it consists of openings (or apertures) formed by longitudinal and transverse ribs or three ribs as shown in Figure 2.25. Uniaxail geogrid has a high strength in the longitudinal direction. Biaxial geogrid has strengths in both longitudinal and transverse directions. Triaxial geogrid has strengths in multiple directions. Geogrids are mostly manufactured by a punched-drawn method, a coated woven method, or a wielded method. High-density polyethylene (HDPE), PET, and PP are three commonly used polymers for geogrids.

Geonet is a grid like material as shown in Figure 2.26, which is made of PE polymer. Different from geogrid, geonet has thickness difference in ribs and is commonly used with nonwoven geotextile to form a geocomposite.

Geomembrane is an impermeable thermoplastic material [mostly PE and sometimes PP or polyvinyl chloride (PVC)] as shown in Figure 2.27, which is manufactured by extrusion. It can have smooth or rough (textured) surfaces.

Geosynthetic clay liner (GCL) consists of a bentonite clay with adhesive bonded with two geotextiles by a needle-punched or stitch-bonded method as shown in Figure 2.28.

Geocell is an expandable three-dimensional honey comb-like polymer structure, which is made of HDPE, PET, or other polymer material as shown in Figure 2.29. It has different pocket sizes (diameter and height).

Geofoam is a block or planar rigid cellular foam polymeric material, which is made of lightweight expanded polystyrene (EPS) or extruded polystyrene (XPS). It has typical density from 11 to 29 kg/m^3.

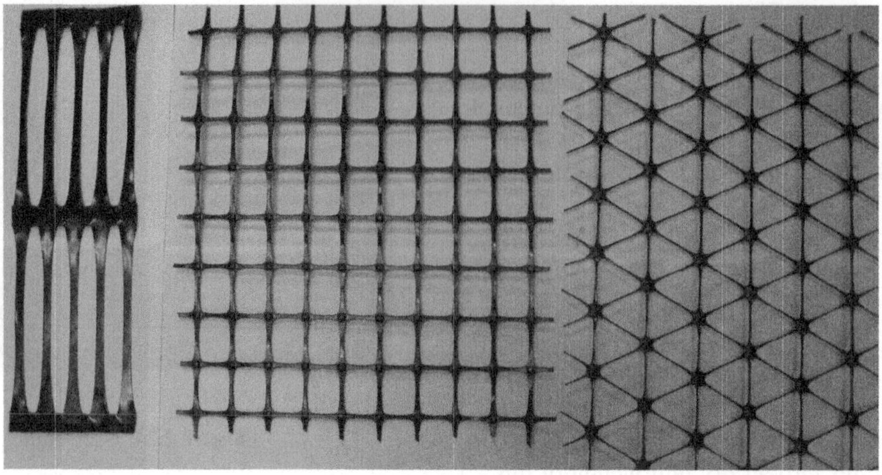

(a) Uniaxial (b) Biaxial (c) Triaxial

Figure 2.25 Geogrid products.

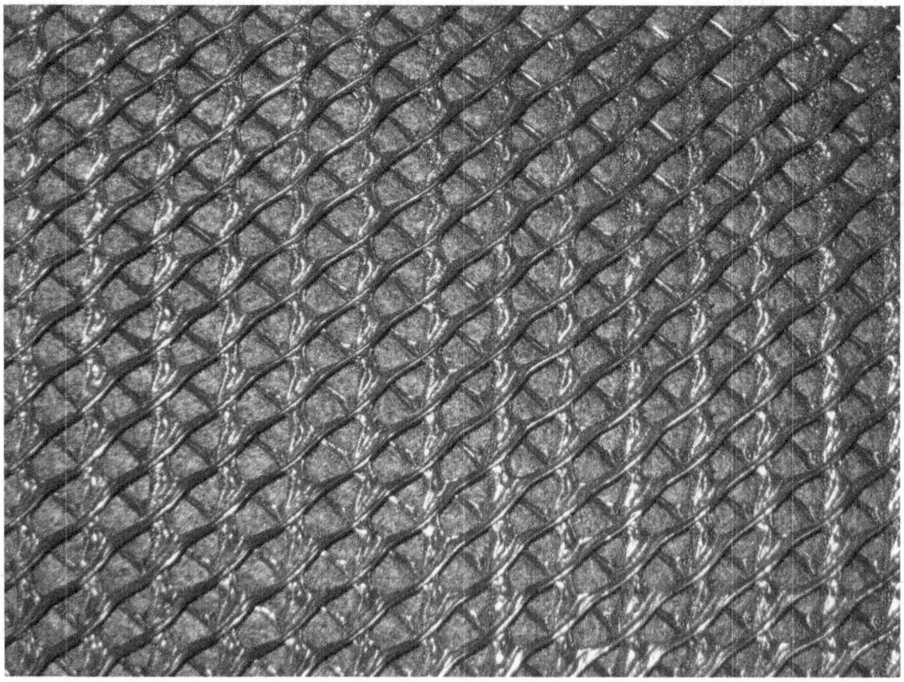

Figure 2.26 Geonet product.

Geocomposite is a composite of two or more materials, which can be a combination of two geosynthetic materials, for example, geotextile and geonet.

2.3.2 Function

The general functions of geosynthetics include separation, filtration, drainage, reinforcement, barrier, and erosion protection. The relationship between geosynthetic type and function is provided in Table 2.6. Geofoam has a unique function by serving as a lightweight fill.

When two dissimilar geomaterials are in contact, they may be intermixed under loading, especially under cyclic loading. Geosynthetics, such as geotextile or geomembrane, can be placed between these two dissimilar materials to keep the integrity and functioning of two dissimilar materials intact and prevent them from being intermixed. This function is often referred to as separation.

(a) Rough (b) Smooth

Figure 2.27 Geomembrane products.

Figure 2.28 Geosynthetic clay liner products.

Similar to natural filter, geotextile can be selected with proper opening sizes to allow for adequate liquid flow and limit soil particle loss across the interface plane. This function is often referred to as filtration.

Since geotextile, geonet, and geocomposite have large porosity, high permeability, and drainage capacity, they allow water to quickly flow through and can help drain water out of geomaterials or pavement structures. To survive in installation and maintain long-term drainage capacity, geosynthetics used for drainage should have sufficient strengths and the functions of separation and filtration. Geosynthetics have been used for drainage inside retaining walls, slopes, roadways, and landfills. The design of geosynthetics for drainage will be discussed in Chapter 6.

Geomaterials are typically strong under compression and shear but weak under tension. Geosynthetics, such as woven geotextiles, geogrids, and geocells, can provide tensile resistance to geomaterials. Geosynthetic reinforcement of geomaterials can be achieved through the mechanisms of anchorage, lateral restraint (interlocking, closed confinement,

Figure 2.29 Geocell product.

Table 2.6 Functions of Geosynthetics

Type	Separation	Filtration	Drainage	Reinforcement	Erosion Protection	Barrier
Geotextile	√	√	√	√	√	√
Geogrid				√		
Geonet			√			
Geomembrane	√				√	√
GCL	√				√	√
Geocell				√	√	
Geocomposite	√	√	√	√	√	√

and friction), and tensioned membrane. It should be pointed out that nonwoven geotextile is generally too weak for reinforcement. Even though all three types of geogrid can be used for reinforcement, they are often used for different applications. For example, uniaxial geogrid is commonly used to reinforce embankments, slopes, and earth-retaining walls due to their two-dimensional nature. Biaxial and triaxial geogrids are commonly used to stabilize unpaved and paved roads and railroad ballast due to three-dimensional traffic loading. Dong et al. (2011) showed that triaxial geogrid provides more uniform tensile resistance than biaxial geogrid. Geocell has been used to stabilize granular bases over weak subgrade (Han et al., 2011). The design of geosynthetics for fill reinforcement will be discussed in Chapter 10.

Barriers can be formed by impermeable or low permeable geomembrane and geosynthetic clay liner, or large-pored geotextile and geocomposite, which lower or stop capillary rise. Geomembranes and GCLs have been commonly used as liners in landfills and river channels and waterproof in dams. Geotextile and geocomposite have been used as capillary barriers to replace coarse sand and gravel capillary barriers to reduce frost heave. The basic idea of the capillary barrier is to "place a large-pored layer between the water table and the freezing front to stop the upward flow of water during freezing" (Henry and Holtz, 2001).

Geosynthetics have been used to protect ground surfaces (especially slopes) from loss of soil particles by running water as a function of erosion protection. Geosynthetics can avoid water drops directly hitting on soil surface and slow down water flow on soil surface so that less soil particles are carried away by running water. Geosynthetics have also been used to protect coastal areas from wave erosion and scour.

Some geosynthetics only have one function but others have multiple functions. For example, geotextile can serve all the functions. Geocell has dual functions (reinforcement and erosion protection).

2.3.3 Properties and Test Methods

Physical Properties Physical properties are mainly used for quality control and assurance, which include polymer type, mass per unit area, thickness (geotextiles and geomembranes), height (geocells), roll length, roll width, roll weight, roll diameter, specific gravity and density, and surface characteristics (geomebrannes). Typical specific gravity of polymer is provided in Table 2.7.

Mass per unit area is one important design and quality assurance parameter. The typical mass per unit area values for most geotextiles, geogrids, and geonets are 150–750, 200–1000 and 700–1000 g/m^2, respectively.

High-density polyethylene (HDPE) has a density higher than 0.941 mg/L while low-density polyethylene (LDPE) has density lower than 0.941 mg/L.

The thickness of a geosynthetic sheet is often measured in a special unit, mil, which is 1/1000 of inch (i.e., 0.025 mm). Typical thickness for geotextile ranges from 0.25 to 7.5 mm. It should be pointed out that the thickness of needle-punched geotextile can decrease significantly under a normal stress. The thickness of the geotextile at the normal stress that corresponds to the field condition should be used for design.

Depending on applications, geomembranes can have smooth or rough surfaces. The geomembranes with rough surfaces have higher interface shear strength with geomaterial; therefore, they are mostly used on slopes. The geomembranes with smooth surfaces are used in a level ground.

Hydraulic Properties Hydraulic properties are important for filtration and drainage design, which include opening characteristics (geotextiles), permeability and permittivity, soil retention ability, and in-plane flow capacity.

Apparent opening size (AOS) is used to define an equivalent opening size of a geotextile, at which 5% or less glass beads pass. This size is also often referred to as O_{95}. The test for AOS is similar to a sieve analysis test. Instead of a sieve, a geotextile sample is placed on a sieve frame. Uniform glass beads are placed on the geotextile and subjected to 10 min

Table 2.7 Typical Specific Gravity of Polymer

Polymer	Specific Gravity
Polyvinyl chloride	1.69
Polyester	1.22–1.38
Nylon	1.05–1.14
Polyethylene	0.90–0.96
Polypropylene	0.91

Source: Koerner (2005).

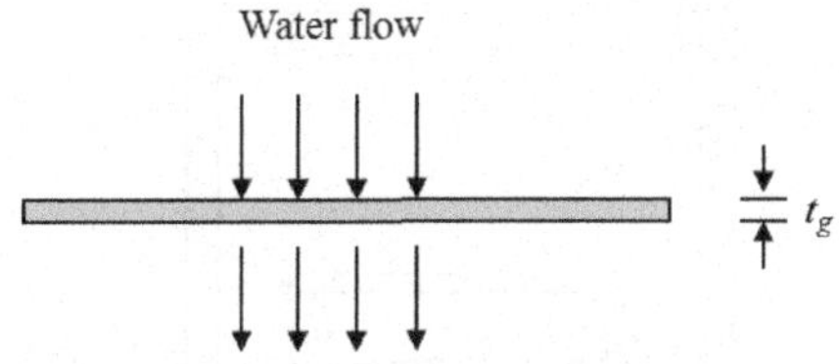

Figure 2.30 Permittivity test.

shaking. The test starts with large glass beads and then reduces the size of glass beads until more than 5% glass beads pass the geotextile.

Permittivity is a measure of the rate of water flow perpendicularly through a geosynthetic sheet (mostly geotextile or geomembrane) as shown in Figure 2.30. Permittivity is defined as

$$\psi_g = \frac{k_n}{t_g} \tag{2.18}$$

where ψ_g = permittivity (s^{-1})
k_n = cross-plane permeability coefficient (m/s)
t_g = thickness at a specific normal stress (m)

Transmissivity is a measure of water flow parallel to the geosynthetic sheet (mostly geotextile, geonet, or geocomposite), as shown in Figure 2.31. Transmissivity is defined as

$$\theta_g = k_p t_g \tag{2.19}$$

where θ_g = transmissivity (m^2/s)
k_p = in-plane permeability coefficient (m/s)
t_g = thickness at a specific normal stress (m)

The measured permittivity and transmissivity of a geosynthetic sheet from laboratory should be reduced by considering the following factors:

- Creep reduction of void space under normal stress in field
- Intrusion of adjacent material into void space
- Soil clogging and blinding
- Chemical clogging
- Biological clogging

Since geomembranes are mostly used as barriers, it is important to determine how slow water vapor can transmit

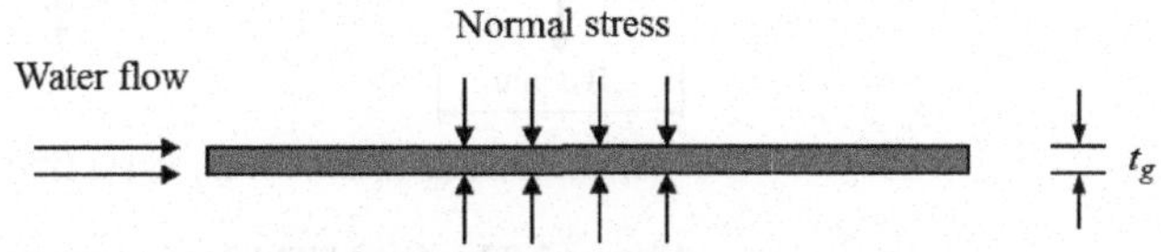

Figure 2.31 Transmissivity test.

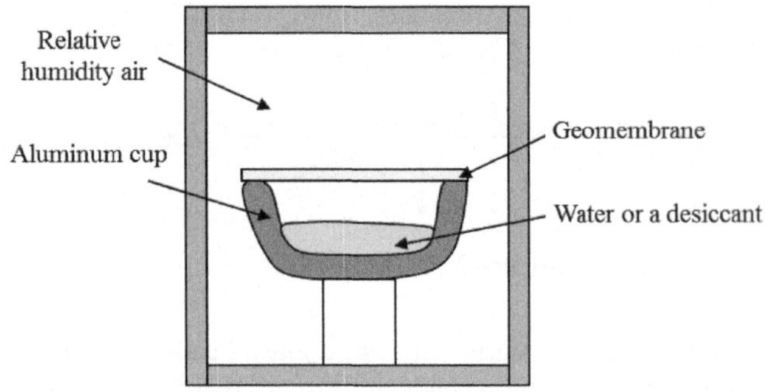

Figure 2.32 Water vapor transmission test.

through them. A water vapor transmission test as shown in Figure 2.32 can be used to determine this parameter as follows:

$$\mathrm{WVT} = \frac{24\,m_l}{At} \tag{2.20}$$

where WVT = water vapor transmission (g/m^2/day)
m_l = mass loss or gain (g)
t = time (h)
A = area of specimen (m^2)

Mechanical Properties Mechanical properties are needed for the evaluation of survivability of geosynthetics during installation and during service when they are used for the reinforcement function. The mechanical properties include grab strength (geotextiles), wide-width tensile strength (geotextiles), single-rib and multi-rib tensile strength (geogrids), narrow strip tensile strength (geomembranes), junction strength (geogrids), creep strength, seam strength, tear strength, burst strength, puncture resistance, and penetration resistance.

Figure 2.33 shows the grab, narrow strip, and wide-width tensile tests of geosynthetics. Figure 2.34 shows an example of tensile stress–strain curve obtained by a short-term tensile test. From this curve, the initial, secant, and tangential moduli can be determined. It is common that the secant moduli are reported at 2 and 5% strains. For geogrids and geocells, the ultimate tensile strengths, T_{ult}, are often determined at 10% strain.

Geosynthetics have limited width and length. At certain locations, geosynthetics (geotextiles and geomembranes) are connected by seams. To evaluate the connection strength, geosynthetic seam tests as shown in Figure 2.35 can be performed. The ratio of the tensile strength with a seam to that without a seam is the efficiency of the seam, which typically ranges from 50 to 70%.

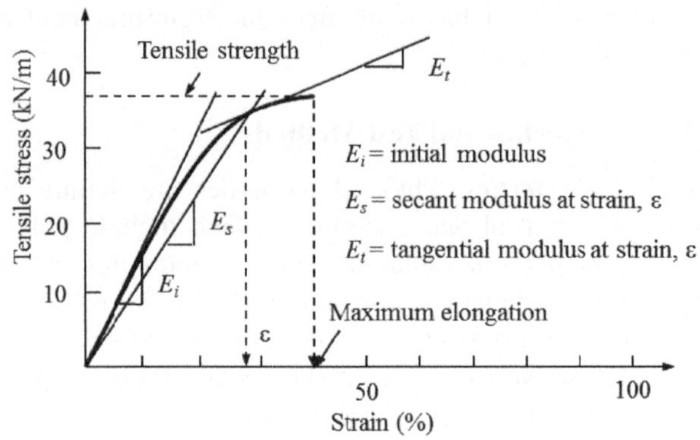

Figure 2.34 Tensile stress–strain curve.

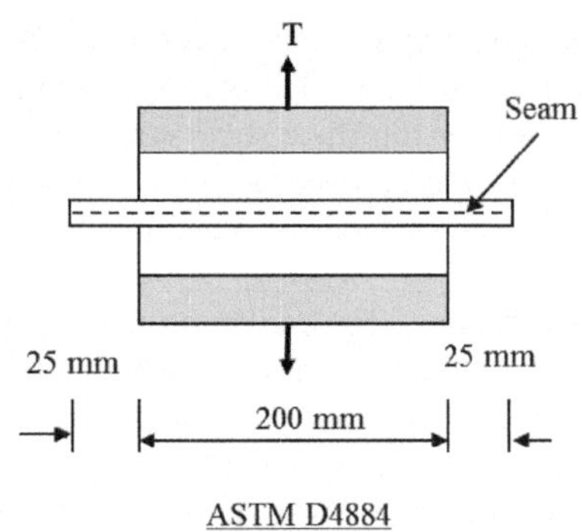

Figure 2.35 Geosynthetic seam test.

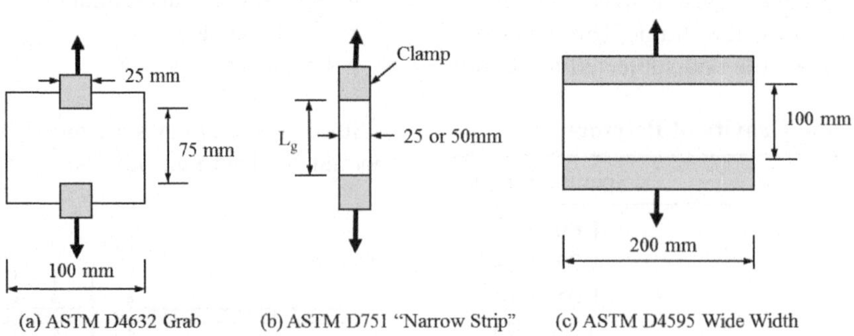

Figure 2.33 (a) ASTM D4632 grab, (b) ASTM S751 narrow strip, and (c) ASTM D4595 wide width tests.

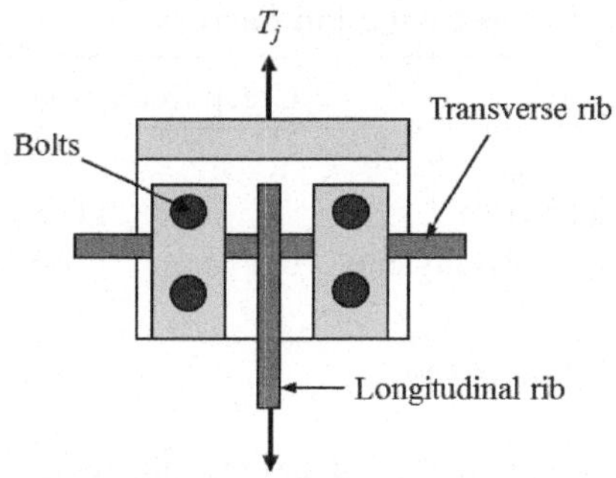

Figure 2.36 Geogrid junction strength test.

Due to different methods of geogrid manufacture, geogrid junction strength can be very different. Figure 2.36 shows the test method to determine the efficiency of geogrid junction strength as follows:

$$E_j = \frac{T_j}{T_s} \times 100\% \quad (2.21)$$

where E_j = junction efficiency
T_j = junction tensile strength
T_s = single-rib tensile strength

The typical junction efficiency of geogrid ranges from 7 to 100%.

Figure 2.37 shows the Mullen burst test from which the burst strength can be determined. Figure 2.38 shows a geosynthetic puncture test in which a piston is pushed into a clamped geosynthetic sample until it breaks. The maximum load is the puncture strength of the geosynthetic.

There are three standard tear tests available to determine the tear strength of a geosynthetic sample: (1) trapezoidal tear test (ASTM D3786), (2) tongue tear test (ASTM D751), and (3) Elmendorf tear test (ASTM D1424). The trapezoidal tear test as shown in Figure 2.39 is applicable to woven and nonwoven geotextiles but not to knitted geotextiles. The tongue tear test is applicable to all geotextiles. The Elmendorf tear test is applicable to woven geotextiles but not to nonwoven geotextiles, and it is commonly used in Europe.

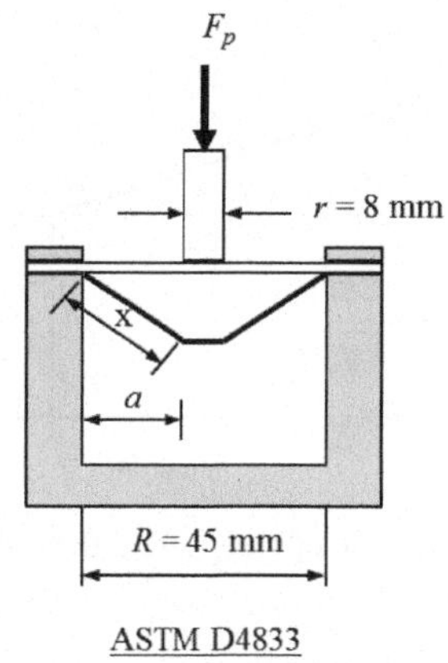

Figure 2.38 Geosynthetic puncture test.

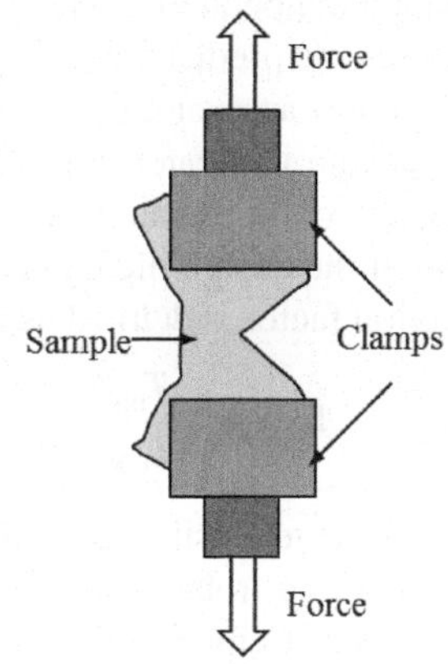

Figure 2.39 Geosynthetic tear tests.

Since geosynthetics are polymer materials, they creep with time under sustained loads. Creep tests can be conducted at different load intensities with time as shown in Figure 2.40. ASTM D5262 requires the minimum duration of a range of load intensities is 10,000 h. The creep (rupture) strength and its corresponding time are defined at 10% strain. However,

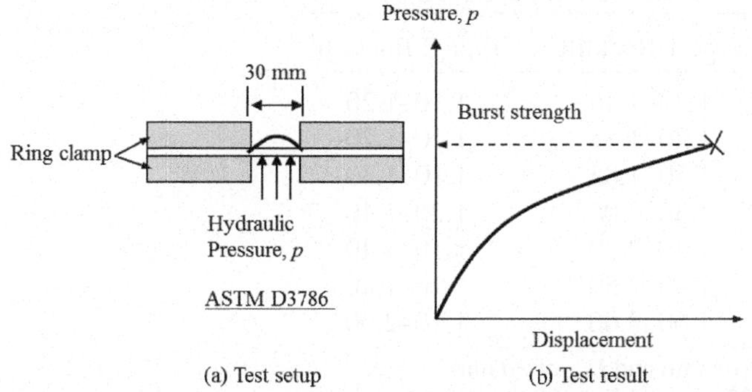

Figure 2.37 Mullen burst test: (a) test setup and (b) test result.

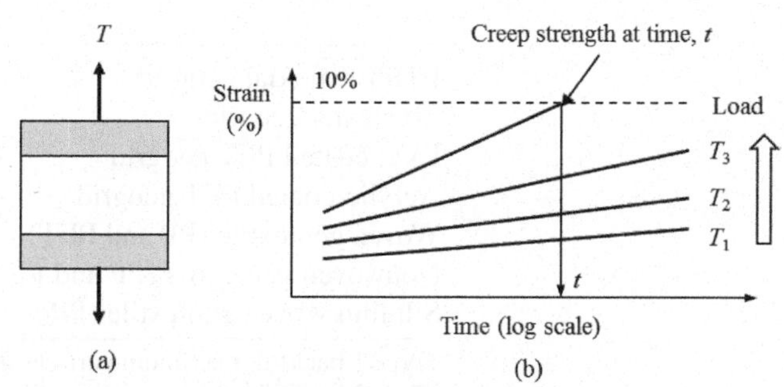

Figure 2.40 Creep test results.

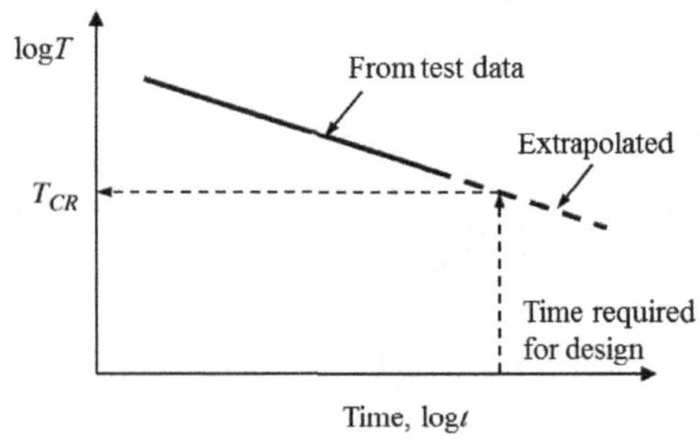

Figure 2.41 Estimation of creep strength at the end of design life.

the time corresponding to the creep strength from one series of tests is often not enough for the required time for design. To accelerate creep deformations, creep tests are often conducted at elevated temperatures. The creep test results at elevated temperatures can predict the creep strength at longer time. Figure 2.41 shows an example of creep test data. The required time for design is greater than that from the tests; therefore, the test curve is extrapolated so that the creep strength at the end of the design life can be determined.

The creep reduction factor is defined as follows:

$$RF_{CR} = \frac{T_{ult}}{T_{CR}} \tag{2.22}$$

where T_{ult} = ultimate tensile strength of the geosynthetic from a short-term tensile test
T_{CR} = creep strength of the geosynthetic

Typical reduction factors for different polymers are shown in Table 2.8.

During installation of geosynthetics in the field, compaction of fill above the geosynthetics may cause damage to the geosynthetics. Installation damage tests can be performed in the field to determine percent retained strength after compaction. During the tests, the same fill material and geosynthetic should be used for the evaluation. The reduction factor due to installation damage can be determined by the following formula:

$$RF_{ID} = \frac{T_{control}}{T_{damaged}} \tag{2.23}$$

where $T_{control}$ = ultimate tensile strength of geosynthetic not subjected to installation damage
$T_{damaged}$ = ultimate tensile strength of geosynthetic subjected to installation damage

Table 2.8 Creep Reduction Factors

Polymer Type	Creep Reduction Factor, RF_{CR}
Polyester (PET)	1.6–2.5
Polypropylene (PP)	4.0–5.0
High-density polyethylene (HDPE)	2.6–5.0

Source: Berg et al. (2009).

Berg et al. (2009) provide typical reduction factors due to installation damage for different geosynthetics under two different fill materials in Table 2.9.

Geosynthetics should also be evaluated for other degradation factors, including sunlight (ultraviolet) degradation, temperature degradation, oxidation degradation, hydrolysis degradation, chemical degradation, radioactive degradation, biological degradation, and others (such as fire). The basic idea is to determine percent retained strength after exposure to these factors for a certain time.

Berg et al. (2009) evaluate the resistance of polymers to specific environments in Table 2.10. Geosynthetics made of PET polymers are questionable and require exposure tests when they are used in soils with pH > 9.

Berg et al. (2009) also suggest the durability (aging) reduction factors for PET as shown in Table 2.11. The default

Table 2.9 Reduction Factors Due to Installation Damage

Geosynthetic	Reduction Factor, RF_{ID}	
	Type 1 Backfill[a]	Type 2 Backfill[b]
HDPE uniaxial geogrid	1.20–1.45	1.10–1.20
PP biaxial geogrid	1.20–1.45	1.10–1.20
PVC-coated PET geogrid	1.30–1.85	1.10–1.30
Acrylic-coated PET geogrid	1.30–2.05	1.20–1.40
Woven geotextile (PP and PET)	1.40–2.20	1.10–1.40
Nonwoven geotextile (PP and PET)	1.40–2.50	1.10–1.40
Silt film woven geotextile (PP)	1.60–3.00	1.10–2.00

[a]Type 1 backfill: maximum particle size of 102 mm and D_{50} of 30 mm.
[b]Type 2 backfill: maximum particle size of 20 mm and D_{50} of 0.7 mm.
Source: Elias et al. (2001).

Table 2.10 Resistance of Polymers to Specific Environment

Soil Environment	Polymer[a]		
	PET	PE	PP
Acid sulfate soils	NE	ETR	ETR
Organic soils	NE	NE	NE
Saline soils, pH < 9	NE	NE	NE
Ferruginous	NE	ETR	ETR
Calcareous soils	ETR	NE	NE
Modified soils/lime, cement	ETR	NE	NE
Sodic soils, pH > 9	ETR	NE	NE
Soils with transition metals	NE	ETR	ETR

[a]NE = no effect and ETR = exposure test required.
Source: Berg et al. (2009).

Table 2.11 Durability Reduction Factor for PET

Product[a]	Reduction Factor, RF_D	
	$5 < pH < 8$	$3 \leq pH \leq 5$ or $8 \leq pH \leq 9$
Geotextiles $M_n < 20,000$ and $40 < CEG < 50$	1.6	2.0
Coated geogrids or geotextiles $M_n > 25,000$ and $CEG < 30$	1.15	1.3

[a]M_n = number-average molecular weight and CEG = carboxyl end group.
Source: Berg et al. (2009).

durability reduction factor for PP and HDPE is 1.3, and certain criteria are required for the use of this default value.

Geosynthetic–Soil/Block Interaction Properties The interaction between geosynthetic and fill is often evaluated by interface shear tests and pullout tests. The interface shear test is more appropriate for fill sliding on a geosynthetic sheet or the geosynthetic sheet sliding on fill. Figure 2.42 shows the illustration of an interface shear test setup. At least three tests at different normal stresses are required to determine the interface cohesion and friction angle as shown in Figure 2.43. Figure 2.43 also shows the peak and residual strengths.

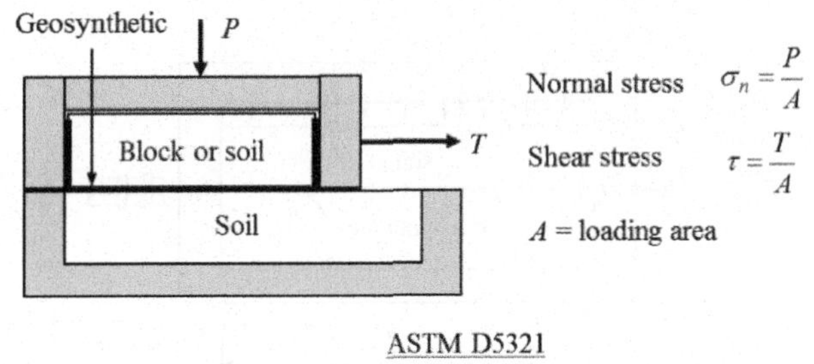

Figure 2.42 Interface shear test (modified from Koerner, 2005).

Interaction coefficient between geosynthetic and soil is often used in practice and defined below:

$$C_i = \frac{\text{Interface strength}}{\text{Soil strength}} = \frac{c_\alpha + \sigma_n \tan\delta}{c + \sigma_n \tan\phi} \tag{2.24}$$

where C_i = interaction coefficient
c_α = interface cohesion
c = fill cohesion
δ = interface friction angle
ϕ = fill friction angle
σ_n = normal stress

If $c = c_\alpha = 0$,

$$C_i = \frac{\tan\delta}{\tan\phi} \tag{2.25}$$

The interaction coefficient, C_i, depends on geosynthetic type and fill type. Smooth geomembranes have low C_i values

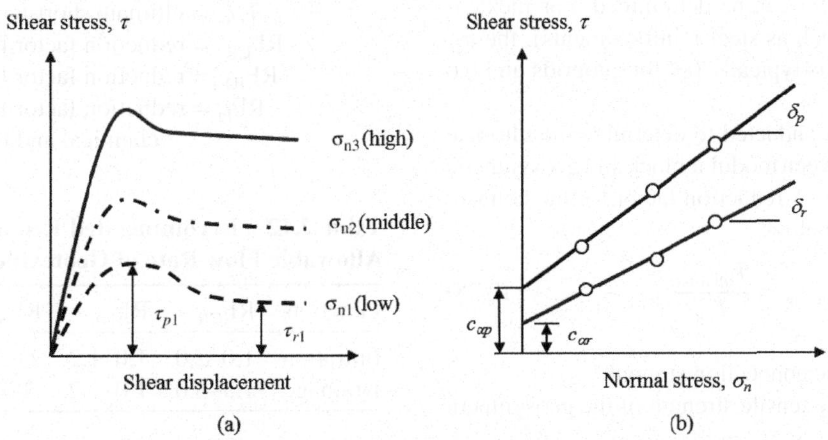

Figure 2.43 Interface shear test results (after Koerner, 2005).

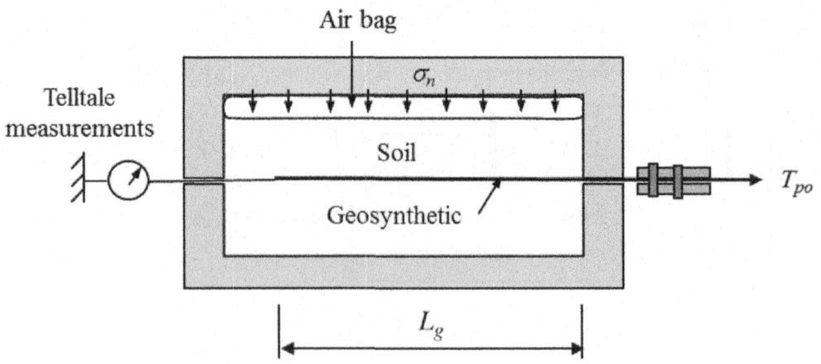

Figure 2.44 Geosynthetic pullout test.

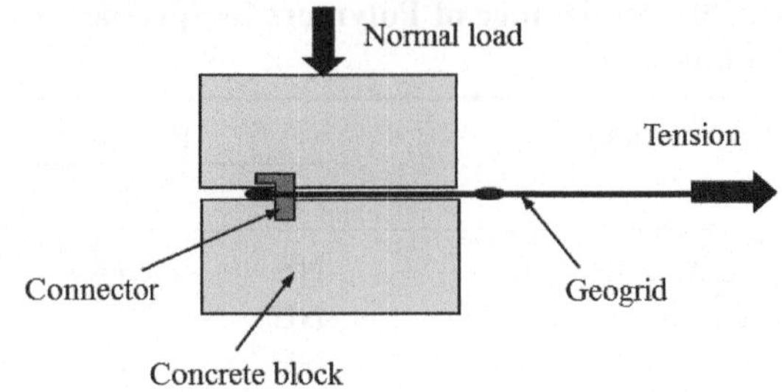

Figure 2.45 Connection test.

while triaxial and biaxial geogrids have high C_i values. For uniaxial geogrids, C_i values mostly range from 0.6 to 0.8.

Interface shear tests can also be performed between two types of geosynthetics (e.g., geotextile over geomembrane), which may be used for landfill applications.

Pullout test is another commonly used interface shear test as shown in Figure 2.44, which is more suitable to evaluate the anchorage capacity of a geosynthetic reinforcement in the fill. The normal stress is typically applied by an air bag. The geoynthetic reinforcement is pulled through a clamp from the front. The pullout capacity can be expressed as follows:

$$T_{po} = 2F^{*}\alpha_{se}\sigma_n L_g = 2C_i \tan(\phi)\alpha_{se}\sigma_n L_g \tag{2.26}$$

$$F^{*} = \tan\delta = C_i \tan\phi \tag{2.27}$$

where F^{*} = pullout friction factor
α_{se} = scale effect correction factor

The scale effect correction factor, α_{se}, depends on the mobilization of the anchorage length. To determine this factor, telltales are connected to the geosynthetic at different locations along the length to measure the mobilized lengths at different loads. Based on the load and mobilized length relationship, the α_{se} factor can be determined. For inextensible reinforcements (such as steel reinforcements), the α_{se} factor is close to 1.0. It is typically 0.8 for geogrids and 0.6 for geotextiles.

A similar test can be conducted to determine the ultimate connection strength between modular block and geosynthetic as shown in Figure 2.45. A reduction factor for the connection strength is defined below:

$$RF_{conn} = \frac{T_{ultconn}}{T_{ult}} \tag{2.28}$$

where $T_{ultconn}$ = ultimate connection strength
T_{ult} = ultimate tensile strength of the geosynthetic

Allowable Properties The allowable long-term flow rate for a geotextile used for filtration and drainage can be determined as follows:

$$q_{wa} = \frac{q_{wu}}{RF_{CR} \times RF_{IN} \times RF_{SCB} \times RF_{CC} \times RF_{BC}} \tag{2.29}$$

where q_{wa} = allowable long-term flow rate
q_{wu} = ultimate short-term flow rate
RF_{CR} = reduction factor for creep reduction of void space
RF_{IN} = reduction factor for adjacent material's intrusion
RF_{SCB} = reduction factor for soil clogging and blinding
RF_{CC} = reduction factor for chemical clogging
RF_{BC} = reduction factor for biological clogging

Table 2.12 provides the recommended reduction factors for the allowable flow rate of a geotextile.

The allowable long-term tensile strength of a woven geotextile or a geogrid can be determined as follows:

$$T_a = \frac{T_{ult}}{RF_{CR} \times RF_{ID} \times RF_D} \tag{2.30}$$

where T_a = allowable long-term tensile strength
T_{ult} = ultimate short-term tensile strength
RF_{CR} = reduction factor for creep
RF_{ID} = reduction factor for installation damage
RF_D = reduction factor for durability including chemical and biological degradation

Table 2.12 Recommended Reduction Factors for Allowable Flow Rate of Geotextile

Function	RF_{CR}	RF_{IN}	RF_{SCB}	RF_{CC}	RF_{BC}
Filtration	1.0–2.0	1.0–1.2	2.0–10.0	1.0–1.5	1.0–5.0
Drainage	1.5–3.0	1.0–1.2	2.0–4.0	1.0–1.5	1.0–1.5

The typical values of these reduction factors are provided in Tables 2.8, 2.9, and 2.11.

Example 2.8

A geotextile specimen (0.3 m × 0.3 m × 0.003 m thick) under a normal stress of 25 kPa is subjected to in-plane flow. The water head is 0.3 m. The quantity of water flow in 1 hour is 3.5 m^3. Determine: (1) hydraulic gradient, (2) in-plane permeability of geotextile, and (3) transmissivity of geotextile.

Solution

The hydraulic gradient is

$$i = \frac{h_w}{L_g} = \frac{0.3}{0.3} = 1.0$$

The in-plane permeability of geotextile is

$$k_p = \frac{Q_w}{iA} = \frac{3.5}{1.0 \times 0.3 \times 0.3} = 38.9\ \mathrm{m/hr}$$
$$= 1.08 \times 10^{-2}\ \mathrm{m/s}$$

The transmissivity of geotextile is

$$\theta_g = k_p t_g = 1.08 \times 10^{-2} \times 0.003$$
$$= 3.24 \times 10^{-5}\mathrm{m^2/s} = 1.94 \times 10^{-3}\mathrm{m^2/\,min}$$

Example 2.9

Given the following set of data from four interface shear tests of woven geotextile/sand: (1) plot the Mohr–Coulomb failure envelope, (2) obtain the interface friction angle, and (3) calculate the interaction coefficient (use the friction angle of sand from Example 2.4.

Example Table 2.2

Normal Stress (kPa)	Peak shear stress (kPa)
17	8.6
35	19
70	37
140	75

Solution

Based on the provided normal stresses and peak shear stresses, the Mohr–Coulomb failure envelope is plotted in Example Figure 2.2. Since it is a sand, the failure envelop passes the origin. The peak interface friction angle is

$$\delta_p = \tan^{-1}(0.5344) = 28.1°$$

The interaction coefficient is

$$C_i = \frac{\tan \delta_p}{\tan \phi_p} = \frac{\tan 28.1°}{\tan 42.0°} = 0.59$$

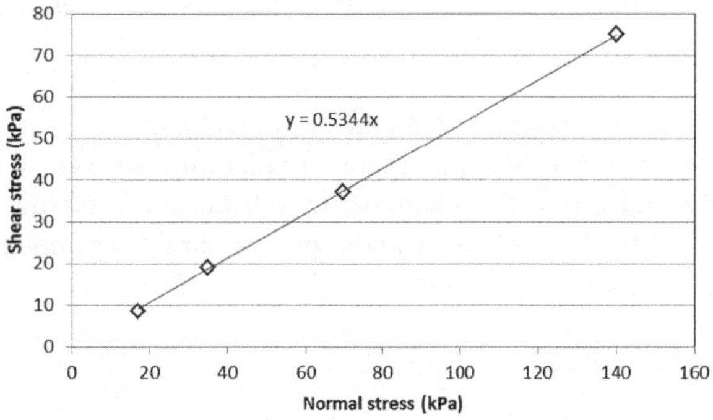

Example Figure 2.2

Example 2.10

Uniaxial HDPE geogrids have an ultimate short-term tensile strength of 300 kN/m from multi-rib tension tests. These geogrids will be used for a geosynthetic-reinforced earth wall project. Backfill material has a maximum particle size of 75 mm and a mean particle size of 25 mm. Estimate the allowable long-term tensile strength of the geogrids considering creep, installation damage, and durability based on the default values.

Solution

The default creep reduction factor for HDPE is 2.6 −5.0. $\mathrm{RF_{CR}} = 3.0$ is selected.

The backfill in this project is close to type I backfill in Table 2.9. The default installation damage reduction factor for uniaxial HDPE geogrid is 1.20 − 1.45. $\mathrm{RF_{ID}} = 1.30$ is selected.

The default durability reduction factor for HDPE is 1.3. $\mathrm{RF}_D = 1.30$ is selected.

The allowable long-term tensile strength of geogrids is

$$T_a = \frac{T_{\mathrm{ult}}}{\mathrm{RF_{CR}RF_{ID}RF_D}} = \frac{300}{3.0 \times 1.3 \times 1.3} = 59.2\,\mathrm{kN/m}$$

2.4 IN SITU TESTING

2.4.1 Standard Penetration Test

Introduction Standard penetration test (SPT) is one of the most commonly used in situ test methods. This test method has a long record of experience, much available test data, and correlation. The test is performed during soil sampling using the split-spoon sampler and is fast and inexpensive. However, this method is crude, has many variants, and does not have a continuous profile.

The basic procedure for the SPT method is shown in Figure 2.46:

- Drill a boring hole to the desired depth.
- Insert an SPT split-spoon sampler into the hole.
- Raise a 63.5-kg hammer to a distance of 760 mm and allow it to fall. Repeat this process until the sampler has penetrated 450 mm.
- Record the number of hammer blows required for each 150-mm interval for three intervals.
- Compute the N value by summing the blow counts for the last 300 mm of penetration.
- Remove the SPT sampler and soil sample.

Measured Parameter The measured SPT N value should be corrected (see Table 2.13) by considering four key factors as follows:

$$N_{60} = \frac{C_E C_B C_S C_R N}{0.60} \tag{2.31}$$

where N_{60} = SPT N value corrected for 60% of the theoretical free-fall hammer energy
C_E = hammer efficiency
C_B = borehole diameter correction
C_S = sampler correction
C_R = rod length correction
N = measured SPT N value

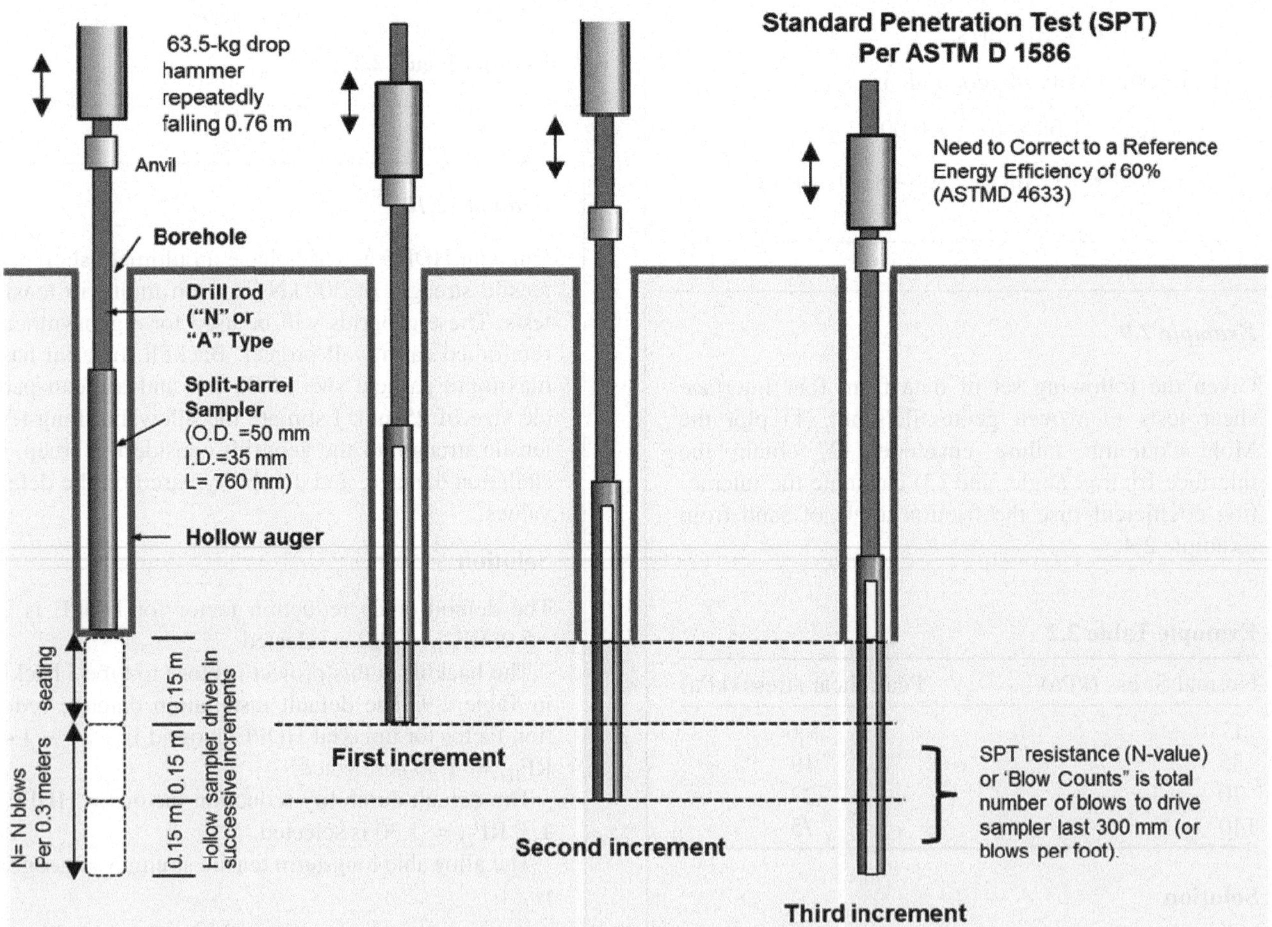

Figure 2.46 SPT test (Mayne et al., 2001).

Table 2.13 Correction Factors for SPT *N* Value

Effect	Variable	Term	Value
Energy ratio	Safety hammer	C_E	0.6–0.85
	Donut hammer		0.3–0.6
	Automatic hammer		0.85–1.0
Borehole diameter	65–115 mm	C_B	1.00
	150 mm		1.05
	200 mm		1.15
Sampling method	Standard sampler	C_S	1.0
	Sampler without liner		1.2
Rod length	3–4 m	C_R	0.75
	4–6 m		0.85
	6–10 m		0.95
	10– 30 m		1.0

Considering the overburden stress effect, N_{60} can be further corrected as follows:

$$(N_1)_{60} = N_{60}\sqrt{\frac{100}{\sigma'_{z0}}} \tag{2.32}$$

where

$(N_1)_{60}$ = corrected N_{60} considering the overburden stress
σ'_{z0} = overburden stress at a depth of interest (kPa)

Correlation Cubrinovski and Ishihara (1999) proposed the following correlation for the relative density of granular soils:

$$D_r = \left[\frac{N_{60}\left(0.23+0.06/D_{50}\right)^{1.7}}{9}\left(\frac{100}{\sigma'_{z0}}\right)^{0.5}\right]^{0.5} \times 100\% \tag{2.33}$$

where σ'_{z0} = effective overburden stress (kPa)
D_{50} = mean grain size (mm)

Kulhawy and Mayne (1990) proposed the following correlation for the effective friction angle of sands:

$$\phi' = \tan^{-1}\left[\frac{N_{60}}{12.2+20.3\left(\sigma'_{z0}/100\right)}\right]^{0.34} \tag{2.34}$$

Wolff (1989) proposed the following correlation for the effective friction angle of sands:

$$\phi' = 27.1° + 0.3(N_1)_{60} - 0.00054(N_1)^2_{60} \tag{2.35}$$

Kulhawy and Mayne (1990) proposed the following correlation for the elastic moduli of sands:

$$E_s = K_E N_{60} \tag{2.36}$$

where E_s = elastic modulus (MPa)
K_E = 0.5 for sands with fines, 1.0 for clean NC sands, and 1.5 for clean OC sands

Example 2.11

A soil boring is conducted in general accordance with the ASTM D1586 procedures with the standard penetration tests and a split barrel sampler at regular interval depths. The 75-mm diameter boring is advanced using a hollow-stem auger. An automatic drop hammer system that has an average efficiency of 83% is used. The SPT at a depth of 9 m within a sand layer results in an SPT *N* value of 12. The mean particle size of the sand is D_{50} = 1.5 mm. The average initial rod length above the existing ground for each SPT is about 1.2 m. The groundwater table is at the depth of 3 m. The unit weight above and below the groundwater is 19.7 kN/m^3. Determine the SPT $(N_1)_{60}$, the relative density of sand, and the friction angle of sand.

Solution

The corrected SPT value can be calculated as follows:

$$N_{60} = \frac{C_E C_B C_S C_R N}{0.60}$$

In this case, C_E = 83% = 0.83, C_B = 1.00, C_S = 1.00, and C_R depends on the depth of the test.

Total rod length to the test point of the sand layer $= 1.2 + 9.0 = 10.2$ m, $C_R = 0.95$, therefore,

$$N_{60} = \frac{C_E C_B C_S C_R N}{0.60} = \frac{0.83 \times 1.00 \times 1.00 \times 0.95 \times 12}{0.60}$$
$$= 15.8$$

The effective overburden stress at this test point is

$$\sigma'_{z0} = 3.0 \times 19.7 + (9.0 - 3.0) \times (19.7 - 9.81)$$
$$= 118 \text{ kN/m}^3$$

$$(N_1)_{60} = N_{60}\sqrt{\frac{100}{118}} = 14.1$$

Cubrinovski and Ishihara (1999) proposed the following correlation for the relative density of granular soils:

$$D_r = \left[\frac{N_{60}\left(0.23 + 0.06/D_{50}\right)^{1.7}}{9}\left(\frac{100}{\sigma'_{z0}}\right)^{0.5}\right]^{0.5} \times 100\%$$

$$= \left[\frac{15.8 \times (0.23 + 0.06/1.5)^{1.7}}{9} \times \left(\frac{100}{118}\right)^{0.5}\right]^{0.5}$$
$$\times 100\% = 41.8\%$$

The friction angle based on Kulhawy and Mayne (1990) is

$$\phi = \tan^{-1}\left[\frac{N_{60}}{12.2 + 20.3\left(\sigma'_{z0}/100\right)}\right]^{0.34}$$

$$= \tan^{-1}\left[\frac{15.8}{12.2 + 20.3\ (118/100)}\right]^{0.34} = 37.0^\circ$$

2.4.2 Cone Penetration Test

Introduction The cone penetration test (CPT) is another commonly used in situ test method in the practice following the ASTM standard D3441. CPT utilizes a cone, which was once known as the Dutch cone. There are two types of CPT cones: mechanical and electric; however, nowadays, most of the CPT cones are electric. More CPT cones have a preinstalled inclinometer to ensure the verticality of the cone during penetration. A typical cone has a 60° apex with a cross-sectional area of 1000 or 1500 mm^2. A typical CPT can measure tip resistance and sleeve friction (also called side resistance). The cone with one or more water pressure sensors is called CPTu or piezocone. The cone as shown in Figure 2.47 is pushed continuously by a hydraulic jack at a speed of 20 mm/s. The readings are taken at every 10 to 50 mm.

The advantages of CPT are: (1) it can obtain more information (two or more parameters) and (2) it can get continuous and more consistent soil profiles. However, it does not have soil sampling and is unreliable for soils containing large particles, such as gravel.

Measured Parameters Figure 2.48 shows the typical profiles of tip resistance, q_t, and sleeve friction f_s from a site.

Soil Classification and Correlation The soil type can be determined based on the design chart developed by Robertson (1990) as shown in Figure 2.49. In this design chart, the normalized cone parameters are used:

$$Q_t = \frac{q_t - \sigma_{zo}}{\sigma'_{zo}} \quad (2.37a)$$

$$B_q = \frac{u_b - u_0}{q_t - \sigma_{zo}} \quad (2.37b)$$

$$F_t = \frac{f_s}{q_t - \sigma_{zo}} \quad (2.37c)$$

where

σ_{zo} = total overburden stress behind the cone tip
σ'_{zo} = effective overburden stress behind the cone tip
u_0 = hydrostatic water pressure behind the cone tip
u_b = measured water pressure behind the cone tip
q_t = corrected tip resistance (approximately equal to measured tip resistance, q_c, if u_b is not too high)

The measured data point within a specific zone represents a specific soil behavior type as shown in Figure 2.49. For example, Zone 2 is a clay or clay to silty clay type.

Kulhawy and Mayne (1990) proposed the following correlation to estimate the relative density of sand:

$$D_r(\%) = \sqrt{\frac{q_c}{30500 Q_c \text{OCR}^{0.18}}\sqrt{\frac{100}{\sigma'_{z0}}}} \quad (2.38)$$

where q_c = CPT tip resistance (kPa)
Q_c = compressibility factor (ranging from 0.9 to 1.1)
σ'_{z0} = effective overburden stress (kPa)

Kulhawy and Mayne (1990) proposed the following correlation to estimate the effective friction angle of sand:

$$\phi' = \tan^{-1}\left[0.1 + 0.38 \log\left(\frac{q_c}{\sigma'_{z0}}\right)\right] \quad (2.39)$$

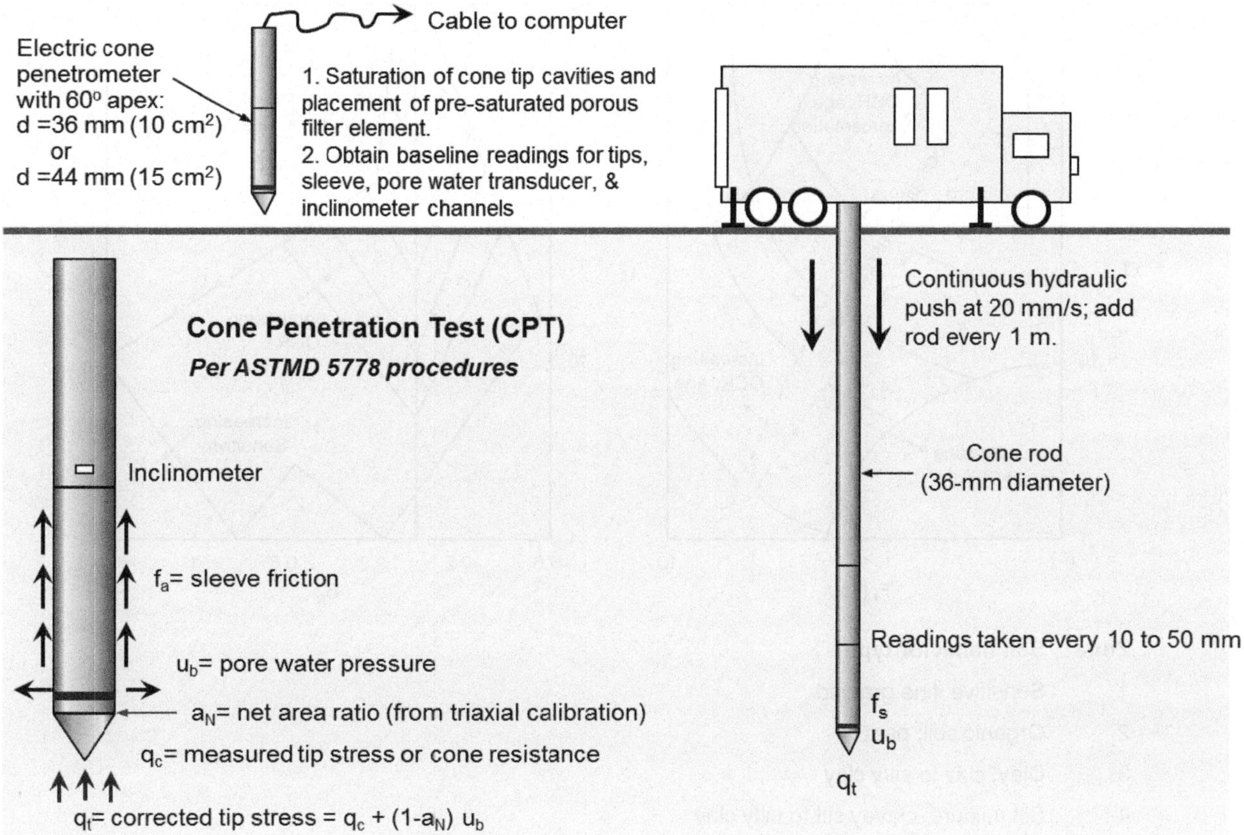

Figure 2.47 Cone penetration test (Mayne et al., 2001).

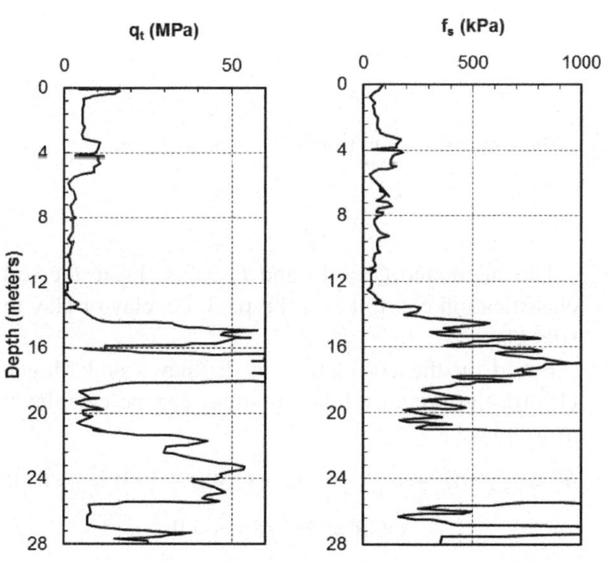

Figure 2.48 CPT profiles (Courtesy of Paul Mayne).

CPT tip resistance can also be used to estimate the undrained shear strength of clay as follows:

$$c_u = \frac{q_c - \sigma_{z0}}{N_k} \tag{2.40}$$

where σ_{z0} = total overburden stress
N_k = cone factor.

Salgado (2006) suggested $N_k = 10 - 12$.

Kulhawy and Mayne (1990) proposed the following correlation to estimate the preconsolidation stress of clay:

$$p_c = 0.33(q_c - \sigma_{z0}) \tag{2.41}$$

The moduli of soils can be estimated based on the CPT tip resistance provided in Table 2.14.

Table 2.14 Soil Elastic Modulus Estimated by CPT Tip Resistance

Soil Type	E_s
Sand (normally consolidated)	$(2–4)q_c$
Sand (overconsolidated)	$(6–30)q_c$
Clayey sand	$(3–6)q_c$
Silty sand	$(1–2)q_c$
Soft clay	$(3–8)q_c$

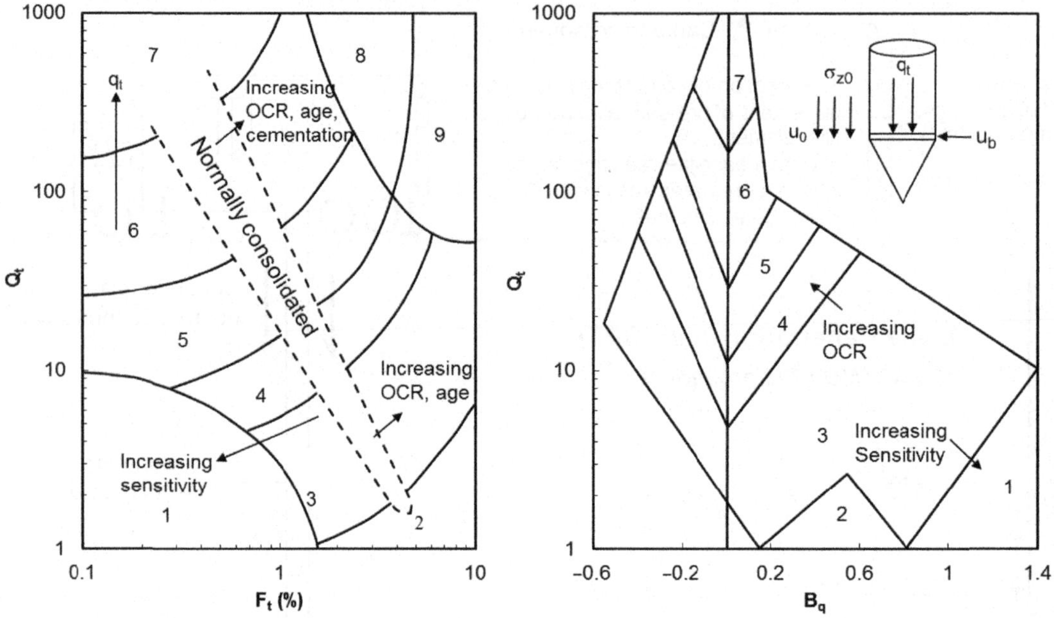

Zone	Soil behavior type
1	Sensitive, fine grained
2	Organic soil; peat
3	Clay; clay to silty clay
4	Silt mixture; clayey silt to silty clay
5	Sand mixture; silty sand to sandy silt
6	Sand; clean sands to silty sand
7	Gravelly sand to sand
8	Very stiff sand to clayey sand
9	Very stiff fine grained

Figure 2.49 CPT soil type classification (after Robertson, 1990, with permission from *Canadian Geotechnical Journal*).

Example 2.12

A CPT is conducted on a site with a groundwater table near the ground surface. The measured tip resistance and sleeve friction at the depth of 10.7 m are 1.5 MPa and 65 kPa, respectively. The unit weight of the soil is 18.1 kN/m^3. Classify this soil and determine the OCR value and the undrained shear strength.

Solution

At the depth of 10.7 m, the total and effective overburden stresses are

$$\sigma_{z0} = 10.7 \times 18.1 = 193\,\text{kPa}$$

$$\sigma'_{z0} = 10.7 \times (18.1 - 9.81) = 88\ \text{kPa}$$

The calculated $Q_t = 15$ and $F_t = 5\%$. From the soil classification chart, it is soil type 3, i.e. clay or clay to silty clay.

Based on the correlation by Kulhawy and Mayne (1990), the preconsolidation stress can be calculated as follows:

$$\sigma'_p = 0.33\,(q_c - \sigma_{z0}) = 0.33 \times (1500 - 193) = 431\ \text{kPa}$$

$$\text{OCR} = 431/88 = 4.9$$

Therefore, it is an overconsolidated soil.

The undrained shear strength of the clay can be calculated using the following formula (use $N_k = 11$):

$$c_u = \frac{q_c - \sigma_{z0}}{N_k} = \frac{1500 - 193}{11} = 119\,\text{kPa}$$

2.4.3 Vane Shear Test

Introduction Vane shear test is the most widely used method for the in situ determination of the undrained shear strengths of soft clays. This method was originally used in Sweden in 1919 and has been extensively employed worldwide since the late 1940s. Vane shear test is a good in situ test to directly determine undrained shear strength of clay, especially soft clay. ASTM standard D2573 is available for this method. The basic test procedure includes:

- Insert a metal vane into the soil.
- Rotate the vane at a rate of 6° per minute until the soil fails in shear.
- Record the maximum torque during the rotation.

Figure 2.50 shows a vane used in a field test.

Measured Value For a flat vane, the undrained shear strength of the soil can be calculated as follows:

$$c_u = \frac{2T_f}{\pi d_v^2 (L_v + 0.33 d_v)} \tag{2.42}$$

where T_f = torque at soil failure
L_v = vane length
d_v = vane diameter.

If $L_v/d_v = 2$, the above equation can be simplified into

$$c_u = \frac{0.86 T_f}{\pi d_v^3} \tag{2.43}$$

For a tapered vane, the undrained shear strength of the soil can be calculated as follows (Bowles, 1996):

$$c_u = \frac{0.3183 T_f}{1.354 d_v^3 + 0.354(d_1 d_v^2 - d_v d_1^2) + 0.2707 d_1^3} \tag{2.44}$$

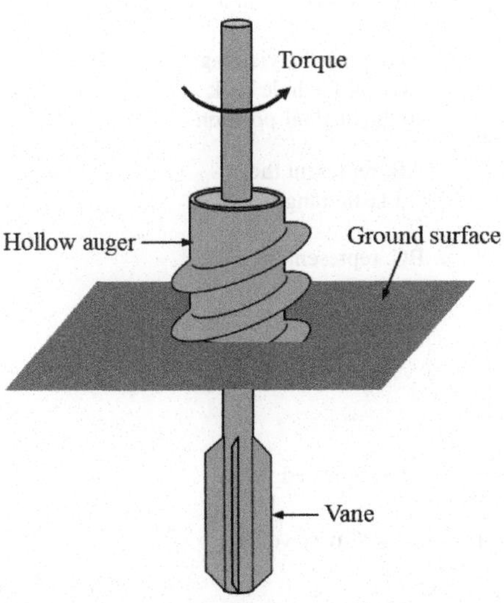

Figure 2.50 Vane shear test.

where d_1 is the diameter of the rod (typically 12 –22 mm).

Bjerrum (1972) found that the vane shear tests overestimated the undrained shear strengths of clays as compared with those back-calculated from failed embankments. Therefore, Bjerrum (1972) suggested the following correction:

$$c_{uc} = \lambda_v c_u \tag{2.45}$$

where c_{uc} = corrected undrained shear strength
c_u = measured undrained shear strength
λ_v = correction factor, which can be estimated by the following correlation with plasticity index of soil:

$$\lambda_v = 1.18 - 0.0107\,\text{PI} + 0.0000513\,\text{PI}^2 \le 1 \tag{2.46}$$

It is unique for a vane shear test that it can be used to determine the soil sensitivity by running the test in undisturbed and remolded soils:

$$S_t = \frac{c_u}{c_{um}} \tag{2.47}$$

where c_u = undrained shear strength of the undisturbed soil
c_{um} = undrained shear strength of the remolded soil

Example 2.13

A vane shear test is performed at a depth of 5.0 m in a soft clay stratum (its plasticity index PI = 50). The measured torque at soil failure is 9 N·m. The flat vane is 60 mm in diameter and 120 mm long. Compute the corrected undrained shear strength of the clay.

Solution

The measured undrained shear strength of the clay can be calculated as follows (for $L_v/d_v = 2$):

$$c_u = 0.86 \frac{T_f}{\pi d_v^3}$$

At a depth of 5.0 m, $T_f = 9 \cdot \text{N} \cdot \text{m}$. The undrained shear strength is

$$c_u = 0.86 \frac{T_f}{\pi d_v^3} = 0.86 \frac{9}{3.14 \times (0.06)^3}$$

$$= 11400\,\text{N/m}^2 = 11.4\,\text{kPa}$$

For PI = 50, the correction factor is $\lambda = 1.18 - 0.0107\,\text{PI} + 0.0000513\,\text{PI}^2 = 1.18 - 0.0107 \times 50 + 0.0000513 \times 50^2 = 0.77$. Therefore, the corrected undrained shear strength is

$$c_{uc} = \lambda_v c_u = 0.77 \times 11.4 = 8.8\ \text{kPa}$$

2.4.4 Pressuremeter Test

Introduction The pressuremeter test was first conceived, designed, constructed, and used by Menard of France in 1955. This method is adopted as ASTM standard D4719. Figure 2.51 shows the illustration of a pressuremeter test in field. The pressuremeter includes one measuring cell and two guard cells. The measuring cell is filled with water while the guard cells are filled with gas. This is a theoretically sound test method in determination of soil parameters. It tests a larger zone of soil mass than other in situ tests. From the test results, a complete stress–strain curve can be developed. However, this test method has a complicated test procedure and requires a high level of expertise in the field. It is time consuming and expensive. Also, the pressuremeter is delicate and easily damaged.

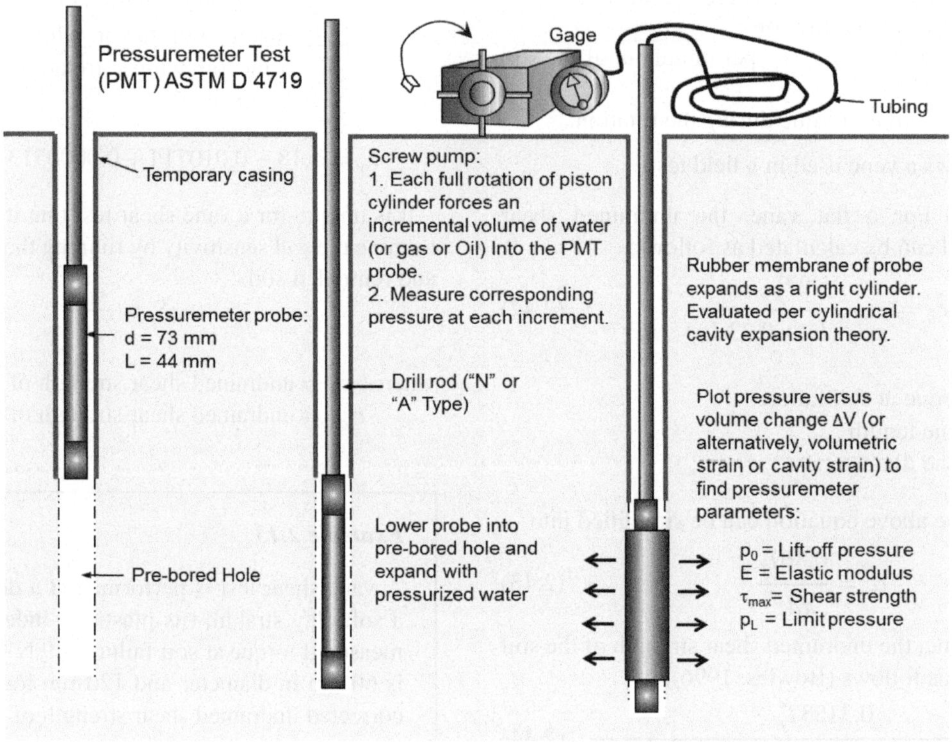

Figure 2.51 Pressuremeter test (Mayne et al., 2001).

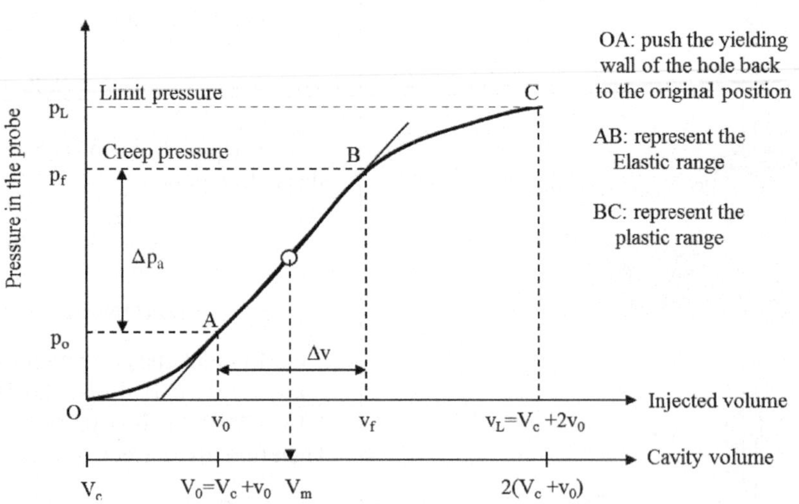

Figure 2.52 Typical pressuremeter curve.

The basic procedure of the pressuremeter test is

- Drill a hole.
- Insert a cylindrical probe.
- Inflate the probe with pressure.
- Measure the pressure and volume change.

Measured Parameters Pressuremeter test is usually carried out at 1-m intervals. The typical pressuremeter result is shown in Figure 2.52. From this curve, the mean volume, V_m, the slope, $\Delta p_a/\Delta v$, and the limit pressure, p_L, can be determined. The mean volume and the slope can be used to calculate the geomaterial pressuremeter modulus, E_{PMT}, as follows:

$$E_{\mathrm{PMT}} = 2(1+v)V_m\frac{\Delta p_a}{\Delta v} \quad (2.48)$$

where v is the geomaterial Poisson ratio (typically 0.33).

Example 2.14

A pressuremeter test is carried out at a site at a depth of 7 m below the ground surface. The water table level is at a depth of 1.5 m. The average unit weight of saturated soil is 17.3 kN/m^3. The deflated volume of the probe is V_c = 53,5000 mm^3. The linear portion of the curve starts at p_0 = 105 kPa and v_0 = 160,000 mm^3 and ends at p_f = 530 kPa and v_f = 200,000 mm^3. Determine the soil pressuremeter modulus, E_{PMT}.

Solution

Since $V_c = 535\,\mathrm{cm}^3$, $v_0 = 160\,\mathrm{cm}^3$, and $v_f = 200\,\mathrm{cm}^3$, the mean volume is

$$V_m = V_c + \frac{v_0 + v_f}{2} = 535{,}000 + (160000 + 200000)/2$$

$$= 715000\,\mathrm{mm}^3 = 7.15 \times 10^{-4}\mathrm{m}^3$$

$$\frac{\Delta p_a}{\Delta v} = \frac{530 - 105}{200{,}000 - 160{,}000} = 0.0106\ \mathrm{kPa/mm}^3$$

$$= 1.06 \times 10^7\,\mathrm{kPa/m}^3$$

Hence,

$$E_{\mathrm{PMT}} = 2.66V_m\frac{\Delta p_a}{\Delta v} = 2.66 \times 7.15 \times 10^{-4} \times 1.06 \times 10^7$$

$$= 20{,}200\,\mathrm{kPa} = 20.2\,\mathrm{MPa}$$

2.4.5 Plate Load Test

Introduction Plate load test is the test method that closely simulates the response of a foundation on soil under loading. This test method can determine the ultimate bearing capacity and elastic modulus of the foundation soil. A test can be performed using a force or displacement rate control method. The force control method is more commonly used in the field. In this method, the load is applied in increments. After each load is applied, it is maintained for a certain time until the preset criteria are met. During this time period, the settlement of the plate is measured. The accumulated settlement from the beginning of loading is plotted against the corresponding applied pressure as shown in Figure 2.53.

Measured Parameters The ultimate bearing capacity is often selected as the applied pressure before plunging or a preset settlement (typically 10% plate width or diameter) reached. It can also be used to calculate the elastic modulus of the foundation soil as follows:

$$E_s = 0.79(1 - v^2)d_f\frac{p}{S} \quad (2.49)$$

where v = Poisson's ratio of soil
d_f = diameter of a circular rigid plate
p = applied bearing pressure
S = settlement

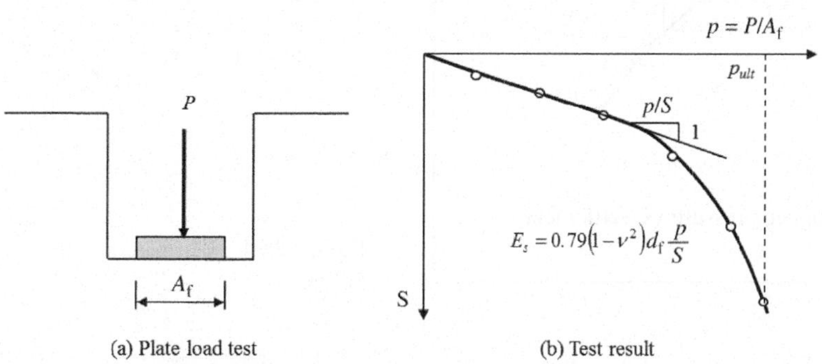

Figure 2.53 (a) Plate load test and (b) test result.

Example 2.15

A plate load test with 1-m diameter circular steel plate is performed on a natural ground. The measured pressures and settlements are provided in Example Table 2.3. Determine the ultimate bearing capacity and the elastic modulus of the foundation soil.

Example Table 2.3

Pressure (kPa)	0	25	50	75	100	125	150	175	200
Settlement (mm)	0	10	20	33	47	64	83	110	155

Solution

Plot the pressure versus settlement curve in Example Figure 2.3. There is a significant increase in settlement from the applied pressure from 175 to 200 kPa; and 200 kPa can be considered as the plunging point; therefore, the ultimate bearing capacity is 175 kPa. Alternatively, based on the criterion for the settlement-to-plate diameter ratio at 10%, the ultimate bearing capacity is 167 kPa. They are close.

The slope of the linear line (pick up any point on the line for calculation) is

$$\frac{P}{S} = \frac{100}{45} = 2.22 \text{ kN/mm} = 2220 \text{ kN/m}$$

Assume the Poisson ratio of the soil is 0.3. The elastic modulus can be calculated:

$$E_s = 0.79(1 - \nu^2)d_f\frac{P}{S} = 0.79 \times (1 - 0.3^2) \times 1.0 \times 2220$$

$$= 1.60 \text{ MPa}$$

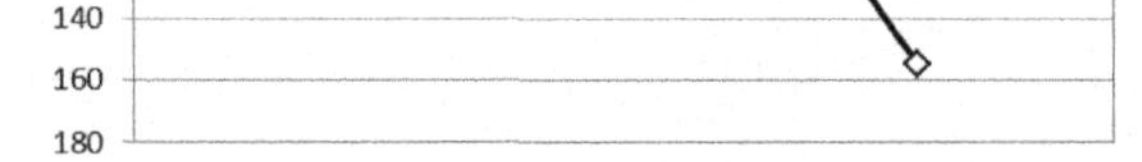

Example Figure 2.3 Applied pressure vs. settlement.

2.5 SHALLOW FOUNDATION DESIGN

2.5.1 Bearing Capacity

When a footing on top of the ground or embedded in the ground is subjected to a load from a superstructure, it transfers the load to the soil underneath the footing. The soil under the load deforms, the soil wedge forms, and shear stresses develop within the soil as shown in Figure 2.54. With an increase of the applied load, the shear stresses continue increasing. When the shear stress is equal to the soil shear strength, the soil yields and slip surfaces develop. When the slip surfaces fully develop, the soil is no longer able to carry any additional load; therefore, the footing collapses. The maximum load the footing is able to carry prior to collapse is often referred to the ultimate bearing capacity. Vesic (1973) found that the possible failure mode in sand, including local, punching, and general shear failure, depends on the relative density of sand and the relative depth. General shear failure is the common failure mode for a shallow foundation.

Several ultimate bearing capacity formulas have been developed for shallow foundations in the past. The commonly used ones include Terzaghi (1943), Meyerhof (1953), Hansen (1970), and Vesic (1973). The Meyerhof bearing capacity formula is mostly used in this book; therefore, it is reviewed here.

The ultimate bearing capacity of a shallow foundation for a general failure can be expressed as

$$q_{\text{ult}} = cN_c s_c d_c + 0.5\gamma' B_f N_\gamma s_\gamma d_\gamma + \sigma'_D N_q s_q d_q \tag{2.50}$$

where B_f = footing width
c = soil cohesion
γ' = soil effective unit weight at the base of the foundation
σ'_D = soil effective overburden stress at the base of the footing

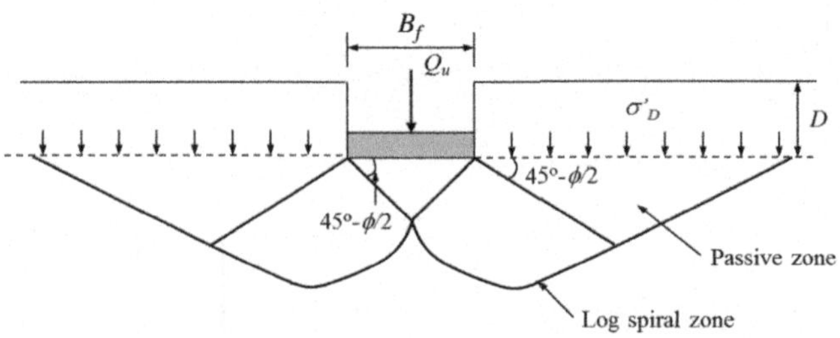

Figure 2.54 Ultimate bearing capacity considering general shear failure.

N_c, N_γ, N_q = bearing capacity factors, which can be calculated as follows:

$$N_q = \tan^2\left(45^\circ + \frac{\phi}{2}\right) e^{\pi \tan \phi} \quad (2.51a)$$

$$N_c = \frac{N_q - 1}{\tan \phi} \text{ if } \phi > 0; N_c = 5.14 \text{ if } \phi = 0 \quad (2.51b)$$

$$N_\gamma = (N_q - 1)\tan(1.4\phi) \quad (2.51c)$$

s_c, s_γ, s_q = shape factors
d_c, d_γ, d_q = depth factors

The shape and depth factors are provided in Table 2.15.

When the saturated, undrained soft soil under the continuous footing has a limited thickness (e.g., underlain by bedrock or a firm soil stratum), the bearing capacity factor, N_c, can be estimated using Figure 2.55.

Table 2.15 Shape Factors and Depth Factors for Bearing Capacity Calculations

Friction Angle	Formula[a]
$\phi = 0$	$s_c = 1 + 0.2(B_f/L_f)$
	$s_q = s_\gamma = 1$
$\phi = 10^\circ$	$s_c = 1 + 0.2K_p(B_f/L_f)$
	$s_q = s_\gamma = 1 + 0.1K_p(B_f/L_f)$
$\phi = 0$	$d_c = 1 + 0.2(D_f/B_f)$
	$d_q = d_\gamma = 1$
$\phi \geq 10^\circ$	$d_c = 1 + 0.2\sqrt{K_p}(D_f/B_f)$
	$d_q = d_\gamma = 1 + 0.1\sqrt{K_p}(D_f/B_f)$

[a] L_f = length of footing and $K_p = \tan^2(45^\circ + \phi/2)$.

Example 2.16

A square concrete footing with 1 m × 1 m is embedded in clay at a depth of 1 m. The undrained shear strength of the clay is 25 kPa. The groundwater table is at the depth of 0.6 m from the ground surface. The unit weights of the clay above and below the groundwater table are 18 and 20 kN/m^3, respectively. Calculate the ultimate bearing capacity of this footing on the clay under an undrained condition.

Solution

The overburden stress at the base of the footing is

$$\sigma'_D = 0.6 \times 18 + 0.4 \times (20 - 9.81) = 14.9 \text{ kPa}$$

Under an undrained condition (i.e., $c_u \neq 0$ and ϕ_u = 0), $N_c = 5.14, N_q = 1.0$, and $N_\gamma = 0$:

$$q_{\text{ult}} = cN_c s_c d_c + 0.5\gamma' B_f N_\gamma s_\gamma d_\gamma + \sigma'_D N_q s_q d_q$$
$$= 5.14 c_u s_c d_c + \sigma'_D s_q d_q$$

$$s_c = 1 + 0.2(B_f/L_f) = 1 + 0.2 \times (1/1) = 1.2$$
$$d_c = 1 + 0.2(D_f/B_f) = 1 + 0.2 \times (1/1) = 1.2$$
$$s_q = 1$$
$$d_q = 1$$

The ultimate bearing capacity under the undrained condition is

$$q_{\text{ult}} = 5.14 c_u s_c d_c + \sigma'_D s_q d_q$$
$$= 5.14 \times 25 \times 1.2 \times 1.2 + 14.9 \times 1 \times 1 = 200 \text{kPa}$$

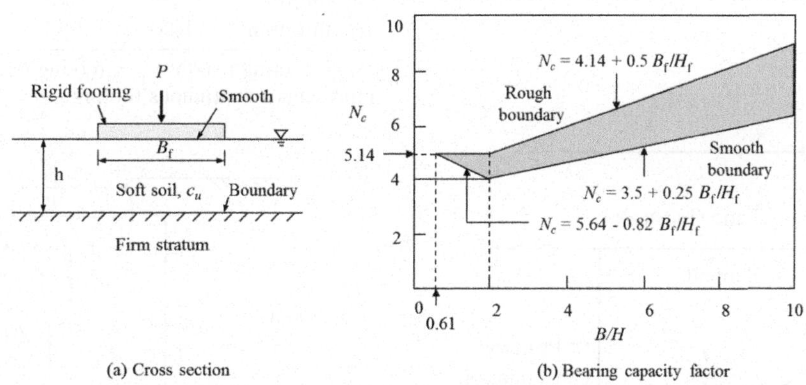

Figure 2.55 Bearing capacity factor considering limited thickness of soft soil (modified from Bonaparte et al., 1978).

2.5.2 Settlement

Foundations settle under vertical static loads. The total settlement of a foundation includes three components as follows:

$$S_t = S_i + S_c + S_s \tag{2.52}$$

where S_t = total settlement
S_i = immediate settlement
S_c = primary consolidation settlement
S_s = secondary settlement or compression

Figure 2.56 shows the settlement–time relationship under a constant load.

Immediate settlement happens under an instantaneous load; therefore, it completes during construction. The applied load generates excess pore water pressure in geomaterial. Primary consolidation settlement develops as excess pore water pressure dissipates with time. Depending on geomaterial type, consolidation settlement can occur quickly for geomaterials with high permeability (such as cohesionless geomaterials) but slowly for geomaterials with low permeability (such as cohesive geomaterials). Consolidation of cohesionless geomaterials typically can finish during construction while consolidation of cohesive geomaterials can take years to complete. The complete dissipation of excess pore water pressure is the end of the consolidation, t_p. Secondary compression, also referred to as creep deformation, is mostly attributed to the change of soil fabric. Mitchell and Soga (2005) suggest that "the mechanism of secondary compression involves sliding at interparticle contacts, expulsion of water from microfabric elements, and rearrangement of adsorbed water molecules and cations into different positions."

Different methods are available to calculate settlement, including the (1) elastic solution, (2) consolidation test-based method, (3) empirical method, (4) numerical method, and (5) observational method. The elastic solution, the method based on consolidation tests, and the empirical method are commonly used in practice and therefore are reviewed here.

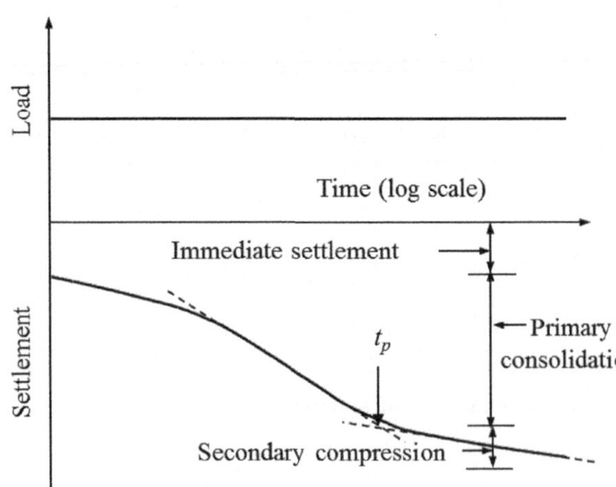

Figure 2.56 Settlement vs. time.

Elastic Solution The basic elastic solution is developed based on the elastic theory considering a footing on a half-space uniform medium as follows:

$$S_e = pB_f\frac{1-\nu_s^2}{E_s}I \tag{2.53}$$

where S_e = elastic settlement
p = applied pressure
B_f = foundation width or diameter (= d_f)
E_s = geomaterial elastic modulus
ν_s = geomaterial Poisson's ratio
I = influence factor, which is provided in Table 2.16

Mayne and Poulos (1999) and Elhakim (2005) modified the preceding equation by considering the flexibility and embedment depth of a circular footing and a Gibson soil as shown in Figure 2.57:

$$S_{\text{center}} = \frac{p_n(1-\nu_s^2)\, d_f I_G I_F I_E}{E_0} \tag{2.54}$$

Table 2.16 Influence Factor for Settlement Calculation

Shape	L_f/B_f[a]	Center	Corner	Average (rigid)
Circular		1.00	0.64	0.79
Square	1.0	1.12	0.56	0.88
Rectangular	2.0	1.53	0.77	1.21
Rectangular	10.0	2.54	1.27	2.10

[a] L_f = footing length; B_f = footing width or diameter; L_f/B_f = 10 represents a continuous footing.

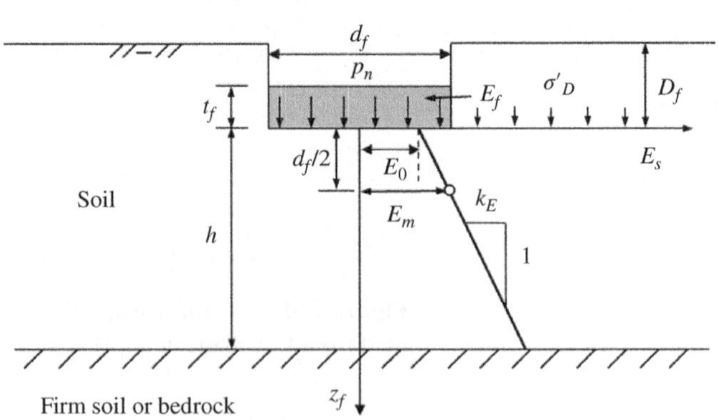

Figure 2.57 Shallow foundation and soil profile.

where S_{center} = settlement at the center of the foundation
p_n = net applied pressure (i.e., $p_n = p - \sigma'_D$)
σ'_D = effective overburden stress at the base of the footing
d_f = diameter of the footing
E_0 = soil elastic modulus at the base of the footing
I_G, I_F, I_E = influences factors considering the Gibson soil, embedment depth, and foundation flexibility can be determined as follows:

$$I_G \approx \frac{1}{(1 + 0.23d_f/h) + 1.6[E_0/(k_E d_f)]^{-0.8}} \tag{2.55a}$$

$$I_E \approx 1 - \frac{1}{3.5[(d_f/D_f) + 1.6]\exp(1.22v_s - 0.4)} \tag{2.55b}$$

$$I_F \approx \frac{\pi}{4} + \frac{1}{4.6 + 10K_f} \tag{2.55c}$$

$$K_f = (1 - v_s^2)\frac{E_f}{E_m}\left(\frac{2t_f}{d_f}\right)^3 \tag{2.56}$$

where D_f = embedment depth of the footing
t_f = thickness of the footing
K_f = flexibility factor ($K_f > 10$, perfectly rigid; $0.01 < K_f < 10$, intermediately rigid; and $K_f < 0.01$, perfectly flexible)
k_E = slope of soil modulus increase
E_f = modulus of the footing.
E_m = representative soil elastic modulus located beneath the footing base (i.e., the soil modulus at the depth equal to the radius of the footing)

The settlement at the edge of the circular footing can be estimated as follows:

$$\frac{S_{edge}}{S_{center}} \approx 1 - \frac{1.533}{4.6 + 10K_f} \tag{2.57}$$

The settlement at the corner of a square or rectangular foundation can be estimated as follows (an equivalent diameter of the foundation should be first obtained based on the equal area concept):

$$\frac{S_{corner}}{S_{center}} \approx 1 - \frac{2.3}{4.6 + 10K_f} \tag{2.58}$$

In the basic elastic solution and the modified solutions by Mayne and Poulos (1999) and Elhakim (2005), the elastic modulus of the soil can be an undrained or drained modulus. When an undrained modulus is used, the calculated settlement is the immediate settlement. However, when a drained modulus is used, the calculated settlement is the consolidation settlement.

Consolidation Test-based Method The primary consolidation settlement can be calculated based on consolidation test results. Figure 2.58 shows the $e - \log \sigma'$ plots. When the total applied stress, σ'_z is lower than the preconsolidation stress, p_c (i.e., the maximum stress the soil has experienced), the following equation can be used to calculate the primary consolidation settlement:

$$S_c = \Sigma \frac{C_r \Delta h}{1 + e_0} \log \frac{\sigma'_{z0} + \Delta\sigma_z}{\sigma'_{z0}} \tag{2.59}$$

where C_r = recompression index
Δh = thickness of geomaterial sublayer
e_0 = geomaterial initial void ratio
σ'_{z0} = overburden stress at midpoint of geomaterial sublayer
$\Delta\sigma_z$ = additional vertical stress at midpoint of geomaterial sublayer

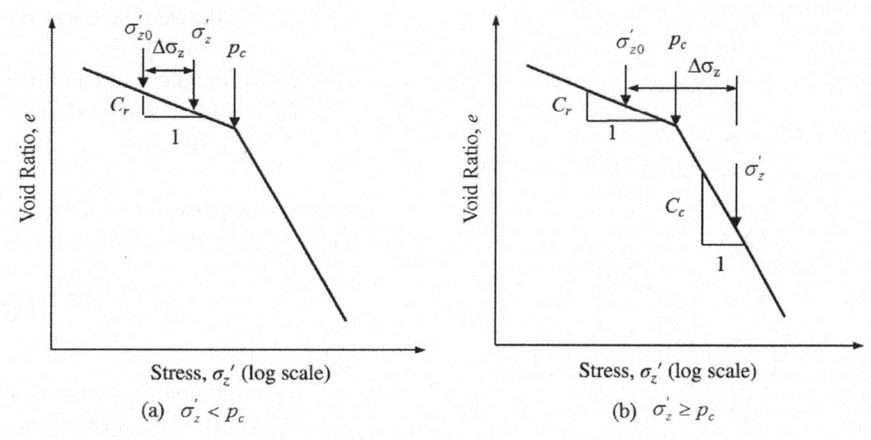

Figure 2.58 e–log σ′ plot.

When the total applied stress, σ'_z, is equal or higher than the preconsolidation stress, p_c, the following equation can be used to calculate the primary consolidation settlement:

$$S_c = \Sigma \left(\frac{C_r \Delta h}{1+e_0} \log \frac{p_c}{\sigma'_{z0}} + \frac{C_c \Delta h}{1+e_0} \log \frac{\sigma'_{z0} + \Delta\sigma_z}{p_c} \right) \quad (2.60)$$

where C_c is the geomaterial compression index.

The preceding equation can be used for both normally consolidated soil ($p_c = \sigma'_{z0}$) and overconsolidated soil when $\sigma'_z \geq p_c$.

The additional vertical stress at the midpoint of a geomaterial sublayer can be calculated by the Boussinesq solution or estimated by the following equation (Coduto, 2001):

$$\Delta\sigma_z = \left\{ 1 - \left[\frac{1}{1 + (0.5B_f/z_f)^{1.38+0.62B_f/L_f}} \right]^{2.60-0.84B_f/L_f} \right\} p_n \quad (2.61a)$$

where L_f and B_f is the length and width of the footing. For a continuous foundation, $B/L = 0$.

In practice, alternatively, the distributed vertical stress along the axis of a rectangular foundation can be estimated using a stress distribution angle method as shown in Figure 2.59 and expressed as follows:

$$\Delta\sigma = \frac{p_n L_f B_f}{(B_f + 2z_f \tan\theta)(L_f + 2z_f \tan\theta)} \quad (2.61b)$$

where θ = stress distribution angle.

Salgado (2008) found that when $\tan\theta = 0.5$ (i.e., a 2[vertical]:1 [horizontal] distribution angle), the stress distribution angle method calculates the distributed vertical stress along the axis close to that by the elastic theory. The method assuming 2:1 distribution angle is often referred to as the 2:1 distribution method in practice.

There are many empirical relationships between C_c or C_r and other soil parameters developed by different researchers. Typically, $C_c = 0.1 - 0.8$ and $C_r = C_c/5 - C_c/10$.

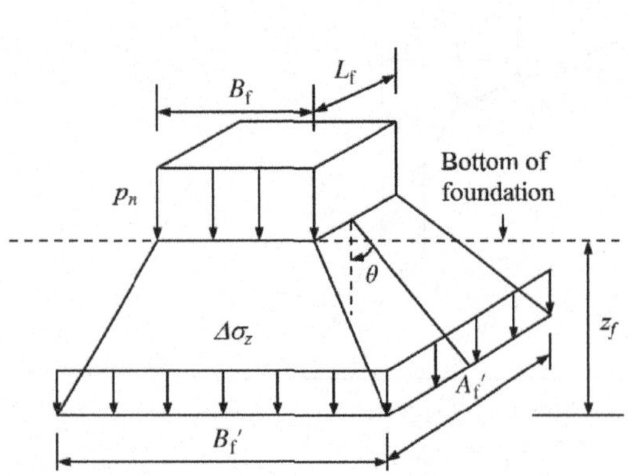

Figure 2.59 Stress distribution in a soil medium.

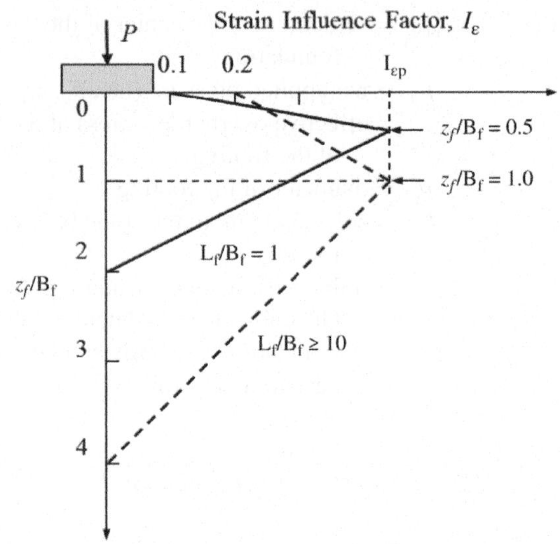

Figure 2.60 Strain influence factor, I_ε, with depth.

Empirical Method The Schmertmann et al. (1978) method, which was developed for shallow foundations with sandy soils based on the strain influence factor concept as shown in Figure 2.60, has been commonly used in practice. The settelement of a shallow foundation can be calculated as follows:

$$S_e = C_1 C_2 C_3 p_n \sum \frac{I_\varepsilon}{E_s} \Delta h \quad (2.62)$$

where C_1 = correction factor for embedment depth of the foundation, i.e., $C_1 = 1 - 0.5(\sigma'_D/p_n)$
C_2 = correction factor for soil creep, i.e., $C_2 = 1 + 0.2\log(t/0.1)$, t is time after the load is applied (years) (t = 50 years has been typically used)
C_3 = correction factor for foundation shape, i.e., $C_3 = 1.03 - 0.03L_f/B_f \geq 0.73$
$I_{\varepsilon p} = 0.5 + 0.1\sqrt{(p_n/\sigma'_{zp})}$
I_ε = influence factor at midpoint of each soil sublayer
Δh = thickness of each soil sublayer
E_s = soil elastic modulus at midpoint of each sublayer

Secondary Compression The secondary compression of a foundation can be estimated using the following formula:

$$S_s = \Sigma \frac{C_\alpha \Delta h}{1+e_0} \log \frac{t}{t_p} \quad (2.63)$$

where C_α = geomaterial secondary compression index
t_p = time at end of primary consolidation
t = time after the pressure is applied ($t > t_p$).

Mesri and Godlewski (1977) investigated a variety of soils for the secondary compression indices and correlated C_α with C_c. They suggested the average C_α/C_c values for inorganic clays and silts, organic clays and silts, and peats are 0.04 ± 0.01, 0.05 ± 0.01, and 0.075 ± 0.01, respectively. Mesri et al. (1990) found that the C_α/C_c values for clean sand range from 0.015 to 0.03.

Example 2.17

A circular concrete mat foundation with a diameter of 12.66 m and a thickness of 1.22 m is to be constructed on uniform silty sand with an elastic modulus of 19.5 MPa, a Poisson ratio of 0.15, and a unit weight of 18 kN/m^3. Groundwater table and bedrock are at great depths. This mat foundation embedded at 1.22 m carries a vertical load of 14.29 MN. Concrete modulus is 35.6 GPa and concrete unit weight is 23.6 kN/m^3. Calculate the settlement of this foundation.

Solution

The area of the mat foundation is

$$A_f = \pi r_f^2 = 3.14 \times \left(\frac{12.66}{2}\right)^2 = 125.9\,\text{m}^2$$

The weight of the mat foundation is

$$W_f = A_f t_f \gamma_f = 125.9 \times 1.22 \times 23.6$$
$$= 3624.9\,\text{kN} = 3.62\,\text{MN}$$

The applied bearing pressure is

$$p = \frac{P + W_f}{A_f} = \frac{14.29 + 3.62}{125.9} = 0.142\,\text{MPa} = 142\,\text{kPa}$$

The net applied pressure is

$$p_n = p - \sigma'_{zD} = 142 - 1.22 \times 18 = 120.4\,\text{kPa}$$

The foundation flexibility factor is

$$K_f = \left(\frac{E_f}{E_m}\right)\left(\frac{t_f}{r_f}\right)^3 (1 - v_s{}^2) = \left(\frac{35{,}600}{19.5}\right)$$
$$\cdot \left(\frac{1.22}{6.33}\right)^3 (1 - 0.15^2) = 12.6 > 10$$

therefore, the mat foundation is rigid.

The settlement of a mat foundation on a half-space medium can be calculated as follows:

$$S = \frac{p_n(1 - v_s{}^2) d_f I_G I_F I_E}{E_0}$$

Since it is a uniform soil, $I_G = 1.0$.

For a rigid foundation, $I_F = 0.793$.

The influence factor considering embedment depth is

$$I_E = 1 - \frac{1}{3.5 \exp(1.22 v_s - 0.4)\,[(d_f/D_f) + 1.6]}$$
$$= 1 - \frac{1}{3.5 \exp(1.22 \times 0.15 - 0.4)[(12.66/1.22) + 1.66]}$$
$$= 0.97$$

Therefore, the calculated settlement is

$$S = \frac{p_n(1 - v_s{}^2) d_f I_G I_F I_E}{E_0}$$
$$= \frac{120.4 \times (1 - 0.15^2) \times 12.66 \times 1 \times 0.793 \times 0.97}{19{,}500}$$
$$= 0.059\,\text{m}$$

Example 2.18

A project site consists of uniform normally consolidated sand. The groundwater table is at a depth of 2.0 m below the ground surface. The elastic modulus of the sand is 10 MPa. The unit weights of the sand above and below the groundwater table are 17 and 20 kN/m^3, respectively. The unit weight of concrete is 23.6 kN/m^3. A 625-kN load is to be supported on a 2.5 m × 2.5 m footing to be founded at a depth of 2.0 m in this soil. Use Schmertmann's method to compute the settlement of this footing soon after construction and the settlement at 50 years after construction.

Solution

The applied pressure at the base of the footing is

$$p = \frac{P + W_f}{A} = \frac{625 + (2.5 \times 2.5) \times 2 \times 23.6}{2.5 \times 2.5}$$
$$= 147.2\,\text{kPa}$$

The effective overburden stress at the depth of the embedment is

$$\sigma'_{zD} = \gamma' D_f = 17 \times 2 = 34\,\text{kPa}$$

The net applied pressure is

$$p_n = p - \sigma'_{zD} = 147.2 - 34 = 113.2\,\text{kPa}$$

The effective overburden stress at the depth of the peak value of the strain influence factor is

$$\sigma'_{zp} = \gamma z_w + (\gamma_{\text{sat}} - \gamma_{\text{w}})(D_f + 0.5B_f - z_w)$$
$$= 17 \times 2 + (20 - 9.8) \times (2 + 0.5 \times 2.5 - 2)$$
$$= 46.8\,\text{kPa}$$

The peak value of the strain influence factor is

$$I_{\varepsilon p} = 0.5 + 0.1\sqrt{\frac{p_n}{\sigma'_{zp}}} = 0.5 + 0.1\sqrt{\frac{113.2}{46.8}} = 0.66$$

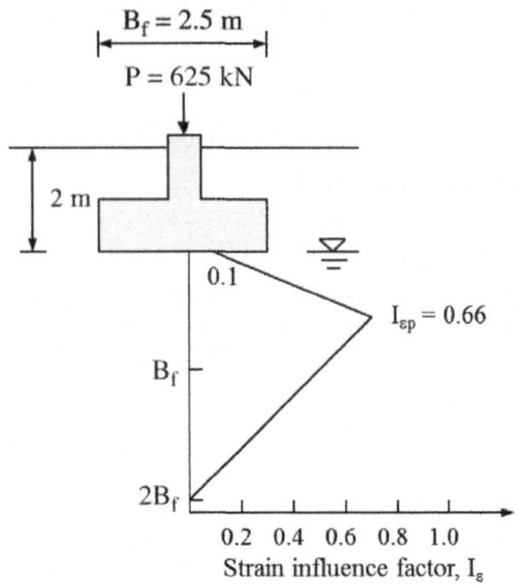

Example Figure 2.4

Example Table 2.4

Layer No.	Depth (m)	$E_s(kPa)$	$z_f(m)$	I_ε	$\Delta h(m)$	$I_\varepsilon \Delta h/E_s$
1	2 to 3	10000	0.5	0.343	1	0.000034
2	3 to 5	10000	2.0	0.514	2	0.000103
3	5 to 6	10000	3.5	0.257	1	0.000026
4	6 to 7	10000	4.5	0.086	1	0.000009
						$\Sigma I_\varepsilon \Delta h/E_s = 0.000171\ \mathrm{m^3/kN}$

$$C_1 = 1 - 0.5\left(\frac{\sigma'_{zD}}{p_n}\right) = 1 - 0.5 \times \left(\frac{34}{113.2}\right) = 0.85$$

$$C_3 = 1.03 - 0.03\frac{L_f}{B_f} = 1.03 - 0.03 \times \left(\frac{2.5}{2.5}\right)$$

$$= 1.0 > 0.73$$

At $t = 0.1$ yr : $C_2 = 1.$

$$S = C_1 C_2 C_3 p_n \Sigma\frac{I_\varepsilon \Delta h}{E_s} = 0.85 \times 1 \times 1 \times 113.2 \times 0.000171$$

$$= 0.016\,\mathrm{m} = 16\,\mathrm{mm}$$

At $t = 50$ yrs : $C_2 = 1 + 0.2\log\left(\frac{t}{0.1}\right)$

$$= 1 + 0.2\log\left(\frac{50}{0.1}\right) = 1.54$$

$$S = C_1 C_2 C_3 p_n \Sigma\frac{I_\varepsilon \Delta h}{E_s}$$

$$= 0.85 \times 1.54 \times 1 \times 113.2 \times 0.000171$$

$$= 0.025\,\mathrm{m} = 25\,\mathrm{mm}$$

2.5.3 Consolidation

Consolidation is dissipation of excess pore water pressure in geomaterial that is generated by an applied load. Figure 2.61 shows a clay layer underlain by a sand layer subjected to a uniform instantaneous pressure, $\Delta\sigma_z$. The excess pore water pressure at the moment of loading, that is, u_0 is equal to $\Delta\sigma_z$. Since the sand layer has high permeability and it can serve as a drainage layer, the water in the clay layer can drain out vertically upward and downward with time (i.e., one-dimensional consolidation problem). This drainage process can be described by a differential equation as follows:

$$\frac{\partial u}{\partial t} = c_v \frac{\partial^2 u}{\partial z^2} \tag{2.64}$$

where u = excess pore water pressure in soil at depth, z and time, t

c_v = coefficient of vertical consolidation of soil

Considering initial and boundary conditions, the above equation was solved by Terzaghi (1943):

$$u = \sum_{m=0}^{\infty} \frac{2\Delta\sigma_z}{M} \sin\left(\frac{Mz}{h_{dr}}\right) e^{-M^2 T_v} \tag{2.65}$$

where

$$M = \frac{(2m+1)\pi}{2} \qquad T_v = \frac{c_v t}{h_{dr}^2}$$

and h_{dr} = longest drainage distance due to vertical flow (if top and bottom drainage surfaces exist, half of the soil thickness between these two surfaces should be used).

To quantify the rate of excess pore water pressure dissipation, degree of consolidation, or rate of consolidation is

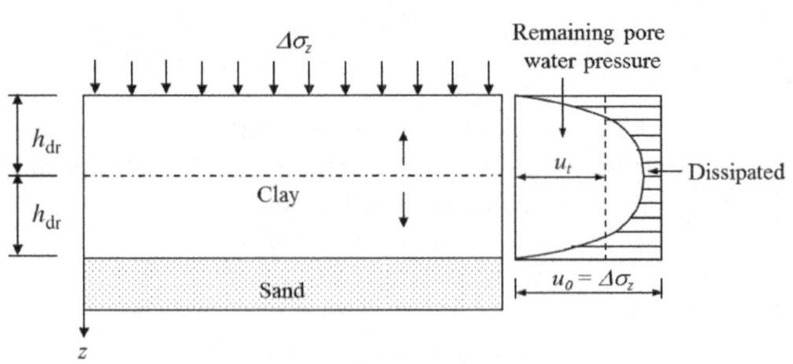

Figure 2.61 Dissipation of excess pore water pressure and one-dimensional consolidation.

defined as follows:

$$U = \frac{u_0 - u_t}{u_0} \times 100\% \qquad (2.66)$$

where u_t is the average excess pore water pressure at time t, which is calculated based on the area of the remaining pore water pressure divided by the thickness of the clay layer in Figure 2.61.

2.6 SLOPE STABILITY ANALYSIS

2.6.1 Introduction

Slopes exist in three different forms: natural slope, fill slope, and cut slope. They can be one side (e.g., a hillside of a mountain or a river bank), two sides (e.g., an embankment), or multiple sides. Figure 2.62 shows a typical embankment, which has the components of foundation, toe, facing, crest, and slope angle. In practice, slope angle is often expressed as a ratio of horizontal distance to vertical distance or a ratio of vertical distance to horizontal distance, depending on the traditions in different countries. To avoid confusing, the slope angle is expressed in either a number or a ratio in a form of m_h (H):1 (V) in this book, in which H stands for horizontal distance and V stands for vertical distance.

As shown in Figure 2.63, steepening a slope toward a wall increases usable space as well as cost but reduces stability. The maximum slope angle of an unreinforced slope is limited by soil strength. To maintain the stability of a steepened slope or wall, internal reinforcements (e.g., geosynthetics or soil nailing) and/or external support (e.g., gravity wall) are needed. In practice, there is an artificial division between

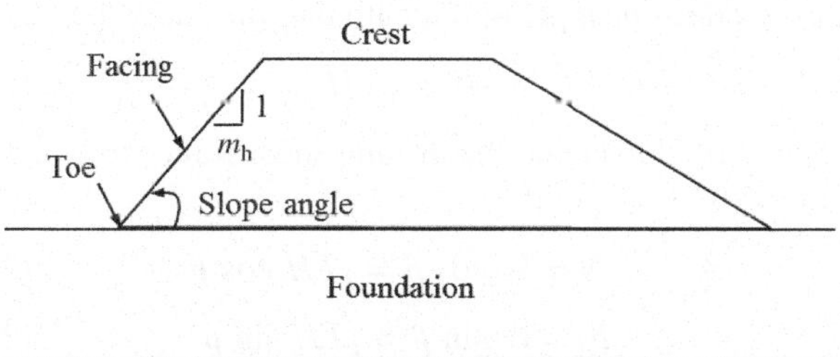

Figure 2.62 Components of a slope.

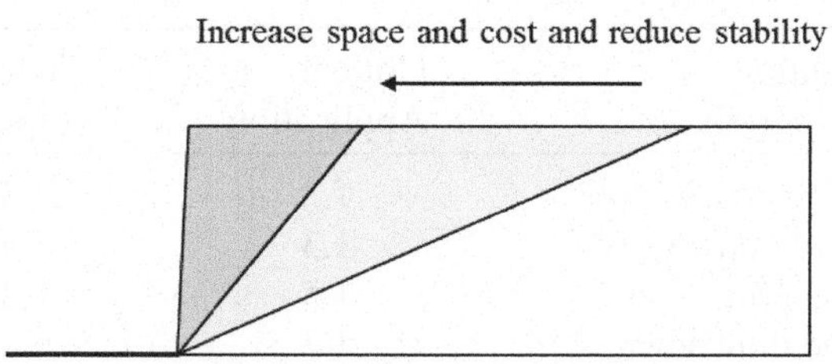

Figure 2.63 Space, cost, and stability.

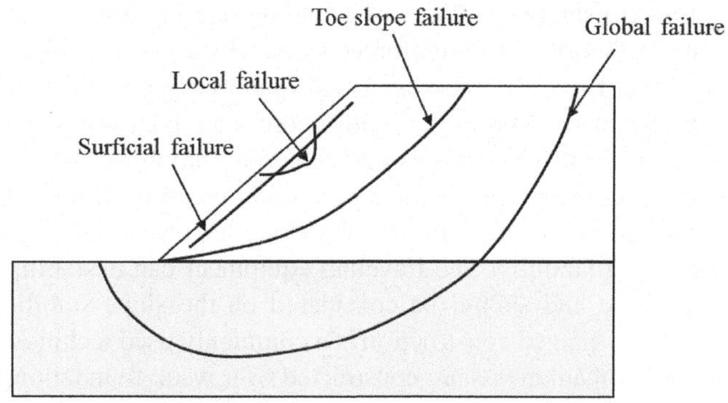

Figure 2.64 Possible failure modes of a slope.

slopes and walls by a slope angle. When the slope angle is less than 70°, the earth structure is considered as a slope. When the slope angle is equal or greater than 70°, the earth structure is considered a wall. Different theories have been proposed and used to analyze slopes and walls.

Figure 2.64 shows possible failure modes of a slope, including local failure, surficial failure, toe slope failure, and global failure. Local failure typically happens at local areas where geomaterial is poorly compacted, weak geomaterial exists, water intrudes, and/or there are other triggering factors. Due to limited sizes, local failure is often not a major concern and can be fixed easily. Surficial failure is the most common failure mode for slopes, especially in areas subjected to frequent rainfall. The causes for surficial failure are poor compaction, low overburden stress, loss of cohesion, saturation, and seepage force. Biological stabilization (such as tree roots) and geosynthetic reinforcement are commonly used to stabilize surficial failure. Toe slope failure happens through the toe of a slope, mostly due to large slope angle, low soil strength, and/or surcharge. Different ground improvement methods can be used to prevent toe slope failure, such as geosynthetic reinforcement, soil nailing, ground anchors, micropiles, and stone columns. Global failure, also referred to as deep-seated failure, happens when a weak foundation exists under the slope. Improvement of the weak foundation is the key to prevent global failure. The above-mentioned ground improvement methods will be discussed in later chapters of this book.

2.6.2 Methods for Slope Stability Analysis

Stability Conditions for Analysis Depending on the nature of a problem, slope stability analysis may be conducted under different conditions: (1) during construction, (2) end of construction, (3) long-term stability, (4) rapid drawdown, (5) rainfall, and (6) earthquake.

During construction, there are three common slope stability problems: (1) excavation and dewatering, (2) construction equipment traveling on a slope, and (3) staged construction of an embankment. Excavation and dewatering often create an unsupported or unreinforced slope with a water head difference. Effective stress analysis should be adopted for the excavated slope because negative pore water pressure is generated during the removal of geomaterial. Undrained analysis cannot consider the negative pore water pressure. Construction equipment traveling on the slope happens when constructing landfills. The traveling equipment can destabilize the slope and should be considered on the slope stability analysis. Staged construction is a commonly used technique when embankments are constructed over weak foundations. Due to low strength of the weak foundation, the embankment sometimes has to be constructed in more than one stage for fill placement. After each stage, the soft soil is allowed to consolidate and gain strength before the next stage fill is placed. For this application, a consolidation analysis needs to be performed first to calculate the effective stress and estimate the soil strength gain. Then a slope stability analysis is performed using the increased undrained shear strength of the soil. The analysis for staged construction will be discussed in Chapter 7.

End of construction is often a critical condition for slope stability. For low permeability soils (such as clays and silts), the analysis is typically done using soil undrained shear strength. For highly permeable soils (such as sands and gravels), soil effective strength can be used for the analysis.

Long-term stability is necessary for the serviceability of a slope. A higher factor of safety is typically required for long-term stability. Effective strength parameters should be used for this application. If there is groundwater flow or seepage, seepage analysis should be first performed.

Rapid drawdown happens along rivers or inside dams. Due to the fast lowering of the water level, except for free-draining geomaterial, the geomaterial does not have enough time to drain. As a result, the geomaterial is in an undrained condition. Therefore, undrained shear strength parameters should be used for the slope stability analysis under rapid drawdown conditions.

Heavy rainfall can saturate the geomaterial near the slope face in a short time. Water flow in the saturated geomaterial will develop seepage force, which destabilizes the slope and causes surficial failure. The failure often happens at the bottom of the saturated geomaterial. A slope stability analysis considering seepage is needed for this problem.

An earthquake can generate uplift and horizontal forces, which will destabilize the slope. These earthquake forces should be considered in the slope stability analysis. Due to the low occurrence rate of earthquakes and short duration of earthquake forces, the required factor of safety is lower for the slope under the earthquake loading than under static loading.

Factor of Safety The factor of safety (FS) for slope stability analysis is typically defined in terms of shear stress, shear force, and moment as follows:

$$\mathrm{FS} = \frac{\tau_f}{\tau_d} \tag{2.67}$$

$$\mathrm{FS} = \frac{T_r}{T_d} \tag{2.68}$$

$$\mathrm{FS} = \frac{M_r}{M_d} \tag{2.69}$$

where τ_f and τ_d = geomaterial shear strength and shear stress, respectively
T_r and T_d = geomaterial resisting and driving forces, respectively
M_r and M_d = geomaterial resisting and driving moments, respectively

Table 2.17 provides typical factors of safety used in practice for slope stability analyses.

There have been a number of studies on three-dimensional (3D) slope stability, for example, Leshchinsky et al. (1985) and Gao et al. (2013); however, most slope stability analyses in practice are still performed in two dimensions. This is because most slope stability problems are in two dimensions (2D) (such as embankments) or close to two dimensions, and 2D slopes are often more critical than 3D slopes. Therefore, only 2D slope stability analyses are discussed below.

Infinite Slope Analysis Figure 2.65 shows the force diagram of an infinite slope. Assume the planar slip surface parallel to the ground surface and at a depth of H. The weight of the wedge *abcd*, per unit length into the paper direction is

$$W = \gamma LH \tag{2.70}$$

The normal load and the driving shear force along the slip plane are

$$N = W \cos\beta = \gamma LH \cos\beta \tag{2.71}$$

$$T_d = W \sin\beta = \gamma LH \sin\beta \tag{2.72}$$

Table 2.17 Required Minimum Factor of Safety for Slope Stability Analysis

Condition for Analysis	Critical Application	Noncritical Application
During construction	1.3	1.2
End of construction	1.3	1.3
Long term	1.5	1.3
Rapid drawdown	1.3	1.1
Earthquake	1.2	1.1

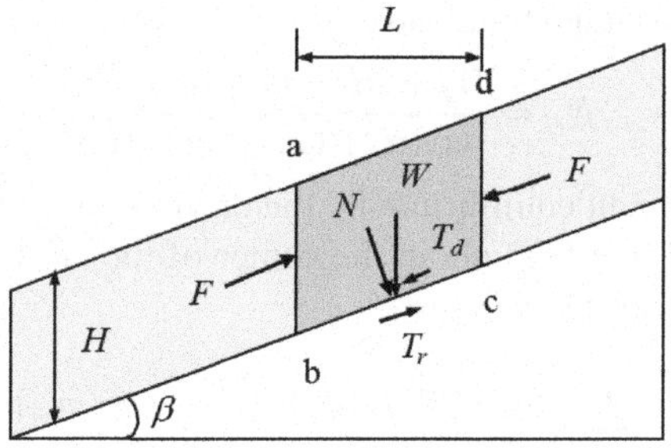

Figure 2.65 Force diagram of infinite slope.

The normal stress and the shear stress are

$$\sigma_n = \frac{N}{bc} = \frac{\gamma LH \cos \beta}{L/\cos \beta} = \gamma H \cos^2 \beta \quad (2.73)$$

$$\tau_d = \frac{T_d}{bc} = \frac{\gamma LH \sin \beta}{L/\cos \beta} = \gamma H \sin \beta \cos \beta \quad (2.74)$$

Based on the Mohr–Coulomb failure criterion, the shear strength along the slip plane is

$$\tau_f = c + \sigma_n \tan \phi = c + \gamma H \cos^2 \beta \tan \phi \quad (2.75)$$

According to the definition of the factor of safety in terms of shear strength versus shear stress, the following equation can be obtained:

$$\mathrm{FS} = \frac{\tau_f}{\tau_d} = \frac{c + \gamma H\cos^2\beta \tan \phi}{\gamma H \sin \beta \cos \beta} = \frac{2c}{\gamma H \sin 2\beta} + \frac{\tan \phi}{\tan \beta} \quad (2.76)$$

If the geomaterial cohesion $c = 0$, the preceding equation can be simplified as follows:

$$\mathrm{FS} = \frac{\tan \phi}{\tan \beta} \quad (2.77)$$

To ensure the stability of the slope, the factor of safety, FS, in Equation (2.77) must be greater than 1. In other words, the slope angle, β, must be less than the geomaterial friction angle, ϕ. When $\beta = \phi, \beta$ is referred to as the repose angle. Typically, fine-grained soils have effective friction angles of approximately 30°. This is why 2(H):1(V) (i.e., 26.7°) is a commonly used slope angle mostly under an unsaturated condition in practice.

When the slope gets fully saturated from water seepage, the flow lines and the slip plane are parallel to the slope face, as shown in Figure 2.66. Based on the basic concept of flow net, equipotential lines (i.e., *fe* is one equipotential line) should be perpendicular to the flow lines; therefore, the water head difference and pore water pressure at point *e* are

$$h_w = H\cos^2\beta \quad (2.78)$$

$$u = \gamma_w h_w = \gamma_w H\cos^2\beta \quad (2.79)$$

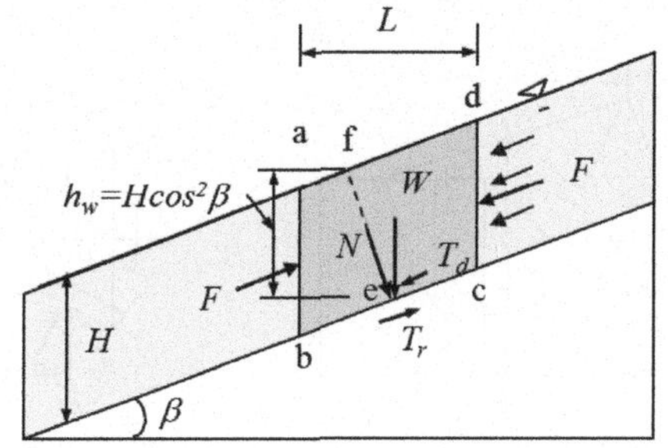

Figure 2.66 Force diagram of infinite slope with seepage.

As a result, the geomaterial shear strength along the slip plane is

$$\begin{aligned}\tau_f &= c' + (\sigma_n - u) \tan \phi' \\ &= c' + (\gamma_{sat} H\cos^2\beta - \gamma_w H\cos^2\beta) \tan \phi' \\ &= c' + \gamma' H\cos^2\beta \tan \phi' \quad (2.80)\end{aligned}$$

Therefore, the factor of safety of the saturated slope with seepage is

$$\mathrm{FS} = \frac{c' + \gamma' H\cos^2\beta \tan \phi'}{\gamma_{sat} H \sin \beta \cos \beta} = \frac{2c'}{\gamma_{sat} H \sin 2\beta} + \frac{\gamma'}{\gamma_{sat}} \frac{\tan \phi'}{\tan \beta} \quad (2.81)$$

If the effective cohesion, $c' = 0$, the preceding equation becomes

$$\mathrm{FS} = \frac{\gamma'}{\gamma_{sat}} \frac{\tan \phi'}{\tan \beta} \quad (2.82)$$

For most geomaterials, $\gamma'/\gamma_{sat} \approx 0.5$, that is, FS(seepage) $\approx$ 0.5FS (no seepage). For typical fine-grained soils (i.e., $\phi' = 30°$), the stable slope under a saturated condition with seepage should have a slope angle less than approximately 16°, which responds to a 4(H):1(V) slope.

Ordinary Method of Slices The ordinary method of slices is also referred to as the Swedish method of slices. It uses a circular slip surface to divide the slope into a moving mass and a stationary mass as shown in Figure 2.67. The moving mass is further divided into vertical slices. The force diagram of each slice is shown in Figure 2.67(b). In this method, it is assumed that the resultant force of T_i and P_i on one side of the slice is equal to that of T_{i+1} and P_{i+1} on another side of the slice and their lines of action coincide. The mobilized shear strength at the base of the slice is as follows:

$$\begin{aligned}T_{ri} &= \frac{\tau_f \Delta l_i}{\mathrm{FS}} = \frac{(c + \sigma_n \tan \phi)\Delta l_i}{\mathrm{FS}} = \frac{c\Delta l_i + N_{ri} \tan \phi}{\mathrm{FS}} \\ &= \frac{c\Delta l_i + W_i \cos \alpha_i \tan \phi}{\mathrm{FS}} \quad (2.83)\end{aligned}$$

The moment equilibrium by the driving force (W_i) and the resisting force (T_{ri}) about the center of the circle, O, for all

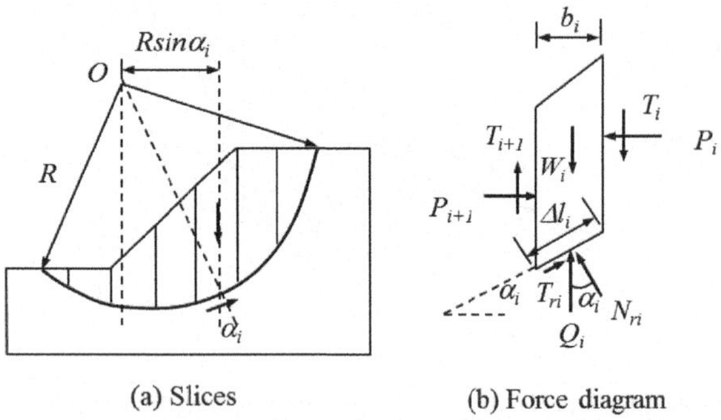

Figure 2.67 Circular slip surface with (a) vertical slices and (b) force diagram.

the slices is

$$\sum_{i=1}^{m}(W_i R \sin \alpha_i) = \sum_{i-1}^{m}(T_{ri} R) \tag{2.84}$$

Solving the above equation yields the following equation:

$$\text{FS} = \frac{\sum_{i=1}^{m}(c\Delta l_i + W_i \cos \alpha_i \tan \phi)}{\sum_{i=1}^{m}(W_i \sin \alpha_i)} \tag{2.85}$$

The above derivation shows that the ordinary method of slices only satisfies the moment equilibrium.

Simplified Bishop's Method Different from the ordinary method of slices, the simplified Bishop method (Bishop, 1955) has a vertical side force difference as follows:

$$\Delta T = T_i - T_{i-1} \tag{2.86}$$

The force equilibrium in the vertical direction is

$$W_i + \Delta T = N_{ri} \cos \alpha_i + \left(\frac{c\Delta l_i + N_{ri} \tan \phi}{\text{FS}} \right) \sin \alpha_i \tag{2.87}$$

The above equation leads to:

$$N_{ri} = \frac{W_i + \Delta T - c\Delta l_i \sin \alpha_i / \text{FS}}{\cos \alpha_i + \tan \phi \sin \alpha_i / \text{FS}} \tag{2.88}$$

The moment equilibrium by the driving force (W_i) and the resisting force (T_{ri}) about the center of the circle, O, for all the slices results in

$$\text{FS} = \frac{\sum_{i=1}^{m}[(cb_i + W_i \tan \phi + \Delta T \tan \phi)/m_{\alpha i}]}{\sum_{i=1}^{m} W_i \sin \alpha_i} \tag{2.89}$$

$$m_{\alpha i} = \cos \alpha_i + \frac{\tan \phi \sin \alpha_i}{\text{FS}} \tag{2.90}$$

For simplicity, let $\Delta T = 0$ and then

$$\text{FS} = \frac{\sum_{i=1}^{m}[(cb_i + W_i \tan \phi)/m_{\alpha i}]}{\sum_{i=1}^{m} W_i \sin \alpha_i} \tag{2.91}$$

The preceding derivation shows that the simplified Bishop method satisfies the moment equilibrium and the vertical force equilibrium.

Spencer's Method The Spencer method is a general method of slices, which is suitable for a circular or planar slip surface. Therefore, it has been used for analysis of a two-part or three-part wedge failure. In this method, a constant relationship between the magnitudes of the interslice shear and normal forces is assumed as follows (Figure 2.68):

$$\tan \theta_i = \frac{T_i}{P_i} = \frac{T_{i+1}}{P_{i+1}} \tag{2.92}$$

The vertical force equilibrium leads to the following equation:

$$W_i - (T_{i+1} - T_i) - N_{ri} \cos \alpha_i - T_{ri} \sin \alpha_i = 0 \tag{2.93}$$

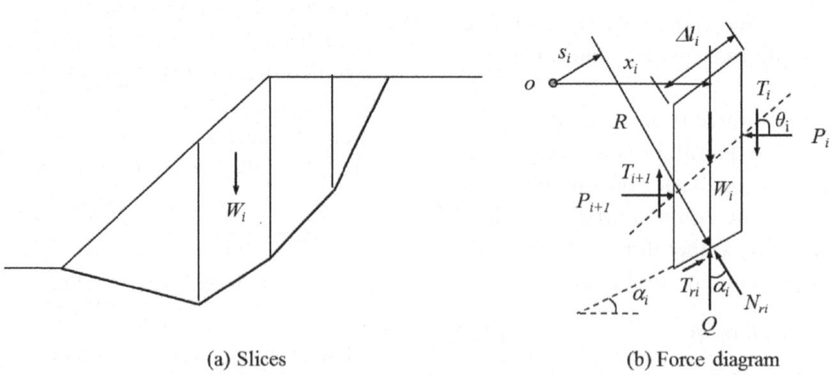

Figure 2.68 Planar slip surface with (a) vertical slices and (b) force diagram.

From the preceding equation, N_{ri} can be solved as follows:

$$N_{ri} = \left[W_i - \left(P_i - P_{i+1}\right)\tan\theta_i - \frac{c\,\Delta l_i \sin\alpha_i}{\text{FS}}\right]\frac{1}{m_{\alpha i}} \quad (2.94)$$

The horizontal force of the slice, F_{hi}, is

$$F_{hi} = (P_{i+1} - P_i) + N_{ri}\sin\alpha_i - T_{ri}\cos\alpha_i \quad (2.95)$$

Assume

$$\sum_{i=1}^{m}(P_{i+1} - P_i) = 0 \quad \sum_{i=1}^{m} F_{hi} = 0$$

leads to

$$\text{FS}_F = \frac{\sum_{i=1}^{m}(c\Delta l_i \cos\alpha_i + N_{ri}\tan\phi\cos\alpha_i)}{\sum_{i=1}^{m} N_{ri}\sin\alpha_i} \quad (2.96)$$

Assume

$$\sum_{i=1}^{m} M_o = 0$$

leads to

$$\sum_{i=1}^{m}(W_i x_i - T_{ri}R - N_{ri}s_i) = 0 \quad (2.97)$$

From the preceding equation, the following equation can be solved:

$$\text{FS}_M = \frac{\sum_{i=1}^{m}(c\Delta l_i R + N_{ri}R\tan\phi)}{\sum_{i=1}^{m}(W_i x_i - N_{ri}s_i)} \quad (2.98)$$

The factors of safety, FS_F and FS_M, depend on the interslice force angle. These two factors of safety at different interslice force angles are plotted in Figure 2.69. The final factor of safety, FS_{FM}, that satisfies both force and moment equilibriums is sought as shown in Figure 2.69.

Circular slip surfaces are not necessarily always most critical, especially when the slope and/or the foundation is not uniform. It is a good idea to check out other possible slip surfaces using two-part or three-part wedge analyses as shown in Figure 2.70.

Minimum Factor of Safety and Safety Map The methods for slope stability analyses discussed above are based on a known slip surface, which is not necessarily the critical one. In order to identify the most critical slip surface, many slip surfaces should be analyzed. The most critical slip surface corresponds to the minimum factor of safety. There are two common ways to search for the critical slip surface: (1) the search center method as shown in Figure 2.71 and (2) the start and exit points method as shown in Figure 2.72. The search center method requires defining the area of search centers and tangential limits while the start and exit points method requires selecting the ranges of start points and end points. It is recommended that initial searches should start with a large area or wide range. After gaining the knowledge of possible critical slip surfaces, the searches can be focused on a smaller area or narrow range.

Baker and Leshchinsky (2001) developed a method to present the spatial distribution of factors of safety in a

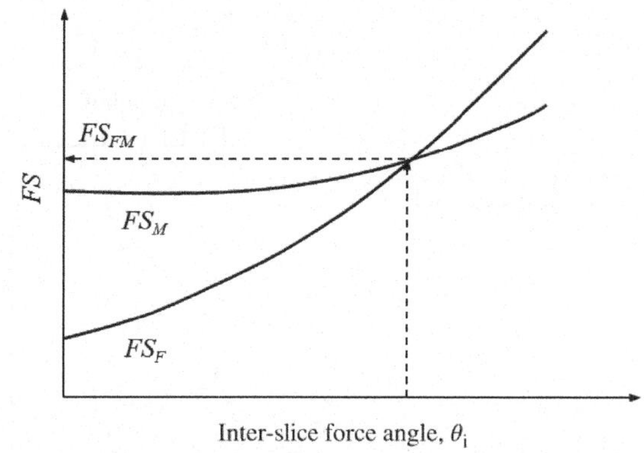

Figure 2.69 Determination of interslice force angle and factor of safety.

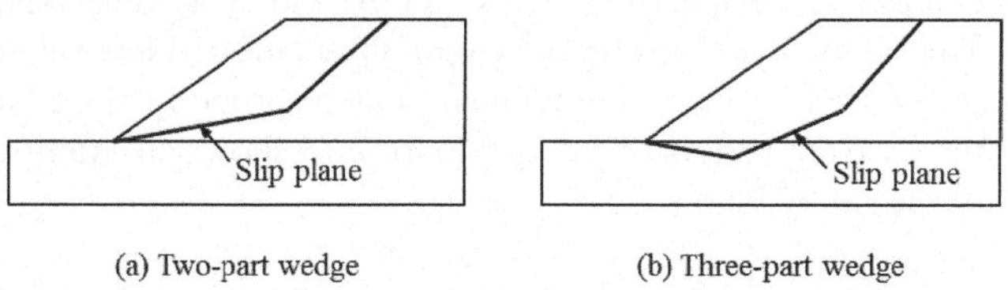

Figure 2.70 (a) Two-part and (b) three-part wedge analyses.

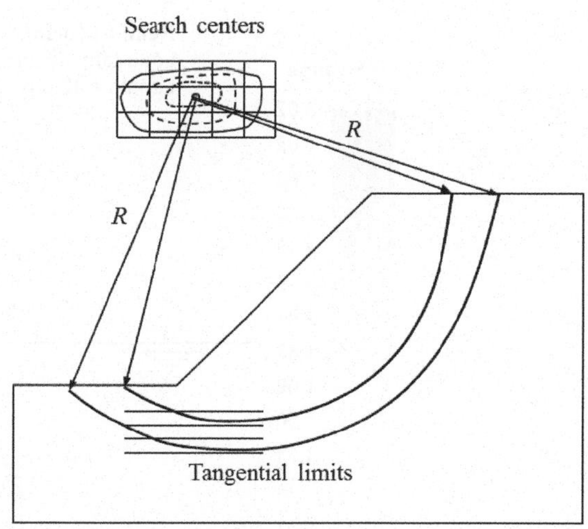

Figure 2.71 Search center method.

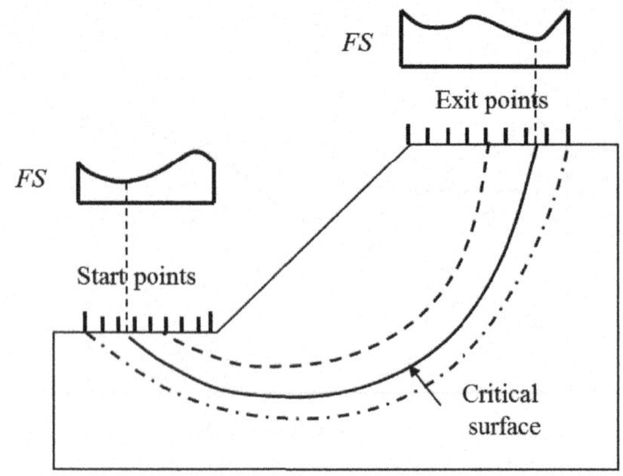

Figure 2.72 Start and exit point method.

color-coded map, which is also referred to as a safety map. The safety map is useful for visual observation of low safety factor areas so that reinforcements can be properly placed to increase the factors of safety. Figure 2.73 shows an example of the safety map.

Numerical Methods Numerical methods, based on finite-difference or finite-element approach, have been increasingly used in recent years to analyze slopes. They adopt a shear strength reduction technique to solve for a factor of safety of slope stability. In this technique, a series of trial factors of safety are used to adjust the cohesion, c, and the friction angle, ϕ, of geomaterial as follows:

$$c_{\text{trial}} = \frac{1}{\text{FS}_{\text{trial}}} c \tag{2.99}$$

$$\phi_{\text{trial}} = \arctan\left(\frac{1}{\text{FS}_{\text{trial}}} \tan\phi\right) \tag{2.100}$$

Adjusted cohesion and friction angle of geomaterial layers are re-inputted in the model for equilibrium analysis. The factor of safety is sought when the adjusted cohesion and friction angle make the slope become unstable from the verge of stability or become stable from the verge of instability.

Cundall (2002) compared the characteristics of numerical solutions and limit equilibrium methods in solving the factors of safety of slopes and concluded that continuum-mechanics-based numerical methods have the following advantages: (1) no predefined slip surface is needed; (2) the slip surface can be of any shape; (3) multiple failure surfaces are possible; (4) no static assumptions are needed; (5) structures (such as footings, tunnels, etc.) and/or structural elements (such as beams, cables, etc.) and interfaces can be included without concerns about compatibility; and (6) kinematics is satisfied. However, a complicated and large size problem may require significant computation time for numerical methods. The inclusion of structural elements and interfaces may create numerical instability leading to questionable solutions. Some specific searches are difficult to perform (e.g., surficial slope instability needs to be prevented in order to study the deep-seated slope stability). Localized and inconsequential failures, which may not be of interest to the study (e.g., locally overstressed soil), may mislead the investigators. Han and Leshchinsky (2004) and Leshchinsky and Han (2004) verified that numerical methods and limit equilibrium methods (such as the simplified Bishop method) mostly resulted in the similar results when they are used to analyze unreinforced slopes and geosynthetic-reinforced slopes and walls. However, they resulted in different results when used to analyze column-supported embankments (Han et al., 2005; Abusharar and Han, 2011). Details about these comparisons can be found in Chapter 5.

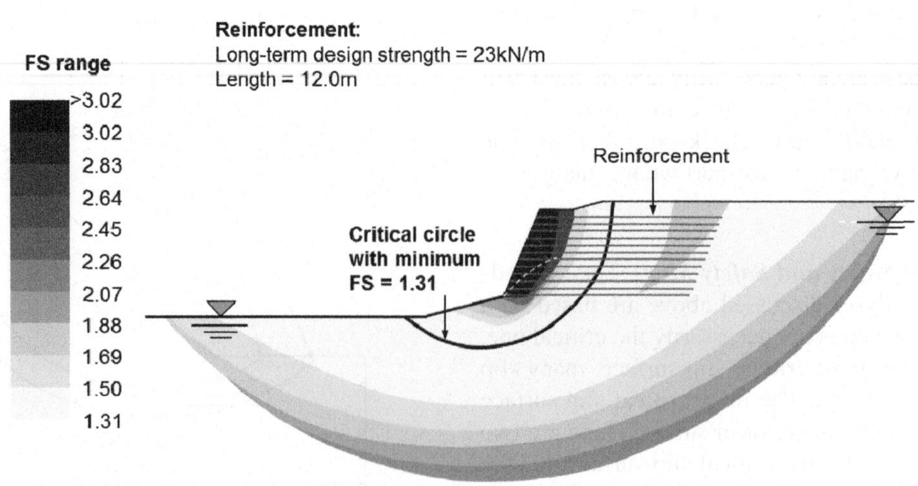

Figure 2.73 Example of safety map (Courtesy of Adama Engineering, Inc.).

Example 2.19

A 30-m-high slope with a slope angle of 23° consists of a silty clay with cohesion of 5 kPa and a friction angle of 28°. The effective cohesion and effective friction angle of the clay are 0kPa and 28°, respectively. During a rainy season, the slope face can be saturated down to 1.2 m vertically. The unit weight of the clay during the dry season is 17 kN/m^3 but during the rainy reason is 19 kN/m^3. Calculate the factors of safety for the slope with a slip surface at the depth of saturation during the dry and rainy seasons.

Solution

Since the depth of the slip surface is much smaller than the height of the slope, it can be considered as an infinite slope problem.

The factor of safety of the slope during the dry season is

$$\begin{aligned} FS &= \frac{2c}{\gamma H \sin 2\beta} + \frac{\tan\phi}{\tan\beta} \\ &= \frac{2\times 5}{17\times 1.2\times \sin(2\times 23°)} + \frac{\tan(28°)}{\tan(23°)} = 1.93 \end{aligned}$$

The factor of safety of the slope during the raining reason is

$$\begin{aligned} FS &= \frac{2c'}{\gamma_{sat} H \sin 2\beta} + \frac{\gamma'}{\gamma_{sat}}\frac{\tan\phi'}{\tan\beta} \\ &= \frac{2\times 0}{19\times 1.2\times \sin(2\times 23°)} + \frac{19-9.81}{19}\frac{\tan(28°)}{\tan(23°)} \\ &= 0.61 \end{aligned}$$

Hence, this slope is unstable during the rainy season.

2.7 EARTH RETAINING WALL ANALYSIS

2.7.1 Type of Wall

Depending on the construction method, there are cut and fill walls as shown in Figure 2.74. Cut walls are formed by excavating part of the existing ground, while fill wall is formed by placing fill on the existing ground.

In terms of rigidity, walls can be classified as rigid, flexible, and hybrid walls, as shown in Figure 2.75. A gravity wall is a typical rigid wall, and its body will not deform but can move or rotate. A sheet pile wall is a typical flexible wall, which will deform under lateral earth pressure. A reinforced wall, such as geosynthetic-reinforced earth wall or soil nailed wall, is a hybrid wall, which is designed as a flexible wall for internal stability but as a rigid wall (i.e., the reinforced mass) for external stability. The design of reinforced walls will be discussed in Chapters 9 and 10.

2.7.2 Lateral Earth Pressure Coefficient

When geomaterial is in the ground or behind a wall, it is subjected to vertical and lateral stresses. The lateral stress is important for the stability of the earth retaining wall. Lateral earth pressure coefficient, K, is used to establish the relationship between the vertical and lateral stresses as follows (also shown in Figure 2.76):

$$K = \frac{\sigma'_x}{\sigma'_z} \tag{2.101}$$

where σ'_z and σ'_x are the vertical and lateral (horizontal) effective stresses in geomaterial, respectively.

The lateral stress at the base of the wall can be expressed as follows:

$$\sigma'_x = K\sigma'_z = K\gamma' H \tag{2.102}$$

where γ' is the effective unit weight of soil and H is the height of the wall. The lateral thrust, which is acted at the height of $1/3\,H$, is

$$P = \frac{1}{2}K\gamma' H^2 \tag{2.103}$$

The magnitude of the lateral stress depends on movement of the wall element. When there is no movement, the lateral stress is often referred to as the lateral earth pressure at rest. The lateral earth pressure coefficient at rest, K_0, in cohesionless soil or cohesive soil under a drained condition can be expressed as follows:

$$K_0 = (1-\sin\phi')\mathrm{OCR}^{\sin\phi'} \tag{2.104}$$

where ϕ' = effective friction angle of soil
OCR = overconsolidation ratio of soil

The active earth pressure develops when the wall moves away from the soil mass while the passive earth pressure develops when the wall moves toward the soil mass. The relationships among the coefficients of at-rest, active, and passive earth pressures, K_0, K_a, and K_p, are illustrated in Figure 2.77 and their magnitudes can be presented in order as follows:

$$K_a < K_0 < K_p \tag{2.105}$$

K_a is mobilized at small movement; however, K_p is mobilized at large movement, which is often too large for practical applications. In addition, scour or freeze–thaw cycles may remove or degrade the passive support. Due to these facts, passive earth pressure is often ignored or reduced for many applications, especially for permanent structures.

2.7.3 Rankine's Theory

The active and passive earth pressures of a geomaterial can be obtained using Rankine's theory. As shown in Figure 2.78,

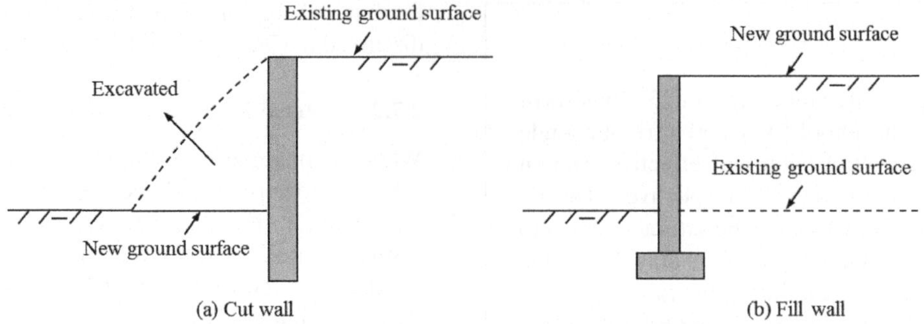

Figure 2.74 Type of wall based on construction.

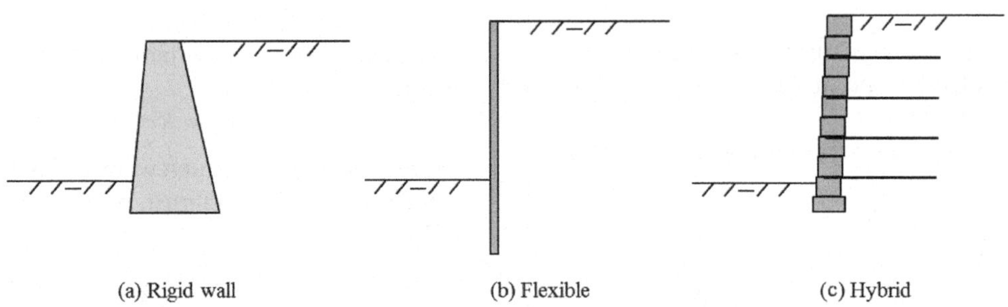

Figure 2.75 Type of wall based on rigidity.

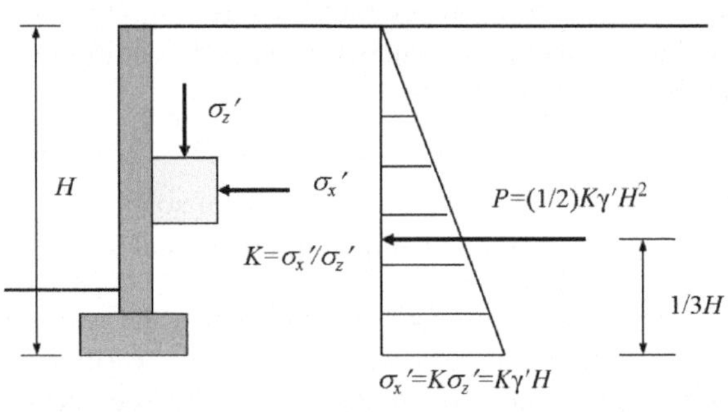

Figure 2.76 Lateral earth pressure.

the initial geomaterial state in the ground can be represented by the Mohr circle, *AB*, in which the vertical stress is σ_z' and the lateral stress is $K_0\sigma_z'$. When the lateral stress decreases, the geomaterial yields until the Mohr circle touches the failure envelope defined by the strength parameters, c' and ϕ'. The Mohr circle AEC represents an active state of the geomaterial in which the lateral active stress is $\sigma_x' = \sigma_a' = K_a\sigma_z'$. However, when the lateral stress increases, the geomaterial yields until the Mohr circle touches the same failure envelope. The Mohr circle *AFD* represents a passive state of the geomaterial in which the lateral passive stress is $\sigma_x' = \sigma_p' = K_p\sigma_z'$.

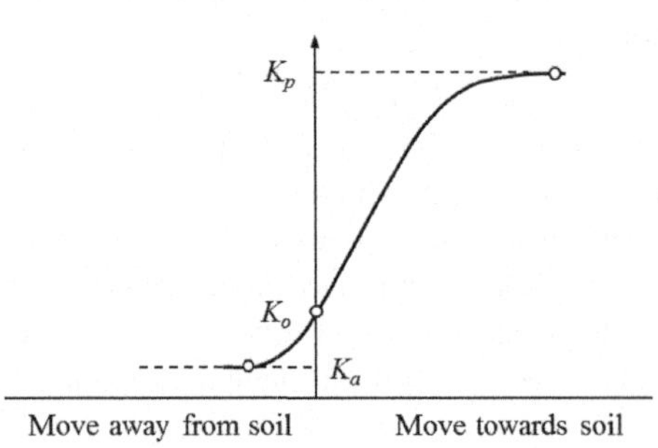

Figure 2.77 Coefficients of lateral earth pressure with movement.

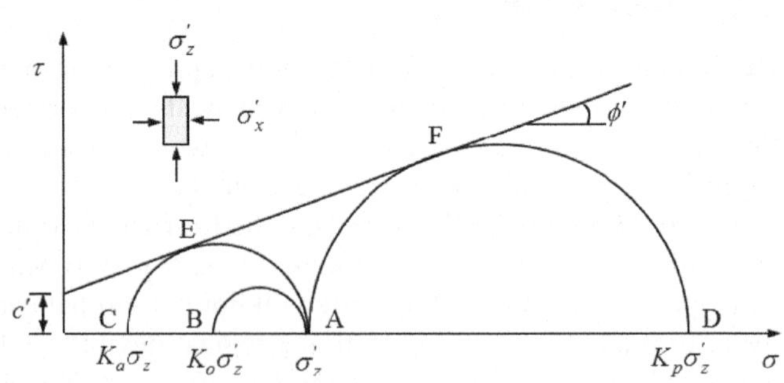

Figure 2.78 Rankine's active and passive earth pressures.

Based on the geometry, the active and passive earth pressures of the geomaterial can be derived as follows:

$$\sigma_a' = \sigma_z' \tan^2\left(45^\circ - \frac{\phi'}{2}\right) - 2c' \tan\left(45^\circ - \frac{\phi'}{2}\right) \quad (2.106)$$

$$\sigma_p' = \sigma_z' \tan^2\left(45^\circ + \frac{\phi'}{2}\right) + 2c' \tan\left(45^\circ + \frac{\phi'}{2}\right) \quad (2.107)$$

For cohesionless or cohesive normally-consolidated geomaterial under a drained condition (i.e., $c' = 0$),

$$\sigma_a' = \sigma_z' \tan^2\left(45^\circ - \frac{\phi'}{2}\right) \quad (2.108)$$

$$\sigma_p' = \sigma_z' \tan^2\left(45^\circ + \frac{\phi'}{2}\right) \quad (2.109)$$

The active and lateral earth pressure coefficients are

$$K_a = \frac{\sigma_a'}{\sigma_z'} = \tan^2\left(45^\circ - \frac{\phi'}{2}\right) \quad (2.110)$$

$$K_p = \frac{\sigma_p'}{\sigma_z'} = \tan^2\left(45^\circ + \frac{\phi'}{2}\right) \quad (2.111)$$

For cohesive geomaterial under an undrained condition (i.e., $c_u \neq 0$ and $\phi_u = 0$),

$$\sigma_a = \sigma_z' - 2c_u \quad (2.112)$$

$$\sigma_p = \sigma_z' + 2c_u \quad (2.113)$$

The preceding earth pressures and their coefficients are based on a vertical wall, a frictionless interface between wall and geomaterial, and a level top ground. If there is a sloping top ground, the active and passive lateral earth pressure coefficients for a geomaterial with $c' = 0$ are

$$K_a = \cos\beta \left(\frac{\cos\beta - \sqrt{\cos^2\beta - \cos^2\phi'}}{\cos\beta + \sqrt{\cos^2\beta - \cos^2\phi'}}\right) \quad (2.114)$$

$$K_p = \cos\beta \left(\frac{\cos\beta + \sqrt{\cos^2\beta - \cos^2\phi'}}{\cos\beta - \sqrt{\cos^2\beta - \cos^2\phi'}}\right) \quad (2.115)$$

where β is the top slope angle.

Example 2.20

A 5-m-high retaining wall has a vertical and frictionless wall back and a top slope with a slope angle of 15°. The backfill is cohesionless soil and has a friction angle of 35°. Calculate the active and passive earth pressure coefficients.

Solution

The active earth pressure coefficient is

$$K_a = \cos\beta \left(\frac{\cos\beta - \sqrt{\cos^2\beta - \cos^2\phi'}}{\cos\beta + \sqrt{\cos^2\beta - \cos^2\phi'}}\right)$$

$$= \cos 15^\circ \left(\frac{\cos 15^\circ - \sqrt{\cos^2 15^\circ - \cos^2 35^\circ}}{\cos 15^\circ + \sqrt{\cos^2 15^\circ - \cos^2 35^\circ}}\right]$$

$$= 0.297$$

The passive earth pressure coefficient is

$$K_p = \cos\beta \left(\frac{\cos\beta + \sqrt{\cos^2\beta - \cos^2\phi'}}{\cos\beta - \sqrt{\cos^2\beta - \cos^2\phi'}}\right)$$

$$= \cos 15^\circ \left(\frac{\cos 15^\circ + \sqrt{\cos^2 15^\circ - \cos^2 35^\circ}}{\cos 15^\circ - \sqrt{\cos^2 15^\circ - \cos^2 35^\circ}}\right]$$

$$= 3.144$$

2.7.4 Coulomb's Theory

Different from Rankine's theory, Coulomb's theory considers an inclined wall back face and interface friction between wall and geomaterial. A planar slip surface is formed within the geomaterial. Figure 2.79 shows the force diagram of the geomaterial wedge above the slip plane. The active thrust corresponds to the minimum force needed to maintain the stability of the wedge when it slides down. The passive thrust corresponds to the maximum force needed to push the wedge upward to failure. The active and passive thrusts can be used to calculate the active and passive earth pressure coefficients as follows:

$$K_a = \frac{\cos^2(\phi' + \omega)}{\cos^2\omega \cos(\omega - \delta)\left[1 + \sqrt{\frac{\sin(\phi'+\delta)\sin(\phi'-\beta)}{\cos(\omega-\delta)\cos(\omega+\beta)}}\right]^2} \quad (2.116)$$

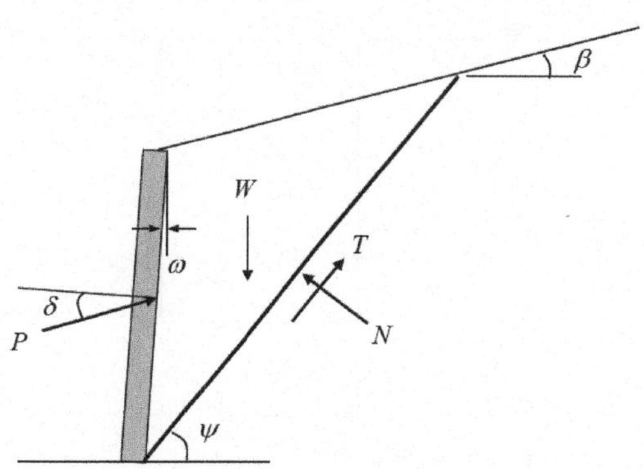

Figure 2.79 Coulomb's earth pressure.

$$K_p = \frac{\cos^2(\phi' - \omega)}{\cos^2\omega\cos(\omega+\delta)\left[1 - \sqrt{\frac{\sin(\phi'-\delta)\sin(\phi'+\beta)}{\cos(\omega-\delta)\cos(\omega-\beta)}}\right]^2} \quad (2.117)$$

where ω = batter of the wall back face
δ = interface friction angle between wall and geomaterial
ψ = angle of failure plane

2.8 LIQUEFACTION ANALYSIS

2.8.1 Liquefaction Potential

Liquefaction occurs when pore water pressure at a depth of cohesionless geomaterial is equal to its total stress, that is, the effective stress is equal to zero. As a result, the geomaterial loses its strength and induces large deformation and even failure of superstructures and substructures. Liquefaction often happens when cohesionless geomaterial is saturated, loose, and subjected to earthquake loading. To evaluate liquefaction potential, the earthquake-induced maximum shear stress in geomaterial needs be first calculated. Figure 2.80 shows a rigid geomaterial column with a unit cross-sectional area subjected to a ground acceleration, a. The shear stress is

$$\tau = \frac{F_e}{A} = \frac{ma}{1} = \frac{W}{g}a = \frac{\gamma z}{g}a = \sigma_{z0}\frac{a}{g} \quad (2.118)$$

where F_e = earthquake-induced force
A = cross-sectional area of the rigid geomaterial column ($A = 1$)
m = mass of the column
W = weight of the column
γ = total unit weight of the column
z = height of the column
a = ground acceleration
g = gravitational acceleration
σ_{z0} = total overburden stress at a depth of z

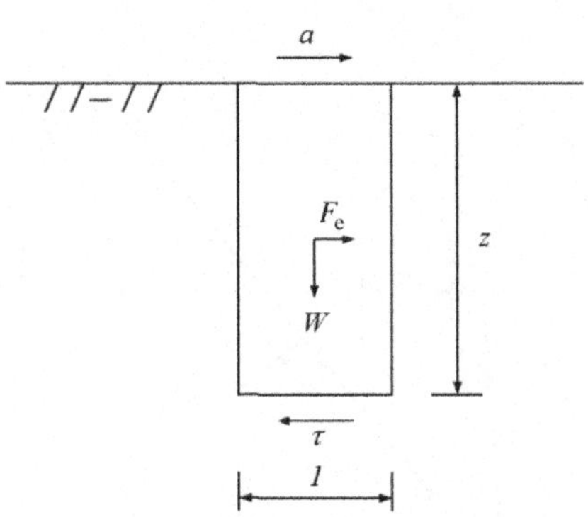

Figure 2.80 Maximum shear stress at the base of a rigid soil column.

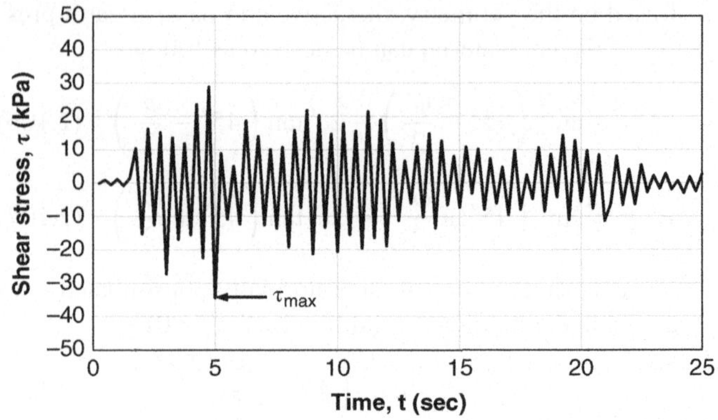

Figure 2.81 Shear stress variation with time.

Under earthquake loading, the shear stress varies with time as shown in Figure 2.81. The maximum shear stress is the most critical to potential liquefaction of geomaterial.

To simplify the problem, Seed and Idriss (1971) converted the maximum shear stress with a nonuniform shear stress variation to a uniform cyclic shear stress as follows:

$$\tau_{cyc} = 0.65\tau_{max} \quad (2.119)$$

Considering the effective overburden stress effect, a cyclic stress ratio (CSR) is defined as follows:

$$\text{CSR} = \frac{\tau_{cyc}}{\sigma'_{z0}} = 0.65 r_d \left(\frac{\sigma_{z0}}{\sigma'_{z0}}\right)\left(\frac{a_{max}}{g}\right) \quad (2.120)$$

where σ'_{z0} = effective overburden stress at a depth of z
a_{max} = maximum ground acceleration
r_d = stress reduction factor

The stress reduction factor was introduced by Seed and Idriss (1971) to account for the fact that the geomaterial is not a rigid body and the shear stress decreases with depth. Later on, Idriss (1999) found that the stress reduction factor depends on the magnitude of earthquake. Figure 2.82 shows the relationship between the stress reduction factor and the depth at different earthquake magnitude. Martin and Lew (1999) suggested that the minimum depth of analysis for liquefaction evaluation should be 15 m below the existing ground or lowest proposed finished grade or 6 m below the lowest expected foundation level (e.g., pile tip).

To evaluate liquefaction potential, it is also necessary to estimate the cyclic resistance ratio. Seed et al. (1985) correlated the cyclic resistance ratio (CRR) to the SPT $(N_1)_{60}$ as shown in Figure 2.83 for clean and silty sands at an earthquake magnitude of 7.5. When the earthquake magnitude is different from 7.5, the cyclic resistance ratio from Figure 2.83 should be multiplied by a magnitude scaling factor (MSF) below to

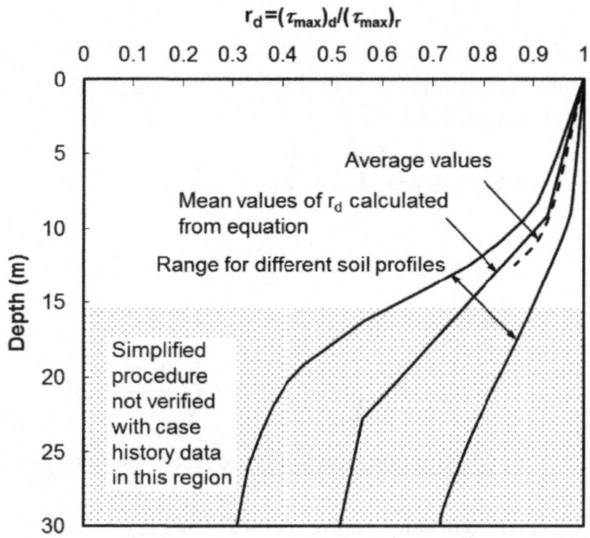

Figure 2.82 Shear stress reduction factor, r_d, versus depth curves developed by Seed and Idriss (1971) with added mean value line (after Youd and Idriss, 1997).

correct the magnitude effect (Idriss, 1999):

$$\text{MSF} = 6.9 \exp\left(-\frac{M_w}{4}\right) - 0.06, \quad \text{for} \quad M_w > 5.2 \tag{2.121}$$

$$\text{MSF} = 1.82, \quad \text{for} \quad M_w < 5.2 \tag{2.122}$$

where M_w is the moment magnitude of the earthquake.

A similar procedure has been developed by Robertson and Wride (1998) to estimate the cyclic resistance ratio based on CPT data.

With CSR and CRR, the factor of safety against liquefaction can be calculated as follows:

$$\text{FS} = \frac{\text{CRR}_M}{\text{CSR}} = \frac{\text{MSF} \cdot \text{CRR}_{M=7.5}}{\text{CSR}} \tag{2.123}$$

where CRR_M is the cyclic resistance ratio at a specific magnitude of earthquake.

Martin and Lew (1999) suggested the factor of safety for liquefaction hazard assessment should be based on consequence of liquefaction and $(N_1)_{60}$ of clean sand as shown in Table 2.18.

Table 2.18 Factor of Safety for Liquefaction Hazard Assessment

Consequence of Liquefaction	$(N_1)_{60}$ (Clean Sand)	Factor of Safety
Settlement	≤ 15	1.1
	≥ 30	1.0
Surface manifestation	≤ 15	1.2
	≥ 30	1.0
Lateral spread	≤ 15	1.3
	≥ 30	1.0

Source: Martin and Lew (1999).

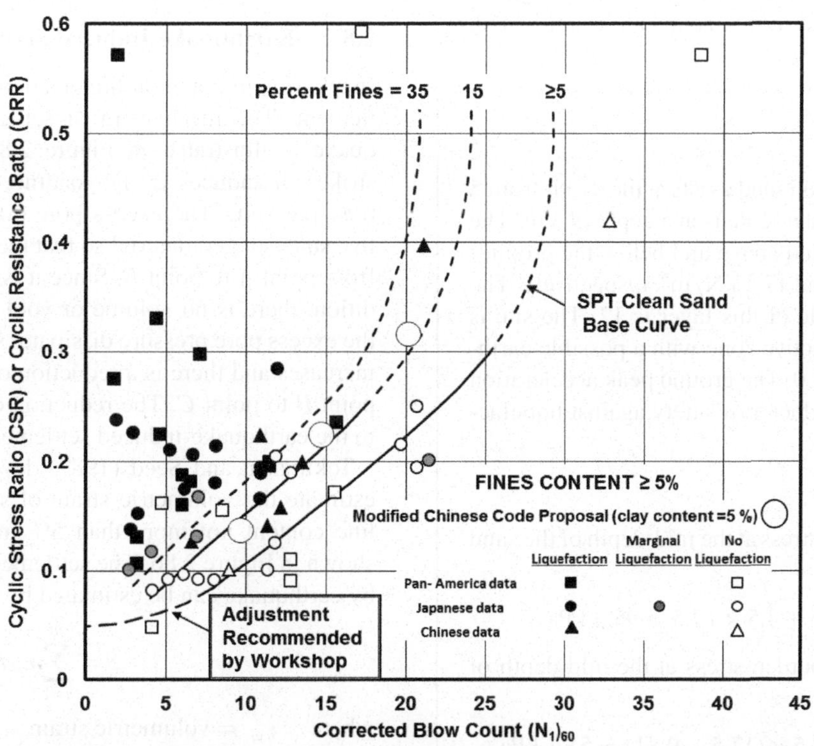

Figure 2.83 Cyclic resistance ratio for clean and silty sand at earthquake magnitude of 7.5 based on SPT data (after Youd and Idriss, 1997).

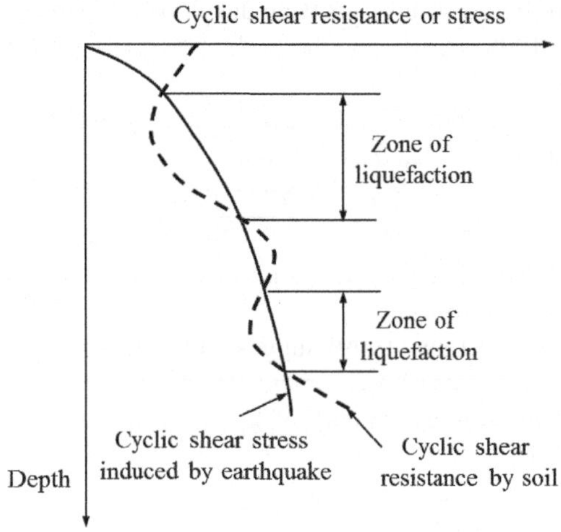

Figure 2.84 Zone of liquefaction.

Surface manifestation of liquefaction includes sand boils, ground fissures, and the like. It can happen during an earthquake with a level ground. Lateral spread is a flow slide, which results from the reduction of undrained shear strength of geomaterial due to liquefaction. Lateral spread can happen with a gently sloping ground or nearby drainage or stream channels.

From the preceding evaluation, a zone of liquefaction can be determined as illustrated in Figure 2.84. If the cyclic shear resistance is less than the cyclic shear stress, the saturated cohesionless geomaterial is liquefiable. The potential liquefaction within this zone can be mitigated by ground improvement.

Example 2.21

A 1-m-thick loose clean sand exists at the depth from 5 to 6 m. The groundwater table is at a depth of 1 m. The unit weights of the sand above and below the groundwater table are 16.5 and 17.5 kN/m^3, respectively. The SPT N_{60} at the middle of this layer is 12. The site is located in a seismic active zone with a possible earthquake magnitude of 7.0. The ground peak acceleration is $0.3g$. Calculate the factor of safety against liquefaction.

Solution

The total overburden stress at the mid-depth of the sand layer is

$$\sigma_{z0} = 1 \times 16.5 + 4.5 \times 17.5 = 95.3 \text{ kPa}$$

The effective overburden stress at the mid-depth of the sand layer is

$$\sigma'_{z0} = 1 \times 16.5 + 4.5 \times (17.5 - 9.81) = 51.1 \text{ kPa}$$

The stress reduction coefficient at the depth of 5.5 m and the earthquake magnitude of 7.0 is $r_d = 0.93$.

The cyclic stress ratio is

$$\text{CSR} = 0.65 r_d \left(\frac{\sigma_{z0}}{\sigma'_{z0}}\right)\left(\frac{a_{\max}}{g}\right)$$

$$= 0.65 \times 0.93 \times \left(\frac{95.3}{51.1}\right) \times 0.3 = 0.338$$

The corrected $(N_1)_{60}$ value is

$$(N_1)_{60} = N_{60}\sqrt{\frac{100}{\sigma'_{z0}}} = 12 \times \sqrt{\frac{100}{51.1}} = 17$$

The CRR value from Figure 2.83 is $\text{CRR}_{M=7.5} = 0.17$.

The magnitude scaling factor (MSF) is

$$\text{MSF} = 6.9 \exp\left(-\frac{M_w}{4}\right) - 0.06$$

$$= 6.9 \times \exp\left(-\frac{7.0}{4}\right) - 0.06 = 1.14$$

The factor of safety against liquefaction is

$$\text{FS} = \frac{\text{MSF} \cdot \text{CRR}_{M=7.5}}{\text{CSR}} = \frac{1.14 \times 0.17}{0.338}$$

$$= 0.57 < 1.0 \qquad \text{liquefaction}$$

2.8.2 Earthquake-Induced Settlement

In addition to potential liquefaction, earthquake induces settlement. The mechanism of settlement induced by earthquake is illustrated in Figure 2.85. When an earthquake strikes, it induces cyclic loading, which generates excess pore pressure. The excess pore pressure reduces the effective stress in geomaterial so that the geomaterial state moves from point A to point B. Since it is under an undrained condition, there is no volume or void ratio change. With time, the excess pore pressure dissipates so that the effective stress increases and there is a reduction of void ratio, that is, from point B to point C. The reduction of void ratio corresponds to the earthquake-induced settlement.

Tokimatsu and Seed (1987) developed a design chart to estimate the volumetric strain of saturated clean sand with fine content not more than 5% induced by earthquake as shown in Figure 2.86. The settlement of free ground induced by earthquake can be estimated by the following equation:

$$S = \sum_{i=1}^{m} \varepsilon_{vi} h_i \tag{2.124}$$

where ε_{vi} = volumetric strain
h_i = thickness of each sublayer

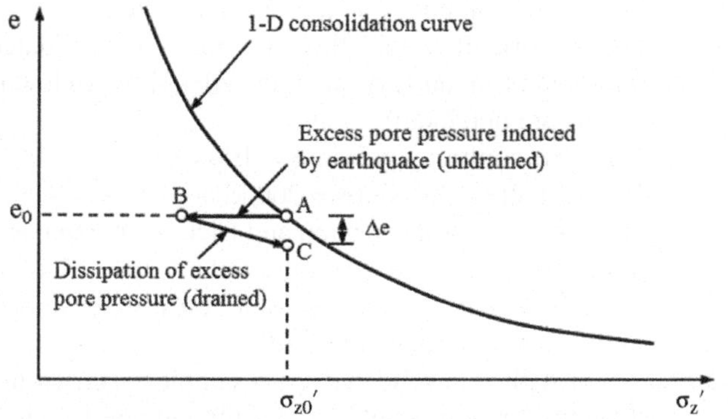

Figure 2.85 Mechanism for earthquake-induced settlement.

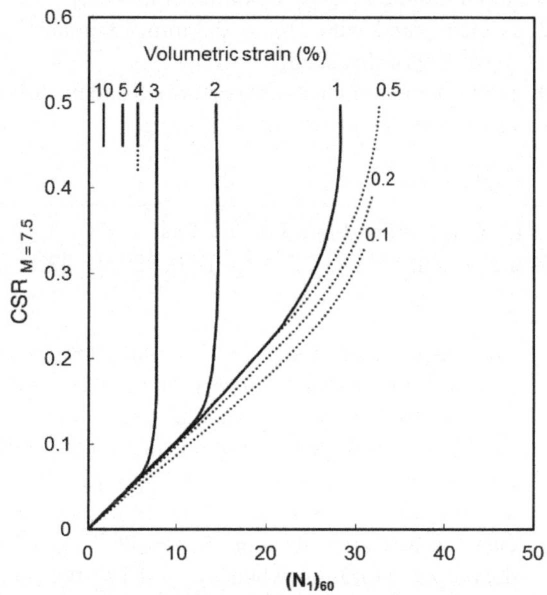

Figure 2.86 Earthquake-induced volumetric strain of saturated clean sand (after Tokimatsu and Seed, 1987, with permission from ASCE).

Dashti and Bray (2013) indicated that earthquake-induced building settlement is contributed not only by volumetric type strain but also by deviatoric type of strain. The volumetric type of strain results from partial drainage during earthquake loading, sedimentation, and consolidation. The deviatoric type of strain is due to partial bearing capacity loss under the static load of the building and soil–structure interaction induced building ratcheting. So far, no simplified method is available to consider all these factors.

PROBLEMS

2.1. Determine D_{10}, D_{15}, D_{30}, D_{50}, D_{60}, and D_{85} from the following gradation curve.

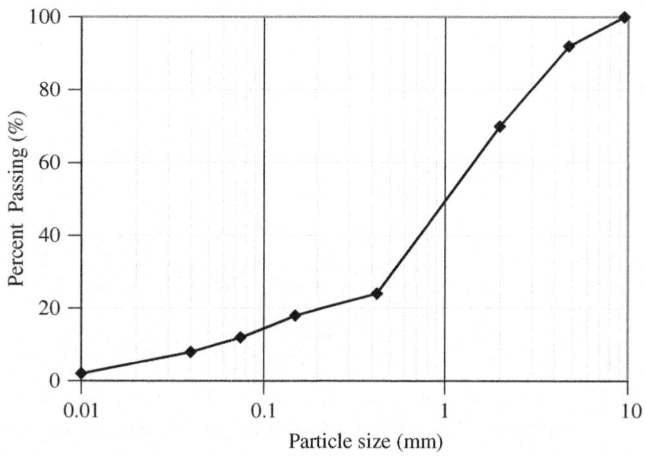

2.2. Determine void ratio, porosity, and degree of saturation of a soil sample with the following known information: moist weight of soil sample = 1000 g, volume of soil sample = 580,000 mm^3, specific gravity = 2.70, and dry weight of soil = 910.0 g.

2.3. The maximum and minimum void ratios of sand are 0.82 and 0.42, respectively. What is the void ratio of the sand corresponding to a relative density of 70%?

2.4. A soil sample has 30% particles retained on the U.S. No. 200 sieve. The plastic limit and liquid limit of this soil are 20 and 60, respectively. Classify this soil according to the USCS method.

2.5. Laboratory gradation analyses and Atterberg limits are performed on a soil sample. Results are summarized below. No organic odor and materials are noted in the sample.

U.S. Sieve Size	Percent Passing
9.8 mm	100
No. 4	95
No. 10	90
No. 40	82
No. 100	75
No. 200	68

Liquid limit (LL) = 51% and plastic limit (PL) = 21%.
Classify this soil according to the USCS system.

2.6. Calculate the total vertical stress σ_z, pore water pressure u, and effective vertical stress σ_z' at A, B, C, and D in the following figure.

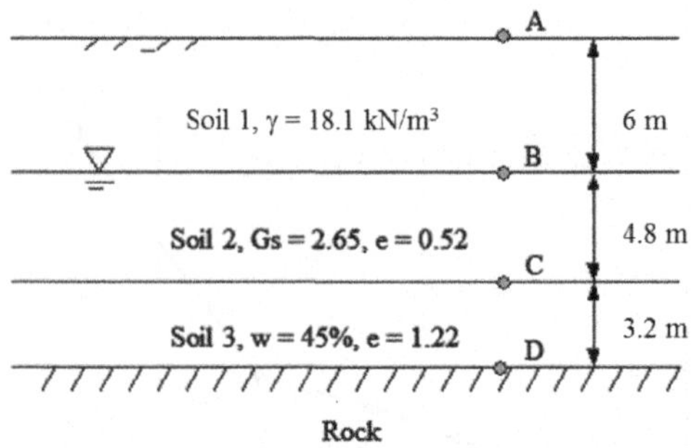

2.7. Consolidation test results of a soil sample taken at a depth of 2.5 m from a field with a groundwater table at the ground surface are shown here:

Stress (kPa)	10	20	40	80	160	320
Void ratio	0.910	0.851	0.760	0.629	0.490	0.352

The saturated unit weight and permeability of the soil sample are 19.0 kN/m^3 and 6.5 × 10^{-7}m/s. Calculate: (1) effective overburden stress in the field; (2) preconsolidation stress; (3) overconsolidation ratio (OCR); (4) coefficient of recompression; (5) coefficient of recompression; (6) coefficient of volumetric compressibility in the stress range between 100 and 200 kPa; (7) coefficient of consolidation.

2.8. A clayey soil has a liquid limit of 60. Estimate: (1) the coefficients of consolidation at virgin compression, reloading, and remolding stages and (2) the ratio of the coefficient of consolidation at virgin compression to that after remolding.

2.9. Three direct shear tests on soil samples of 63.6 mm in diameter are conducted. Test results are shown here:

Test Number	Normal load (N)	Shear force (N)
1	45.4	36.3
2	90.8	52.2
3	136.1	70.8

Determine soil cohesion and friction angle.

2.10. For a normally consolidated soil, a consolidated-undrained triaxial test is conducted and the following results are obtained:
Chamber confining pressure = 100 kPa
Deviator stress at failure = 200 kPa
(1) Determine soil cohesion and friction angle under a total stress condition; (2) if the soil effective frictional angle is 36°, determine pore water pressure developed in the clay specimen at failure.

2.11. A normally consolidated clay sample taken from a depth of 2.5 m in a field has a saturated unit weight of 18.7 kN/m^3 and a plasticity index of 25. Estimate the undrained shear strength, effective peak friction angle, and undrained elastic modulus of this clay.

2.12. A clean sand with D_{10} = 0.3 mm. Estimate the permeability of this sand.

2.13. Four compaction tests are performed and the following test results are obtained:

Test Number	1	2	3	4
Mass of moist soil and mold (kg)	3.81	3.94	3.93	3.82
Moisture content (%)	17.0	20.0	22.1	23.9

The mass of the model is 2.13 kg and the volume of mold is 943,000 mm^3. The specific gravity of the soil is 2.68. Plot the dry unit weight versus water content curve, determine the maximum dry unit weight and optimum moisture content, plot the zero air void curve, and plot the dry unit weight versus moisture content curve at a degree of saturation of 80%. What is the degree of saturation when the soil has the optimum moisture and maximum dry unit weight?

2.14. Given the following constant-head data for planar flow of water in a 1.50-mm-thick geotextile that is 300 mm wide × 600 mm long:

Head Difference (mm)	90	180	270	360
Rate of Flow (mm³ / min)	21,000	41,000	60,000	79,000

Calculate the transmissivity in m^3/ min -m of the geotextile and then the planar coefficient of permeability in mm/s.

2.15. A tensile test is conducted on a wide-width geotextile sample with a width of 200 mm. Test results are shown below:

Strain (%)	2	4	6	8	10	12
Tensile force (kN)	2.16	4.32	5.80	6.56	7.24	7.72

Determine (1) initial modulus, (2) secant modulus at 2% strain, and (3) tensile stress at 10%.

2.16. Four soil–geogrid interface friction tests are conducted and the test results are provided as follows:

Normal Stress (kPa)	17	35	70	140
Shear Strength (kPa)	8.6	20	36	75

(a) Plot the Mohr-Coulomb failure envelope, (b) obtain the interface friction angle, and (c) calculate the coefficient of interaction based on a soil friction angle of 34°.

2.17. A steel mesh of 0.3 m wide and 1.2 m long is placed within sand for a pullout test. The applied normal stress is 50 kPa. The friction angle of the sand is 35°. The interaction coefficient of steel mesh and sand is 0.6. Calculate the pullout capacity of the steel mesh in the sand.

2.18. A nonwoven geotextile with ultimate permittivity of 1.2 s^{-1} is used for a filtration purpose. Estimate the allowable permittivity of the geotextile after considering creep, intrusion, clogging, and blinding.

2.19. An HDPE geogrid is designed for a reinforced slope. The short-term ultimate tensile strength is 100 kN/m. The reduction factors for creep, installation damage, and durability are 2.70, 1.25, and 1.10, respectively. Calculate the allowable tensile strength of this geogrid.

2.20. A standard penetration test is performed in a 150-mm-diameter borehole. The borehole is 11.0 m below the ground surface. An automatic hammer with an efficiency of 92% and a standard SPT sampler are used to perform the SPT. The actual blow count (N) is found to be 21.The soil is overly consolidated silty sand with an OCR of 1.5. The unit weights of the soil above and below the groundwater table are 17.8 and 18.5 kN/m^3, respectively with D_{50} of 0.35 mm. The groundwater is found at a depth of 5 m below the ground surface. Determine N_{60}, $(N_1)_{60}$, D_r, and ϕ' of this soil.

2.21. A CPT is conducted in field with a groundwater table at 1.0 m below the ground surface. The measured tip resistance at a depth of 4.5 m is 5.1 MPa and the sleeve friction is 0.3 MPa. The unit weights of the soil above and below the groundwater table are 17.5 and 18.5 kN/m^3, respectively. Classify the type of the soil. If it is a sandy soil, determine the relative density and friction angle. If it is a clayey soil, determine the undrained shear strength and OCR.

2.22. A soft clay stratum is found to have liquid and plastic limits of 110 and 35, respectively. A series of vane shear tests with a vane in a dimension of 60 mm in diameter and 120 mm long are performed at different depths. The measured torques at depths are provided below:

Depth (m)	4.0	5.0	6.0	7.0	9.0
T_f(N-m)	8.3	9.4	10.5	12.3	15.2

Calculate the soil uncorrected and corrected undrained shear strengths at each depth and then develop a profile of corrected undrained shear strength versus depth.

2.23. A plate loading test with a square rigid plate in a dimension of 1 m × 1 m is run on a uniform soil. The applied load–settlement data are provided here:

Load (kN)	20	40	60	80	100	120	140	160
Settlement (mm)	5	10	18	25	35	47	60	110

Plot the applied pressure–settlement curve, determine the ultimate bearing capacity, and calculate the elastic modulus of the soil (assume Poisson's ratio = 0.3).

2.24. A square footing in a dimension of 1.8 m × 1.8 m is embedded in a uniform soil at a depth of 1.0 m. The unit weight, effective cohesion, and effective friction angle of the soil are 18.0 kN/m^3, 5 kPa, and 25°, respectively. No groundwater is encountered. Calculate the ultimate bearing capacity of this footing.

2.25. A spread footing is designed to support a column load of 390 kN. The proposed design is a 1.6 m × 1.6 m square footing with an embedment depth of 0.5 m. A site investigation is performed for the site. It is found that the underlying soil has unit weights of 18.4 and 19.0 kN/m^3 above and below the groundwater table, respectively. The groundwater is found at a depth of 1.8 m below the ground surface. Compute the effective overburden stress and additional vertical stress beneath the center of the footing at a point 0.8 m below the bottom of the footing.

2.26. A spread footing with a dimension of 2.0 m × 2.0 m is embedded in uniform clay at a depth of 1.0 m to support a vertical downward load of 300 kN. The preconsolidation stress of the soil is 50 kPa and their compression indices are $C_r/(1 + e_0)$ and $C_c/(1 + e_0)$ of

0.025 and 0.11, respectively. The soil has unit weights of 16.8 and 17.8 kN/m^3 above and below the groundwater table, respectively. The soil stratum extends to a great depth and the groundwater table is at a depth of 1.5 m below the ground surface. Estimate the consolidation settlement of this footing.

2.27. A spread footing with a dimension of 2.25 m × 2.25 m is embedded in silty sand at a depth of 2.0 m to support a 500-kN column load. A CPT is done for the site and the results are shown here:

Layer Number	Depth (m)	Average Cone Tip Resistance, q_c (MPa)
1	2.0–3.0	2.4
2	3.0–4.0	3.2
3	4.0–5.0	4.1
4	5.0–7.0	6.5
5	7.0–9.0	8.5
6	9.0–10.0	6.7
7	10.0–12.0	11.0

The groundwater table is at a depth of 1.8 m and the unit weights of the sand are 17.5 and 18.1 kN/m^3 above and below the groundwater table, respectively. Compute the settlement of the footing right after construction and the settlement at 30 years after construction using the Schmertmann method.

2.28. A 1.5-m-thick clay layer starts at a depth of approximately 2.4 m below the ground surface and is overlain by a permeable sandy soil and underlain by gravel and cobble. The clay layer has a vertical coefficient of consolidation of 6.2 × 10^{-8}m^2/s. Calculate the number of days required for the clay layer to achieve 90% consolidation by a large area of fill placed instantly.

2.29. An infinite slope with a slope angle of 18.6° consists of a 1-m-thick clayey sand above bedrock. The soil has undrained cohesion of 30 kPa, total cohesion of 10 kPa and friction angle of 23° (for of the most soil), and effective cohesion of 0 kPa and effective friction angle of 30°. The soil moist and saturated unit weights are 17.5 and 18.8 kN/m^3, respectively. The interface shear strength between soil and bedrock is equal to the soil strength. The slope has chances to be subjected to heavy rainfall. Evaluate the stability of the slope under undrained, drained, and regular (moist) conditions.

2.30. Calculate the factor of safety for the slope along the slip surface shown below using the ordinary method of slices and the simplified Bishop method.

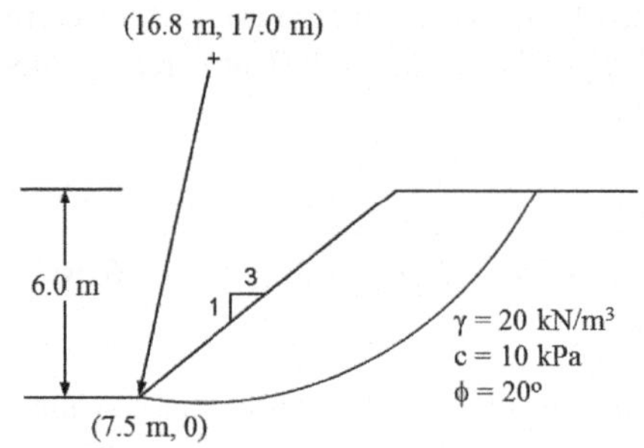

2.31. The retaining wall is shown below. Determine the active and passive Rankine earth pressure distributions, the resultant forces, and the locations of the resultant forces.

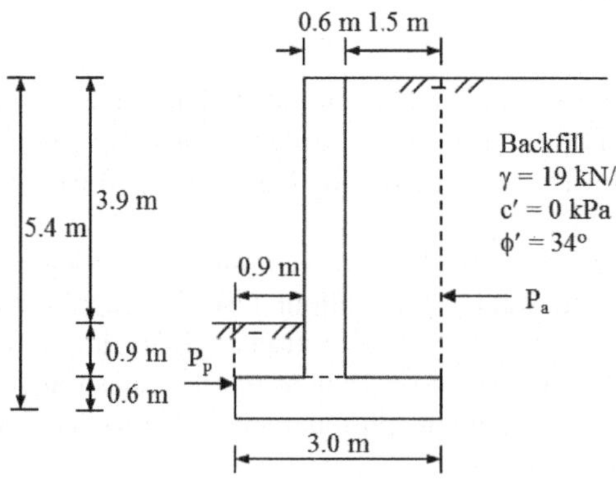

2.32. A 2.0-m-thick loose sand layer with 15% fines exists at the depth starting from 3.0 m. The groundwater table is at 0.5 m from the ground surface. The unit weights of the sand above and below the groundwater are 17.4 and 18.1 kN/m^3, respectively. The SPT N_{60} values at the depths of 3.5 and 4.5 m are 10 and 11.5, respectively. The site is located at a seismically active area with a possible maximum magnitude earthquake of 7.8 and a peak ground acceleration of 0.35g. Calculate the factor of safety of this sand against liquefaction and the seismically induced settlement.

REFERENCES

Abusharar, S. and Han, J. (2011). "Two-dimensional deep-seated slope stability analysis of embankments over stone columns." *Eng. Geol.*, 120: 103–110.

Andrus, R.D. and Stokoe, K.H. (2000). "Liquefaction resistance of soils from shear-wave velocity." *J. Geotech. Geoenviron. Eng.*, 126(11): 1015–1025.

Baker, R. and Leshchinsky, D. (2001). "Spatial distribution of safety factors." *J. Geotech. Geoenviron. Eng.*, 127(2): 135–145.

Berg, R.R., Christopher, B.R., and Samtani, N.C. (2009). *Design of Mechanically Stabilized Earth Walls and Reinforced Soil Slopes*. FHWA-NHI-10-024, FHWA, Washington, DC.

Bishop, A.W. (1955). "The use of the slip circle in the stability analysis of slopes." *Geotechnique*, 5(1): 7–17.

Bjerrum, L. (1972). "Embankments on soft ground." *Proceedings of the Specialty Conference, American Society of Civil Engineers* (2): 1–54.

Bonaparte, R., Holtz, R. D., and Giroud, J. P. (1987). "Soil reinforcement design using geotextiles and geogrids." Geotextile Testing and the Design Engineer, ASTM, STP No. 952, 69–116.

Bowles, J.E. (1996). *Foundation Analysis and Design* (5th ed.). McGraw-Hill, New York.

Budhu, M. (2000). *Soil Mechanics and Foundations*. Wiley, New York.

Conduto, D.P. (2001). *Foundation Design: Principles and Practices* (2nd ed.).Prentice Hall, Upper Saddle River, NJ.

Cubrinovski, M. and Ishihara, K. (1999)."Empirical correlation between SPT N-value and relative density for sandy soils." *Soils and Foundations* 39(5): 61–71.

Cundall, P.A. (2002). "The replacement of limit equilibrium methods in design with numerical solutions for factor of safety." Powerpoint presentation, Itasca Consulting Group. *Minneapolis*, Minnesota.

Dashti, S. and Bray, J.D. (2013). "Numerical simulation of building response on liquefiable sand." *J. Geotech. Geoenviron. Eng.*, 139(8): 1235–1249.

Dong, Y.L., Han, J., and Bai, X.H. (2011). "Numerical analysis of tensile behavior of geogrids with rectangular and triangular apertures." *Geotext. Geomembr.*, 29(2): 83–91.

Elias, V., Christopher, B.R., and Berg, R.R. (2001). *Mechanically Stabilized Earth Walls and Reinforced Slopes Design and Construction Guidelines*. Publication no. FHWA-NHI-00-043.US Department of Transportation, Federal Highway Administration, Washington, DC. Available at http://isddc.dot.gov/OLPFiles/FHWA/010567.pdf.

Elhakim, A.F. (2005). Evaluation of Shallow Foundation Displacements. Ph.D. Dissertation, the Georgia Institute of Technology, 396p.

Freeze, R.A. and Cherry, J.A. (1979). *Groundwater*. Prentice-Hall, Englewood Cliffs, N.J.

Gao, Y., Zhang, F., Lei, G.H., Li, D., Wu, Y., and Zhang, N. (2013). "Stability charts for 3D failures of homogeneous slopes." *J. Geotechn. Geoenviron. Eng.*, 139(9): 1528–1538.

Han, J. and Leshchinsky, D. (2004). "Limit equilibrium and continuum mechanics-based numerical methods for analyzing stability of MSE walls." Proceedings of the 17th Engineering Mechanics Conference, ASCE, University of Delaware, Newark, DE, June 13–16.

Han, J., Parsons, R.J., Sheth, A.R., and Huang, J. (2005). "Factors of safety against deep-seated failure of embankments over deep mixed columns." Deep Mixing 2005 Conference, Sweden, Vol. 1.2, May 23–25, 231–236.

Han, J., Pokharel, S.K., Yang, X., Manandhar, C., Leshchinsky, D., Halahmi, I., and Parsons, R.L. (2011). "Performance of geocell-reinforced RAP bases over weak subgrade under full-scale moving wheel loads." *J. Mat. Civil Eng.*, 23(11): 1525–1534.

Hansen, J.B. (1970). *A Revised and Extended Formula for Bearing Capacity*, Bulletin 28, Danish Geotechnical Institute, Copenhagen.

Henry, K.S.and Holtz, R.D. (2001). "Geocomposite capillary barriers to reduce frost heave in soils."*Canad. Geotech. J.*, 38: 678–694.

Holm, T.A. and Valsangkar, A.J. (1993). "Lightweight aggregate soil mechanics:Properties and applications." *Transportation Research Record* 1422, 7–13.

Idriss, I. M. (1999). "Presentation notes: An update of the Seed-Idriss simplified procedure for evaluating liquefaction potential." *Proc., TRB Workshop on New Approaches to Liquefaction Anal.*, Publ. No. FHWARD-99-165, Federal Highway Administration, Washington, D.C.

Joint Departments of the Army and Air Force, USA (1983), Technical Manual TM 5-818-1/AFM 88-3, Chapter 7, Soils and Geology Procedures for Foundation Design of Buildings and Other Structures (Except Hydraulic Structures), 21 October 1983.

Koerner, R.M. (2005). *Designing with Geosynthetics*, 5th ed., Prentice Hall, 796. Upper Saddle River, NJ.

Kulhawy, F.H., and Mayne, P.W. (1990). *Manual on Estimating Soil Properties for FoundationDesign*. Report EL-6800, Electric Power Research Institute, Palo Alta, CA.

Ladd, C.C. (1971). *Settlement Analysis for Cohesive Soils*. Research Report R71-2, Soils Publication 272, Massachusetts Institute of Technology, Cambridge, MA.

Lambe, T.W. (1958). "The structure of compacted clay." *J. Soil Mech. Found. Div.*, 84(SM2): 1654-1–1654-34.

Leshchinsky, D. and Han, J. (2004). "Geosynthetic reinforced multitiered walls." *J. Geotech. Geoenviron. Eng.*, 130(12): 1225–1235.

Leshchinsky, D., Baker, R., and Silver, M.L. (1985). "Three dimensional analysis of slope stability." *Int. J. Numerical Anal. Methods Geomech.*, 9: 199–223.

Louden, A.G. (1952). "The computation of permeability from simple soil tests." *Geotechnique*, 3: 165–183.

Lukas, R.G. (1995). *Dynamic Compaction*. Geotechnical Engineering Circular No. 1, FHWA-SA-95-037, 105, FHWA, Washington, DC.

Martin, G.R. and Lew, M. (1999). *Recommendation Procedures for Implementation of DMG Special Publication 117 Guidelines for Analyzing and Mitigating Liquefaction Hazards in California*. The Southern California Earthquake Center, Los Angeles.

Mayne, P.W., Christopher, B.R., and DeJong, J. (2001). *Subsurface Investigations: Geotechnical Site Characterization*. Publication no. FHWA-NHI-01-031.US Department of Transportation, Federal Highway Administration, Washington, DC. Available at http://isddc.dot.gov/OLPFiles/FHWA/012546.pdf.

Mayne, P.W. and Poulos, H.G. (1999). "Approximate displacement influence factors for elastic shallow foundations." *J. Geotech. Geoenviron. Eng.*, 125(6): 453–460.

Mesri, G. and Godlewski, P.M. (1977). "Time and stress—compressibility interrelationships." *J. Geotech. Eng. Div.*, 103(GT5): 417–430.

Mesri, G., Feng, T.W., and Benak, J.M. (1990). "Postdensification penetration resistance of clean sands." *J. Geotech. Eng.*, 116(7): 1095–1115.

Meyerhof, G.G. (1953). "The bearing capacity of foundations under eccentric and inclined loads." *Proceedings, the Third International Conference on Soil Mechanics and Foundation Engineering*, Zurich, 1, 440–445.

Mitchell, J.K. and K. Soga (2005). *Fundamentals of Soil Behavior*(3rd ed.). Wiley, New York.

NAVFAC (1982). *Foundations and Earth Structures Design Manual 7.2*. Department of the Navy, Naval Facilities Engineering Command, Alexandria, VA.

Oweis, I.S. and Khera, R.P. (1998). *Geotechnology of Waste Management*, 2nd ed., PWS Publishing, Boston, MA.

Peck, R.B., Hanson, W.E., and Thornburn, T.H. (1974). *Foundation Engineering*, 2nd ed., Wiley, New York.

Robertson, P.K. and Wride, C.E. (1998). "Evaluating cyclic liquefaction potential using the cone penetration test." *Canad. Geotech. J.*, 35(3): 442–459.

Sabatini, P.J., Bachus, R.C., Mayne, P.W. Schneider, J.A., and Zettler, T.E. (2002). *Geotechnical Engineering Circular No. 5: Evaluation of Soil and Rock Properties*. Publication no. FHWA-IF-02-034.

Salgado, R. (2006). *The Engineering of Foundations*. McGraw-Hill Higher Education, Boston, MA.

Schmertmann, J.H., Hartman, J.P., and Brown, P.R. (1978). "Improved strain influence pactor diagrams." *J. Geotech. Eng. Div.*, 119(4): 662–674.

Seed, H.B. and Chan, C.K. (1959). "Structure and strength characteristics of compacted clays." *J. Soil Mech. Found. Div.*, 85(SM5), 87–129.

Seed, H.B. and Idriss, I.M. (1971). "Simplified procedure for evaluating soil liquefaction potential." *J. Soil Mech. Found. Div.*, 97(SM9): 1249–1273.

Seed, H.B., Tokimatsu, K., Harder, L.F., and Chung, R.M. (1985). "The influence of SPT procedures in soil liquefaction resistance evaluations." *J. Geotech. Eng.*, 111(12): 1425–1445.

Spencer, E. (1981). "Slip circles and critical shear planes." *J. Geotech. Eng. Div.*, 107(GT7): 929–942.

Terzaghi, K. (1943). *Theoretical Soil Mechanics*. Wiley, New York.

Tokimatsu, K. and Seed, H.B. (1987)."Evaluation of settlements in sands due to earthquake shaking." *Journal of Geotechnical and Geoenvironmental Engineering* 113(8), 661–678.

Vesic, A.S. (1973). "Analysis of ultimate loads of shallow foundations." *J. Soil Mech. Found. Div.*, 99(SM1): 45-73.

Wolff, T. F. (1989). "Pile Capacity Prediction Using Parameter Functions."*Predicted and Observed Axial Behavior of Piles, Results of a Pile Prediction Symposium*, sponsored by the Geotechnical Engineering Division, ASCE, Evanston, IL (June) 1989, ASCE Geotechnical Special Publication 23, 96–106.

Youd, T.L. and Idriss, I.M. (1997). *Proceedings of the NCEER Workshop on Evaluation of Liquefaction Resistance of Soils*. Technical Report NCEER-97-0022, National Center for Earthquake Engineering Research, Buffalo, NY.

CHAPTER 3

Shallow and Deep Compaction

3.1 INTRODUCTION

Shallow and deep compaction methods have been commonly used to improve geomaterial properties near surface and at depth through a densification process by vibration, pressure, kneading, and/or impact on ground surface. This technology is effective to improve cohesionless geomaterial or cohesive geomaterial with low plasticity. Conventional plate or roller compaction has been used for many years, and it densifies geomaterial to a shallow depth by repeated passing of a vibratory plate or a roller on a relatively thin lift. Intelligent compaction is a new technology and has evolved in the past few years through research and implementation. In addition to providing the same compaction capabilities as conventional compaction, intelligent compaction includes sensors, which provide feedback on the location, stiffness, and machine driving power on geomaterial on a color-coded map. With this map, an engineer or operator can identify areas that require more or less compaction to create a uniform foundation. Deep dynamic compaction extends the depth of geomaterial densification to a greater depth by applying high-energy impact through repeated dropping of a large and heavy weight on ground surface. Rapid impact compaction is an intermediate compaction technology from shallow to deep compaction, and it rapidly applies impact on ground surface using a hydraulic hammer. Vibro-compaction densifies cohesionless soil by driving a vibrating probe into the ground to apply lateral vibratory forces which rearrange particles into a dense state. For saturated cohesionless soil or when water is injected into the ground, vibration can also cause liquefaction to the soil and the soil is densified after the dissipation of excess pore water pressure. There is no well-accepted definition of shallow to deep compaction. Shallow, intermediate, and deep compaction is mostly effective for densification of geomaterials up to depths of 1 m, 6 m, and greater than 6 m, respectively.

3.2 DENSIFICATION PRINCIPLES

The basic principle of densification is the rearrangement of particles into a denser state (i.e., the void ratio of geomaterial decreases). As a result, the modulus, strength, and resistance to liquefaction of the geomaterial are increased while the permeability and collapsibility are reduced. Chapter 2 describes the principles of soil compaction and influence factors, which are mostly based on shallow compaction of unsaturated disturbed geomaterial. Intermediate compaction can be used for disturbed and in situ geomaterials. Deep compaction, however, mostly improves in situ geomaterials, which can be unsaturated or saturated. Therefore, their densification principles may be different. In general, there are four principles of densification: (1) densification under static or kneading pressure, (2) densification under dynamic loading (vibration or impact), (3) densification due to liquefaction, and (4) densification due to consolidation. Static or kneading pressure is effective for densifying unsaturated cohesive geomaterial. Vibration or impact is effective for densifying unsaturated cohesionless or collapsible geomaterial. Vibration or impact can also be used to densify saturated cohesionless geomaterial by inducing liquefaction to the geomaterial. Saturated cohesive geomaterial is difficult to be densified immediately under pressure, vibration, and/or impact; however, saturated cohesive geomaterial at low plasticity can be densified by consolidation, which dissipates excess pore water pressure induced by impact. The densification of saturated cohesive geomaterial by consolidation will be discussed in Chapter 7.

3.3 CONVENTIONAL COMPACTION

3.3.1 Introduction

Basic Concept Conventional compaction is to use rollers or plates to repeatedly apply static pressure, kneading action, or vibration on ground surface to densify geomaterials, as shown in Figure 3.1. This is one of the most commonly used ground improvement methods in practice for earthwork. To achieve better densification, proper compaction equipment should be chosen, geomaterials should be prepared at appropriate lift thickness and moisture content, which is close to an optimum moisture content, and sufficient compactive energy should be applied.

Suitability Rollers are larger and heavier than plate compactors; therefore, they are more commonly and efficiently used for large-area compaction than plate compactors. However, in constraint areas or unstable edges, such as inside

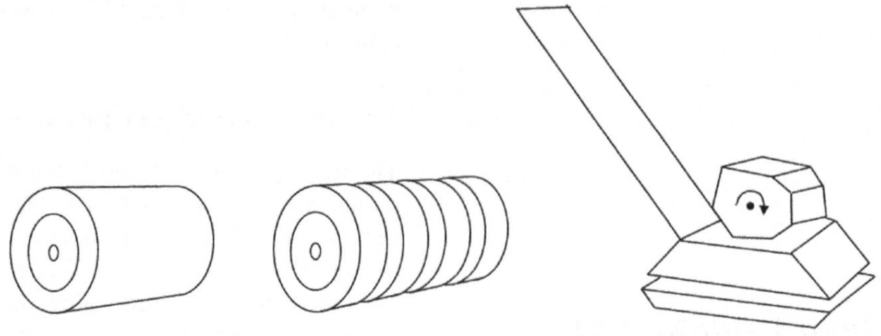

Figure 3.1 Different compaction equipment; (a) roller, (b) rubber tire compactor, and (c) vibrating plate compactor.

trenches or close to a slope or wall face, rollers are often not suitable. Under such conditions, plate compactors are used.

Conventional compaction is used to densify a wide range of cohesionsless and cohesive geomaterials in lifts, but mainly for fill under an unsaturated condition. The lift thickness is typically limited to 300 mm. Depending on geomaterial type, different types of equipment may be selected.

Applications Conventional compaction has been used for earthworks, such as roads, embankments, dams, slopes, walls, parking lots, and sports fields.

Advantages and Limitations Construction equipment is readily available. It is a well-established ground improvement method that has a long history and extensive knowledge in the industry.

The main limitations of this method are: (1) the depth of improvement is limited, (2) it is mainly used for fill and not for in situ natural geomaterials, (3) geomaterials should be within the moisture content close to the optimum content to be more effective, and (4) it is challenging to achieve uniform compaction of geomaterials in a large area.

3.3.2 Principles

Compaction Curve As discussed in Chapter 2, most geomaterials have a maximum dry unit weight with a corresponding optimum moisture content. When the moisture content is lower than the optimum moisture content, the geomaterial has a lower dry unit weight because the friction between particles makes densification difficult. When the moisture content is higher than the optimum moisture content, most of the voids are occupied by water so that limited voids can be compressed. Figure 3.2 shows typical compaction curves. Compaction curves are always below the zero air void line, which represents a fully saturated condition. High-energy compaction results in a high maximum dry unit weight and a low optimum moisture content.

Specifications often require a specific dry unit weight in terms of the percentage of the maximum dry unit weight determined by laboratory standard or modified Proctor tests to be achieved during field compaction. Figure 3.2 shows that the compaction curve for the low energy compaction is below the required unit weight line; therefore, this low-energy compaction will result in a unit weight that never meets the specification requirement. However, high-energy compaction can result in the dry unit weights at the moisture contents ranging from w_1 to w_2 to meet the specification requirement. When intermediate energy is applied, the compaction still can result in the dry unit weights to meet the specification requirement, but within a narrow moisture content range (w_3 to w_4).

The degree of saturation of a geomaterial sample can be calculated by the moisture content on the compaction curve divided by the saturated moisture content at the same dry unit weight. For example, the degree of saturation at the moisture content, w_2, is w_2/w_{sat}.

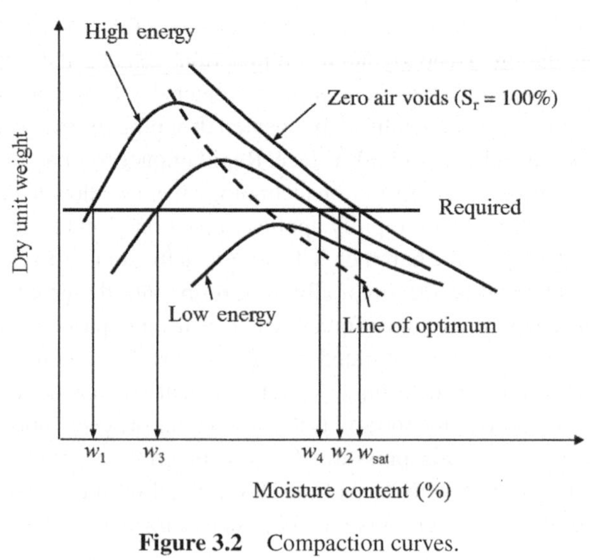

Figure 3.2 Compaction curves.

Relative Compaction Relative compaction, RC, which is defined below, has been commonly used in the field to control and assure the quality of compaction:

$$RC = \frac{\gamma_d}{\gamma_{d,\max}} \times 100\% \quad (3.1)$$

where γ_d = dry unit weight in field
$\gamma_{d,\max}$ = maximum dry unit weight determined by standard or modified Proctor tests in laboratory

For a granular geomaterial, the degree of compaction is better presented in terms of relative density because compaction tests are difficult to do on granular geomaterials in compaction molds in the laboratory. However, some specifications still specify required relative compaction for granular geomaterials rather than relative density. The relative density, D_r, can be converted from relative compaction using the following formula:

$$D_r = \frac{RC - \gamma_{d,\min}/\gamma_{d,\max}}{RC(1 - \gamma_{d,\min}/\gamma_{d,\max})} \times 100\% \quad (3.2)$$

where $\gamma_{d,\min}$ is the minimum dry unit weight determined by minimum index density tests in the laboratory and maximum dry unit weight determined by maximum index density tests in the laboratory.

Lee and Singh (1971) developed an empirical relationship between RC and D_r based on 47 different soils as follows:

$$D_r = \frac{RC - 0.80}{0.20} \times 100\% \quad (3.3)$$

When $\gamma_{d,\min}/\gamma_{d,\max} = 0.80$ and RC is not less than 80% (most projects in the field require 95% relative compaction), Equation (3.2) is approximately equal to Equation (3.3). In other words, the empirical relationship in Equation (3.3) is approximately accurate only when the $\gamma_{d,\min}/\gamma_{d,\max}$ is close to 0.80 and RC is not less than 80%.

Schroeder et al. (2004) pointed out that the maximum dry unit weight determined by the maximum index density tests (ASTM D4253) is close to that determined by the modified Proctor tests (ASTM 1557).

One-Point Method The one-point method, as shown in Figure 3.3, is a simple and approximate way to estimate the compactive energy used in the field and the maximum dry unit weight and its corresponding optimum moisture content of the fill under compaction. To use the one-point method, two previously developed compaction curves in the laboratory should be plotted first. These two curves can be conveniently selected as those from standard and modified Proctor tests. After compaction, field dry unit weight and moisture content are determined and then they are plotted

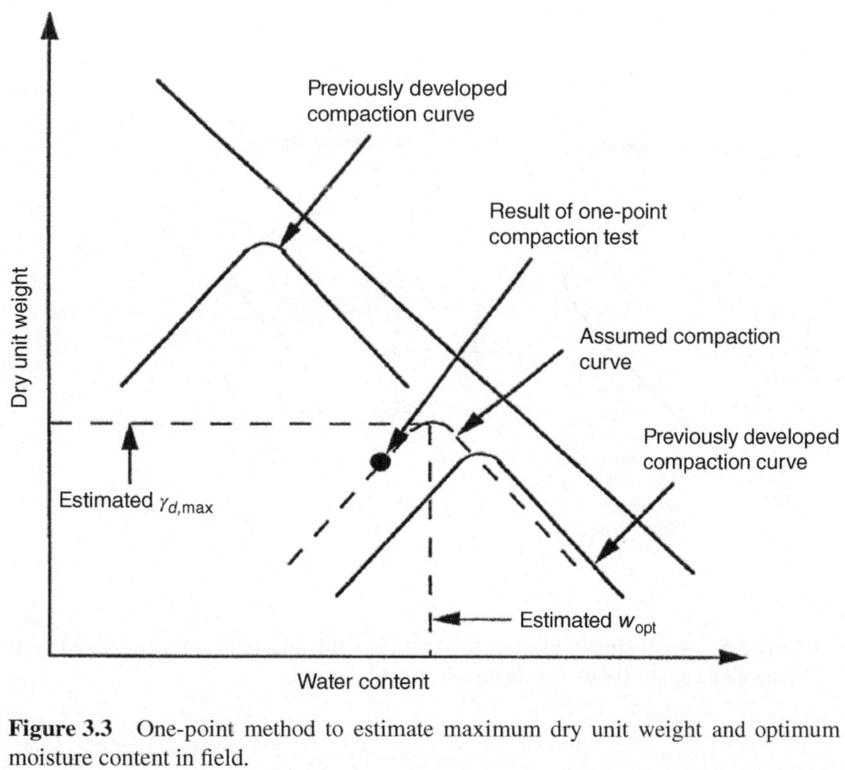

Figure 3.3 One-point method to estimate maximum dry unit weight and optimum moisture content in field.

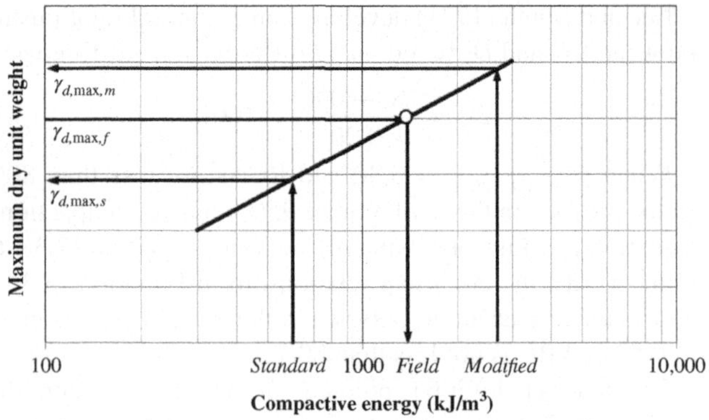

Figure 3.4 Maximum dry unit weight and compactive energy relationship (modified from Hausmann, 1990).

on the same figure. The compaction curve should be drawn through the test data point and "parallel" to those two existing curves. The maximum dry unit weight should be on the line of optimum. As a result, the maximum dry unit weight and its corresponding optimum moisture content of the fill onsite can be determined. Both Hausmann (1990) and Schroeder et al. (2004) indicated that the relationship between the maximum dry unit weight and the compactive energy is approximately linear in a semilog plot, as shown in Figure 3.4. When the maximum dry unit weights from standard and modified Proctor compaction tests, $\gamma_{d,\max,s}$ and $\gamma_{d,\max,m}$ are known, this linear relationship can be established. Therefore, the field compactive energy can be estimated based on the known maximum dry unit weight in the field, $\gamma_{d,\max,f}$.

Influence Factors The relative compaction in the field depends on the following factors:

- Geomaterial type
- Moisture content
- Compaction method, such as static pressure, kneading, vibration, and impact
- Compactive effort including applied energy, compactor size, lift thickness, and number of passes
- Relative layer stiffness (i.e., upper layer over lower layer)

Influence Depth The lift thickness can be determined based on the influence depth of a roller on a specific fill. Figure 3.5 shows that for a large lift thickness, the relative density increases with the depth and then decreases with depth after reaching the maximum value. The low-density value near the surface results from low overburden stress, which makes the fill be disturbed easily. With an increase of the depth, the confining stress increases so that the fill can be easily densified. However, with a further increase of the depth, the applied stress decreases so that its applied compactive energy decreases. D'Appolonia et al. (1969) proposed a method to determine a lift thickness to achieve a desired relative density in Figure 3.5(b). During the compaction of each lift, there is a similar relative density profile with the depth. When the relative density at the overlapping point between two adjacent lifts is higher than the desired density, the lift thickness is appropriate.

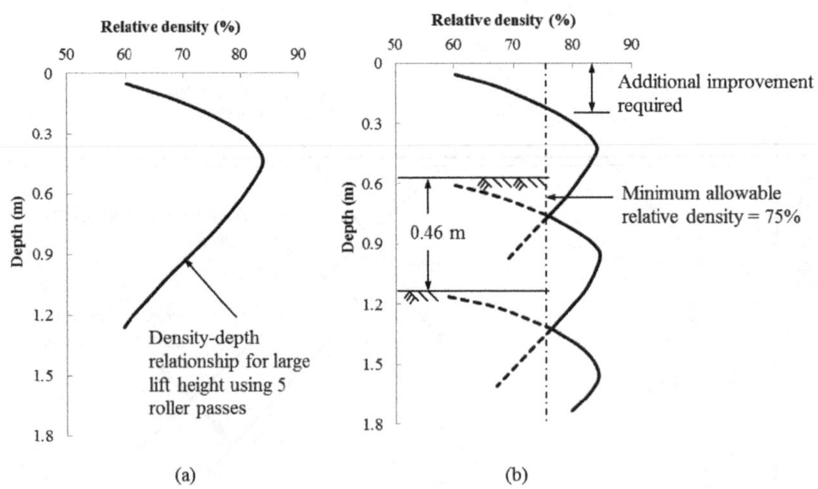

Figure 3.5 Approximate method to determine lift thickness: lifts (modified from D'Appolonia et al., 1969): (a) single lift and (b) multiple.

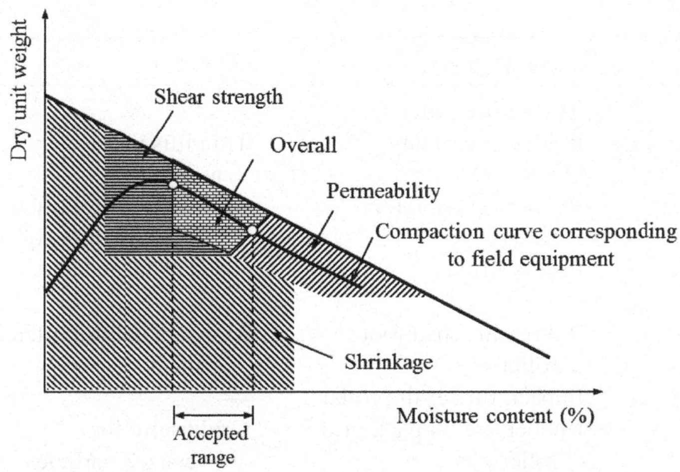

Figure 3.6 Accepted zone of improvement based on different performance requirements (modified from Daniel and Wu, 1993).

3.3.3 Design Considerations

Performance Requirements The relative compaction can be determined based on a specific application. For example, if compacted soil is used for slopes, shear strength is one of the important parameters for slope stability. Assume a relative compaction first (typically 95%), prepare fill samples at this relative compaction, and then test for the strength values. If the strength value is satisfactory for slope stability, this relative compaction is used for construction specifications. However, if this strength value is unsatisfactory, samples at a higher relative compaction should be retested until a satisfactory result is achieved.

However, when a project has more than one performance requirement, the acceptable zone of compaction should be determined on the compaction curve based on different requirements as shown in Figure 3.6. A zone, which satisfies all the requirements, should be selected. Based on the compaction curve corresponding to field equipment, the accepted range of moisture content to be used in field can be determined as shown in Figure 3.6. Table 3.1 provides typical compaction requirements for different applications.

Hausmann (1990) suggested that as a rough guide, the maximum dry unit weights of sand and clay determined by standard Proctor tests are about 95 and 90% those determined by modified Proctor tests, respectively.

Selection of Compaction Equipment Different compaction equipment can produce different action on fill. Smooth drum rollers can apply uniform pressure on fill while sheepsfoot rollers can apply highly concentrated pressure on fill. Pneumatic rubber tire rollers can apply pressure and kneading action on fill, while vibratory rollers can apply vibration on fill. Since granular fills have high friction between particles, they can be more effectively densified by vibration. On other hand, cohesive fills have cohesion and capillary force between particles. These forces can be more effectively broken by high pressure; therefore, rollers with high contact pressure should be used for cohesive fills. Table 3.2 provides the guidance for this selection. Vibratory rollers are more suitable for sands and gravels, while sheepsfoot or pad foot rollers are more suitable for clays and silts.

Optimum Moisture Content and Maximum Dry Unit Weight The *U.S. Navy Design Manual* (1962) suggested the following relationships to estimate the optimum moisture content and the maximum dry unit weight determined in standard Proctor tests based on liquid limit (LL) and

Table 3.1 Typical Compaction Requirements

Compaction Fill for	% Modified Maximum Dry Unit Weight	Moisture Range about Optimum Moisture Content (%)
Roads		
Depth = 0–0.5 m	90–105[a]	−2 to +2
Depth > 0.5 m	90–95[a]	−2 to +2
Small earth dam	90–95	−1 to +3
Large dam	95	−1 to +2
Embankment	95	−2 to +2
Foundation for structure	95	−2 to +2
Wall	90	−2 to +2
Trench	90	−2 to +2
Clay liner	90	0 to +4

[a]Depending on fill type, traffic loading, and function of fill.
Source: Modified from Hausmann (1990).

Table 3.2 Recommended Type of Compaction Equipment

Geomaterial Type	First Choice	Second Choice	Comment
Rock fill	Vibratory roller	Rubber tire roller	—
Plastic soils—CH, MH	Sheepsfoot or pad foot roller	Rubber tire roller	Thin lifts usually needed
Low plasticity soils—CL, ML	Sheepsfoot or pad foot roller	Rubber tire vibratory roller	Moisture control often critical for silty soils
Plastic sands and gravels—GC, SC	Vibratory, pneumatic roller	Pad foot roller	—
Silty sands and gravels—SM, GM	Vibratory roller	Rubber tire, pad foot roller	Moisture control often critical
Clean sands—SW, SP	Vibratory roller	Impact, rubber tire roller	—
Clean gravels—GW, GP	Vibratory roller	Rubber tire, impact, grid roller	Grid useful for oversized particles

Source: Modified from Rollings and Rollings (1996).

plasticity index (PI):

$$w_{\text{opt}} = 6.77 + 0.43\ \text{LL} - 0.21\ \text{PI} \qquad (3.4)$$

$$\gamma_{d,\max} = 20.48 - 0.13\ \text{LL} + 0.05\ \text{PI} \qquad (3.5)$$

where w_{opt} = optimum moisture content (%)
$\gamma_{d,\max}$ = maximum dry unit weight (kN/m^3)

Alternatively, the one-point method can be used to estimate the optimum moisture content and the maximum wet unit weight in a design chart as shown in Figure 3.7. When measured moisture content and wet unit weight of a natural geomaterial are known, their values are plotted on Figure 3.7. The closest curve or the interpolated curve between two curves is used to determine the optimum moisture content and the maximum wet unit weight.

The relationship between the maximum dry unit weight and the optimum moisture content in Figure 3.7 can be approximately expressed as follows:

$$\gamma_{d,\max} = 33.7 - 5.9 \ln(w_{\text{opt}}) \qquad (3.6)$$

where $\gamma_{d,\max}$ = maximum dry unit weight (kN/m^3)
w_{opt} = optimum moisture content (%)

Please note the relationships by the *U.S. Navy Design Manual* (1962), and the Ohio compaction curves should only be used for preliminary design. The estimated parameters should be verified by laboratory or field tests for actual projects.

When fill contains oversized particles (>4.75 mm for 100-mm-diameter mold or >19 mm for 150-mm-diameter mold), maximum dry unit weight and optimum moisture content should be corrected based on the ASTM test

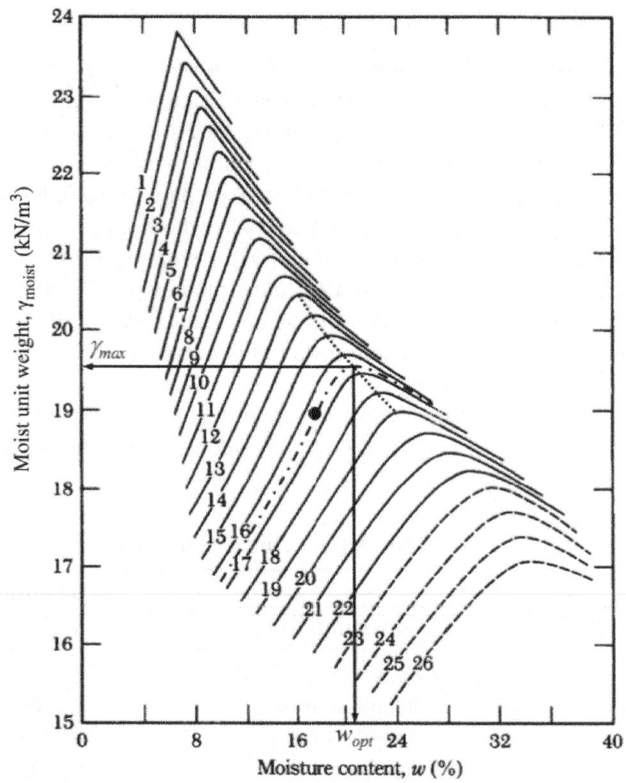

Figure 3.7 Ohio compaction curves (modified from Johnson and Sallberg, 1960).

designation D4718 as follows (valid for 30% or less oversized particles of the total fill material by weight):

$$\gamma_{d,\max,c} = \frac{\gamma_w}{\frac{P_{\text{os}}}{G_{\text{os}}} + \frac{\gamma_w(1-P_{\text{os}})}{\gamma_{d,\max}}}$$

Table 3.3 Lift Thickness and Number of Passes for Different Compaction Equipment

Equipment Type	Applicability	Compacted Lift Thickness (mm)	Number of Passes
Sheepsfoot rollers	For fine-grained fills or coarse-grained fills with more than 20% fines	150	4–6 for fine-grained fills 6–8 for coarse-grained fills
Rubber tire roller	For clean, coarse-grained fills with 4–8% fines	250	3–5
	For fine-grained fills or well-graded coarse-grained fills with more than 8% fines	150–200	4–6
Smooth wheel rollers	Appropriate for subgrade or base course compaction of well-graded sand-gravel mixtures	200–300	4
	May be used for fine-grained fills other than earth dams	150–200	6
Vibrating sheeps-foot rollers	For coarse-grained fills and sand-gravel mixtures	200–300	3–5
			3–5
Vibrating smooth drum rollers	For coarse-grained fills and sand-gravel mixtures—rock fills	200–300 (soil) to 900 (rock)	4–6
Vibrating plate compactors	For coarse-grained fills with less than 4–8% fines, placed thoroughly wet	200–250	3–4
Crawler tractor	Best suited for coarse-grained fills with less than 4–8% fines, placed thoroughly wet	150–250	3–4
Power tamper or rammer	For difficult access, trench backfill. Suitable for all inorganic fills	100–150 for silt or clay, 150 for coarse-grained fills	2

Source: Modified from U.S. Navy (1986).

where $\gamma_{d,\max,c}$ = corrected maximum dry unit weight
$\gamma_{d,\max}$ = maximum dry unit weight without oversized particles determined in laboratory
γ_w = unit weight of water
G_{os} = specific gravity of over-sized particles
P_{os} = percent of oversized particles by weight (in decimal)

The corrected optimum moisture content is

$$w_{\text{opt},c} = w_{\text{opt}}(1 - P_{\text{os}}) + w_{\text{SSD}} P_{\text{os}}$$

where $w_{\text{opt},c}$ = corrected optimum moisture content
w_{opt} = optimum moisture content without oversized particles determined in laboratory
w_{SSD} = saturated surface dry (SSD) moisture content of oversized particles

Lift Thickness and Number of Passes Table 3.3 provides the recommended lift thickness and number of passes for different compaction equipment.

Borrow Volume Since the unit weight of geomaterial on a borrow site is often different from that on a construction site after compaction, a conversion factor is used to make this conversion. The convention factor also considers weight loss in striping, waste, oversize, and transportation. The total borrow volume required for compacted fill can be estimated as follows (U.S. Navy, 1986):

$$V_b = \frac{\gamma_{d,f}}{\gamma_{d,b}} V_{f,r} + \frac{W_l}{\gamma_{d,b}} \tag{3.7}$$

where V_b = borrow volume
$V_{f,r}$ = required fill volume
$\gamma_{d,f}$ = dry unit weight of fill
$\gamma_{d,b}$ = dry unit weight of borrow
W_l = dry weight loss in striping, waste, oversize, and transportation

For a rough estimation, the borrow volume can be estimated by the following formula:

$$V_b = \zeta V_{f,r} \tag{3.8}$$

where ζ is the volume conversion factor considering "shrinkage" or "swell" from the borrow geomaterial to the compacted fill. Borrow soil typically has a conversion factor of 1.10–1.15 while rock has a conversion factor of 0.7–0.9.

3.3.4 Design Parameters and Procedure

Design Parameters The design parameters for conventional compaction may include the following parameters:

- Project requirement(s)
- Relative compaction
- Area and thickness of compacted fill
- Type and gradation of fill
- Type of equipment
- Optimum moisture content and maximum dry unit weight or minimum and maximum void ratios
- Borrow volume
- Thickness and number of lifts
- Number of passes

Design Procedure The following procedure may be adopted to design a compaction project:

1. Collect fill material.
2. Conduct laboratory tests (including sieve analysis test, Atterberg limit tests, and specific gravity test) to classify soil type.
3. Run standard and/or modified Proctor compaction tests in laboratory to obtain the compaction curve, including the zero air void line and determine optimum moisture content and maximum unit weight. For a preliminary design, the optimum moisture content and the maximum dry unit weight may be estimated based on Equations (3.4) and (3.5). For cohesionless fill, minimum and maximum void ratio tests should be performed.
4. Based on the project requirements (e.g., required shear strength, modulus, permeability, shrinkage potential, and/or swell potential), run laboratory tests at a trial dry unit weight (typically 95% of the maximum dry unit weight) or different compaction levels to determine the required dry unit weight by meeting the project requirements so that the required relative compaction can be calculated. Use Table 3.1 if there is no available test data.
5. Based on fill type and gradation, select compaction equipment.
6. Based on the selected compaction equipment and fill type, select lift thickness and number of passes.
7. Based on the area and thickness of compacted fill, calculate fill volume and borrow volume.
8. For a large and important project, a field trial is recommended to verify/adjust the above design parameters.

3.3.5 Design Example

A 200 m × 200 m parking lot needs to be backfilled with 0.4-m-thick well-graded coarse-grained fill as a base course before being paved by 75-mm-thick asphalt surface. The available fill material has 15% fines and 20% oversized particles (G_{os} = 2.68 and w_{SSD} = 2.6%). The granular fill is mainly used to support the pavement; therefore, the fill strength and stiffness are the key performance parameters. Standard Proctor tests are performed on the fill without oversized particles, which result in the test results shown in Example Table 3.1. The specific gravity of the fill without oversized particles is 2.65. Determine the parameters needed for field compaction.

Example Table 3.1 Compaction Test Results

Moisture content (%)	6	8	10	12	14
Moist unit weight (kN/m^3)	17.6	18.5	20	19.5	19

Solution

1. Collect fill material: Done.
2. Conduct laboratory tests for fill classification: Done.
3. Run compaction tests and determine maximum dry unit weight and optimum moisture content: Tests are done. Since moist unit weights are determined from the laboratory tests, dry unit weights should be calculated for the compaction curve. At the moisture content of 6%, the dry unit weight can be calculated as follows:

$$\gamma_d = \frac{\gamma_{moist}}{1+w} = \frac{17.6}{1+0.06} = 16.6 \text{ kN/m}^3$$

Other dry unit weight results are shown in Example Table 3.2. The zero air void line can be determined using the following formula (using w = 6% as an example):

$$\gamma_{d,\,zav} = \frac{\gamma_w}{w+(1/G_s)} = \frac{9.81}{0.06+(1/2.65)} = 22.4 \text{ kN/m}^3$$

Example Table 3.2 Compaction Test Results

Moisture content (%)	6	8	10	12	14
Moist unit weight (kN/m^3)	17.6	18.5	20	19.5	19
Dry unit weight (kN/m^3)	16.6	17.1	18.2	17.4	16.7
Dry unit weight at zero air void line (kN/m^3)	22.4	21.4	20.6	19.7	19.0

The compaction curve and the zero air void line are plotted in Example Figure 3.1. The compaction curve is below the zero air void line; therefore, the test data are valid. From the compaction curve, it can be determined that the maximum dry unit weight is 18.2 kN/m^3 and the optimum moisture content is 10.0%.

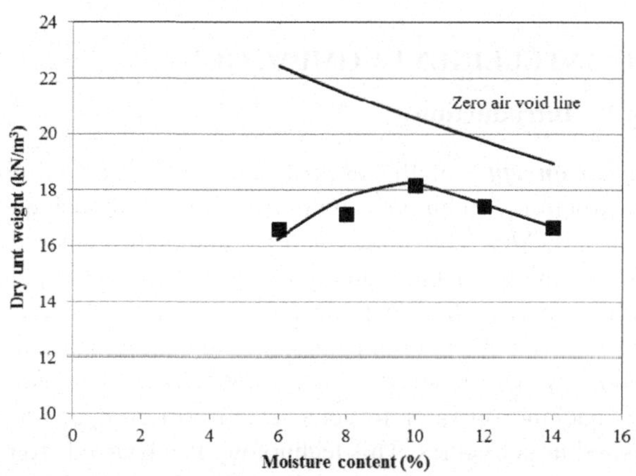

Example Figure 3.1 Compaction curve.

Since this fill contains oversized particles, its maximum dry unit weight and the optimum moisture content should be corrected as follows:

$$\gamma_{d,\max,c} = \frac{\gamma_w}{\frac{P_{os}}{G_{os}} + \frac{\gamma_w(1-P_{os})}{\gamma_{d,\max}}} = \frac{9.81}{\frac{0.20}{2.68} + \frac{9.81\times(1-0.20)}{18.2}}$$

$$= 19.4 \text{ kN/m}^3$$

$$w_{\text{opt},c} = w_{\text{opt}}(1 - P_{os}) + w_{\text{SSD}} P_{os}$$

$$= 0.10 \times (1 - 0.20) + 0.026 \times 0.20$$

$$= 0.085 = 8.5\%$$

4. Since there is no available test data, use Table 3.1 for the relative compaction. Since the total thickness (fill + asphalt surface) is 0.475 m (< 0.5 m), the required relative compaction is 90–105%. Considering a parking lot will not be subjected to highway truck loads, 90% relative compaction is selected. Since this relative compaction is based on the modified Proctor compaction tests, the relative compaction corresponding to the standard Proctor compaction tests is

$$\text{RC} = \frac{90\%}{0.95} = 95\%$$

5. Since the fill is a well-graded coarse-grained fill with 15% fines, a vibratory roller or rubber tire roller may be used based on Tables 3.2 and 3.3. Select the rubber tire roller.
6. Based on the selected equipment (i.e., the rubber tire roller), the compacted lift thickness is 150–200 mm and the number of passes is 4–6. Considering the base course thickness is 400 mm, it is convenient to choose the lift thickness of 200 mm. For the thicker lift thickness, the higher number of passes at 6 is selected.
7. The total compacted fill volume is 200 × 200 × 0.4 = 16,000 m^3. Considering the "shrinkage" from the borrow fill to the compacted fill (use a factor of 1.15), the required borrow fill volume is

$$V_b = \zeta V_{f,r} = 1.15 \times 16000 = 18400 \text{ m}^3$$

8. Since a parking lot is typically not a large project, no field trial is necessary.

3.3.6 Construction

Conventional compaction involves transportation of fill material from a borrow site, placement of the fill, watering or drying of fill if needed, and compaction.

It is likely that the fill transported from a borrow site onto the project is too wet or dry. If the fill is too wet, it can be spread out on the site for drying. When appropriate moisture content is achieved, compaction can start. However, when the fill is dry, water can be spread on the fill by a water tank. After water is spread, it is necessary to wait a certain amount of time (depending on fill type) to allow water to penetrate into fill and have uniform moisture distribution. Silty fill can lose moisture quickly; therefore, compaction should start immediately after appropriate moisture content is reached.

During compaction, rollers should travel at a constant speed in a regular pattern. Sudden break, acceleration, and turning, which may disturb compacted fill and subgrade, should be avoided.

Fill should not be overcompacted. After the fill has a degree of saturation close to full saturation (say 95% or higher),

Table 3.4 Recommended Number of Tests

Earth Structure	Volume of Fill per Test (m^3)
Embankment	500–2000
Impermeable liner	200–1000
Subgrade	500–1500
Base course	500–1000
Backfill in trench or around structure	100–200

Source: U.S. Navy (1986) and Hausmann (1990).

continued compaction may not densify the fill, instead it may disturb the fill. The additional compactive energy may create shear stresses in the fill that remold the compacted fill and smear fill particles. As illustrated by Seed and Chan (1959), compaction at the wet side of optimum can change soil fabric from a flocculated to a dispersed one. A fill with a dispersed fabric has a low shear strength. Obvious ground heave or movement is an indication of overcompaction. To fix the overcompaction problem, the fill should be excavated and dried for recompaction.

For sandy and clayey fills, the surface after compaction may not be smooth, especially after the compaction by sheepsfoot rollers. Smooth wheel rollers can be used for finish operation of fill compaction to smooth out the ground surface. Smooth wheel rollers have also been commonly used for proof rolling. The purpose of proof rolling is to identify areas of fill that are undercompacted or too wet. Proof rolling may not be able to detect the problem for the areas compacted at dry side of optimum because they are relatively strong and stiff due to capillary action.

3.3.7 Quality Control and Assurance

Quality control and assurance are important parts of conventional compaction because it often involves large-area construction. Quality control typically includes the following:

- Quality of fill material, such as type, gradation, and Atterberg limits
- Type and weight of equipment
- Lift thickness
- Moisture content at compaction
- Number of passes

Quality assurance is achieved through field tests after each lift compaction and/or completion of the overall compaction. Table 3.4 provides recommended number of field tests to be performed on a project site.

There are several field test methods available for quality assurance of compaction as provided in Table 3.5. Moisture content, density, and stiffness are three commonly measured parameters. Most of the current specifications for compaction only require moisture content and density. Nuclear gauge and sand cone tests are two of the most commonly used methods in the field. Stiffness is a performance index, which has gained more attention by the transportation community in recent years. It is a trend that fill stiffness will be included as a performance index in future specifications. Details of these test methods can be found in the ASTM standards.

3.4 INTELLIGENT COMPACTION

3.4.1 Introduction

Basic Concept Intelligent compaction (IC) is a continuous construction technique with quality assurance/quality control (QA/QC) processes incorporated in the construction. This technique includes automatic measurements of geomaterial properties through preinstalled sensors on the machine and adjustment and optimization of the vibration amplitude, frequency, and/or speed using a feedback control system by the machine operator to achieve uniform compaction and desired target values. This technology is advanced from the continuous compaction control (CCC) system initiated in Europe in 1970s. Figure 3.8 shows the main components of the CCC/IC compaction system, which consists of a roller, a controlling unit, machine-integrated sensors, an onboard computer, and a global positining system (GPS). The computer monitor displays operation parameters,

Table 3.5 Field Tests for Quality Assurance of Compaction

Test Method	Measurement	Standard
Sand cone	Density	ASTM D1556
Rubber balloon	Density	ASTM D2167
Nuclear gauge	Moisture content and density	ASTM D6938
Dynamic cone penetrometer	Penetration index	ASTM D6951
Soil stiffness gauge	Stiffness	ASTM D6758
Falling weight deflectometer	Stiffness	ASTM D4694
Light weight deflectometer	Stiffness	ASTM E2583
Electrical density gauge	Density	ASTM D7830
Time domain reflectometry	Moisture content	ASTM D6565

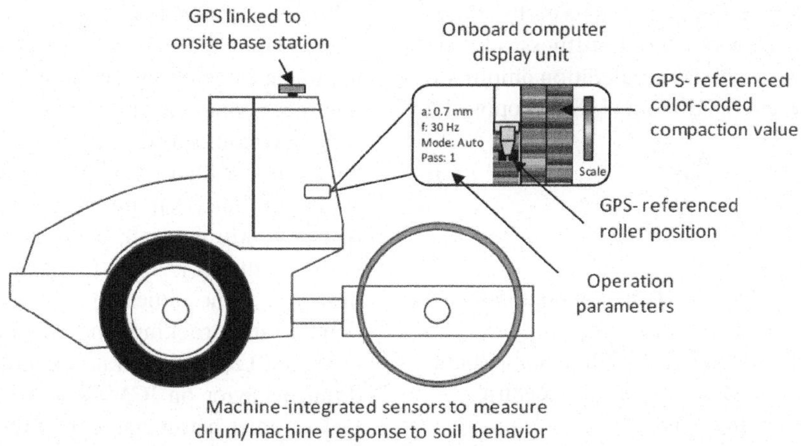

Figure 3.8 Main components of CCC/IC compaction system (Chang et al., 2011).

Table 3.6 Available Single IC Drum Rollers[a]

Vendor	Ammann Case	Bomag	Caterpillar	Dynapac	Sakai
Model	ACEplus	VarioControl	NA	DCA-S (GPS)	CIS
Model number	SV	BW213-4BVC	NA	CA 152-702	SV505/SV510
Auto-feedback	Yes	Yes	Yes	Yes	No
Measurement system	Yes	Yes	Yes	Yes	Yes
Measurement value	K_b	E_{vib}	CMV	CMV	CCV
Measurement unit	MN/m	MN/m^2	Dimensionless	Dimensionless	Dimensionless
GPS capability	Yes	Yes	Yes	Yes	Yes
Documentation system	ACEPlus	BCM05 Office and Mobile	AccuGrade	PCA	AithonMT-R

[a]K_b = stiffness or ground bearing capacity; E_{vib} = vibration modulus; CMV = compaction meter value; and CCV = compaction control value.
Source: Modified from Chang et al. (2011).

GPS-referenced roller position, and GPS-referenced color-coded compaction value. The color-coded compaction value allows the operator to identify areas of weak spots or poor compaction and make necessary adjustments in rolling operations. The GPS mounted on the machine is linked to an onsite base station to determine the location of the roller. There are several IC rollers available in practice. Table 3.6 lists five commonly used single IC drum rollers. Drums can be smooth or padfoot. Double drum rollers are also available but are mainly used to compact asphalt pavement materials.

Suitability Similar to conventional compaction rollers, IC rollers can be used to compact natural subgrade and fill ranging from cohesive to cohesionless geomaterials. IC rollers are even more suitable for nonuniform subgrade conditions than conventional compaction rollers.

Applications The IC method has the same applications as the conventional compaction method.

Advantages and Limitations The IC method has the measurement and feedback system, which helps the operator identify weak spots for further compaction or treatment and QC/QA, avoid overcompaction, and reduce the number of passes. The IC method can serve as full coverage QC/QA and minimize the need for field inspection. Therefore, it can result in more efficient and uniform compaction.

Currently, the equipment is more expensive than conventional compaction equipment. It requires managing and analyzing a significant amount of data. This method often requires test strips for calibration of target compaction values prior to production compaction.

3.4.2 Principles

Intelligent Compaction Measurement Value Intelligent compaction has compaction passes and measuring passes. During the measuring pass, the machine takes measurements of drum–soil interaction. Several IC roller manufacturers develop different intelligent compaction measurement values (ICMVs) for their machines. The commonly used ICMVs in the United States include: (1) compaction meter value (CMV), (2) machine drive power (MDP), (3) compaction control value (CCV), (4) stiffness, K_b, and (5) vibration modulus, E_{vib}.

1. Compaction meter value (CMV) Based on early studies on the relationship between geomaterial stiffness and the drum acceleration amplitude and the acceleration amplitude of its harmonics, Thurner and Sandstrom (1980) proposed a CMV as follows:

$$\mathrm{CMV} = C\frac{A_{2\Omega}}{A_{\Omega}} \tag{3.9}$$

where C = constant (i.e., 300)
A_{Ω} = amplitude of the fundamental component of vibration (i.e., operating)
$A_{2\Omega}$ = amplitude of the first harmonic component of vibration (i.e., twice the eccentric excitation frequency)

Mooney and Adams (2007) suggested that the $A_{2\Omega}/A_{\Omega}$ ratio is an indicator of whether geomaterial and drum have a nonlinear interaction. The nonlinear interaction results from nonlinear behavior of geomaterial and varying contact area between geomaterial and drum. When geomaterial and drum have a linear interaction, the $A_{2\Omega}/A_{\Omega}$ ratio is equal to zero. When a roller is on an uncompacted loose geomaterial, the $A_{2\Omega}/A_{\Omega}$ ratio is small. With an increase of the geomaterial stiffness, the $A_{2\Omega}/A_{\Omega}$ ratio increases as shown in Figure 3.9; however, at the same time, the chance for the drum to have double jump, rocking, and chaotic motion increases. Adam and Kopf (2004) illustrate the influence of soil stiffness and drum behavior on ICMVs in Figure 3.10. To consider this fact, another parameter, called resonant meter value (RMV)

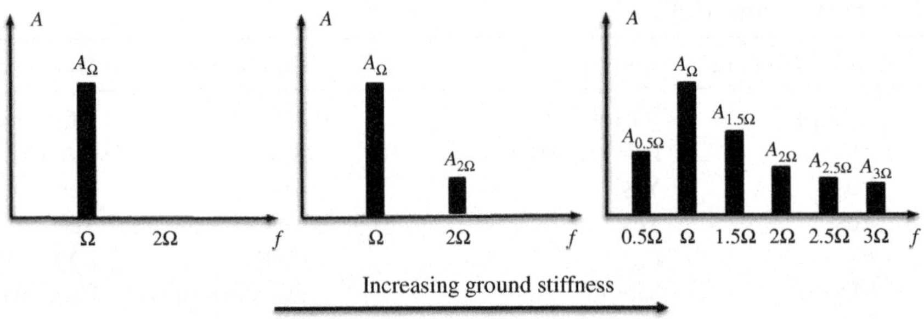

Figure 3.9 Changes in amplitude spectrum with an increase of ground stiffness (Chang et al., 2011).

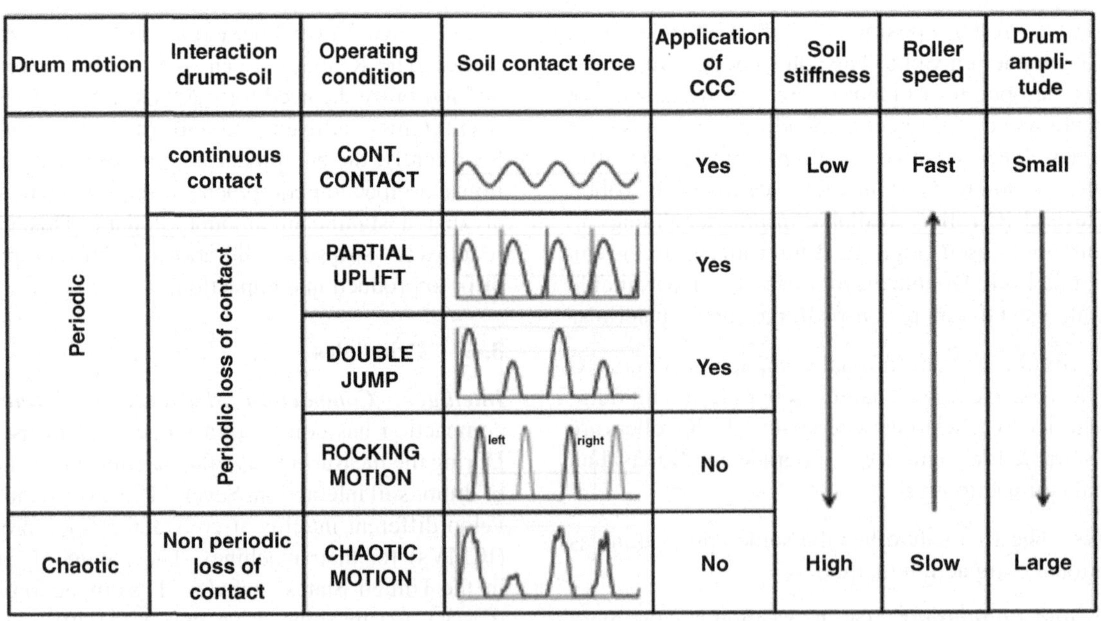

Drum motion	Interaction drum-soil	Operating condition	Soil contact force	Application of CCC	Soil stiffness	Roller speed	Drum ampli-tude
Periodic	continuous contact	CONT. CONTACT		Yes	Low	Fast	Small
	Periodic loss of contact	PARTIAL UPLIFT		Yes	↓	↑	↓
		DOUBLE JUMP		Yes			
		ROCKING MOTION	left right	No			
Chaotic	Non periodic loss of contact	CHAOTIC MOTION		No	High	Slow	Large

Figure 3.10 Influence of soil stiffness and drum behavior on ICMVs (Adam and Kopf, 2004).

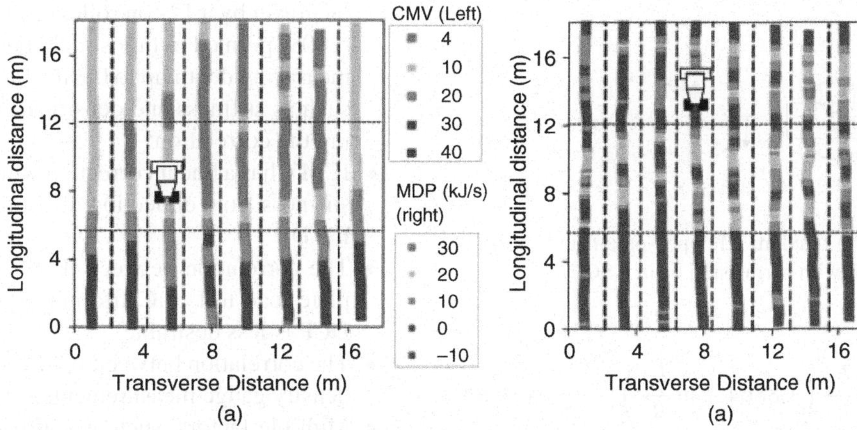

Figure 3.11 Example of color-coded CMVs and MDPs in one case study (Vennapusa et al., 2010, with permission from ASCE).

or bouncing value (BV), is suggested as follows:

$$\text{RMV or BV} = C\frac{A_{0.5\Omega}}{A_{\Omega}} \quad (3.10)$$

where $A_{0.5\Omega}$ is the subharmonic amplitude caused by jumping (i.e., the drum skips every other cycle).

When RMV = 0, there is continuous contact or partial uplift between the drum and the geomtareial. Theoretically speaking, the drum starts to double jump when RMV > 0. Based on one field study, however, Vennapusa et al. (2010) indicated a value of RMV > 2 considered as a practical cutoff value for further analysis.

2. Compaction control value (CCV) Similar to CMV, CCV is determined from the measured acceleration data based on the harmonic frequency. Figure 3.9 shows that stiff ground results in a "jumping" motion with vibration accelerations at different frequency components. CCV is defined as

$$\text{CCV} = \frac{A_{0.5\Omega} + A_{1.5\Omega} + A_{2\Omega} + A_{2.5\Omega} + A_{3\Omega}}{A_{0.5\Omega} + A_{\Omega}} \quad (3.11)$$

3. Machine drive power (MDP) Machine drive power (MDP) is defined below by relating the power needed to drive the machine during compaction to the properties of compacted geomaterial (Chang et al., 2011):

$$\text{MDP} = P_g - W_r v_r \left(\sin\beta + \frac{a_m}{g}\right) - (m_m v_r + b_m) \quad (3.12)$$

where MDP = machine drive power (kJ/s)
P_g = gross power needed to drive the machine (kJ/s)
W_r = roller weight (kN)
β = slope angle
a_m = machine acceleration (m/s^2)
g = acceleration of gravity (m/s^2)
v_r = roller velocity
m_m (kJ/m) and b_m (kJ/s) = machine-dependent internal loss coefficients

MDP is a relative value calibrated against a hard compacted surface, which is set as MDP = 0. A positive MDP value implies a less compacted material than the hard surface while a negative value implies a more compacted material.

Figure 3.11 shows color-coded CMVs and MDPs in one case study.

4. Roller-integrated stiffness (K_b) The roller–ground interaction can be modeled using a one-degree-freedom lumped-parameter model as shown in Figure 3.12. In this model, the drum is modeled as a rigid body with a mass, m_d; m_0e_0 is the eccentric moment of the unbalanced mass generated by the roller; m_f is the mass of the frame; z_d and $\ddot{z}_d$ are the vertical displacement and the acceleration of the drum, respectively; Ω is the excitation frequency (i.e., $2\pi f$); and F_s is the drum–ground contact force. The ground stiffness, K_b, is defined as the slope of the unloading portion of the contact force with the drum vertical displacement, that is, $\Delta F_s/\Delta z_d$, and can be expressed as follows when the drum velocity is equal to zero (Andergg and Kaufmann, 2004):

$$K_b = 4\pi^2 f^2 \left(m_d + \frac{m_0 e_0 \cos\Delta_p}{z_d}\right) \quad (3.13)$$

where Δ_p is the phase angle.

5. Vibratory modulus, E_{vib} The vibratory modulus, E_{vib}, can also be obtained using the one-degree-freedom lumped-parameter model with the analytical solution developed by Lundberg (1939) for a rigid cylinder on an elastic half-space

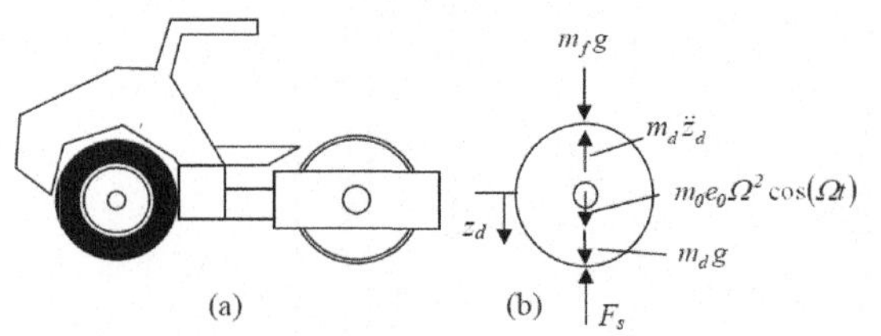

Figure 3.12 One-degree freedom lumped-parameter model (Mooney and Adam, 2007, with permission from ASCE).

as follows:

$$z_d = \frac{2(1 - V_s^2)}{\pi E_{\text{vib}}} \frac{F_s}{L_d} \left(1.8864 + \ln \frac{L_d}{B_d}\right) \tag{3.14}$$

where V_s = Poisson's ratio of the geomaterial
L_d = length of the drum
B_d = contact width of the drum

According to Hertz's contact theory (Hertz, 1895), the following relationship can be established:

$$B_d = \sqrt{\frac{16}{\pi} \frac{R_d(1 - V_s^2)}{E_{\text{vib}}} \frac{F_s}{L_d}} \tag{3.15}$$

where R_d is the radius of the drum. E_{vib} in Equations (3.14) and (3.15) can be solved by iterations. Kröber et al. (2001) provided a detailed description about the E_{vib} measurement.

Field Correlation Intelligent compaction measurement values (ICMVs) as discussed above are recorded during compaction. They can be used by the operator to identify possible weak spots, adjust compaction parameters, and/or complete the compaction after reaching target values. How the ICMVs are relevant to geomaterial properties determined by traditional test methods is a practical question. Researchers have conducted a few studies to correlate the ICMVs with dry unit weight, strength, and modulus or stiffness of geomaterial. The recent study by Chang et al. (2011) for the field correlation with ICMVs evaluated 13 in situ QC/QA test methods, including nuclear gauge, electrical density gauge, lightweight deflectometer (LWD), soil stiffness gauge, static plate load test, falling weight deflectometer (FWD), Briaud compaction device, seismic pavement analyzer, Clegg hammer, Shelby tube sampling, dynamic cone penetrometer (DCP), static cone penetrometer, and heavy test rolling. Below are a few key findings:

- Different test methods have different measuring depths. The measuring depth and field moisture content are two main factors affecting the correlation between ICMVs and field point test methods. ISSMGE (2005) suggested that the measuring depth of a roller depends on the weight of the roller and the maximum measuring depth on gravel by a 12-ton roller can reach 0.8–1.5 m. White (2008) pointed out most field point test methods have a measuring depth up to 0.3 m. Target material properties such as stiffness and density are other factors influencing the correlation.
- ICMV has a linear correction with back-calculated layer stiffness from deflection measurements (i.e., LWD and FWD).
- The correlation between ICMV and the stiffness from plate load test or California bearing ratio (CBR) from DCP is less desirable.
- The correlation between ICMV and nuclear/nonnuclear density gauge measurements is least desirable.
- Multiple factors, such as soil moisture content and vibration frequency and amplitude, affect ICMV.

3.4.3 Design Considerations

Intelligent compaction is a construction-based ground improvement method. In addition to the typical design for the conventional compaction method, the design of IC is mostly related to the planning of test strips and the selection or determination of ICMV and compaction parameters. ISSMGE (2005) developed detailed specifications for the IC operation while Chang et al. (2011) provided sample specifications.

Rollers Rollers can be smooth or padfoot single-drum vibratory rollers. They should be self-propelled and have accelerometers or other sensors mounted on the machine to enable compaction measurements. The output of the measurements is the ICMV.

Correlation for Preliminary Design Researchers have developed different relationships between ICMV and field test measurements. However, most of these relationships are site specific. Table 3.7 lists a few relationships with some consistent results from different sites and locations, which may be used for preliminary design. The preliminary design can be used as a start point to prepare for a test section.

Test Section ISSMGE (2005) suggests selecting a test section of 100 m long with entire road width within a construction site. Chang et al. (2011) suggested a test section of 75 m long and 8 m wide. The test section should be representative in terms of the properties of geomaterial layers on the site and be divided into several roller tracks with the maximum track width corresponding to the drum width. The overlap between tracks should not exceed 10% of the drum width. If different materials exist or are used on the project site, all of them should be evaluated. Dynamic passes should be performed as measuring passes with the same machine settings (speed, amplitude, and frequency)

Table 3.7 Correlation between ICMV and Average Field Test Measurements

Correlation	Soil Type	Notes	Reference
$\gamma_d(\text{kN/m}^3) = -a\log(\text{MDP}) + b$, $a = 1.7$–4.6, $b = 20.2$–22.7, $R^2 = 0.90$–0.96	Granular soil	Combine three soil types at different locations	White et al. (2008)
$\gamma_d(\text{kN/m}^3) = a\,\text{CMV} + b$, $a = 0.1$–0.2, $b = 12.0$–17.2, $R^2 = 0.83$–0.96	Granular soil	Combine three soil types at different locations	White et al. (2008)
$\gamma_d(\text{kN/m}^3) = a - b\,\text{MDP} + cw$, $a = 10.6$–15.3, $b = 0.2$, $c = 0.1$–0.3, $R^2 = 0.60$–0.93	Cohesive soil	Combine three soil types at different locations	White et al. (2008)
$\gamma_d(\text{kN/m}^3) = 15.47 - 0.37\,\text{MDP} + 0.16w$, $R^2 = 0.78$	Cohesive soil	Padfoot roller, $A_m = 0.8$ mm, $f = 33$ Hz, $v_r = 4$ km/h	Mooney et al. (2010)
$E_{\text{LWD}}(\text{MPa}) = a - b\,\text{MDP}$, $a = 41.4$–48.5, $b = 1.6$–1.3, $R^2 = 0.46$ to 0.78	Cohesive soil	Combine three soil types at different locations; E_{LWD} determined using Keros lightweight deflectometer	White et al. (2008)

Note: w = moisture content of soil; A_m = amplitude; E_{LWD} = elastic modulus of soil determined by a lightweight deflectometer (LWD).

throughout the test section. During the measuring pass the ICMVs are recorded. After each pass, nondestructive devices should be used to measure the density and/or stiffness of the geomaterial at different locations. Chang et al. (2011) suggested 10 uniformly spaced locations on the test section for density or stiffness tests. ISSMGE (2005) suggested 9 locations for static plate load tests and 36 locations for LWD tests. Based on the recorded ICMVs, an IC compaction curve can be established between the ICMVs and the number of passes.

Selection of Target ICMV Different countries or organizations have different specifications for the selection of target ICMV and acceptance criteria. Two such specifications are discussed herein.

ISSMGE (2005) recommended the use of the target initial load secant modulus, E_i (either E_{v1}, a modulus from a static plate load test or E_{LWD}, a modulus from LWD) to establish the target ICMV. A linear regression method as shown in Figure 3.13 is used. The target mean ICMV corresponds to $1.05E_i$ (E_{v1} or E_{LWD}), the target minimum ICMV corresponds to $0.95E_i$, and the target maximum ICMV corresponds to 1.5 times the minimum ICMV.

Chang et al. (2011) suggested a two-step procedure to determine the production target ICMV based on the density requirement. In the first step, the target ICMV is determined

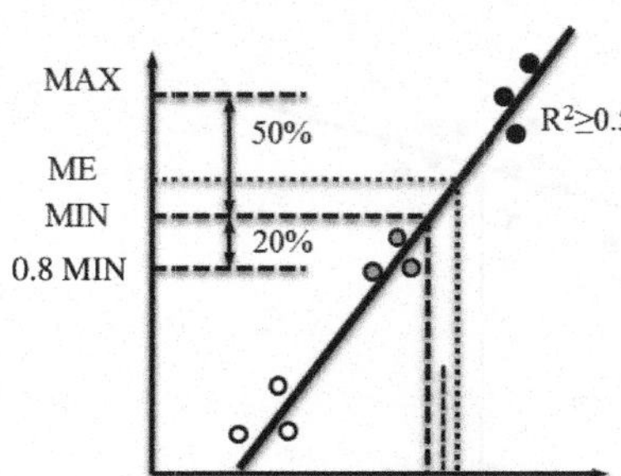

Level	E_{v1} (MN/m²)
1-m Below Subgrade*	15 (cohesive); 20 (cohesionless)
Top of Subgrade	25 (cohesive); 35 (cohesionless)
Top of Subbase	60 (rounded); 72 (angular)
Top of Base	75 (rounded); 90 (angular)
Level	**E_{LWD} (MN/m²)**
1-m Below Subgrade*	18 (cohesive); 24 (cohesionless)
Top of Subgrade	30 (cohesive); 38 (cohesionless)
Top of Subbase	58 (rounded); 68 (angular)
Top of Base	70 (rounded); 82 (angular)

* If fill section is to be constructed

Figure 3.13 Establishment of target values (ISSMGE, 2005).

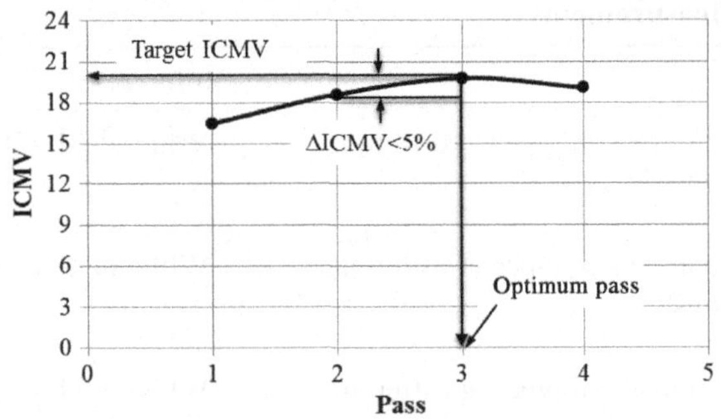

Figure 3.14 IC compaction curve and target ICMV (Chang et al., 2011).

as the point where the increase in the ICMV of the geomaterial between passes is less than 5% on the compaction curve as shown in Figure 3.14. The corresponding number of passes is the optimum pass. In the second step, an adjoining section is compacted using the same roller settings and number of passes for the established target ICMV to verify the compaction with the same density device after the final pass. A linear regression method as shown in Figure 3.15 is used to determine the production target ICMV based on the required density, in which G_{mm} is the maximum specific gravity of a mix.

3.4.4 Construction

Most of the construction procedures for conventional compaction apply to the IC operation. Some special IC procedures are provided below:

- Construction site should be divided into measuring sections with a size similar to the test section to cover the entire area.
- No intersection is permitted.
- Measuring passes should be performed in a forward direction while reverse travel should be performed statically. The roller's travel speed should be kept constant between 2 and 6 km/h.
- If a measuring pass encounters significant jump, the testing sequence should be interrupted. The measuring pass should be repeated with lower amplitude and/or higher travel speed.
- After the measuring pass, the roller should be moved to the next track.

3.4.5 Quality Control and Assurance

In addition to typical acceptance processes for materials and construction operations that are the same as those for conventional compaction, ISSMGE (2005) and Chang et al. (2011) have additional requirements specific to the IC technology.

ISSMGE (2005) suggests the following ICMV acceptance criteria in a production area for quality control and assurance:

- Production mean ICMV ≥ target mean ICMV.
- One hundred percent of production ICMV ≥ 0.8 target minimum ICMV.
- Ninety percent production ICMV ≥ target minimum ICMV.
- Compaction must be performed until production mean ICMV (current pass) > production mean ICMV (preceding pass) by less than 5%.
- If 100% of production ICMV > target minimum ICMV, the coefficient of variation (COV) for the entire area ≤ 20%.
- If 0.8 target minimum ICMV < production mean ICMV < target minimum ICMV, 100% of production ICMV ≤ maximum target ICMV.

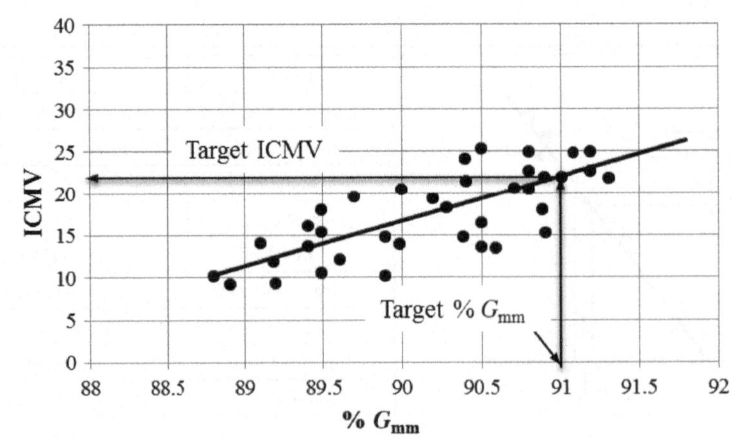

Figure 3.15 Linear regression method to determine production target ICMV (Chang et al., 2011).

Chang et al. (2011) suggest the following quality control and assurance procedure:

- Ninety percent or more of the individual construction area shall meet the optimum pass requirement and 70% of the target ICMV determined from the test sections.
- Rework and reevaluation are necessary for the construction area not meeting the ICMV criterion prior to continuing with the production in that area.

3.5 DEEP DYNAMIC COMPACTION

3.5.1 Introduction

Basic Concept Deep dynamic compaction is to repeatedly drop a weight ("tamper") freely from a height onto the ground surface in a pattern to compact problematic geomaterial to a deep depth as shown in Figure 3.16. Repeated impacts reduce voids, densify the geomaterial, and induce ground movement. A tamper typically has a weight of 5–40 tons and drops from a height of 10–40 m. Different from shallow compaction, deep dynamic compaction can compact problematic geomaterial down to a depth of 10 m. The concept of dynamic compaction can be traced back to Roman times. The modern technology has been credited to the French engineer, Louis Menard since 1960s. The article published by Menard and Broise in 1975 provided theoretical bases for dynamic consolidation of fine-grained soils by heavy tamping.

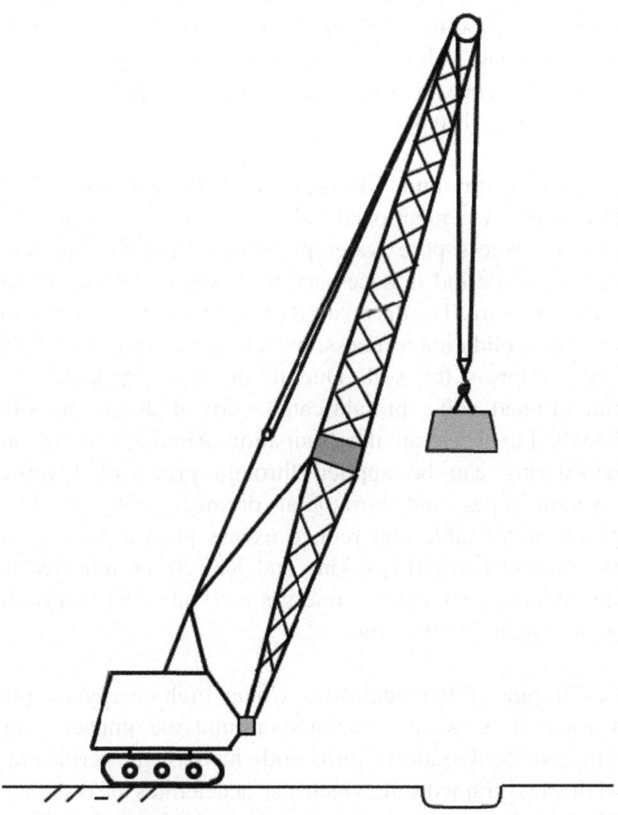

Figure 3.16 Dynamic compaction.

Suitability Deep dynamic compaction is suitable for the following conditions:

- Loose and partially saturated fills
- Saturated free-drained soils
- Silts with plasticity index less than 8
- Clayey soil with a low degree of saturation (moisture content lower than plastic limit)

Deep dynamic compaction is generally not recommended for clayey soil with high plasticity index (greater than 8) and high degree of saturation. However, this method has been used to improve clayey soils in some countries (Han, 1998; Liang and Xu, 2011). Drainage and/or dewatering are often required to reduce excess pore water pressure in clayey soil generated by deep dynamic compaction. A certain waiting period is necessary for the dissipation of excess pore water pressure. High groundwater table (within 2 m of the starting level) minimizes the effectiveness of dynamic compaction. Under such a condition, dewatering may be necessary. Deep dynamic compaction is more economic when the area of a site is larger than 5000 m^2. Due to the size of a crane for deep dynamic compaction operation, certain clearance is necessary. Table 3.8 lists adverse situations for dynamic compaction.

Applications Deep dynamic compaction has been used to improve problematic geomaterials by increasing bearing capacity, reducing settlement, minimizing collapsible potential, and mitigating liquefaction for commercial and residential buildings, storage tanks, highways and railways, airports, and harbors.

Advantages and Limitations Deep dynamic compaction can improve a large area of geomaterials in a relatively short time at low cost. It is very effective to densify loose and partially saturated fill with less than 15% fines. This method often can detect weak or loose areas during operation so that they can be treated properly, such as overexcavation and replacement. Dynamic compaction can change a heterogeneous geomaterial to a more uniform, denser, and stronger material. The major equipment needed for this method is a crane and a tamper, which are readily available from many contractors.

Deep dynamic compaction is generally less to not effective to improve saturated clayey soils. Special measures have to be taken for this method to be reasonably effective for

Table 3.8 Adverse Situations for Dynamic Compaction

Adverse Situation	Possible Difficulty
Soft clays (undrained shear strength less than 30 kPa)	Insufficient resistance to transmit tamper impulse
High groundwater level	Need to dewater and to consider possible effects of subsequent recovery in water level
Vibration effects (may be worse if groundwater level is high)	Distance from closest structure to be of the order of 30 m or more
Clay surface	May be inadequate for heavy cranes and unsuitable for imprint backfilling
Clay fills	May be subject to collapse settlement if inundated later
Flying debris	Precautions for site and public safety
Voided ground or Karst features below treated ground	Treatment may not reach the voided zone or may make it less stable
Biologically degrading material	Compaction may create anaerobic conditions and regenerate or change the seat of the biological degradation

Source: Mitchell and Jardine (2002).

these soils, such as providing drainage and/or dewatering and having a long waiting period for dissipation of excess pore water pressure. Impact by deep dynamic compaction induces noise, vibration, and lateral movement, which may cause problems to nearby buildings, substructures, and utility lines. This method often requires instrumentations to monitor vibration, noise level, and ground movement. When it is used in saturated clayey soils, piezometers are needed to monitor generation and dissipation of excess pore water pressure. The tamping work may cause flying debris, which poses danger to workers onsite. The mobilization cost may be high when large crane and tamper are used.

3.5.2 Principles

Dynamic Densification When dynamic compaction is used on unsaturated granular geomaterial, the impact by a heavy tamper immediately displaces particles to a denser state, compresses or expels air out of voids, and reduces the volume of voids. Under such a condition, typically there is ground depression without any ground heave. A hard plug is formed under the tamper (Moseley and Kirsch, 2004).

Dynamic Consolidation The theory of dynamic consolidation was proposed by Menard and Broise (1975) to explain why saturated fine-grained soil can also be improved by repeatedly dropping a heave tamper. They attributed dynamic consolidation to four main mechanisms: (1) compressibility of saturated soil, (2) liquefaction, (3) change of permeability, and (4) thixotropic recovery. Repeated impacts do not necessarily always liquefy fine-grained soils, instead, they generate excess pore water pressure, which can accumulate under repeated loading. The accumulated excess pore water pressure starts to dissipate once tamping stops. Therefore, it is more appropriate to refer this mechanism to the generation and dissipation of excess pore water pressure.

1. Compressibility of Saturated Soil It is common knowledge that saturated fine-grained soil is incompressible and cannot have volume change under immediate loading (i.e., an undrained condition). Menard and Broise (1975) attributed the immediate volume change of saturated fine-grained soil to the existence of microbubbles in most quaternary soils ranging from 1 to 4%.

2. Generation and Dissipation of Excess Pore Water Pressure As mentioned above, dynamic compaction induces excess pore water pressure during the operation. A waiting period is necessary to dissipate the excess pore water pressure. The dissipation of excess pore water pressure is a consolidation process, which can induce settlement and compress the soil. Due to the low permeability of fine-grained soils, prefabricated vertical drains are often installed to accelerate the dissipation. Alternatively, vacuum dewatering can be applied through preinstalled vertical vacuum pipes and horizontal drainage pipes to lower groundwater table and reduce excess pore water pressure (Liang and Xu, 2011). Liang and Xu (2011) indicated that this method is suitable for fine-grained soil with permeability greater than 5×10^{-9} m/s.

3. Change of Permeability Under high-energy tamping, vertical fissures are generated around the impact points. These vertical fissures significantly increase the permeability of the fine-grained soil, which also accelerates the dissipation of excess pore water pressure and consolidation.

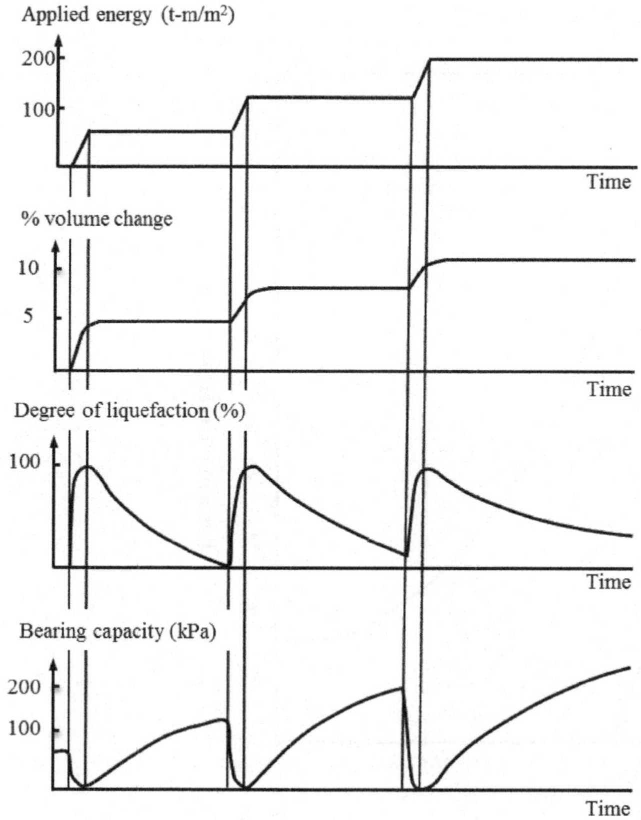

Figure 3.17 Variations of volume, excess pore water pressure, and soil strength during and after the tamping process (after Menerd and Broise, 1975).

4. Thixotropic Recovery Due to the disturbance of fine-grained soil caused by tamping, it degrades and reduces its strength. This strength regains with time due to the thixotropic recovery. This is also the reason why fine-grained soils should be evaluated at least 30 days after tamping.

The changes of volume, excess pore water pressure, and soil strength during and after tamping are illustrated in Figure 3.17.

Dynamic Replacement When a clayey soil is too soft and has too low permeability, it is not effective to be densified or consolidated during and after tamping. Instead of improving the soil, the soil can be displaced by tamping and replaced by stones or coarse aggregates. The process of dynamic replacement involves tamping, backfilling, and continued tamping until stone columns are formed, as shown in Figure 3.18. The design of dynamic replacement is similar to that for stone columns installed by a vibro-probe or casing method to be discussed in Chapter 5. Therefore, this method will not be discussed further in this chapter.

3.5.3 Design Considerations

Site Investigation Before the design of deep dynamic compaction, a geotechnical investigation is required to evaluate the site conditions, which include:

- Geomaterial profiles including geomaterial type, particle size, fine content, degree of saturation, and Atterberg limits
- Relative density of cohesionless geomaterial
- Groundwater level
- Possible voids
- Possible presence of hard lenses within the depth of improvement
- Possible sensitive soil

Influence Factors The design of deep dynamic compaction should consider the following influence factors:

- Geomaterial type
- Depth and area of improvement
- Tamper geometry and weight
- Drop height and energy
- Pattern and spacing of drops
- Depth of crater
- Number of drops and passes
- Degree of improvement
- Induced settlement
- Environmental impact (vibration, noise, and lateral ground movement)

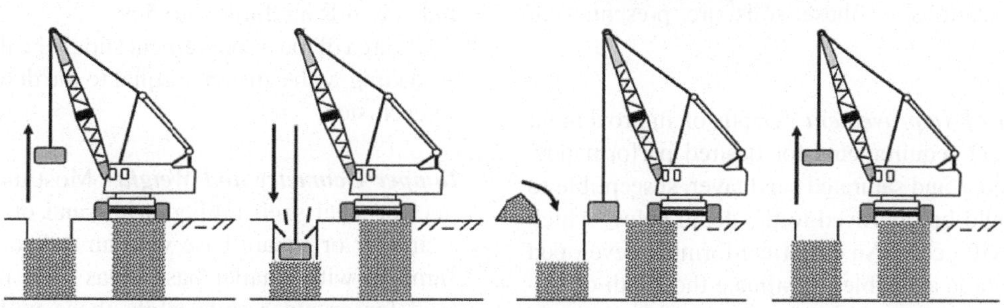

Figure 3.18 Dynamic replacement (after Yee and Ooi, 2010).

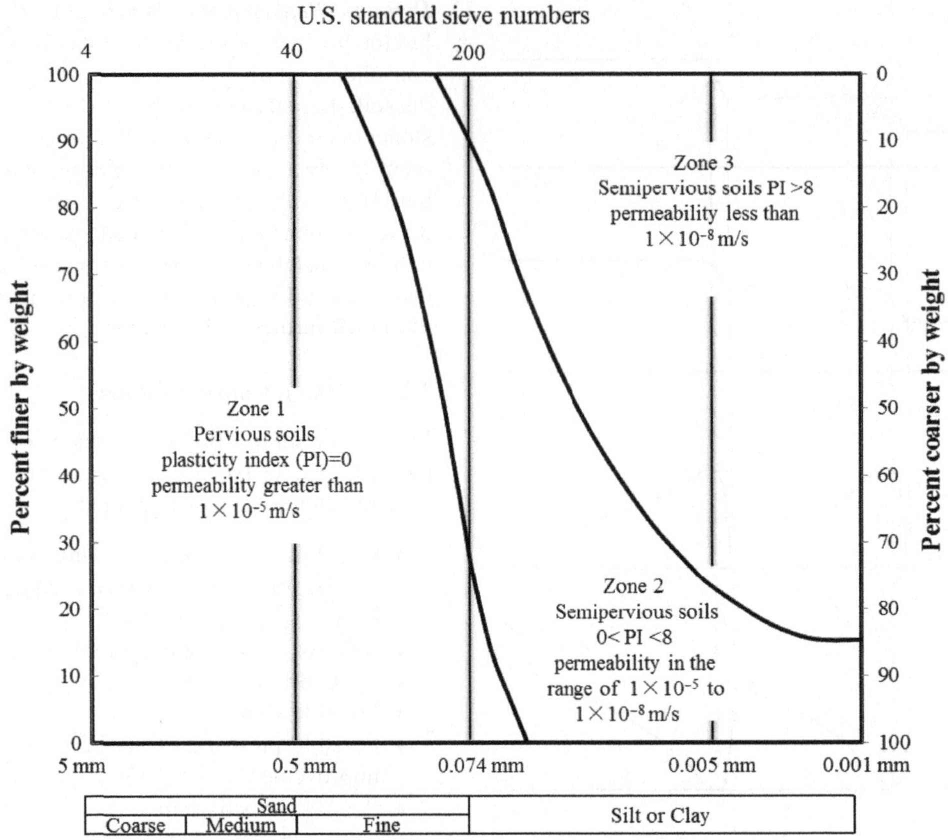

Figure 3.19 Soil types for dynamic compaction (Lukas, 1995).

- Presence of soft layer
- Presence of hard layer
- High groundwater table
- Elapsed time
- Pilot trial

Soil Type Lukas (1995) defined three types of soil that are suitable for dynamic compaction: (1) previous soil deposits—granular soil, (2) semipervious deposit—primary silts with plasticity index less than 8, and (3) semipervious deposit—primary clayey soil with plasticity index greater than 8. The gradations of these soils are presented in Figure 3.19.

Depth and Area of Improvement Depth of improvement depends on project requirements for desired performance. For example, a loose and saturated sand layer, susceptible to liquefaction, should be improved to the depth below which no liquefaction will occur. An empirical formula developed based on field data is available to estimate the depth of improvement as follows:

$$D_i = n_c\sqrt{W_t H_d} \qquad (3.16)$$

where D_i = depth of improvement (m)
W_t = weight of tamper (ton)
H_d = height of drop (m)
n_c = constant, depending on soil type, degree of saturation, and speed of drop

It should be pointed out that the above formula is units dependent. The specific units as noted in the definitions should be used. Table 3.9 provides the recommended n_c values.

Field data show that the depths of improvement for granular soils are mostly up to 10 m while those for cohesive soils and clay fills are limited to 5 m.

The area of the improvement should be that beyond the area of loading with a distance equal to the depth of improvement on each side.

Tamper Geometry and Weight Most tampers are made of steel or steel shell infilled with sand or concrete and have a circular or square base with an area of 3–6 m^2 or larger. Tampers with smaller base areas (3–4 m^2) are commonly used for granular soils while those with large base areas (larger than 6 m^2) are used for cohesive soils. The weight of a tamper typically ranges from 5 to 40 tons.

Table 3.9 Recommended n_c Value

Soil Type[a]	Degree of Saturation	n_c
Pervious soil deposits—granular soils	High	0.5
	Low	0.5–0.6
Semipervious deposits—primary silts with PI < 8	High	0.35–0.4
	Low	0.4–0.5
Semipervious deposits—primary clayey soils with PI > 8	High	Not recommended
	Low (w < PL)	0.35–0.4

[a]PI = plasticity index, w = moisture content, and PL = plastic limit. For W_tH_d = 1–3 MJ/m^2 and a tamper drop using a single cable.
Souce: Lukas (1995).

Figure 3.20 Layout of drop points.

Drop Height and Energy The height of tamper drop is typically 10–40 m. Based on Mayne et al. (1984), the energy per drop in practice mostly ranges from 800 to 8000 kN·m. Mayne et al. (1984) also provided a chart of relationship between weight of tamper and drop height based on field data. This relationship can be approximately expressed as follows:

$$H_d = (W_tH_d)^{0.54} \tag{3.17}$$

where W_tH_d = energy per drop of tamper (ton-m), which is determined from Equation (3.16) based on the required depth of improvement.

The calculated drop height may be adjusted based on the available tamper owned by the contractor.

Pattern and Spacing of Drops Square and triangular patterns of drops are commonly used. Often both patterns are used on the same job to accommodate different passes of compaction in the same phase. Figure 3.20 shows a typical layout of drop points in two primary phases (phase 1 and phase 2). Phase 1 has two passes (also called as phase 1-1 and phase 1-2 in Figure 3.20). Compaction often starts with drop points at larger spacing (e.g., phase 1-1 and phase 1-2), which are to densify deeper soil layers. Drop points at smaller spacing (e.g., phase 2) are to densify shallower soil layers. The secondary phase of compaction uses a lower energy tamper to cover the whole site. This compaction technique is also called ironing compaction. The depth of densification by this dynamic compaction sequence and the change of geomaterial property are illustrated in Figure 3.21. During phase 1-1, only the deeper geomaterial is densified. Phase 1-2 further densifies the deeper geomaterial. Phase 2 densifies the geomaterial within the intermediate depth. During the two phases of compaction, surface deposit is often loosened to the depth of crater penetration due to low overburden stresses. Ironing phase with lower energy is to densify the loosened deposit. Lukas (1995) indicated that the maximum improvement usually occurs between $D_i/3$ to $D_i/2$ (D_i is the maximum depth of improvement). Spacing of drop points (s_1 or s_2) is commonly selected to be 1.5–2.5 times the diameter or width of a tamper (s_1 and s_2 are often equal to create uniform compaction).

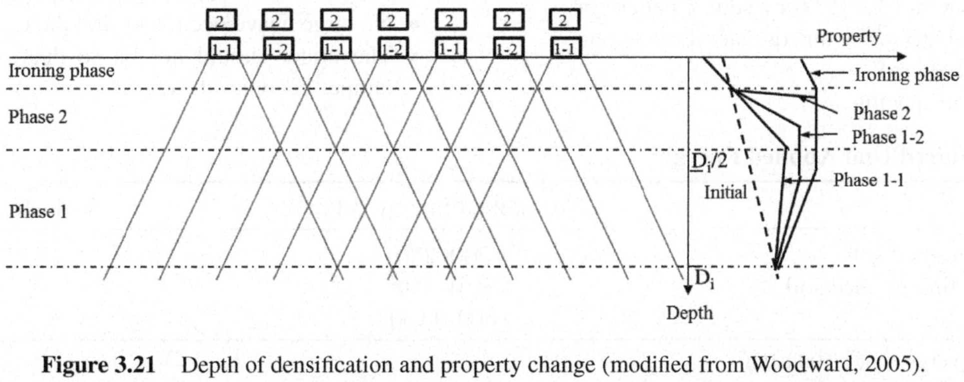

Figure 3.21 Depth of densification and property change (modified from Woodward, 2005).

Depth of Crater A crater is formed under each tamper drop and its depth increases with the number of drops. High-energy compaction can induce a crater of 1.0–1.5 m deep. The crater depth should be limited to the height of a tamper plus 0.3 m to ensure the safety and ease of compaction operation. When the crater depth gets too deep, the compaction operation should be divided into two or multiple phases. Rollins and Kim (2010) proposed empirical formulas to estimate crater depth, d_{cd}, in soils with a low degree of saturation after dynamic compaction:

For a rough estimate

$$d_{cd} = 0.028 N_d^{0.55} \sqrt{W_t H_d} \tag{3.18}$$

For a more accurate estimate

$$\log d_{cd} = -1.42 + 0.553 \log N_d + 0.213 \log H_d + 0.873 \log W_t - 0.435 \log\left(\frac{s_d}{d_t}\right) - 0.118 \log p \tag{3.19}$$

where H_d = drop height (m)
W_t = tamper weight (tons)
N_d = number of drops
s_d = drop spacing (m)
d_t = tamper width or diameter (m)
p = contact pressure in (t /m^2)

Dynamic compaction on soil with a high degree of saturation would result in deeper crater depth.

Number of Drops and Passes The number of drops and passes can be estimated based on applied energy on a site. Applied energy (AE) at each drop point location can be calculated as follows:

$$\text{AE} = \frac{N_d W_t H_d}{A_e} \tag{3.20}$$

where N_d = number of drops by one pass at each drop location (typically 5–10 drops)
W_t = weight of tamper
H_d = drop height
A_e = influence (equivalent) area of each impact point ($A_e = s^2$ for a square pattern or 0.867 s^2 for an equilateral triangular pattern)
s = drop spacing

Total applied energy is the sum of the energy applied during high-energy passes plus ironing pass. Unit applied energy (UAE) is defined based on the depth of improvement as follows:

$$\text{UAE} = \frac{\text{AE}_{\text{total}}}{D_i} = \frac{\text{AE}_{\text{HEP}} N_p + \text{AE}_{\text{IP}}}{D_i} \tag{3.21}$$

where AE_{HEP} = applied energy by a high-energy pass
AE_{IP} = applied energy by an ironing pass
N_p = number of passes
D_i = depth of improvement

Lukas (1995) provided the guidelines for required UAE based on soil type as shown in Table 3.10.

Ironing pass is mainly used to compact loosened soil within the depth of craters. The required applied energy for ironing compaction is estimated as follows:

$$\text{AE}_{\text{IP}} = \text{UAE} \cdot d_{cd} \tag{3.22}$$

where d_{cd} is the depth of the crater. The number of drops for ironing pass can be determined using Equation (3.20) if the weight and drop height of the tamper and the area of the tamper (i.e. the influence area of each impact point) are known.

From the required UAE in Table 3.10 and the AE_{IP}, in Equation (3.22) the total applied energy required for high-energy compaction can be calculated using Equation (3.21) and the number of drops can be calculated using Equation (3.20) if one pass is assumed. When the number of drops at one location in a single pass is too large (greater than 10 passes) or the crater depth is too deep, the operation of compaction should be divided into two or multiple passes.

The number of drops can be determined through trial tamping work onsite. The Chinese *Ground Improvement Technical Code* (China Academy of Building Research, 2000) sets the following criteria to determine the number of drops from trial tamping work:

- The average vertical displacement induced by the last two drops is not greater than 50 mm. When high drop energy is used, it should not be greater than 100 mm.
- No large heave occurs around the crater.
- The crater should not be so deep that lifting of the tamper becomes difficult.

Table 3.10 Required Unit Applied Energy[a]

Soil Type	Unit Applied Energy (kJ/m^3)	% Standard Proctor Energy
Pervious coarse-grained soil	200–250	33–41
Semi-impervious fine-grained soil	250–350	41–60
Landfill	600–1100	100–180

[a]Standard Proctor energy equals 600 kJ/m^3.
Source: Lukas (1995).

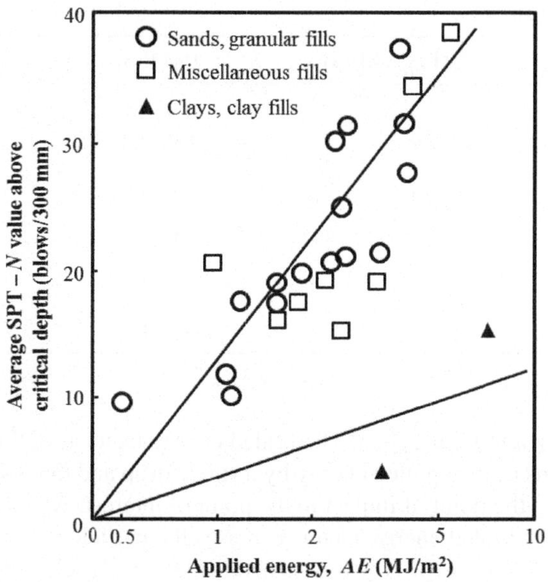

Figure 3.22 Average SPT N value after improvement (after Lukas, 1995).

Degree of Improvement The degree of improvement depends on geomaterial type, fine content, groundwater table, applied energy, drop layout, and time. Figures 3.22, 3.23, and 3.24 show the average SPT N values, CPT tip resistance, and pressuremeter (PML) limit pressure above the improvement depth. Table 3.11 provides upper bound test values after dynamic compaction. These figures and table can be used as target values for dynamic compaction preliminary design. The actual degree of improvement should be evaluated by in situ testing after compaction.

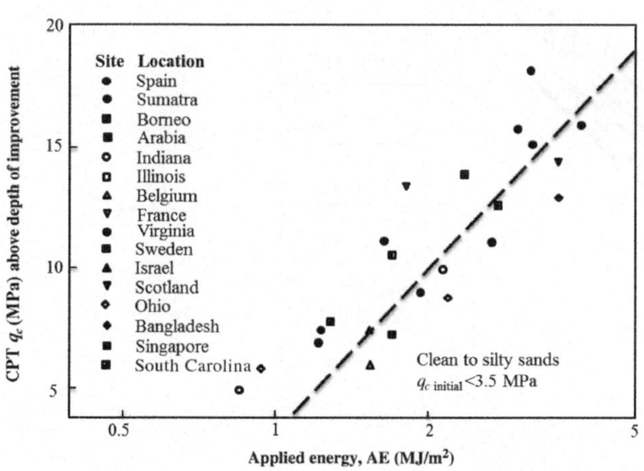

Figure 3.23 Average CPT q_c value after improvement (Lukas, 1995).

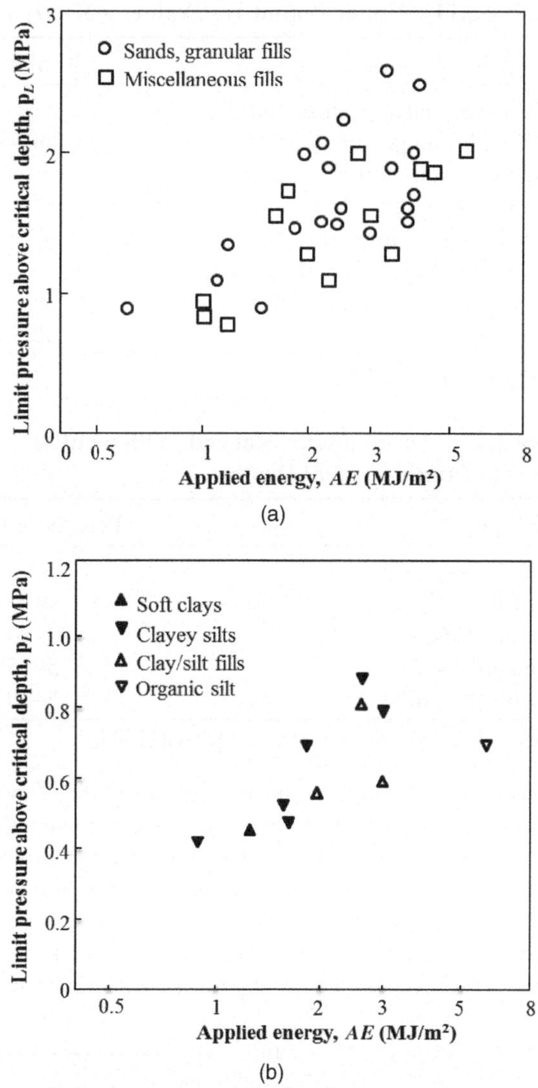

Figure 3.24 Average PMT p_L value after improvement (Lukas, 1995).

Induced Settlement After each pass of dynamic compaction, construction equipment, most commonly bulldozers, is used to level the ground surface. Ground settlement (also called subsidence) is measured based on the current ground elevation as compared with the initial elevation. In unsaturated soil, the settlement occurs immediately after compaction. In saturated soil, however, the settlement increases gradually with time after the initial compression under each compaction. Most of the settlement results from filling large craters induced by tampers. The approximate induced settlement as percent of improvement depth is provided in Table 3.12.

The induced settlement can also be estimated based on the crater depth (see the design example in Section 3.5.5).

Table 3.11 Upper Bound Test Values after Dynamic Compaction

	SPT N (blows/0.3 m)	CPT q_c (MPa)	PMT p_L (MPa)
Previous coarse-grained soil:			
Sands and gravels	40–50	19–29	1.9–2.4
Semipervious soil:			
Sands and gravels	40–50	19–29	1.9–2.4
Silts and clayey silts	25–35	10–13	1.0–1.4
Partially saturated impervious deposits			
Clay fill and mine spoil	30–40	NA	1.4–1.9
Landfills	20–40	NA	0.5–1.0

Source: Lukas (1995).

Table 3.12 Approximate Induced Settlement as Percent of Improvement Depth

Soil Type	Percent of Depth
Natural clays	1–3
Clay fills	3–5
Natural sands	3–10
Granular fills	5–15
Uncontrolled fills	5–20

Source: Modified from Moseley and Kirsch (2004).

Environmental Impact It is expected that applying high-energy impact on ground induces environmental impact, mostly vibration, noise, and lateral ground movement. This fact has to be considered in the selection of a suitable ground improvement technique.

Field measurements show that particle velocity depends on the scaled energy factor and the geomaterial density as shown in Figure 3.25. The scaled energy factor is defined in terms of the applied energy by a single drop and the distance from the point of impact to the point of interest. An increase of the scaled energy factor increases the particle velocity. A loose soil or fill typically generates lower particle velocity. Lukas (1995) indicated that the frequency of ground vibrations induced by dynamic compaction ranges from 6 to 10 Hz. Mayne et al. (1984) provided the following formula to estimate the upper limit of peak particle velocity (PPV) in terms of applied single-drop energy and distance to the drop point:

$$\mathrm{PPV} = 70\left(\frac{\sqrt{W_t H_d}}{x_{dp}}\right)^{1.4} \tag{3.23}$$

where PPV = peak particle velocity (mm/s)
W_t = tamper weight (ton)
H_d = drop height (m)
x_{dp} = distance to the drop point (m)

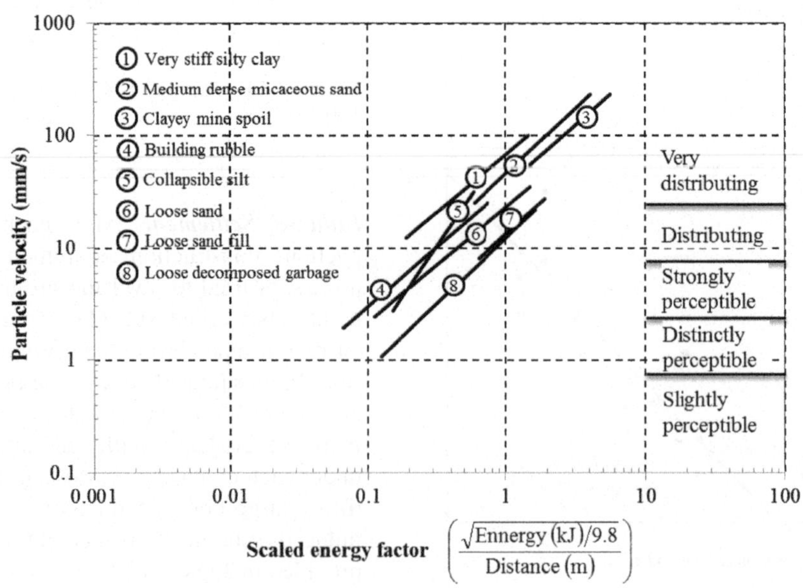

Figure 3.25 Scaled energy factor versus particle velocity (FHWA, 1986).

Table 3.13 Typical Threshold Particle Velocity

Structural Type	Velocity (mm/s)
Commercial, industrial	20–40
Residential	5–15
Sensitive	3–5

Siskind et al. (1980) suggested that threshold particle velocity depends on type of structure and frequency of vibration. Different countries have established different guidelines for threshold particle velocity for vibration-induced damage to buildings. Table 3.13 provides typical values of threshold particle velocity based on the typical vibration frequency generated by deep dynamic compaction. Moseley and Kirsch (2004) suggested that the particle velocity at 2.5 mm/s would annoy occupants of buildings. To reduce particle velocity induced by dynamic compaction, isolation trenches have been commonly used around a construction site to cut off or minimize surface waves (i.e., the Rayleigh waves). Trenches should be 2–3 m deep and at least 1 m wide at the bottom.

Dynamic compaction also induces lateral ground movement, which may damage existing underground utility lines or buried structures. The Federal Highway Administration (FHWA) (1986) documented that lateral ground movement at a magnitude of 19–76 mm was measured at a distance of 6.1 m from the point of impact. Field monitoring of ground vibrations and movement is necessary if there are nearby buildings and/or underground utility lines and structures.

Presence of Soft Layer When a soft layer exists near the ground surface, it may not be able to support the equipment for dynamic compaction operation or absorb applied energy so that limited energy is transmitted to soil at depth. Under such a condition, this soft layer should be excavated or stabilized by a stabilizing layer (typically 0.3–1.2 m thick) to provide a stable working platform for dynamic compaction equipment and to limit crater depth. The most favorable material for the stabilizing layer is a coarse-grained geomaterial, such as gravel, crushed stone, or building rubble. An extra-thick stabilizing layer reduces the depth of improvement below the stabilizing layer; therefore, it should be avoided.

Presence of Hard Layer When a hard layer to a certain thickness (1–2 m) exists near the ground surface, it distributes the applied energy over a wide area so that the energy transmitted to the depth is greatly reduced. As a result, the depth and degree of improvement are reduced. Under such a condition, the hard layer should be removed or loosened. When a hard layer is thin, however, a tamper may penetrate this layer and deliver proper energy to the underlying layer.

High Groundwater Table It is a general requirement for dynamic compaction that the groundwater table should be 2 m below the ground surface. When the groundwater table is within 2 m, dynamic compaction likely encounters some difficulties. Typically, a crater depth ranges from 1.0 to 1.5 m. Dynamic compaction generates excess pore water pressure so that the groundwater rises and enters the craters. The geomaterial and water can be intermixed during compaction. To avoid such a problem, the groundwater table should be lowered by dewatering or additional fill should be added to increase the distance from the ground surface to the groundwater table.

Elapsed Time Dynamic compaction induces excess pore water pressure if the geomaterial is saturated. The excess pore water pressure can accumulate under multiple drops of impact if the geomaterial is not pervious. The accumulated excess pore water pressure reduces geomaterial strength, destabilizes the ground, and minimizes the densification. Under such a condition, the number of drops in each phase or pass should be limited. An elapsed time is needed between two phases or passes to allow the dissipation of the excess pore water pressure. For sandy soil, the dissipation of the excess pore water pressure is rapid and can complete within a few minutes. However, the geomaterial with fine contents may take a few days to weeks to dissipate the excess pore water pressure. To shorten the time for pore water pressure dissipation, prefabricated vertical drains or vacuum dewatering have been used. The details about the function and design of prefabricated vertical drains and vacuum dewatering can be found in Chapters 7 and 6, respectively.

3.5.4 Design Parameters and Procedure

The influence factors discussed above are the design parameters for deep dynamic compaction. The following procedure may be followed for design of deep dynamic compaction:

1. Based on geotechnical profile and potential problem, select the depth of improvement.
2. Based on geomaterial type and degree of saturation, select the n_c value from Table 3.9.
3. Calculate the required energy per blow for the high-energy impact using Equation (3.16) based on the required depth of improvement.
4. Estimate the drop height using Equation (3.17) and then the tamper weight.
5. Based on the applied energy guidelines, the unit applied energy can be selected based on the geomaterial type using Table 3.10.
6. Calculate the required total applied energy using Equation (3.21).
7. Based on the geomaterial type and degree of saturation near the ground surface, the required unit applied energy for the ironing pass can be selected using Table 3.10.

8. Calculate the required applied energy for the ironing pass using Equation (3.22) with an assumed crater depth (typically 1.0–1.5 m).
9. Calculate the required total applied energy for high-energy compaction by subtracting the required applied energy for the ironing pass from the total required applied energy.
10. Based on the tamper diameter, estimate the spacing of drops.
11. Based on the required total applied energy for high-energy compaction and the spacing of drops, calculate the required number of drops (round up to an integer number). If the required number of drops on one location is greater than 10, multiple passes or phases are required.
12. Estimate the crater depth using Equation (3.18) or (3.19).
13. Select target performance values after improvement.
14. Estimate the settlement after improvement based on Table 3.12 or crater depth.

3.5.5 Design Example

A 5-m-high highway embankment is to be constructed over a landfill that has a fine-grained soil cover underlain by a soil mixture with the total thickness ranging from 5.0 to 8.2 m. This soil mixture includes primarily silts and clays with construction waste (concrete blocks, brick fragments, etc.). At certain locations, there are voids and loose pockets within the landfill. Standard penetration tests performed prior to ground improvement indicate SPT values ranging from about 5 to 20 with an average of 10. The predicted settlement ranged from 140 to 274 mm. Dynamic compaction is selected to reduce the anticipated total and differential settlements. The required SPT *N* value after improvement should be at least 20. The surface of the landfill is strong enough to support the dynamic compaction equipment. Leachate inside the landfill was at a relatively shallow depth (approximately 2.5 m from the existing surface). To minimize the generation of excess pore water pressure, multiple pass construction may be needed. The contractor has an 18.2-ton tamper that has the diameter of 1.5 m and the height of 1.5 m.

You are requested to provide a preliminary design for the dynamic compaction project and estimate the settlement after compaction.

Solution

Considering the thickness of the landfill typically ranging from 5.0 to 8.2 m, the depth of improvement is selected as 8.2 m. Based on the composition of the landfill, it can be considered as a semipervious soil deposit. Since the landfill has a high degree of saturation, the n_c value is selected as 0.35. As a result, the required energy per blow can be computed as follows:

$$W_t H_d = \left(\frac{8.2}{0.35}\right)^2 = 550\ t\text{-}m$$

The contractor provided an 18.2-t tamper, therefore, the required drop height is 550 t-m/18.2 t = 30.2 m. Based on Equation (3.17), the estimated drop height is also 30.2 m.

Based on the applied energy guidelines, the unit applied energy for landfills ranges from 600 to 1100 kJ/m^3. The average unit applied energy is 850 kJ/m^3, therefore, the required total applied energy is AE_{total} = 850 kJ/m^3 × 8.2 m = 6970 kJ/m^2 = 6.97 MJ/m^2.

Ironing passes are typically used to compact the geomaterial near the surface, which is close to the depth of the craters. Typically, the crater depth ranges from 1.0 to 1.5 m. The geomaterial above the landfill is most likely fine grained. Since the geomaterial near the surface is above the groundwater table, the unit applied energy for the semipervious fine-grained soils of 300 kJ/m^3 may be used for the ironing passes. Therefore, the required total applied energy for ironing passes is AE_{IP} = 300 kJ/m^3 × 1.5 m = 450 kJ/m^2 = 0.45 MJ/m^2. The required total applied energy for high-energy compaction is $AE_{HEP}\ N_p$ = 6.97 MJ/m^2 – 0.45 MJ/m^2 = 6.52 MJ/m^2.

To allow for pore water pressure dissipation during energy application, multiple passes are needed. Assume two passes are adopted. The required applied energy for each pass is AE_{HEP} = 6.52 MJ/m^2/2 = 3.26 MJ/m^2.

Typical drop spacing is 1½–2½ times the tamper diameter. The factor of 2.0 is selected for this site, that is, drop spacing = 2.0 × 1.5 m = 3.0 m (assuming a square pattern). The number of drops at each specific drop point location can be computed by

$$N_d = \frac{AE_{HEP} \times (A_e)}{W_t H_d}$$

$$= \frac{3260\,\text{kJ}/\text{m}^2 \times (3.0\,\text{m})^2}{18.2\text{t} \times 30.2\,\text{m} \times 10\,\text{m}/\text{s}^2} = 5.3$$

Select the number of the drops for each pass at 6. Example Figure 3.2 depicts the layout of tamper drops.

For the number of drops at one location at 6 for each pass, the crater depth can be estimated as follows:

$$d_{cd} = 0.075\sqrt{W_t H_d} = 0.075\sqrt{550} = 1.75\,\text{m}$$

The allowable crater depth for construction is 1.5 + 0.3 = 1.8 m, which is the same as the estimated crater depth expected in the field; therefore, it is OK.

Based on the FHWA guidelines, the upper bound of SPT N value after dynamic compaction ranges from 20 to 40.

The induced settlement for uncontrolled fill ranges from 5 to 20%. If the average percentage (i.e., 13%) is considered, the possible induced settlement is $0.13 \times 8.2\ \text{m} = 1.10\ \text{m}$. However, based on the estimated crater depth, the expected settlement may be estimated as follows (assume the crater diameter is the same as the tamper diameter and no heave). The area ratio of improvement, defined as the area of each crater to the influence area of each tamping point, is:

$$\text{Area ratio of improvement} = \frac{3.14 \times (1.5/2)^2}{3.0^2} = 0.20$$

the induced settlement by two passes of dynamic compaction $= 2 \times 0.20 \times 1.75 = 0.70$ m. If heave is considered, the induced settlement will be smaller.

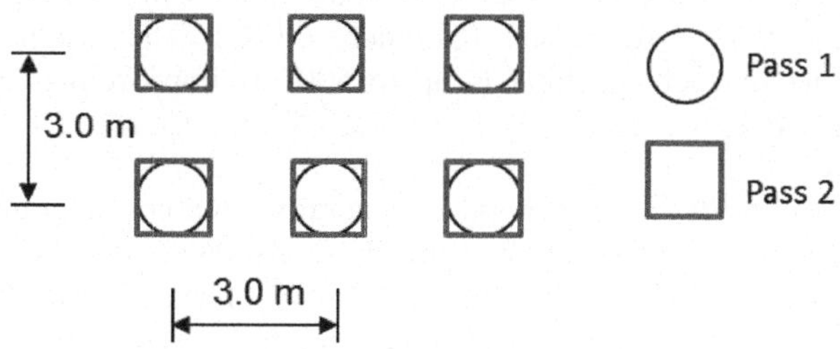

Example Figure 3.2 Layout of tamper drops.

3.5.6 Construction

Before construction, the equipment used for lifting and dropping a tamper should be selected based on the weight of the tamper. FHWA (1986) provided a guideline for the selection of equipment for different tamper weights as shown in Table 3.14. A conventional crawler crane with a single cable and a free spool is typically sufficient for a tamper weighing up to 220 kN. For heavier tampers, the crawler crane should be reinforced with stronger components.

A typical flow of tamping work is shown in Figure 3.26 and its detailed steps are as follows:

Table 3.14 Required Equipment for Different Tamper Weights

Tamper Weight (kN)	Crawler Crane Capacity (kN)	Cable Size (mm)
50–70	360–440	19–22
70–130	440–890	22–25
130–160	890–1100	25–29
160–220	1300–1600	32–38

Source: Modified from FHWA (1986).

1. Prepare a site by removing large objects (e.g., trees), leveling the ground, dewatering, and filling existing ponds and local depressed area. If the groundwater is within 2 m from the ground surface, it should be lowered by dewatering or additional fill is placed. If the surface soil is too weak to support equipment, a construction platform should be constructed first.
2. If there are nearby existing structures or utility lines, an isolation trench is required to minimize vibration and lateral movement. Trench should be at least 2–3 m deep and 1 m wide at the bottom of the trench.
3. Place stakes at the locations for the centers of all the drop points for each pass and survey the ground elevations.
4. Position the equipment and move the tamper right above the drop point.
5. Survey the top elevation of the tamper on the ground.
6. Lift the tamper to the desired height and then let it drop freely onto the ground. Survey the top elevation while the tamper is still in the crater. Alternatively, measure the dimensions of the crater after removing the tamper. If the tamper is tilted after reaching the ground, level the base of the crater after removing the tamper.
7. Repeat step 6 until the number of drops on one tamping point reaches the target value and other criteria are met. Move to the next tamping point.
8. Repeat steps 4–7 until all the tamping points are complete for the first pass.
9. Use bulldozers to level the ground and measure the ground elevation. The difference between the current elevation and the previous elevation is the induced settlement.
10. After an elapsed time depending on geomaterial and groundwater conditions, repeat steps 3–8 until all the tamping points are complete for the next passes if needed.
11. Apply ironing tamping over the whole compaction area.

3.5.7 Quality Control and Assurance

Before any tamping work, the height of drop and locations of drop points should be verified. During field tamping operation, it is important to have monitoring and close visual observations. Adjustments may be made based on monitoring and observations. For example, if one drop location has a much deeper crater depth than other locations, this is an indication that much weaker geomaterial exists at that location. Special measures may be necessary to improve this area, such as overexcavation and replacement. If additional tamping induces large heave around the crater, this is an indication that further densification is not effective so that the tamping should be suspended or terminated at this location. Common field monitoring includes piezometers in

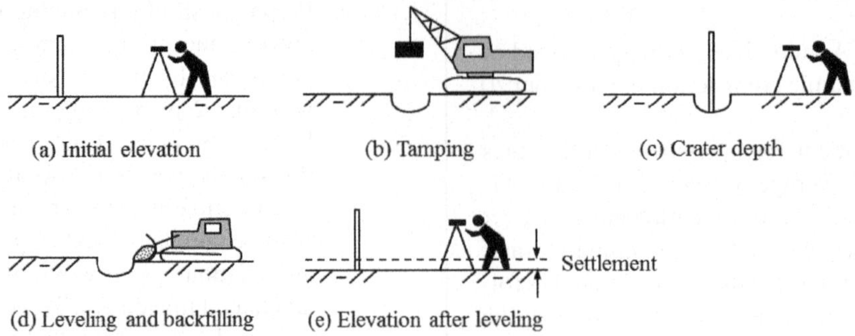

Figure 3.26 Flow of tamping work: (a) initial elevation, (b) tamping, (c) crater depth, (d) leveling and backfilling, and (e) elevation after leveling.

saturated fine-grained geomaterial, inclinometer casings for lateral movement, and accelerometers for ground vibrations.

After the completion of the tamping work, field explorations should be conducted to evaluate or verify the degree and depth of improvement. Depending on geomaterial type and groundwater level, for coarse-grained geomaterial, the field evaluation should be performed at least in 1–2 weeks after the completion of tamping; for fine-grained geomaterial, the evaluation should be performed at least in 3–4 weeks after the completion of tamping. Field explorations include sampling for laboratory tests, SPT, CPT, or PMT. The depth of the test should be below the design depth of improvement. Static plate load tests may be performed in a large project site. Since PMT and plate load tests are more sensitive to the change of soil stiffness than SPT and CPT, they are good methods to be adopted for this purpose.

3.6 RAPID IMPACT COMPACTION

3.6.1 Introduction

Basic Concept Rapid impact compaction is an intermediate compaction method between conventional shallow compaction and deep dynamic compaction. It densifies geomaterial by repeatedly dropping a hydraulic hammer mounted on an excavator at a fast rate as shown in Figure 3.27. The weight of hammer is typically 5–12 tons, which is dropped freely from a height of 1.2 m on a circular steel foot with a diameter of 1.0–1.5 m (the most common one is 1.5 m in diameter). The rapid impact compaction machine can generate 40–60 blows per minute, which is much faster than the deep dynamic compaction machine. There is a monitoring system on the machine to record impact energy and foot penetration. The depth of improvement depends on geomaterial type and properties, groundwater table level, and applied energy. This technology has been adopted since its initial use in the United Kingdom in 1990s (Adam and Paulmichl, 2007). The production rate of each machine is up to 500 m^2 improvement area per day.

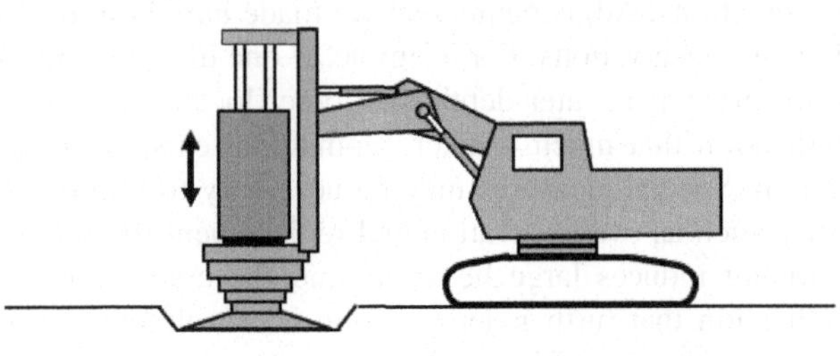

Figure 3.27 Rapid impaction compaction.

Suitability This method is generally suitable for granular geomaterials, including gravel, sands, silts, uncontrolled fills (i.e., a mixture of sand, silt, and clay), and industrial and mine wastes. It has also been successfully used to minimize collapsible potential of loess. This method generally can improve geomaterials up to a depth of 6 m deep (mostly 3–4 m).

Applications Rapid impact compaction has been used for increasing bearing capacity and stiffness of geomaterial to support building foundations, floor slabs, tanks, highways, railways, parking lots, and airport runways, mitigating liquefaction, and reducing waste volume and collapsible potential. It has also been used to compact granular fills in large lifts (up to 3 m).

Advantages and Limitations The operation of rapid impact compaction is fast and under a much controlled manner as compared with deep dynamic compaction. It induces smaller vibrations than deep dynamic compaction due to low impact energy; therefore, it can be operated at closer distances to existing structures. Because the impact foot is always in contact with the ground, it eliminates the risk of generating flying debris. Similar to deep dynamic compaction, it can detect weak areas during the construction. It has better mobility and works in areas with difficult access.

The key limitation for this technology is that the depth of improvement is smaller than that of deep dynamic compaction.

3.6.2 Principles

Rapid impact compaction is mainly used for granular geomaterials. Its general principle is dynamic densification, which is similar to that for deep dynamic compaction when used to densify granular geomaterials. Since rapid impact compaction has lower single-drop energy with a larger number of drops, the densification of geomaterial is progressive and accumulated. BRE (2003) and Serridge and Synac (2006) explained the progressive densification by a concept of soil plug, which is different from that of deep dynamic compaction. In deep dynamic compaction, deeper geomaterial is densified first by tamping at larger spacing and then shallow geomaterial is densified by tamping at smaller spacing. Serridge and Synac (2006) referred to this densification process as a "bottom-up" process. In rapid impact compaction, however, a dense soil plug is first formed under the steel foot by the first few blows. Further blows push the dense geomaterial plug as a rigid block deeper to densify the underlying geomaterial until no or little further penetration can be achieved. This process is referred to as the "top-down" process by Serridge and Synac (2006). This densification process can be used to explain why rapid impact compaction can densify geomaterials at deeper depths even though its single-drop energy is much lower than that by deep dynamic compaction. It is important to point out that in rapid impact compaction, the accumulated impact energy has more influence on the depth of improvement than the single-drop energy. Since rapid impact compaction can densify the geomaterial directly below the steel foot, impact points at close spacing are necessary.

3.6.3 Design Considerations

Depth of Improvement Due to the different densification processes by deep dynamic compaction and rapid impact compaction, the formula to estimate the depth of improvement for deep dynamic compaction [i.e., Equation (3.16)] cannot be used for rapid impact compaction. BRE (2003) and SAICE (2006) provide the guidelines to estimate the depth of improvement for rapid impact compaction as shown in Tables 3.15 and 3.16.

Table 3.15 Typical Improvement Depth with Rapid Impact Compaction

Geomaterial	Applied Energy (ton-m/m^2)	Depth of Improvement (m)
Loose building waste	150	4.0
Ash fill	150	3.5
Select granular fill	150	4.0
Sandy silt	80	2.0
Silty sand	190	3.0

Source: BRE (2003).

Table 3.16 Test Results of Rapid Impact Compaction by 9-Ton Hammer

Geomaterial	SPT *N* Value after Improvement	Typical Improvement Depth (m)
Sand	20–30	6
Silty sand	15	4.5
Sandy silt	10–15	3.5–4.5
Uncontrolled fill	>10	3 to 5

Source: SAICE (2006).

For typical impact spacing (1.5 m × 1.5 m), 30 blows of 9-ton hammer with 1.2-m drop height generate about 150 ton-m/m^2 applied energy.

Patterns of Impact Points Rapid impact compaction typically adopts three patterns of impact points:

1. Arc pattern, that is, primary impact points are arranged in the arc around the center as shown in Figure 3.28. Secondary impact points are arranged between primary impact points.
2. Square pattern, that is, primary impact points are arranged within a 6 m × 6 m or 9 m × 9 m area for each impact grid as shown in Figure 3.29. Within each impact zone, secondary and tertiary impact points are uniformly distributed between primary impact points.
3. Triangular pattern.

For rapid impact compaction, no ironing pass is needed because rapid impact compaction is similar to ironing compaction with low energy and close spacing. However, surface compaction with rollers is often needed to densify shallow geomaterial and level the ground surface.

Number of Blows Number of blows on each point can be estimated using Equation (3.20) based on the required applied energy, weight of hammer, height of drop, and spacing of impact points. Number of blows typically ranges from 10 to 40.

Groundwater A high degree of saturation of geomaterial near ground surface affects the effectiveness of rapid impact compaction. Watts and Cooper (2011) reported that this method was still effective when the groundwater table was at a depth as close as 1.1 m. The depth of groundwater table at 1 m is the minimum requirement for rapid impact compaction. If the groundwater table is too close to the ground surface, dewatering or additional fill should be implemented prior to compaction.

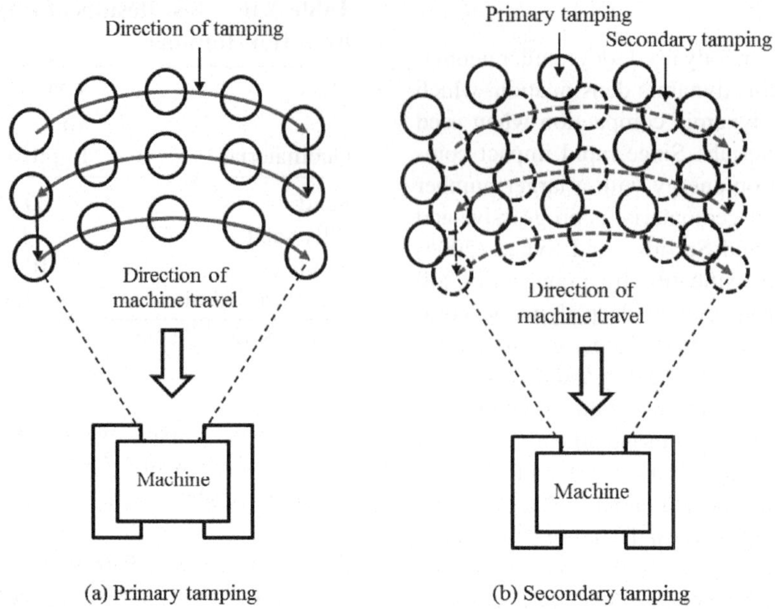

Figure 3.28 Arc pattern of impact points: (a) primary and (b) secondary tamping (modified from Braithwaite and du Preez, 1997).

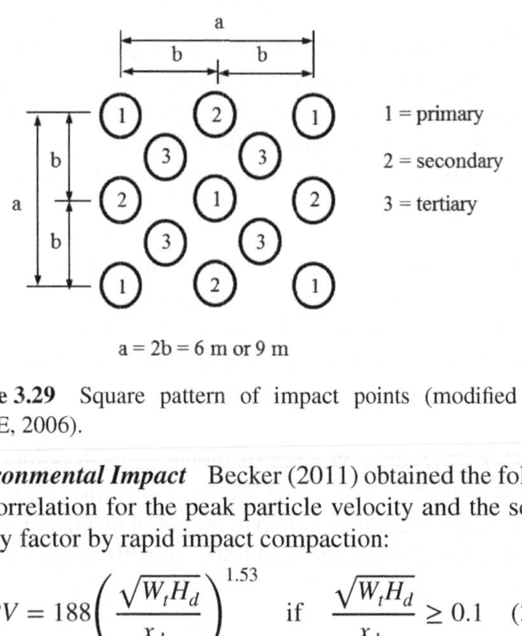

Figure 3.29 Square pattern of impact points (modified from SAICE, 2006).

Environmental Impact Becker (2011) obtained the following correlation for the peak particle velocity and the scaled energy factor by rapid impact compaction:

$$PPV = 188\left(\frac{\sqrt{W_t H_d}}{x_{dp}}\right)^{1.53} \quad \text{if} \quad \frac{\sqrt{W_t H_d}}{x_{dp}} \geq 0.1 \qquad (3.24)$$

$$PPV = 36\left(\frac{\sqrt{W_t H_d}}{x_{dp}}\right)^{0.79} \quad \text{if} \quad \frac{\sqrt{W_t H_d}}{x_{dp}} < 0.1 \qquad (3.25)$$

where PPV = peak particle velocity (mm/s)
W_t = weight of tamper (ton)
H_d = height of drop (m)
x_{dp} = distance from the center of the impact point (m)

For most rapid impact compaction, the scaled energy factor is greater than 0.1 (ton-m)$^{0.5}$/m. The comparison between Equations (3.23) and (3.24) shows that rapid impact compaction produces greater peak particle velocity than deep dynamic compaction (DDC) at the same scaled energy factor. Tara and Wilson (2004) pointed out that the greater peak particle velocity induced by rapid impact compaction results from the fact that the steel foot is always in contact with the ground. The greater peak particle velocity transfers the impact energy to the ground more efficiently. Despite rapid impact compaction produces the greater peak particle velocity than deep dynamic compaction at the same scaled energy factor, the minimum allowable distance to existing structures for rapid impact compaction is typically larger than that for deep dynamic compaction because the energy per blow by rapid impact compaction is lower.

Allen (1996) reported that rapid impact compaction induced the vibration frequencies ranging from 9 to 15 Hz, which are higher than those by deep dynamic compaction. Based on the vibration frequency and threshold particle velocity for different structures, Becker (2011) recommended the minimum allowable distance of rapid impact compaction to structures as shown in Table 3.17.

Kristainsen and Davies (2004) reported the use of rapid impact compaction in a distance of 5 m to a buried utility line and 6 m to a residential building without any evidence of problem. However, in their project, shallow trenches were excavated to minimize the vibration from the source.

Table 3.17 Minimum Allowable Distance of Rapid Impact Compaction to Structures

Structure Type	Threshold Particle Velocity (mm/s)	Minimum Allowable Distance (m)
Drywall	19	14.5
Plaster	13	19.0
All others	51	7.5

Source: Modified from Becker (2011).

Field Trial Test As compared with deep dynamic compaction, rapid impact compaction is still relatively new. Limited research results and case histories are available. As a result, trial test sections are often needed to determine the optimum grid spacing, number of blows per point, and a minimum final set (vertical displacement in mm/blow), or a maximum total depth of penetration before full-scale production work. Watts and Cooper (2011) reported trial areas at different projects ranged from 15 m × 15 m to 40 m × 35 m. A compaction trial is typically done with various spacing and numbers of blows. The resulting degree and depth of improvement are evaluated using in situ tests, such as SPT and CPT, by comparing the properties of geomterials before and after compaction.

3.6.4 Design Parameters and Procedure

Design parameters for rapid impact compaction include:

- Geomaterial type
- Depth of groundwater table
- Weight of hammer
- Height of drop
- Diameter of steel foot
- Depth of improvement
- Pattern and spacing of impact points
- Number of blows
- Distance to existing structures or utility lines

The following procedure may be followed for design of rapid impact compaction:

1. Based on geotechnical profile and potential problem, determine whether rapid impact compaction is suitable.
2. Select the depth of improvement.
3. Determine the required applied energy for primary pass using Table 3.15.
4. Select a pattern and spacing of impact points.
5. Based on the required applied energy and the pattern and spacing of impact points, determine the number of blows using Equation (3.20).
6. Use Table 3.16 or other related results to evaluate possible improvement.
7. Based on the single-drop energy and the closest distance to existing structures, calculate the peak particle velocity using Equation (3.24) or (3.25) and then compare it against the threshold particle velocity in Table 3.17. Alternatively, compare the distance against the minimum allowable distance in Table 3.17.

3.6.5 Design Example

A newly reclaimed land with 3-m-thick loose hydraulic fill (mostly silty sand, SPT N value = 5) needs to be improved to an SPT N value of 15 for building foundations. The groundwater table is at 1.5 m deep. An existing warehouse is located at the close distance to the project site about 10 m. A contractor has rapid impact compaction with a hammer of 9 tons and a drop height of 1.2 m. Provide a preliminary design using rapid impact compaction.

Solution

1. Based on the geotechnical information, the problematic geomaterial is silty sand and the groundwater table is at 1.5 m deep; therefore, the soil type and the groundwater condition are suitable for rapid impact compaction.
2. The required depth of improvement is 3 m.
3. Based on Table 3.15, the required applied energy for primary pass is 190 ton-m/m^2.
4. Assume a square pattern of impact points at spacing of 1.5 m.
5. Based on Equation (3.20), the required number of blows can be calculated as follows:
$$N_d = \frac{AE \cdot A_e}{W_t H_d} = \frac{190 \times 1.5 \times 1.5}{9 \times 1.2} = 39.6$$
Therefore, the number of blows at 40 is selected.
6. Based on Table 3.16, the SPT N value after improvement is 15, which meets the requirement.
7. The scaled energy factor can be calculated as follows:
$$\frac{\sqrt{W_t H_d}}{x_{dp}} = \frac{\sqrt{9 \times 1.2}}{10} = 1.0 > 0.1$$
$$PPV = 188\left(\frac{\sqrt{W_t H_d}}{x_{dp}}\right)^{1.53} = 188 \times \left(\frac{\sqrt{9 \times 1.2}}{10}\right)^{1.53}$$
$$= 34 \text{ mm/s} < 51 \text{ mm/s for other structures (OK)}$$

Alternatively, x_{dp} = 10 m > the minimum allowable distance of 7.5 m (OK).

This is a preliminary design. Trial compaction is necessary to verify or adjust the design parameters and the degree and depth of improvement.

3.6.6 Construction

The following procedure may be followed for the rapid impact compaction work:

1. Prior to equipment mobilization, prepare a site by removing large objects (e.g., trees), leveling ground, dewatering, and filling existing ponds and local depressed area. If the groundwater is within 1 m from the ground surface, it should be lowered by dewatering or additional fill is placed. The site shall be graded such that the water will not pond. Any large debris or rubble uncovered during grading that may interfere with compaction shall be removed and replaced with loose granular fill.
2. If there are nearby existing structures or utility lines, an isolation trench is required to minimize vibration and lateral movement. Trench should be at least 2–3 m deep and 1 m wide at the bottom of the trench.
3. Place stakes based on the pattern of impact points to layout the area to be compacted.
4. Position the steel foot and the hammer on the point to be compacted.
5. Perform compaction until preset criteria, such as the number of blows and a minimum final set, or a maximum total depth of penetration are met. Compaction starts from the outside with large spacing as the primary pass.
6. After each pass, relevel the work area and reestablish survey points for the next pass.
7. If areas are found to be excessively hard or soft during compaction, they should be overexcavated and replaced with granular fill.
8. After the final pass, level all craters and apply surface compaction by vibratory rollers.
9. Take final surveys to estimate the settlement after compaction.

3.6.7 Quality Control and Assurance

During rapid impact compaction, quality control can be performed by utilizing a data acquisition system built into the equipment. The data acquisition system displays operating parameters for each impact point during compaction including total number of blows, total energy input, set (vertical displacement in mm/blow), and total depth of penetration of steel foot. These parameters are monitored by the data acquisition system during operation and used as termination criteria for each impact point. The termination criteria are determined from field trial tests.

Quality assurance tests should be performed after compaction of the entire project site to verify whether rapid impact compaction achieves the required degree and depth of improvement. The common in situ test methods are SPT and CPT. Plate load tests are sometimes used.

3.7 VIBRO-COMPACTION

3.7.1 Introduction

Basic Concept Vibro-compaction drives a vibrating probe into the ground, which generates lateral vibratory forces to rearrange particles into a dense state as shown in Figure 3.30. The rearrangement of particles becomes possible only when the induced forces are higher than the interparticle friction. In saturated cohesionless geomaterial, vibration can generate excess pore water pressure, which reduces interparticle contact forces (i.e., effective stresses) so that the interparticle friction (i.e., shear strength) is reduced. As a result, the rearrangement of particles becomes easier. In dry cohesionless geomaterial, water can be injected to make the compaction easier. Water or air is often used to assist the penetration and densification. Backfill is also often used to improve the degree of densification. This technique, called the vibro-flotation method, was first developed in Germany in 1930s and has been successfully used worldwide. The probe for vibro-flotation is commonly referred to as a vibro-flot. During the 1989 Loma Prieta earthquake, the hydraulic fill densified by vibro-compaction at Emeryville on Treasure Island and Bay Farm Island in California in the United States did not liquefy (Seed et al., 1990). However, some untreated areas close to the treated areas had excessive settlement, lateral spreading, and sand boils. This case history demonstrates the effectiveness of vibro-compaction in mitigating liquefaction of cohesionless soil.

In addition to vibro-flotation, there are other types of equipment used for vibro-compaction as shown in Figure 3.31. Vibro-flotation has a vibrator at the bottom of the probe (also called bottom vibrator), which generates vibration by the rotation of an eccentric weight at the bottom of the probe. Other types of equipment shown in Figure 3.31 have a vibrator or hammer on the top of the probe (also called top vibrator). Most of the theory and design methods are developed based on the vibro-flotation; therefore, it is focused herein.

Suitability Vibro-compaction is suitable for densifying deep deposits of cohesionless geomaterial with up to 20% fines (preferably less than 10%) but less than 2–3% clay particles.

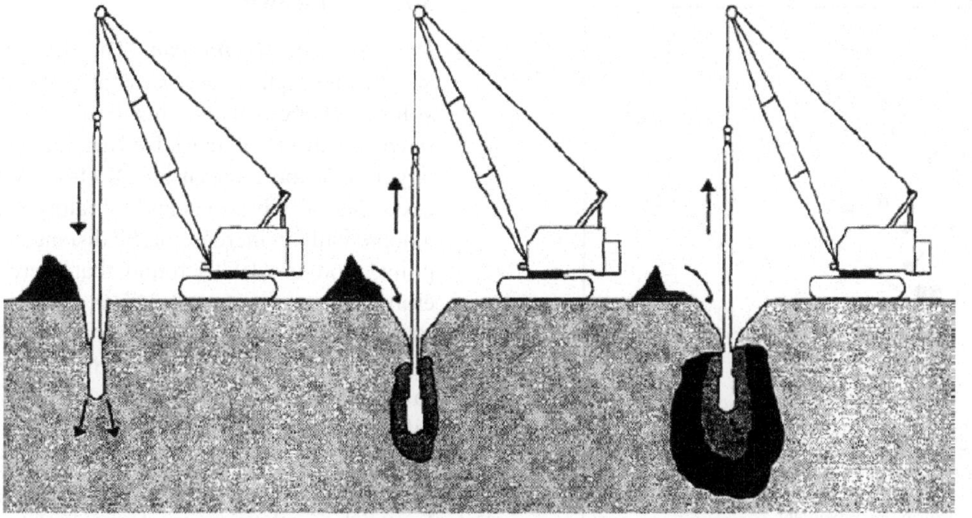

Figure 3.30 Vibro-compaction (Hayward Baker, Inc.).

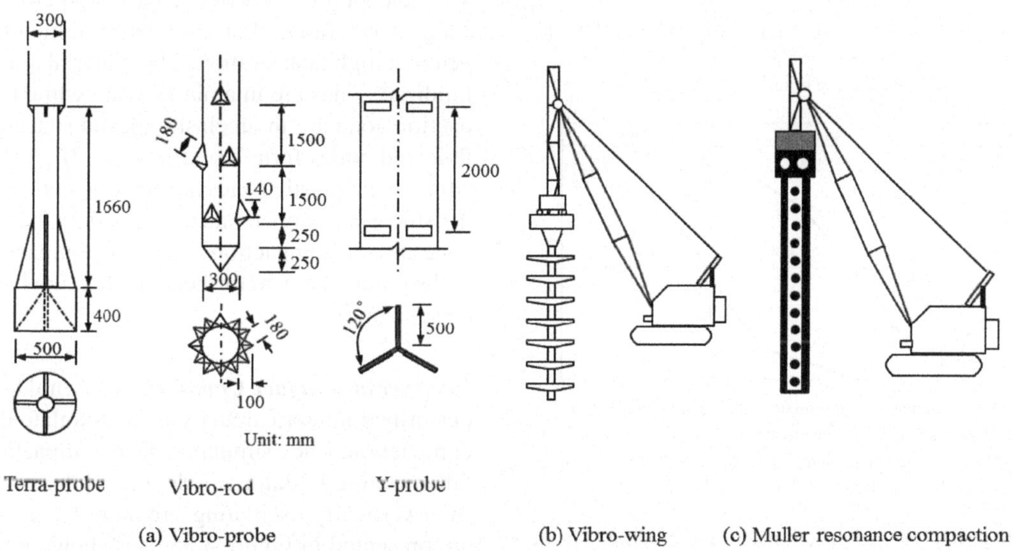

Figure 3.31 Other vibro-compaction equipment: (a) vibro-probe, (b) vibro-wing, and (c) Muller resonance compaction (modified from Massarsch and Fellenius, 2005).

The 20% limit was developed based on the field evaluation before and after vibro-compaction by Saito (1977) as shown in Figure 3.32. It is clearly shown that at a low fine content, the increase of the SPT *N* value is more significant.

Figure 3.33 shows the suitable or unsuitable soils for vibro-compaction. The most suitable zone is zone B, which ranges from fine sand to fine gravel. Zone A is suitable for vibro-compaction but may be difficult due to large particles of gravels. Zone C may be feasible but requires longer time for densification due to apparent cohesion for unsaturated soil or relatively low permeability for saturated soil. Fine-grained soil in zone D makes densification impossible; therefore, a deep replacement method, such as stone columns, should be used. Deep replacement methods will be discussed in Chapter 5. The increase of fineness and plasticity of soil reduces the effectiveness of vibro-compaction. This technique works well for saturated loose cohesionless geomaterial. If dry geomaterial is encountered, flushing water may be used or the whole site is even flooded prior to vibro-compaction. Vibro-compaction method has been used to densify loose cohesionless soil up to a depth of 40 m (mostly within 20 m).

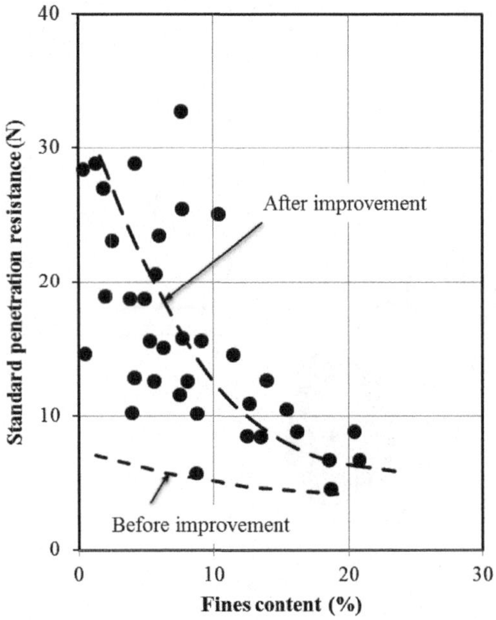

Figure 3.32 Effect of fine content on SPT N value (Saito, 1977).

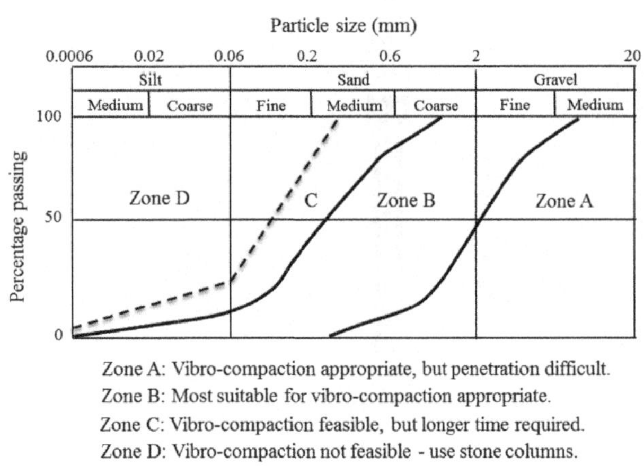

Figure 3.33 Suitability for vibro-compaction (modified from Woodward, 2005).

Applications Vibro-compaction has been mostly used to increase bearing capacity, reduce settlement, and mitigate liquefaction for a variety of projects when loose cohesionless geomaterial exists. The examples of these projects are storage tanks, buildings, roadways, dams, and dikes or levees.

Advantages and Limitations Vibro-compaction is a quick, easy, and economical construction process. It is a proven ground improvement method for many successful projects around the world. This method can be used to improve geomaterials above and below the groundwater table. However, this method is limited to cohesionless geomaterial with a low clay content (i.e., less than 3%). Installation induces vibration and possible ground subsidence.

3.7.2 Principles

Densification Mechanism The densification of cohesionless geomaterials is a process of particle rearrangement and volume change, which result in a denser state. To rearrange positions of particles, the induced forces should be larger than the interparticle friction. Vibro-compaction induces lateral vibrations and vibratory forces. The forces attenuate with an increase of the distance from the compaction point. Rodger (1979) found that there is a critical acceleration of approximately 0.5 g, above which the dynamic stresses induced by dynamic compaction destroy the structure of granular soils. When the acceleration is increased to more than $1.5g$, the shear strength of the soil is significantly reduced and the soil is fluidized. A further increase of acceleration exceeding $3.0g$ causes soil dilation. Rodger (1979) proposed idealized response of cohesionless soil to vibration as shown in Figure 3.34. There are four zones around the vibrating probe: (1) dilatory zone, (2) fluidized zone, (3) compaction zone, and (4) elastic zone. The fluidization is referred to the dry soil losing its stability at a high acceleration. For saturated soil, liquefaction may happen at a high acceleration. The materials in the dilatory and fluidized zones are in a plastic state while those in the compaction zone are in an elastic-plastic state. The dilatory and fluidized zones form a plastic zone. High shear stresses existing in the plastic zone cause plastic deformations. The best densification happens at the boundary between the plastic zone and the compaction zone. The compaction zone extends to the point where no further densification happens (i.e. in the elastic zone).

Deep versus Surface Compaction Arnold and Herle (2009) performed numerical analyses to simulate deep and surface compaction. They simulated deep compaction by a horizontally rotating vibrator at a depth and the surface compaction by a vertically oscillating vibrator. Their numerical results are presented in Figure 3.35. It is shown that the deep compaction has a wider radial influence distance than the surface compaction. The deep compaction also generates more uniform volume change than the shallow compaction. Arnold and Herle (2009) attributed the better performance of the deep compaction to the multidirectional shearing mode induced by the combined vertical and rotational movement of the vibrator.

Installation Process To minimize probe shaft resistance, penetration and extraction should be done at a high frequency. Massarsch and Fellenius (2005) suggested a frequency higher than 30 Hz for penetration and extraction. At a high frequency, there is low ground vibration, and less energy is transferred to the surrounding soil so that the penetration and extraction become easier. During the compaction,

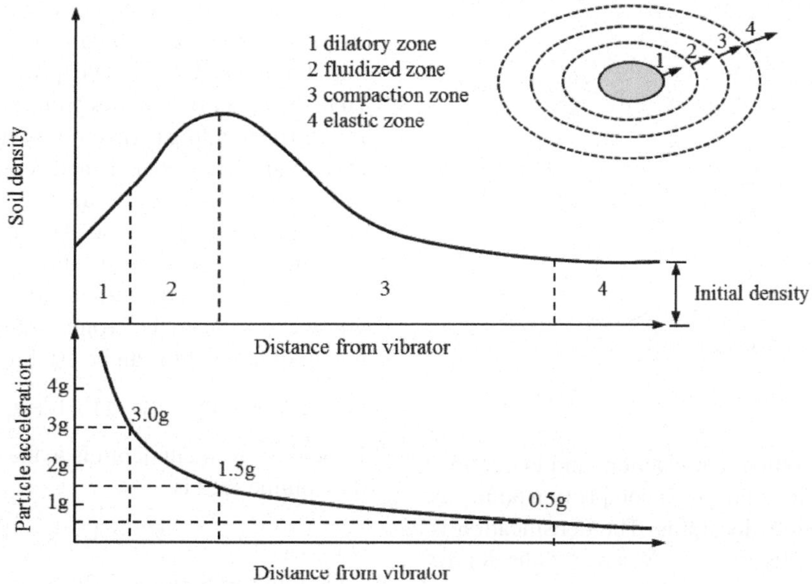

Figure 3.34 Idealized response of cohesionless soil around a vibrating probe (modified from Rodger, 1979).

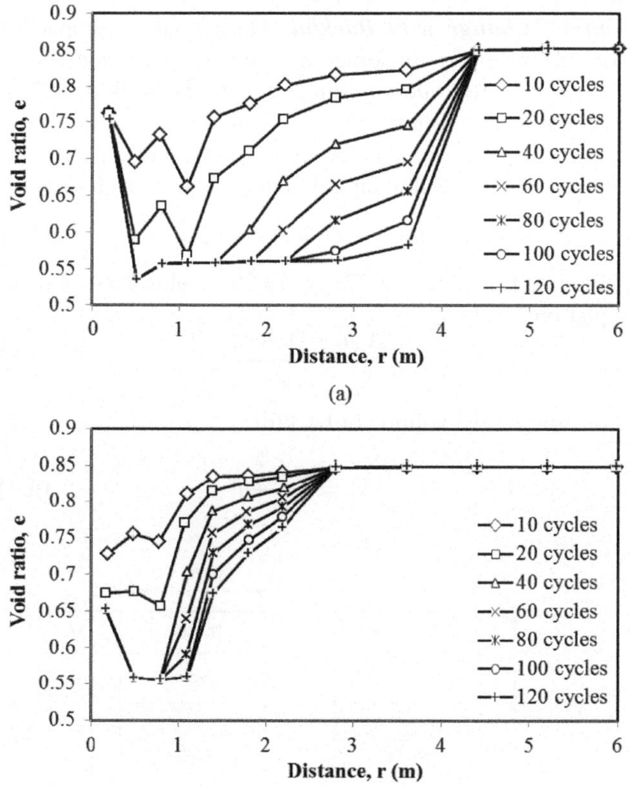

Figure 3.35 Evolution of void ratio along radial distance by: (a) deep vibration and (2) surface vibration (after Arnold and Herle, 2009, with permission from John Wiley and Sons).

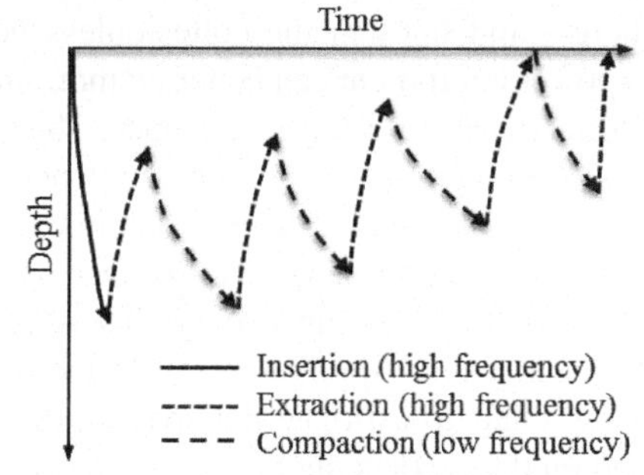

Figure 3.36 Penetration, compaction, and extraction process (Massarsch and Fellenius, 2005).

however, the preferable frequency is close to the resonance of the geomaterial mass so that more energy is transferred to the surrounding geomaterial to make the compaction efficient. Typical frequency of resonance of the soil is around 15–20 Hz (Massarsch and Fellenius, 2005). During this process, it is slow for the probe to penetrate. Duration of compaction is an important parameter, which depends on initial and target geomaterial properties, horsepower (HP) of the equipment, and efficiency of energy transfer from the probe to the surrounding geomaterial. It is often controlled by the maximum amperage preset from trial compaction test or past experience. Figure 3.36 shows the penetration and extraction process for the vibro-compaction.

Degree and Distance of Influence Figure 3.34 shows the soil density change during vibro-compaction under a fixed point vibration. Figure 3.36 shows the construction process

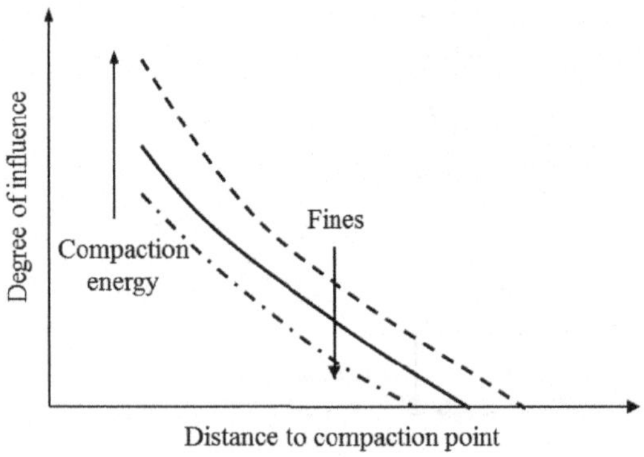

Figure 3.37 Influence of vibro-compaction.

involving repeated penetration, compaction, and extraction. In addition, the construction of the next compaction point has an influence on the previous locations. The densification is accumulated as shown in Figure 3.35. As a result, the degree of influence decreases with the radial distance to the compaction point. At a certain distance, the vibro-compaction has no influence on the soil density. Figure 3.37 shows that the degree of influence depends on the distance, the compaction energy, and fines in the cohesionless geomaterial. High horsepower equipment can generate more influence on the geomaterial density at a farther distance. The existence of fines reduces the effectiveness of vibro-compaction in terms of degree of influence and distance of influence. Mitchell (1981) showed that the influence radius decreased from approximately 1.8 m for clean sands to 0.6–0.9 m for sands with more than 25% fines. D'Appolonia (1953) and others developed a design chart to estimate the spacing of compaction points based on these concepts.

Volume Change without Backfill Vibro-compaction without backfill often induces ground subsidence. The volume change can be explained using a phase diagram as shown in Figure 3.38. Under an initial condition, a geomaterial mass consists of solid and void and the total volume, V_0, is equal to the sum of the solid volume, V_s, and the void volume, V_v. The initial state of the geomaterial can be described by an initial void ratio, e_0, which is the ratio of V_v to V_s. The total initial volume, V_0, can be expressed as $V_s\ (1 + e_0)$. Assume vibro-compaction induces ground subsidence, S. The volume loss corresponding to the ground subsidence within the influence (equivalent) area of each compaction probe, A_e, is equal to $V_f = A_eS$. As a result, the void volume after improvement is V_v' and the total volume after improvement, V_1, is $V_s\ (1 + e_1)$, where e_1 is the void ratio after improvement. The volume change from the initial state to the improved state is the volume loss due to the ground subsidence. Therefore, the following relationship can be established:

$$A_eS = V_s(1 + e_0) - V_s(1 + e_1) = V_s(e_0 - e_1) \tag{3.26}$$

If the improvement depth is h, the total initial volume within the volume area is

$$V_0 = A_eh = V_s(1 + e_0) \tag{3.27}$$

Dividing Equations (3.26) by (3.27) yields the following equation:

$$\frac{S}{h} = \frac{e_0 - e_1}{1 + e_0} \tag{3.28}$$

Volume Change with Backfill During vibro-compaction, backfill materials are sometimes added to help densify surrounding cohesionless geomaterial. If the backfill volume is equal to the volume change from the initial state to the improved state (i.e., no ground subsidence) as shown in Figure 3.39, the following relationship can be established:

$$V_f = V_s(1 + e_0) - V_s(1 + e_1) = V_s(e_0 - e_1) \tag{3.29}$$

Dividing Equations (3.27) by (3.29) yields the following equation:

$$\frac{A_eh}{V_f} = \frac{1 + e_0}{e_0 - e_1} \tag{3.30}$$

Considering the volume of backfill

$$V_f = \frac{\pi d_{cl}^{\ 2}}{4}h \tag{3.31}$$

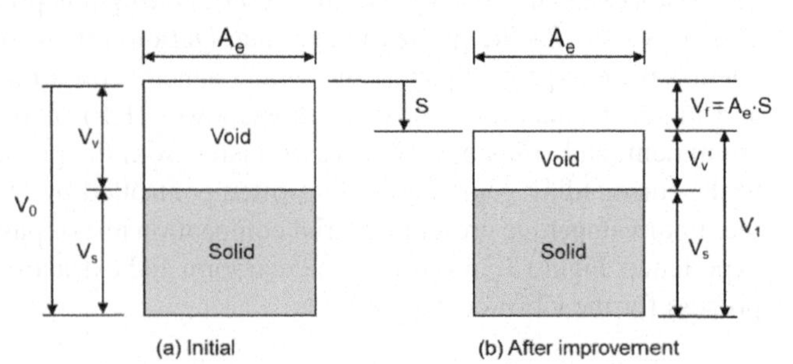

Figure 3.38 Volume changes during densification without backfill: (a) initial and (b) after improvement.

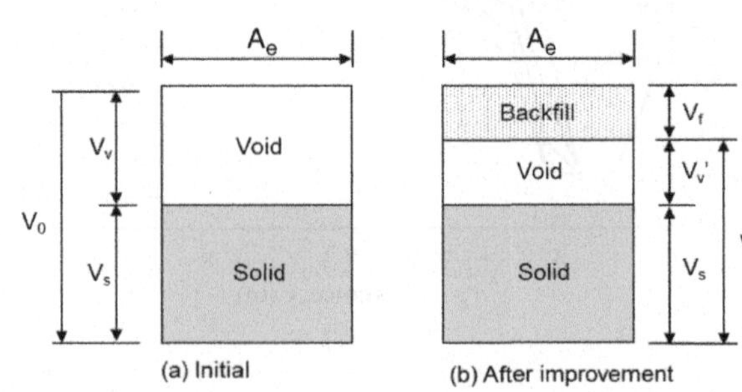

Figure 3.39 Volume changes during densification with backfill: (a) initial and (b) after improvement.

where d_{cl} is the diameter of the column with backfill. The improvement area by an individual column can be calculated as follows:

$$A_e = s^2 = \frac{\pi d_{cl}^2}{4} \cdot \frac{1 + e_0}{e_0 - e_1} \quad (3.32)$$

The spacing of columns, s, can be determined as follows:

$$s = 0.89 d_{cl} \sqrt{\frac{1 + e_0}{e_0 - e_1}} \quad \text{(square pattern)} \quad (3.33a)$$

$$s = 0.95 d_{cl} \sqrt{\frac{1 + e_0}{e_0 - e_1}} \quad \text{(triangular pattern)} \quad (3.33b)$$

If there is ground subsidence or heave after vibro-compaction, the required volume of backfill is

$$V_f = A_e H \frac{e_0 - e_1}{1 + e_0} \pm A_e S = s^2 \left(h \frac{e_0 - e_1}{1 + e_0} \pm S \right) \quad (3.34)$$

in which the + sign represents a ground heave condition while the − sign represents a ground subsidence condition.

Combining equations (3.31) and (3.34) yields the following equations:

$$s = 0.89 d_{cl} \sqrt{\frac{(1 + e_0)h}{(e_0 - e_1)h \pm (1 + e_0)S}} \quad \text{(square pattern)} \quad (3.35a)$$

$$s = 0.95 d_{cl} \sqrt{\frac{(1 + e_0)h}{(e_0 - e_1)h \pm (1 + e_0)S}} \quad \text{(triangular pattern)} \quad (3.35b)$$

It should be pointed out that the volume changes without and with backfill material presented above are based on the assumption that the soil is uniformly compacted. This is not necessarily always the case in the field. Therefore, the above equations can only be used for preliminary design and should be verified through field trial tests.

The use of backfill material in vibro-compaction is mainly to assist the compaction. Backfill material is also used to replace weak soil for stiffness, strength, and permeability purposes, which will be discussed in Chapter 5 for deep replacement.

3.7.3 Design Considerations

Performance Criteria For most vibro-compaction projects, the following performance criteria should be considered (Elias et al., 2004):

- Relative density of geomaterial, $D_r \geq 60\%$ for floor slabs, flat bottom tanks, and embankments
- $D_r \geq 70$–75% for column footings and bridge foundations
- $D_r \geq 80\%$ for machinery and mat foundations.

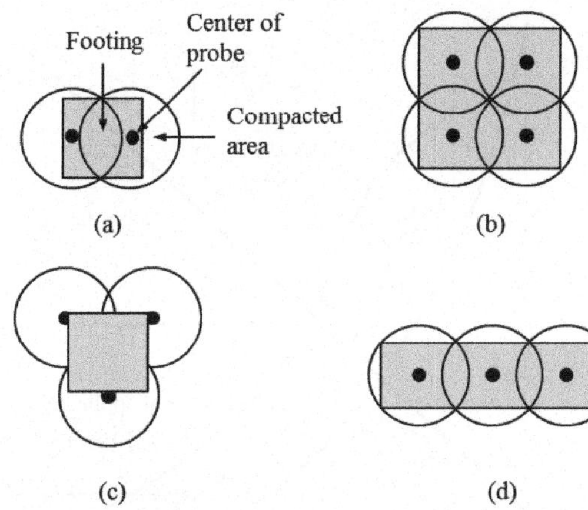

Figure 3.40 Typical arrangements of compaction probe points below isolated and strip footings (modified from Kirsch and Kirsch, 2010).

Area and Depth of Improvement In general, the area of improvement should be larger than footprints of foundations. Kirsch and Kirsch (2010) suggested typical arrangements of compaction probe points below isolated and strip footings as shown in Figure 3.40. Under a general condition, one to two rows of compaction points may be installed outside of a footing. On a liquefiable soil site, two to four rows of compaction points may be installed outside of a footing. The depth of improvement should be deep enough to eliminate all potential problems for problematic geomaterials.

Grid Pattern and Spacing Grid points for vibro-compaction can be in a square, rectangular or triangular pattern. Typical spacing for vibro-compaction ranges from 1.5 to 3.5 m, depending on type, initial density, and target density of the geomaterial and horsepower of the vibrator. Engineers have developed design charts to estimate the spacing of compaction points. Figure 3.41 is one such design chart. This design chart was first developed by D'Appolonia in 1953 based on a single 30-HP vibro-flot in clean sand with a triangular compaction point pattern. In the 1970s, Brown added his curve based on a larger horsepower vibro-flot (i.e., 100 HP) in fine to coarse sands. This is why Brown's curve is above D'Appolonia's curve. About the same time, Webb and Hall (1969) produced a similar curve using a large vibro-flot but in sand with fines up to 15%; therefore, their curve is below Brown's curve. Glover (1982) presented these results in his study. The dash line in the middle between the lines by D'Appolonia and Brown corresponds to a typical application with a 100-HP vibro-flat in sand with 5–10 fines (Yee et al., 2013). The influence coefficient corresponds to the centroid point of the triangular pattern and is the accumulated effect by three compaction points. Yee (2013) provided an example to illustrate how

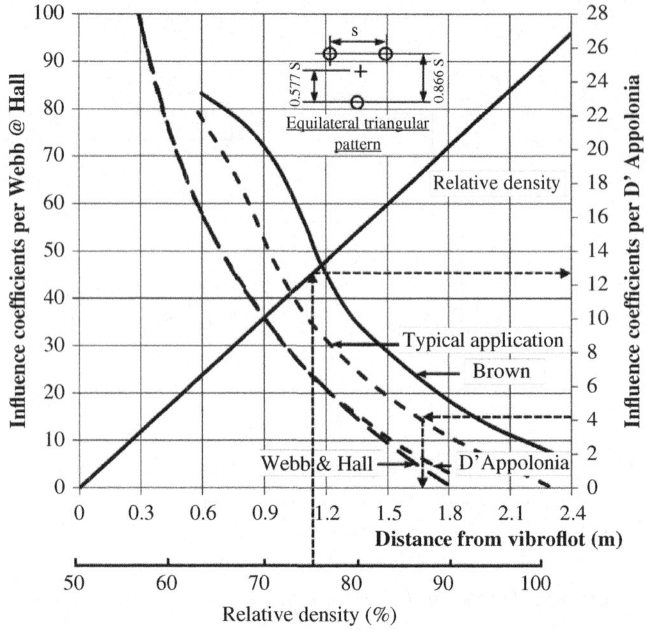

Figure 3.41 Design chart for compaction point spacing and relative density (Modified from Yee, 2013; Glover, 1982).

this design chart should be used. The general procedure is as follows:

- From a target relative density, D_r, draw a vertical line up to intersect the relative density line.
- Draw a horizontal line from the intersect point to the right (for all the methods except the Webb & Hall method) or left (for the Webb & Hall method) depending on which method is selected to find the influence coefficient.
- Divide the influence coefficient by a factor of 3 to account for the accumulated contributions from three compaction points (i.e., the divided coefficient is for a single compaction point).
- Based on the influence coefficient for a single compaction point, draw a horizontal line to intersect the selected curve (D'Appolonia, Brown, Webb and Hall, or the typical curve).
- From the intersected point, draw a vertical line to find the distance to the vibro-flot (i.e., the spacing of compaction points).

Another design chart as shown in Figure 3.42 has also been used in practice. Based on the soil type and the target relative density, the tributary area for each compaction point can be estimated from this figure.

After the spacing of compaction points is determined, the average site subsidence can be estimated using Equation (3.28) without any backfill. If the ground subsidence is too large, backfill can be added to minimize ground subsidence.

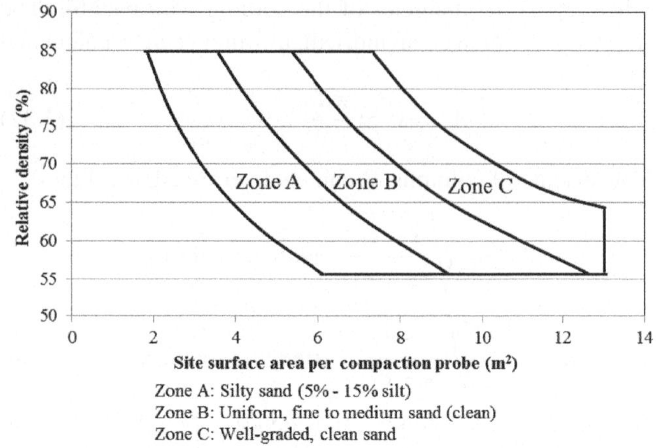

Figure 3.42 Tributary area of compaction point versus relative density of soil (Hayward Baker).

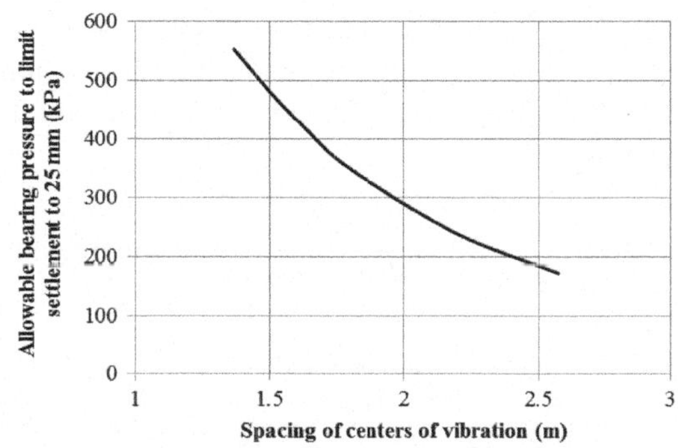

Figure 3.43 Allowable bearing capacity versus spacing of compaction points for footing width of 1–3 m (Thorburn, 1975).

If backfill is used, Equation (3.35a) or (3.35b) can be used to estimate the required spacing of compaction points. If the spacing is fixed, the ground subsidence after adding backfill can be estimated.

Based on a required allowable bearing capacity, the spacing of compaction points can also be estimated as shown in Figure 3.43. Figure 3.43 shows that the allowable bearing capacity of the soil after improvement ranges from 200 to 550 kPa.

3.7.4 Design Parameters and Procedure

Design parameters for vibro-compaction include:

- Geomaterial type, fine content, and percent of clay particles
- Thickness and depth of problematic geomaterial
- Depth of groundwater table
- Initial void ratio or relative density of geomaterial
- Target void ratio or relative density of geomaterial
- Pattern and spacing of compaction points
- Area of improvement

- Equipment type and horsepower
- Frequency of penetration, compaction, and extraction
- Duration of compaction
- Ground subsidence
- Diameter of column if backfill is used

The following procedure can be followed for the design of vibro-compaction:

1. Based on geotechnical profile and potential problems, determine whether vibro-compaction is suitable. The most important parameters are fine content and percent of clay particles.
2. If the geomaterial is deemed suitable for vibro-compaction, estimate the initial void or relative density.
3. Based on the performance requirement (e.g., bearing capacity, settlement, and liquefaction), select a target void ratio or relative density.
4. Based on the required relative density or allowable bearing capacity, estimate the required spacing of compaction points.
5. Calculate a potential ground subsidence.
6. If the ground subsidence is greater than the requirement, backfill should be used to minimize the ground subsidence.

3.7.5 Design Example

Example 1: Vibro-Compaction without Backfill

A 5-m-thick loose sand exists on a site that has a fine content of 8% without any clay particles. The minimum and maximum void ratios of this sand are 0.456 and 0.950, respectively. This sand has an initial void ratio of 0.673. The design requires the sand to be densified to a target relative density of 75%. A large horsepower vibrator (such as HP100) is available for vibro-compaction. Estimate the required spacing of compaction points in a triangular pattern and the average ground subsidence after vibro-compaction without backfill.

Solution

Figure 3.41 can be used for the design of vibro-compaction. Based on the target relative density, D_r = 75%, the required influence coefficient as shown in Figure 3.41 is 12.5. Since this influence coefficient at the centroid of a triangular pattern is contributed by three compaction points. The influence coefficient by a single compaction point is 12.5/3 = 4.2. The influence coefficient of 4.2 is used with the dash line (a typical application) to find the required spacing is 1.7 m. This procedure is illustrated in Figure 3.41.

The relative density, $D_r = 75\%$, corresponds to a void ratio:

$$e_1 = e_{\max} - D_r(e_{\max} - e_{\min})$$
$$= 0.950 - 0.75 \times (0.950 - 0.456) = 0.580$$

The average ground subsidence after vibro-compaction without backfill is

$$S = h\frac{e_0 - e_1}{1 + e_0} = 5 \times \frac{0.673 - 0.580}{1 + 0.673} = 0.279 \text{ m}$$

Example 2: Vibro-Compaction with Backfill

A site consists of uniform medium sand with 5% fine content. The thickness of this sand layer is 12 m starting from the ground surface. The minimum and maximum void ratio values of the sand are 0.45 and 0.98, respectively. The groundwater table is at 1.5 m from the ground surface. The SPT N_{60} value at a depth of 6.0 m is 5. This site is located in a seismic-active area. The potential earthquake magnitude can be 7.0. The peak ground acceleration can reach as high as 0.3g. Vibro-compaction backfilled with granular material is proposed to densify this possible liquefiable soil to prevent potential liquefaction. Assume the average diameter of granular columns can reach 0.8 m, the length of granular columns is 12 m, and the ground subsides 50 mm after the installation. You are requested to determine the required spacing of granular columns in a square pattern to eliminate the liquefaction potential of this sand.

Solution

Assume the unit weights of the sand above and below the groundwater table are 19 and 20 kN/m^3. Based on ASTM, particle size of medium sand ranges from 0.425 mm (No. 40) to 2.00 mm (No. 10). Assume D_{50} = 1.2 mm.

At a depth of 6 m, the total and effective overburden stresses are

$$\sigma_{z0} = 19 \times 1.5 + 20 \times 4.5$$
$$= 119 \text{ kPa}$$
$$\sigma'_{z0} = 19 \times 1.5 + (20 - 9.81) \times 4.5$$
$$= 74 \text{ kPa}$$

The corrected SPT N value is

$$(N_1)_{60} = N_{60}\sqrt{\frac{100 \text{ kPa}}{\sigma'_{z0}}} = 5 \times \sqrt{\frac{100}{74}} = 5.8$$

The initial relative density is

$$D_r = \left[\frac{N_{60}(0.23 + 0.06/D_{50})^{1.7}}{9} \times \left(\frac{100}{\sigma'_{z0}}\right)^{0.5}\right]^{0.5} \times 100\% = 29\%$$

The initial void ratio is

$$e_0 = e_{max} - D_r(e_{max} - e_{min})$$
$$= 0.98 - 0.29 \times (0.98 - 0.45) = 0.826$$

Based on the chart from Seed et al. (1985) in Figure 2.83 using $(N_1)_{60}$, the cyclic stress ratio (CSR) under magnitude 7.5 earthquake, $CSR_{M=7.5}$, can be determined as 0.06. The cyclic resistance ratio (CRR) under magnitude 7.0 earthquake $CSR_{M=7.0}$ can be estimated as follows:

$$CRR_{M=7.0} = MSF\ CRR_{M=7.5} = [6.9 \exp(-7.0/4) - 0.06] \times 0.06 = 0.068$$

The stress reduction factor $r_d = 0.96$ is based on the Seed and Idriss (1971) chart in Figure 2.82.

The cyclic shear stress induced by earthquake and the corresponding CSR value are

$$\tau_{cyc} = 0.65 \frac{a_{max}}{g} \sigma_{z0} r_d = 0.65 \times \frac{0.3g}{g} \times 119 \times 0.96 = 22 \text{ kPa}$$

$$CSR = \frac{\tau_{cyc}}{\sigma'_{z0}} = \frac{22}{74} = 0.297$$

The factor of safety of the sand against liquefaction before improvement is a

$$FS = \frac{CRR_{M=7.0}}{CSR} = \frac{0.068}{0.297} = 0.23 < 1$$

Therefore, the sand is liquefiable and improvement is needed.

Let's design the improved ground with a factor of safety equal to 1.2. Therefore, the required $CRR_{M=7.0} = 1.2 \times 0.297 = 0.356$ and the required $CRR_{M=7.5} = CRR_{M=7.0}/MSF = 0.356/1.14 = 0.312$. From the chart in Seed et al. (1985), the corresponding $(N_1)_{60} = 26$ and N_{60} can be calculated as follows:

$$N_{60} = (N_1)_{60} / \sqrt{\frac{100 \text{ kPa}}{\sigma'_{z0}}} = 26 / \sqrt{\frac{100}{74}} = 22.4$$

Therefore, the required relative density of the sand after ground improvement is

$$D_r = \left[\frac{N_{60}(0.23 + 0.06/D_{50})^{1.7}}{9} \times \left(\frac{100}{\sigma'_{z0}} \right)^{0.5} \right]^{0.5} \times 100\% = 58\%$$

The final void ratio of the sand is

$$e_1 = e_{max} - D_r(e_{max} - e_{min})$$
$$= 0.98 - 0.58 \times (0.98 - 0.45) = 0.672$$

Considering the use of granular columns in a square pattern and the ground subsidence, the required spacing of columns is as follows:

$$s = 0.89 d_{cl} \sqrt{\frac{(1 + e_0)h}{(e_0 - e_1)h - (1 + e_0)S}}$$

$$= 0.89 \times 0.8 \sqrt{\frac{(1 + 0.827) \times 12}{(0.827 - 0.672) \times 12 - (1 + 0.827) \times 0.05}}$$

$$= 2.5 \text{ m}$$

3.7.6 Construction

The equipment used for vibro-compaction typically includes a crane, a hydraulic or electric vibrator, a probe, and a water pump or an air compressor. Table 3.18 provides specifications of several vibrators.

Vibrators have been commonly used to penetrate into cohesionless geomaterial and densify the geomaterial by vibration. During the penetration and densification, water is sometimes introduced to help liquefy the geomaterial. A certain vibration time is needed at each depth to densify the geomaterial within a desired range. Following is a typical installation procedure using a vibro-flotation with water (a similar procedure can be followed for other equipment):

1. Position the probe to the desired location, turn on water and power, and check water pressure and voltage to ensure the equipment functions well.
2. Sink the probe into the ground at a rate of 0.3–0.5 m/min and observe the variation of electrical current (reduce the rate of penetration if the current is too high).
3. Reduce water pressure when the probe is close to the desired depth to minimize disturbance.
4. Continue the probe to the desired depth and maintain the vibration for at least 30 s.
5. Withdraw the probe at a rate of 0.3–0.5 m/min and pause for at least 30 s after each withdrawal of 0.3–0.5 m until the ground surface (sometimes repenetration and reextraction are needed to densify the geomaterial).
6. Turn off the power and water and move the probe to the next location.

Table 3.18 Specifications of Common Vibrators

Manufacturer	Bauer	Bauer	Keller	Keller	Keller	ICE	Vibro	Vibro
Name	TR13	TR85	M	S	A	V180	V23	V32
Length (m)	3.13	4.2	3.3	3.0	4.35	4.96	3.57	3.57
Diameter (mm)	300	420	290	400	290	360	350	250
Mass (kg)	1000	2090	1600	2450	1900	2580	2200	2200
Motor (kW)	105	210	50	120	50	395	130	130
Speed (rpm)	3250	1800	3000	1800	2000	1800	1800	1800
Amplitude (mm)	6	22	7.2	18	13.8	20	23	32
Dynamic force (kN)	150	330	150	280	160	195	300	450

Source: Modified from Layne Christiansen Company.

3.7.7 Quality Control and Assurance

During vibro-compaction, quality control can be implemented to ensure vibro-compaction to achieve desired performance. The quality control items may include the location of compaction points, depth of penetration, vibratory power consumption or electric current amperage, frequency and duration of penetration, compaction, and extraction, depth of obstruction encountered if any, and ground subsidence. If water is used, it is important to have well-planned water flow channels to direct water from compaction points into settling ponds. If backfill is used, the quantity and rate of adding backfill should be well controlled according to the design. Trial compaction tests may be performed to verify design parameters.

Quality assurance tests should be performed after compaction of the entire project site to verify whether vibro-compaction achieves the required degree and depth of improvement. The common in situ test methods are SPT and CPT. Plate load tests are sometimes used. These tests should be performed at the centroid of the pattern of the compaction points.

PROBLEMS

3.1. What is the maximum depth conventional compaction can densify?

3.2. Is it possible for a data point above the zero air line on a compaction curve? Why?

3.3. For a cohesionless geomaterial, which parameter, relative density or relative compaction, is more appropriate for conventional compaction specification in the field? Why?

3.4. Is it possible for relative compaction greater than 100%? Why?

3.5. Describe how to use the one-point method to estimate the maximum dry unit weight, optimum moisture content, and applied compactive energy in the field.

3.6. How to determine the accepted zone of improvement using conventional compaction if there are different performance requirements?

3.7. If the required relative compaction in terms of the maximum dry unit weight based on the standard Proctor tests is 97%, what is the approximate equivalent relative compaction based on the modified Proctor tests?

3.8. If the geomaterial is a low plasticity clay (CL), which roller is best suited for compaction of this geomaterial?

3.9. A fine-grained soil has LL = 30 and PI = 10, estimate the maximum dry unit weight and the optimum moisture content using the *U.S. Navy Design Manual* method.

3.10. A compacted fine-grained soil has moisture content of 16% and moist unit weight of 17.5 kN/m^3, estimate the maximum dry unit weight and the optimum moisture content using the Ohio compaction curves.

3.11. A fill material has 10% oversized particles. The compaction tests on the fill without oversized particles result in a maximum dry unit weight of 18.7 kN/m^3 and its corresponding optimum moisture content of 12%. The oversized particles have specific gravity of 2.67 and saturated surface dry (SSD) moisture content of 3%. Calculate the corrected maximum dry unit weight and optimum moisture content including the oversized particles.

3.12. A contractor excavates 10,000 m^3 soil at moist unit weight of 17.5 kN/m^3 and moisture content of 10% from a borrow pit and transports it to a project site. The project has an area of 20,000 m^2 to be filled with this compacted soil. If the required dry unit weight and moisture content of the compacted soil are 18.3 kN/m^3 and 12.5% (assume there is no soil loss during transportation and compaction), what is the thickness of the compacted soil and how much water needs to be added?

3.13. In Problem 3.12, if there are 2% dry soil loss and 3% moisture loss, what is the thickness of the compacted soil and how much water needs to be added to maintain the compaction at the required moisture content?

3.14. What are the major differences between conventional and intelligent compaction?

3.15. An intelligent compaction machine on a compacted fill generates the following amplitudes of vibration: $A_{0.5\Omega}$ = 0.6 mm, A_{Ω} = 2.0 mm, $A_{1.5\Omega}$ = 1.0 mm, $A_{2\Omega}$ = 0.5 mm, $A_{2.5\Omega}$ = 0.4 mm, and $A_{3\Omega}$ = 0.3 mm. Calculate CMV, BV, and CCV.

3.16. An intelligent compaction machine with machine acceleration, $a_m = 3$ m/s^2 requires a gross power, P_g = 100 kJ/s to drive a roller with a weight, W_r = 20 kN, at a roller velocity, v_r = 5 km/h on a slope with an angle, $\beta = 1\%$. The machine-dependent internal loss coefficients are m_m = 5 kJ/m and b_m = 2 kJ/s. Calculate the machine drive power (MDP).

3.17. An intelligent compaction machine with a length of 2.1 m and radius of 0.75 m is compacted on a geomaterial with a Poisson's ratio of 0.3. When the machine applies an excitation force of 175 kN, the vertical displacement of the drum is 1.0 mm. Calculate vibratory modulus, E_{VIB}.

3.18. If the machine drive power for an intelligent machine on a granular soil is 110 kJ/s, use a correlation to estimate the dry unit weight of the soil.

3.19. If the compaction meter value for an intelligent machine on a cohesive soil is 50, use a correlation to estimate the dry unit weight of the soil.

3.20. A test section on a cohesionless subgrade results in the following intelligent compaction meter values (CMVs) and the initial load secant moduli, E_{v1}:

E_{v1} (MPa)	10	15	22	28	35	38	43	45	55
CMV	15	20	34	38	45	50	54	58	63

The target initial load secant moduli, E_{v1} = 35 MPa. Determine the target CMV and develop acceptance criteria for production.

3.21. Explain why deep dynamic compaction is less effective for saturated fine-grained soil.

3.22. Discuss the main differences among dynamic densification, dynamic consolidation, and dynamic replacement.

3.23. What measures should be taken prior to deep dynamic compaction if the groundwater table is at 0.5 m below the ground surface?

3.24. A 6-m-thick granular fill with a groundwater table at 9 m deep needs to be improved by deep dynamic compaction. Determine the required single tamper drop energy, the tamper mass, and the drop height.

3.25. Dynamic compaction with a tamper weight of 15 ton and a drop height of 12 m applies 7 drops on fill with low degree of saturation. Estimate the crater depth.

3.26. An unsaturated fill with a CPT cone resistance of 2.0 MPa is subjected to dynamic compaction with applied energy of 2.5 MJ/m^2. Estimate the CPT cone resistance within the improvement depth after deep dynamic compaction.

3.27. Estimate the approximate induced settlement after deep dynamic compaction on a 5-m-thick unsaturated clay fill.

3.28. Estimate the peak particle velocity at a distance of 10 m from the tamping point of dynamic compaction with tamper weight of 15 ton and drop height of 15 m.

3.29. A three-story structure is to be constructed over an 8000-m^2 site. The initial subsurface exploration indicates the presence of sinkholes and voids due to dissolution of the limestone formation. The predominant soil type is a silty fine sand grading to a fine sand with seams of sandy clay. The design indicates that shallow foundations can be used for this project provided the soils were made more homogeneous as far as load support and no voids were present within the depth up to 7.6 m below the ground surface. Assume groundwater is not a concern. Dynamic compaction is proposed to improve the ground. The local contractor doing dynamic compaction has a 15-ton tamper with the diameter of 2.0 m and the height of 1.4 m. You are requested to conduct the preliminary design for this dynamic compaction project including drop height, spacing, number of drops, number of passes, estimated crater depth, and settlement.

3.30. What is the main difference in dynamic densification between deep dynamic compaction and rapid impact compaction?

3.31. What are the favorable applications for rapid impact compaction?

3.32. Explain why rapid impact compaction with low single-drop energy can improve soil up to 6 m deep.

3.33. A rapid impact compaction machine equipped with a 9-ton hammer with 1.2-m drop height and a 1.5-m-diameter steel foot is used to compact a loose building waste area with a groundwater table at 3 m below the ground surface. The impact points are arranged in a triangular pattern with spacing of 2.0 m. Estimate the applied energy, the number of drops on each impact point, and the depth of improvement.

3.34. A rapid impact compaction machine equipped with a 9-ton hammer with 1.2-m drop height and a 1.5-m diameter steel foot is used to compact loose silty sand. Estimate the peak particle velocity at a distance of 12 m from the impact point.

3.35. A site consists of a problematic soil, which has the following gradation:

U.S. Sieve Size	Percent of Passing
No. 40	100
No. 60	95
No. 100	64
No. 200	6

Plastic index = 5.4.
Evaluate whether vibro-compaction is suitable.

3.36. A site consists of 3 m thick loose sand with an in-situ void ratio of 0.75. The laboratory tests show that the maximum and minimum void ratios of the sand are 0.85 and 0.50, respectively. Based on the field SPT value of 5, this soil will be liquefiable at an earthquake magnitude of 6.5. Vibrocompaction method with granular backfill is proposed to densify this soil. Assume the diameter of granular columns is 0.75 m. To minimize the liquefaction potential, the densified soil should have an SPT N value of 20, which corresponds to the relative density of 70%. Assume the ground settles 40 mm after the densification. Find the required spacing of granular columns.

REFERENCES

Adam, D. and Kopf, F. (2004). "Operational devices for compaction optimization and quality control (Continuous Compaction Control & Light Falling Weight Device)." *Proc. of the Intl. Seminar on Geotechnics in Pavement and Railway Design and Construction*, December, Athens, Greece (Invited paper), 97–106.

Adam, D. and Paulmichl, I. (2007). "Rapid impact compactor - an innovative dynamic compaction device for soil improvement." Improvement of Soil Properties, Bratislava, June 4–5.

Allen, S. (1996). The Low Energy Dynamic Compaction of Soil. Ph.D. Dissertation. University of Wales. Cardiff, Wales, UK.

Anderegg R. and Kaufmann, K. (2004). "Intelligent compaction with vibratory rollers—feedback control systems in automatic compaction and compaction control," Transportation Research Record No. 1868, *J. Transport. Res. Board*, 124–134.

Arnold, M. and Herle, I. (2009). "Comparison of vibrocompaction methods by numerical simulations." *Int. J. Numerical Anal. Methods Geomech.*, 33(16): 1823–1838.

Becker, P.J. (2011). Assessment of Rapid Impact Compaction for Transportation Infrastructure Applications. Master's Thesis, Iowa State University.

Braithwaite, E.J., and du Preez, R.W. (1997). "Rapid impact compaction in Southern Africa." Presented at Geology for Engineering, Urban Planning, and the Environment, Midrand, South Africa, November 12–14.

BRE (2003). Specifying Dynamic Compaction. Building Research Establishment, BRE Report, BR458.

Chang, G., Xu, Q., Rutledge, J., Horan, B., Michael, L., White, D., and Vennapusa, P. (2011). *Accelerated Implementation of Intelligent Compaction Technology for Embankment Subgrade Soils, Aggregate Base, Asphalt Pavement Material*. Final Report No. FHWA-IF-12-002, FHWA, Washington, DC.

China Academy of Building Research (2000). *Ground Improvement Technical Code*. The China Planning Press, Beijing.

Daniel, D.E. and Wu, Y.K. (1993). "Compacted clay liners and covers for arid sites." *J. Geotech. Eng. Div.*, 119(2).

D'Appolonia, E. (1953) "Loose sands – Their compaction by vibroflotation." *Proc. Symp. on Dynamic Testing of Soils*, STP 156, ASTM, Philadelphia, PA.

D'Appolonia, D.J., Whitman, R.V., and D'Appolonia, E.D. (1969). "Sand compaction with vibratory rollers." *J. Soil Mech. Found. Div.*, 95(SM1): 263–284.

Elias, V., Welsh, J., Warren, J., Lukas, R., Collin, J.G., and Berg, R.R. (2004). *Ground Improvement Methods*. FHWA-NHI-04-001, 1022 pp.

FHWA (1986). *Dynamic Compaction for Highway Construction, Vol. 1: Design and Construction Guidelines*. Report FHWA/RD-86/133. FHWA, Washington, DC.

Glover, J. C. (1982). *Sand Compaction and Stone Columns by the Vibroflotation Process*. Symposium on Recent Developments in Ground Improvement Techniques, Bangkok, 3–15.

Han, J. (1998). "Ground modification by a combination of dynamic compaction, dynamic consolidation, and dynamic replacement." *Proceedings of Fourth International Conference on Case Histories in Geotechnical Engineering*, St, Louis, MO, March 8–15.

Hertz, H. (1895). *Über die Berührung fester elasticscher Körper, Gesammelte Werke*, Bd. 1. Aufbau-Verlag, Leipzig.

Huasmann, M.R. (1990). *Engineering Principles of Ground Modification*. McGraw-Hill, New York.

ISSMGE (2005). "Roller-integrated continuous compaction control (CCC): Technical contractual provisions, recommendations, TC3: Geotechnics for pavements in transportation Infrastructure." International Society for Soil Mechanics and Geotechnical Engineering, London.

Johnson, A.W. and Sallberg, J.R. (1960). "Factors that influence field compaction of soils." Bulletin No. 262, Highway Research Board.

Kirsch, K. and Kirsch, F. (2010). *Ground Improvement by Deep Vibratory Methods*. Spon Press, Abingdon, Oxfordshire, UK.

Kristiansen, H. and Davies, M. (2004). "Ground improvement using rapid impact compaction." *Proceedings of the 13th World Conference on Earthquake Engineering*, Vancouver, B.C., Canada, August 1–6.

Kröber, W., Floss, E., and Wallrath, W. (2001). "Dynamic soil stiffness as quality criterion for soil compaction." In *Geotechnics for Roads, Rail Tracks and Earth Structures*. A.A.Balkema, Lisse, 189–199.

Liang, R.Y. and Xu, S. (2011). "Innovative soft clay improvement technique using vacuum and dynamic compaction (HVDM)." 2011 Pan-Am CGS Geotechnical Conference. Toronto, Ontario, Canada.

Lee, K.L. and Singh, A. (1971). "Relative density and relative compaction." *J. Soil Mech. Found. Div.*, 97(SM7): 1049–1052.

Lukas, R.G. (1995). *Dynamic Compaction*. Geotechnical Engineering Circular No. 1, FHWA-SA-95-037, 105, FHWA, Washington, DC.

Lundberg, G. (1939). Elastische Berührung zweier Halbräume, Forschung, auf dem Gebiete des Ingenieurwesens, Band 10, Göteborg, 201–211.

Massarsch, K.R. and Fellenius, B.H. (2005). "Deep vibratory compaction of granular soils." In *Ground Improvement-Case Histories*. B. Indranatna and C. Jian (eds.), Elsevier, Amsterdam, 633–658.

Mayne, P.W., Jones, Jr.,, J.S., and Dumas, J.C. (1984). "Ground response to dynamic compaction." *J. Geotech. Eng.*, 110(6): 757–774.

Menard, L. and Broise, Y. (1975). "Theoretical and practical aspects of dynamic compaction." *Geotechnique*, 25(1): 3–18.

Mitchell, J.K. (1981). "Soil improvement—State-of-the-art report." *Proceedings of the 10th International Conference on Soil Mechanics and Foundation Engineering*, Stockholm, 4, 509–565.

Mitchell, J.M. and Jardine, F.M. (2002). *A Guide to Ground Treatment*. CIRIA Publication, London C573, CIRIA.

Mooney, M. and Adam, D. (2007). "Vibratory roller integrated measurement of earthworks compaction: an overview." *Proceedings of the Seventh International Symposium on Field Measurements in Geomechanics*, September 24–27, Boston.

Mooney, M., Einehart, R., Facas, N., Musimbi, O., White, D., and Vennapusa, P. (2010). *Intelligent Soil Compaction Systems*. NCHRP Report 676, Transportation Research Board, Washington, DC.

Moseley, M.P. and Kirsch, K. (2004). *Ground Improvement*, 2nd ed. Spon Press, London and New York.

Rodger, A.A. (1979). *Vibrocompaction of Cohesionless Soils*. Cementation Research Limited, Int. Rep., R. 7/79, summarized in Greenwood, D.A. and Kirsch, K (1983), "State of the art report on specialist ground treatment and vibratory and dynamic methods". *Proceedings of the Conference on Piling and Ground Treatment*. London, Thomas Telford Ltd. pp 17–45.

Rollings, M. P. and Rollings, R.R. (1996). *Geotechnical Materials in Construction*. McGraw-Hill, New York.

Rollins, K. and Kim, J. (2010). "Dynamic compaction of collapsible soils based on U.S. case histories." *J. Geotech. Geoenviron. Eng.*, 136(9): 1178–1186.

SAICE (2006). "Innovative new ground improvement method uses controlled dynamic compaction." *Civil Eng.*, 14(5): 3–6.

Saito, A. (1977). "Characteristics of penetration resistance of a reclaimed sandy deposit and their changes through vibratory compaction." *Soils and Foundations* 17(4), 31–43.

Schroeder, W.L., Dickenson, S.E., and Warrington, D.C. (2004). *Soils in Construction*, 5th edition, Pearson/Prentice Hall, Upper Saddle River, NJ.

Seed, H.B. and Chan, C.K (1959). "Structure and strength characteristics of compacted clays." *J. Soil Mech. Found. Div.*, 85(SM5): 87–128.

Seed, R.B., Dickenson, S.E., Riemer, M.F., Bray, J.D., Sitar, N., Mitchell, J.K., Idriss, I.M., Keyen, R.E., Kropp, A., Harder, L.F., and Power, M.S. (1990). *Preliminary Report on the Principal Geotechnical Aspects of the October 17, 1989 Loma Prieta Earthquake*. UCB/EERC-90/05, Berkeley, CA.

Seed, H.B. and Idriss, I.M. (1971). "Simplified procedure for evaluating soil liquefaction potential." *Journal of Soil Mechanics and Foundation Division* 97(SM9), 1249–1273.

Seed, H.B., Tokimatsu, K., Harder, L.F., and Chung, R.M. (1985). "The influence of SPT procedures in soil liquefaction resistance evaluations." *Journal of Geotechnical Engineering* 111(12), 1425–1445.

Serridge, C.J. and Synac, O. (2006). "Application of the rapid impact compaction (RIC) technique for risk mitigation in problematic soils." IAEG2006, Nottingham, UK, September 6–10.

Siskind, D.E., Stagg, M.S., Kopp, J.W., and Dowding, C.H. (1980). *Structure Response and Damage Produced by Ground Vibrations from Surface Mine Blasting*. Bureau of Mines, Department of Investigation, Washington, DC. RI8507.

Tara, D. and Wilson, P. (2004). *Rapid Impact Compactor Ground Improvement, Vibration Monitoring and Densification Assessment*. Final report submitted to Rapid Impact Compactors Ltd., Thurber Engineering Ltd., Delta, BC.

Thorburn, S. (1975). "Building structures supported by stabilized ground." *Geotechnique*, 25(1).

Thurner, H. and Sandstrom, A. (1980). "Continuous compaction control, CCC." *Proceedings of International Conference on Compaction*, Paris, 237–245.

U.S. Navy (1986). *Foundations and Earth Structures*. Design Manual 7.02. Naval Facilities Engineering Command, September.

U.S. Navy (1962). *Soil Mechanics, Foundations, and Earth Structures*. NAVFAC Design Manual DM-7, Washington DC.

Vennapusa, P.K.R., White, D.J., and Morris, M.D. (2010). "Geostatistical analysis for spatially referenced roller-integrated compaction measurements." *J. Geotech. Geoenvrion. Eng.*, 136(6): 813–822.

Watts, K.S. and Cooper, A. (2011). "Compaction of fills in land reclamation by rapid impact." *Geotechnica l Engineering, Proceedings of the Institute of Civil Engineers*, 164(GE3), 181–193.

Webb, D.L. and Hall, R.I.V. (1969). "Effects of vibroflotation on clayey sands." *Journal of the Soil Mechanics and Foundations Division* 95(SM6).

White, D.J., Vennapusa, P.K., and Gieselman, H. (2008). "Roller-integrated compaction monitoring technology: Field evaluation, spatial visualization, and specifications." *Proceedings of the 12th International Conference of International Association for Computer Methods and Advances in Geomechanics* (IACMAG), 1-6 October, Goa, India, 4424–4431.

Woodward, J. (2005). *Introduction to Geotechnical Processes*. CRC Press, Boca Raton, FL.

Yee, K. (2013). Personal communication.

Yee, K. and Ooi, T.A. (2010). "Ground improvement—A green technology towards a sustainable housing, infrastructure and utilities developments in Malaysia." *Geotechnical Engineering Journal of the SEAGS & AGSSEA* 41(2), 1–20.

CHAPTER 4

Overexcavation and Replacement

4.1 INTRODUCTION

4.1.1 Basic Concept

Overexcavation and replacement is one of the traditionally but still commonly used ground improvement methods in practice. The basic concept of this method is to remove a problematic geomaterial and replace it with nonproblematic fill. Replacing fills are often rock, gravel, and sand. Chemically stabilized soil, such as lime or cement-stabilized soil, can be used as well. Onsite geomaterial may be excavated and then recompacted back to the original location without or with addition of lime or cement. Figure 4.1 shows a typical design section of overexcavation and replacement for a shallow foundation. Depending on geotechnical condition and depth of excavation, the excavation can be vertical or sloped. The replaced area is typically larger than the area of the footing. The thickness of the replaced zone is limited to a certain value to be economical. When the bottom of the excavation is below groundwater table, dewatering is necessary but increases the cost of the project. Design of dewatering will be discussed in Chapter 6.

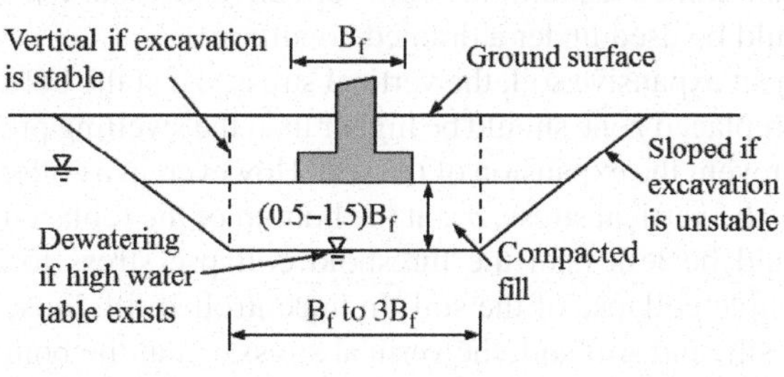

Figure 4.1 Typical design section of overexcavation and replacement (modified from Lawton, 2001).

4.1.2 Suitability

The overexcavation and replacement method is suitable for improving shallow problematic geomaterials. Problematic geomaterials include uncontrolled fill, loose sand and silt, soft soil, expansive soil, collapsible soil, liquefiable soil, and frozen soil that may have excessive deformation and/or potential bearing failure due to low strength during service. This method is often used for the following conditions:

- The area of overexcavation is limited.
- The depth of excavation is less than 3 m.
- No or limited temporary shoring and dewatering are required.
- No existing structure is close to the overexcavation area.
- Removed soil can be easily disposed or reused.
- Fill material is readily available.

4.1.3 Applications

The purpose of overexcavation and replacement depends on the issue(s) associated with the problematic geomaterial. This method can be used to:

- Increase bearing capacity
- Reduce settlement
- Eliminate expansion/shrinkage of expansive soil
- Eliminate the freeze–thaw of frozen soil

This method has been commonly used to improve geomaterials under continuous (strip) and isolated (square or rectangular) footings. It has also been used for highway and railway construction when problematic geomaterials are encountered within limited areas and depths.

4.1.4 Advantages and Limitations

This method is often cost-effective to improve problematic geomaterials when their area and depth are limited and fill materials are readily available. This method is simple, reliable, and well established. It does not require specialty contractors and special equipment except excavators and rollers if no temporary shoring and dewatering are not required.

Depending on site conditions, this method may be limited by

- Deep excavation required
- High groundwater table
- Onsite or nearby existing structures and utility lines
- Limited truck access to the site

- Long distance for hauling fill material and disposing of problematic geomaterial
- Time

4.2 PRINCIPLES

4.2.1 Stress Distribution

The basic principle of overexcavation and replacement is to eliminate potential problems by removing a problematic geomaterial and replacing it with nonproblematic fill. A complete replacement of the problematic geomaterial is preferred; however, a partial replacement is also acceptable and more economical as long as the performance of a structure on the replaced fill meets requirements. For expansive soil, collapsible soil, and frozen soil, the depth of the excavation should be equal or greater than the active depth of the problematic geomaterial, while the width of the excavation should ensure no influence of the problematic geomaterial outside the replaced zone on the performance of the structure. For uncontrolled fill, loose sand and silt, and soft soil, the depth and width of the excavation should be equal or greater than those needed for the replaced foundation to meet bearing capacity and settlement requirements. Figure 4.2 shows the stress distribution of an applied load from the base of the footing to the bottom of the replaced zone. The net pressure applied at the base of the footing is

$$p_n = \frac{P + W_f}{A_f} - \sigma'_D \tag{4.1}$$

where P = column load applied on the footing
W_f = weight of the footing
A_f = cross-sectional area of the footing
σ'_D = effective overburden stress at the base of the footing

The additional vertical stress at the center and the bottom of the replaced zone induced by the net pressure at the base of

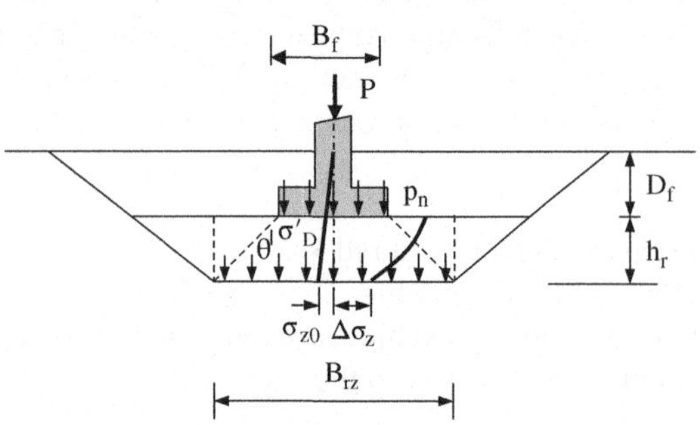

Figure 4.2 Stress distribution through the replaced zone (modified from Ye et al., 1994).

the footing can be estimated by a stress distribution method as follows:

For a rectangular footing:

$$\Delta\sigma_z = \frac{p_n A_f}{A'_f} = \frac{p_n L_f B_f}{L'_f B'_f} \tag{4.2a}$$

$$L'_f = L_f + 2h_r \tan\theta \tag{4.2b}$$

$$B'_f = B_f + 2h_r \tan\theta \tag{4.2c}$$

For a circular footing:

$$\Delta\sigma_z = \frac{p_n A_f}{A'_f} = \frac{p_n d_f^2}{(d'_f)^2} \tag{4.3a}$$

$$d'_f = d_f + 2h_r \tan\theta \tag{4.3b}$$

For a continuous footing:

$$\Delta\sigma_z = \frac{p_n B_f}{B'_f} \tag{4.4}$$

where A_f, L_f, B_f, d_f = area, length, width, diameter of the footing, respectively
$A'_f L'_f, B'_f, d'_f$ = area, length, width, diameter of the distributed foundation (the area, length, width, and diameter of the replaced zone, A_{rz}, L_{rz}, B_{rz}, and d_{rz} should be greater than those of the distributed foundation)
h_r = thickness of the replaced zone
θ = distribution angle

The total and effective vertical stresses at the bottom of the replaced zone are

$$\sigma_z = \sigma_{z0} + \Delta\sigma_z \tag{4.5a}$$

$$\sigma'_z = \sigma'_{z0} + \Delta\sigma_z \tag{4.5b}$$

where σ_{z0} and σ'_{z0} are total and effective overburden stresses of the geomaterial at the bottom of the replaced zone, respectively (total overburden stress should be used under an undrained condition, while effective overburden stress should be used under a drained condition).

In an expansive soil, the vertical stress, σ_z, at the bottom of the replaced zone should be higher than the swelling pressure to prevent the expansion of the soil. However, in a collapsible soil, the vertical stress, σ_z, at the bottom of the replaced zone should be less than the threshold collapse stress to avoid possible collapse of the soil. In uncontrolled fill, loose sand and silt, and soft soil, the vertical stress, σ_z, at the bottom of the replaced zone should be less than the allowable bearing capacity of the underlying soil. In addition, the design should ensure that the fill material in the replaced zone has sufficient bearing capacity and the total settlement of the footing is smaller than a tolerable value.

4.2.2 Failure Modes

Figure 4.3 shows possible failure modes of an improved foundation with a replaced zone. The mode for the general failure within the replaced zone [Figure 4.3(a)] likely develops under at least one of the following conditions: (1) the fill is too weak, (2) the area of the footing is too small, (3) the embedment depth of the footing is too shallow, and (4) the applied load is too high. This failure mode can be used to determine or verify the required strength of the fill if other parameters are fixed, or to determine the area and/or depth of the footing if the fill is selected.

The mode of a possible punching failure through a replaced zone [Figure 4.3(b)] likely occurs when the thickness of the replaced zone is too thin and the underlying soil is too weak. This failure mode can be used to determine the required thickness of the replaced zone. Use of higher strength fill or geosynthetic(s) can also minimize the chance of this failure. The use of geosynthetic(s) to prevent punching failure will be discussed in Chapter 10.

The distributed failure through a replaced zone [Figure 4.3(c)] is controlled by the strength of the underlying soil. Use of thicker and higher modulus fill or geosynthetic(s) helps distribute loads to a larger area and minimizes the chance of the failure of the underlying soil.

The punching failure of the replaced zone into the underlying soil happens when the area of the replaced zone is too small and the underlying soil is too weak. This failure mode is mostly dominated by the area of the replaced zone.

The ultimate bearing capacities of the footing on the replaced zone for all the possible failure modes should be calculated based on the short-term and long-term strengths of the soil and fill. The minimum ultimate bearing capacity among all the calculated values should be selected for the design. If the minimum allowable bearing capacity is less than the applied pressure, the design parameters for the dimensions of the footing and dimensions and properties of the replaced zone should be adjusted until meeting the bearing capacity requirement.

4.3 DESIGN CONSIDERATIONS

The design of the replaced zone should consider bearing capacity and settlement requirements as presented by Lawton (2001) in addition to other requirements (such as swelling, collapsible, liquefaction, and freezing–thaw weakening potentials) for special problematic geomaterials. The following parameters should be determined during design:

- Depth of replaced zone
- Length and width of replaced zone
- Thickness of replaced zone
- Fill quality including strength and modulus of fill

In addition to settlement calculations, the design should examine all possible failure modes as shown in Figure 4.3

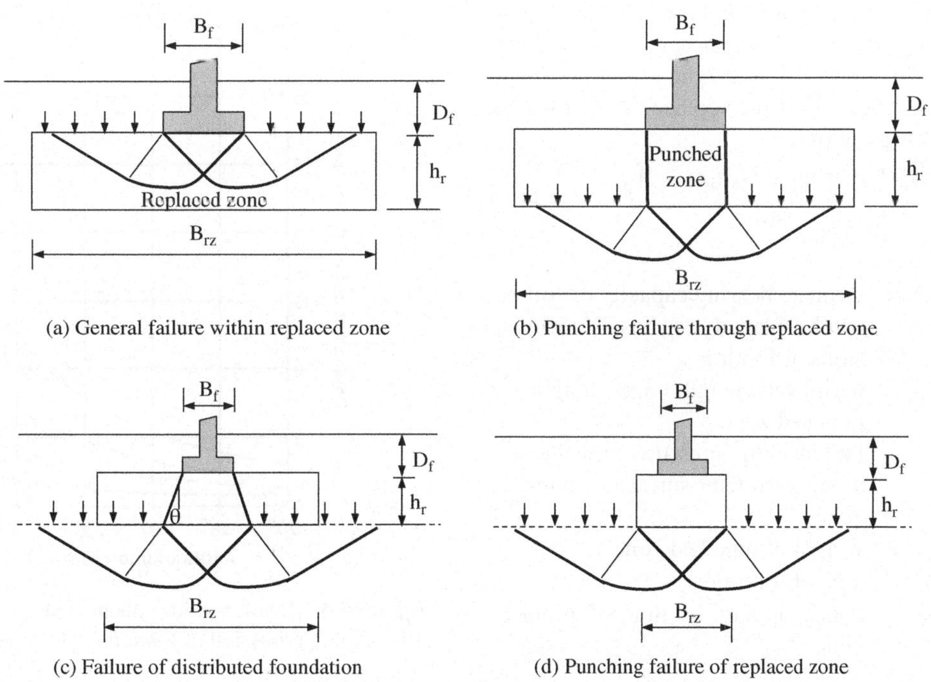

Figure 4.3 Possible failure modes of foundation with replaced zone (modified from Lawton, 2001).

by calculating their ultimate bearing capacities and factors of safety.

4.3.1 General Shear Failure within Replaced Zone

A bearing capacity formula for a homogeneous soil can be used to estimate the ultimate bearing capacity of the fill:

$$q_{\text{ult}} = N_c c s_c d_c + 0.5\gamma' B_f N_\gamma s_\gamma d_\gamma + \sigma'_D N_q s_q d_q \tag{4.6}$$

where
- c = cohesion of fill
- γ'_1 = effective unit weight of fill
- σ'_D = effective overburden stress at the base of the footing
- N_c, N_γ, N_q = bearing capacity factors of fill
- s_c, s_γ, s_q = shape factors
- d_c, d_γ, d_q = depth factors

The equations for all the factors are provided in Chapter 2. Since the methods for punching failure to be discussed in the next sections were developed by Meyerhof and Hanna (1978), it is recommended that the Meyerhof formulas for bearing capacity factors, shape factors, and depth factors in Chapter 2 be used to be consistent. This bearing capacity is mainly dominated by the quality of fill and dimensions of the footing.

4.3.2 Punching Failure through the Replaced Zone

Meyerhof and Hanna (1978) proposed a method to calculate the ultimate bearing capacity of one strong soil layer over a weak soil layer. This method can also be used to estimate the ultimate bearing capacity of the replaced zone over its underlying soil layer as follows:

$$q_{\text{ult}} = q_b + \frac{U_p P_h \tan\phi_1 + U_p h_r c_1 - W_{\text{pz}}}{A_f} \tag{4.7}$$

where
- q_b = ultimate bearing capacity of soil beneath replaced zone based on dimensions of footing
- U_p, h_r = perimeter length and height of punched zone
- P_h = lateral earth pressure thrust acting along perimeter surface of punched zone
- W_{pz} = weight of punched zone
- A_f = area of footing
- c_1 and ϕ_1 = cohesion and friction of replaced zone

The lateral earth pressure thrust acting along the perimeter surface of the punched zone can be estimated by

$$P_h = \int_{D_f}^{D_f+h_r} K_s \sigma'_{z0} dz \tag{4.8}$$

where
- K_s = coefficient of punching shear
- σ'_{z0} = effective overburden stress

If the coefficient of punching shear is constant within the replaced zone, Equation (4.8) can be simplified as follows:

$$P_h = K_s\left(\gamma'_1 D_f h_r + 0.5\gamma'_1 h_r^2\right) \tag{4.8a}$$

where γ'_1 is the unit weight of the fill within and above replaced zone.

The coefficient of punching shear, K_s, can be estimated based on the chart developed by Meyerhof and Hanna (1978) in Figure 4.4.

The approximate bearing capacity ratio can be determined based on the following equations:

$$q_1 = c_1 N_{c1} + 0.5\gamma'_1 B_f N_{\gamma 1} \tag{4.9}$$

$$q_2 = c_2 N_{c2} + 0.5\gamma'_2 B_f N_{\gamma 2} \tag{4.10}$$

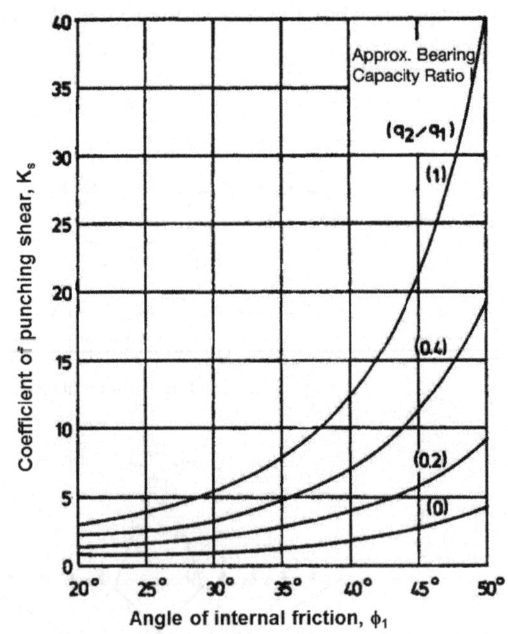

Figure 4.4 Coefficient of punching shear (Meyerhof and Hanna, 1978, with permission of *Canadian Geotechnical Journal*).

where γ'_2 = unit weight of in situ soil
q_1, q_2 = ultimate bearing capacities for a continuous footing with a width B_f under a vertical load on the surfaces of homogeneous thick deposits of the fill in the replaced zone and the underlying soil, respectively
c_1, c_2 = cohesion of the fill and the underlying soil, respectively
$N_{c1}, N_{\gamma 1}$ = bearing capacity factors of the fill
$N_{c2}, N_{\gamma 2}$ = bearing capacity factors of the underlying soil

4.3.3 Failure of Distributed Foundation

The applied pressure at the base of the footing can be distributed to the bottom of the replaced zone. The distributed area can be calculated using Equations (4.2b), (4.2c), and/or (4.3b) depending on the shape of the footing. The ultimate bearing capacity of the distributed foundation as a rigid footing on the underlying soil can be calculated as follows:

$$q_b = N_c c_2 s'_c d'_c + 0.5\gamma'_2 B'_f N_{\gamma 2} s'_\gamma d'_\gamma + \sigma'_{z0} N_{q2} s'_q d'_q \quad (4.11)$$

where γ'_2 = effective unit weight of the underlying soil
s'_c, s'_γ, s'_q = shape factors of the distributed foundation
d'_c, d'_γ, d'_q = depth factors of the distributed foundation
σ'_{z0} = effective overburden stress at the bottom of the replaced zone

The distribution angle, θ, depends on the modulus ratio of the replaced zone and the underlying soil, the width ratio of the replaced zone at the bottom to the footing, and the ratio of the replaced zone thickness to the footing width. It should be noted that if the distributed area is greater than the actual area of the replaced zone at the bottom, the actual dimensions of the replaced zone should be used. Based on the Burmister (1958) solution, Giroud and Han (2004) obtained an approximate formula to estimate the stress distribution angle from the upper layer to the lower layer based on the equivalent maximum stress at center as follows:

$$\tan\theta = \tan\theta_0 \left[1 + 0.204\left(\frac{E_1}{E_2} - 1\right)\right] \quad (4.12)$$

where θ_0 is the stress distribution angle for a reference uniform medium (typically 26.7°) and E_1 and E_2 are the moduli of the upper layer and lower layer, respectively. This formula is only valid when the width of the upper layer is much larger than the width of the footing. With a limited width of the replaced zone, the stress distribution angle in Equation (4.12) must be adjusted by a factor as follows:

$$\tan\theta = f_{\mathrm{lw}} \tan\theta_0 \left[1 + 0.204\left(\frac{E_1}{E_2} - 1\right)\right] \geq \tan\theta_0 \quad (4.13)$$

where f_{lw} is distribution angle reduction factor considering the limited area of the replaced zone. More research is needed to determine this distribution angle reduction factor.

The Chinese Ground Improvement Technical Code (JGJ79-91) (China Academy of Building Research, 2000) recommended the stress distribution angles as listed in Table 4.1.

It should be pointed out that the ultimate bearing capacity calculated by Equation (4.11) is at the base of the distributed foundation. This capacity should be converted to that under the footing. The following formula can be used to calculate the ultimate bearing capacity at the base of the footing:

$$q_{\mathrm{ult}} = \frac{A'_f}{A_f}(q_b - \gamma'_1 h_r) \quad (4.14)$$

where A_f, A'_f = areas of the footing and the distributed foundation, respectively
γ'_1 = effective unit weight of fill

4.3.4 Punching Failure of Replaced Zone into In Situ Soil

When the width of the replaced zone is too small, there is a possibility that the replaced zone as a rigid footing punches into the underlying soil. The ultimate bearing capacity of the punching failure of the replaced zone into the underlying soil can be calculated as follows (Lawton, 2001):

$$q_{\mathrm{ult}} = \frac{A_{\mathrm{rz}}}{A_f} q_b + \frac{Q_U - W_{\mathrm{rz}}}{A_f} \quad (4.15)$$

$$Q_U = U_1 P_{h1} \tan\phi_i + U_1 H c_i \quad (4.15a)$$

Table 4.1 Stress Distribution Angle, θ

	Fill Material		
h_r/B_f[a]	Medium Sand, Coarse Sand, Gravel, Cobble, Aggregate	Clay and Silt (8 < PI < 14)	Lime-Stabilized Soil
0.25	20	6	30
≥0.50	30	23	30

[a]When $h_r/B_f < 0.25$, use $\theta = 0$ except for lime-stabilized soil (θ = 30°); when $0.25 < h_r/B_f < 0.50$, interpolate between the values for $h_r/B_f = 0.25$ and $h_r/B_f \geq 0.50$.

Source: China Academy of Building Research (2000).

where q_b = ultimate bearing capacity of the soil beneath the replaced zone based on the dimensions of the replaced zone
Q_U = shear resistance along the perimeter surface of the replaced zone
U_1 = perimeter length of the replaced zone
P_{h1} = lateral earth pressure thrust acting along the perimeter surface of the replaced zone
W_{rz} = weight of the replaced zone
A_{rz} = area of the replaced zone
c_i, ϕ_i = cohesion and friction angle of the replaced zone (c_1 and ϕ_1) or the in situ soil (c_2 and ϕ_2)

The lateral earth pressure thrust can be estimated by the following equation:

$$P_{h1} = K_{s1}\left(\gamma_2' D_f h_r + 0.5\gamma_2' h_r^2\right) \tag{4.16}$$

where K_{s1} is the coefficient of punching shear around the perimeter surface of the replaced zone, which can be estimated by assuming $q_1 = q_2$ and using the friction angle of the in situ soil, ϕ_2 in Figure 4.4. However, K_{s1} should be equal or greater than K_0. For a normally consolidated in situ soil, $K_0 = 1 - \sin\phi_2$. Q_U should be the lesser of the calculated shear resistance using c_1 and ϕ_1 and c_2 and ϕ_2.

4.3.5 Minimum Bearing Capacity and Factor of Safety

The minimum ultimate bearing capacity is the least of all the bearing capacities calculated based on the failure modes shown in Figure 4.3 under undrained and drained conditions. The minimum bearing capacity and the corresponding failure mode control the design in terms of the bearing capacity. The factor of safety (FS) against the bearing failure can be calculated as follows:

$$\mathrm{FS} = \frac{q_{\mathrm{ult(min)}}}{p} \tag{4.17}$$

where $q_{\mathrm{ult(min)}}$ is the minimum ultimate bearing capacity considering all failure modes under undrained and drained conditions and p is the applied pressure under the footing.

The minimum factor of safety should be greater than the required factor of safety. Otherwise, quality of fill material and/or dimensions of the replaced zone should be adjusted until the requirement is met.

4.3.6 Settlement of a Footing on Layered Soils of Infinite Width

Settlement of a footing on layered soils of infinite width can be calculated using an elastic layered theory, which is often used in design of airfields and highways because the lengths and widths of pavement layers are much larger than those of a tire contact area. The settlement of a circular footing on a two-layer soil system as shown in Figure 4.5 is as follows:

$$S = \rho_r \frac{p_n d_f}{E_2} I_s \tag{4.18}$$

where S = settlement at center if the footing is flexible or average settlement if the footing is rigid
ρ_r = rigidity factor of the footing (1.0 for a flexible footing or 0.79 for a rigid footing)
p_n = net applied pressure
d_f = diameter of the footing
E_2 = elastic modulus of the lower soil layer
I_s = settlement influence factor for a circular, uniform pressure on the surface of a two-layer elastic medium (Poisson's ratios, $\nu_1 = 0.2$ and $\nu_2 = 0.4$), which is provided in Figure 4.6

Figure 4.6 shows that the increase of the modulus ratio reduces the settlement influence factor and thus the settlement. Even though it is possible for an upper soil layer to have an elastic modulus of 20 times higher than the lower layer, it is suggested that the modulus ratio should be limited to 20 unless more field test data and experience justify a higher ratio. Clearly, the increase of the thickness of the upper layer, which has a higher modulus than the lower layer, can reduce the settlement as well.

4.3.7 Settlement of a Footing on a Replaced Zone with Limited Area

Elastic Solution for Circular Footing on Replaced Zone It is clear that a replaced zone has a limited area to be economic; therefore, Equation (4.18), which was developed based on infinite area, may not be accurate for calculating the settlement of a footing on a replaced zone if the area of the replaced zone is small. To develop the full reduction in the settlement as that with the infinite area, Lawton (2001) suggested the area of the replaced zone should be large enough to cover 95% the applied load within the bottom area of the replaced zone.

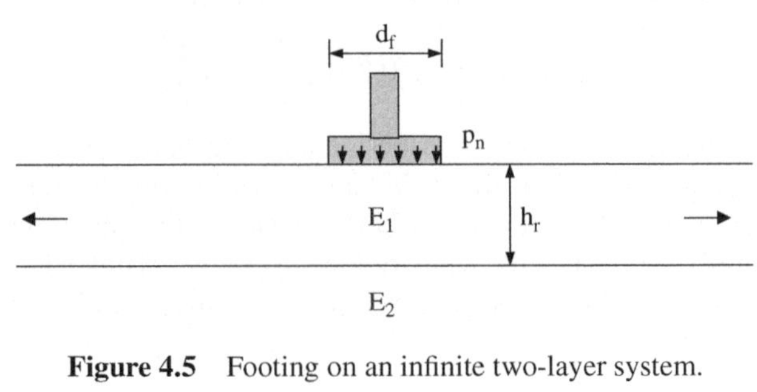

Figure 4.5 Footing on an infinite two-layer system.

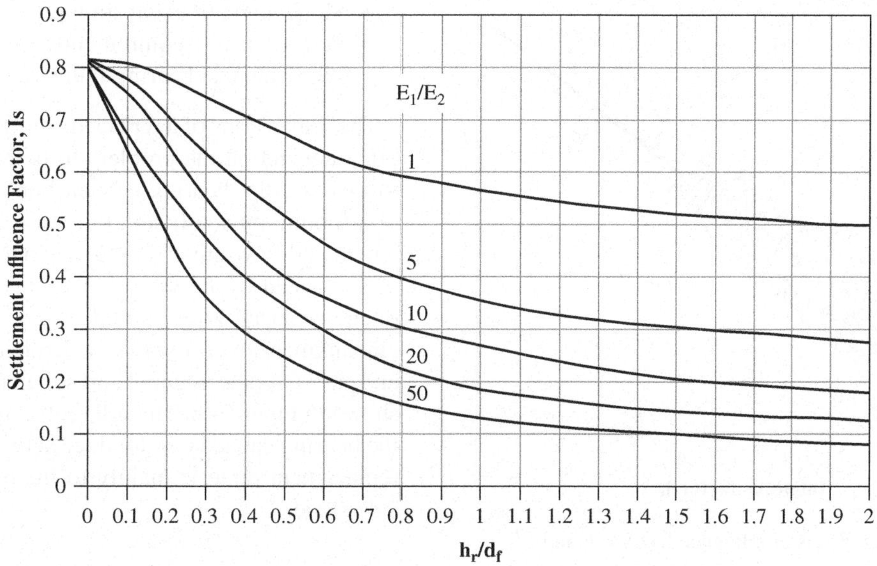

Figure 4.6 Settlement Influence Factors (Burmister, 1958).

Based on this suggestion, Lawton (2001) proposed the following condition for the full reduction in the settlement:

For a rectangular footing:

$$L_{rz} \geq L_f + 2h_r \tan \theta_s \quad (4.19a)$$

$$B_{rz} \geq B_f + 2h_r \tan \theta_s \quad (4.19b)$$

For a circular footing:

$$d_{rz} \geq d_f + 2h_r \tan \theta_s \quad (4.19c)$$

where θ_s is the distribution angle corresponding to the area to cover 95% applied load. Based on the design chart provided in Lawton (2001), the following approximate relationship can be obtained:

$$\tan \theta_s = 0.67 \left(\frac{E_1}{E_2} \right)^{0.45} \quad (4.20)$$

It should be pointed out that the stress distribution angle in Equation (4.20) is different from that in Equation (4.12). The relationship in Equation (4.12) was developed based on the maximum stress at the center of the interface between the upper and lower layers while the relationship in Equation (4.20) was developed based on 95% applied load within the distributed area.

If the dimension of the replaced zone does not meet the requirement in Equation (4.19a) and (4.19b) or (4.19c), the following modification factors should be used to adjust the settlement influence factor, I_s in Equation (4.18) from Figure 4.6:

$$I'_s = I_s I_D I_h I_{drz} I_v \quad (4.21)$$

$$I_D = 1 - \frac{1}{1.1(d_f/D_f + 1.6)} \quad (4.22)$$

$$I_h = 1 + 2\frac{h_r}{d_f}\left[\left(\frac{E_1}{E_2}\right)^{0.1} - 1\right] \quad (4.23)$$

$$I_{drz} = \left(\frac{E_1}{E_2}\right)^{0.05} \left(\frac{d_{rz}}{d_f}\right)^{-0.2} \geq 1.0 \quad (4.24)$$

$$I_v = \frac{1 - \nu_2}{1 - 0.4} \quad (4.25)$$

where
I_D = modification factor for the embedment of the footing
I_h = modification factor for the thickness of the replaced zone
I_{drz} = modification factor for the width of the replaced zone

Except Equation (4.25) suggested by Lawton (2001), the preceding modification factors were developed by the author based on the tabulated settlement influence factors for a flexible, uniform, circular pressure applied onto the replaced zone provided in Lawton (2001). The accuracy of the approximate formula is examined by the comparisons shown in Figure 4.7, in which I'_s is calculated using Eq. (4.21) and I_s is from Lawton (2001).

Empirical Method The settlement of a footing on a replaced zone may also be estimated using an empirical method. In the empirical method, a distributed foundation is first determined by the stress distribution method mentioned in Section 4.2. The compression of the replaced zone from the base of the footing to the bottom of the replaced zone

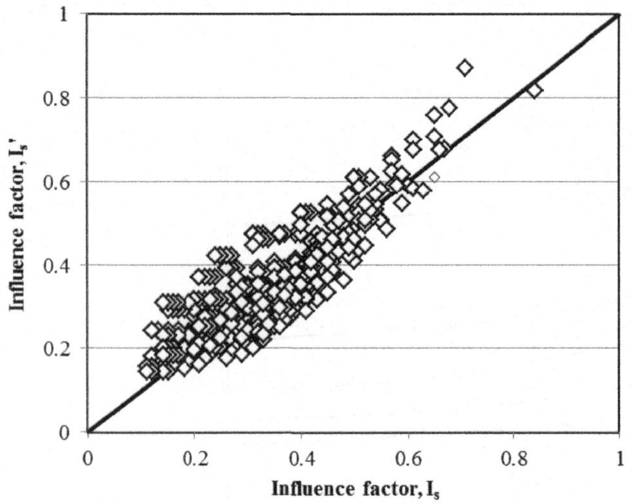

Figure 4.7 Comparisons of influence factors, I_s and I_s'.

can be estimated as follows:

$$S_{rz} = \frac{p_n + \Delta\sigma_z}{2E_1} h_r \tag{4.26}$$

Then the distributed foundation is treated as a rigid footing. The settlement below the distributed foundation can be estimated using the consolidation index method for clays or the Schmertmann method for sands (Schmertmann, 1970) as described in Chapter 2. The total settlement of the footing is the sum of the compression of the replaced zone and the settlement below the distributed foundation.

The calculated settlement by the elastic solution or empirical method should be less than the allowable settlement to meet the performance requirement. Otherwise, quality of fill material and/or dimensions of the replaced zone should be adjusted until the requirement is met.

4.4 DESIGN PARAMETERS AND PROCEDURE

4.4.1 Design Parameters

Typical design parameters for overexcavation and replacement include:

- Shape and dimensions of footing (such as width, length or diameter, and embedment depth)
- Applied load on the footing
- In situ geomaterial conditions, including unit weight, undrained and drained shear strengths, and the groundwater table
- Dimensions of the replaced zone, including width, length or diameter, and thickness
- Quality of replacing fill including unit weight and shear strength
- Slope angle of excavation pit
- Performance requirements, including the factor of safety against bearing failure and allowable settlement

The dimensions of a replaced zone and properties of the fill often depend on the problem to be mitigated. For example, to increase the bearing capacity and reduce the settlement of a footing on a soft soil, the replaced zone often needs its length and/or width of B_f–$3B_f$ and a thickness of $0.5B_f$–$1.5B_f$ as suggested by Lawton (2001). In addition, the thickness of the replaced zone should be greater than 0.5 m. For an embankment over a soft soil, Broms (1979) suggested full and partial replacement of soft soil under the embankment, as shown in Figure 4.8. The full replacement is done to increase the bearing capacity and reduce settlement, while the partial replacement done is mainly to increase the stability of the side slope.

4.4.2 Design Procedure

The following procedure should be followed for design of overexcavation and replacement:

1. To gather information about geomaterial conditions, including soil types, thicknesses, and properties and depth of groundwater table.
2. To gather information about foundation type, dimensions, and loading conditions.
3. To gather information about required performance.
4. To evaluate potential problem(s) without replacement; such as low bearing capacity, excessive settlement, shrinkage and heave, soil collapse, freez–thaw, and liquefaction.
5. To develop a trial design section of a replaced zone including the area and thickness of the replaced zone and type and properties of fill.

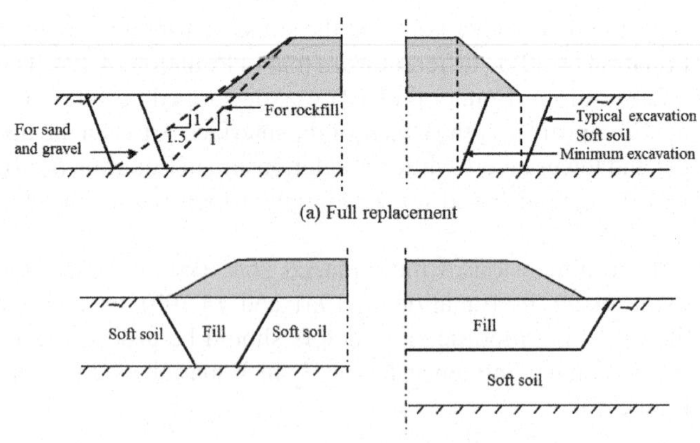

Figure 4.8 (a) Full and (b) partial replacement under an embankment (modified from Broms, 1979).

6. Depending on the nature of the problem, the following steps may be different. Using a foundation on soft soil as an example, the factors of safety against possible bearing failures should be examined following the calculations in Section 4.3.
7. To calculate the settlement of the footing on the replaced zone following the calculations in Section 4.3.
8. To compare the calculated performance results in steps 6 and 7 with the required performance. If the results meet the required performance, the design is complete. Otherwise, go back to step 5, adjust the design section, and repeat steps 6 and 7 until satisfactory results are obtained.
9. Calculate the volumes of excavation and fill material.

4.5 DESIGN EXAMPLE

A 1.0-m-wide square concrete footing is designed to be embedded at a depth of 1.0 m. The site has 9-m-thick soft clayey soil from the ground surface and underlain by a very dense sand layer. The soft soil is a normally consolidated clay, which has undrained shear strength of 25 kPa and effective friction angle of 25°. The undrained elastic modulus of this clay is 200 times the undrained shear strength, and the effective elastic modulus is approximately half of its undrained elastic modulus. The groundwater table is at the depth of 0.6 m. The unit weight of the soil above the groundwater table is 18 kN/m^3, while that below the groundwater table is 20 kN/m^3. The column load applied on this footing is 150 kN and the concrete unit weight is 23.6 kN/m^3. Overexcavation and replacement are adopted to increase the bearing capacity and reduce the short-term and long-term settlements. The backfill material for the replaced zone is a well-graded gravel, which has a unit weight of 20 kN/m^3, friction angle of 38°, and elastic modulus of 50,000 kPa after compaction. The designed replaced zone is 1.0 m thick, 2.0 m long, and 2.0 m wide. The required factor of safety against bearing failure is 3.0, and the tolerable settlement including immediate and consolidation settlements is 25 mm. You are required to: (1) calculate the ultimate bearing capacity and factor of safety of the foundation without replacement; (2) calculate the ultimate bearing capacity and factor of safety of the foundation after replacement; and (3) calculate the settlement of the foundation after replacement. See Example Figure 4.1 for the cross section and parameters.

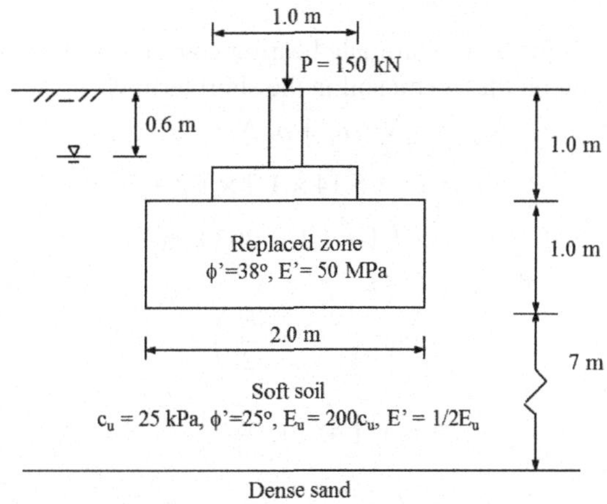

Example Figure 4.1 Design cross section and parameters.

Solution

1. Applied bearing pressure: This example has a high groundwater table, which is not necessarily favorable for overexcavation and replacement. However, it is sometimes done in practice and requires dewatering and/or shoring. This example was prepared to provide a comprehensive design with a complicated problem.
   Applied pressure at the base of the footing is

$$p = \frac{P + W_f}{A_f} - u_D = \frac{150 + 1 \times 1 \times 1 \times 23.6}{1 \times 1} - 0.4 \times 9.81 = 170\,\text{kPa}$$

2. Bearing capacity and factor of safety without replacement: The ultimate bearing capacity of the footing on the in situ soil is

$$q_{\text{ult}} = cN_c s_c d_c + 0.5\gamma' B_f N_\gamma s_\gamma d_\gamma + \sigma'_D N_q s_q d_q$$

   The effective overburden stress at the base of the footing is

$$\sigma'_D = 0.6 \times 18 + 0.4 \times 20 - 0.4 \times 9.81 = 15\,\text{kPa}$$

   In the short term, $\phi_u = 0$; therefore, $N_c = 5.14$, $N_\gamma = 0$, $N_q = 1$, and $K_p = 1$. The shape factors and depth factors can be calculated as follows:

$$s_c = 1 + 0.2K_p \frac{B_f}{L_f} = 1 + 0.2 \times 1 \times \frac{1}{1} = 1.2$$

$$s_q = 1 + 0.1K_p \frac{B_f}{L_f} = 1 + 0.1 \times 1 \times \frac{1}{1} = 1.1$$

$$d_c = 1 + 0.2\sqrt{K_p} \frac{D_f}{B_f} = 1 + 0.2 \times 1 \times \frac{1}{1} = 1.2$$

$$d_q = 1 + 0.1\sqrt{K_p} \frac{D_f}{B_f} = 1 + 0.1 \times 1 \times \frac{1}{1} = 1.1$$

Hence, the ultimate bearing capacity of the footing on the in situ soil in the short term is

$$q_{ult} = c_u N_c s_c d_c + \sigma'_D N_q s_q d_q$$
$$= 25 \times 5.14 \times 1.2 \times 1.2 + 15 \times 1$$
$$\times 1.1 \times 1.1 = 203 \text{ kPa}$$

In the long term,

$$N_q = \tan^2\left(45° + \frac{\phi'}{2}\right) e^{\pi \tan \phi'}$$
$$= \tan^2\left(45° + \frac{25°}{2}\right) e^{\pi \tan 25°}$$
$$= 10.66$$
$$N_\gamma = (N_q - 1)\tan(1.4\phi') = 6.77$$
$$K_p = \tan^2\left(45° + \frac{25°}{2}\right) = 2.46$$

The shape factors and depth factors are

$$s_\gamma = s_q = 1 + 0.1K_p\frac{B_f}{L_f} = 1 + 0.1$$
$$\times \tan^2\left(45° + \frac{25°}{2}\right) \times \frac{1}{1} = 1.25$$
$$d_\gamma = d_q = 1 + 0.1\sqrt{K_p}\frac{D_f}{B_f} = 1 + 0.1$$
$$\times \tan\left(45° + \frac{25°}{2}\right) \times \frac{1}{1} = 1.16$$

Hence, the ultimate bearing capacity of the footing on the natural soil in the long term is

$$q_{ult} = 0.5\gamma' B_f N_\gamma s_\gamma d_\gamma + \sigma'_D N_q s_q d_q$$
$$= 0.5 \times (20 - 9.81) \times 1 \times 6.77 \times 1.25 \times 1.16$$
$$+ 15 \times 10.66 \times 1.25 \times 1.16 = 278 \text{ kPa}$$

q_{ult}(long term) > q_{ult}(short term); therefore, the short term controls. The factor of safety is

$$FS = \frac{203}{170} = 1.20 < 3.0$$

therefore, overexcavation and replacement are necessary.

3. Bearing capacity and factor of safety after replacement:

   General failure within the replaced zone. Since the replaced zone consists of gravel, only a drained analysis is needed. The ultimate bearing capacity based on this failure mode (i.e., $c'_1 = 0$, $\phi'_1 = 38°$) can be calculated as follows:

$$N_{q1} = \tan^2\left(45° + \frac{\phi'_1}{2}\right) e^{\pi \tan \phi'_1}$$
$$= \tan^2\left(45° + \frac{38°}{2}\right) e^{\pi \tan 38°} = 48.9$$
$$N_{\gamma 1} = (N_{q1} - 1)\tan(1.4\phi'_1)$$
$$= (48.9 - 1) \times \tan(1.4 \times 38°) = 64.1$$
$$s_\gamma = s_q = 1 + 0.1K_p\frac{B_f}{L_f} = 1 + 0.1$$
$$\times \tan^2\left(45° + \frac{38°}{2}\right) \times \frac{1}{1} = 1.42$$
$$d_\gamma = d_q = 1 + 0.1\sqrt{K_p}\frac{D_f}{B_f} = 1 + 0.1$$
$$\times \tan\left(45° + \frac{38°}{2}\right) \times \frac{1}{1} = 1.21$$
$$q_{ult} = 0.5\gamma'_1 B_f N_{\gamma 1} s_\gamma d_\gamma + \sigma'_D N_{q1} s_q d_q$$
$$= 0.5 \times (20 - 9.81) \times 1 \times 64.1 \times 1.42 \times 1.21$$
$$+ 15 \times 48.9 \times 1.42 \times 1.21 = 1805 \text{ kPa}$$

The factor of safety is

$$FS = \frac{1805}{170} = 10.61 > 3.0 \quad \text{(OK)}$$

Punching failure through the replaced zone. Since the punching failure through the replaced zone into the soft soil, undrained (i.e., short term) and drained (i.e., long term) analyses are needed.

For the short term:

Replaced zone : $N_{q_1} = 48.9$, $N_{\gamma 1} = 64.1$

Soft soil : $N_{c2} = 5.14$ $N_{q_2} = 1.0$

The bearing capacity of a continuous footing on the surface of the replaced zone is

$$q_1 = c_1 N_{c1} + 0.5\gamma'_1 B_f N_{\gamma 1} = 0 + 0.5$$
$$\times (20 - 9.81) \times 1 \times 64.1 = 326 \text{ kPa}$$

The bearing capacity of a continuous footing on the surface of the soft soil is

$$q_2 = c_2 N_{c2} = 25 \times 5.14 = 129 \text{ kPa}$$
$$\frac{q_2}{q_1} = \frac{129}{326} = 0.39 \quad \phi_1 = 38° \rightarrow K_s = 7.0$$

from Figure 4.4. The ultimate bearing capacity of the punching portion on the underlying soft soil can be calculated as follows:

$$s_c = 1 + 0.2K_p\frac{B_f}{L_f} = 1 + 0.2 \times 1 \times \frac{1}{1} = 1.2$$

$$s_q = 1 + 0.1K_p\frac{B_f}{L_f} = 1 + 0.1 \times 1 \times \frac{1}{1} = 1.1$$

$$d_c = 1 + 0.2\sqrt{K_p}\frac{D_f + h_r}{B_f} = 1 + 0.2 \times 1 \times \frac{2}{1} = 1.40$$

$$d_q = 1 + 0.1\sqrt{K_p}\frac{D_f + h_r}{B_f} = 1 + 0.1 \times 1 \times \frac{2}{1} = 1.20$$

$$\sigma'_{z0} = 0.6 \times 18 + 1.4 \times 20 - 1.4 \times 9.81 = 25 \text{ kPa}$$

$$q_b = c_u N_{c2} s_c d_c + \sigma'_{z0} N_{q2} s_q d_q = 25 \times 5.14 \times 1.2 \times 1.40 + 25 \times 1 \times 1.1 \times 1.20 = 249 \text{ kPa}$$

$$P_h = K_s(0.5\gamma'_1 h_r^2 + \gamma'_2 D_f h_r) = 7.0 \times [0.5 \times (20 - 9.81) \times 1^2 + 18 \times 0.6 \times 1 + (20 - 9.81) \times 0.4 \times 1] = 140 \text{ kN/m}$$

$$W_{pz} = (20 - 9.81) \times 1 \times 1 \times 1 = 10 \text{kN}$$

$$q_{ult} = q_b + \frac{U_p P_h \tan \phi_1 + U_p h_r c_1 - W_{pz}}{A_f} = 249 + \frac{4 \times 1 \times 140 \times \tan 38° + 0 - 10}{1 \times 1} = 676 \text{ kPa}$$

For the long-term:

$$q_1 = 326 \text{kPa}$$

$$N_{q2} = 10.66$$

$$N_{\gamma 2} = 15.89$$

$$q_2 = 0.5\gamma_2 B_f N_{\gamma 2} = 0.5 \times (20 - 9.81) \times 1 \times 15.89 = 81 \text{kPa}$$

$$\frac{q_2}{q_1} = \frac{81}{326} = 0.25 \quad \phi_1 = 38^o \rightarrow K_s = 4$$

$$s_\gamma = s_q = 1 + 0.1K_p\frac{B_f}{L_f} = 1 + 0.1 \times \tan^2\left(45° + \frac{25°}{2}\right) \times \frac{1}{1} = 1.25$$

$$d_\gamma = d_q = 1 + 0.1\sqrt{K_p}\frac{D_f + h_r}{B_f} = 1 + 0.1 \times \tan\left(45° + \frac{25°}{2}\right) \times \frac{2}{1} = 1.31$$

$$q_b = 0.5\gamma_2 B_f N_{\gamma 2} s_\gamma d_\gamma + \sigma'_{z0} N_{q2} s_q d_q = 0.5 \times (20 - 9.81) \times 1 \times 15.89 \times 1.25 \times 1.31 + 25 \times 10.66 \times 1.25 \times 1.31 = 570 \text{ kPa}$$

$$P_h = K_s(0.5\gamma'_1 h_r^2 + \gamma'_1 D_f h_r) = 4 \times [0.5 \times (20 - 9.81) \times 1^2 + 18 \times 0.6 \times 1 + (20 - 9.81) \times 0.4 \times 1] = 80 \text{ kN/m}$$

$$q_{ult} = q_b + \frac{U_1 P_h \tan \phi_1 + U_1 h_r c_1 - W_{pz}}{A_f} = 570 + \frac{4 \times 1 \times 80 \times \tan 38° + 0 - 10}{1 \times 1} = 809 \text{ kPa}$$

q_{ult}(long term) > q_{ult}(short term), therefore, the short-term controls, that is, $q_{ult} = 676$ kPa.

The factor of safety is

$$FS = \frac{676}{170} = 3.98 > 3.0, \text{ (OK)}$$

Failure of distributed foundation. For a well-graded gravel and $h_r/B_f \geq 0.5$, the distribution angle is 30° based on the Chinese Ground Improvement Technical Code (JGJ79-91). The distributed width and length at the base of the replaced zone are

$$L'_f = B'_f = B_f + 2h_r \tan \theta = 1 + 2 \times 1 \times \tan 30° = 2.15 \text{m} > B_{rz}$$

Use

$$L'_f = B'_f = 2 \text{m}$$

The effective overburden stress at the base of the distributed foundation is

$$\sigma'_{z0} = 0.6 \times 18 + (1.0 + 1.0 - 0.6) \times (20 - 9.81) = 25 \text{kPa}$$

For the short term:

$$s'_c = 1 + 0.2K_p\frac{B'_f}{L'_f} = 1 + 0.2 \times 1 \times \frac{2}{2} = 1.2$$

$$s'_q = 1 + 0.1K_p\frac{B'_f}{L'_f} = 1 + 0.1 \times 1 \times \frac{2}{2} = 1.1$$

$$d'_c = 1 + 0.2\sqrt{K_p}\frac{D_f + h_r}{B'_f} = 1 + 0.2 \times 1 \times \frac{2}{2} = 1.20$$

$$d'_q = 1 + 0.1\sqrt{K_p}\frac{D_f + h_r}{B'_f} = 1 + 0.1 \times 1 \times \frac{2}{2} = 1.10$$

The ultimate bearing capacity at the base of the distributed foundation is

$$q_b = c_u N_{c2} s'_c d'_c + \sigma'_{z0} N_{q2} s'_q d'_q = 25 \times 5.14 \times 1.2 \times 1.2 + 25 \times 1 \times 1.1 \times 1.1 = 216\,\text{kPa}$$

The ultimate bearing capacity at the base of the footing is

$$q_{\text{ult}} = \frac{L'_f B'_f}{L_f B_f}(q_b - \gamma'_1 h_r) = \frac{2.0 \times 2.0}{1.0 \times 1.0} \times (216 - 10.89 \times 1.0) = 819\,\text{kPa}$$

For the long term:

$$s'_\gamma = s'_q = 1 + 0.1 K_p \frac{B'_f}{L'_f} = 1 + 0.1 \times \tan^2\left(45^\circ + \frac{25^\circ}{2}\right) \times \frac{2}{2} = 1.25$$

$$d'_\gamma = d'_q = 1 + 0.1\sqrt{K_p}\frac{D_f + h_r}{B'_f} = 1 + 0.1 \times \tan\left(45^\circ + \frac{25^\circ}{2}\right) \times \frac{2}{2} = 1.16$$

The ultimate bearing capacity at the base of the distributed foundation is

$$q_b = 0.5\gamma'_2 B'_f N_{\gamma 2} s'_\gamma d'_\gamma + \sigma'_{z0} N_{q2} s'_q d'_q$$
$$= 0.5 \times (20 - 9.81) \times 2.0 \times 15.89 \times 1.25 \times 1.16 + 25 \times 10.66 \times 1.25 \times 1.16 = 619\text{ kPa}$$

The ultimate bearing capacity at the base of the footing is

$$q_{\text{ult}} = \frac{L'_f B'_f}{L_f B_f}(q_b - \gamma'_1 h_r) = \frac{2.0 \times 2.0}{1.0 \times 1.0} \times (619 - 10.89 \times 1.0) = 2435\,\text{kPa}$$

$$q_{\text{ult}}(\text{long term}) > q_{\text{ult}}(\text{short term})$$

therefore, the short-term controls, that is, $q_{\text{ult}} = 819$ kPa.

The factor of safety is

$$\text{FS} = \frac{819}{170} = 4.83 > 3.0\ (\text{OK})$$

Punching of replaced zone into the in situ soil. The length and width of the replaced zone are used as those of the footing for the bearing capacity calculation, $L_{rz} = 2m$ and $B_{rz} = 2m$.

For the short term:

$$\frac{q_2}{q_1} = 1 \quad \phi_2 = 0^\circ \rightarrow K_s \geq 1 - \sin\phi_2 = 1$$

$$s_c = 1 + 0.2K_p\frac{B_{rz}}{L_{rz}} = 1 + 0.2 \times 1 \times \frac{2}{2} = 1.2$$

$$s_q = 1 + 0.1K_p\frac{B_{rz}}{L_{rz}} = 1 + 0.1 \times 1 \times \frac{2}{2} = 1.1$$

$$d_c = 1 + 0.2\sqrt{K_p}\frac{D'_f}{B_{rz}} = 1 + 0.2 \times 1 \times \frac{2}{2} = 1.2$$

$$d_q = 1 + 0.1\sqrt{K_p}\frac{D'_f}{B_{rz}} = 1 + 0.1 \times 1 \times \frac{2}{2} = 1.1$$

$$q_b = c_u N_{c2} s_c d_c + \sigma'_{z0} N_{q2} s_q d_q = 25 \times 5.14 \times 1.2 \times 1.13 + 25 \times 1 \times 1.2 \times 1.1 = 215\,\text{kPa}$$

$$P_{h1} = K_s(0.5\gamma'_2 h_r^2 + \gamma'_2 D_f h_r) = 1 \times [0.5 \times (20 - 9.81) \times 1^2 + 18 \times 0.6 \times 1 + (20 - 9.81) \times 0.4 \times 1]$$
$$= 20\text{ kN/m}$$

$$Q_{U1} = U_1 P_{h1}\tan\phi_1 + U_1 h_r c_1 = 4 \times 2 \times 20 \times \tan 38^\circ + 0 = 125\,\text{kN}$$

$$Q_{U2} = U_1 P_{h1}\tan\phi_2 + U_1 h_r c_2 = 0 + 4 \times 2 \times 1 \times 25 = 200\,\text{kN}$$

$$Q_U = \min(Q_{U1}, Q_{U2}) = 125\,\text{kN}$$

$$W_{rz} = (20 - 9.81) \times 2 \times 2 \times 1 = 41\,\text{kN}$$

$$q_{\text{ult}} = q_b\frac{A_{rz}}{A_f} + \frac{Q_U - W_{rz}}{A_f} = 215 \times \frac{2 \times 2}{1 \times 1} + \frac{125 - 41}{1 \times 1} = 946\text{ kPa}$$

For the long term:

$$\frac{q_2}{q_1} = 1 \quad \phi_1 = 25^\circ \rightarrow K_s = 3.7$$

$$s_\gamma = s_q = 1 + 0.1K_p \frac{B_{rz}}{L_{rz}} = 1 + 0.1 \times \tan^2\left(45^\circ + \frac{25^\circ}{2}\right) \times \frac{2}{2} = 1.25$$

$$d_\gamma = d_q = 1 + 0.1\sqrt{K_p}\frac{D_f'}{B_{rz}} = 1 + 0.1 \times \tan\left(45^\circ + \frac{25^\circ}{2}\right) \times \frac{2}{2} = 1.16$$

$$q_b = 0.5\gamma_2' B_{rz} N_{\gamma 2} s_\gamma d_\gamma + \sigma_{z0}' N_q s_q d_q$$
$$= 0.5 \times (20 - 9.81) \times 2 \times 15.89 \times 1.25 \times 1.16 + 25 \times 10.66 \times 1.25 \times 1.16 = 619 \text{ kPa}$$

$$P_{h1} = K_s(0.5\gamma_2' h_r^2 + \gamma_2' D_f h_r) = 3.7 \times [0.5 \times (20 - 9.81) \times 1^2 + 18 \times 0.6 \times 1 + (20 - 9.81) \times 0.4 \times 1]$$
$$= 74 \text{ kN/m}$$

$$Q_{U1} = U_1 P_{h1} \tan \phi_1 + U_1 h_r c_1 = 4 \times 2 \times 74 \times \tan 38^\circ + 0 = 462 \text{ kN}$$

$$Q_{U2} = U_1 P_{h1} \tan \phi_2 + U_1 h_r c_2 = 4 \times 2 \times 74 \times \tan 25^\circ + 0 = 276 \text{ kN}$$

$$Q_U = \min(Q_{U1}, Q_{U2}) = 276 \text{ kN}$$

$$q_{ult} = q_b \frac{A_{rz}}{A_f} + \frac{Q_U - W_{rz}}{A_f} = 619 \frac{2 \times 2}{1 \times 1} + \frac{276 - 41}{1 \times 1} = 2710 \text{ kPa}$$

q_{ult}(long term) > q_{ult}(short term), therefore, the short-term controls, that is, $q_{ult} = 946$ kPa.

The factor of safety against this failure mode is

$$\text{FS} = \frac{946}{170} = 5.57 > 2.0 \quad \text{(OK)}$$

In summary, the minimum ultimate bearing capacity of the footing on the replaced zone, $q_{ult} = 676$ kPa, due to possible punching failure through the replaced zone and its corresponding factor of safety is FS = 4.77, which is higher than the required factor of safety of 3.0; therefore, the design is acceptable based on the bearing capacity requirement.

4. Settlement calculation after replacement: The net applied pressure at the bottom of the footing for settlement calculation is

$$p_n = p - \sigma_D' = 170 - 15 = 155 \text{ kPa}$$

The equivalent diameter of the footing and the replaced zone is

$$d_f = \sqrt{\frac{4A_f}{\pi}} = \sqrt{\frac{4 \times 1 \times 1}{3.14}} = 1.13 \text{ m}$$

$$d_{rz} = \sqrt{\frac{4A_{rz}}{\pi}} = \sqrt{\frac{4 \times 2 \times 2}{3.14}} = 2.26 \text{ m}$$

The following dimension ratios can be obtained:

$$\frac{D_f}{d_f} = \frac{1}{1.13} = 0.89 \quad \frac{h_r}{d_f} = \frac{1}{1.13} = 0.89$$

$$\frac{d_{rz}}{d_f} = \frac{2.26}{1.13} = 2.0$$

For immediate (short-term) settlement

$$E_{u2} = 200 \times 25 = 5000 \text{ kPa}, \ \nu_2 = 0.5$$

$$\frac{E_1}{E_{u2}} = \frac{50,000}{5000} = 10 \rightarrow \tan \theta_s = 0.67\left(\frac{E_1}{E_{u2}}\right)^{0.45} = 0.67(10)^{0.45} = 1.89$$

therefore, the required width of the replaced zone for full reduction in settlement is $d_f + 2h_r \tan \theta_s = 1.13 + 2 \times 1 \times 1.89 = 4.91 \text{ m} > 2 \text{ m} \rightarrow$ the solution for limited width of a replaced zone is needed:

$$\frac{E_1}{E_2} = 10 \quad \text{and} \quad \frac{h_r}{d_f} = 0.89 \rightarrow I_s = 0.29$$

(center) from Figure 4.6 for an infinite width condition. Due to the limited width of the replaced zone, the following modification factors can be determined:

$$I_D = 1 - \frac{1}{1.1(d_f/D_f + 1.6)} = 1 - \frac{1}{1.1(1/0.89 + 1.6)} = 0.67$$

$$I_h = 1 + 2\left(\frac{h_r}{d_f}\right)\left[\left(\frac{E_1}{E_2}\right)^{0.1} - 1\right] = 1 + 2 \times (0.89) \times (10^{0.1} - 1) = 1.46$$

$$I_{\text{drz}} = \left(\frac{E_1}{E_2}\right)^{0.05} \left(\frac{d_{\text{rz}}}{d_f}\right)^{-0.2} = (10)^{0.05} \times 2^{-0.2}$$

$$= 0.98 < 1.0 \quad (I_{\text{drz}} = 1.0 \text{ is used})$$

$$I_v = \frac{1 - \nu_2}{1 - 0.4} = \frac{1 - 0.5}{1 - 0.4} = 0.83$$

The adjusted settlement influence factor considering the above factors is

$$I_s' = I_s I_D I_h I_{\text{drz}} I_v$$

$$= 0.29 \times 0.67 \times 1.46 \times 1.0 \times 0.83 = 0.24$$

The immediate settlement is

$$S_i = \frac{p_n d_f}{E_{u2}} I_s' = \frac{155 \times 1.13}{5000} \times 0.24 = 0.008 \text{ m} = 8 \text{ mm}$$

For consolidation (long-term) settlement

$$E_2 = \frac{E_{u2}}{2} = 2500 \text{ kPa}, \ \nu_2 = 0.3 \text{ (assumed)}$$

$$\frac{E_1}{E_2} = \frac{50,000}{2500} = 20 \rightarrow \tan\,\theta_s$$

$$= 0.67\left(\frac{E_1}{E_2}\right)^{0.45} = 0.67(20)^{0.45} = 2.58$$

therefore, the required width of the replaced zone for full reduction in settlement is $d_f + 2h_r \tan\,\theta_s = 1.13 + 2 \times 1 \times 2.58 = 6.29 \text{ m} > 2 \text{ m} \rightarrow$ the solution for limited width of a replaced zone is needed:

$$\frac{E_1}{E_2} = 20 \quad \text{and} \quad \frac{h_r}{d_f} = 0.89 \rightarrow I_s = 0.21$$

(center) from Figure 4.6 for an infinite width condition. Due to the limited width of the replaced zone, the following modification factors can be determined:

$$I_D = 1 - \frac{1}{1.1(d_f/D_f + 1.6)}$$

$$= 1 - \frac{1}{1.1(1/0.89 + 1.6)} = 0.67$$

$$I_h = 1 + 2\left(\frac{h_r}{d_f}\right)\left[\left(\frac{E_1}{E_2}\right)^{0.1} - 1\right]$$

$$= 1 + 2 \times (0.89) \times (20^{0.1} - 1) = 1.62$$

$$I_{\text{drz}} = \left(\frac{E_1}{E_2}\right)^{0.05} \left(\frac{d_{\text{rz}}}{d_f}\right)^{-0.2} = (20)^{0.05} \times 2^{-0.2}$$

$$= 1.01 > 1.0 \quad (I_{\text{drz}} = 1.01 \text{ is used})$$

$$I_v = \frac{1 - \nu_2}{1 - 0.4} = \frac{1 - 0.3}{1 - 0.4} = 1.17$$

The adjusted settlement influence factor considering the above factors is

$$I_s' = I_s I_D I_h I_{\text{drz}} I_v$$

$$= 0.21 \times 0.67 \times 1.62 \times 1.01 \times 1.17 = 0.27$$

The consolidation settlement is

$$S_c = \frac{p_n d_f}{E_2} I_s' = \frac{155 \times 1.13}{2500} \times 0.27 = 0.019 \text{ m} = 19 \text{ mm}$$

The total settlement without including the secondary settlement

$$S_t = S_i + S_c = 8 + 19 = 27 \text{ mm}$$

Assuming the footing is rigid, the average settlement of the footing is

$$S = \rho_r S_t = 0.79 \times 27 = 21 \text{ mm} \quad \text{(OK)}$$

4.6 CONSTRUCTION

4.6.1 Selection of Fill

Fill material for replacement should be better than in situ geomaterial in terms of strength and modulus. It should not contain organic matter and should not be problematic (such as expansive soil and frozen soil in a cold region). Commonly used fills include sand, gravel, rockfill, lime or cement-stabilized soil, and slag. Low-plasticity clay is sometimes used to replace collapsible (e.g., loess) or expansive soil. For any selected fill, its physical and mechanical properties should be determined for design and QC/QA purposes. The key parameters include optimum moisture content, maximum dry unit weight, cohesion, friction angle, and the elastic modulus or compression index. For the mitigation of collapsible soil, collapsible potential of the selected fill should be evaluated. For the mitigation of expansive soil, the expansion and shrinkage potentials of the selected fill should be evaluated. For the mitigation of frozen soil, frozen potential of the selected fill should be evaluated. In situ liquefiable soil may be excavated and reused as fill after compaction. If slag is used, its long-term durability and possible environmental impact should be evaluated.

4.6.2 Excavation

Excavation is an important part of this ground improvement method. A well-developed excavation plan should include at least but not limited to the following aspects:

- Ensure least amount of in situ soil is removed.
- Maintain stability of excavation pit.
- Cause the least disturbance to existing soil and surrounding substructures (e.g., pipes and utility lines) or superstructures (e.g., buildings).
- Have a well-thought-out plan for disposal of excavated soil and delivery of fill material.
- Remove water, especially inside the excavation pit, under a high groundwater table condition. Design of dewatering will be discussed in Chapter 6.

When open land is available, excavation with slope is preferable. Typical slope angles ranging from 2(H):1(V) to 4(H) to 1(V) are used, which depend on geomaterial conditions, groundwater conditions, and excavation depths. However, where there is limited space, vertical excavation with internal support, such as timber shoring with struts, sheet piles, and the like, is commonly used.

Base soil before the placement of a foundation should not be overexposed to water, which will reduce soil strength and increase soil compressibility. If the excavation pit will be exposed for a certain time (e.g., 6 h or longer), the last layer of soil (at least 0.3 m thick) should be removed right before the placement of the foundation.

4.6.3 Placement and Compaction

Appropriate compaction equipment should be selected based on the type of fill material and the condition of the site (e.g., space available for compaction may be limited.) Fill material should be placed and compacted in lifts. The lift thickness and number of passes depend on geomaterial type and properties and compaction equipment. Details on fill compaction can be found in Chapter 3.

4.7 QUALITY CONTROL AND ASSURANCE

4.7.1 Locations and Dimensions

The locations, elevations, and dimensions of an excavation trench and a replaced zone should be verified during excavation and before placement of fill. Any deviation should be within a tolerable limit; otherwise, it should be corrected.

4.7.2 Compacted Fill

Fill material received at a site should be verified. For granular fill, grain size distribution tests should be conducted. For cement or lime-stabilized fill, a mix design should be verified by making and curing specimens in at least 7 days and conducting unconfined compressive tests.

During compaction, moisture content and dry unit weight of compacted fill should be verified for each lift against specifications. Nuclear or nonnuclear gauge should be used to make these measurements. For granular fill, sand cone test is often useful for determining the fill density. For cement or lime-stabilized fill, specimens can be cored after at least 7 days curing. The cored specimens are sent to laboratory for determining their unconfined compressive strengths. Dynamic cone penetrometer (DCP) test may be performed to evaluate the strength and stiffness of fill material.

4.7.3 Performance Evaluation

To evaluate the performance of overexcavation and replacement, plate loading test may be performed. For a replaced zone over soft soil, the plate loading test can determine the ultimate bearing capacity and the elastic modulus of the replaced zone over soft soil. To evaluate the benefit of the replacement, a plate loading test on in situ soil without replacement should be performed as well. For a replaced zone over collapsible or expansive soil, a plate loading test should be performed first to a design load and continue after flooding with water to observe potential plate movement.

PROBLEMS

4.1. List three possible geotechnical problems that are suitable for overexcavation and replacement.

4.2. What are the benefits by replacing soft soil with higher strength and stiffness fill?

4.3. Explain why overexcavation and replacement method is suitable for shallow ground improvement.

4.4. What are possible problems if the depth of excavation is below the groundwater table?

4.5. If the replaced zone has the same width as the footing, will this replaced zone have any benefits? Why?

4.6. Under what kind of condition will the punching failure through the replaced zone likely happen?

4.7. Explain why an increase of the replaced zone modulus can reduce the vertical stress applied on top of the soft soil.

4.8. What is the main limitation of the layered elastic theory used to analyze a replaced foundation?

4.9. What are the benefits of widening the replaced zone?

4.10. What are the benefits of deepening the replaced zone?

4.11. A 3-m-long and 2-m-wide rectangular concrete footing embedded at 1.5 m deep in a uniform soft soil is subjected to a 600-kN column load. The unit weight of the soil above or below a groundwater table is 19 kN/m^3 and the groundwater table is at a depth of 2 m. What are the total and net applied pressures at the base of the footing?

4.12. In Problem 4.11, if the groundwater table rises to a depth of 1 m, what are the total and net applied pressures at the base of the footing?

4.13. In Problem 4.12, if the soft soil has an undrained shear strength of 20 kPa, what is the factor of safety against bearing failure? If the required factor of safety is 3.0, does the calculated factor of safety meet the requirement?

4.14. A circular concrete footing of 1.5 m in diameter has an embedment depth of 0.6 m in a uniform loose sand. The column load on the footing is 300 kN. The loose sand has a unit weight of 16 kN/m^3 and a friction angle of 26°. The groundwater table is 10 m below the ground surface. The required factor of safety is 3.0. Calculate the ultimate bearing capacity and factor of safety against bearing failure. If a replaced zone of 2.5 m in diameter and 1.0 m thick granular fill with unit weight of 20 kN/m^3, cohesion of 0, and friction angle of 40° is constructed under the footing, what are the ultimate bearing capacity and factor of safety against bearing failure?

4.15. In problem 4.14, if the loose sand and the granular fill have elastic moduli of 10 MPa and 50 MPa, respectively, calculate the settlements of the footing on the loose sand without and with replacement using the elastic layered theory with settlement influence factors.

4.16. Use the empirical method based on stress distribution and the Schmertmann approach to calculate the settlement in the design example. Assume a distribution angle of 30°.

REFERENCES

Broms, B.B. (1979). "Problems and solutions to construction in soft clays." *Proceedings of the 6th Regional Conference on Soil Mechanics and Foundation Engineering*, Singapore, 2, 27–36.

Burmister, D.M. (1958). "Evaluation of pavement systems of the WASHO road test by layered systems method." *Bulletin 177*, Highway Research Board, 26–54.

China Academy of Building Research (2000). *Ground Improvement Technical Code*. The China Planning Press, Beijing 156p.

Giroud, J.P. and Han, J. (2004). "Design method for geogrid-reinforced unpaved roads—Part I: theoretical development." *ASCE J. Geotech. Geoenviron. Eng.*, 130(8): 776–786.

Lawton, E.C. (2001). "Section 6A Nongrouting techniques." In *Practical Foundation Engineering Handbook*, R.W. Brown (ed.). McGraw-Hill, New York.

Meyerhof, G.G. and Hanna, A.M. (1978). "Ultimate bearing capacity of foundtaions on layered soils under inclined load." *Canad. Geotech. J.*, 15(4): 565–572.

Schmertmann, J.H. (1970). "Static cone to compute settlement over sand." *J. Soil Mech. Found. Div.*, 96(SM3): 1011–1043.

Ye, S.L., Han, J., and Ye, G.B. (1994). *Ground Improvement and Underpinning Technologies*, 2nd ed. China Building Industry Press, Beijing.

CHAPTER 5

Deep Replacement

5.1 INTRODUCTION

5.1.1 Basic Concepts

Deep replacement methods improve the ground to a great depth by partially excavating or displacing problematic geomaterials, which are replaced with better-quality and/or densified fill or concrete in a column form, as shown in Figure 5.1. The columns and the surrounding geomaterial form a composite ground or foundation to carry vertical loads and/or shear forces. Since both the fill in the columns and the material around the columns are geomaterials, the term "soil" is mostly used for the surrounding material in this chapter to avoid confusion.

Excavation of geomaterials from the ground can be accomplished in two ways: (1) by injecting water into the ground, turning the geomaterial into slurry, and flushing it out from the hole [Figure 5.1(a)] and (2) by drilling a hole in the ground [Figure 5.1(g)]. Displacement of soils in the ground can be achieved in at least four ways: (1) by injecting air into the ground [Figures 5.1(b) and 5.1(c)], (2) by driving a steel casing into the ground [Figure 5.1(e) and 5.1(f)], (3) by driving a reverse flight displacement auger into the ground [Figure 5.1(d)], and (4) by dropping a tamper to penetrate into the ground [Figure 5.1(h)]. The excavation method has little effect on the surrounding soil while the displacement method has more significant effect on the surrounding soil, depending on soil type. For cohesionless and/or unsaturated soils, the displacement process may densify the surrounding soil. However, for saturated cohesive soils, the displacement process may cause disturbance to the surrounding soil, including heaving and lateral movement. The displacement method is not suitable for sensitive soils. The vibro-compaction method, which densifies in situ soils by driving a vibrating probe into the ground, is considered as a deep compaction method and has been discussed in Chapter 3. In current practice, more vibro-compaction involves using backfill materials like a deep replacement method. However, the key difference is that vibro-compaction relies on the densified in situ soils, while deep replacement relies on using columns together with the surrounding soil to form a composite foundation. Different installation procedures are available for deep replacement, as shown in Figure 5.1. Displacement of soils and densification of columns may involve the densification of surrounding soil, which is similar to vibro-compaction. An excavation method is mostly suitable for cohesive soils, while a displacement method is suitable for both cohesive and cohesionless soils.

The technologies involving excavation include vibro-replacement by vibro-flotation and rammed aggregate columns. The columns installed by vibro-replacement (also called the wet method) are commonly referred to as stone columns or granular piles. Rammed aggregate columns are installed by removing soil with an auger and then backfilling aggregate followed by tamping. During the tamping process, there is some degree of displacement toward the surrounding soil. These technologies backfill granular fill from the top of holes; therefore, their installation procedure is often referred to as top feeding and the columns installed using this procedure are called top-fed columns.

The technologies involving displacement include vibro-casing, vibro-probe, reverse flight displacement, and dynamic replacement. The vibro-casing method, commonly referred as the sand compaction column or pile method, drives a steel casing down into the ground with a vibrator to displace the soil, backfills granular fill (mainly sand but sometimes aggregate) through the casing, and then densifies the fill by a repeated extraction and penetration process. The vibro-probe technology (also called the dry method) can be used with pressurized air to displace the soil, introduce granular fill through a central tube or a side tube, and then densify the fill by vibration of the probe. Instead of granular fill, a mixture of cement, fly ash, and gravel or low-strength concrete is sometimes used. The columns formed by the cement, fly ash, and gravel mixture are often called as CFG columns, while those formed by concrete are called vibro-concrete columns. The vibro-casing or vibro-probe method introduces fill from the bottom of holes; therefore, this installation procedure is often referred to as bottom feeding. The reverse flight displacement method displaces soils by a specially made auger and then replaces the displaced soils by grouting. The grout can be designed and prepared at different strength and modulus based on the needs for a project; therefore, these types of columns are referred to as controlled modulus (stiffness) columns or auger displacement columns. Dynamic replacement, even though the word "replacement" is used, is a special displacement method. Granular fill is pushed into the ground by repeated dropping of a tamper to displace soft soil. In this

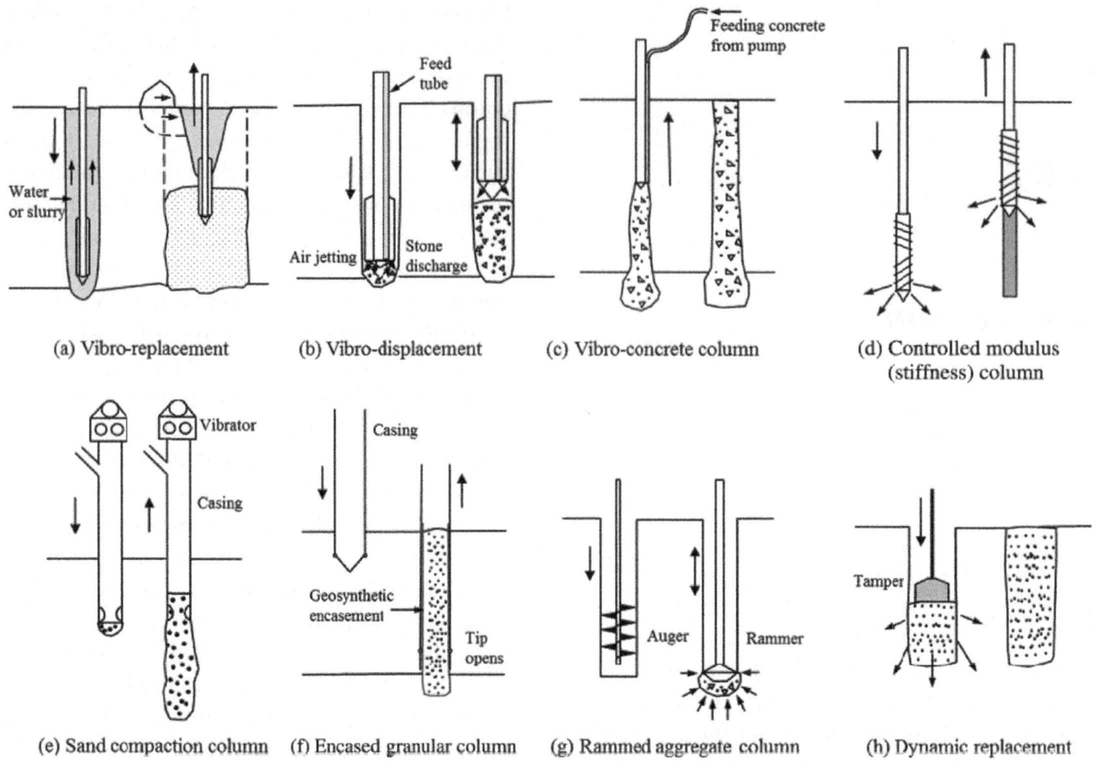

Figure 5.1 Various deep replacement methods.

construction, granular fill is backfilled into a crater generated by the deep dynamic compaction. Details about the deep dynamic compaction can be found in Chapter 3.

The deep replacement methods form columns after installation. Granular fill or concrete has been mostly used as the backfill material. Therefore, the columns installed by deep replacement methods are either granular or concrete columns. "Granular column" is a general term, which includes stone columns, rammed aggregate columns, and sand compaction columns in this book. Since columns have higher strength and stiffness, they carry more loads than the surrounding soils. In addition, granular fill has higher permeability than the surrounding soil. As a result, the columns can increase bearing capacity, reduce settlement, increase stability, and accelerate consolidation of soft foundations. When columns are formed by granular fill, their load capacities highly depend on the strength of the fill and the confining stress of the surrounding soil. In soft soil, granular columns may not have sufficient capacities to support loads. Concrete columns or CFG columns may be used instead. Alternatively, geosynthetic encasement may be used outside the columns to provide lateral confinement and increase their capacities as shown in Figure 5.1(f). This technology has drawn more attention in recent years and will be discussed in this chapter.

In recent years, new types of deep replacement columns have been developed, such as hollow concrete columns (Liu et al., 2003), multiple stepped columns (Liu, 2007a), X-shape (Liu, 2007b) or Y-shape (Chen et al., 2010) concrete columns, and grouted stone columns (Liu, 2007a). Most of these new columns are concrete columns, which have higher strength and stiffness, but different shapes have been used to reduce the amount of concrete to create more efficient and economic solutions.

Based on the stiffness, Han and Ye (1991) and Han (2012) classified the columns into three types: (1) flexible columns, (2) rigid columns, and (3) semirigid columns. Flexible columns (e.g., stone columns and sand compaction columns) have relatively lower load capacities and stiffness while rigid columns (e.g., vibro-concrete columns) have higher load capacities and stiffness. Semirigid columns have the load capacities and stiffness between flexible and rigid columns. An important characteristic of semirigid columns is that the capacities and stiffness of these columns vary with the type and properties of the replacing material in or around the columns, such as the stiffness of geosynthetic reinforcement for geosynthetic-encased granular columns and the mixture ratio for controlled modulus (stiffness) columns.

5.1.2 Suitability

Since deep replacement has a variety of installation technologies and column types, it can be used to improve almost all types of geomaterials. However, for individual technologies, they have different suitability.

The vibro-replacement technology is mostly used for cohesive soils with undrained shear strength higher than 15 kPa. The same method used for cohesionless soils is so-called vibro-compaction, which is discussed in Chapter 3. Figure 5.2 shows the soils suitable for vibro-compaction and vibro-replacement. The typical depth of vibro-replacement is 10–15 m.

The vibro-displacement technology is suitable for insensitive cohesive soils with undrained shear strength ranging from 15 to 60 kPa when it is used to install stone columns. It can also be used to install vibro-concrete columns (VCC). Vibro-concrete columns are most suitable for very soft clays and organic soils that may not be able to provide sufficient lateral confinement to stone columns. The typical depth of vibro-concrete columns is 10–15 m.

Controlled modulus (stiffness) column technology is suitable for various geomaterial conditions, such as loose sands, uncontrolled fill, soft clays and silts, and organic soils including peat. Typical depth of controlled modulus (stiffness) columns is 10–20 m.

Sand compaction column technology is suitable for cohesionless and cohesive soils. However, their functions for ground improvement are different. When sand compaction columns are installed in cohesionless soils, they function as vibro-compaction. When sand compaction columns are installed in cohesive soils, they function as deep replacement. This technology has been used to a depth of up to 70 m.

Encased granular column technology is used for very soft soils and organic soils with undrained shear strength as low as 5 kPa. The typical depth of the columns is 5–10 m.

Rammed aggregate column technology is suitable for soft to stiff clays, loose silt and sand to dense sand, and uncontrolled fill. The favorable condition for soft clays is to have an undrained shear strength higher than 15 kPa. High-groundwater table and cohesionless soils may make construction difficult. Soft clays and loose sands may require casing for hole stability. Rammed aggregate columns have been mostly used to improve geomaterials within a depth of 10 m.

Dynamic replacement is suitable to improve saturated cohesive soils and soft organic soils. Granular columns installed by dynamic replacement can reach a depth up to 8 m, but the diameter of the columns is rather large, depending on the diameter of a tamper. This technology has been mostly on land; however, it is used off shore to improve shear resistance of soft seabed (Chu et al., 2009).

5.1.3 Applications

Columns with surrounding soils to form a composite ground or foundation can increase bearing capacity, reduce and accelerate settlements, increase shear strength for slope stability, and increase resistance to liquefaction. Most of the deep replacement columns can be used to support industrial, residential, retail buildings, storage tanks, embankments and walls, bridge abutments, roadway widening, wind turbines, and utilities and pipelines.

5.1.4 Advantages and Limitations

Sand compaction columns have the following advantages (Barksdale, 1987): (1) use of lower cost material (often less expensive than stone), (2) fast construction, (3) fully supported hole by a casing during construction, and (4) limited intrusion by surrounding soil. However, sand compaction columns have the following disadvantages: (1) friction angle

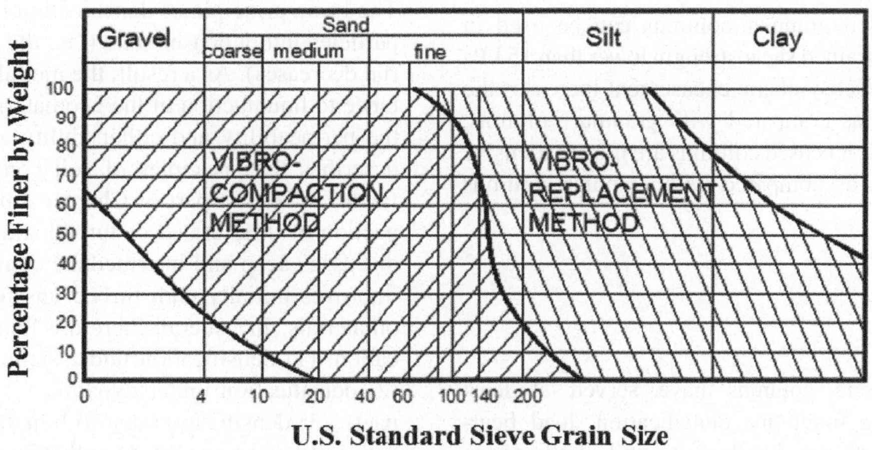

Figure 5.2 Soils suitable for vibro-compaction and vibro-replacement. (Bauman and Bauer, 1974, with permission from *Canadian Geotechnical Journal*).

of sand lower than that of stone, (2) casing penetration and extraction inducing smear to surrounding soil, and (3) insufficient permeability as vertical drains during earthquake.

Stone columns have been widely used worldwide for several decades. This technology has a long track record of successful applications. Most design methods for granular columns were developed based on stone columns. Installation of stone columns is fast and easy. They have higher strength and stiffness than sand compaction columns but lower values than concrete columns. Installation of stone columns by bottom feeding does not generate spoil; however, that by top feeding generates spoil, which is not environmentally friendly.

Rammed aggregate columns have a relatively rapid installation procedure if holes are stable, have a high level of compaction, can prestress surrounding soil, can provide uplift force with preinstalled anchor bars, and have an easy QC/QA procedure. However, rammed aggregate columns have limited improvement depth and are difficult to be installed in clean sands with a high groundwater table.

All the granular columns can provide drainage. However, they are not suitable for very soft soil with undrained shear strength lower than 15 kPa due to excessive bulging at low confining stresses near ground surface.

Vibro-concrete columns have higher strength and modulus because of the use of concrete. The installation process is quick and does not produce spoil because it is a displacement method by pressured air. They are more expensive than granular columns.

Controlled modulus (stiffness) columns can have strength and modulus according to project needs. Their strengths and moduli are higher than those of granular columns. They are installed fast with real-time monitoring and without any vibration and spoil. There is no issue of hole collapse. It has a low mobilization cost. However, they are more expensive than granular columns. Installation may be difficult in soils with rocks and boulders.

Geosynthetic-encased granular columns can be used in very soft soil with undrained shear strength lower than 15 kPa and as low as 5 kPa. Geosynthetic encasement increases the stiffness of columns as compared with granular columns. However, geosynthetic-encased columns are more expensive and slower to install as compared with granular columns without geosynthetic.

5.2 PRINCIPLES

5.2.1 Functions

Granular and concrete columns have served at least one of the following functions: densification, load bearing, reinforcement, stress distribution, and drainage in geotechnical applications. The function of densification is mostly done during the installation of columns and not significant when the soil is saturated and cohesive. This function has been discussed in Chapter 3 for vibro-compaction and a brief discussion is included in this section. Other functions are served by the columns during their service.

Due to the higher strength and stiffness (or modulus) of granular and concrete columns as compared with surrounding soils, columns often serve as load-bearing elements carrying a large portion to most of the load from superstructures and transmit it to a deep competent layer or through side friction as piles; therefore, it is also referred to as a pile effect. This effect increases the bearing capacities and reduces the settlements of the soft soils.

Similar to steel-reinforced concrete, columns can also serve as reinforcements to soft soils in the composite foundation in which soft soils are the matrix. In the composite foundation, columns and soils work together to share the applied load and provide shear resistance against sliding.

Column-reinforced composite foundations have higher equivalent strengths and stiffness, which help distribute vertical loads to a wider area and reduce the distributed vertical stresses onto the soft soils. This is the function of stress distribution. As a result, possible failure of the underlying soft soil is prevented and its deformation is reduced.

It is obvious that granular columns can serve as drainage paths to soft soils. Concrete columns can also accelerate the dissipation of excess pore water pressure in soft soils (Zheng et al., 2011). Han and Ye (2001) found that the acceleration of consolidation of stone column-reinforced composite foundations is attributed to the drainage of the columns and the reduction of the vertical stresses on soft soils. The acceleration of consolidation of the composite foundations by low permeable columns is mainly attributed to the reduction of the vertical stresses on soft soils.

5.2.2 Densification

The basic principle of densification is the rearrangement of particles into a denser state (i.e., the void ratio of geomaterial decreases). As a result, the modulus, strength, and resistance to liquefaction of the geomaterial are increased, while the permeability and collapsibility are reduced. Chapter 2 describes the principles of soil compaction and influence factors. The first part of Chapter 3 discusses shallow compaction of unsaturated disturbed soil or fill. Chapter 3 also discusses deep and intermediate compaction, which mostly improves initially undisturbed unsaturated or saturated geomaterials. In general, there are four principles of densification: (1) densification under static or kneading pressure, (2) densification under dynamic loading (vibration or impact), (3) densification due to liquefaction, and (4) densification due to consolidation. Static or kneading pressure is effective for densifying cohesive geomaterial. Vibration or impact is effective for densifying unsaturated cohesionless or collapsible geomaterial. Vibration or impact can also be used to densify saturated cohesionless geomaterial by inducing liquefaction to the material. Saturated cohesive geomaterial is difficult to be densified immediately under pressure, vibration, and/or impact; however, saturated cohesive geomaterial can be improved by consolidation that dissipates excess pore water pressure.

5.2.3 Load Transfer Mechanisms

Equal Stress versus Equal Strain In geotechnical analyses, there are two ideal loading and displacement conditions: equal strain and equal stress. The equal strain condition exists under rigid loading (e.g., rigid footing), while equal stress exists under flexible loading (e.g., tire pressure). In a column-reinforced soft foundation, columns carry a higher stress than the soil under an equal strain (also equal settlement on the columns, S_{cl} and the soil, S_{sl}) condition due to the modulus difference between columns and the surrounding soil (Figure 5.3). The ratio of the stress on the column ($\Delta\sigma_c$) to that on the soil ($\Delta\sigma_s$) is defined as the stress concentration ratio (n), which is often used to describe the load transfer between columns and soft soils. However, under an equal stress condition, the columns and the soil carry the same stress (i.e., the stress concentration ratio is equal to 1) but have different settlements (i.e., $S_{sl} > S_{cl}$). As a result, there is a differential settlement between the columns and the soil. A column-supported embankment not only has a stress concentration ratio greater than 1.0 but also has a differential settlement (e.g., Han and Gabr, 2002; Huang et al., 2009). Therefore, a column-supported embankment has a condition between equal strain and equal stress.

Unit Cells and Stress Concentration Ratio Unit cell, which consists of one column and its surrounding soil, is often used to analyze a column-reinforced soft foundation. One-dimensional (1D) unit cell, which does not allow lateral deformation of the column as shown in Figure 5.4(a), has been mostly assumed by many researchers in their theoretical developments and analyses in the past (e.g., Han and Ye, 2001, 2002; Xie et al., 2009a, 2009b). Han and Ye (2001) pointed out that an increase of a lateral stress from the column affected the variation of the excess pore water pressure in the surrounding soil. Castro and Sagaseta (2011) and Jiang et al. (2013) found that the unit cell allowing the column to have lateral deformation, δ_h as shown in Figure 5.4(b) affected the settlement and the consolidation rate of the column-reinforced soft foundation.

When a 1D unit cell is under an equal vertical strain condition, it has the following relationship:

$$\varepsilon_z = \frac{\Delta\sigma_c}{D_c} = \frac{\Delta\sigma_s}{D_s} \tag{5.1}$$

where ε_z = vertical strain at a depth of z
$\Delta\sigma_c$ = vertical stress on the column
$\Delta\sigma_s$ = vertical stress on the soil
D_c = constrained modulus of the column
D_s = constrained modulus of the soil

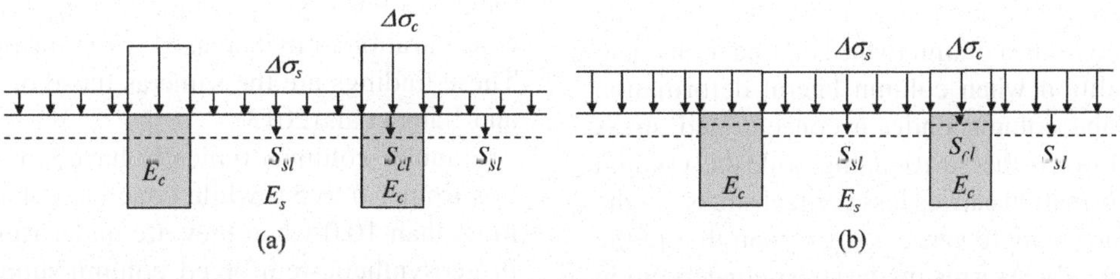

Figure 5.3 (a) Equal strain-rigid loading versus (b) equal stress-flexible loading.

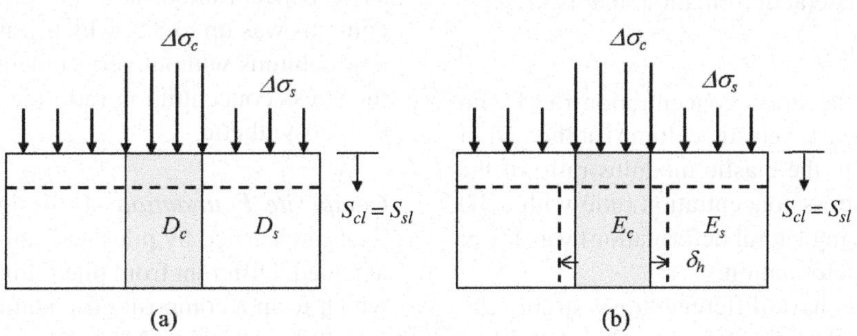

Figure 5.4 (a) One-dimensional unit cells without and (b) with lateral deformation of column.

Stress concentration ratio is defined as the ratio of the stress on the column to that on the soil, that is,

$$n_{1D} = \frac{\Delta\sigma_c}{\Delta\sigma_s} = \frac{D_c}{D_s} \tag{5.2}$$

where n_{1D} is the stress concentration ratio under a 1D condition.

Therefore, the stress concentration ratio is equal to the constrained modulus ratio of the column to the soil under an equal vertical strain condition.

When a unit cell under an equal vertical strain condition allows the column to have lateral deformation, it has the following relationship:

$$\varepsilon_z = \frac{\Delta\sigma_{cz} - \nu(\Delta\sigma_{cx} + \Delta\sigma_{cy})}{E_c} = \frac{\Delta\sigma_{sz} - \nu(\Delta\sigma_{sx} + \Delta\sigma_{sy})}{E_s} \tag{5.3}$$

$$n_{3D} \neq \frac{E_c}{E_s} \tag{5.4}$$

where

ε_z = vertical strain at a depth of z
$\Delta\sigma_{cx}, \Delta\sigma_{cy}, \Delta\sigma_{cz}$ = stresses on the column in the x, y, z directions, respectively
$\Delta\sigma_{sx}, \Delta\sigma_{sy}, \Delta\sigma_{sz}$ = stresses on the soil in the x, y, z directions, respectively
E_c = elastic modulus of the column
E_s = elastic modulus of the soil
n_{3D} = stress concentration ratio considering lateral deformation of the column (i.e., a 3D condition)

Therefore, the stress concentration ratio is not equal to the elastic modulus ratio of the column to the soil under an equal vertical strain condition when column lateral deformation is allowed. Since the column under a concentrated stress deforms laterally toward the surrounding soil, the column with the lateral deformation carries less vertical stress but the surrounding soil carries more vertical stress than that in the 1D unit cell, respectively. As a result, the stress concentration ratio considering column lateral deformation is lower than that without considering lateral deformation, that is,

$$n_{3D} < n_{1D} \tag{5.5}$$

Figure 5.5 shows that the stress concentration ratio with no lateral deformation (i.e., 1D elastic column) at the end of consolidation was equal to the elastic modulus ratio of the column to the soil. The stress concentration ratio with a 3D elastic column (i.e., allowing lateral deformation) was lower than that with no lateral deformation.

Since columns and soils have different stress–strain relationships, the stress concentration ratio, n, is not constant and depends on the properties of columns and soils and the stress or strain level as shown in Figure 5.6. Figure 5.6 shows that the stress concentration ratio first increases with the strain and then decreases with the strain after the column yields. The increase of the stress concentration ratio indicates the stress transfer from the soil to the column while the decrease of the ratio indicates the stress transfer from the column to the soil. For flexible columns, the columns and the soft soil may yield a similar strain level. For semirigid and rigid columns, however, the columns often fail first before the soil mobilizes its strength. Figure 5.5 shows that the foundation with elastic-plastic columns had a much lower stress concentration ratio than that with elastic columns because the columns yield at a lower stress. Figure 5.5 also shows that the stress concentration ratio increased with time due to consolidation. These findings are the same as those of Han and Ye (2001) and Jiang et al. (2013).

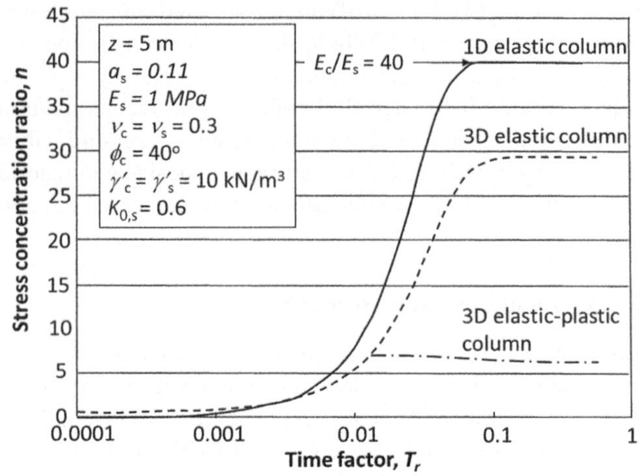

Figure 5.5 Stress concentration ratio under different conditions. (after Castro and Sagaseta, 2011, with permission from Elsevier).

Granular columns typically have stress concentration ratios from 1.0 to 5.0 while concrete columns have the ratios more than 10.0 when they are under rigid loading plates or in geosynthetic-reinforced column-supported (GRCS) embankments (Han, 2012). Castro et al. (2013) showed that the stress concentration ratio for geosynthetic-encased granular columns was up to 8.5, which was higher than that for granular columns without geosynthetic encasement. In addition, the stress concentration ratio increases with the stiffness of the geosynthetic.

Composite Foundation In design of pile foundations, all loads are carried by piles and surrounding soils do not share any load. Different from pile foundations, columns and soils, which form a composite foundation, share the applied loads and deform together. Most theories developed so far for composite foundations are based on rigid loading and equal strain conditions.

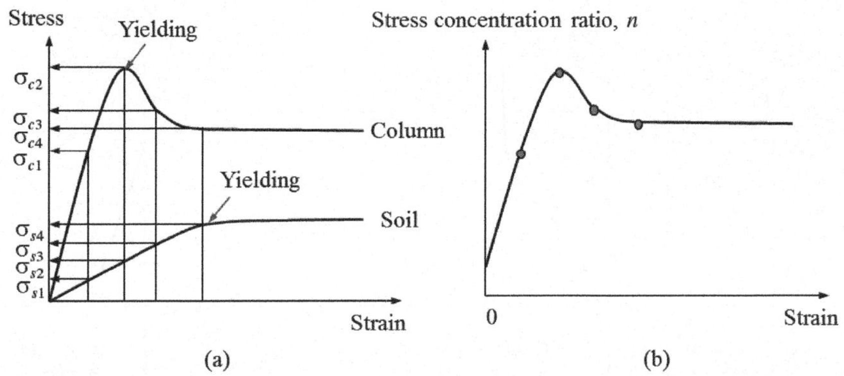

Figure 5.6 Stress concentration ratio at different strain: (a) stress–strain relationship and (b) stress concentration ratio.

Under rigid loading, the stress distribution on the columns and the soil can be simplified as shown in Figure 5.7. Based on the force equilibrium, the following relationship can be established:

$$\Delta\sigma_z\, A = \Delta\sigma_s(A_e - A_c) + \Delta\sigma_c\, A_c \quad (5.6)$$

where $\Delta\sigma_z$ = average vertical stress applied on the composite foundation
A_e = influence area (also called effective or tributary area) of one column
$\Delta\sigma_s$ = vertical stress on the soil
$\Delta\sigma_c$ = vertical stress on the column
A_c = cross sectional area of the column

Dividing Equation (5.6) by A_e on both sides yields

$$\Delta\sigma_z = \Delta\sigma_s(1 - a_s) + \Delta\sigma_c a_s \quad (5.7)$$

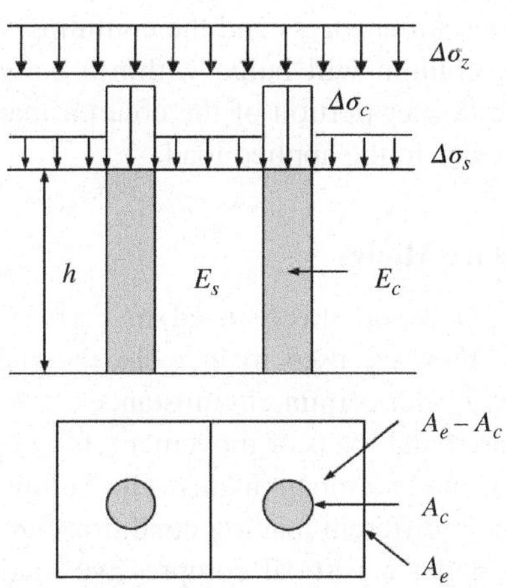

Figure 5.7 Stress distribution model.

where a_s is the area replacement ratio, defined as the ratio of the column cross-section area to the influence area.

Considering the stress concentration ratio (i.e., $n = \Delta\sigma_c/\Delta\sigma_s$), Equation (5.7) can be rewritten as

$$\Delta\sigma_z = [(1 - a_s) + na_s]\,\Delta\sigma_s = [1 + (n-1)a_s]\,\Delta\sigma_s \quad (5.8)$$

The stress on the soil is

$$\Delta\sigma_s = \mu\,\Delta\sigma_z \quad (5.9)$$

$$\mu = \frac{1}{1 + (n-1)a_s} \quad (5.10)$$

where μ is the stress reduction factor.

Equation (5.10) shows that the stress reduction factor is less than 1, and an increase of the stress concentration ratio and/or the area replacement ratio reduces this factor. In other words, less stress is applied on the soil.

Assume the deformations of the column and the soil are one dimensional and equal, that is,

$$\varepsilon_c = \varepsilon_s = \varepsilon_z \quad (5.11)$$

where ε_c = vertical strain of the column
ε_s = vertical strain of the soil
ε_z = average vertical strain

Dividing Equation (5.7) by Equation (5.11) results in

$$E_{eq} = E_s(1 - a_s) + E_c a_s \quad (5.12)$$

where E_{eq} = equivalent modulus of the composite foundation
E_s = soil modulus
E_c = column modulus

The above equation can also be expressed as

$$E_{eq} = [1 + (n-1)a_s]E_s \quad (5.13)$$

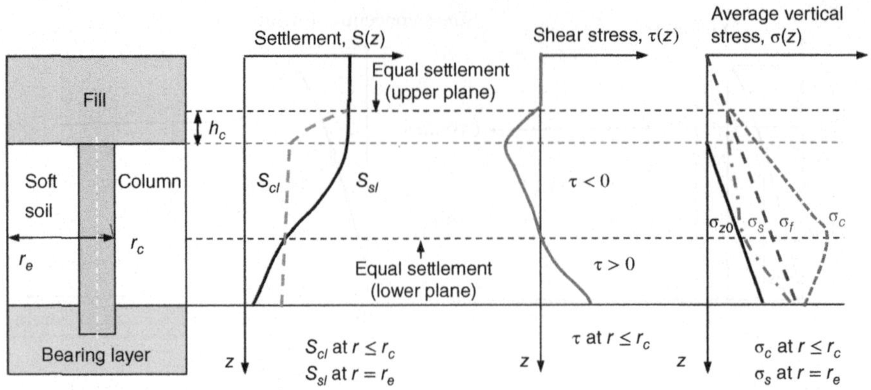

Figure 5.8 Stress transfer in column-supported embankment (modified from Simon and Schlosser, 2006).

Stress Transfer Stress transfer between columns and soil in the composite foundation depends on the rigidity of loading, the rigidity of columns relative to soil, and the end-bearing condition. Simon and Schlosser (2006) clearly illustrated the load transfer and deformations in column-supported embankments in Figure 5.8, which is considered an intermediate condition between flexible loading and rigid loading. In this illustration, the column is relatively rigid but deformable and the bearing layer is not firm. There exist two equal settlement planes in the system, one in the embankment fill and one in the soft soil, if the embankment height is higher than the critical height (h_c). Based on the field measurements, Chen et al. (2010) reported the critical heights in the embankments were in the range of 1.0–1.5 times the clear spacing of the columns. Based on the model and field tests, Filz et al. (2012) developed a relationship of the critical height with the diagonal clear spacing and the diameter of the columns. Due to the relative difference between the column settlement (S_{cl}) and the soft soil settlement (S_{sl}), the negative shear stress (τ) (often referred as negative skin friction) develops along the column between the upper and lower equal settlement planes, but the positive shear stress develops below the lower equal settlement plane. The negative shear stress increases the average vertical stress (σ_c) in the column and reduces the average vertical stress (σ_s) in the soft soil, which is higher than the initial overburden stress (σ_{z0}) but lower than the average vertical stress (σ_f) with fill surcharge. Figure 5.8 also shows that the stress concentration ratios at different depths are different. The vertical stresses above the column and the soil at the upper equal settlement plane are equal, while the highest stress concentration on the column exists at the lower equal settlement plane. If the bearing layer is firm, the lower equal settlement plane will be at the top of the bearing layer. Under such a condition, all the shear stress will be negative. The above load transfer mechanism was included in the analytical model proposed by Chen et al. (2008). If the column is flexible, low stress concentration occurs on the column, the column may bulge, and less shear stress will develop between the column and the soil to a limited depth.

Under flexible loading, the vertical stresses on the top of the column and the soil are equal. However, the soil deforms more than that of the column because of their difference in the moduli. This behavior is similar to that under fill loading. Therefore, the stress transfer between rigid columns and soil is similar to that below the fill in Figure 5.8.

Under rigid loading, the column and the soil deform equally in the vertical direction on the top; therefore, there is no shear stress between the column and the soil. If the bearing layer is firm, the differential settlement between the column and the soil will be minimal; therefore, the vertical stresses in the column and the soil remain constant with depth. However, if the bearing layer is soft, a rigid column may penetrate into the soft bearing layer. As a result, positive shear stress will develop between the column and the soil. If the column is flexible, the stress transfer is dominated by lateral confinement. Near the ground surface, since the overburden stress and the confining stress are low, the flexible column will bulge within a certain depth. As a result, the deeper portion of the column may not provide much resistance to the applied load.

5.2.4 Failure Modes

Columns have been mostly used to carry vertical loads. Sometimes, they are used to increase the shear resistance of soft soils. Under certain circumstances, columns are subjected to horizontal loads or movement, for example, during excavation or under embankments. The failure modes of the columns under different loading conditions are different.

Columns under a vertical compressive load transmit the load through the surrounding soil by side friction or lateral confinement. The possible failure modes are illustrated in

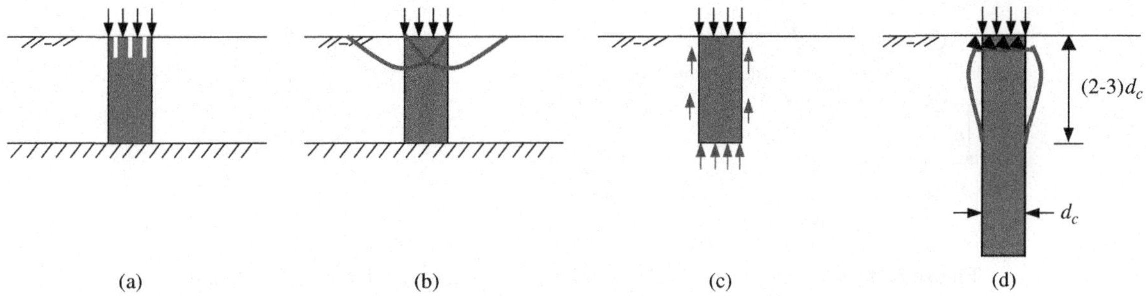

Figure 5.9 Possible failure modes of individual columns subjected to vertical loads: (a) crushing, (b) shear, (c) punching, and (d) bulging (modified from Barksdale and Bachus, 1983; Han and Ye, 1991).

Figure 5.9. Columns may crush when the applied load is higher than the strengths of the concrete columns. This failure more likely happens to relatively brittle concrete columns with an end-bearing condition. The shear failure may happen to granular columns. The punching failure may happen to short granular or concrete columns without an end-bearing layer. Bulging failure more likely happens to granular columns in soft soils within the top portion of two to three times the diameter of the column. Rammed aggregate piers have been used to provide uplift resistance by preinstalling a steel plate at the bottom of the pier (Farrell et al., 2008).

Several researchers (e.g., Hughes and Withers, 1974; Mckelvey et al., 2004; Najjar et al., 2010) investigated the column critical length, below which the column length does not provide any additional bearing capacity. From these studies, it can be concluded that the critical lengths of flexible columns (i.e., granular columns) mostly range from four to six times the column diameter.

Columns under a horizontal load or movement provide shear resistance or bending moment and may fail under shear, bending, or rotation. Shear failure may happen to granular columns. Bending failure may happen to concrete columns especially when no steel reinforcement is included. Rotation may happen to short concrete columns in soft soils. Rigid inclusions can be installed inside columns to increase the shear and bending resistance of columns (Zheng et al., 2009). For geosynthetic-encased columns, another possible failure mode is bursting of geosynthetic reinforcement.

5.3 DESIGN CONSIDERATIONS

In general, granular column-reinforced foundations are designed as composite foundations while concrete columns are designed as piles. In the composite foundation design, a unit cell concept is often used for simplification. This section addresses the following design issues: (1) general rules (such as backfill material, area of improvement, pattern, area replacement ratio, and depth of improvement, and diameter of column), (2) bearing capacity, (3) settlement, (4) consolidation, (5) stability, and (6) liquefaction.

5.3.1 General Rules

Backfill A rating system has been developed by Brown (1977) to judge the suitability of backfill material for vibro-replacement based on the settling rate of the backfill in water and project experience using the suitability number Table 5.1:

$$S_N = 1.7\sqrt{\frac{3}{(D_{50})^2} + \frac{1}{(D_{20})^2} + \frac{1}{(D_{10})^2}} \tag{5.6c}$$

Where D_{50}, D_{20}, and D_{10} are particle sizes of 50%, 20%, and 10% finer, respectively, in a unit of mm.

Patterns Three patterns of columns as shown in Figure 5.10 have been commonly used in practice. When the center-to-center spacing (s_1) is equal to s_2 in Figure 5.10(a), it becomes a square pattern. When s_1 is equal to s_2 in Figure 5.10(b), it becomes an equilateral triangular pattern. Rectangular and triangular patterns are commonly used for most foundations while the radial pattern is most suitable for circular or ring foundations (e.g., tank foundations).

Diameter of Column The diameter of columns depends on the equipment used to install the columns. Table 5.2 provides the typical column diameters used in practice.

Area Replacement Ratio When columns are installed, the area replacement ratio is defined as the ratio of the cross-sectional area of a column to the tributary area of the column, as shown in Figure 5.10, that is,

$$a_s = \frac{A_c}{A_e} = C\left(\frac{d_c}{s}\right)^2 \tag{5.7}$$

Table 5.1 Suitability of Backfill

Suitability Number	0–10	10–20	20–30	30–40	>50
Rating	Excellent	Good	Fair	Poor	Unsuitable

Source: Brown (1977).

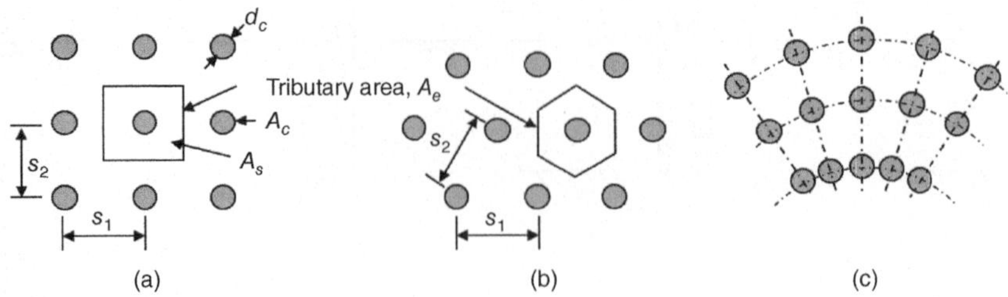

Figure 5.10 Typical patterns of compaction probe points or columns: (a) rectangular, (b) triangular, and (c) radial.

Table 5.2 Typical Diameters of Columns

Column Type	Equipment for Installation	Diameter (m)
Sand compaction columns	Casing	0.6–0.8
Stone columns	Vibro-probe	0.5–1.2
Rammed aggregate columns	Auger and rammer	0.7–0.9
Vibro-concrete columns	Vibro-probe	0.5–0.6 (shaft) and 0.6–0.9 m (head and base)
Controlled modulus (stiffness) columns	Reverse flight auger	0.3–0.5
Geosynthetic-encased columns	Closed casing or open casing with auger	0.7–0.9

where a_s = area replacement ratio
A_c = cross sectional area of the column
A_e = tributary area of the column
d_c = diameter of the column
s = center-to-center spacing between columns in a square or equilateral triangular pattern
C = constant ($\pi/4$ or 0.785 for a square pattern or $\pi/(2\sqrt{3})$ or 0.907 for an equilateral triangular pattern)

Area replacement ratios for granular columns without geosynthetic encasement typically range from 0.1 to 0.4. Larger ratios are used for very soft or loose soil. Geosynthetic-encased granular columns are typically designed with area replacement ratios from 0.1 to 0.2 (Alexiew and Thomson, 2013). Concrete columns are typically installed with area replacement ratios of 0.05–0.15, based on the cross sectional area of column shafts.

Depth of Improvement Depth of improvement should be determined based on site conditions, soil properties, and performance requirements. The following rules may be followed in the determination:

- When a firm stratum exists at a relatively shallow depth, the depth of improvement should reach this stratum.
- When a firm stratum exists at a relatively deep depth, the depth of improvement should be determined to meet performance requirements, such as bearing capacity, settlement, slope stability, and liquefaction.
- Typical depth of improvement ranges from 5 to 15 m.

Area of Improvement The area of improvement should be determined based on site conditions and importance of superstructures. In general, the area of improvement should be larger than footprints of footings. Under a general condition, one to two rows of columns may be installed outside of a footing. On a liquefiable soil site, two to four rows of columns may be installed outside of a footing.

5.3.2 Densification Effect

The densification effect by vibro-compaction is discussed in Chapter 3. The method for determining the required spacing of granular columns to reduce the void ratio of a soil from its initial value to the target value is presented in Section 3.7 of Chapter 3. Readers are encouraged to review that section to understand the principles and design of granular columns for this application. Sand compaction columns can also be used to densify cohesionless soil by driving a vibratory casing to displace the surrounding soil. The method for volume change by vibro-compaction with backfill can be used to analyze the densification effect on the surrounding soil. In addition, Figure 5.11 can be used to estimate the SPT N values midway between sand compaction columns (N_1) and in the center of

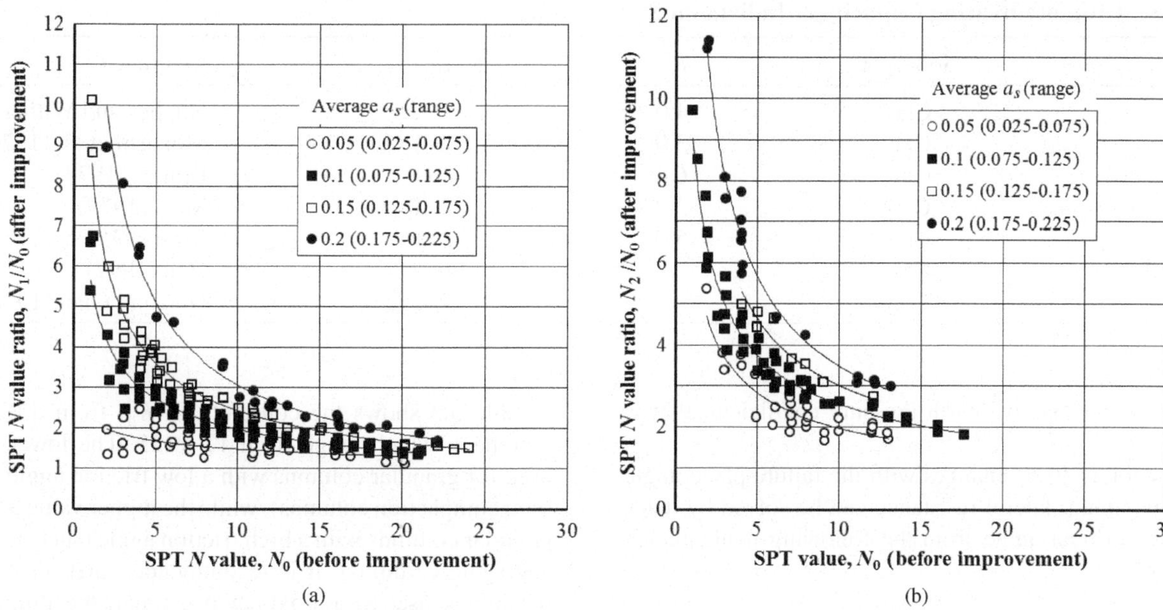

Figure 5.11 SPT N values after installing sand compaction columns: (a) midway between columns and (b) center of columns (modified from Kensetsu Kikai Chosa Co.).

sand compaction columns (N_2). The SPT N values depend on the initial SPT N values (N_0), the area replacement ratio (a_s), and the location (between or in the center of columns). It is shown that an increase of the initial SPT N value (N_0) or the area replacement ratio (a_s) increases the SPT N value after installation (N_1 or N_2). In addition, the SPT N value in the center of the columns (N_2) is higher than that midway between columns (N_1).

Kensetsu Kikai Chosa Co. suggested the average weighted SPT N value including the sand compaction column and the surrounding soil as follows:

$$N_{eq} = a_s N_2 + (1 - a_s) N_1 \tag{5.8}$$

where N_{eq} = average weighted (equivalent) SPT N value
N_1 = SPT N value in the surrounding soil
N_2 = SPT N value in the sand compaction column

5.3.3 Bearing Capacity

Granular Columns Brauns (1978) proposed a simplified method to estimate the ultimate bearing capacity of an individual stone column in saturated soft soil under an undrained condition based on an axisymmetric model as shown in Figure 5.12. This method considers a passive shear failure from the column to the surrounding soil. It is assumed: (1) the interface between the column and the soil is smooth, (2) no circumferential (hoop) stress exists, and (3) no self-weight is considered for the column and the soil. Based

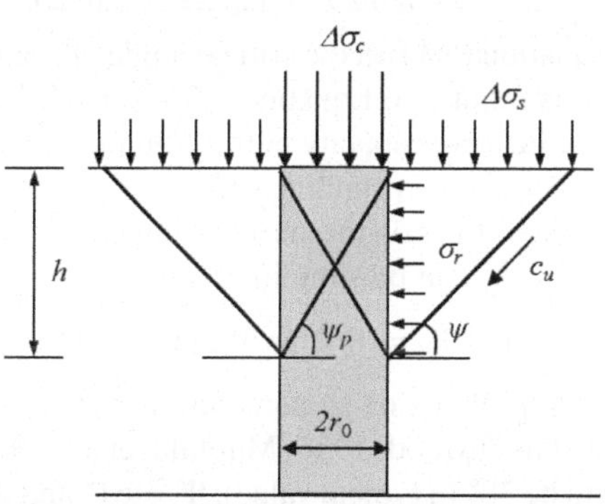

Figure 5.12 Individual column failure mode.

on the force equilibrium, it results in the following formula:

$$\sigma_r = \left(\Delta\sigma_s + \frac{2c_u}{\sin 2\Psi}\right)\left(1 + \frac{\tan \Psi_p}{\tan \Psi}\right) \tag{5.9}$$

where σ_r = lateral stress from the surrounding soil
$\Delta\sigma_s$ = vertical stress on the soil
c_u = undrained shear strength of the soil
Ψ_p = passive failure plane angle within the column (i.e., $\Psi_p = 45° + \phi_c/2$)
Ψ = failure plane angle within the soil
ϕ_c = friction angle of the column

Considering the passive shear failure of the column, it has

$$\Delta\sigma_c = \sigma_r K_p \tag{5.10}$$

Table 5.3 Ultimate Bearing Capacity of Individual Stone Column

c_u (kPa)	Soil Type	K'	$K' K_p$	Reference
19.4	Clay	4.0	25.2	Hughes and Withers (1974)
19.0	Clay	3.0	15.8–18.8	Mokashi et al. (1976)
—	Clay	6.4	20.8	Brauns (1978)
20.0	Clay	5.0	20.0	Mori (1979)
—	Clay	5.0	25.0	Broms (1979)
15.0 to 40.0	Clay	—	14.0–24.0	Han (1992)
—	Clay	—	12.2–15.2	Guo and Qian (1990)

Source: Ye et al. (1994).

where K_p is the passive earth pressure coefficient, that is, $\tan^2\Psi_p$.

In Equation (5.9) σ_r changes with the failure plane angle, Ψ. The maximum lateral resistance can be solved by determining the critical angle from the following mathematical operation:

$$\frac{\partial \sigma_r}{\partial \Psi} = 0 \tag{5.11a}$$

$$\frac{\Delta\sigma_s}{2c_u}\tan\Psi_p = -\frac{\tan\Psi}{\tan 2\Psi} - \frac{\tan\Psi_p}{\tan 2\Psi} - \frac{\tan\Psi_p}{\sin 2\Psi} \tag{5.11b}$$

From Equation (5.11b), the critical angle, Ψ, can be solved, and then it is input into Equations (5.9) and (5.10) to obtain the ultimate bearing capacity of the column with the vertical stress ($\Delta\sigma_s$) on the surrounding soil.

If the vertical stress on the surrounding soil is equal to zero, Equation (5.11b) can be simplified into

$$\tan\Psi_p = 1/2\tan\Psi(\tan^2\Psi - 1) \tag{5.12}$$

The friction angles of stone columns mostly range from 35° to 45° (Greenwood, 1970; Munfakh et al., 1983). If ϕ_c is assumed to be 38° (a typical value), $\Psi = 61°$ and the ultimate bearing capacity of the individual stone column is

$$q_{ult,c} = 20.75c_u \tag{5.13}$$

Other solutions were also obtained by different researchers. From most of the solutions, below is a common equation for the lateral soil resistance:

$$\sigma_r = \sigma_{r0} + K_l c_u \tag{5.14}$$

where σ_{r0} = lateral soil stress induced by the overburden stress
K_l = constant obtained by different researchers

The ultimate bearing capacity of the individual stone column can be expressed as follows:

$$q_{ult,c} = (\sigma_{r0} + K_l c_u)K_p = K' K_p c_u \tag{5.15}$$

where K' is a constant considering the effect of σ_{r0}, which is not significant at a shallow depth.

Table 5.3 shows the constants ($K'K_p$) from different researchers, which range from 12 to 25. The lower value is used for granular columns with a low friction angle (such as sand compaction columns), while the higher value is used for granular columns with a high friction angle (such as rammed aggregate columns). It is recommended that the following formula be used to approximately estimate the ultimate bearing capacity of an individual stone column:

$$q_{ult,c} = 20c_u \tag{5.16}$$

Since granular columns and the surrounding soil mobilize their strengths at a similar strain level, the ultimate bearing capacity (q_{ult}) of a granular column-reinforced composite foundation can be estimated as follows:

$$q_{ult} = q_{ult,c}a_s + q_{ult,s}(1 - a_s) \tag{5.17}$$

where $q_{ult,s}$ is the ultimate bearing capacity of the surrounding soil, which can be estimated as $5c_u$ as suggested by Barksdale (1987).

Concrete Columns The ultimate load capacity ($Q_{ult,c}$) of a single concrete column can be estimated using the same method for a single rigid pile, that is,

$$Q_{ult,c} = f_s A_s + q_t A_t \tag{5.18}$$

where f_s = side friction
q_t = toe resistance
A_s = surface area of the column
A_t = toe cross-sectional area of the column

For some concrete columns, for example, VCC, the toe area is larger than the column shaft area.

The side friction under an undrained condition can be estimated by the α method, that is,

$$f_s = \alpha c_u \tag{5.19a}$$

where α is the interface reduction factor, mostly ranging from 0.5 to 1.0. A smaller value may be used for a replacement method but a larger value may be used for a displacement method.

The side friction under a drained condition can be estimated by the β method, that is,

$$f_s = \beta_i \sigma'_{z0} \tag{5.19b}$$

where β_i = $K \tan \delta_i$
σ'_{z0} = effective overburden stress
K = coefficient of lateral earth pressure, typically ranging from 0.5 to 1.0 K_0 (a smaller constant may be used for a replacement method but a larger value may be used for a displacement method)
K_0 = coefficient of lateral earth pressure at rest
δ_i = interface friction angle, mostly equal to 0.8 to 1.0 times the soil effective friction angle, ϕ'

The toe resistance under an undrained condition is

$$q_t = N_c^* c_u \approx 9c_u \tag{5.20a}$$

The toe resistance under a drained condition is

$$q_t = 0.5 d_c N_\gamma^* + \sigma'_D N_q^* \tag{5.20b}$$

where d_c = diameter of column
σ'_D = effective overburden stress at the depth of the column toe

N_γ^*, N_q^* are provided in Table 5.4

The ultimate load capacity of an individual column should be limited by the unconfined compressive strength of the concrete, that is,

$$Q_{\text{ult},c} \le q_u A_c \tag{5.21}$$

where q_u = unconfined compressive strength of the concrete (typically 20 MPa at 28 days)
A_c = cross-sectional area of the column head

The ultimate load capacity (q_{ult}) of a concrete column-reinforced composite foundation can be estimated using the method proposed by Poulos (2001) for a piled raft foundation as shown in Figure 5.13, which is the lesser of the following two load capacities: (1) the sum of the load capacities of the raft and all the concrete columns and (2) the sum of the load capacities of the equivalent pier and the raft outside the periphery of the pier (the capacity of the equivalent pier can be estimated as an individual column using the methods discussed above):

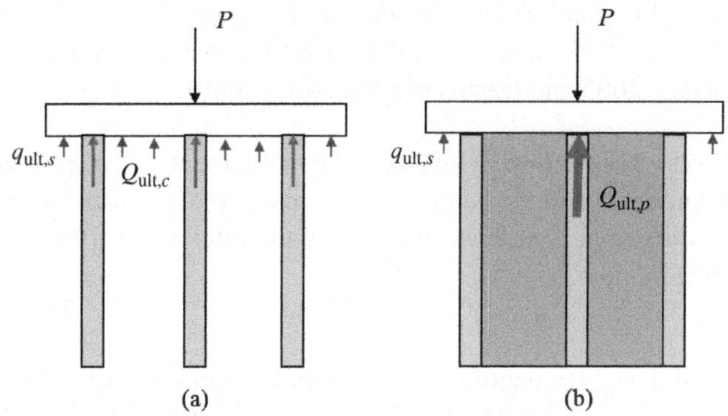

Figure 5.13 Concrete column-reinforced composite foundation.

$$Q_{\text{ult}} = \sum Q_{\text{ult},ci} + q_{\text{ult},s} A_{\text{sr}} \tag{5.22a}$$

$$Q_{\text{ult}} = Q_{\text{ult},p} + q_{\text{ult},s} A_{\text{so}} \tag{5.22b}$$

where $Q_{\text{ult},ci}$ = individual column ultimate load capacity
$q_{\text{ult},s}$ = ultimate bearing capacity of soil
A_{sr} = area of soil under the raft
$Q_{\text{ult},p}$ = ultimate load capacity of the equivalent pier (the same above procedure used to calculate the ultimate load capacity of the pier by treating the pier as an individual column)
A_{so} = area of soil outside the periphery of the pier

In the preceding calculations, the α method should be used to calculate the side friction under an undrained condition ($\alpha = 1$ for an equivalent pier) while the β method should be used under a drained condition. Undrained shear strength should be used to calculate the toe resistance under an undrained condition while effective friction angle should be used under a drained condition.

5.3.4 Settlement

Granular Columns Han (2010) summarized the methods for calculating the settlement of granular column-reinforced foundations, including (1) the stress reduction method

Table 5.4 N_γ^* **and** N_q^*

ϕ'(°)	20	22	24	26	28	30	32	34	36	38	40	42	44	46
N_γ^*	3.3	4.3	5.8	7.4	10	14	19	24	32	48	66	90	120	180
N_q^*	13	16	21	26	33	42	54	69	90	130	170	220	290	400

Sources: Vesic (1975) and Vesic (1977).

(Aboshi et al., 1979), (2) the improvement factor method (Priebe, 1995), and (3) the elastic-plastic method (Pulko and Majes, 2005 and Castro and Sagaseta, 2009).

Stress Reduction Method The settlement of a natural foundation under a large loading area (i.e. the width of the loading area is at least three times the thickness of the soft soil) is

$$S = m_{v,s} \Delta\sigma_z h \tag{5.23}$$

where $m_{v,s}$ = coefficient of volume compressibility natural soil
$\Delta\sigma_z$ = additional vertical stress
h = thickness of soil layer

When the loading area is relatively small (i.e., the width of the loading area is less than three times the thickness of the soft soil), Equation 2.53 or 2.54 should be used to estimate the settlement instead.

The coefficient of volume compressibility of soil can be determined from consolidation tests as discussed in Chapter 2 or by the following relationship:

$$m_{v,s} = \frac{1}{D_s} = \frac{(1+v_s)(1-2v_s)}{E_s(1-v_s)} \tag{5.24}$$

where D_s = constrained modulus of soil
E_s = elastic modulus of soil
v_s = Poisson's ratio of soil

The constrained modulus of soil can also be estimated for a normally consolidated soil by

$$D_s = \frac{2.30(1+e_0)\sigma'_{z0}}{C_c} \tag{5.25}$$

where e_0 = initial void ratio of the soil
σ'_{z0} = effective overburden stress
C_c = compression index of the soil

Near the ground surface, the soil is likely overconsolidated. Under such a condition, the recompression index, C_r, should be used instead of C_c if the total stress is less than the preconsolidation stress.

The settlement of a composite foundation based on the compression of the soil is

$$S' = m'_{v,s} \Delta\sigma_s h = m'_{v,s} \mu \Delta\sigma_z h \tag{5.26}$$

where $m'_{v,s}$ = coefficient of volume compressibility of soil after column installation
μ = stress reduction factor, as defined in Equation (5.10)
$\Delta\sigma_z$ = additional vertical stress
h = thickness of soil layer

The settlement ratio of the composite foundation to the natural foundation is

$$\frac{S'}{S} = \frac{m'_{v,s}}{m_{v,s}} \mu \tag{5.27}$$

For soft soils, their coefficients of volume compressibility before and after column installation do not change much. Assume $m'_{v,s} = m_{v,s}$ and then Equation (5.27) can be simplified into

$$\frac{S'}{S} = \mu \tag{5.28}$$

or

$$S' = \frac{1}{1 + a_s(n-1)} S \tag{5.29}$$

Barksdale and Bachus (1983) developed an empirical design chart as shown in Figure 5.14 to determine the stress concentration ratio, which can be approximated as follows for the average ratio (Han, 2010):

$$n = 1 + 0.217\left(\frac{E_c}{E_s} - 1\right) \tag{5.30}$$

where E_c = elastic modulus of the column
E_s = elastic modulus of the soil

Based on field data, the modulus ratio (E_c/E_s) should be limited to 20. The typical stress concentration ratio for granular columns under rigid loading ranges from 2 to 4; however, that under embankment loading ranges from 1 to 2.

Improvement Factor Method Priebe (1995) proposed a basic improvement factor method to calculate the settlement of stone column-reinforced soft foundations by a

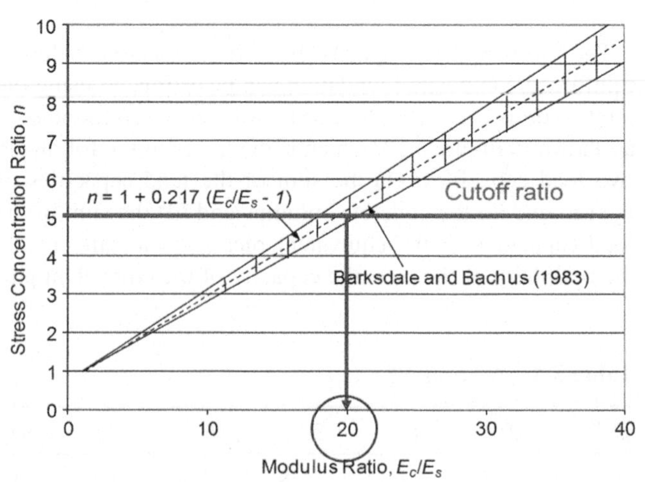

Figure 5.14 Stress concentration ratio versus modulus ratio. (modified from Barksdale and Bachus, 1983).

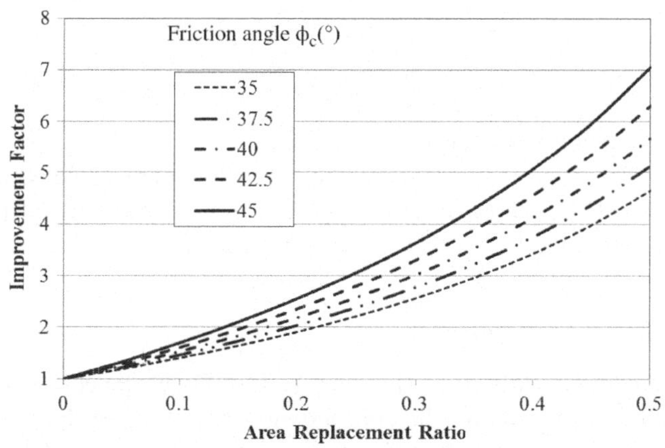

Figure 5.15 Improvement factor versus area replacement ratio. (drawn from Priebe, 1995).

vibro-replacement method considering rigid and incompressible columns with a bulging over the column length as follows:

$$S' = \frac{1}{I_f}, \quad I_f = 1 + a_s \left[\frac{5 - a_s}{4\left(1 - a_s\right)\tan^2(45° - \phi_c/2)} - 1 \right] \tag{5.31}$$

where I_f = improvement factor, which is also shown in Figure 5.15
ϕ_c = friction angle of the column material

Priebe (1995) also suggested the consideration of the column compressibility and overburden stress in addition to the basic improvement factor. The formulas and design charts for such consideration, referred to as a modified improvement factor method, can be found in Priebe (1995). However, the basic improvement factor method is more commonly used in practice.

Elastic-Plastic Method Pulko and Majes (2005) and Castro and Sagaseta (2009) proposed the methods to calculate the settlement of the granular column-reinforced soft foundation based on elastic-plastic constitutive models. In their methods, the soft soil is assumed to be linearly elastic while the granular columns are assumed to be linearly elastic–perfectly plastic following the Mohr–Coulomb failure criterion with a constant dilatancy angle. The plasticity starts with the upper portion of the column and can extend deeper to the whole length of the column with an increase of the applied load. This method is theoretically sound but more complicated for use in practice.

Among all the preceding methods, Priebe (1995) is still a favorable method to estimate the settlement of stone column-reinforced soft foundations in practice (McCabe and Egan, 2010).

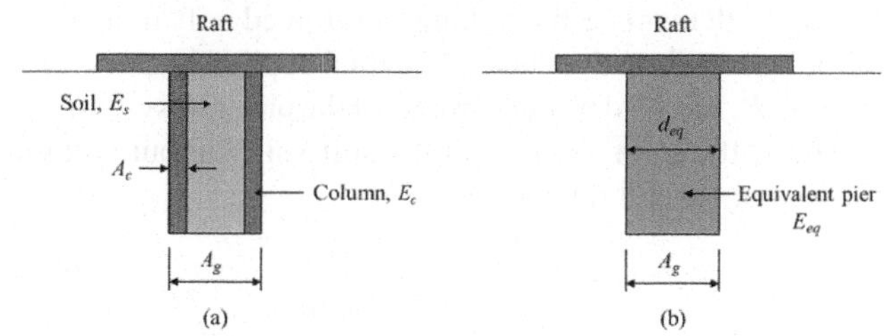

Figure 5.16 Equivalent pier in the piled raft method: (a) column group and (b) equivalent pier.

Concrete Columns The settlement of soft foundations improved by end-bearing concrete columns should not be an issue in most cases. However, if concrete columns are not end bearing, the settlement may become an issue. The settlement of concrete column-reinforced soft foundations can be estimated using the method for piled rafts or pile groups. Horikoshi and Randolph (1999) and Poulos (2001) proposed simplified design methods to calculate the settlement of piled rafts, which are based on pile–raft interaction. Horikoshi and Randolph (1999) used the equivalent pier concept for the piled raft method as shown in Figure 5.16. To ensure the suitability of this method, a R_f factor should be calculated first, which is $R_f = (N_{cl}s/L_c)^{0.5}$, where N_{cl} is the number of columns, s is the center-to-center spacing of columns, and L_c is the column length). If the R_f factor is less than 2, the equivalent pier approach is suitable. In this method, the equivalent modulus can be calculated by

$$E_{eq} = E_s + (E_c - E_s)\frac{A_{tc}}{A_g} \tag{5.32}$$

where E_{eq} = equivalent modulus of the pier
A_{tc} = total cross-sectional area of all individual columns
A_g = total area of the column group, including all individual columns and the soil between columns

The stiffness of the pier–raft system in Figure 5.16(b) can be calculated by

$$K_{pr} = \frac{P_{eq} + P_r}{S_{pr}} = \frac{K_{eq} + K_r(1 - 2\alpha_{pr})}{1 - (K_r/K_{eq})\alpha_{pr}^2} \tag{5.33}$$

where K_{pr} = stiffness of the pier–raft system
P_{eq} = load carried by the equivalent pier
P_r = load carried by the raft
S_{pr} = settlement of the pier–raft system
K_{eq} = stiffness of the equivalent pier
K_r = stiffness of the raft on the soil
α_{pr} = pier–raft interaction factor

The settlement of the column-reinforced soft foundation, S' or S_{pr}, can be calculated as the total applied load (i.e., $P = P_{eq} + P_r$) divided by the stiffness of the pier-raft system, K_{pr}.

The stiffness of the raft on the soft soil without columns (assuming a rigid raft) is

$$K_r = \frac{P_r}{S} = \frac{E_s A_r}{(1 - v_s^2) B_r I_s} \tag{5.34}$$

where P_r = load carried by the raft
S = settlement of the raft
A_r = area of the raft
B_r = width or diameter of the raft
I_s = influence factor (0.88 for a square raft or 0.79 for a circular raft)
v_s = Poisson's ratio of the soil

The pier–raft interaction factor is

$$\alpha_{pr} = 1 - \frac{\ln(r_r/r_{eq})}{\ln(r_m/r_{eq})} \tag{5.35}$$

where r_r = radius of the equivalent raft in a circular shape
r_{eq} = radius of the equivalent pier
r_m = maximum influence radius of an individual pile, which can be determined as follows:

$$r_m = Cr_c + \{0.25 + \xi[2.5\rho(1 - v_s) - 0.25]\}L_c \tag{5.36}$$

where ξ = G_L/G_b
ρ = G_{avg}/G_L
G_L = shear modulus of the soil at the base of columns
G_b = shear modulus of the soil below the base of columns
G_{avg} = average shear modulus of the soil within the length of columns
L_c = length of the columns
r_c = radius of the column
C = 0 if $L_c/r_c > 5$ or 5 if $L_c/r_c \leq 5$

The stiffness of the equivalent pier is

$$K_{eq} = G_L r_{eq} \left[\frac{\dfrac{4\eta}{(1 - v_s)\,\xi} + \rho \cdot \dfrac{2\pi}{\zeta} \cdot \dfrac{\tanh(\mu_L L_{eq})}{\mu_L L_{eq}} \cdot \dfrac{L_{eq}}{r_{eq}}}{1 + \dfrac{1}{\pi\lambda} \cdot \dfrac{4\eta}{(1 - v_s)\xi} \cdot \dfrac{\tanh(\mu_L L_{eq})}{\mu_L L_{eq}} \cdot \dfrac{L_{eq}}{r_{eq}}} \right] \tag{5.37}$$

where η = $r_{eq,b}/r_{eq}$
$r_{eq,b}$ = radius of the equivalent pier at the base
L_{eq} = length of the equivalent pier (= L_c)
ζ = $\ln(r_{m,p}/r_{eq})$, $r_{m,p}$ is the maximum influence radius of the pier calculated by Eq. (5.36) using r_{eq} instead of r_c.

$$\mu_L L_{eq} = \sqrt{2/(\xi\lambda)}(L_{eq}/r_{eq})$$
$$\lambda = E_{eq}/G_L$$

For a uniform equivalent pier, $\eta = 1$.

The stress distribution method for pile groups included in AASHTO (2006) may also be used for this purpose. Readers are referred to AASHTO (2006) for details.

5.3.5 Consolidation

Granular Columns Han and Ye (2001) developed a simplified solution for the consolidation rate of a free-draining stone column-reinforced soft foundation unit cell as shown in Figure 5.17. The excess pore water pressure in the soft soil dissipates due to vertical and radial drainages. When the water reaches the free-draining stone column, it is assumed to dissipate immediately. It is also assumed that the stone column and the soil deform one dimensionally at the same strain, that is,

$$\frac{\partial e_s}{1 + e_s} = \frac{\partial e_c}{1 + e_c} \tag{5.38}$$

where e_s is the void ratio of the soil and e_c is the void ratio of the column.

The pressure $\Delta\sigma_z$ is applied instantaneously on the top of the cell and shared by the column and the soil as follows:

$$\Delta\sigma_s A_s + \Delta\sigma_c A_c = \Delta\sigma_z A_e \tag{5.39}$$

where $\Delta\sigma_s$ = average vertical stress in the soil
$\Delta\sigma_c$ = average vertical stress in the column

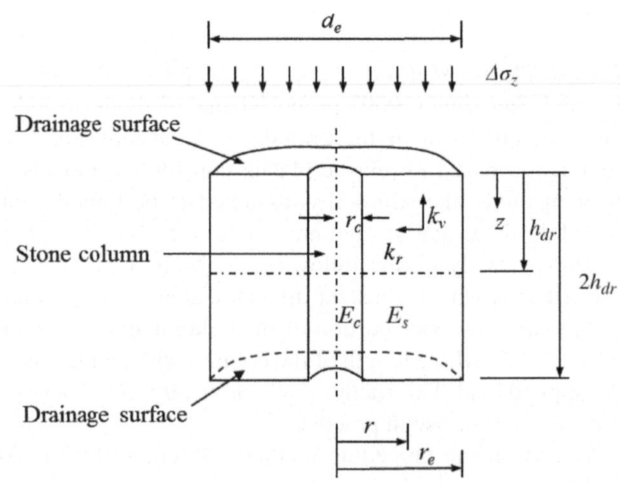

Figure 5.17 Consolidation model with a free-draining stone column.

From Equation (5.38), the following relationship can be obtained:

$$\partial\sigma'_s = \frac{m_{v,c}}{m_{v,s}}\partial\sigma'_c \tag{5.40}$$

where σ'_s = average effective stress in the soil
σ'_c = average effective stress in the column
$m_{v,c}$ = volumetric compressibility of the column in the vertical direction, and
$m_{v,s}$ = volumetric compressibility of the soil in the vertical direction

Combining Equations (5.39) and (5.40) and considering the total stress in the soil $\sigma_s = \sigma'_s + \overline{u}$ result in the following equation:

$$\frac{\partial\sigma'_s}{\partial t} = -\frac{m_{v,c}}{m_{v,s}}\frac{A_s}{A_c}\frac{\partial\sigma_s}{\partial t} = -\frac{m_{v,c}A_s}{m_{v,c}A_s + m_{v,s}A_c}\frac{\partial\overline{u}}{\partial t} \tag{5.41}$$

where $\overline{u}$ is average excess pore water pressure in the soil.

Based on the basic concept that the volume change of the soil is equal to the water discharge from the soil, the following equation can be obtained.

$$\frac{k_r}{\gamma_w}\left(\frac{1}{r}\frac{\partial u}{\partial r} + \frac{\partial^2 u}{\partial r^2}\right) + \frac{k_v}{\gamma_w}\frac{\partial^2 u}{\partial z^2} = \frac{m_{v,s}m_{v,c}(1-a_s)}{m_{v,c}(1-a_s)+m_{v,s}a_s}\frac{\partial\overline{u}}{\partial t} \tag{5.42}$$

$$c_{rm}\left(\frac{1}{r}\frac{\partial u}{\partial r} + \frac{\partial^2 u}{\partial r^2}\right) + c_{vm}\frac{\partial^2 u}{\partial z^2} = \frac{\partial\overline{u}}{\partial t} \tag{5.43}$$

where

$$c_{rm} = c_r\left(1 + n\frac{1}{N_D^2 - 1}\right) = c_r\left(1 + n\frac{a_s}{1-a_s}\right)$$

$$c_{vm} = c_v\left(1 + n\frac{1}{N_D^2 - 1}\right) = c_v\left(1 + n\frac{a_s}{1-a_s}\right)$$

c_r = the coefficient of consolidation of soft soil in the radial direction, i.e., $k_r/(\gamma_w m_{v,s})$
c_{rm} = the modified coefficient of consolidation of soft soil in the radial direction
c_v = the coefficient of consolidation of soft soil in the vertical direction, i.e., $k_v/(\gamma_w m_{v,s})$
k_v = permeability of the soil in the vertical direction
k_r = permeability of the soil in the radial direction
γ_w = unit weight of water
u = excess pore water pressure at a distance r
N_D = the diameter ratio, defined as the ratio of the influence diameter (d_e) to the column diameter ($d_c = 2r_c$) in a unit cell, and
n = stress concentration ratio at the end of consolidation. This stress concentration ratio can be estimated using Figure 5.14 but should be limited to not more than 5.0.

Equation (5.43) is the same as that presented in Chapter 7 in terms of format for vertical drains; therefore, the Terzaghi one-dimensional solution can be used for the vertical flow while the Barron solution (Barron, 1948) can be used for the radial flow.

Terzaghi (1943) suggested the following approximate relationships for the degree of soil consolidation under one-dimensional vertical flow:

$$\text{For } U_v = 0 \text{ to } 52.6\%,\ U_v = \sqrt{\frac{4T_v}{\pi}} \tag{5.44a}$$

$$\text{For } U_v > 52.6\%,\ U_v = 1 - 0.81 \times 10^{-1.07T_v} \tag{5.44b}$$

where U_v = average degree of consolidation due to vertical flow, and
T_v = time factor due to vertical flow (i.e., $T_v = c_v t/h_{\text{dr}}^2$, t is time and h_{dr} is longest drainage distance due to vertical flow)

Barron (1948) proposed the following solution for the degree of consolidation due to radial flow towards a free-draining sand drain:

$$U_r = 1 - e^{-\frac{8}{F(N_D)}T_r},\quad F(N_D) = \frac{N_D^2}{N_D^2 - 1}\ln(N_D) - \frac{3N_D^2 - 1}{4N_D^2} \tag{5.45}$$

where U_r = average degree of consolidation due to radial flow
N_D = diameter ratio (i.e., $N_D = d_e/d_c$)
d_e = equivalent diameter of a unit cell
d_c = diameter of a sand drain; and
T_r = time factor due to radial flow, (i.e., $T_r = c_r t/d_e^2$).

The overall degree of consolidation of soil due to vertical and radial flows can be calculated as follows (Carillo, 1942):

$$U_{vr} = 1 - (1 - U_v)(1 - U_r) \tag{5.46}$$

where U_{vr} = overall degree of consolidation
U_v = degree of consolidation in a vertical direction
U_r = degree of consolidation in a radial direction

For granular column-reinforced foundations, the modified coefficient of consolidation of soft soil in the vertical direction, c_{vm}, can be used to calculate the time factor for Terzaghi's solution for the degree of one-dimensional consolidation, as $T_{vm} = c_{vm}t/h_{\text{dr}}^2$. The modified coefficient of consolidation of soft soil in the radial direction, c_{rm}, can be used to calculate the time factor for Barron's solution (Barron, 1948) for the degree of radial consolidation as $T_{rm} = c_{rm}t/d_e^2$.

The modified coefficients of consolidation of soft soil accounts for the contribution of the stress concentration on the

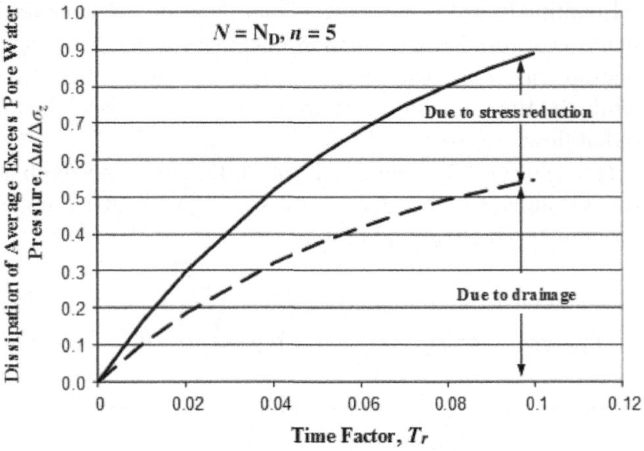

Figure 5.18 Dissipation of excess pore water pressure (after Han and Ye, 2001, with permission from ASCE).

columns due to the modulus difference between column and soil. Figure 5.18 shows that the rate of pore water pressure dissipation or consolidation is contributed not only by the drainage of the column but also by the stress reduction on the soft soil. During the consolidation, the stress on the soil is transferred to the column so that the stress on the soil decreases thus the excess pore water pressure in the soil is released.

Han and Ye (2002) also developed a simplified solution for consolidation degree of a stone column-reinforced soft foundation unit cell with smear and well resistance, as shown in Figure 5.19, as follows:

$$U = 1 - \exp\left(-\frac{8}{F'_m} T_{\mathrm{rm}}\right) \tag{5.47}$$

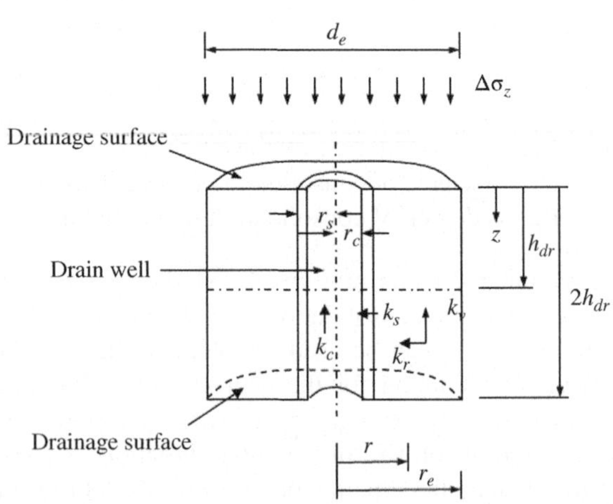

Figure 5.19 Unit cell model for stone column-reinforced soft foundation (after Han and Ye, 2002, with permission from ASCE).

$$F'_m = \frac{N_D^2}{N_D^2 - 1}\left(\ln\frac{N_D}{N_S} + \frac{k_r}{k_s}\ln N_S - \frac{3}{4}\right) + \frac{N_S^2}{N_D^2 - 1}\left(1 - \frac{k_r}{k_s}\right)\left(1 - \frac{N_S^2}{4N_D^2}\right) + \frac{k_r}{k_s}\frac{1}{N_D^2 - 1}\left(1 - \frac{1}{4N_D^2}\right) + \frac{32}{\pi^2}\left(\frac{k_r}{k_c}\right)\left(\frac{h_{\mathrm{dr}}}{d_c}\right)^2 \tag{5.48}$$

where N_S = d_s/d_c, the diameter ratio of the smeared zone to the stone column
d_s = diameter of the smeared zone (= $2r_s$)
d_c = diameter of the column (= $2r_c$)
k_r = permeability of the undisturbed surrounding soil in the radial direction
k_s = permeability of the smeared soil in the radial direction

Baez and Martin (1995) and Boulanger et al. (1998) indicated that the intrusion of native soil into stone columns during field installation could reach 20% by weight. Field injection tests indicated the ratio of the permeability of stone columns to that of the native soil ranging from 15 to 40 while laboratory tests indicated this ratio from 40 to 100. Han (2010) used the formula for the permeability of a granular drain with fine contents in the FHWA *Highway Subdrainage Design* manual (Moulton, 1980) to estimate the permeability of granular columns as follows:

$$k = \frac{2.19(D_{10})^{1.478} n_p^{6.654}}{(P_{200})^{0.597}} \tag{5.49}$$

where k = permeability of the granular column (unit: m/s)
D_{10} = effective grain size corresponding to 10% passing the sieve size (unit: mm)
P_{200} = percentage of particles passing U.S. No. 200 sieve
n_p = porosity of the stone column = $1 - \frac{\gamma_d}{\gamma_w G_s}$
γ_d = dry unit weight of the stone column
γ_w = unit weight of water
G_s = specific gravity

This formula was originally developed for granular bases and subbases for roadway applications and should be further verified for stone columns.

Han (2010) used D_{10} = 5μm for clay particles when stone columns are installed in clay. Due to the insufficient field data to separate smear and well resistance effects, Han (2010) combined these two effects into the well resistance. As a result, Equation (5.48) can be simplified into:

$$F'_m = \frac{N_D^2}{N_D^2 - 1}\left(\ln N_D - \frac{3}{4}\right) + \frac{1}{N_D^2 - 1}\left(1 - \frac{1}{4N_D^2}\right) + \frac{32}{\pi^2}\left(\frac{k_r}{k_c}\right)\left(\frac{h_{dr}}{d_c}\right)^2 \quad (5.50)$$

Han (2010) found that the solution considering the well resistance effect better predicted the rate of settlement as compared with the field data than the solution without any well resistance.

Xie et al. (2009b) derived a solution for the degree of consolidation of the stone column-reinforced foundation considering the variation of the permeability of the surrounding soil from the interface between the column and the surrounding soil to the boundary of a unit cell. Determination of such variations in the field is a challenging task. Xie et al. (2009a) also considered a step or ramp loading situation, which is useful to simulate the construction load, such as the filling of an embankment. In addition to a step or ramp loading, Wang (2009) developed a solution for the degree of consolidation for the stone column-reinforced foundation subjected to a cyclic loading, which may be useful to simulate traffic loading. Both Xie et al. (2009a) and Wang (2009) compared their solutions to that of Han and Ye (2002) under an instantaneous load and resulted in good agreement. All the above-mentioned solutions were developed based on the assumptions of equal strain and one-dimensional vertical elastic deformation in the column and the soil.

Castra and Sagaseta (2009) developed their solutions considering lateral deformation of stone columns under a vertical load. The lateral deformation of the column reduces the load carried by the column and slows down the rate of consolidation. In addition, Castra and Sagaseta (2009) treated the stone column as a linearly elastic, perfectly plastic material. The upper portion of the column can yield under a certain load and the plastic zone can extend from the top to the bottom of the column when the applied load is increased. They also developed modified coefficients of consolidation based on elastic and plastic deformations. However, the Castra and Sagaseta (2009) solutions did not consider smear and well resistance effects on the degree of consolidation of the stone column-reinforced foundation.

Concrete Columns The rate of consolidation of a concrete column-reinforced foundation with an end-bearing condition should not be a design concern because the concrete columns carry a majority of the load and the settlement happens immediately. When concrete columns partially penetrate into the soft soil, the rate of consolidation can be an issue. Zheng et al. (2011) showed that the excess pore water pressure in the soft clay between CFG rigid columns dissipated rapidly after each embankment loading. Research is needed to develop a solution for the rate of consolidation of a concrete column-reinforced foundation with partial column penetration.

5.3.6 Stability

Granular Columns The most likely failure mode of granular columns under embankments is shear failure. In practice, it is common to use equivalent parameters (cohesion, c_{eq}, and friction angle, ϕ_{eq}) for granular column-reinforced soft foundations in stability analysis. The equivalent parameters for the composite foundation are estimated based on the area average of these parameters from granular columns and the soft soil (e.g., Cooper and Rose 1999, Abusharar and Han, 2011) as follows:

$$c_{eq} = c_s(1 - a_s) \quad (5.51)$$

$$\phi_{eq} = \arctan[a_s \tan\phi_c + (1 - a_s)\tan\phi_s] \quad (5.52)$$

where c_s = soil cohesion
ϕ_s = soil friction angle
ϕ_c = column friction angle
a_s = area replacement ratio
c_{eq} = equivalent cohesion
ϕ_{eq} = equivalent friction angle

Under embankment loading as shown in Figure 5.20, numerical analysis using the strength reduction method showed that the factor of safety of an embankment over an individual granular column-reinforced foundation was approximately 90% that of the embankment over a composite foundation with the equivalent parameters under an undrained condition (Abusharar and Han, 2011).

Zhang et al. (2013) confirmed Abusharar and Han's (2011) finding for an undrained condition; however, they found that under a drained condition, the factors of safety calculated by the individual column and equivalent area methods are almost the same.

Priebe (1978) proposed a method to account for the effect of stress concentration ratio, n, on the equivalent friction angle:

$$\phi_{eq} = \arctan[\omega_f \tan\phi_c + (1 - \omega_f)\tan\phi_s] \quad (5.53)$$

$$\omega_f = \frac{a_s n}{1 + a_s(n - 1)} \quad (5.54)$$

Typically, the value of ω_f is within a range of 0.4–0.6. However, Zhang et al. (2013) found that the consideration of the effect of stress concentration ratio is not conservative

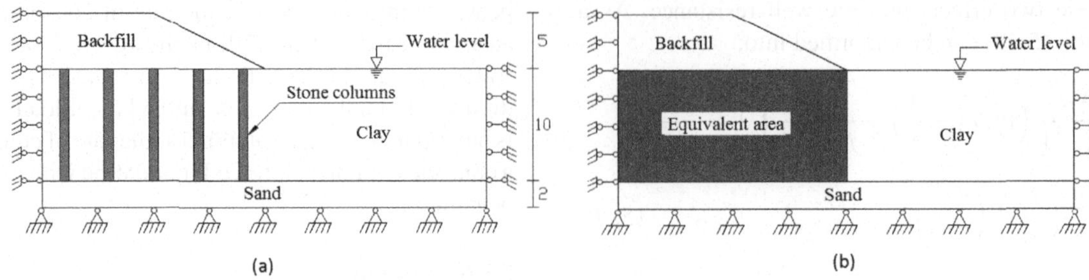

Figure 5.20 Stability of granular column-supported embankments: (a) individual column and (b) equivalent area (modified from Abusharar and Han, 2011 with permission from Elsevier).

because the stress concentration ratio approaches 1.0 at the state of failure.

Concrete Columns The most likely failure modes of concrete columns under embankments are rotational and bending failure. Zheng et al. (2010) investigated numerically the stability of embankments over rigid column-reinforced foundations. They found that the bending moments and shear forces in the rigid columns increased with the increase of the embankment load. The bending failure happened progressively on the rigid columns. In the soft soil, after the bending failure of the column at a certain depth, the portion of the column above the failure location would rotate and still provide resistance to the stability of the embankment. In the relatively firm soil, the second bending failure might occur in the upper portion of the column. Further research is needed to develop simplified solutions for analyzing the stability of rigid column-supported embankments.

5.3.7 Liquefaction

The columns installed in liquefiable soil may have the following three effects: (1) increase cyclic resistance ratio (CRR) by densfying the soil, (2) reduce cyclic stress ratio (CSR) by stiffer columns, and (3) dissipate excess pore water pressure quickly by granular columns.

The increase of CRR can be considered by increasing the SPT N value as shown in Figure 5.11 or back-calculated from the increased relative density, D_r.

Seed and Booker (1977) proposed a simple radial consolidation analytical model to analyze the dissipation of excess pore water pressure through gravel drains. The governing equation for free-draining gravel drains is as follows:

$$\frac{k_r}{\gamma_w m_{v,s}}\left(\frac{\partial^2 u}{\partial r^2}+\frac{1}{r}\frac{\partial u}{\partial r}\right)=\frac{\partial u}{\partial t}-\frac{\partial u_g}{\partial N_{\rm cyc}}\frac{\partial N_{\rm cyc}}{\partial t} \tag{5.55}$$

where u = excess pore water pressure
u_g = pore water pressure generated by the alternating shear stresses
$N_{\rm cyc}$ = number of cycles of alternating shear stress

The limitations of this method are: (1) it assumes infinite column permeability and (2) $m_{v,s}$ is constant. Research shows that $m_{v,s}$ remains constant until the ratio of excess pore water pressure to effective stress reaches 0.6 (Baez, 1995). The number of cycles, $N_{\rm 1cyc}$, for liquefaction can be obtained by solving Equation (5.55).

At the same shear strain, columns carry more shear stresses than the soil due to their stiffness difference as shown in Figure 5.21. Under the same shear force, the soil is subjected to lower shear stress. As a result, the chance for the soil to be liquefied is reduced.

Based on the shear strain compatibility, Baez (1995) suggested a shear stress reduction factor K_G for a granular column-reinforced foundation as follows:

$$K_G=\frac{\tau_s}{\tau}=\frac{1}{R_G[a_s+(1-a_s)/R_G]} \tag{5.56}$$

where K_G = shear stress reduction factor (i.e., ratio of shear stress in soil with granular columns to that in soil without granular columns)
τ_s = shear stress in soil between granular columns
τ = average shear stress in the granular column composite foundation
R_G = shear modulus ratio (i.e., G_c/G)
G_c = shear modulus of granular columns
G = shear modulus of soil

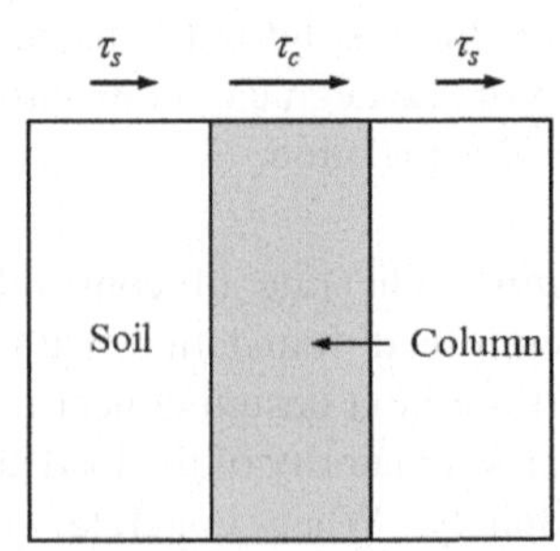

Figure 5.21 Shear stresses in the column and the soil.

Baez and Martin (1993) reported that R_G ranges from 2 to 7. Before the liquefaction analysis, the CSR of the soil induced by earthquake should be reduced as follows:

$$\mathrm{CSR}_s = \mathrm{CSR} \cdot K_G \tag{5.57}$$

However, recent research shows that the deformation of columns and soil may not necessarily be compatible during the motion induced by earthquake. Based on numerical results, Rayamajhi et al. (2014) proposed a revised shear stress reduction factor, K_{Gr}, as follows:

$$K_{\mathrm{Gr}} = \frac{1}{R_G[a_s R_\gamma + (1 - a_s)/R_G]} \tag{5.58}$$

where R_γ is the shear strain ratio, which is defined as the ratio of the shear strain in the column to that in the soil.

Rayamajhi et al. (2014) developed an approximate relationship between the shear strain ratio and the shear modulus ratio as follows:

$$R_\gamma = 1.04 R_G^{-0.65} - 0.04 \tag{5.59}$$

The comparisons of the calculated shear stress reduction factors before and after the revision with the numerical results are shown in Figure 5.22. It is shown that the revised shear stress reduction factor is higher than the original factor.

5.3.8 Design of Geosynthetic-encased Granular Columns

The key for the design of geosynthetic-encased granular columns is how to consider the benefit of geosynthetic encasement. Raithel and Kempfert (2000) proposed a method to calculate the settlement of a geosynthetic-encased granular column-reinforced foundation under a drained condition while Murugesan and Rajagopal (2010) proposed a method to calculate the bearing capacity of an individual geosynthetic-encased soil column in soft soil under an undrained condition.

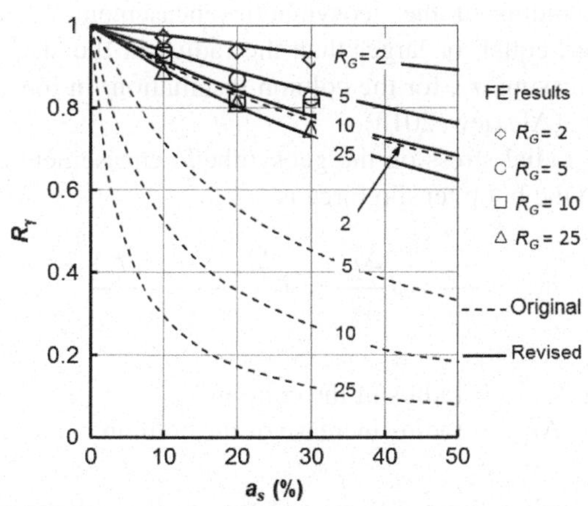

Figure 5.22 Comparison of shear stress reduction factor. (modified from Rayamajhi et al., 2014).

Bearing Capacity In addition to the lateral confinement by surrounding soil, Murugesan and Rajagopal (2010) considered additional confinement by geosynthetic encasement for the ultimate bearing capacity of an individual column due to possible bulging failure with a bulging length of four times the column diameter (i.e., 4*d*) as follows:

$$q_{\mathrm{ult},c} = (\sigma_{r0} + 4c_u + \sigma_{r,g})K_p \tag{5.60}$$

where σ_{r0} = lateral soil stress induced by the overburden stress at the middle point of the column bulging length (i.e., 2*d*, *d* is the column diameter)

$\sigma_{r,g}$ = additional confining stress by geosynthetic encasement

K_p = coefficient of passive earth pressure of the column

The additional confining stress can be calculated by considering the hoop tensile force in the geosynthetic encasement as follows:

$$\sigma_{r,g} = \frac{2T_g}{d_c} \tag{5.61}$$

where T_g is the hoop tensile force of the geosynthetic.

The hoop tensile force may be controlled by the geosynthetic elongation at failure (typically 5%) or by the vertical strain of the column, that is,

$$\varepsilon_g = \frac{1 - \sqrt{1 - \varepsilon_a}}{\sqrt{1 - \varepsilon_a}} \tag{5.62}$$

where ε_g = hoop strain of the gesoynthetic and ε_a = vertical compressive strain of the column (i.e., the vertical compression divided by the column bulging length). The vertical compression is typically limited to 10% the diameter of the column.

The hoop tensile force can be calculated as $T_g = J\varepsilon_g$, where J = tensile stiffness of the geosynthetic. Reduction factors considering creep, installation damage, and chemical and biological degradation should be considered for allowable tensile strength of geosynthetic encasement.

The ultimate bearing capacity of an individual column due to possible punching failure should also be checked if the column is not end-bearing. The method for the ultimate bearing capacity of concrete columns discussed in the earlier section may be used for this calculation.

Once the ultimate bearing capacity of individual geosynthetic-encased columns is known, the ultimate bearing capacity of the composite foundation can be calculated using Equation (5.17).

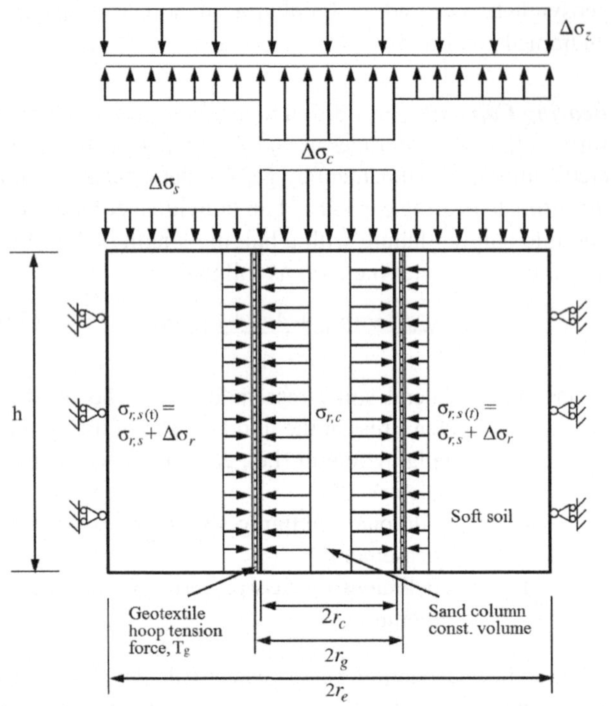

Figure 5.23 Unit cell model for a geosynthetic-encased column. (Raithel and Kempfert, 2000).

Settlement The proposed method by Raithel and Kempfert (2000) for the settlement of a geosynthetic-encased granular column-reinforced foundation is based on a unit cell concept as shown in Figure 5.23. Different from a unit cell for a column-reinforced foundation without geosynthetic, this model considers the contribution of geosynthetic encasement by providing additional confinement to the column.

Following are the assumptions for this development:

- The loading size is much larger than the thickness of the soft soil; therefore, the applied additional stress does not decrease with depth.
- The settlements on the top of the column and the soft soil are equal.
- No settlement is below the toe of the column.
- The column is at an active earth pressure state.
- Before loading, the soil is at an at-rest state. For an excavation installation method, the earth pressure coefficient of the soil is $K_{o,s} = 1 - \sin\phi_s$ (ϕ_s = friction angle of the soil). For a displacement installation method, a larger earth pressure coefficient, $K^*_{o,s}$, should be used.
- The geosynthetic encasement has linearly elastic behavior.
- The granular column is incompressible (i.e., its volume is constant).
- The design is based on a drained condition.

The radial stresses in the column and the soil are contributed by the overburden stresses of the column and the soil and the additional stresses on them as follows:

$$\sigma_{r,c} = \Delta\sigma_c K_{a,c} + \sigma_{z0,c} K_{a,c} = \left(\frac{1}{a_s}\Delta\sigma_z - \frac{1-a_s}{a_s}\Delta\sigma_s\right) K_{a,c} + \sigma_{z0,c} K_{a,c} \quad (5.63)$$

$$\sigma_{r,s} = \Delta\sigma_s K_{0,s} + \sigma_{z0,s} K_{o,s} \quad (5.64)$$

where $\sigma_{z0,c}$ = overburden stress of the column
$\sigma_{z0,s}$ = overburden stress of the soil
$\Delta\sigma_c$ = additional vertical stress in the column
$\Delta\sigma_s$ = additional vertical stress in the soil
$K_{a,c}$ = active earth pressure coefficient in the column
$K_{0,s}$ = at-rest earth pressure coefficient in the soil ($K^*_{0,s}$, should be used if a displacement installation method is used)

Raithel and Kempfert (2000) assume that the geosynthetic encasement has linearly elastic behavior with tensile stiffness, J. The hoop tensile force is

$$T_g = J\frac{\Delta r_g}{r_g} \quad (5.65)$$

where Δr_g = radius increase of the geosynthetic encasement
r_g = radius of the geosynethtic encasement

The compatibility of the radial deformation requires

$$\Delta r_c = \Delta r_g + (r_g - r_c) \quad (5.66)$$

The radius of the geosynthetic encasement, r_g, can be smaller, equal, or larger than the radius of the steel installation casing, r_c, for the column installation. In most cases, $r_g = r_c$ (Alexiew, 2013).

The radial stress on the geosynthetic encasement equivalent to the hoop tensile force is

$$\sigma_{r,g} = \frac{T_g}{r_g} = J\frac{\Delta r_g}{r_g^2} = J\frac{\Delta r_c - (r_g - r_c)}{r_g^2} \quad (5.67)$$

where r_c = radius of the column
Δr_c = radius increase of the column

The radial stress difference between the column and the soil is

$$\Delta\sigma_r = \sigma_{r,c} - \sigma_{r,s} - \sigma_{r,g} \quad (5.68)$$

The radial displacement, Δr_c, can be calculated based on Ghionna and Jamiolkowski (1981) for a radially and axially loaded hollow cylinder:

$$\Delta r_c = \frac{\Delta\sigma_r}{E^*}\left(\frac{1}{a_s} - 1\right) r_c \tag{5.69}$$

$$E^* = \left(\frac{1}{1-\nu_s} + \frac{1}{1+\nu_s}\frac{1}{a_s}\right) E_s \tag{5.70}$$

$$E_s = \frac{(1+\nu_s)(1-2\nu_s)}{1-\nu_s} D_s \tag{5.71}$$

where D_s = constrained modulus of the soil, which is equal to $1/m_{v,s}$
$m_{v,s}$ = coefficient of soil volumetric compressibility
E_s = elastic modulus of the soil
ν_s = Poisson's ratio of the soil

Substituting Equations (5.67) and (5.68) into Equation (5.69) results in the following equation:

$$\Delta r_c = \frac{\sigma_{r,c} - \sigma_{r,s} + \dfrac{(r_g - r_c)J}{r_g^2}}{\dfrac{a_s E^*}{(1-a_s) r_c} + \dfrac{J}{r_g^2}} \tag{5.72}$$

Since the radial stress difference results in the column expansion, the settlement of the soft soil, S_{sl}, can be obtained according to Ghionna and Jamiolkowski (1981) as follows:

$$S_{sl} = \left[\frac{\Delta\sigma_s}{D_s} - \frac{2}{E^*}\left(\frac{\nu_s}{1-\nu_s}\right)\Delta\sigma_r\right] h \tag{5.73}$$

where h is the thickness of the soil or length of the column.

Based on the constant volume assumption, the following equation for the settlement of the column can be obtained:

$$S_{cl} = \left[1 - \frac{r_c^2}{(r_c + \Delta r_c)^2}\right] h \tag{5.74}$$

Based on the equal strain assumption for the column and the soil (i.e., $S_{cl} = S_{sl} = S'$), the following equation can be established:

$$\frac{\Delta\sigma_s}{D_s} - \frac{2}{E^*}\left(\frac{\nu_s}{1-\nu_s}\right)\Delta\sigma_r = 1 - \frac{r_c^2}{(r_c + \Delta r_c)^2} \tag{5.75}$$

Substituting Equation (5.72) into Equation (5.75) results in an equation with only one unknown, $\Delta\sigma_s$, which can be solved mathematically. Raithel and Kempfert (2000) pointed out that D_s is stress dependent; therefore, the above equation should be solved by iterations. The settlement of the geosynthetic-encased column foundation, S', can be calculated using Eqs. (5.72) and (5.74). However, this procedure is complicated and requires the development of a program. Approximately, a constant D_s value may be used for a soil sublayer at the corresponding vertical stress level as shown in Equation (5.25) (for a normally consolidated soil). Near the ground surface, the soil is likely overconsolidated. Under such a condition, the recompression index, C_r, should be used instead of C_c, if the total effective stress is less than the preconsolidation stress.

The improvement factor for a geosynthetic-encased column foundation is

$$I_f = \frac{\Delta\sigma_z h}{D_s S'} \tag{5.76}$$

Based on measured data, Raithel et al. (2005) obtained the improvement factors for granular columns with or without geosynthetic encasement as shown in Figure 5.24. It is shown that the gesoynthetic encasement increases the improvement factor and the geosynthetic encasement with higher stiffness has a higher improvement factor.

From the field-measured settlement, Raithel and Kirchner (2008) found that the geosynthetic-encased columns reduced the secondary settlement after primary consolidation. They proposed a reduction factor to estimate the secondary settlement of the geosynthetic-encased column as follows:

$$S_\alpha = R_\alpha C_{\alpha\varepsilon} h_p \log\left(\frac{t}{t_p}\right) \tag{5.77}$$

where R_α = creep reduction factor
$C_{\alpha\varepsilon}$ = coefficient of secondary volumetric compression, i.e., $C_{\alpha\varepsilon} = C_\alpha/(1+e_p)$
C_α = coefficient of secondary compression
e_p = void ratio of soil at the end of primary consolidation
h_p = soil thickness at the end of primary consolidation
t_p = time of primary consolidation

Based on the field data, Raithel and Kirchner (2008) suggested that R_α ranges from 0.25 to 0.50.

Selection of geosynthetic encasement Alexiew et al. (2012) suggested that the geosynthetic encasement should have the following characteristics:

- No joint or seam
- High tensile stiffness
- Low creep reduction factor
- High permeability
- Limited installation damage
- High chemical and biological resistance

Typical tensile stiffness of geosynthetic encasement ranges from 1500 to 6000 kN/m and typical ultimate hoop tensile strength ranges from 100 to 400 kN/m.

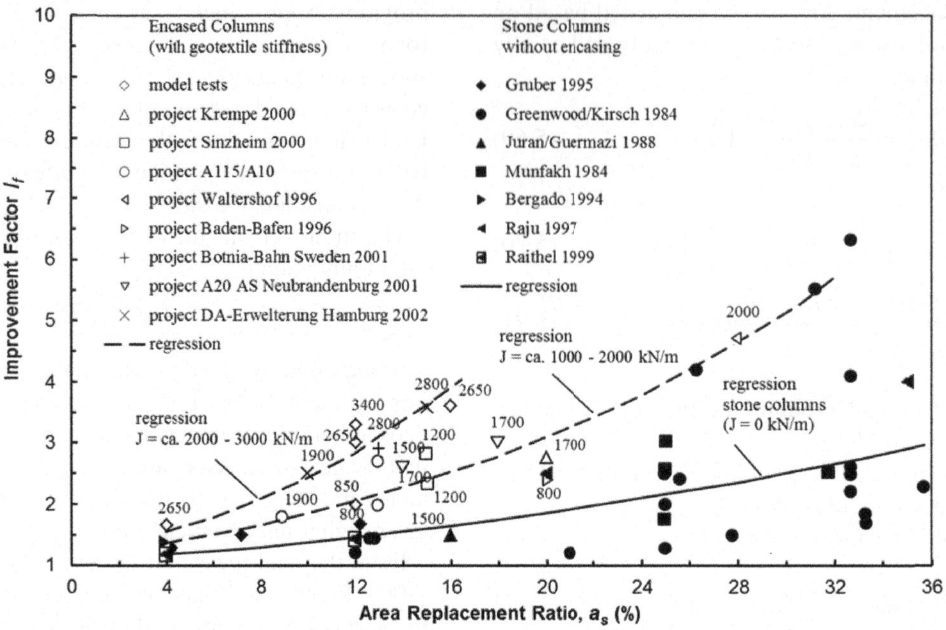

Figure 5.24 Improvement factors for geosynthetic-encased columns. (Raithel et al., 2005).

5.4 DESIGN PARAMETERS AND PROCEDURE

5.4.1 Granular Columns

Design Parameters Depending on applications, geotechnical problems, and installation methods, the following design parameters are typically needed:

- Soil type (cohesionless or cohesive soil)
- Thickness and depth of problematic soil
- Depth of groundwater table
- Initial void ratio or relative density of cohesionless soil
- Undrained shear strength for short-term design and effective friction angle for long-term design for cohesive soil
- Consolidation coefficient of cohesive soil
- Coefficient of volume compressibility of soil
- Coefficient of volume compressibility of column
- Magnitude and area of load or magnitude of earthquake
- Target void ratio or relative density of soil for cohesionless soil
- Required allowable bearing capacity
- Tolerable settlement
- Gradation of backfill for columns
- Friction angle of backfill for columns
- Pattern and spacing of columns
- Diameter of columns
- Area of improvement
- Type of installation method (replacement or displacement)
- Frequency of penetration, compaction, and extraction (sand compaction column)
- Duration of compaction for cohesionless soil
- Ground subsidence for cohesionless soil

Design Procedure The design procedure for granular columns in cohesionless soil mostly involves the density increase of the soil after installation of columns. The following procedure may be followed for design of granular columns in cohesionless soil in addition to the design of vibro-compaction discussed in Chapter 3:

1. Based on geotechnical profile and potential problem, determine whether densification is possible. The most important parameters are fine content and percent of clay particles.
2. If the soil is deemed suitable for densification, estimate the initial void ratio or relative density.
3. Assume the diameter of columns based on installation equipment or typical value.
4. Based on the performance requirement (e.g., bearing capacity, settlement, liquefaction, and global slope stability), select a target void ratio, relative density, or SPT *N* value.
5. Based on the required relative density or SPT value, estimate the required spacing of columns as discussed in Chapter 3 or area replacement ratio using Figure 5.11 for sand compaction columns.

6. Considering the strengthening or stiffness effect of the columns, calculate the reduced shear stress inducing liquefaction using Equations (5.57) and (5.58).
7. The procedures for the increased allowable bearing capacity, the reduced settlement, and the increased factor of safety of slope stability are similar to those for granular columns in cohesive soil, which are presented below.

The design procedure for granular columns in cohesive soils involves bearing capacity, settlement, rate of consolidation, and global slope stability. The typical procedure is as follows:

1. Based on undrained shear strength of cohesive soil, calculate the ultimate bearing capacity of natural soil as $5c_u$ (c_u is undrained shear strength of soil).
2. Based on undrained shear strength of cohesive soil, calculate the ultimate bearing capacity of individual columns using Equation (5.15) or (5.16).
3. Calculate the ultimate bearing capacity of the composite foundation by considering ultimate bearing capacities of natural soil and individual columns, and the area replacement ratio using Equation (5.17).
4. Based on the coefficient of volumetric compression, calculate the settlement of the natural soil using Equation (5.23).
5. Based on the column–soil modulus ratio or column friction angle, calculate the stress reduction factor or improvement factor and then calculate the settlement of the composite foundation using Equation (5.29) or (5.31).
6. The degree of consolidation of a composite foundation can be estimated using the Han and Ye solution considering well resistance using Equation (5.47). In this calculation, the permeability of the column should be estimated first, possibly using the FHWA highway drainage method Equation (5.49).
7. The factor of safety against global slope failure can be calculated using the individual column method or equivalent area method. When the equivalent area method is used for undrained analysis, a reduction factor of 0.9 should be applied to the calculated value.

5.4.2 Concrete Columns

Concrete columns are mostly used for cohesive soils. The design of concrete columns is similar to pile–raft or pile group design.

Design Parameters The design parameters for concrete columns typically include:

- Thickness and depth of problematic soil
- Depth of groundwater table
- Soil undrained shear strength for short-term and effective friction angle for long-term design
- Elastic modulus of soil
- Elastic modulus of column
- Properties of end-bearing soil
- Magnitude and area of load
- Required allowable bearing capacity
- Tolerable settlement
- Pattern and spacing of columns
- Diameter and length of columns
- Area of improvement
- Type of installation method (excavation or displacement)

Design Procedure The design procedure for concrete columns is mostly focused on bearing capacity and settlement. Their methods are similar to those for pile–raft systems or pile groups. The typical design procedure is as follows:

1. Based on undrained shear strength of cohesive soil, calculate the ultimate bearing capacities of individual columns under an undrained condition using the α method for the side friction, i.e., Equation (5.19a) and the bearing capacity factor of 9 for the toe bearing capacity, i.e., Equation (5.20a).
2. Based on effective friction angle of cohesive soil or sand, calculate the ultimate bearing capacities of individual columns under a drained condition using the β method for the side friction, i.e., Equation (5.19b) and the friction angle-dependent bearing capacity factor for the toe bearing capacity Equation, i.e., (5.20b).
3. Calculate the ultimate bearing capacities for an equivalent pier under undrained and drained conditions.
4. Calculate the ultimate bearing capacities for a pier–raft system under undrained and drained conditions using Equation (5.22).
5. Use the piled raft method to calculate the settlement of the concrete column-reinforced foundation using Equations from (5.32) to (5.37).

5.4.3 Geosynthetic-encased Granular Column

Design Parameters The design parameters for geosynthetic-encased granular columns typically include:

- Thickness and depth of problematic soil
- Depth of groundwater table
- Undrained shear strength of soil
- Effective friction angle of soil
- Elastic modulus and Poisson's ratio of soil
- Elastic modulus of column
- Magnitude and area of load
- Tolerable settlement

- Pattern and spacing of columns
- Diameter and length of columns
- Area of improvement
- Tensile strength and stiffness of geosynthetic encasement
- Type of installation method (excavation or displacement)

Design Procedure

1. Based on the undrained shear strength of the soil, calculate the ultimate bearing capacity of the natural soil under an undrained condition.
2. Based on the undrained shear strength of the soil and the diameter of the column, the ultimate bearing capacity of the individual column can be calculated by considering the lateral confinement by the soil and the geosynthetic encasement using Equation (5.60).
3. Calculate the ultimate bearing capacity of the composite foundation based on the ultimate bearing capacities of the soil and the individual column, and the area replacement ratio using Equation (5.17).
4. Based on the soil and column properties, the column radius, area replacement ratio, and length, the geosynthetic encasement properties, and the applied pressure, calculate the radius change of the column and the settlement of the composite foundation using Equations (5.63) to (5.76).
5. Based on the coefficient of secondary compression of the soil and the time for service after the completion of primary consolidation, estimate the reduced secondary compression using Equation (5.77).

5.5 DESIGN EXAMPLES

Example 5.1 Bearing Capacity of Granular Columns

Problem

A site consists of 20-m-thick soft clay with an undrained shear strength of 20 kPa. The groundwater table is at a depth of 1.0 m. The unit weights of the soil above and below the groundwater table are 18 and 19 kN/m^3. A square footing with a width of 2.0 m will be constructed on this site with an embedment depth of 1.0 m. The applied load of this footing (including the weight of the footing) is 400 kN. The required factor of safety against bearing failure is 2.5. Granular columns of 10 m long are designed in an equilateral triangular pattern with column spacing of 1.5 m and column diameter of 0.8 m. Evaluate whether this design meets the bearing capacity requirement.

Solution

The applied bearing pressure at the base of the footing is

$$p = \frac{P + W_f}{A_f} = \frac{400}{2.0 \times 2.0} = 100\ \text{kPa}$$

The ultimate bearing capacity of a single granular column is

$$q_{\text{ult},c} = Kc_u = 20 \times 20 = 400\ \text{kPa}$$

The ultimate bearing capacity of a square footing embedded at a depth of 1.0 m in a natural ground under an undrained condition is

$$\begin{aligned} q_{\text{ult},s} &= cN_c s_c d_c + 0.5\gamma' B_f s_\gamma d_\gamma + \sigma_D{'} N_q s_q d_q \\ &= cN_c s_c d_c + \sigma_D{'} N_q \\ s_c &= 1 + 0.2(B_f/L_f) = 1 + 0.2(2.0/2.0) = 1.2 \\ d_c &= 1 + 0.2(D_f/B_f) = 1 + 0.2(1.0/2.0) = 1.1 \\ q_{\text{ult},s} &= cN_c s_c d_c + \sigma_D{'} N_q \\ &= 20 \times 5.15 \times 1.21 \times 1.1 + 18 \times 1.10 = 154\ \text{kPa} \end{aligned}$$

The area replacement ratio can be calculated as follow:

$$a_s = 0.907\left(\frac{d_c}{s}\right)^2 = 0.907 \times \left(\frac{0.8}{1.5}\right)^2 = 0.26$$

According to Equation (5.17), the ultimate bearing capacity of the composite foundation is

$$q_{\text{ult}} = 400 \times 0.26 + 154 \times (1 - 0.26) = 218\ \text{kPa}$$

The factor of safety against bearing failure is

$$\text{FS} = \frac{q_{\text{ult}}}{p} = \frac{218}{100} = 2.2 < 2.5$$

Therefore, this design does not meet the bearing capacity requirement. A closer column spacing should be used or other column type may be used to increase the factor of safety.

Example 5.2 Bearing Capacity of Concrete Columns

Problem

A 5.5 m × 5.5 m square bridge footing is supported by a group of nine vibro-concrete columns in 10-m-thick soft soil, which is underlain by a dense sand. The soft soil has a saturated unit weight of 16 kN/m^3 and an undrained shear strength of the soil of 20 kPa. The groundwater table is at the ground surface. The dense sand has a saturated unit weight of 18.6 kN/m^3 and an effective friction angle of 36°. The vibro-concrete columns are 10 m long and have a shaft diameter of 0.5 m and a toe diameter of 0.7 m. The unconfined compressive strength of the concrete is 20 MPa. Example Figure 5.1 shows the layout of the columns. Calculate the ultimate load capacity of an individual column and the ultimate bearing capacity of the foundation (ignoring the foundation embedment due to possible scour) under an undrained condition.

Solution

1. Area of columns and foundations: The cross-sectional area of the column shaft is

$$A_c = 3.14 \times \left(\frac{0.5}{2}\right)^2 = 0.196\ \text{m}^2$$

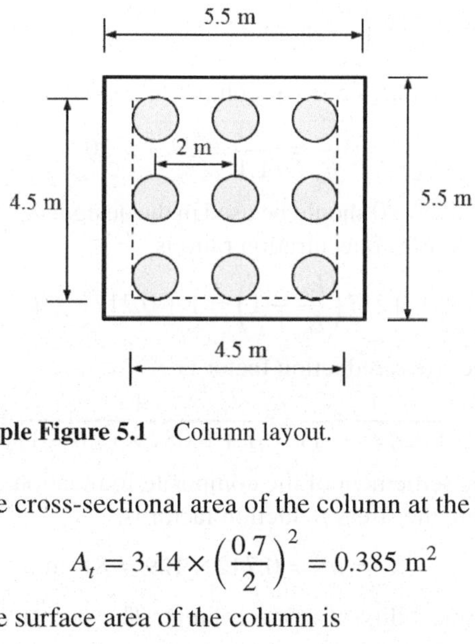

Example Figure 5.1 Column layout.

The cross-sectional area of the column at the toe is

$$A_t = 3.14 \times \left(\frac{0.7}{2}\right)^2 = 0.385\ \text{m}^2$$

The surface area of the column is

$$A_s = 3.14 \times 0.5 \times 10 = 15.71\ \text{m}^2$$

2. Ultimate load capacity for the individual column: The α method is used for the side friction. Considering a displacement method is used for the installation of vibro-concrete columns. Select $\alpha = 0.9$. The side friction is

$$f_s = \alpha c_u = 0.9 \times 20 = 18\ \text{kPa}$$

Since the end-bearing soil is sand, the toe resistance can be calculated as follows:

$$q_t = 0.5 d_c N_\gamma^* + \sigma_D' N_q^*$$

According to $\phi' = 36^\circ$, $N_\gamma^* = 32$ and $N_q^* = 90$. The effective overburden stress at the toe is

$$\sigma_D' = (16 - 9.81) \times 10 = 61.9\ \text{kPa}$$

The toe resistance is

$$q_t = 0.5 d_c N_\gamma^* + \sigma_D' N_q^* = 0.5 \times 0.7 \times 32 + 61.9 \times 90$$
$$= 5582\ \text{kPa}$$

The ultimate load capacity of the column is

$$Q_{\text{ult},c} = f_s A_s + q_t A_t = 18 \times 15.71 + 5582 \times 0.385$$
$$= 2431\ \text{kN}$$

The ultimate bearing capacity of the column is

$$q_{\text{ult},c} = \frac{Q_{\text{ult},c}}{A_c} = \frac{2431}{0.196} = 12381\ \text{kPa} < 20000\ \text{kPa}$$

(compressive strength of concrete) (OK)

3. Ultimate bearing capacity of the foundation: The ultimate bearing capacity of a shallow foundation on the undrained soft soil without an embedment is

$$q_{\text{ult},s} = cN_c s_c d_c + 0.5\gamma' B_f N_\gamma s_\gamma d_\gamma + \sigma_D' N_q s_q d_q = cN_c s_c$$

$$s_c = 1 + 0.2\left(\frac{B_f}{L_f}\right) = 1 + 0.2\left(\frac{5.5}{5.5}\right) = 1.2$$

$$q_{ult,s} = cN_c s_c = 20 \times 5.14 \times 1.2 = 123\ \text{kPa}$$

The area of the soil under the foundation is

$$A_{\text{sr}} = A_r - N_{cl} A_c = 5.5 \times 5.5 - 9 \times 0.196 = 28.5\ \text{m}^2$$

The ultimate load capacity of the foundation considering the combined contribution of the individual columns and the soil is

$$Q_{\text{ult1}} = \sum Q_{\text{ult},ci} + q_{\text{ult},s} A_{\text{sr}} = 9 \times 2431 + 123 \times 28.5$$
$$= 25393\ \text{kN}$$

Considering an equivalent pier that includes nine columns, the cross-sectional area and the surface area are

$$A_{\text{pt}} = 4.5 \times 4.5 = 20.25\ \text{m}^2$$

$$A_{\text{ps}} = 4 \times 4.5 \times 10 = 180\ \text{m}^2$$

The side friction of the equivalent pier (select $\alpha = 1.0$ because the shear mostly occurs between the soil) is

$$f_{ps} = \alpha c_u = 1.0 \times 20 = 20 \text{ kPa}$$

The toe resistance of the equivalent pier is

$$q_{pt} = 0.5B_p N_\gamma^* + \sigma_D' N_q^* = 0.5 \times 4.5 \times 32 + 61.9 \times 90$$
$$= 5643 \text{ kPa}$$

The ultimate load capacity of the equivalent pier is

$$Q_{ult,p} = f_{ps}A_{ps} + q_{pt}A_{pt} = 20 \times 180 + 5643 \times 20.25$$
$$= 117871 \text{ kN}$$

The ultimate load capacity of the foundation considering the combined contribution of the equivalent pier and the soil is

$$Q_{ult2} = Q_{ult,p} + q_{ult,s}A_{so} = 117871$$
$$+123 \times (5.5 \times 5.5 - 4.5 \times 4.5)$$
$$= 119104 \text{ kN}$$

Since $Q_{ult1} < Q_{ult2}$, $Q_{ult} = Q_{ult1} = 25393$ kN

The ultimate bearing capacity of the foundation is

$$q_{ult} = \frac{Q_{ult}}{A_f} = \frac{25393}{5.5 \times 5.5} = 839 \text{ kPa}$$

Example 5.3 Consolidation Settlement of Granular Columns

Problem

A 40-m-wide (on the crest) and 1.8-m-high embankment with the fill unit weight of 18 kN/m^3 is constructed over a 5-m-thick soft clay underlain by a stiff clay. Above the soft soil, there is a 0.3-m-thick fill, which is provided as a construction platform and drainage layer. The groundwater table is 1 m below the ground surface. The soft clay has the following properties: $\gamma = 15$ kN/m^3, $E_s = 1.1$ MPa, $k_r = 3.47 \times 10^{-9}$ m/s, and $k_v = 1.16 \times 10^{-9}$ m/s. The stone columns installed in this project have a diameter of 0.8 m and a length of 6 m with a square pattern at spacing of 2.4 m. The dry unit weight, the specific gravity, the elastic modulus, and Poisson's ratio of the columns are 15.7 kN/m^3, 2.70, 30 MPa, and 0.3, respectively. In the stone columns, there are 20% clay particles with $D_{10} = 0.005$ mm. Calculate the consolidation settlements with and without stone columns and the settlement of the stone column-reinforced foundation at one month after the construction of the embankment (assuming instantaneous placement of the embankment).

Solution

1. Settlement of natural ground: The constrained modulus of the soft soil is

$$m_{v,s} = \frac{1}{D_s} = \frac{(1+v_s)(1-2v_s)}{E_s(1-v_s)}$$
$$= \frac{(1+0.3) \times (1-2 \times 0.3)}{1.1 \times (1-0.3)} = 0.675 \text{ MPa}^{-1}$$
$$= 0.000675 \text{ kPa}^{-1}$$

The settlement of the natural ground is

$$S = m_{v,s}\Delta\sigma_z h = 0.000675 \times (1.8 \times 18) \times 5$$
$$= 0.109 \text{ m} = 109 \text{ mm}$$

2. Settlement of composite foundation: Since the stone columns are arranged in a square pattern at spacing of 2.4 m, the area replacement ratio can be calculated as follows:

$$a_s = C\left(\frac{d_c}{s}\right)^2 = 0.785 \times \left(\frac{0.8}{2.4}\right)^2 = 0.087$$

The modulus ratio of column to soil is

$$\frac{E_c}{E_s} = \frac{30}{1.1} = 27.3 > 20$$

$E_c/E_s = 20$ should be used in the design. As a result, the stress concentration ratio is

$$n = 1 + 0.217\left(\frac{E_c}{E_s} - 1\right) = 1 + 0.217 \times (20 - 1) = 5.0$$

The stress reduction factor is

$$\mu = \frac{1}{1 + a_s(n-1)} = \frac{1}{1 + 0.087 \times (5.0 - 1)} = 0.742$$

The settlement of the composite foundation considering the stress reduction factor is

$$S' = \mu S = 0.742 \times 109 = 81 \text{ mm}$$

3. Permeability of stone column: The porosity of the stone column is

$$n_p = 1 - \frac{\gamma_d}{\gamma_w G_s} = 1 - \frac{15.7}{9.81 \times 2.7} = 0.407$$

The permeability of the stone column is

$$k_c = \frac{2.19(D_{10})^{1.478} n_p^{6.654}}{(P_{200})^{0.597}} = \frac{2.19 \times 0.005^{1.478} \times 0.407^{6.654}}{20^{0.597}} = 3.67 \times 10^{-7} \text{m/s}$$

4. Modified coefficients of consolidation of soft soil: The coefficients of consolidation of soft soil in the vertical and radial directions are

$$c_v = \frac{k_v}{\gamma_w m_{v,s}} = \frac{1.16 \times 10^{-9}}{9.81 \times 0.000675} = 1.75 \times 10^{-7} \text{m}^2/\text{s}$$

$$c_r = \frac{k_r}{\gamma_w m_{v,s}} = \frac{3.47 \times 10^{-9}}{9.81 \times 0.000675} = 5.24 \times 10^{-7} \text{m}^2/\text{s}$$

The equivalent diameter of the unit cell is

$$d_e = \sqrt{\frac{4s^2}{\pi}} = \sqrt{\frac{4 \times 2.4^2}{3.14}} = 2.7 \text{ m}$$

The diameter ratio is

$$N_D = \frac{d_e}{d_c} = \frac{2.7}{0.8} = 3.4$$

The modified coefficients of consolidation of soft soil in the vertical and radial directions are

$$c_{\text{vm}} = c_v \left(1 + n\frac{a_s}{1 - a_s}\right) = 1.75 \times 10^{-7} \left(1 + 5.0 \times \frac{0.087}{1 - 0.087}\right) = 2.61 \times 10^{-7} \text{m}^2/\text{s}$$

$$c_{\text{rm}} = c_r \left(1 + n\frac{a_s}{1 - a_s}\right) = 5.24 \times 10^{-7} \left(1 + 5.0 \times \frac{0.087}{1 - 0.087}\right) = 7.80 \times 10^{-7} \text{m}^2/\text{s}$$

5. Time factors: The time factors in the vertical and radial directions are

$$T_{\text{vm}} = \frac{c_{\text{vm}} t}{h_{\text{dr}}^2} = \frac{2.61 \times 10^{-7} \times (30 \times 24 \times 60 \times 60)}{5^2} = 0.027$$

$$T_{\text{rm}} = \frac{c_{\text{rm}} t}{d_e^2} = \frac{7.80 \times 10^{-7} \times (30 \times 24 \times 60 \times 60)}{2.7^2} = 0.277$$

6. Degree of consolidation: The degree of consolidation due to the vertical flow according to one-dimensional Terzaghi consolidation theory is

$$U_{\text{vm}} = \sqrt{\frac{4T_{\text{vm}}}{\pi}} = \sqrt{\frac{4 \times 0.027}{3.14}} = 0.185$$

For the radial flow, considering the smear zone as part of the well resistance

$$\begin{aligned} F'_m &= \frac{N_D^2}{N_D^2 - 1}\left(\ln\frac{N_D}{N_S} + \frac{k_r}{k_s}\ln N_S - \frac{3}{4}\right) \\ &+ \frac{N_S^2}{N_D^2 - 1}\left(1 - \frac{k_r}{k_s}\right)\left(1 - \frac{N_S^2}{4N_D^2}\right) \\ &+ \frac{k_r}{k_s}\frac{1}{N_D^2 - 1}\left(1 - \frac{1}{4N_D^2}\right) + \frac{32}{\pi^2}\left(\frac{k_r}{k_c}\right)\left(\frac{h_{\text{dr}}}{d_c}\right)^2 \\ &= \frac{N_D^2}{N_D^2 - 1}\left(\ln N_D - \frac{3}{4}\right) + \frac{1}{N_D^2 - 1}\left(1 - \frac{1}{4N_D^2}\right) \\ &+ \frac{32}{\pi^2}\left(\frac{k_r}{k_c}\right)\left(\frac{h_{\text{dr}}}{d_c}\right)^2 \\ &= \frac{3.4^2}{3.4^2 - 1}\left(\ln 3.4 - \frac{3}{4}\right) + \frac{1}{3.4^2 - 1}\left(1 - \frac{1}{4 \times 3.4^2}\right) \\ &+ \frac{32}{3.14^2} \times \left(\frac{3.47 \times 10^{-9}}{3.67 \times 10^{-7}}\right) \times \left(\frac{5}{0.8}\right)^2 \\ &= 1.81 \end{aligned}$$

The degree of consolidation due to the radial flow is

$$U_{\text{rm}} = 1 - \exp\left(\frac{-8}{F'_m}T_{\text{rm}}\right) = 1 - \exp\left(\frac{-8}{1.81}0.277\right) = 0.706$$

The degree of consolidation due to combined vertical and radial flows is

$$U_{\text{vr}} = 1 - (1 - U_{\text{vm}})(1 - U_{\text{rm}}) = 1 - (1 - 0.185) \times (1 - 0.706) = 0.761$$

7. Consolidation settlement at one month after construction: The consolidation settlement of the composite foundation at one month after construction is

$$S_t = U_{\text{vr}} S = 0.761 \times 81 = 61 \text{ mm}$$

Example 5.4 Settlement of Geosynthetic-Encased Granular Columns

Problem

A naturally deposited soft soil site is proposed for land development. The site consists of a 5-m-thick soft clay underlain by a firm soil. This soft soil has the following properties: normally consolidated, constraint modulus, $D_s = 1658$ kPa, saturated unit weight, $\gamma_s = 15.0\ \text{kN/m}^3$, Poisson's ratio, $v_s = 0.3$, and effective friction angle, $\phi'_s = 20°$. The groundwater table is at the ground surface. The elevation of this site will be raised by 3-m fill, which has a unit weight of $\gamma_f = 18.0\ \text{kN/m}^3$. Geosynthetic-encased sand columns of 0.5 m in diameter are proposed in a triangular pattern with spacing of 1.2 m to minimize the settlement. The sand columns have a unit weight, $\gamma_c = 19.0\ \text{kN/m}^3$ and an effective friction angle, $\phi'_c = 34°$. The tensile stiffness of the geosynthetic encasement is $J = 3000$ kN/m. The diameter of the geosynthetic encasement has the same diameter as the installation pipe. Calculate the settlement without ground improvement, the settlement with ground improvement by geosynthetic-encased sand columns, and the improvement factor.

Stress and Modulus of Soft Soil

1. Modulus of soft soil: The effective overburden stress of the soft soil is estimated at the middepth of the soil layer, that is, $z = 2.5$ m.

$$\sigma'_{z0,s} = (\gamma_s - \gamma_w)z = (15.0 - 9.81) \times 2.5 = 13.0 \text{ kPa}$$

The elastic modulus of the soil is

$$E_s = \frac{(1+v_s)(1-2v_s)}{1-v_s} D_s = \frac{(1+0.3)\times(1-2\times 0.3)}{1-0.3} \times 1658 = 1232 \text{ kPa}$$

2. Settlement of the natural ground: The applied stress by the fill is

$$\Delta\sigma_z = \gamma_f H = 18 \times 3 = 54 \text{ kPa}$$

The settlement of the natural ground is

$$S = \frac{\Delta\sigma_z}{D_s} h = \frac{54}{1658} \times 5 = 0.163 \text{ m} = 163 \text{ mm}$$

3. Settlement of the foundation improved by geosynthetic-encased columns: The area replacement ratio is

$$a_s = C\left(\frac{d_c}{s}\right)^2 = 0.907 \times \left(\frac{0.5}{1.2}\right)^2 = 0.16$$

The elastic modulus of the soil considering the effect of the columns is

$$E^* = \left(\frac{1}{1-v_s} + \frac{1}{1+v_s}\frac{1}{a_s}\right) E_s = \left(\frac{1}{1-0.3} + \frac{1}{1+0.3} \times \frac{1}{0.16}\right) \times 1232 = 7776 \text{ kPa}$$

The coefficients of lateral earth pressure in the soil and the column are

$$K_{0,s} = 1 - \sin\phi'_s = 1 - \sin 20° = 0.658$$

$$K_{a,c} = \tan^2\left(45° - \frac{\phi'_c}{2}\right) = \tan^2\left(45° - \frac{34°}{2}\right) = 0.283$$

The effective overburden stress in the column at the mid-depth of the soil layer is

$$\sigma'_{z0,c} = (\gamma_c - \gamma_w)z = (19 - 9.81) \times 2.5 = 23.0 \text{ kPa}$$

The lateral stress in the column is

$$\sigma_{r,c} = \left(\frac{1}{a_s}\Delta\sigma_z - \frac{1-a_s}{a_s}\Delta\sigma_s\right) K_{a,c} + \sigma'_{z0,c}K_{a,c} = \left(\frac{1}{0.16} \times 54 - \frac{1-0.16}{0.16}\Delta\sigma_s\right) \times 0.283 + 23.0 \times 0.283 = 103.4 - 1.51\Delta\sigma_s$$

The lateral stress by the soil is

$$\sigma_{r,s} = \Delta\sigma_s K_{o,s} + \sigma'_{z0,s}K_{o,s} = \Delta\sigma_s \times 0.658 + 13.0 \times 0.658 = 0.658\Delta\sigma_s + 8.54$$

The radius change of the column after subjected to the fill loading is

$$\Delta r_c = \frac{\sigma_{r,c} - \sigma_{r,s} + \frac{(r_g - r_c)J}{r_g^2}}{\frac{a_s E^*}{(1-a_s)r_c} + \frac{J}{r_g^2}} = \frac{103.4 - 1.51\Delta\sigma_s - 0.658\Delta\sigma_s - 8.54 + 0}{\frac{0.16 \times 7776}{(1-0.16)\times 0.25} + \frac{3000}{0.25^2}}$$

$$= \frac{94.9 - 2.17\Delta\sigma_s}{53813} = 0.00176 - 0.0000403\Delta\sigma_s$$

The lateral resistance provided by the geosynthetic encasement is

$$\sigma_{r,g} = J\frac{\Delta r_c - (r_g - r_c)}{r_g^2}$$

$$= 3000 \times \frac{0.00176 - 0.0000403\Delta\sigma_s}{0.25^2}$$

$$= 84.7 - 1.94\Delta\sigma_s$$

The lateral stress difference between the column and the soil is

$$\Delta\sigma_r = \sigma_{r,c} - \sigma_{r,s} - \sigma_{r,g} = 103.4 - 1.51\Delta\sigma_s$$
$$-0.658\Delta\sigma_s - 8.54 - 84.7 + 1.94\Delta\sigma_s$$
$$= 10.3 - 0.234\Delta\sigma_s$$

Based on the equal settlement of the column and the soil

$$\frac{\Delta\sigma_s}{D_s} - \frac{2}{E^*}\left(\frac{\nu_s}{1-\nu_s}\right)\Delta\sigma_r = 1 - \frac{r_c^2}{(r_c + \Delta r_c)^2}$$

$$\frac{\Delta\sigma_s}{1658} - \frac{2}{7776} \times \left(\frac{0.3}{1-0.3}\right) \times (10.3 - 0.234\Delta\sigma_s) =$$
$$1 - \frac{0.25^2}{(0.25 + 0.00176 - 0.0000403\Delta\sigma_s)^2}$$

$$0.000603\Delta\sigma_s - 0.00113 + 0.000026\Delta\sigma_s =$$
$$1 - \frac{0.0625}{(0.25176 - 0.0000403\Delta\sigma_s)^2}$$

$$1.00113 - 0.000577\Delta\sigma_s - \frac{0.0625}{\begin{pmatrix}0.25176 - \\ 0.0000403\Delta\sigma_s\end{pmatrix}^2} = 0$$

From the above equation, $\Delta\sigma_s$ can be solved as 16.9 kPa.

The radius change of the column is

$$\Delta r_c = 0.00176 - 0.0000403\Delta\sigma_s$$
$$= 0.00176 - 0.0000403 \times 16.9 = 0.0011 \text{ m}$$

The settlement of the foundation after the ground improvement is

$$S' = \left[1 - \frac{r_c^2}{(r_c + \Delta r_c)^2}\right] h$$

$$= \left[1 - \frac{0.25^2}{(0.25 + 0.0011)^2}\right] \times 5 = 0.043 = 43 \text{ mm}$$

4. Improvement factor: The improvement factor is

$$I_f = \frac{S}{S'} = \frac{163}{43} = 3.8$$

From Figure 5.24, the improvement factors, I_f, for J = 1000 to 2000 kN/m and 2000 to 3000 kN/m are 2.6 and 3.9, respectively. The improvement factor for J = 3000 kN/m is 3.9, which is almost the exact value as calculated in this example.

5.6 CONSTRUCTION

5.6.1 Sand Compaction Columns

Steel pipe casing has been commonly used to install sand compaction columns. Typically, the equipment includes a 4.5- to 6-ton hydraulic or electric vibrator attached on the top of a 0.4- to 0.6-m-diameter steel pipe. The casing should be slightly longer than the desired depth of the sand compaction column and have a foldable base, which can be closed during the penetration but be open during the filling of backfill. Following is a typical installation procedure (also shown in Figure 5.25):

1. Position the pipe casing to the desired location, turn on the vibrator, and check to ensure the equipment functions well.
2. Drive the casing into the ground to the desired depth (the casing is usually filled with sand to minimize construction time).
3. Fill the casing if not filled prior to being driven.
4. Withdraw the casing with a stroking motion (typically pulled up 1.8–3 m and then vibrate it back down 1.0–2.0 m) until reaching the top elevation of the column. During this process, air pressure (typically 250–500 kPa) is applied to the top of the casing to force the sand out and prevent soft soil from flowing into the casing.
5. Turn off the power and move the casing to the next location.

This installation method can also be used to install stone columns if stone instead of sand is used as the backfill material.

5.6.2 Stone Columns

In addition to the casing method, top and bottom feed methods are more commonly used to install stone columns. The top-feed method requires water for the installation; therefore, it is often referred to as a wet method. The top-feed method for stone column installation is similar to that for vibro-compaction; however, water is mainly used to excavate soil and maintain the stability of the hole in cohesive soil instead of inducing liquefaction of cohesionless soil.

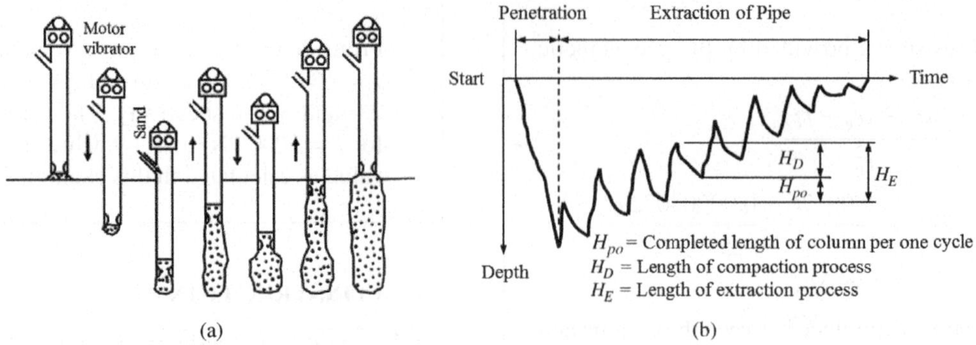

Figure 5.25 Construction of sequence for sand compaction column: (a) installation and (b) casing tip movement. (after Tanimoto, 1973), Courtesy of CSIRU Publishing).

The bottom-feed method usually uses air instead of water; therefore, it is often referred to as a dry method. Both methods use a vibratory probe. The installation procedure of stone columns using the top-feed wet method is as follows:

1. Position the probe to the desired location, turn on water and power, and check water pressure and voltage to ensure the equipment functions well.
2. Penetrate the probe into the ground at a rate of 0.3–0.5 m/min and observe the variation of electrical current (reduce the rate of penetration if the current is too high).
3. Reduce water pressure when the probe is close to the desired depth to minimize disturbance.
4. Continue the probe to the desired depth and then move the probe up and down twice without complete removal from the hole to flush out the hole.
5. Withdraw the probe gradually by adding stone from the top to fill the hole in 0.6–1.2 m lifts, maintain the flow of water from the bottom jet to prevent carving or collapsing of the hole, and then repenetrate the probe into the stone to densify and force the stone radially into the surrounding soil until the predetermined amperage level is reached for each lift or the top elevation of the column is reached.
6. Turn off the power and water and move the probe to the next location.

During the installation, spoil is generated. It is important to properly manage the spoil on site by directing the spoil to a slurry pond.

The installation procedure of stone columns using the bottom-feed dry method is similar to that using the top-feed wet method. The following are two exceptions:

1. Instead of water, air is used to help the penetration of the probe and maintain the stability of the hole.
2. The stone is supplied through a feeding tube on the side of the probe to the bottom of the hole.

In this installation, limited spoil is generated. Therefore, the construction site using the bottom-feed method is much cleaner than that using the top-feed method.

5.6.3 Rammed Aggregate Columns

Rammed aggregate columns are installed through a drilling, backfilling, and ramming process. The ramming equipment consists of a hydraulic excavator, a hydraulic break rammer, and a specially designed 45° beveled ram (Farrell and Taylor, 2004). The typical installation procedure is as follows:

1. Position the excavator and drill a hole to the desired depth. Steel casing is sometimes used if the hole cannot maintain its stability.
2. Fill crushed rock into the hole in lifts. It is important to form a proper "bottom bulb" prior to constructing the column. In weak soil, a certain quantity of rock may be used to stabilize the bottom bulb. Less amount of rock is used in the later lifts.
3. Apply a high-frequency ramming action of the beveled ram on each rock lift and densify the rock and force the rock into the surrounding soil until the top elevation of the column is reached.

5.6.4 Vibro-Concrete Columns

Vibro-concrete columns are installed by an electrically driven bottom-feed vibratory probe (Hussin, 1994), which is the same as that used for installing stone columns. The probe penetrates into the ground under its weight and vibrations. The following installation procedure is typically adopted:

1. Penetrate the vibrating probe to a desired depth by displacing or densify surrounding soils.
2. Pump concrete through a supply tube to the tip of the probe.
3. Repeatedly raise and lower the probe to form a bulb of concrete at the base.

4. Continuously pump concrete until the column is formed up to the ground surface.

5.6.5 Controlled Modulus (Stiffness) Columns

The installation of controlled modulus (stiffness) columns or auger displacement columns uses a hollow-stem displacement auger with a high torque pull-down rig. The displacement auger includes three parts from the bottom to the top: (1) constant volume flight to expel cuts upward during penetration, (2) middle displacement part of the same diameter as the auger and the hole to prevent spoils from escaping and push them laterally, and (3) flight in the opposite direction to that in the lower part, which prevents the spoils from collapsing down from the hole to the displacement part.

Below is the typical installation procedure for controlled modulus (stiffness) columns:

1. Drive the auger into the ground to a desired depth by applying high torque and downward thrust to displace in situ soil.
2. After the desired depth is reached, pump grout down the hole through the hollow stem at moderate pressure (typically lower than 10 bars) to overcome gravity and lateral soil resistance at the tip of the auger.
3. Extract the auger gradually by maintaining the rotation in the direction as that during the penetration to avoid loss of grout around the hole and the Kelly bar, until reaching the ground surface.

Grout is typically sand-mix mortar or pea-gravel concrete. The strength of the concrete, typically ranging from 7 to 21 MPa, is controlled by the amount of cement used in the mix. Depending on the required strength, the grout typically has slump ranging from 200 to 300 mm. The rate of pumping grout is controlled based on the volume. The diameter of the column is the same or larger than the auger diameter. Steel reinforcements can be placed into the grouted hole before grout is hardened to form a steel-reinforced column.

5.6.6 Geosynthetic-encased Granular Columns

Geosynthetic-encased granular columns can be installed in two methods: the excavation method and the vibro-displacement method (Raithel et al., 2005). In the excavation method, an open steel pipe is used while in the vibro-displacement method, a steel pipe with two adjustable base flaps is used. The base flaps are closed when the pipe is vibrated down but open when it is withdrawn under vibration. The following installation sequence is generally followed:

1. Drive a steel pipe in the ground to a desired depth.
2. Excavate soil inside the pipe (for the excavation method only).
3. Place a premanufactured geosynthetic tube inside the hole.
4. Place fill (mostly sand or gravel) into the geosynthetic encasement.
5. Withdraw the pipe under vibration to densify the soil inside the geosynthetic encasement.

The excavation method is more suitable when the pipe penetrates through a high resistance soil or displacement and vibration are of concern to adjacent structures. The vibro-displacement method is faster, more economic, and can densify or prestress surrounding soil without soil disposal.

5.7 QUALITY CONTROL AND ASSURANCE

5.7.1 Locations and Dimensions

Field installation should follow design drawings in terms of locations, diameters, and lengths of columns. Any deviations may affect the performance of the composite foundations. A specification should be developed to define tolerable values of deviations prior to installation. Column top elevations should be within 75 mm of the design elevation. Elias et al. (2004) specifies that the offset of column heads from the center design location should not be more than 100 mm and the lateral offset of the column axis should not exceed 1.5%. For any 50 consecutively installed columns, the average diameter over the total length should not be smaller than the design diameter. The diameter of the column at different location and depth should be calculated based on the amount of fill material placed and the in-place density. Excavation of a few randomly selected columns may be conducted to verify actual column diameters.

5.7.2 Fill Material

For granular columns, maximum particle sizes of fill materials are typically limited to 100 mm. Larger particles (not larger than 150 mm) are sometimes used at the bottom of the column to stabilize the base, especially when the installation is in very soft soil. Elias et al. (2004) suggested four possible fill materials for stone columns as shown in Table 5.5. The fill material should be clean, hard, durable stones free from organics, trash, or other deleterious materials.

For sand compaction columns, well-graded fine to medium sand with fines up to 15% but without any particles smaller than 0.05 mm is commonly used.

Representative samples of grout for controlled modulus (stiffness) columns or concrete mixes for vibro-concrete columns should be obtained at the project site for QA/QC testing. Tests are performed to evaluate workability of fresh concrete or grout and strength of hardened samples. Workability is commonly evaluated using slump tests for concrete and slump or flow cone tests for grout. Strength testing is performed in a laboratory after samples from the field are cured

Table 5.5 Gradations for Possible Fill Materials

Sieve Size (mm)	% Passing			
	Fill Type 1	Fill Type 2	Fill Type 3	Fill Type 4
100	100	—	—	—
88	90–100	—	—	—
75	—	90–100	—	—
63	25–100	—	100	—
50	—	40–90	65–100	100
38	0–60	—	—	—
25	—	—	20–100	2
19	0–10	0–10	10–55	—
13	0–5	0–5	0–5	—

Source: Elias et al. (2004).

for certain time periods. Typically, five samples are made from a single batch of 50 m^3 of concrete or grout from each day's work. Among five samples, two samples are tested at an age of 7 days, two at 28 days, and one sample is reserved for further testing if necessary. Strength testing of concrete utilizes conventional 150-mm-diameter and 300-mm-high cylinders. Typical required concrete unconfined compressive strengths at 7 and 28 days are at least 10 and 20 MPa, respectively. For sand–cement grout, small cylinders of 50 or 75 mm in diameter and two times the diameter in height, but mostly 50 × 50 × 50 mm^3 cubic samples are used. There is no consensus internationally on which method is preferred, but cubic samples are easier to prepare and transport.

For geosynthetic-encased columns, in addition to fill material, geosynthetic encasement should be evaluated for its ultimate tensile strength, tensile stiffness, and their associated reduction factors for creep, installation, and chemical and biological degradation. When geosynthetic-encased columns are used as drainage, the geosynthetic encasement should also be evaluated against filtration and drainage criteria.

5.7.3 Installation Parameters

Installation parameters are highly dependent on the method of installation, the type of equipment, and the type of fill material.

Sand Compaction Columns The key installation parameters for sand compaction columns include:

- Steel casing diameter
- Penetration rate
- Extraction rate
- Compressed air pressure
- Frequency and time of vibration
- Power of vibrator (amps)
- Volume of sand supplied and discharged
- Casing tip elevation

Stone Columns The key installation parameters for stone columns by the wet or dry method are:

- Power of vibrator (at least 60 kW and 150 kN eccentric force)
- Penetration rate
- Extraction rate (typically 300–500 mm) increments to allow fill placement
- Frequency and time of vibration
- Water pressure by wet method or air pressure by dry method
- Quantity and rate of water supply (for the wet method only)
- Required amperage for each fill increment
- Volume of stone supplied
- Top and bottom elevations

Rammed Aggregate Columns The key installation parameters for rammed aggregate columns are:

- Tamper energy
- Tamping duration
- Number of passes
- Volume of aggregate each lift
- Drill depth
- Top elevation of column

Vibro-concrete Columns The key installation parameters for vibro-concrete columns by the dry method are similar to those for stone columns by the dry method. The main differences are related to the supply of concrete and densification of concrete in the hole. Following are the unique installation parameters for vibro-concrete columns:

- Rate of concrete supply
- Pumping pressure
- Frequency and time of vibration

Controlled Modulus (Stiffness) Columns During the controlled modulus (stiffness) column installation, an onboard computerized quality control system in the rig has a real-time continuous monitoring of the installation parameters, including (Masse et al., 2004, 2012)

- Speed of rotation
- Rate of advancement and withdrawal of auger
- Torque, pull-down, down-pressure, drilling energy
- Depth of column
- Time of installation
- Pressure and volume of grout

A sensor is mounted to the mast drilling head at the top of the concrete line to monitor grout pressure. Any loss of grout pressure during installation may be an indication of soft or loose soil existing in the ground.

Geosynthetic-encased Columns The key installation parameters for geosynthetic-encased columns are:

- Steel casing diameter
- Penetration rate
- Extraction rate
- Frequency and time of vibration
- Power of vibrator
- Volume of fill supplied and discharged
- Casing tip elevation
- Diameter and length of geosynthetic encasement

5.7.4 Performance Evaluation

Time for Field Evaluation Performance evaluation of granular columns in cohesionless soil can be performed right after the completion of installation. However, performance evaluation of granular columns in cohesive soil should be done in 2 (for low plasticity soil) to 4 weeks (for high plasticity soil) after the completion of installation to allow the dissipation of excess pore water pressure and the recovery of soil strength.

Performance evaluation of concrete columns should be done at 4 weeks after the completion of installation to allow the curing of concrete or grout.

Soil Sampling and Penetration Tests Soil sampling and penetration tests (SPT and CPT) have been commonly used to evaluate surrounding soils before and after ground improvement. These tests can determine the degree of improvement or disturbance caused to the surrounding soils during the installation. Soil densification increases the density, strength, and modulus of the soil. However, disturbance reduces the density, strength, and modulus of the soil. In the surrounding soil, soil sampling, SPT, and CPT should be performed at the centroid point of a column pattern as shown in Figure 5.26. CPT may be used in the center of sand compaction columns but should not be used in the center of stone columns or rammed aggregate columns. SPT may be used in all types of granular columns. SPT and CPT should not be used for concrete columns. Cored samples are sometimes taken in concrete columns to evaluate the integrity, strength, and modulus of the columns.

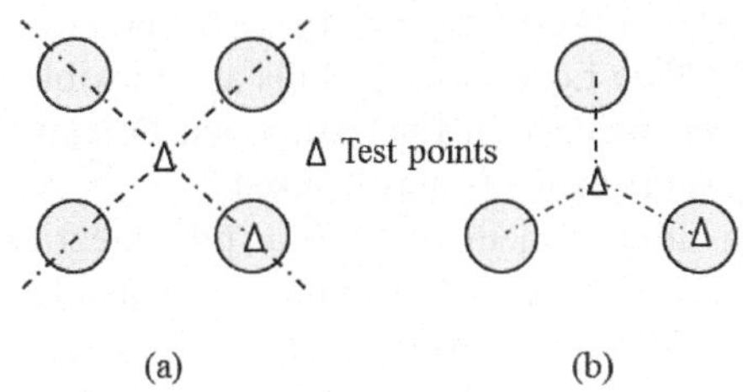

Figure 5.26 Locations for penetration tests: (a) square pattern and (b) triangular pattern.

Plate Loading Tests Plate loading tests are typically required to determine the bearing capacity and settlement of an individual column and the composite foundation as shown in Figure 5.27. Three types of loading tests may be performed: (1) individual column test, (2) individual column composite foundation test, and (3) multiple column composite foundation test. The individual column test should be performed with a rigid loading plate having the same size as the column. The loading plate for an individual column composite foundation test should have the same tributary area as a unit cell. For a multiple column composite foundation test, the area of the loading plate should be equal to the number of columns under the plate multiplied by the tributary area of the unit cell. For all the tests, the test column(s) should be surrounded by at least a row of columns to be representative.

For an individual column test, the ASTM D1143 Standard Test Methods for Deep Foundations under Static Axial Compressive Load may be followed. If a test column is evaluated, the load should be applied until the failure of the column. If a project column is evaluated, the applied load is limited to 1.5 times the required service load. For composite foundation

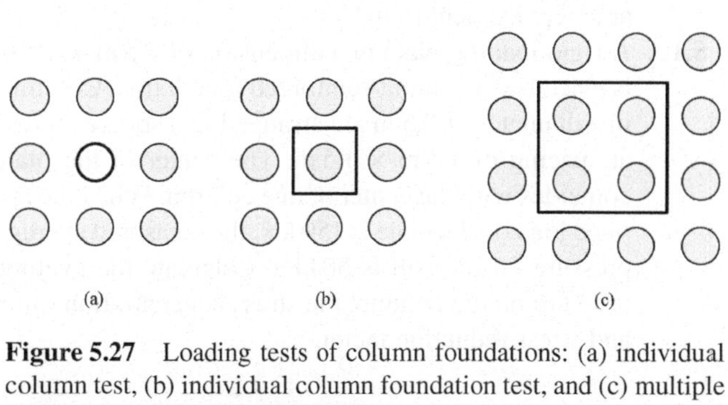

Figure 5.27 Loading tests of column foundations: (a) individual column test, (b) individual column foundation test, and (c) multiple column composite foundation test.

tests, the ASTM D1196 Standard Test Method for Nonrepetitive Static Plate Load Tests of Soils and Flexible Pavement Components, for Use in Evaluation and Design of Airport and Highway Pavements may be used.

Other in situ test methods, such as pile integrity tests and dynamic load tests (e.g., Statnamic) have also been used to evaluate concrete columns in practice.

PROBLEMS

5.1. List the key differences between the vibro-replacement and vibro-displacement methods to install stone columns.

5.2. A site has a soft silty clay with an undrained shear strength of 19 kPa. Is this site suitable for the vibro-replacement method?

5.3. Explain the main reasons why granular columns likely bulge near the ground surface.

5.4. What are the main functions of columns in ground installed by deep replacement methods?

5.5. What are the differences for a column-reinforced soft foundation under plane strain and plane stress conditions? What condition is for columns under an embankment?

5.6. What is the definition of unit cell? Why is it commonly used in practice?

5.7. Discuss the differences between 1D unit cell and unit cell with lateral column deformation.

5.8. Define stress concentration ratio in a column-reinforced foundation.

5.9. If soil and columns are both elastic and their constrained moduli are 3 and 30 MPa, what is the 1D stress concentration ratio? If columns can deform laterally, will the stress concentration ratio change? How?

5.10. Is stress concentration ratio constant? If not, how does the ratio change with load and time?

5.11. What is the composite foundation?

5.12. Columns with a diameter of 0.8 m are arranged in a square pattern at spacing of 2.0 m, what is the area replacement ratio? If the same columns are arranged in a triangular pattern at the same spacing, what is the area replacement ratio?

5.13. A rigid loading plate in a dimension of 1.5 m × 1.5 m is placed on a column-reinforced foundation. Columns in a diameter of 0.5 m are arranged in a square pattern at spacing of 1.5 m × 1.5 m. The center of the plate coincides with the center of one column. When the applied load on the plate is 150 kN, the measured average pressure on the soil is 50 kPa. Calculate the average pressure on the column, the stress concentration ratio, and stress reduction factor.

5.14. In Problem 5.13, if the elastic modulus of the soil is 4.2 MPa, calculate the equivalent modulus of the composite foundation.

5.15. Explain why columns under an embankment may be subjected to down-drag forces. Is the stress concentration ratio of the column-reinforced foundation under the embankment constant with depth? If not, explain why.

5.16. List possible failure modes of single columns in soil under vertical loading. Will they depend on strength or stiffness of the columns?

5.17. A contractor has a backfill material with the following gradation:

Particle Size (mm)	Percent of Passing
25	100
19	65
12.5	30
4.75	10

Evaluate the suitability rating of this backfill material.

5.18. A natural ground consists of an 8-m-thick loose cohesionless deposit over bedrock. The initial void ratio of this deposit is 0.845 and the required average void ratio after improvement is 0.650. Granular columns are selected for this improvement, and they are arranged in an equilateral triangular pattern at column spacing of 2.0 m. All the columns are seated on the bedrock. Assume the ground surface remains the same after ground improvement. Determine the required column diameter. Hint: use the method discussed in Section 3.7 of Chapter 3.

5.19. In Problem 5.18, if there is 40 mm heave after ground improvement, what is the required column diameter?

5.20. A site consists of a cohesionless soil, which has an initial SPT N value of 6. Sand columns with an area replacement ratio of 0.15 are proposed for ground improvement. Estimate the SPT N values midway between columns and in the center of columns after ground improvement.

5.21. Stone columns are installed in saturated clay with an undrained shear strength of 25 kPa. The friction angle of the backfill is 40°. Use the Brauns (1978) method to estimate the ultimate bearing capacity of a single stone column without any vertical stress on the surrounding soil.

5.22. A 3 m × 3 m square concrete footing is constructed on a granular column-reinforced soft clay. The undrained shear strength of the clay is 20 kPa, the saturated unit weight of the clay is 17.2 kN/m^3, and the groundwater table is at the ground surface. Under the footing on the ground surface, there are four 6-m-long granular columns with a diameter of 0.7 m, which are arranged at spacing of 2.0 m in a square pattern. The friction angle of the granular backfill is 38°. Estimate the ultimate bearing capacity of the granular column-reinforced composite foundation.

5.23. A 5.0-m-long vibro-concrete column (VCC) with an average diameter of 0.55 m is installed in a saturated soft clay with an undrained shear strength of 15 kPa, a saturated unit weight of 17.4 kN/m^3, and a groundwater table at the ground surface. The VCC is seated on a saturated stiff clay with an undrained strength of 75 kPa and a saturated unit weight of 18.5 kN/m^3. Estimate the ultimate load capacity of the VCC under an undrained condition.

5.24. In Problem 5.23, if the effective friction angles of the saturated soft clay and the stiff clay are 22° and 34°, estimate the ultimate load capacity of the VCC under a drained condition.

5.25. A 3 m × 3 m square concrete footing with an embedment depth of 1 m is constructed on a VCC-reinforced soft clay, which has a groundwater table at 1 m from the ground surface. The undrained shear strength of the clay is 15 kPa and the unit weights of the clay above and below the groundwater table are 17.5 and 18.5 kN/m^3, respectively. Under the footing, there are four 5-m-long VCC with an average diameter of 0.55 m, which are seated on a stiff clay and arranged at spacing of 2.0 m in a square pattern. The stiff clay has an undrained shear strength of 75 kPa and a saturated unit weight of 19.1 kN/m^3. Estimate the ultimate bearing capacity of the VCC-reinforced composite foundation under an undrained condition.

5.26. A rigid circular footing with a diameter of 5 m is embedded at a depth of 1.5 m in a 10-m-thick soft soil. This soft soil is underlain by a dense sand layer (assume no settlement in the sand layer). The groundwater table is located at a depth of 1.5 m. The drained elastic modulus of the soil is 4 MPa. The footing is subjected to a vertical column load of 2500 kN. Granular columns of 8.5 m long and 1.0 m in diameter are installed in a square pattern with a spacing of 2.0 m under and outside the footing as shown in the figure below. The elastic modulus of the columns is 40 MPa. Calculate the settlement of the footing on the ground before and after improvement using the stress reduction method.

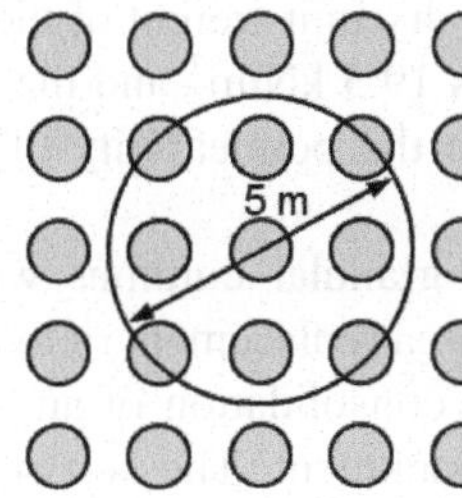

5.27. Granular columns are installed in saturated soft soil. The elastic moduli of the soil and the columns are 3.5 and 50 MPa. Use the Barksdale and Bachus (1983) design chart to estimate the stress concentration ratio.

5.28. A large steel-reinforced mat foundation, in a dimension of 30 m × 30 m, is to be constructed on 6.5-m-thick uniform loose sand, underlain by bedrock. The thickness of the mat foundation is 0.6 m and it has an embedment of 0.6 m. The total load applied on the mat is 30 MN. The average elastic modulus of the sand is 7.5 MPa and the elastic modulus of the mat is 30 GPa. Calculate the maximum settlement of the mat foundation on the natural ground. To reduce the settlement by 50%, stone columns are proposed to be installed to the top of the bedrock. The friction angle of the stone column is 40°. Use the improvement factor method to determine the required area replacement ratio of stone columns.

5.29. A large steel-reinforced mat foundation, in a dimension of 10 m × 10 m, is to be constructed on 15-m-thick uniform soft clay, underlain by dense sand. The mat foundation is 0.5 m thick and is placed on the ground surface. The applied load on top of the mat foundation is 10 MN. The drained elastic modulus of the soft clay increases linearly with 1–5 MPa from the base of the mat foundation to the bottom of the soft clay. The Poisson's ratio of the soft clay is 0.3. The elastic modulus of the dense sand is 40 MPa. Nine vibro-concrete columns (0.5 m in diameter) are installed in a square pattern at spacing of 3 m from the center under the mat foundation. The elastic modulus of the columns is 15 GPa, while the elastic modulus of the mat is 30 GPa. Calculate the settlement of the mat foundation using the piled raft method.

5.30. What are the key differences between granular columns and sand drains in accelerating consolidation of soft soil?

5.31. The backfill material for granular columns has 5% particles passing U.S. No. 200 sieve. The effective grain size corresponding to 10% passing the sieve size is

4.75 mm. The dry unit weight of the compacted granular column is 19.5 kN/m^3 and the specific gravity is 2.67. Estimate the permeability of the granular column.

5.32. Free-draining granular columns with a diameter of 0.8 m and an area replacement ratio of 0.15 are used to accelerate the consolidation of an 8-m-thick soft soil under an embankment. The soft soil has initial void ratio of 0.95, drained elastic modulus of 2 MPa, Poisson's ratio of 0.3, and vertical and radial permeability of 1.0×10^{-9} m/s and 2.0×10^{-9} m/s, respectively. This soft soil is underlain by a highly permeable sand layer. The groundwater is at the ground surface. The granular columns fully penetrate the soft soil into the sand layer. The steady-state stress concentration ratio is 2.5. Calculate the degrees of consolidation without and with granular columns in three months.

5.33. In Problem 5.32, the installation of granular columns induces a smear zone, which is 1.0 m in diameter and has permeability equal to 50% that of the natural soil. The permeability of the stone columns is 3.0×10^{-6} m/s. Calculate the degree of consolidation of the granular column-reinforced soft foundation considering smear and well resistance in three months.

5.34. A site consists of a soft soil with an undrained cohesion of 20 kPa and an effective friction angle of 23°. Granular columns with an area replacement ratio of 0.20 are installed to increase the stability of an embankment against deep-seated failure. The granular columns have zero cohesion and an effective friction angle of 38°. Calculate the equivalent undrained shear strength and the effective friction angle of the column-reinforced soft foundation.

5.35. A site consists of uniform medium sand with 5% fine content. The thickness of this sand layer is 12 m from the ground surface. The minimum and maximum void ratio values are 0.45 and 0.98, respectively. The groundwater table is at 1.5 m from the ground surface. The SPT N_{60} value at a depth of 6 m is 5. This site is located in a seismic-active area. The potential earthquake magnitude can be 7.0. The peak ground acceleration can reach as high as 0.3g. Granular columns are proposed to densify this possible liquefiable soil (please verify the liquefaction potential first) to prevent potential liquefaction. Assume the average diameter of stone columns can reach 0.75 m, the length of stone columns is 12 m, and the ground subsides 50 mm after the installation. The shear modulus ratio of column to sand is 10. Determine the spacing of granular columns in a square pattern without and with considering the shear stress reduction factor.

5.36. Explain how geosynthetic-encased granular columns are different from traditional granular columns.

5.37. Geosynthetic-encased granular columns are installed in normally consolidated soft soil with an undrained shear strength of 10 kPa. The diameter of the columns is 0.5 m. The total unit weight of the soil is 16 kN/m^3 and its effective friction angle is 20°. The friction angle of the granular column is 36°. The geosynthetic encasement has an allowable tensile strength at 5% strain is 20 kN/m. Calculate the ultimate bearing capacity of a single geosynthetic-encased column considering possible bulging failure.

5.38. A large oil tank of 75 m in diameter is planned to be constructed on a normally consolidated soft soil. The maximum tank load (including self-weight) is 600 MN. The site consists of an 8.5-m-thick soft clay underlain by a bedrock (no deformation is considered). This soft soil has the following properties: saturated unit weight, $\gamma_s = 15.0$ kN/m^3, drained elastic modulus, $E_s = 2.5$ MPa, Poisson's ratio, $\nu_s = 0.3$, and effective friction angle, $\phi'_s = 20°$. The tank is embedded at 1.0 m deep and the groundwater table is at 1.0 m below the ground surface. Geosynthetic-encased sand columns of 0.55 m in diameter are proposed in a square pattern with spacing of 1.2 m to minimize the settlement. The sand columns have a unit weight, γ_c = 19.0 kN/m^3, and an effective friction angle, $\phi'_c = 34°$. The tensile stiffness of the geosynthetic casing is J = 2250 kN/m. The initial diameter of the geosynthetic encasement is 0.55 m. Calculate the settlement of the tank without and with geosynthetic-encased sand columns.

REFERENCES

AASHTO (2006). *AASHTO LRFD Bridge Design Specifications*, 2006 interim revisions.

Aboshi, H., Ichimoto, E., Enoki, M., and Harada, K. (1979). "The compozer – a method to improve characteristics of soft clays by inclusion of large diameter sand columns." *Proceedings of the International Conference on Soil Reinforcement*, Vol. 1. E.N.P.C., Paris, 211–216.

Abusharar, S. and Han, J. (2011). "Two-dimensional deep-seated slope stability analysis of embankments over stone columns." *Eng. Geol.*, 120: 103–110.

Alexiew, D. (2013). Personal communications.

Alexiew, D. and Thomson, G. (2013). "Foundations on geotextile encased granular columns: Overview, experience, and perspectives." *Proceedings of International Symposium on Advances in Foundation Engineering* (ISAFE 2013), K.K. Phoon, T.S. Chua, H.B. Yang, and W.M. Cham (ed.). December 5–6, Singapore.

Alexiew, D., Raithel, M., Kuster, V., and Detert, O. (2012). "15 years of experience with geotextile encased granular columns as foundation system." ISSMGE - TC 211 International Symposium on Ground Improvement, IS-GI Brussels, May 31 to June 1.

Baez, J.I. (1995). *A Design Method for the Reduction of Soil Liquefaction by Using Vibro-Stone Columns*. Ph.D. dissertation, the University of Southern California.

Baez, J.I. and Martin, G.R. (1993). "Advances in the design of vibro systems for the improvement of liquefaction resistance." *Proc. 7th Ann. Symp. Ground Improv.*, 1–16.

Baez, J.I. and Martin, G.R. (1995). "Permeability and shear wave velocity of vibro-replacement stone columns." *Soil Improve. Earthquake Haz. Mitig., GSP*, 49: 66–81.

Barksdale, R.D. (1987). *State of the Art for Design and Construction of Sand Compaction Piles*. Technical Report REMR-GT-4, Georgia Institute of Technology, prepared for Department of the Army, US Army Corps of Engineers, Washington, DC.

Barksdale, R.D. and Bachus, R.C. (1983). *Design and Construction of Stone Columns*, FHWA/RD-83/026.

Barron, R.A. (1948). "Consolidation of fine-grained soils by drain wells." *Transaction* 113, 718–742.

Baumann, V. and Bauer, G.E.A. (1974). "The performance of foundation on various soils stabilized by vibrocompaction method." *Canadian Geotechnical Journal*, 11, 509–530.

Boulanger, R.W., Idriss, I.M., Stewart, D.P., Hashash, Y., and Schmidt, B. (1998). "Drainage capacity of stone columns or gravel drains for mitigating liquefaction." *Proceedings of Geotechnical Earthquake Engineering and Soil Dynamics III*, 678–690.

Brauns, J. (1978). "Initial bearing capacity of stone columns and sand piles." Symposium on Soil Reinforcing and Stabilizing Techniques, Sydney, Australia, 477–496.

Broms, B. (1979). "Problem and solution to construction on soft clay." *Proc. 6th Asian Regional Conference on Soil Mechanics and Foundation Engineering*, vol. 2, Singapore, 3–40.

Brown, R.E. (1977). "Vibroflotation compaction of cohesionless soils." *Journal of Geotechnical Engineering* 103(GT12).

Carrillo N. (1942). "Simple two- and three-dimensional cases in the theory of consolidation of soils." *Journal of Mathematics and Physics* 21, 1–5.

Castro, J. and Sagaseta, C. (2009). "Consolidation around stone columns—influence of column deformation." *International Journal for Numerical and Analytical Methods in Geomechanics* 33(7), 851–877.

Castro, J. and Sagaseta, C. (2011). "Consolidation and deformation around stone columns: Numerical evaluation of analytical solutions." *Compu. Geotech.*, 8: 354–362.

Castro J., Sagaseta C., Cañizal J., Da Costa A., Miranda M. (2013). "Foundations of embankments using encased stone columns." *Proceedings of the 18th International Conference on Soil Mechanics and Geotechnical Engineering*, Paris, 2445–2448.

Chen, R.P., Chen, Y.M., Han, J., and Xu, Z.Z. (2008). "A theoretical solution for pile-supported embankments on soft soil." *Canad. Geotech. J.,* 45(5): 611–623.

Chen, R.P., Xu, Z.Z, Chen, Y.M., Ling, D.S., and Zhu, B. (2010). "Field tests of pile-supported embankments on soft ground." *J. Geotech. Geoenviron. Eng.*, 136(6): 777–785.

Chu, J., Varaksin, S., Klotz, U., and Menge, P. (2009). "Construction processes—State of the art report." *Proceedings of the 17th International Conference on Soil Mechanics and Geotechnical Engineering*, Alexandria, Egypt, 5–9 Oct., 3006–3135.

Cooper, M.R. and Rose, A.N. (1999). "Stone column support for an embankment on deep alluvial soils." *Proceedings of the Institution of Civil Engineers, Geotechnical Engineering* 37(1), 15–25.

Elias, V., Welsh, J., Warren, J., Lukas, R., Collin, J.G., and Berg, R.R. (2004). Ground Improvement Methods. FHWA-NHI-04-001, 1022p.

Farrell, T., FitzPatrick, B., and Kenney, W. (2008). "Uplift testing of rammed aggregate pier systems." *Geotechnical Earthquake Engineering and Soil Dynamics IV*, ASCE Geo. Inst., May 18–22, 2008, Sacramento, CA.

Farrell, T. and Taylor, A. (2004). "Rammed aggregate pier design and construction in California—Performance, constructability, and economics." *SEAOC 2004 Convention Proceedings*, 147–154.

Filz, G., Sloan, J., McGuire, M.P., Collin, J., and Smith, M. (2012). "Column-supported embankments: settlement and load transfer." *Geotechnical Engineering State of the Art and Practice,* GSP No. 226: 54–77.

Ghionna, V., Jamiolkowski, M. (1981): "Colonne di ghiaia". X Ciclo di conferenze dedicate ai problemi di meccanica dei terreni e ingegneria delle fondazioni metodi di miglioramento dei terreni. Politecnico di Torino Ingegneria, atti dell'istituto di scienza delle costruzioni, 507.

Greenwood, D.A. (1970). "Mechanical improvement of soils below the ground surface." *Proc. Conf. on Ground Engineering, Inst. of Civil Engineers*, London.

Guo, W. D. and Qian, H. J. (1990), "New methods for calculating bearing capacity of granular pile foundations." *Ground Improvement* (1):1, 38–46 (in Chinese).

Han, J. (1992). "Stone column techniques—general report." *Proc. of 3rd Chinese Soil Improvement Conference*, Qengwangdao, China.

Han, J. (2010). "Consolidation settlement of stone column-reinforced foundations in soft soils." Invited paper, *New Technologies on Soft Soils, Proceedings of Symposium on New Techniques for Design and Construction on Soft Clays*, M. Almeida (ed.), Brazil, May 22–23, 167–179.

Han, J. (2012). "Recent advances in column technologies to improve soft soils." Invited keynote lecture, *Proceedings of International Conference on Ground Improvement and Ground Control*, Vol. 1, B. Indraratna, C. Rujikiatkamjorn, and J. Vinod (eds.), Wollongong, Australia, 30 October to 2 November, Research Publishing, Singapore 99–113

Han, J. and Gabr, M.A. (2002). "Numerical analysis of geosynthetic-reinforced and pile-supported earth platforms over soft soil." *J. Geotech. Geoenviron. Eng.,* 128(1): 44–53.

Han, J. and Ye, S. L. (1991). "Analysis of characteristics to composite grounds." *Proc. of First Young Asian Geotechnical Engineers Conference*, January 7–11, Bankok, Thailand, 197–206.

Han, J. and Ye, S. L. (2001). "A simplified method for computing consolidation rate of stone column reinforced foundations." *J. Geotech. Geoenviron. Eng.,* 127(7): 597–603.

Han, J. and Ye, S. L. (2002). "A theoretical solution for the rate of consolidation of a stone column reinforced foundation accounting for smear and well resistance." *Int. J. Geomech.,* 2(2): 135–151.

Huang, J., Han. J., and Oztoprak, S. (2009). "Coupled mechanical and hydraulic modeling of geosynthetic-reinforced column-supported embankments." *J. Geotech. Geoenviron. Eng.*, 135(8): 1011–1021.

Hughes, J.M.O. and Withers, N.J. (1974). "Reinforcing soft cohesive soil with stone columns." *Ground Eng.*, 7(3): 42–49.

Hussin, J.D. (1994). "Ground modification with vibro concrete columns." *ASCE South Florida Section Meeting*, September 30 to October 1, Ft. Lauderdale, FL.

Horikoshi, K. and Randolph, M.F. (1999). "Estimation of overall settlement of piled rafts." *Soils and Foundations* 39(2), 59–68.

Jiang, Y., Han, J., and Zheng, G. (2013). "Numerical analysis of consolidation of soft soils fully-penetrated by deep-mixed columns." *KSCE J. Civil Eng.*, 17(1): 96–105.

Liu, H.L. (2007a). "New piling techniques for soil improvement in China." *Proc 13 Asian Regional Conf on Soil Mech and Geot Eng.*, Kolkata.

Liu, H.L. (2007b). "Construction method of X shaped pile." Chinese Patent 2007100203063 (in Chinese).

Liu, H.L., Fei, K., Ma, X.H., and Gao, Y.F. (2003). "Large-diameter driven cast-in-place concrete thin-wall pipe pile (I) research and development." *J. Rock Soil Mech.,* 24: 164–198 (in Chinese).

Masse, F., Brockbank, B., and Pearlman, S. (2004). "CMC (controlled modulus columns): Potential application to Canadian soils with a new trend in ground improvement." *Proceedings of the 57th Canadian Geotechnical Conference*, Session 4F, 32–39.

Masse, F., Pearlman, S.L., Walker, M.P., and Swift, S.S. (2012). "Use of controlled modulus columns in retail and industrial development projects." 2012 DFI Annual Conference on Deep Foundations, Houston, TX.

McCabe, B. and Egan, D. (2010). "A review of the settlement of stone columns in compressible soils." *Ground Improvement and Geosynthetics, GSP No.* 207, 197–204.

McKelvey, D., Sivakumar, V., Bell, A., and Graham, J. (2004). "Modelling vibrated stone columns in soft clay." *Proc. Inst. Civil Eng. Geotech. Eng.*, 157(GE3): 137–149.

Mokashi, S.L., Paliwal, S.T., and Bapaye, D.R. (1976). "Use of stone columns for strengthening soft foundation clay." *Proc. Central Board of Irrigation and Power, 45th Annual Research Session* (June 8–11). Vol. III—Soils and Concrete, 61–68.

Mori, H. (1979). "Some case records of stability and settlement of embankment on the soft ground." *Proc. 6th Asian Regional Conference on Soil Mechanics and Foundation Engineering*, Singapore, vol. 2, 169–189.

Moulton, L.K. (1980). *Highway Subdrainage Design*, Report FHWA-TS-80-224, Offices of Research and Development, Federal Highway Administration, Washington DC.

Munfakah, G.A., Sarkar, S.K., and R.J. Castelli, R.J. (1983). "Performance of a test embankment founded on stone columns." *Proceedings of the International Conference on Advances in Pilings and Ground Treatment for Foundations*, London, 259–265.

Murugesan, S. and Rajagopal, K. (2010). "Studies on the behavior of single and group of geosynthetic encased stone columns." *Journal of Geotechnical and Geoenvironmental Engineering* 136(1), 129–139.

Najjar, S.S., Sadek, S., and Maakaroun, T. (2010). "Effect of sand columns on the undrained load response of soft clays." *J. Geotech. Geoenviron. Eng.*, 136(9): 1263–1277.

Poulos, H.G. (2001). "Piled raft foundations—Design and applications." *Geotechnique,* 50(2): 95–113.

Priebe, H. (1978). "Abschatzung des Scherwiderstandes eines durgh Stopfverdichtung verbesserten Baugrundes." *Die Bautechnik* 55 (9), 281–284.

Priebe, H.J. (1995). "The design of vibro replacement." *Ground Engineering*, December: 31–37.

Pulko, B. and Majes, B. (2005). "Simple and accurate prediction of settlements of stone column reinforced soil." *Proceedings of the 16th International Conference on Soil Mechanics and Geotechnical Engineering*, Osaka, Japan, 1401–1404.

Raithel, M. and Kempfert, H. G. (2000). "Calculation models for dam foundations with geosynthetic coated sand columns." *Proceedings of International Conference on Geotechnical and Geological Engineering*, GeoEng 2000, Melbourne, Australia, 347–352.

Raithel, M. and Kirchner, A. (2008). "Calculation techniques and dimensioning of encased columns—Design and state of the art." *Proceedings of the 4th Asian Regional Conference on Geosynthetics*, June 17–20, Shanghai, China.

Raithel, M., Kirchner, A., Schade, C., and Leusink, E. (2005). "Foundation of constructions on very soft soils with geotextile encased columns—State of the art." *Innovations in Grouting and Soil Improvement*, Geotechnical Special Publication, No. 136, V.R. Schaefer, D.A. Bruce, and M.J. Byle (eds.), ASCE, Reston, VA, 1–11.

Rayamajhi, D.R., Nguyen, T.V., Ashford, S.A., Boulanger, R.W., Lu, J., and Elgamal, A. (2014). "Numerical study of shear stress distribution for discrete columns in liquefiable soils." *Journal of Geotechnical and Geoenvironmental Engineering* 140(3), 04013034.

Seed, H.B. and Booker, J.R. (1977). "Stabilization of potentially liquefiable sand deposits using gravel drains." *J. Geotech. Eng. Divis.* 103, 757–768.

Simon, B. and Schlosser, F. (2006). "Soil reinforcement by vertical stiff inclusions in France." *Symposium on Rigid Inclusion in Difficult Subsoil Conditions*. Mexico, 1–22.

Tanimoto, K. (1973). *Introduction to the Sand Compaction Pile Method as Applied to Stabilization of Soft Foundation Grounds*. Division of Applied Geomechanics, GSIRO, Technical Report No. 16, Australia.

Vesic, A.S. (1975). "Bearing Capacity of Shallow Foundations," Chapter 3, In *Foundation Engineering Handbook*, Van Nostrand Reinhold, New York, N.Y., 121–147.

Vesic, A.S. (1977). *NCHRP Synthesis of Highway Practice 42: Design of Pile Foundations*, Transportation Research Board, National Research Council, Washington, D.C., 68.

Wang, G. (2009). "Consolidation of soft clay foundations reinforced by stone columns under time-dependent loadings." *Journal of Geotechnical and Geoenvironmental Engineering* 135(12), 1922–1931.

Xie, K.-H., Lu, M.-M., Hu, A.-F., and Chen, G.-H. (2009a). "A general theoretical solution for the consolidation of a composite foundation." *Comp. Geotechn.*, 36: 24–30.

Xie, K.-H., Lu, M.-M., and Liu, G.-B. (2009b). "Equal strain consolidation for stone columns reinforced foundation." *Int. J. Num. Anal. Methods Geomechan.*, 33(15): 1721–1735.

Ye, S.-L., Han, J., and Ye, G.-B. (1994). *Soil Improvement and Underpinning*, 2nd ed. The Chinese Building Industry Press Beijing, (in Chinese).

Zhang, Z., Han, J., and Ye, G. (2014). "Numerical investigation on factors for deep-seated slope stability of stone column-supported embankments over soft clay." *Eng. Geol.* 168C, 104–113.

Zheng, G., Liu, S.-Y., and Chen, R.-P. (2009). "State of advancement of column-type reinforcement element and its application in China." In *Advances in Ground Improvement*, Geotechnical Special Publication No. 188, J. Han, G. Zheng, V.R. Schaefer, and M.S. Huang (eds.), *Proceedings of the US-China Workshop on Ground Improvement Technologies*, March 14, Orlando, FL, 12–25.

Zheng, G., Liu, L., and Han, J. (2010). "Stability of embankment on soft subgrade reinforced by rigid inclusions (II) – group pile analysis". *Chinese J. Geotech. Eng.*, 32(12): 1811–1820 (in Chinese).

Zheng, G., Jiang, Y., and Han, J. (2011). "Performance of cement-fly ash-gravel pile-supported high-speed railway embankments over soft marine clay." *J. Marine Georesour. Geotechn.*, 29(2): 145–161.

CHAPTER 6

Drainage and Dewatering

6.1 INTRODUCTION

Water can come from the following sources (Figure 6.1):

1. Precipitation
2. Snow melting
3. Seepage from higher ground (e.g., from water inside dams, lakes, and rivers)
4. Rising groundwater
5. Vapor movements due to temperature and/or humidity difference
6. Capillary action

Climate change affects the sources from 1 to 5. Capillary action happens most in fine-grained materials. Movement of water in geomaterial can happen by gravity (fast in coarse-grained materials but slow in fine-grained materials), capillary action (mainly in fine-grained materials), vapor pressure, or a combination of any of the above actions.

Water is one of the most common causes for failures in geotechnical engineering. Detrimental effects of water include but are not limited to:

- Reduction of geomaterial strength and stiffness
- Increase of geomaterial weight
- Generation of excess pore water pressure
- Necessary condition for liquefaction
- Development of uplift force
- Development of seepage force
- Increase of lateral earth pressure
- Expansion of geomaterial
- Collapse of geomaterial
- Erosion of soil particles and rock
- Migration of fines
- Freeze–thaw
- Stripping of asphalt pavement
- Durability cracking of concrete

Removal of water from geomaterials, pavements, or earth structures can prevent or minimize the above problems. One easy solution is to properly manage surface water by designing and constructing slopes, ditches, and impermeable surface layers. Table 6.1 shows the typical requirements for transverse slopes of pavements, shoulders, and ditches. However, management of surface water is not always enough to solve all the problems. Therefore, it is also important to properly control groundwater.

Different methods are available to control groundwater (Figure 6.2):

- Drainage
- Dewatering
- Barrier

Both drainage and dewatering remove groundwater from geomaterials. Drainage removes water by gravity or water head through permeable layers as shown in Figure 6.2(a). Dewatering removes water by induced water head difference from pumps [Figure 6.2(b)] or electric gradients from electroosmosis. Barrier prevents water from entering geomaterials by installing a low permeable geomembrane [Figure 6.2(c)], clay liner, or cutoff wall. To prevent or minimize capillary rise of water, a large pore granular layer or geotextile layer may be placed to break the capillary action. Drainage promotes water movement while barrier prevents water movement. Therefore, barrier is opposite to drainage.

To ensure long-term drainage, a soil filter may be used between the drainage layer and the fine-grained geomaterial. The soil filter allows water to run through but does not allow soil particles to go through. This function is often referred to as filtration.

In addition to granular layers for drainage, soil filters for filtration, and clay layers for barrier, geosynthetics have been used for the same purposes. The types, functions, and properties of geosynthetics are discussed in Chapter 2. Nonwoven and woven geotextiles have apparent advantages for separation and filtration between fine-grained geomaterials and coarse-grained geomaterials. Geocomposites, formed by a geonet and a geotextile, can provide a drainage path to remove water entering geomaterials. Geomembranes are mostly used as barriers for moisture-sensitive geomaterials, such as expansive soils, to prevent water from entering the soil mass. Geomembranes have been placed vertically, horizontally, or in a combination around expansive soil to prevent water intrusion into the expansive soil. This application has been used in limited locations due to the nature of the problems. A technical book by Steinberg (1998) well documented this application; therefore, no further discussion will be provided in this book. Geotextiles and geocomposites utilize their large pores to break capillary rise in fine-grained geomaterials to prevent frost heave (Henry, 1996; Han and Jiang, 2013).

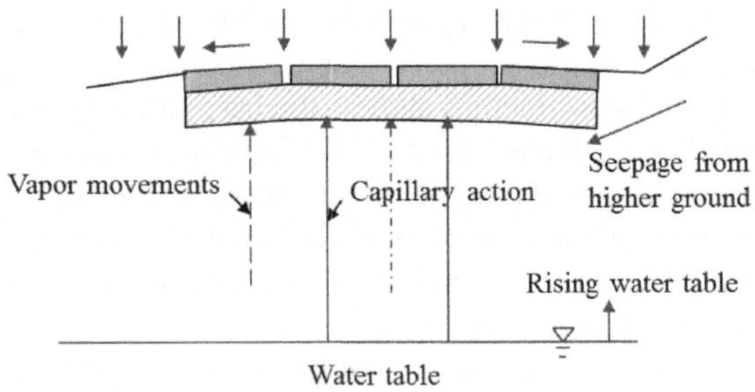

Figure 6.1 Sources of water (FHWA, 1992).

Table 6.1 Typical Transverse Slopes of Pavements, Shoulders, and Ditches

Feature	Pavement	Shoulder	Ditch
Slope (%)	1.5–4	4–6	25–50

Source: modified from Huang (2004).

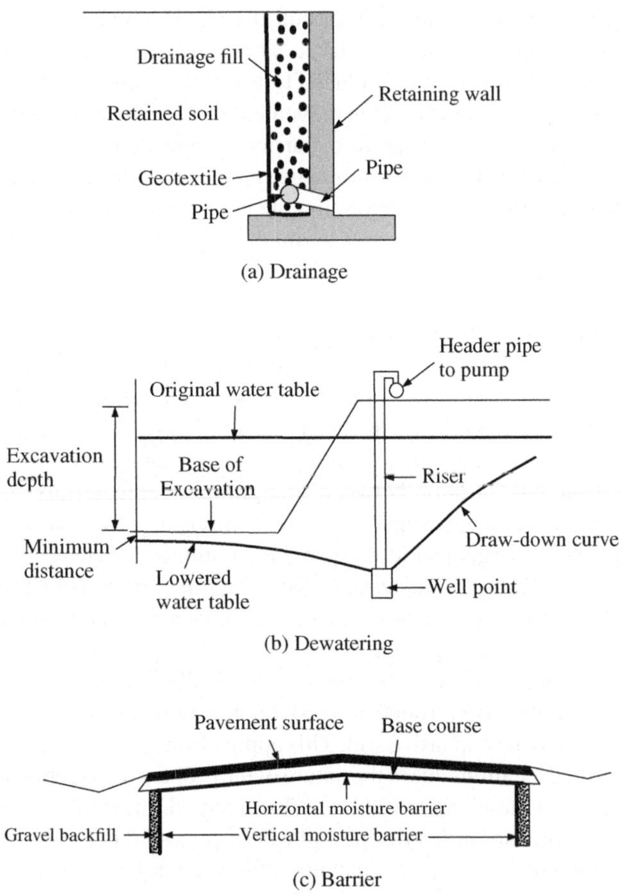

Figure 6.2 (a) Drainage, (b) dewatering, and (c) barrier.

In the following sections, principles of water flow in geomaterials, filtration, drainage, and dewatering will be discussed.

6.2 PRINCIPLES OF WATER FLOW IN GEOMATERIAL

6.2.1 Bernoulli's Equation

From fluid mechanics, the water head in geomaterial can be expressed as Bernoulli's equation as follows:

$$h_w = \frac{u}{\gamma_w} + \frac{v^2}{2g} + Z \tag{6.1}$$

where

h_w = total water head
u = pore water pressure
γ_w = unit weight of water
v = seepage velocity of water
g = acceleration of gravity
Z = water elevation head

The first, second, and third terms in Equation (6.1) are referred to as pressure head, velocity head, and elevation head, respectively.

Since the seepage velocity, v, in geomaterial is often relatively small, the velocity head can be neglected. Hence, Equation (6.1) can be simplified into

$$h_w = \frac{u}{\gamma_w} + Z \tag{6.2}$$

Figure 6.3 shows the water heads at different locations of a soil sample. The hydraulic gradient between positions 1 and 2 is defined as follows:

$$i = \frac{h_{w1} - h_{w2}}{L} = \frac{\Delta h_w}{L} \tag{6.3}$$

where

h_{w1} = water head at position 1
h_{w2} = water head at position 2
L = distance between positions 1 and 2

At datum 1, the total water heads at points A, B, and C are h_{w1}, h_{w3}, and h_{w2}, respectively. Assume the water heads, h_{w1} and h_{w2} are known, the water head, h_{w3}, can be solved below.

For an equal rate of water flow from positions 1 to 2 and positions 1 to 3,

$$Q_{w1} = Q_{w2} \tag{6.4a}$$

$$k\frac{h_{w1} - h_{w2}}{L}A = k\frac{h_{w1} - h_{w3}}{L/2}A \tag{6.4b}$$

$$h_{w3} = h_{w1} - \frac{1}{2}(h_{w1} - h_{w2}) = \frac{h_{w1} + h_{w2}}{3} \tag{6.4c}$$

where

Q_{w1} = quantity of water flow from positions 1 to 2
Q_{w2} = quantity of water flow from positions 1 to 3
k = soil permeability
A = cross-sectional area of soil sample

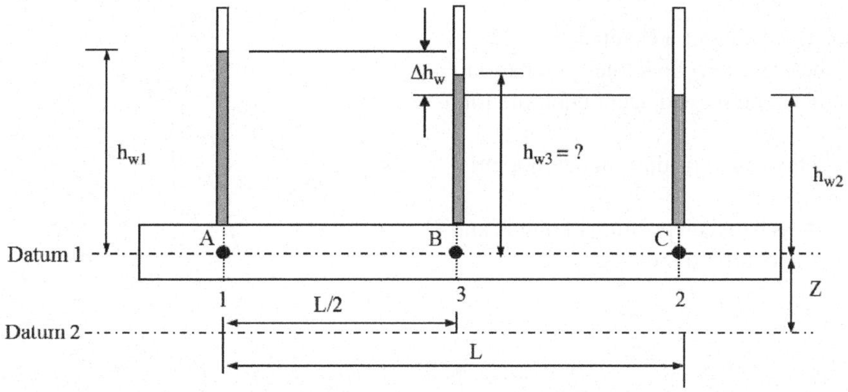

Figure 6.3 Water heads at different locations.

At datum 2, the total water heads at points A, B, and C are $h_{w1} + Z$, $h_{w3} + Z$, and $h_{w2} + Z$, respectively. Based on the same procedure, the water head, h_{w3}, can be solved below:

$$k\frac{(h_{w1}+Z)-(h_{w2}+Z)}{L}A = k\frac{(h_{w1}+Z)-(h_{w3}+Z)}{L/2}A \tag{6.5a}$$

$$h_{w3} = h_{w1} - \frac{1}{2}(h_{w1} - h_{w2}) = \frac{h_{w1}+h_{w2}}{2} \tag{6.5b}$$

The important conclusion from the above exercise is that the total head depends on the elevation, but the pressure head does not depend on the elevation.

6.2.2 Flow Net

Flow net is a simplified method to compute the rate of water flow. Figure 6.4 shows that the soil sample is divided by a flow net, which consists of flow lines and equipotential lines.

A flow net has the following characteristics:

- Flow lines or streamlines represent flow paths of particles of water.
- The area between two flow lines is called a flow channel.
- The rate of flow in a flow channel is constant.
- Flow cannot occur across flow lines.
- The velocity of flow is normal to the equipotential line.
- Flow lines and equipotential lines are orthogonal (perpendicular) to each other.
- The difference in head between two equipotential lines is called the potential drop or head loss.
- The boundary of an impervious layer is a flow line.

For computation convenience, a flow net element is often drawn in a square shape, in which a circle can fit in as shown in Figure 6.4. For each channel, the rate of flow is

$$Q_{wi} = A_i ki = (B_i \cdot 1)k\frac{\Delta h_{wi}}{L_i} = k\frac{\Delta h_w}{N_{hd}}\frac{B_i}{L_i} \tag{6.6}$$

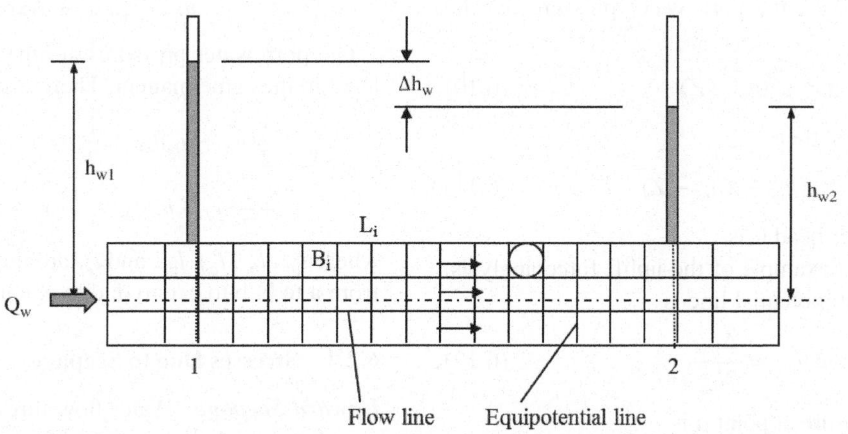

Figure 6.4 Flow net of soil sample.

where

A_i = cross-sectional area of a channel
B_i = distance between two flow lines at location i
L_i = distance between two adjacent equipotential lines,
Δh_{wi} = head loss between two adjacent equipotential lines
Δh_w = total head loss between positions 1 and 2, and
N_{hd} = number of equipotential drops

Since $B_i/L_i = 1$, then

$$Q_{wi} = k\frac{\Delta h_w}{N_{hd}} \quad (6.7)$$

The total rate of flow for all N_f channels is

$$Q_w = N_f Q_{wi} = k\,\Delta h_w \frac{N_f}{N_{hd}} \quad (6.8)$$

Depending on the nature of a problem, a flow net may be inclined or curved as shown in Figure 6.5. However, the characteristics of the flow net as listed above still remain. The "square" shape of an element in a curved flow net is imaginary, as shown in Figure 6.5(b), in which a circle is fitted in.

When a soil has different permeability values in the vertical and horizontal directions, the horizontal scale should be scaled by a factor of $\sqrt{k_v/k_h}$ and then the flow net is drawn. The rate of flow can be calculated as follows:

$$Q_w = \sqrt{k_v k_h}\frac{\Delta h_w N_f}{N_{hd}} \quad (6.9)$$

where

k_v = soil permeability in the vertical direction
k_h = soil permeability in the horizontal direction

6.2.3 Pore Water Pressure and Uplift Force

Based on Equation (6.2), the pore water pressure can be calculated as follows:

$$u = \gamma_w(h_w - Z) \quad (6.10)$$

If there is a head loss, then

$$u = \gamma_w(h_w - \Delta h_w - Z) \quad (6.11)$$

where Δh_w is the water head loss.

Figure 6.6 shows an example of the uplift force analysis. Head loss per each equipotential line is

$$\Delta h_{w1} = \frac{\Delta h_w}{N_{hd}} \quad (6.12)$$

The pore water pressure at point a is

$$u_a = \gamma_w(h_w - \Delta h_{w1} - Z) \quad (6.13a)$$

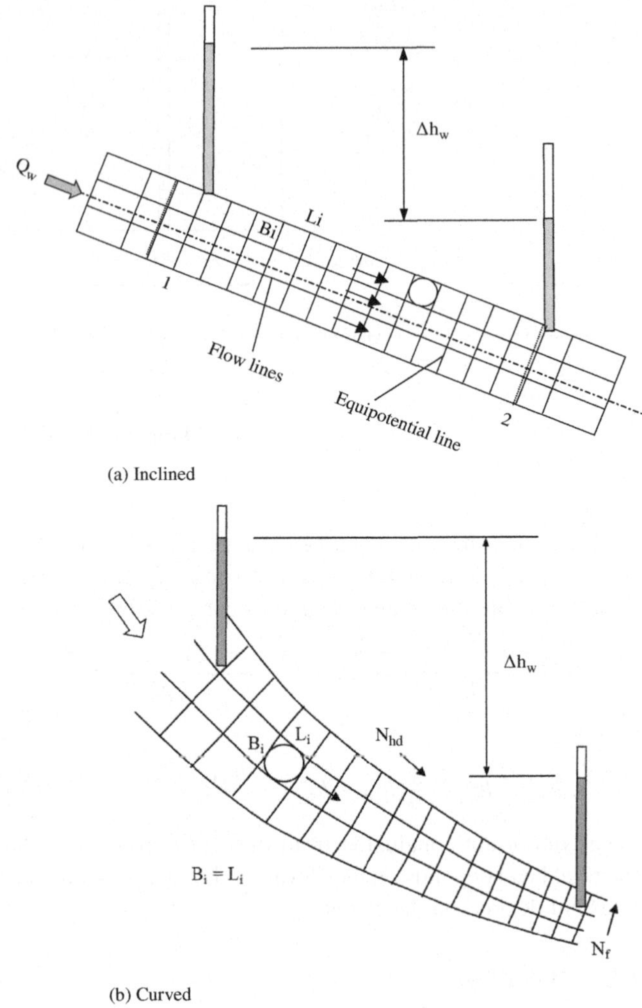

Figure 6.5 Flow nets in different inclinations and shapes: (a) inclined and (b) curved.

The pore water pressure at point b is

$$u_b = \gamma_w(h_w - 2\Delta h_{w1} - Z) \quad (6.13b)$$

The pore water pressures at other locations can be calculated in the same manner. Therefore, the total uplift force is

$$P_u = \frac{1}{2}(u_a + u_b)l_{ab} + \frac{1}{2}(u_b + u_c)l_{bc} + \frac{1}{2}(u_c + u_d)l_{cd} + \frac{1}{2}(u_d + u_e)l_{de} + \frac{1}{2}(u_e + u_f)l_{ef} \quad (6.14)$$

where l_{ab}, l_{bc}, l_{cd}, l_{de}, and l_{ef} are the lengths of the segments from a to b, b to c, c to d, d to e, and e to f.

6.2.4 Stresses Due to Seepage

Upward Seepage Water flow through a soil mass is called seepage, which induces stresses in the soil mass. For an upward seepage (i.e., the water head at point B is higher than

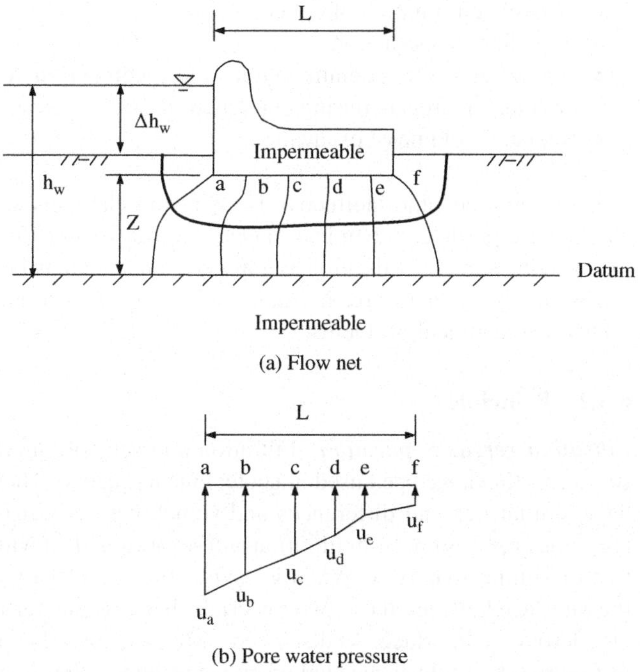

Figure 6.6 Example of uplift force analysis: (a) flow net and (b) pore water pressure.

that at point A) as shown in Figure 6.7, the total vertical stress, the pore water pressure, and the effective vertical stress at point C are

$$\sigma_C = h_w\gamma_w + z\gamma_{sat} \tag{6.15}$$

$$u_C = \left(h_w + z + \frac{z}{L}\Delta h_w\right)\gamma_w \tag{6.16}$$

$$\sigma'_C = \sigma_C - u_C$$

$$= (\gamma_{sat} - \gamma_w)z - \frac{\Delta h_w}{L}z\gamma_w = \gamma' z - iz\gamma_w \tag{6.17}$$

If $\sigma'_C = \gamma' z - iz\gamma_w = 0$, the soil loses stability. This situation is generally referred to as boiling, or a quick condition. It is also called static liquefaction, which is similar to the liquefaction due to dynamic loads, such as earthquake. This situation developing under dams or excavations is often referred to as piping. The hydraulic gradient at the zero effective stress is

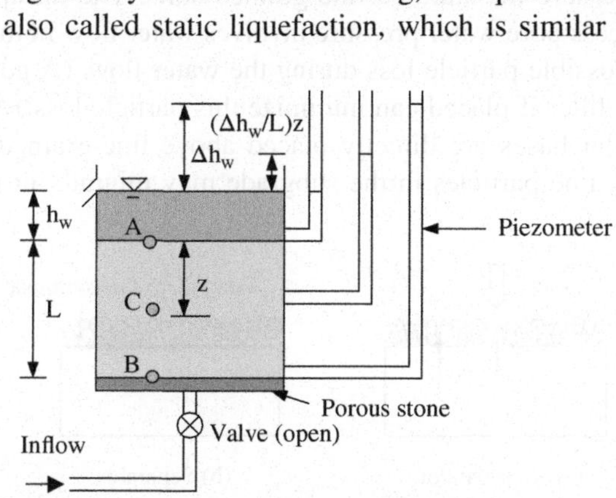

Figure 6.7 Upward seepage through a soil sample.

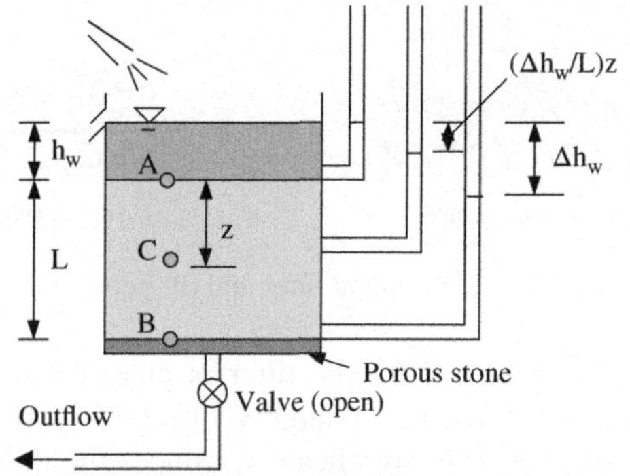

Figure 6.8 Downward seepage through a soil sample.

$$i_{cr} = \frac{\gamma'}{\gamma_w} \tag{6.18}$$

where i_{cr} is the critical hydraulic gradient. For most soils, $i_{cr} =$ 0.9–1.0. To avoid this instability, the hydraulic gradient in the soil should be less than i_{cr}.

Downward Seepage The downward seepage is opposite to the upward seepage (i.e., the water head at point A is higher than that at point B as shown in Figure 6.8). The total vertical stress, the pore water pressure, and the effective vertical stress at point C are

$$\sigma_C = h_w\gamma_w + z\gamma_{sat} \tag{6.19}$$

$$u_C = (h_w + z - iz)\gamma_w \tag{6.20}$$

$$\sigma'_C = \sigma_C - u_C = z\gamma' + iz\gamma_w \tag{6.21}$$

Therefore, the downward seepage increases the effective stress in the soil.

Seepage Force The seepage force due to upward or downward seepage can be expressed as follows:

$$P_s = (iz\gamma_w)A = (i\gamma_w)V \tag{6.22}$$

where V is the volume of soil.

6.3 FILTRATION

6.3.1 Introduction

Basic Concept Filtration is a function of a filter to allow for adequate liquid flow but limit soil particle loss across an interface plane. The liquid can be water, leachate from landfill, or another fluid. Since water is the most common fluid encountered in geotechnical engineering, it is discussed in this chapter. Filters can be a granular material or geotextile

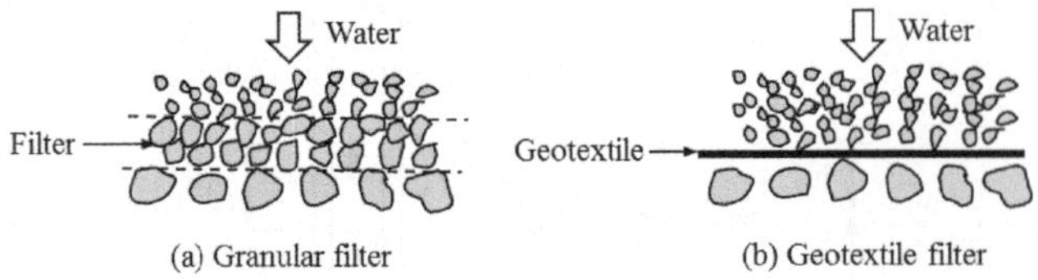

Figure 6.9 (a) Granular filter and (b) geotextile filter.

as shown in Figure 6.9. The filter is placed between two dissimilar geomaterials of largely different particle sizes. The use of geotextile significantly reduces the thickness of the filter as compared with the granular filter. Opening size of the filter is the key to the filtration performance. Two basic criteria for a filter are permeability and retention. Three types of retention are used for different applications (Giroud, 2010):

- Total retention: complete prevention of particle loss with low permeability (e.g., at a downstream side of a dam to prevent piping)
- Optimum retention: a balance between adequate liquid flow and limited particle loss (e.g., for most filter applications)
- Partial retention: slow and progressive loss of particles (e.g., on a waterfront bank subjected to turbulent and multidirectional flow)

Most filtration criteria have been developed based on the optimum retention for filters subjected to nonturbulent, one-directional flow. Giroud (2010) suggested two additional criteria for geotextile filters: porosity and thickness criteria.

Suitability Granular and geotextile filters are mostly used between fine and coarse cohesionless geomaterials. They are sometimes used between clays and cohesionless soils. However, most filter criteria have been developed for cohesionless geomaterials.

Applications Granular and geotextile filters are mostly used in or for dewatering trenches, wall facing and internal drainage, roadway base and edge drainage, landfill drainage, dam toe drainage or drainage behind clay core, liquid reservoir with clay liner, silt fence, and rip-rap erosion protection.

Advantages and Limitations Granular and geotextile filters are easy and inexpensive to install. They are effective to allow for adequate liquid flow and retain particles if properly designed and installed. For most filtration and drainage applications, geotextile filters have shown cost advantages over granular filters because of (Holtz et al., 2008):

- Use of lower-cost drainage aggregates
- Possible use of smaller-sized drains
- Possible elimination of collector pipes
- Expedient construction
- Lower risk of contamination and segregation of drainage aggregate during construction
- Reduced volume of excavation

Filters may become ineffective if their void channels are clogged by particles, chemical, and biological matter during the service life. Filtration is only beneficial when liquid is involved; therefore, it cannot solve any non-liquid-related geotechnical problems.

6.3.2 Principles

Filtration versus Separation Filtration and separation are two terms that have been used in geotechnical practice. They have similarities and differences and sometimes coexist in the same application. Both filtration and separation deal with two dissimilar materials. When geosynthetics are placed at the interface between these two materials they prevent particles across the interface. As discussed earlier and also shown in Figure 6.9, a filter is needed when water flows from finer soils to coarser soils. A separator is needed when the interface is subjected to loading, especially repeated loading, for example, wheel loading. The separator prevents larger particles from moving down and smaller particles from moving up, as shown in Figure 6.10.

When a geosynthetic is used as a filter, it is required to have proper opening sizes. When a geosynthetic is used as separator, it is required to have sufficient tensile strengths. However, when a geosynthetic is used as a filter as well as a separator, it is required to have proper opening sizes and sufficient tensile strengths.

When granular particles are directly in contact with the subgrade, they would penetrate or sink into the subgrade, especially for a soft subgrade, under wheel loading. On the other hand, repeated loading often induces excess pore water pressure in saturated fine-grained soils. The dissipation of excess pore water pressure involves water flow in the soil and possible particle loss during the water flow. (A geosynthetic filter if placed can minimize this particle loss.) When granular bases are directly placed above fine-grained subgrade, fine particles in the subgrade may migrate into base

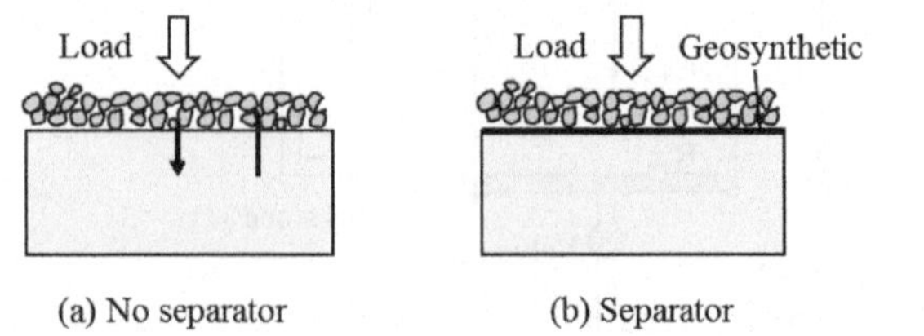

Figure 6.10 Soil separation: (a) no separator and (b) separator.

courses and continue moving up to the roadway surface under wheel loading. The loss of fine particles in the subgrade would weaken its support for the pavement structure and create space for base courses to settle in so that depressions form on the pavement surface. Both repeated loading and water flow would cause the intermixing of base and subgrade, which deteriorate the quality of the base courses and eventually affect the performance of the roadway. Under such a condition, both separation and filtration are needed to mitigate this intermixing problem. In summary, filtration is needed when there is water flow between two dissimilar materials, while separation is needed when there is a load applied on two dissimilar materials.

Nonwoven or woven geotextile can be used as a separator as well as a filter. However, geomembrane can only be used as a separator but not a filter because no or minimum water can flow through the geomembrane. Geogrid can also be used as a separator for coarse aggregate but not a filter because geogrid aperture sizes are too large to retain fine particles. The requirements for a geotextile used as a separator or a filter are different. To ensure the geotextile functions as a separator, it must first survive construction by having sufficient tensile strengths. The serviceability of the geotextile during the construction depends on the subgrade condition, quality of base material, and construction equipment. To ensure the geotextile functions as a filter, it must meet the permeability, retention, porosity, and thickness criteria as suggested by Giroud (2010). To ensure the geotextile functions as a separator as well as a filter, it should meet all the above requirements.

Permeability To allow for adequate water flow, a filter must have sufficient permeability. Giroud (2010) suggested that the permeability criterion include pore water pressure and flow rate requirements. Figure 6.11(a) shows that the inclusion of a filter may change the distribution of excess pore water pressure. The higher pore water pressure in the soil may cause disturbance to the soil, which is not desired. Therefore, the filter should be designed to avoid the buildup of the

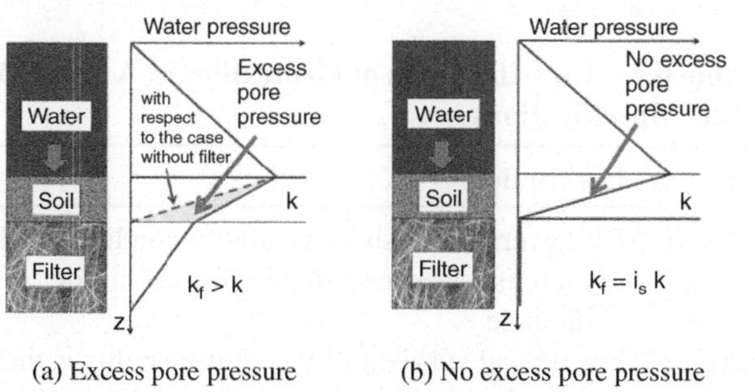

Figure 6.11 Pore water pressure distribution with a filter: (a) excess and (b) no excess pore pressure (Giroud, 2010).

Table 6.2 Typical Hydraulic Gradient in Soil Next to a Filter

Application	Hydraulic Gradient, i_s
Dewatering trench	≤1.0
Vertical wall drainage	1.5
Road edge design	≤1.0
Inland waterway protection	≤1.0
Landfill drainage layer	1.5
Dam toe drain	2.0
Drain behind dam clay core	3 to >10
Liquid reservoir with clay liner	> 10

Source: Giroud (2010).

excess pore water pressure in the soil. To achieve this objective, the permeability of the filter should meet the following condition (Giroud, 2010):

$$k_f \geq i_s k \tag{6.23}$$

where

k_f = permeability of the filter
k = permeability of the soil
i_s = hydraulic gradient in the soil

Typical hydraulic gradient in soil next to a filter is provided in Table 6.2.

Retention To limit loss of particles, a granular filter or geotextile filter should have a limited maximum opening size. During water flow, particles can move and arrange themselves around the openings of the filter. When particles are too small as compared with the opening size of the filter, they will pass through the openings. When particles are sufficiently large, they can build up a stable structure to bridge over the openings as shown in Figure 6.12. Even though a geotextile is shown in the figure, the same principle applies

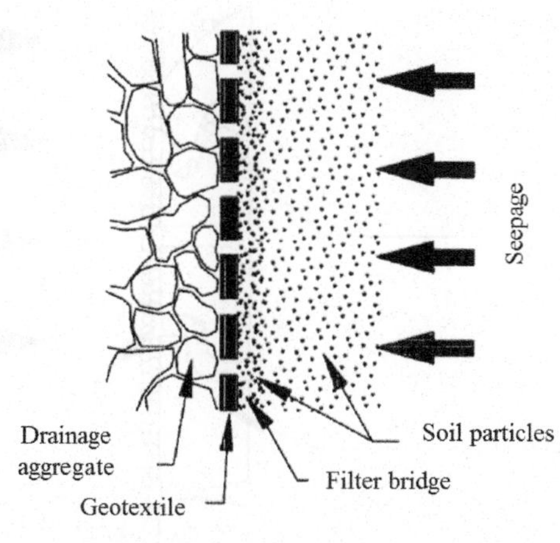

Figure 6.12 Filter bridge formation (Holtz et al., 2008).

to a granular filter. The particles behind the filter bridge are retained by the stable structure. This retention is often referred to as internal stability of the soil. Therefore, the relative dimension of the maximum particle size to the opening size of the filter and the internal stability of the soil are the retention criterion for the filter design. Giroud (2010) suggested that the retention criterion depends on soil density and coefficient of uniformity.

Porosity and Thickness As mentioned earlier, the optimum retention allows the loss of limited soil particles. In other words, a small amount of particles is allowed to pass the opening. However, it is not desired if the particles are trapped in the channel to block water flow, as shown in Figure 6.13. The blockage of water flow by particles is often referred to as clogging. Similar to clogging, the openings may also be blocked by larger particles as shown in Figure 6.13, which is referred to as blinding. Soil clogging and blinding should be limited in the filter design to ensure the long-term effectiveness of a filter. In other words, there should be a sufficient number of openings in the filter. The number of openings is directly related to the porosity of the filter. Giroud (2010) pointed out that the opening size of the nonwoven geotextile also depends on its thickness if there are fewer than 25 constrictions (i.e., the size of passage) in the filtration paths. The constriction is illustrated in Figure 6.14, in which the size of a constriction is the diameter of the largest sphere passing through the constriction. The nonwoven geotextile filter becomes reliable after the number of constrictions is larger than 25 in any given filtration path within the thickness. For granular filters, both porosity and thickness criteria are almost always met; therefore, there is no need to check these criteria for granular filters.

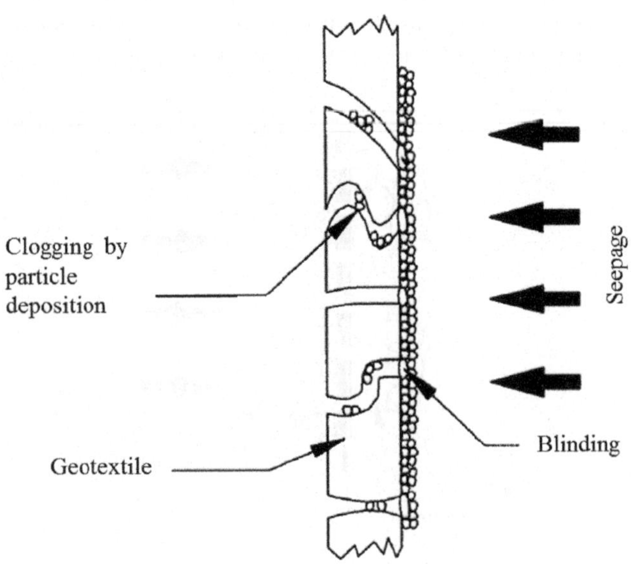

Figure 6.13 Clogging and blinding (Bell and Hicks, 1980).

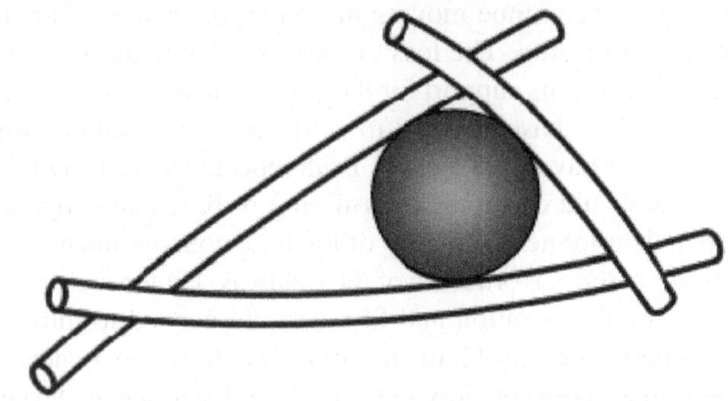

Figure 6.14 Constriction (i.e., passage between fibers) (Giroud, 2010).

In addition to soil clogging, there is possible chemical and biological clogging. Chemical clogging develops when chemical residues accumulate inside the channels or on the opening. Bacteria and/or roots can build up inside the channels as well. These clogging effects on the reduction of flow rate should be considered in the filter design.

6.3.3 Design Considerations

Survivability Requirements To provide proper separation and/or filtration between two dissimilar materials when they are in contact, geotextiles should have sufficient strength to survive during construction. AASHTO M288 (AASHTO, 2006) defines three classes of geotextile products based on their survivability during the construction as presented in Table 6.3.

The evaluation of the geotextile survivability considers its separation function. For the separation of subgrade soil with soaked CBR > 3 or undrained shear strength > 90 kPa, which is a typical roadway construction condition, Class 2 geotextile should be used. When the subgrade soil has a CBR of 1–3 or undrained shear strength of 30–90 kPa, reinforcement of the subgrade is needed. As a result, Class 1 geotextile or geogrid should be used. The design of geosynthetics for subgrade improvement will be discussed in Chapter 10. The required properties of Class 2 geotextiles are provided in Table 6.4.

Table 6.3 Classifications of Geotextiles in AASHTO M288 Specifications

Class	Description
Class 1	For severe or harsh survivability conditions where there is a greater potential for geotextile damage
Class 2	For typical survivability conditions; this is the default classification to be used in the absence of site specific information
Class 3	For mild survivability conditions

Table 6.4 AASHTO M288 Class 2 Geotextile Strength Requirements

Requirements	Test Methods	Units	$\varepsilon_g < 50\%$	$\varepsilon_g \geq 50\%^a$
Grab strength	ASTM D4632	N	1100	700
Sewn seam strength	ASTM D4632	N	990	630
Tear strength	ASTM D4533	N	400	250
Puncture strength	ASTM D6241	N	2200	1375

[a] ε_g = strain of geotextile.

The required properties of other classes can be found in the AASHTO M288 specifications (AASHTO, 2006). For rough estimates, the required strengths of Class 1 geotextiles are approximately 1.3 to 1.4 times those of Class 2 geotextiles, while the required strengths of Class 2 geotextiles are 1.3 to 1.4 times those of Class 3 geotextiles. These strength values ensure the geotextile would not be damaged during the construction, thus affecting its effectiveness as a separator or filter. Geotextiles typically have large tolerable tensile strain before rupture. Strains less and more than 50% have been used to define the strength requirements of woven and nonwoven geotextiles, respectively. Details of these test methods can be found in the ASTM standards.

Lawson (1992) suggested the following equation to ensure sufficient tear strength for a geotextile to survive during construction in terms of the maximum stone size (not more than 100 mm):

$$T_t \geq 750(D_{max})^{0.45}$$

where T_t is the trapezoidal tear strength of the geotextile measured according to ASTM D4533 (N) and D_{max} is the maximum stone diameter of the granular material placed on the geotextile (m).

Lawson (1992) also proposed a relationship between the required geotextile mass per unit area and the stone size and the stone drop height as follows:

$$m_g \geq 1200\sqrt{H_g D_{max}}$$

where

m_g = geotextile mass per unit area (g/m^2) (ASTM D5261)

H_g = height from which the granular material is dropped onto the geotextile (m)

D_{max} = maximum stone diameter of the granular material dropped onto the geotextile (m)

Terzaghi's Filter Criteria The early filter criteria were proposed by Terzaghi, which include permeability and retention criteria. The permeability criterion requires

$$D_{15f} \geq CD_{15} \tag{6.24}$$

where

D_{15f} = filter particle size corresponding to 15% passing

D_{15} = soil particle size corresponding to 15% passing

C = constant (4 or 5 but 5 is commonly used)

The retention criterion requires

$$D_{15f} \leq CD_{85} \tag{6.25}$$

where D_{85} is the soil particle size corresponding to 85% passing. Terzaghi's filter criteria are only suitable for granular filters.

FHWA Filter Criteria Figure 6.15 shows typical filter materials used for roadway constructions. The permeability of the filter material ranges from 0.6 m/day to 30 m/day. The maximum particle size for the filter material is 19 mm.

FHWA (Holtz et al., 2008) developed a comprehensive filter design procedure for geotextile filters as shown in Figure 6.16. This design procedure considers steady-state flow, dynamic flow, and unstable soils. For most applications, the steady-state flow is considered. The dynamic flow condition exists in coastal applications, which involves wave actions. This FHWA filter design procedure includes four criteria: (1) retention, (2) permeability, (3) clogging resistance, and (4) survivability and endurance.

Under the steady-state flow condition, the soil to be protected can be coarse grained or fine grained. For a coarse-grained soil, the coefficient of uniformity is used for the retention criterion, while for a fine-grained soil only D_{85} is needed. It is a general requirement that the permeability of the geotextile filter should be 1.0–10.0 times that of the soil, depending on the applications and site conditions. However, the required permittivity of the geotextile depends on the percent of fines in the soil. Permittivity as discussed in Chapter 2, is a parameter to evaluate the rate of water flow through the geotextile perpendicular to the plane of the geotextile. This parameter is important for the permeability criterion of filtration. O_{95} stands for an apparent opening size (AOS) of a geotextile. This parameter is important for the geotextile to retain fine particles but still allow water flow through the geotextile. To minimize the risk of clogging, it is recommended that the geotextile with the largest available opening size be used to satisfy the retention criterion. For woven geotextiles, the percent opening area (POA) should be greater than 4%. For nonwoven geotextiles, the porosity of the geotextile should be greater than 50% under the actual stress condition. The porosity of the geotextile is defined as follows:

$$n_g = 1 - \frac{m_g}{\rho_f t_g} \tag{6.26}$$

where

m_g = geotextile mass per unit area

t_g = geotextile thickness

ρ_f = density of filaments

In addition to retention, permeability, and clogging resistance criteria, FHWA (Holtz et al., 2008) suggests survivability and endurance criteria. The survivability requirements for

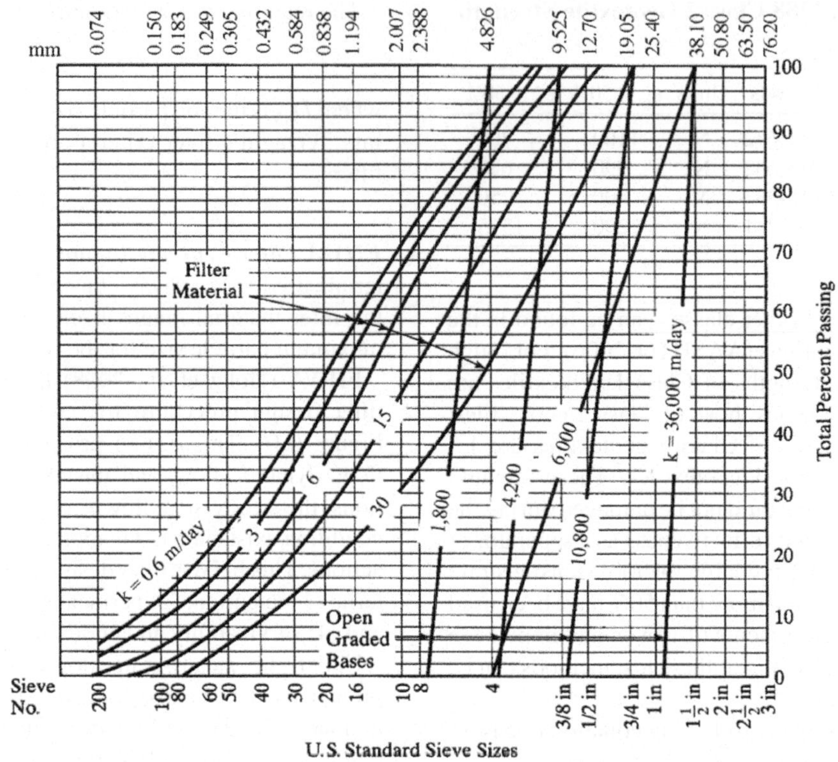

Figure 6.15 Typical granular filter and drainage materials (Cedergren et al., 1972).

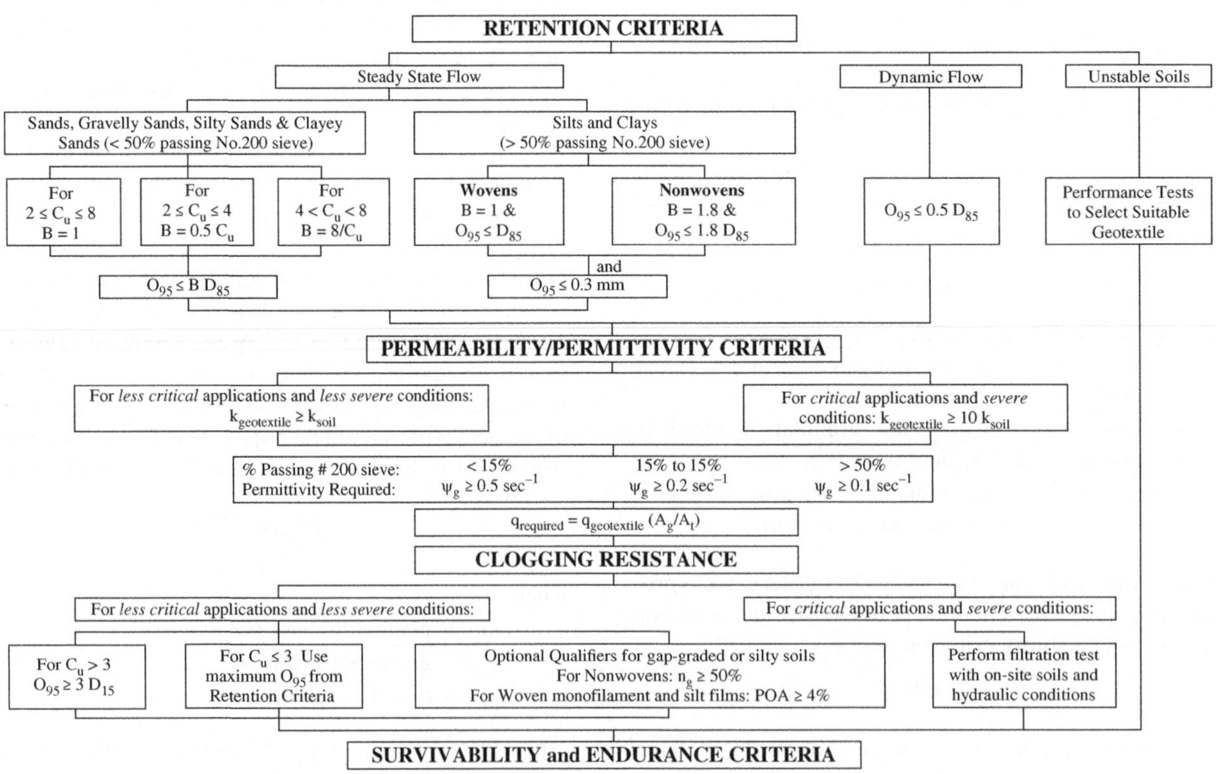

Figure 6.16 Flowchart summary of the FHWA filter design procedure (Holtz et al., 2008).

Table 6.5 AASHTO M288 Subsurface Filtration Geotextile Requirements

	Test Methods	Units	Requirements (% in situ soil passing No. 200)		
			<15	15–50	>50
Geotextile Class				Class 2	
Permittivity	ASTM D4491	s^{-1}	0.5	0.2	0.1
AOS	ASTM D4751	mm	0.43	0.25	0.22
UV stability	ASTM D4355	%	% retained strength after 500 h of exposure		

geosynthetics to survive construction are discussed earlier in this section. Ultraviolet (UV) stability is a parameter to ensure the long-term stability of the geotextile exposed to UV. Typically, the tensile strengths of geotextile products would decrease after exposed to UV.

The minimum requirements for a geotextile used as a separator are provided as follows:

- Class 2 geotextile
- Permittivity = 0.02 s^{-1}
- O_{95} (AOS) = 0.60 mm
- 50% retained strength after 500 h of exposure to UV

The requirements for a geotextile used as a filter are provided in Table 6.5.

Giroud's Filter Criteria Giroud (2010) developed the new criteria for geotextile and granular filters based on the mechanisms of filtration, theoretical analysis, and experimental data. The new criteria are applicable to cohessionless soils. In this development, he considered the following four criteria:

- Permeability criterion
- Retention criterion
- Porosity criterion
- Thickness criterion

The permeability criterion was developed based on two requirements: pore water pressure and flow rate requirements. From these requirements, the following two equations are obtained:

For geotextile filters:

$$k_f \geq \max(i_s k,\ k) \tag{6.27}$$

For granular filters:

$$k_f \geq \max(i_s k,\ 25k) \tag{6.28}$$

The first value in the parenthesis was established based on the pore water pressure requirement while the second value was based on the flow rate requirement. Giroud (2010) pointed out that the Terzaghi permeability criterion was based on the flow rate requirement. Pore water pressure requirement should also be considered for the permeability criterion.

Giroud (2010) suggested that the retention criteria depend on the soil density and the linear coefficient of uniformity. The linear coefficient of uniformity is defined below based on the linear line of the central portion of the actual particle size distribution curve as shown in Figure 6.17:

$$C'_u = \frac{D'_{60}}{D'_{10}} = \sqrt{\frac{D'_{100}}{D'_0}} \tag{6.29}$$

where the preceding parameters are defined in Figure 6.17.

Giroud (2010) used the concept of opening size for a filter. For a granular filter, the opening size can be approximated as follows:

$$O_f \approx \frac{D_{15f}}{5} \tag{6.30}$$

For a geotextile filter, the opening size is

$$O_f = O_{95} \tag{6.31}$$

Based on mathematical analyses on the effects of the soil density and the soil linear coefficient of uniformity, Giroud (2010) obtained the following equations.

For $C'_u \leq 3$:

For loose soils (relative density $D_r \leq 35\%$),

$$O_f \leq (C'_u)^{0.3} D'_{85} \tag{6.32}$$

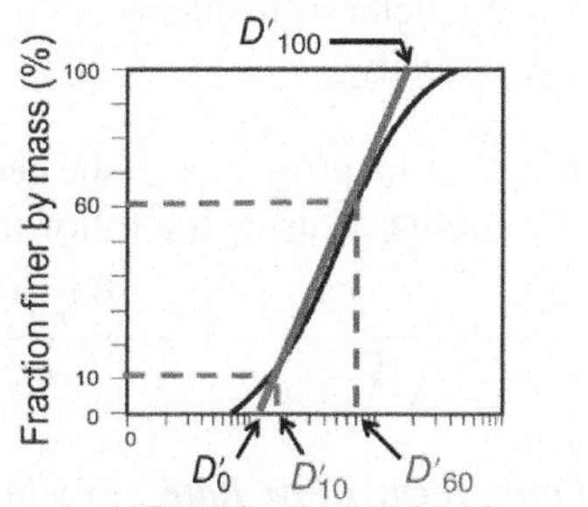

Figure 6.17 Linear coefficient of uniformity (Giroud, 2010).

For medium dense soils (35% $< D_r <$ 65%),

$$O_f \leq 1.5(C'_u)^{0.3} D'_{85} \tag{6.33}$$

For dense soils ($D_r \geq$ 65%),

$$O_f \leq 2(C'_u)^{0.3} D'_{85} \tag{6.34}$$

For $C'_u > 3$:

For loose soils ($D_r \leq 35\%$),

$$O_f \leq \frac{9D'_{85}}{(C'_u)^{1.7}} \tag{6.35}$$

For medium dense soils (35% $< D_r <$ 65%),

$$O_f \leq \frac{13.5D'_{85}}{(C'_u)^{1.7}} \tag{6.36}$$

For dense soils ($D_r \geq$ 65%),

$$O_f \leq \frac{18D'_{85}}{(C'_u)^{1.7}} \tag{6.37}$$

To minimize disturbance of water flow from the soil to the filter and ensure sufficient channels for water flow even if some channels are clogged by small particles, Giroud (2010) suggested geotextile filters should have sufficient porosity as follows:

For woven geotextiles the percent opening area (POA) should be

$$\text{POA} \geq 10\% \tag{6.38}$$

For nonwoven geotextiles, the porosity should be

$$n_g \geq 0.55 \tag{6.39}$$

To ensure the reliability of a nonwoven geotextile filter, the filter should have sufficient thickness, which has the following constriction number requirement:

$$N_{\text{const}} = \frac{m_g}{\rho_f d_{fm}\sqrt{1-n_g}} \geq 25 \tag{6.40}$$

where m_g = geotextile mass per unit area
ρ_f = density of filaments
d_{fm} = diameter of filaments
n_g = porosity

Based on the material properties of the geotextile, its opening size can be estimated using the following equation:

$$\frac{O_f}{d_{fm}} \approx \frac{1}{\sqrt{1-n_g}} - 1 + \frac{10n_g(\rho_f d_{fm})}{m_g} \tag{6.41}$$

Allowable Long-Term Flow Rate In addition to the filter criteria, and the survivability and endurance criteria, a design should evaluate the allowable long-term flow rate for a geotextile used for filtration. The formula for the allowable long-term flow rate and the reduction factors considering creep reduction of void space, adjacent material's intrusion, soil clogging and blinding, chemical clogging, and biological clogging are presented in Chapter 2.

6.3.4 Design Parameters and Procedure

Design Parameters Depending on the criteria used for the filter design, different design parameters may be needed. The Giroud filter criteria are most comprehensive and require most design parameters. For granular filters, the following design parameters are needed for the Giroud filter criteria:

- Soil gradation
- Soil density
- Soil permeability
- Hydraulic gradient in soil
- Granular filter gradation
- Granular filter permeability.

For geotextile filters, the following parameters are needed for the Giroud filter criteria:

- Soil gradation
- Soil density
- Soil permeability
- Hydraulic gradient in soil
- Geotextile permeability
- Geotextile apparent opening size, O_{95}
- Geotextile percent opening area, POA
- Geotextile mass per unit area
- Density of filaments
- Diameter of filaments

In addition to the parameters related to the filter design, the strength parameters and UV stability of geotextiles are needed for the survivability and endurance considerations.

Design Procedure The design of granular filters and geotextile filters can be accomplished in one of the two options: (1) select a filter candidate and then verify its parameters against the criteria (if not satisfactory, reselect the candidate and repeat the procedure) and (2) according to the soil parameters, determine the required filter parameters based on the filter criteria and then select an appropriate filter product. The design procedure for granular filters in terms of the second option using the Giroud criteria includes the following steps:

1. Based on the permeability of the soil, determine the minimum permeability of the filter to meet pore water pressure and flow rate requirements (a typical hydraulic gradient value in Table 6.2 may be used).
2. Based on the soil gradation and density, determine the maximum opening size of the granular filter.

The design procedure for geotextile filters using the Giroud method includes one or two more steps in addition to the first two steps for the granular filter:

1. Determine the minimum percent opening area for woven geotextiles or the minimum porosity for nonwoven geotextiles.
2. Determine the required nonwoven geotextile thickness based on the number of constrictions greater than 25.

A similar procedure can be adopted using the FHWA filter criteria. However, the FHWA filter criteria do not have the thickness requirement.

6.3.5 Design Example

Problem

A nonwoven geotextile is needed as a separator as well as a filter between base course and subgrade soil for a local roadway construction. The subgrade soil is a CL clay ($D_{85} = 0.2$ mm) and its CBR value is 3.5%. This separator should also meet the filter criterion (FHWA filter criteria) to prevent fine particles migrating into the base course. Please provide the specifications of this geotextile.

Solution

Based on the AASHTO M288 geotextile specifications, class 2 geotextile should be used for separation of subgrade when CBR > 3 (CBR = 3.5 for this case). Since the product used in this project is nonwoven geotextile, the elongation at failure is greater than 50%. The strength property requirements based on AASHTO M288 are grab strength, $T_{grab} = 700$ N, sewn seam strength, $T_{seam} = 630$ N, tear strength, $T_{tear} = 250$ N puncture strength, and $T_{puncture} = 1375$ N.

Based on the AASHTO M288 separation geotextile property requirements, other requirements for a geotextile separator are permittivity, $\Psi_g = 0.02\ s^{-1}$, maximum apparent opening size, $O_{95} = 0.60$ mm, and ultraviolet stability (retained strength) = 50% after 500 h of exposure.

A special requirement for the geoextile in this project is to meet the filter criterion of the FHWA filter criteria. Since the subgrade is a CL soil, the percent of fine is greater than 50%. For a nonwoven geotextile, $O_{95} \leq 1.8D_{85}$, that is, $O_{95} \leq 0.36$ mm. In addition, $O_{95} \leq 0.30$ mm is required by the FHWA design flowchart. Among all the opening size requirements for the retention criteria, $O_{95} \leq 0.30$ mm controls.

Considering a local road construction, which is a less critical application, $k_g > k$ and $\psi_g \geq 0.1\ s^{-1}$. This required permittivity is higher than that for separation; therefore, $\psi_g \geq 0.1\ s^{-1}$ should be used.

For clogging resistance, the required porosity for the nonwoven geotextile is $n_g \geq 50\%$.

6.3.6 Construction

Granular and geotextile filters are often used together with drainage materials for drainage purposes. Therefore, the construction using these filters will be discussed in the next section on drainage.

6.3.7 Quality Control and Assurance

For the same reason as that for the construction using granular and geotextile filters, their quality control and assurance will also be discussed in the next section.

6.4 DRAINAGE

6.4.1 Introduction

Basic Concept It is well known that water is detrimental to performance and life of roadways and earth structures in geotechnical engineering. Removal of water would prevent damage and failure, maintain performance, and prolong the life of these structures. Highly-permeable granular, geosynthetic, or combined layers (often called drains) can be installed to remove water from roadways, earth structures, and ground as shown in Figure 6.18. Drains are often protected by filters or have filters as part of the drains to maintain their long-term effectiveness. In Figure 6.18(a), the subbase below the granular base serves as a filter. Geotextile filters have been increasingly used around granular drains due to their cost and volume advantages as shown in Figure 6.18(b). Further volume reduction can be achieved by using geocomposites to replace granular drains as shown in Figure 6.18(c). Geocomposites typically consist of a drainage core (such as a geonet) covered by a geotextile sheet as a filter on one side or two sides as shown in Figure 6.19. The design of filters has been discussed in the previous section.

There are four types of drains commonly used in practice:

- Drainage layer or blanket
- Longitudinal drain
- Transverse drain
- Vertical drain

In granular drains, drainable pipes are often included to collect and discharge water. Vertical drains are often used in

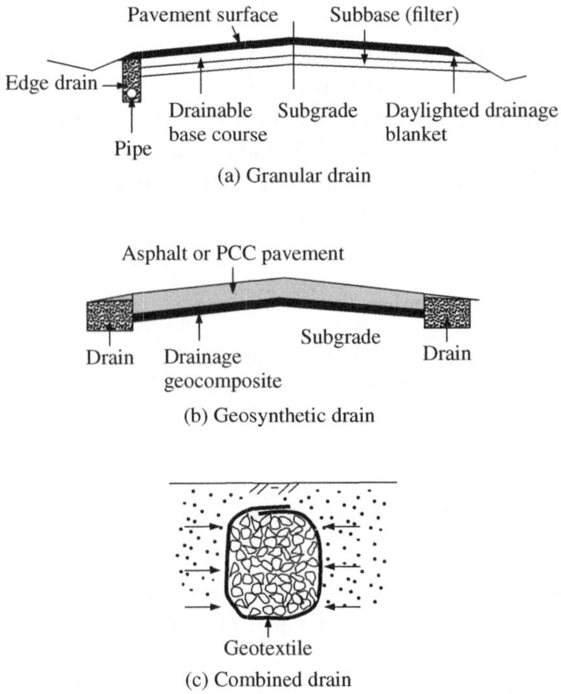

Figure 6.18 Different types of drains: (a) granular, (b) geosynthetic, and (c) combined drains.

Figure 6.19 Geocomposite drain.

the ground to dissipate excess pore water pressure induced by dynamic loading (such as deep dynamic compaction) or preloading. The use of vertical drains for deep dynamic compaction is discussed in Chapter 3, while the use of vertical drains for preloading is discussed in Chapter 7.

The benefits of good drainage have been recognized in many geotechnical designs. For example, free-draining fill is required for mechanically stabilized earth walls, which will be discussed in Chapter 10. The 1993 *AASHTO Guide for Design of Pavement Structures* includes a drainage coefficient to calculate pavement structural capacity. Drainable aggregate layers have been successfully used in rigid and flexible pavements. AASHTO (1993) defined the quality of pavement drainage from excellent (i.e., water is removed within 2 h) to very poor (water does not drain) based on the time for water to be removed within the pavement system. However, to provide enough drainage capacity necessary for good to excellent drainage, a thick drainable granular base (typically 100–300 mm thick) is needed. FHWA (1992) recommended that a drainable base should have a minimum permeability of approximately 300 m/day, which can remove water from the roadway system within a few hours. The open-graded bases in Figure 6.15 are good drainage materials.

Suitability Drains are mostly used above, below, behind, or inside low permeable materials. For example, drainage bases are often placed above fine-grained subgrade and below asphalt or concrete pavements to remove water from the pavement system. Drainage layers are installed behind the facing of retaining walls to reduce lateral earth pressure. Vertical drains are installed in soft clays to accelerate consolidation.

Applications Drains have been used for many different applications. Figure 6.20 shows a few examples of these applications. Trench or edge drains [Figure 6.20(a)] have been commonly used to remove surface runoff water from roadway systems. Pavement drains [Figure 6.20(b)] are to remove water entering into pavement foundations through pavement cracks and joints. Retaining wall drains [Figure 6.20(c)] are to reduce lateral earth pressures on wall facing. Chimney drains [Figure 6.20(d)] are often installed in dams or levees to lower water heads between upper stream and lower stream. Interceptors or toe drains [Figure 6.20(e)] are used to minimize seepage forces on slopes. Vertical drains [Figure 6.20(f)] are to accelerate the dissipation of excess pore water pressure in soft soil.

Advantages and Limitations Drains are easy and often inexpensive to install. They are effective to minimize or prevent many water-related geotechnical problems. They can be used to:

- Reduce soil moisture content, increase soil strength, and increase soil modulus.
- Accelerate consolidation and settlement.
- Reduce lateral earth pressure.
- Reduce uplift force.
- Reduce the rate of pavement deterioration.
- Minimize or eliminate soil expansion.
- Minimize or eliminate soil collapse.

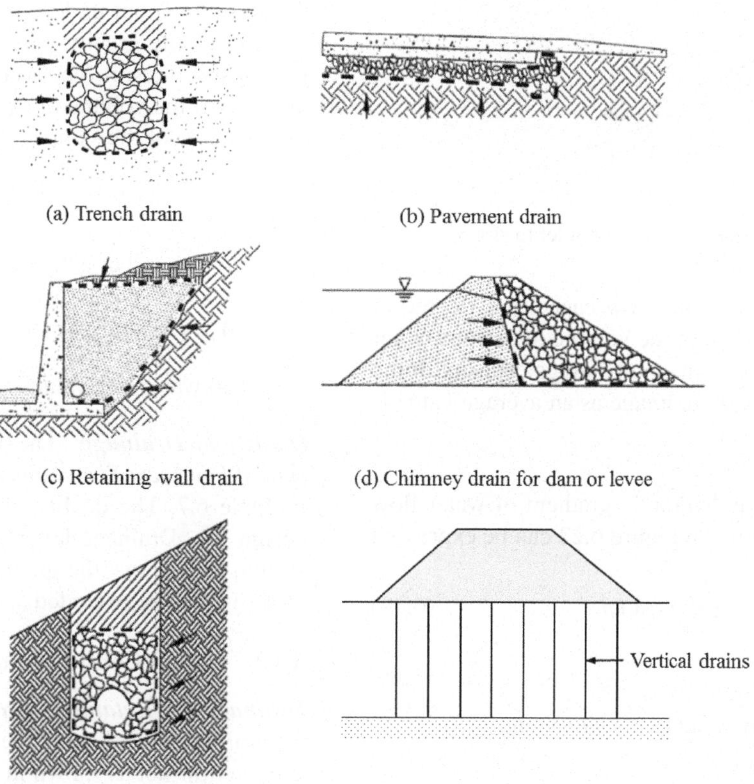

Figure 6.20 Examples of drainage applications (modified from Holtz et al., 2008).

- Minimize or eliminate frost heave and freeze–thaw problems.
- Enhance stability of dams, levees, and slopes.

Vertical drains in soft soils are effective only if there is surcharge or excess pore water pressure induced by dynamic loading. Drains cannot be used to solve non-water-related geotechnical problems. Drains may become ineffective if they are clogged.

6.4.2 Principles

Steady-State Flow versus Unsteady-State Flow Steady-state flow is referred to as the condition when all the flow lines are parallel and they do not interfere with each other. This condition exists when there is a uniform drainage layer and the water head difference between the upper and bottom flow lines is the same at all the locations. Unsteady-state flow is referred to as the condition when the flow lines are not parallel and they may interfere with each other. This condition exists when the drainage layer is not uniform and water head difference is different at different locations. The unsteady-state flow occurs when the drainage layer does not have a full storage capacity as shown in Figure 6.21. Under such a condition, there is a drawdown line, which has a maximum height of water. The design of a drainage layer should ensure that the thickness of the drainage layer should be larger than the maximum height of water.

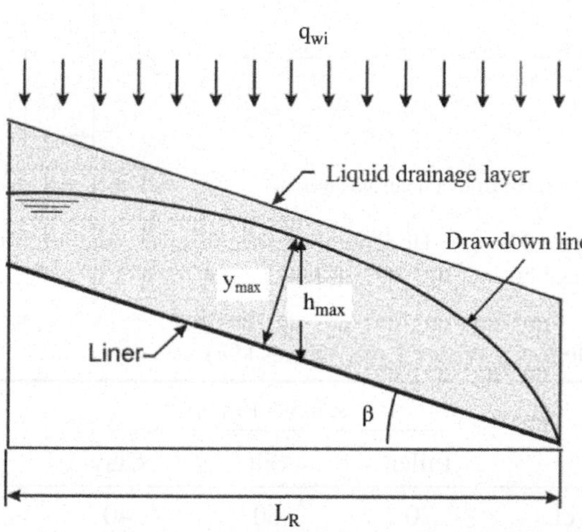

Figure 6.21 Unsteady-state flow and maximum height of water.

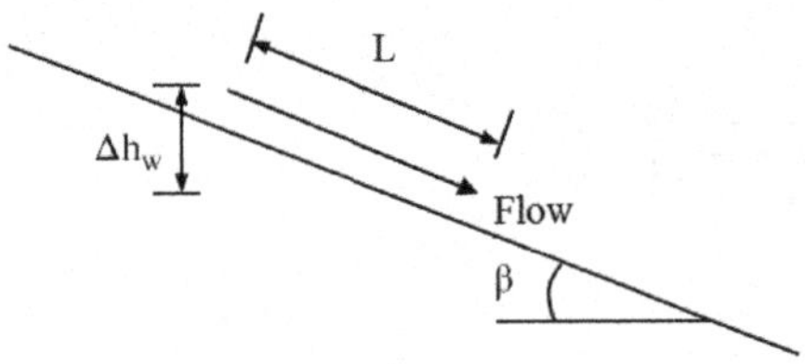

Figure 6.22 Hydraulic gradient of flow parallel to slope.

To describe the unsteady-state flow capacity, a degree of drainage is defined as a ratio of the volume of drained water since the infiltration stops to the total storage capacity of the drainage layer. Fifty percent drainage as an average value is commonly used in practice.

Hydraulic Gradient The hydraulic gradient of water flow parallel to a slope as shown in Figure 6.22 can be expressed as follows:

$$i = \frac{\Delta h_w}{L} = \sin \beta \quad (6.42)$$

where

Δh_w = head difference
L = distance parallel to the slope, and
β = slope angle

When the slope angle is equal to 90° (i.e., a vertical flow as shown in Figure 6.23), the hydraulic gradient $i = 1$.

Effective Porosity Different soils have different water holding capacities under gravity. Not necessarily all the water in the soil drains out under gravity. Effective porosity is used to describe the gravity-induced drainage potential of water from a specific soil. The soil effective porosity is defined as the ratio of the volume of drained water under gravity to the total volume of the sample, that is,

$$n_e = n_p W_L \quad (6.43)$$

where

n_e = soil effective porosity
n_p = soil porosity
W_L = water loss (%)

Typical water loss values are provided in Table 6.6.

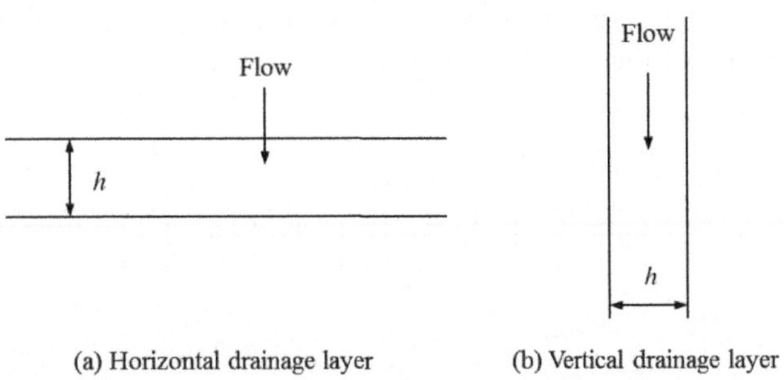

Figure 6.23 (a) Horizontal drainage layer and (b) vertical drainage layer.

Quality of Drainage The 1993 AASHTO *Guide for Design of Pavement Structures* specifies the quality of drainage in Table 6.7. The quality of drainage affects the pavement design life. Drainage layers can be installed in a pavement system to improve the quality of drainage so that the pavement life can be extended.

6.4.3 Design Considerations

Drainage for Retaining Wall The rate of water flow into the drainage layer behind the retaining wall can be estimated using a flow net as shown in Figure 6.24:

$$Q_{wi} = k\Delta h_w \frac{N_f}{N_{hd}} \quad (6.44)$$

The rate of flow out of the drainage layer per unit width is

$$Q_{wo} = kih_{dr} \quad (6.45a)$$

where

h_{dr} = thickness of the drainage layer
k = permeability of the drainage layer, and
i = hydraulic gradient in the drainage layer ($i = 1$ for a vertical drainage layer)

If a geosynthetic (geotextile or geocomposite) layer is used as a drainage layer only as shown in Figure 6.24(a), Equation (6.45a) can be rewritten as follows:

$$Q_{wo} = \theta_g i \quad (6.45b)$$

where θ_g = transmissivity of the geosynthetic.

If a geosynthetic layer is used as a filter, as shown in Figure 6.24(b), the rate of flow through the geosynthetic layer

Table 6.6 Water Loss Values (%)

	<2.5% Fines			5% Fines			>5% Fines		
	Filler	Silt	Clay	Filler	Silt	Clay	Filler	Silt	Clay
Gravel	70	60	40	60	40	20	40	30	10
Sand	57	50	35	50	35	15	25	18	8

Source: FHWA (1992).

Table 6.7 Quality of Drainage

Quality of Drainage	Water Removed Within
Excellent	2 h
Good	1 day
Fair	1 week
Poor	1 month
Very poor	Water will not drain

Source: AASHTO (1993).

into the drainage layer per unit width is

$$Q_w = kiH = (\psi_g t_g)\frac{H}{t_g}(H) = \psi H^2 \tag{6.46}$$

where ψ_g is permittivity of the geosynthetic and t_g is the thickness of the geosynthetic

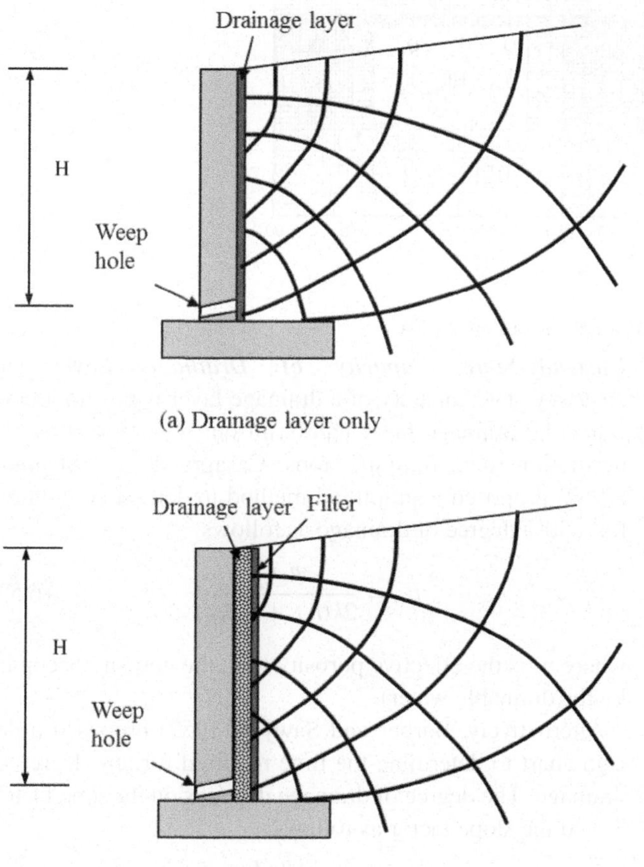

Figure 6.24 Drainage design for retaining wall.

Pavement Infiltration There are different methods available to estimate the infiltration rate of surface water through a pavement. The simplest and rough method is based on an infiltration ratio (Cedergren et al., 1973), that is,

$$q_{wi} = 24C_{in}R_R \tag{6.47}$$

where

q_{wi} = pavement infiltration (m/day)
C_{in} = infiltration ratio (0.33–0.50 for asphalt pavements and 0.50–0.67 for concrete pavements)
R_R = rainfall ratio (m/h)

The infiltration ratio, C_{in}, depends on pavement joints and cracks. A value of 0.5 may be used for a preliminary design. The typical rainfall ratio in the United States ranges from 5 to 66 mm/h.

For a pavement section, the rate of water infiltration per unit width of the pavement in the traffic direction is

$$Q_{wi} = q_{wi}L_R \tag{6.48}$$

where

L_R = drainage length of the pavement section in the transverse (width) direction

Groundwater Inflow In addition to surface infiltration, sometimes there is groundwater inflow as shown in Figure 6.25. The groundwater inflow can be estimated using the design chart in Figure 6.25. The influence distance, L_{R1}, of the drain can be estimated as follows:

$$L_{R1} = 3.8(h_w - h) \tag{6.49}$$

The inflow above the bottom of the drainage layer can be determined as follows:

$$Q_{w1} = \frac{k(h_w - h)^2}{2L_{R1}} \tag{6.50}$$

The inflow below the drainage layer, Q_{w2}, can be determined from Figure 6.25. Therefore, the lateral flow in a symmetrical section (i.e., each side of the section has a drain) is

$$Q_{wi} = Q_{w1} + Q_{w2} \tag{6.51}$$

Steady-State Capacity of Drainage Layer Baber and Sawyer (1952) proposed the following formula to calculate the steady-state discharge capacity of the drainage layer as shown in Figure 6.26:

$$Q_{wo} = kh_{dr}\left(S_R + \frac{h_{dr}}{2L_R}\right) \tag{6.52}$$

where

k = permeability of the drainage layer
h_{dr} = vertical thickness of the drainage layer

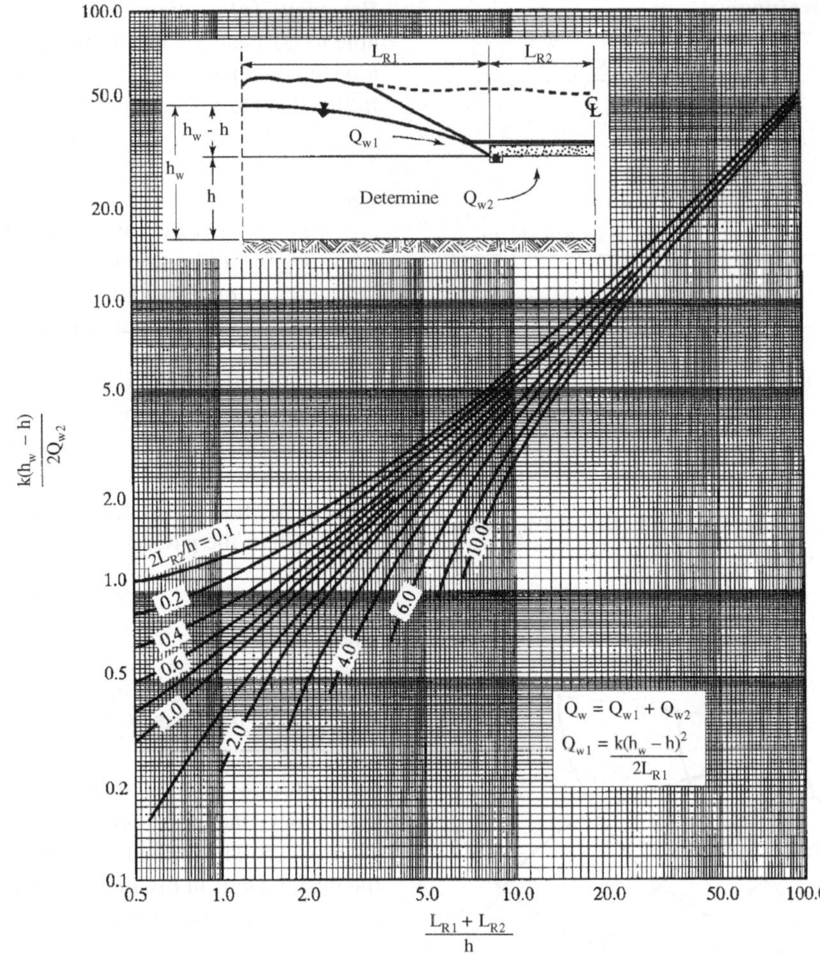

Figure 6.25 Inflow of groundwater (Moulton, 1980).

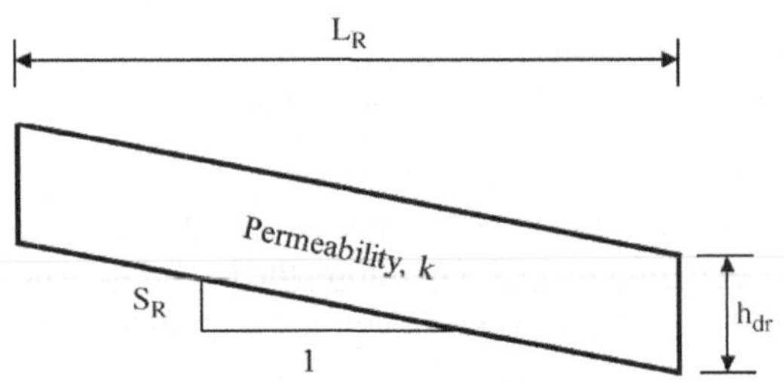

Figure 6.26 Drainage layer.

S_R = slope of the drainage layer
L_R = lateral distance of the drainage layer

The steady-state capacity of a drainage layer is the maximum amount of water the drainage layer can drain out. The design discharge capacity of the drainage layer should be greater than the water inflow. If there is only surface water infiltration, the discharge capacity should be greater than $q_{wi}L_R$.

Unsteady-State Capacity of Drainage Layer The unsteady-state capacity of a drainage layer is the amount of water the drainage layer can drain out with time after the infiltration (e.g., rainfall) stops. Casagrande and Shannon (1952) proposed a simplified method to determine the time for a 50% degree of drainage as follows:

$$t_{50} = \frac{n_e L_R^2}{2k(h_{dr} + S_R L_R)} \tag{6.53}$$

where n_e is the effective porosity (i.e., the porosity occupied by the drainable water).

Alternatively, Barber and Sawyer (1952) proposed a design chart to determine the time required for any degree of drainage. The degree of drainage depends on the time factor, T_t and the slope factor as follows:

$$T_t = \frac{k h_{dr} t}{n_e L_R^2} \tag{6.54}$$

$$S_1 = \frac{L_R S_R}{h_{dr}} \tag{6.55}$$

The 50% drainage approach recommended by FHWA (1992) can be used to determine the time to drain as follows:

$$t = T_{50} \cdot m_F \cdot 24 \tag{6.56}$$

$$m_F = \frac{n_e L_R^2}{k h_{dr}} \tag{6.57}$$

where

t = time to drain in hours
T_{50} = time factor
m_F = m factor

The time factor, T_{50}, can be determined from Figure 6.27 based on the slope factor, S_1, which depends on the resultant slope, S_R, lateral distance, L_R, and thickness, h_{dr}, of the drainage layer.

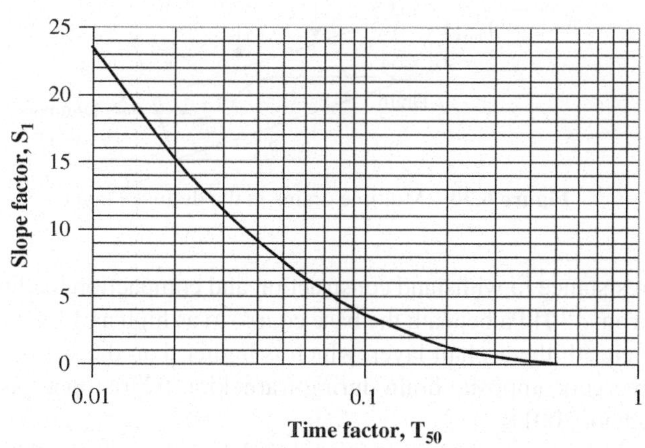

Figure 6.27 Time factor for 50% drainage (FHWA, 1992).

Maximum Height of Flow In the drainage design, the maximum height of flow should be limited to less than the vertical thickness of the drainage layer. In landfill design, the maximum height of liquid flow should be less than 0.3 m. There are several methods available to estimate this maximum height of flow. Three of these methods are discussed below: (1) FHWA (1992), (2) Giroud et al. (1992), and (3) McEnroe (1993).

Figure 6.28 shows the FHWA (1992) design chart to estimate the maximum height of flow. Based on the ratio of filtration rate to soil permeability and the slope, the $L_R/h_{\max}$ ratio can be determined. With a known L_R, the maximum height of flow, $h_{\max}$, can be calculated.

Giroud et al. (1992) developed a solution to estimate the maximum height of flow as follows:

$$h_{\max} = jL_R \frac{[(4q_{wi}/k + S_R^2)^{1/2} - S_R]}{2\cos\beta} \tag{6.58}$$

$$j = 1 - 0.12\exp\left\{-\left[\log\left(\frac{1.6q_{wi}}{kS_R^2}\right)^{5/8}\right]^2\right\} \tag{6.59}$$

Both FHWA (1992) and Giroud et al. (1992) methods assumed a free drainage condition, that is, the drainage range is beyond the lateral distance, L_R. However, for some applications, the drainage range may be smaller than the lateral distance, L_R, as shown in Figure 6.29. McEnroe (1993) developed the solutions based on different drainage range as follows:

$$R_D = \frac{q_{wi}}{k\sin^2\beta} \tag{6.60}$$

$$A = (1 - 4R_D)^{0.5} \tag{6.61}$$

$$B = (4R_D - 1)^{0.5} \tag{6.62}$$

For $R_D < \frac{1}{4}$:

$$h_{\max} = L_R S_R (R_D - R_D S_R + R_D^2 S_R{}^2)^{0.5} \left[\frac{(1 - A - 2R_D)(1 + A - 2R_D S_R)}{(1 + A - 2R_D)(1 - A - 2R_D S_R)}\right]^{0.5/A} \tag{6.63}$$

For $R_D = \frac{1}{4}$:

$$h_{\max} = L_R S_R R_D (1 - 2R_D S_R)/(1 - 2R_D)\exp\{2R_D(S_R - 1)/[(1 - 2R_D S_R)(1 - 2R_D)]\}/(1 - 2R_D) \tag{6.64}$$

For $R_D > \frac{1}{4}$:

$$h_{\max} = L_R S_R (R_D - R_D S_R + R_D^2 S_R^2)^{0.5} \exp\{(1/B)\tan^{-1}[(2R_D S_R - 1)/B] - (1/B)\tan^{-1}[(2R_D - 1)/B]\} \tag{6.65}$$

If the calculated maximum height of flow is greater than the required, more drains can be added as shown in Figure 6.30.

Geocomposites Instead of granular drains, geocomposites, formed by a drainable core (commonly a geonet) wrapped around by a nonwoven geotextile layer, can be used for the same drainage purpose. Since geocomposite has higher drainage capacity than an aggregate layer, a thin geocomposite can serve equally or better than a thick aggregate layer. The geocomposite drain can be placed between the subgrade and the base to shorten the drainage path for the base. Since the geocomposite drain is highly permeable, the effective drainage distance for the water to leave the pavement system is the thickness of the base course rather than the width of the road lane. This shortened drainage distance significantly reduces the time to drain. The geocomposite drain can also be placed between a concrete pavement and a subgrade, a frost

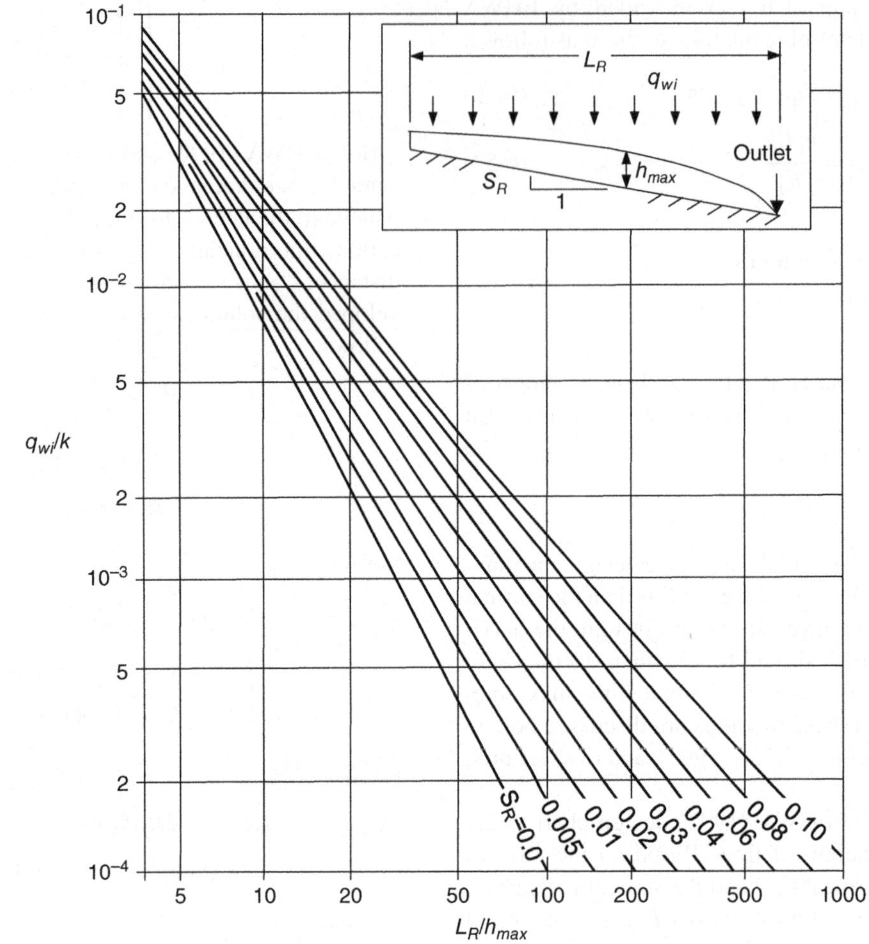

Figure 6.28 Maximum height of flow (FHWA, 1992).

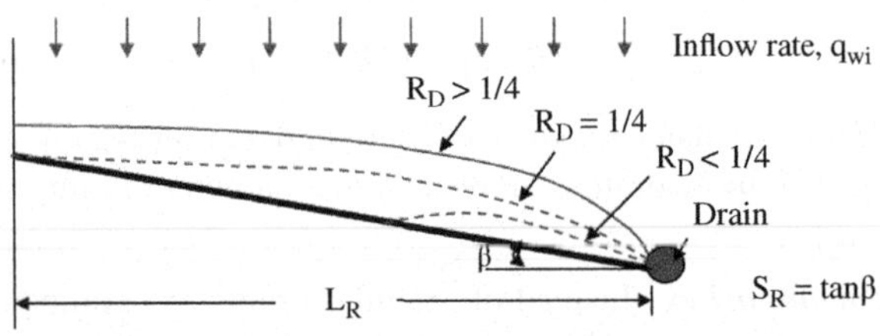

Figure 6.29 Maximum height of flow at different drainage range.

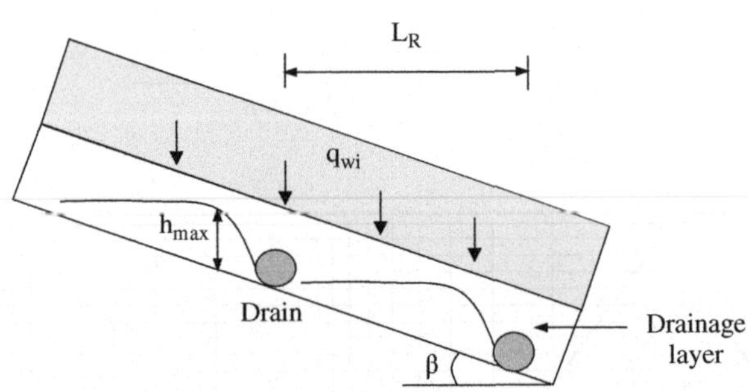

Figure 6.30 Multiple drains in the drainage layer.

susceptible soil and a subgrade, or behind a retaining wall facing or reinforced fill.

To provide equivalent drainage capacity to a 100-mm free-draining base layer with a transmissivity (i.e., permeability multiplied by the thickness) of approximately 30–100 m^2/day, the required transmissivity for a geocomposite drain is approximately 90–300 m^2/day due to unconfined, partially filled flow within the drainage layer (Giroud et al., 2000; Christopher and Zhao, 2001). In addition, the geocomposite drain should have enough crushing resistance to withstand construction and compaction loading from 480 kPa beneath the base course to as high as 1450 kPa beneath the asphalt layer, which is higher than the stress on the geocomposite drain during trafficking (Christopher and Zhao, 2001).

If the geocomposite drain is placed directly underneath the pavement and above the subgrade, the drainage only happens in the geocomposite. The design parameters, S_R, L_R, and h_{max} discussed earlier refer to those of the geocomposite drain. The slope and lateral distance of the drainage layer, S_R and L_R, are equal to the slope and the drainage width of the pavement section. Parameter h_{dr} is equal to the thickness of the geocomposite drain. However, if the geocomposite drain is placed between the base and the subgrade, the time to drain includes the time needed for water to drain through the base and drain along the geocomposite out of the pavement. The time needed for water to drain through the base can be calculated using $S_R = 1$ and L_R and h_{dr} both equal to the thickness of the geocomposite. The time for water to drain along the geocomposite drain is the same as that for the geocomposite directly underneath the pavement.

The m_F factor for a geocomposite can be determined by the following equation:

$$m_F = \frac{n_e L_R^2}{k t_g} = \frac{n_e L_R^2}{\theta_g} \quad (6.66)$$

where

n_e = effective porosity of the geocomposite layer
k = permeability of the geocomposite layer
θ_g = transmissivity of the geocomposite layer

The effective porosity is the ratio of the volume of the drainable water in the geocomposite to the total volume of the geocomposite.

In order for the geocomposite drain to be effective in a long term, the nonwoven geotextile around the drainage core should meet the requirements for filtration. The filter design is discussed in the previous section.

Pipe Drains Collector and outlet pipes as shown in Figure 6.31 should be designed to discharge the water collected from the drainage layer.

The flow capacity of a pipe subjected to gravity flow can be estimated using the Manning formula (Manning, 1891):

$$Q_{wo} = \frac{A}{n_r} R_h^{2/3} S_R^{1/2} \quad (6.67)$$

where

Q_{wo} = flow capacity of the pipe, m^3/s
A = cross-sectional area of the water stream, m^2
R_h = hydraulic radius = A/WP, m
WP = wetted perimeter, m
S_R = slope of pipe in m/m
n_r = roughness coefficient for the drain product, which is provided in Table 6.8

If the flow in the pipe is full, $A = \pi d_p^2/4$ and WP $= \pi d_p$ (d_p is the diameter of the pipe).

Lateral inflow to each outlet pipe can be calculated as follows:

$$Q_w = Q_{wi} L_o \quad (6.68)$$

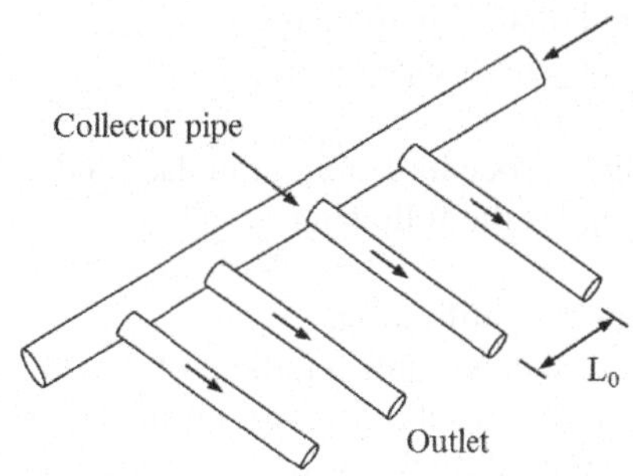

Figure 6.31 Collector and outlet pipes.

where Q_{wi} is the lateral flow per unit width in $m^3/s/m$ (may include infiltration and/or groundwater inflow as calculated by Eq. (6.51)) and L_0 is the distance between outlets in m.

The required diameter of the pipe is

$$d_p = \left(\frac{n_r Q_{wi} L_0}{0.312 S_R^{0.5}} \right)^{0.375} \quad (6.69)$$

Filter Criteria for Drainage Layer When a granular filter is used, the filter criteria for the drainage layer should also be examined. It is undesired that the particles from the granular filter are migrated into the drainage layer. These particles may be drained out or trapped in the drainage layer to clog the drain. In this case, the drainage layer is treated as a filter layer while the granular filter is treated as a soil to be protected by the drainage layer. The same filter criteria as discussed in the previous section can be used for this evaluation.

6.4.4 Design Parameters and Procedure

The design parameters for granular and geosynthetic drainage layers include:

- Soil type and permeability
- Infiltration rate
- Geometry of the system to be drained including drainage width, length, and slope
- Thickness and properties of drainage layer

Table 6.8 Typical Roughness Coefficient for Underdrain

Pipe Type	n_r
Clay drain pipe	0.014–0.018
Concrete drain pipe	0.011–0.015
Asbestos cement pipe	0.011–0.015
Corrugated metal pipe	0.017–0.024
Corrugated plastic pipe	0.020–0.024
Smooth plastic pipe	0.008–0.015

Source: modified from Huang (2004).

- Required quality of drainage
- Type and size of pipe

The design procedure for granular and geosynthetic drainage includes the following:

1. Estimate the infiltration.
2. Calculate the hydraulic gradient of the drainage layer.
3. Calculate the discharge capacity of the drainage layer or the required thickness of the drainage layer.
4. Calculate the required time for drainage.

6.4.5 Design Examples

Problem 1

A geotextile filter is being considered to protect the stone aggregate drain behind a cantilever retaining wall, as shown in Example Figure 6.1. The wall stem is 7.5 m high, retaining a low-plasticity silt with $k = 2.5 \times 10^{-4}$ m/s. The candidate geotextile is a heat-bonded nonwoven geotextile with a permittivity $\psi_g = 0.01\ \text{s}^{-1}$. What is the factor of safety against water flow behind the wall?

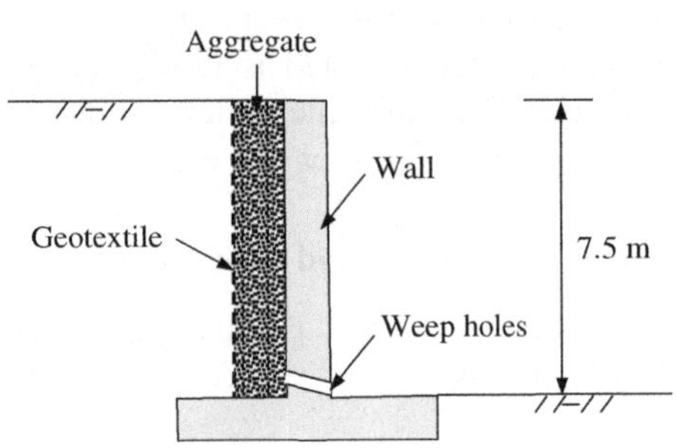

Example Figure 6.1

Solution

The flow nets are drawn in Example Figure 6.2. In this example, the geotextile layer serves as a filter while the stone aggregate layer serves a drain.

The rate of inflow can be calculated as follows:

$$Q_{wi} = k\Delta h_w \frac{N_f}{N_{hd}} = 2.5 \times 10^{-4} \times 7.5 \times \frac{4}{3} = 0.0025\ \text{m}^2/\text{s}$$

The required permittivity of the geotextile is:

$$\psi_{\text{reqd}} = \frac{Q_{wi}}{H^2} = \frac{0.0025}{(7.5)^2} = 4.44 \times 10^{-5}/\text{s}$$

The reduction factors (RF) for retaining wall filters based on Table 2.12 in Chapter 2 are:

$$\text{RF}_{\text{SCB}} = 3.0,\ \text{RF}_{\text{CR}} = 1.75,\ \text{RF}_{\text{IN}} = 1.1,$$
$$\text{RF}_{\text{CC}} = 1.1,\ \text{RF}_{\text{BC}} = 1.15$$

The allowable permittivity is

$$\psi_{\text{allow}} = \frac{\psi_{ult}}{RF_{\text{SCB}}RF_{\text{CR}}RF_{IN}RF_{\text{CC}}RF_{\text{BC}}} = \frac{0.01}{3.0 \times 1.75 \times 1.1 \times 1.1 \times 1.15} = 1.37 \times 10^{-3}/\text{s}$$

The factor of safety against flow is

$$FS = \frac{\psi_{\text{allow}}}{\psi_{\text{reqd}}} = \frac{1.37 \times 10^{-3}}{4.44 \times 10^{-5}} = 30.9 \qquad \text{(acceptable)}$$

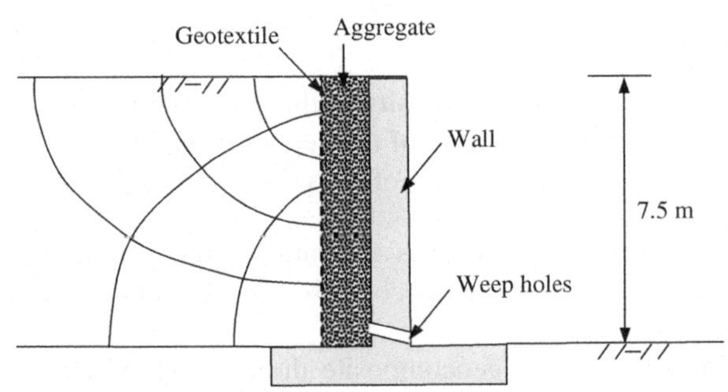

Example Figure 6.2

Problem 2

A 7.3 m wide pavement section with a 2.5% grade (i.e., half of a crowned four lane road) contains 0.30-m–thick base course. The effective porosity and the permeability of the base course are 20% and 0.30 m/day, respectively. A geocomposite material with a thickness of 6.1 mm, effective porosity of 69%, and transmissivity of 618 m^2/day is available to be placed beneath the base course. Calculate the time to drain without or with this geocomposite.

Solution

1. Without geocomposite:

The slope factor is

$$S_1 = \frac{L_R S_R}{h_{dr}} = \frac{7.3 \times 0.025}{0.30} = 0.60$$

The time factor, T_{50}, can be estimated from the chart (Figure 6.27) as 0.3.

The m_F factor is

$$m_F = \frac{n_e L_R^2}{k h_{dr}} = \frac{0.20 \times 7.3^2}{0.30 \times 0.30} = 115.2 \text{ days}$$

The time to drain is

$$t = T_{50} \times m_F \times 24 = 0.3 \times 115.2 \times 24$$

$$= 829 \text{ hours} = 35 \text{ days}$$

Or

$$t_{50} = \frac{n_e L_R^2}{2k(h_{dr} + S_R L_R)}$$

$$= \frac{0.2 \times 7.3^2}{2 \times 0.3 \times (0.30 + 0.025 \times 7.3)} = 37 \text{ days}$$

2. With geocomposite
   (a) Time to drain for the base course
   The slope factor is

$$S_1 = \frac{L_R S_R}{h_{dr}} = \frac{0.30 \times 1}{0.30} = 1$$

The time factor, T_{50}, can be estimated from the chart as 0.25.

The m_F factor is

$$m_F = \frac{n_e L_R^2}{k h_{dr}} = \frac{0.20 \times 0.30^2}{0.30 \times 0.30} = 0.20 \text{ day}$$

The time to drain is

$$t_1 = T_{50} \times m_F \times 24 = 0.25 \times 0.20 \times 24 = 1.2 \text{ hours}$$

   (b) Time to drain for the geocomposite
   The slope factor for the geocomposite layer ($h_{dr} = t_g = 0.0061$ m) is

$$S_1 = \frac{L_R S_R}{h_{dr}} = \frac{7.3 \times 0.025}{0.0061} = 30$$

The time factor, T_{50}, can be estimated from the chart as 0.01.

The m factor is

$$m_F = \frac{n_e L_R^2}{\theta_g} = \frac{0.69 \times 7.3^2}{618} = 0.088 \text{ day}$$

The time to drain is

$$t_2 = T_{50} \times m_F \times 24 = 0.01 \times 0.088 \times 24$$

$$= 0.02 \text{ hour}$$

   (c) The total time to drain

$$t = t_1 + t_2 = 1.20 + 0.02 = 1.22 \text{ hour}$$

6.4.6 Construction

The installation of a drainage layer and filter involves grading, trenching, placement of granular or geosynthetic layers and/or pipes, backfill, and compaction. The following construction procedure is typically adopted:

1. Grade the ground or subgrade to a certain slope based on design drawings. Sloping is important for effective drainage of water out of the system.
2. For some applications, trenches are needed, for example, for edge drains. Trenches should have sufficient widths to accommodate the placement of pipes and backfill materials. When geotextiles are used as filters, they should be installed first before backfill materials are placed.
3. For roadway construction, granular subbase may be used as a filter between subgrade and base course. The granular filter can be replaced by a geotextile or geocomposite. When the geotextile or geocomposite is placed, it shall be placed in contact with the soils without wrinkles or folds and anchored on a smoothly graded surface. The geotextile or geocomposite shall be placed with the machine direction perpendicular to the direction of water flow. The geotextile or geocomposite should be overlapped or seamed at joints. A certain overlap (typically ranging from 0.3 to 1.0 m) is necessary depending on the subgrade condition and soft subgrade requires larger overlaps. If a sewn seam is to be used for the seaming of the geotextile, the thread used shall consist of high-strength polypropylene or polyester.
4. Backfill should be placed carefully in a manner that will not damage or displace geosynthetics and pipes. It should extend a minimum of 150 mm over the geosynthetic or top of the pipe.
5. Light compaction equipment may be used on top of the geosynthetic or pipe with a 150 mm cover. Heavy compaction equipment or traffic is only allowed with a fill cover of at least 300 mm above the geosynthetic or pipe.

6.4.7 Quality Control and Assurance

Quality control and assurance should be followed at each step from construction stakeout, excavation, installation, to backfilling. ASCE (1998) suggested the following items for quality control and assurance:

- Inspection of materials and equipment onsite
- Conformance to line and grade
- Longitudinally stretched pipe or geosynthetics having corrugated core or separated joints
- Deformed or fractured pipes or geosynthetics

- Proper placement of envelope materials
- Backfill including compaction where required
- Proper placement of structures
- Testing of pipe capacity after backfill, by use of deflection test
- Site restoration

6.5 DEWATERING

6.5.1 Introduction

Basic Concept Dewatering is to lower an existing groundwater table by open pumping (sumps, trenches, and pumps), a well system (well points or deep wells), and the electroosmosis method. The most common purpose for dewatering is for construction excavations. Dewatering for construction excavations is mostly temporary. There are structures and highways constructed with permanent dewatering systems, but they are far less than temporary or construction dewatering systems. Permanent dewatering systems require continuous operation without interruption; therefore, they should be conservatively designed and maintained.

Dewatering and drainage have similarities but also have differences. Both methods remove water from soil. Dewatering is to lower a groundwater table from the existing level to a lower level. However, drainage is to remove water that enters into a system from the surface, side, or bottom of the drainage layer. Without a drainage system, the water level may rise. Drainage mostly relies on gravity while dewatering mostly relies on pumping.

Sumps, trenches, and pumps as shown in Figure 6.32 are used in open excavation, can handle a small amount of water inflow in a small area or relatively impermeable soil, and lower the groundwater table by a limited depth (mostly 1.5 m or less). Low power pumps are often used for this operation.

Well points as shown in Figure 6.33 are formed by multiple closely spaced wells, which are connected by pipes to a powerful pump. A typical well point system includes well points, riser pipes, header pipes, and pumps. Well points can be a single stage or multiple stages, depending on a required drawdown depth. Multiple stages of well points can lower a groundwater table by more than 5–6 m.

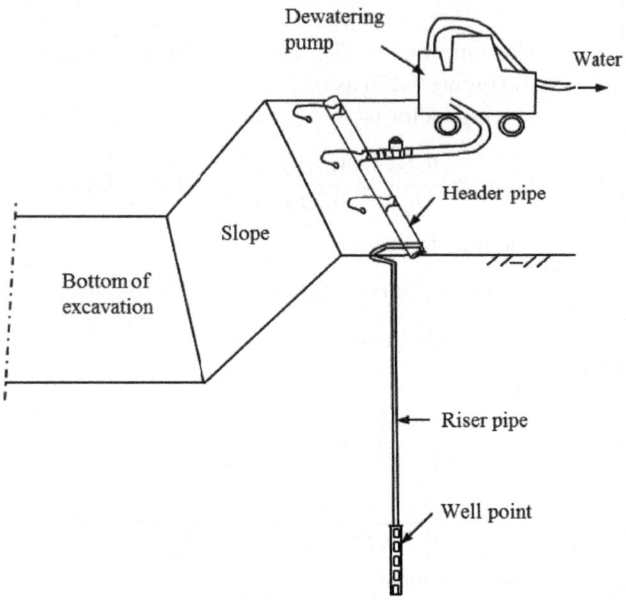

Figure 6.33 Well points and their components.

In deep wells, pumps are located at the bottom of the wells. The water is pumped up from the bottom through a pipe to a discharge point as shown in Figure 6.34. The pumps used for deep wells are more powerful than those for well points. Since deep wells can lower the groundwater table at a greater depth and range, fewer deep wells are installed in larger spacing. Deep wells have been used alone or combined with well points.

Electroosmosis creates electric gradients in geomaterial by installing anode and cathode to induce water flow as shown in Figure 6.35. The cathode is made in a form of a well point to collect and discharge the water.

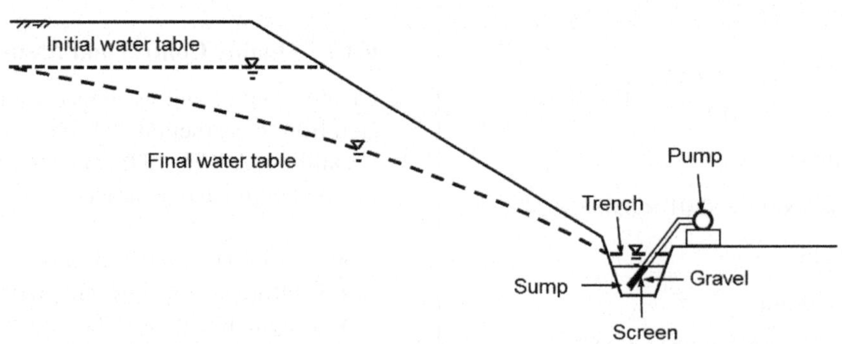

Figure 6.32 Sumps, trenches, and pumps.

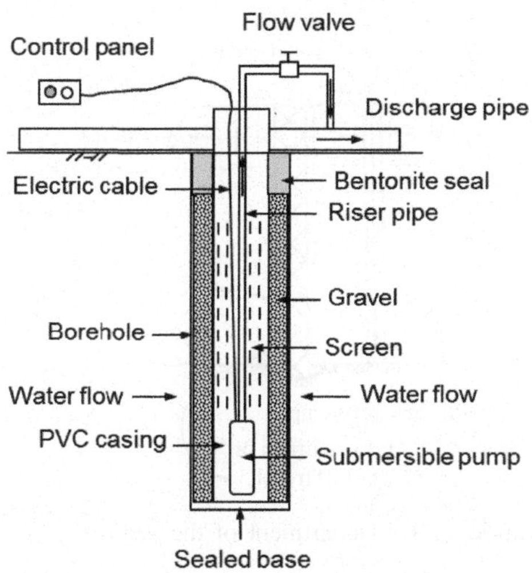

Figure 6.34 Deep well details.

The discharge of a cathode well point can be estimated by the following equation (Department of the Army, 1983):

$$Q_e = k_e i_e s z \tag{6.70}$$

where

k_e = coefficient of electro-osmotic permeability assume 0.98×10^{-4} m/s/volt/m),
i_e = electrical gradient between electrodes (volts/m)
s = effective spacing of well points (m); and
z = depth of soil being stabilized (m).

Typical electric current requirements for electroosmosis operation range from 15 to 30 amperes per well.

Curtain or cutoff walls are sometimes used to stop or minimize seepage into an excavation as shown in Figure 6.36. Examples of such cutoff walls include deep mixed walls, grouted walls, and slurry walls. Deep mixing and grouting methods will be discussed in Chapter 8 of this book.

Suitability The suitability of a dewatering technique depends on:

- Location, type, size, and depth of excavation
- Thickness, stratification, and permeability of geomaterials
- Required depth of the groundwater to be lowered
- Potential damage resulting from failure of the dewatering system
- Cost of installation and operation

Figure 6.37 shows the suitability of different dewatering techniques. Deep wells and well points are suitable for sandy soils. Vacuum is needed for deep wells and well points to dewater in silty soils. The electroosmosis technique can be used for silts, clayey silts, and clayey silty sands, but not suitable for saline soils (requiring high current) and organic soils (having possible adverse environmental impact).

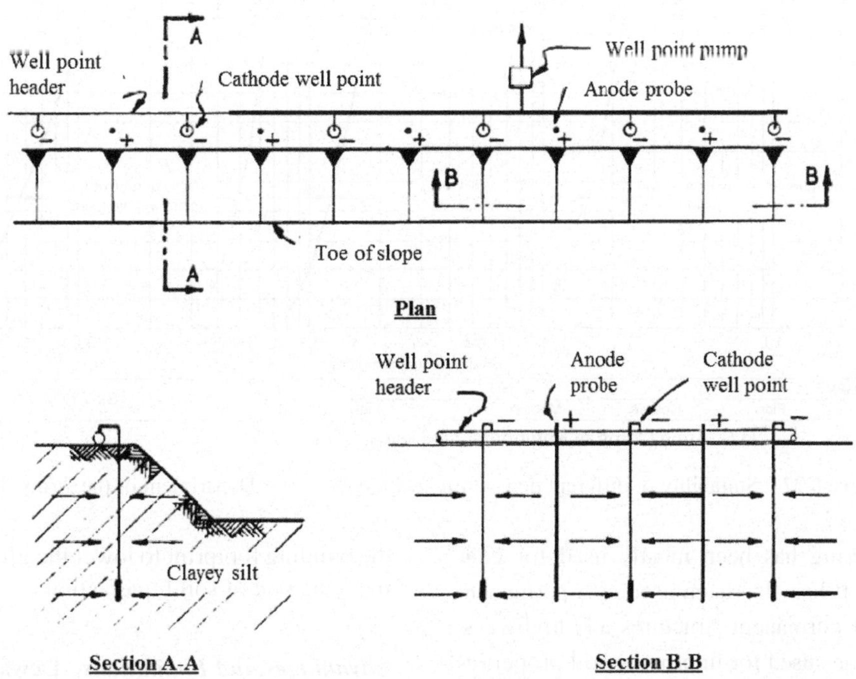

Figure 6.35 Dewatering using electroosmosis (after Department of the Army, 1983).

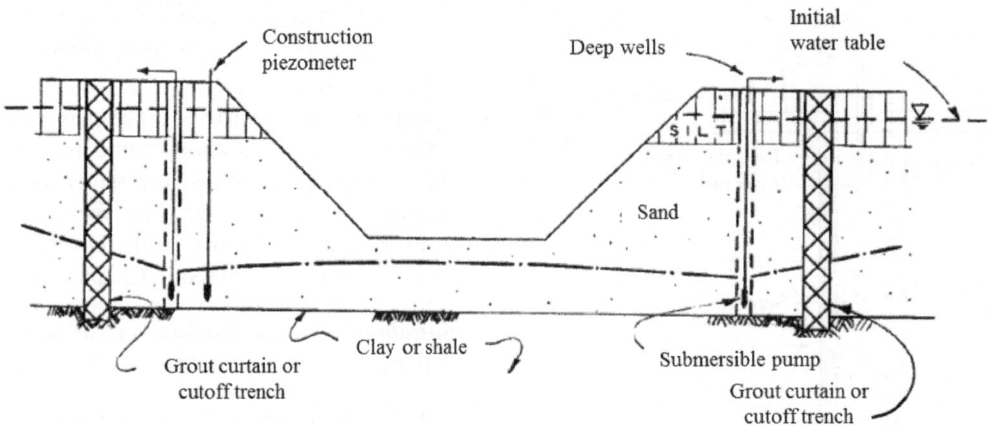

Figure 6.36 Grout curtain or cutoff trench around and excavation (after Department of the Army, 1983).

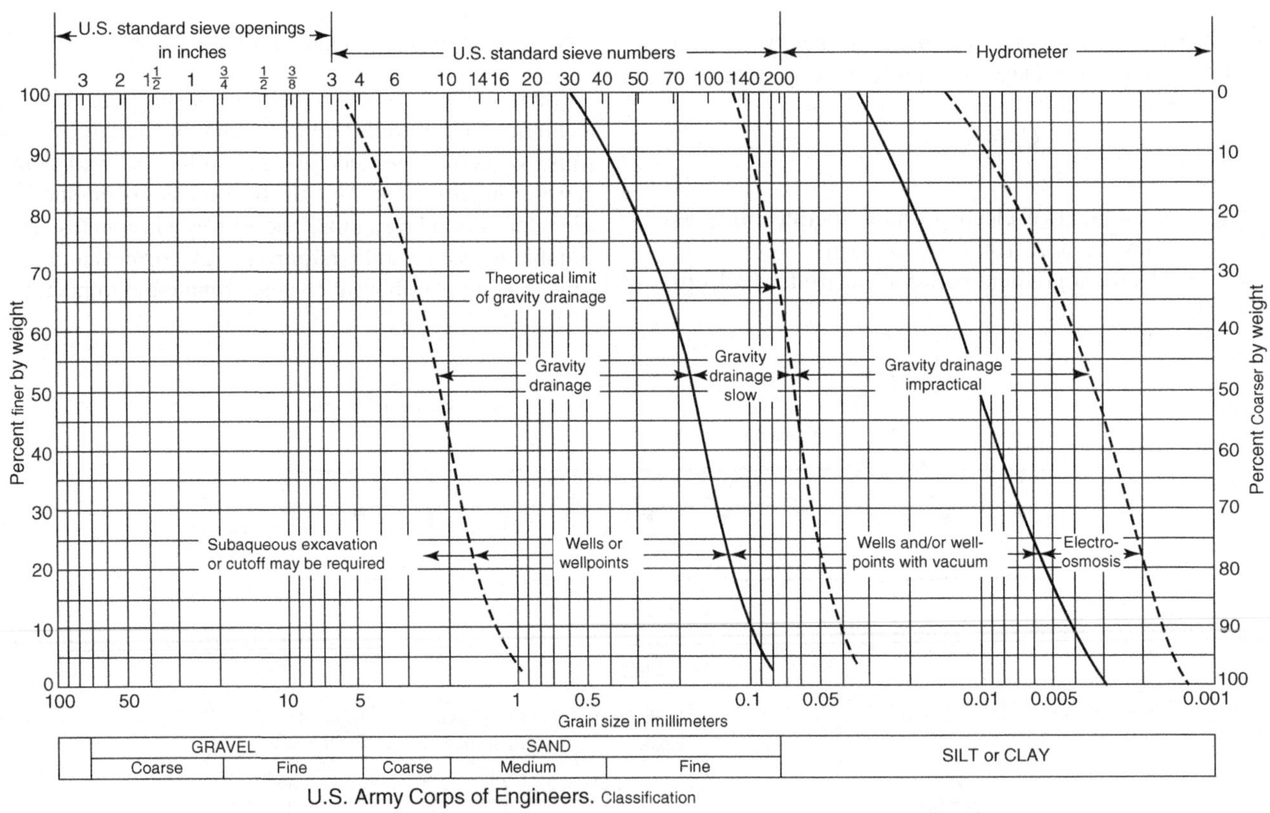

Figure 6.37 Suitability of different dewatering techniques (after Department of the Army, 1983).

Applications Dewatering has been mostly used for construction excavations. It has also been used as a permanent dewatering system for permanent structures and highways. Dewatering is sometimes used for improving soil properties and resistance to liquefaction. Chen and Jensen (2013) reported a case history using permanent pumping wells within the building footprint to lower the groundwater table to eliminate the risk of soil liquefaction.

Advantages and Limitations Dewatering systems are easy to install and can eliminate possible problems associated with water during excavations and permanent uses. Dewatering may induce ground subsidence and cause damage to adjacent structures. It should be used with caution when there are nearby existing structures and utilities. Dewatering requires disposal or recycling of water removed from the ground and continuous power supplies.

6.5.2 Principles

Theoretical Solutions for Water Flow into a Well The theoretical solution for rate of water flow into a well with an unconfined or confined permeable layer can be derived based on Darcy's law. Figure 6.38 shows two simplified models (i.e., Dupuit-Thiem approximation) for a single well fully penetrating in a ground with unconfined and confined permeable layers. They are symmetrical problems with the well in the center. The hydraulic gradient of the phreatic level after pumping (i.e., the drawdown curve) at the height of z (i.e., point A) is

$$i = \frac{dz}{dr} \tag{6.71}$$

Assume the hydraulic gradient below the drawdown curve is the same as that on the drawdown curve. Under the unconfined condition, according to Darcy's law, the rate of water flow into the surface, AB, with the height of z is

$$Q_w = kiA = k\left(\frac{dz}{dr}\right)(2\pi rz) \tag{6.72}$$

A differential equation can be obtained from the preceding equation:

$$\frac{dr}{r} = \left(\frac{2\pi k}{Q_w}\right) z\,dz \tag{6.73}$$

From r_0 to R_i,

$$\int_{r_0}^{R_i} \frac{dr}{r} = \left(\frac{2\pi k}{Q_w}\right) \int_{h_{w1}}^{h_{w0}} z\,dz \tag{6.74}$$

Thus, the rate of water flow (or discharge) can be solved as follows:

$$Q_w = \frac{\pi k(h_{w0}^2 - h_{w1}^2)}{\ln(R_i/r_0)} = 1.364k\frac{(h_{w0}^2 - h_{w1}^2)}{\log(R_i/r_0)} \tag{6.75}$$

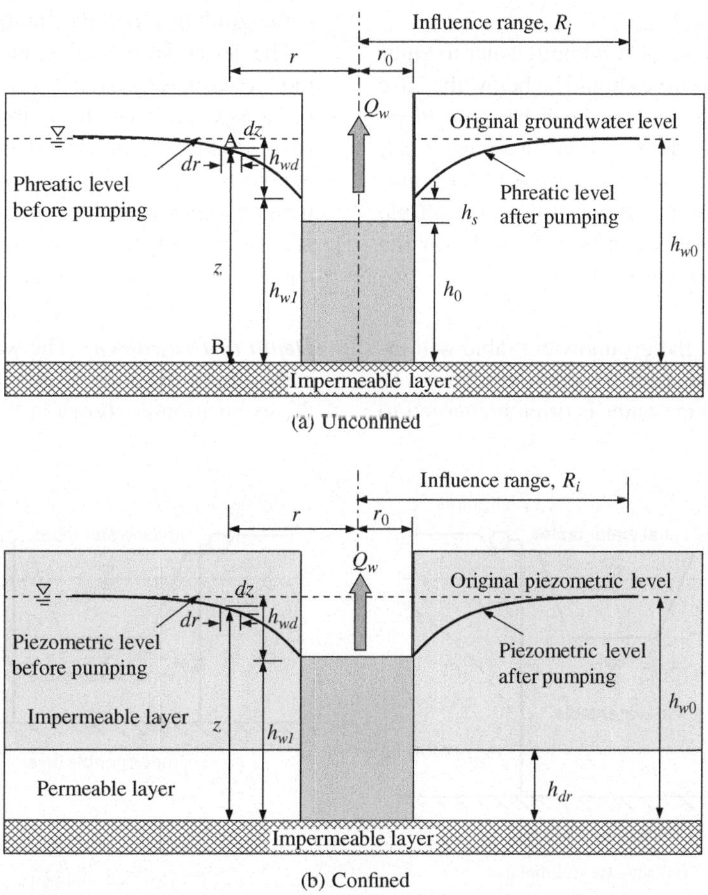

Figure 6.38 Flow of water into a well with an unconfined or confined permeable layer underlain by an impervious layer.

The height of the phreatic level after pumping can be obtained as follows:

$$z^2 - h_{w1}^2 = \frac{Q_w}{\pi k}\frac{\ln(r/r_0)}{} \tag{6.76}$$

Under the confined condition, water only flows into the well from the permeable layer with a thickness of h_{dr}, hence

$$Q_w = kiA = k\left(\frac{dz}{dr}\right)(2\pi r h_{dr}) \tag{6.77}$$

$$\int_{r_0}^{R_i}\frac{dr}{r} = \left(\frac{2\pi k h_{dr}}{Q_w}\right)\int_{h_{w1}}^{h_{w0}} dz \tag{6.78}$$

$$Q_w = \frac{2\pi k h_{dr}(h_{w0} - h_{w1})}{\ln(R_i/r_0)}$$

$$= 2.729 k h_{dr}\frac{h_{w0} - h_{w1}}{\log(R_i/r_0)} \tag{6.79}$$

It should be pointed out that under the unconfined condition, there is a phreatic level, while under the confined condition, there is a piezometric level. The piezometric level is the water head of the permeable layer, and it does not represent the actual groundwater table in the impermeable layer.

Drawdown Curve To ensure no or minimum water to enter the excavation, the drawdown curve should be below the base of the excavation. The drawdown curve can be induced by one-side wells or two-side wells as shown in Figure 6.39, which often depends on the depth of the permeable stratum.

When a single stage of well points is not sufficient, multiple staged well points or a deep well can be used to lower the drawdown curve as shown in Figure 6.40.

Recharge The drawdown of the groundwater table will induce the settlement of the adjacent existing structures. To avoid or minimize this problem, water is often recharged to the ground close to the existing structures so that the groundwater table close to the structures can be maintained, as shown in Figure 6.41.

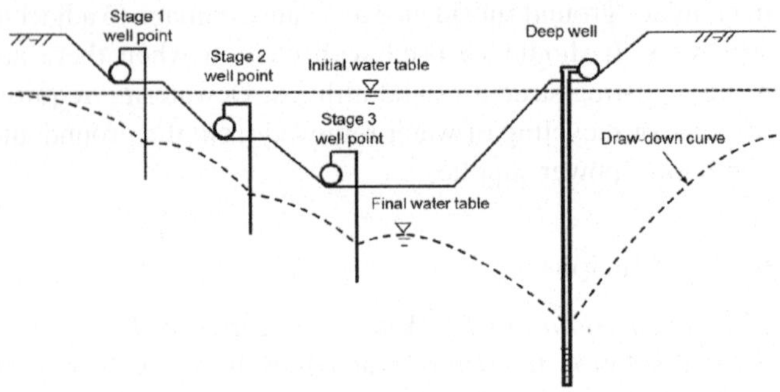

Figure 6.40 Drawdown curve by multiple staged well points or a deep well.

6.5.3 Design Considerations

Selection of Dewatering Technique Selection of a dewatering technique depends on several factors. Table 6.9 provides some guidelines, which can be used for this selection.

The jet-eductor well point system is also referred to as the "jet-eductor system," "ejector system," or "eductor well point system," which is similar to the well point system. Instead of using vacuum to draw water to the well points, the eductor system uses high-pressure water and riser units. The main advantage of the eductor system is that a single pump station can control many well points so that the water table can be lowered in one stage as much as 30 m.

Depth of Drawdown The water level should be lowered to a minimum distance of about 0.5–1.5 m below the base of the excavation as shown in Figure 6.42.

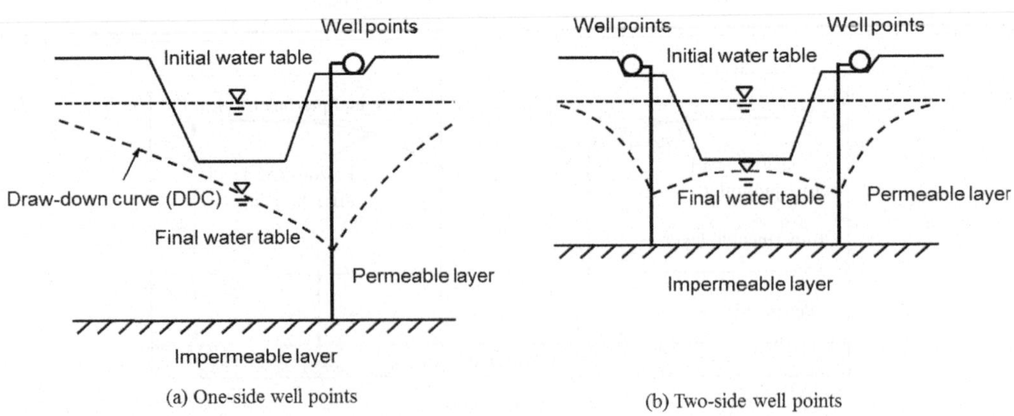

Figure 6.39 Drawdown curves by (a) one-side and (b) two-side well points.

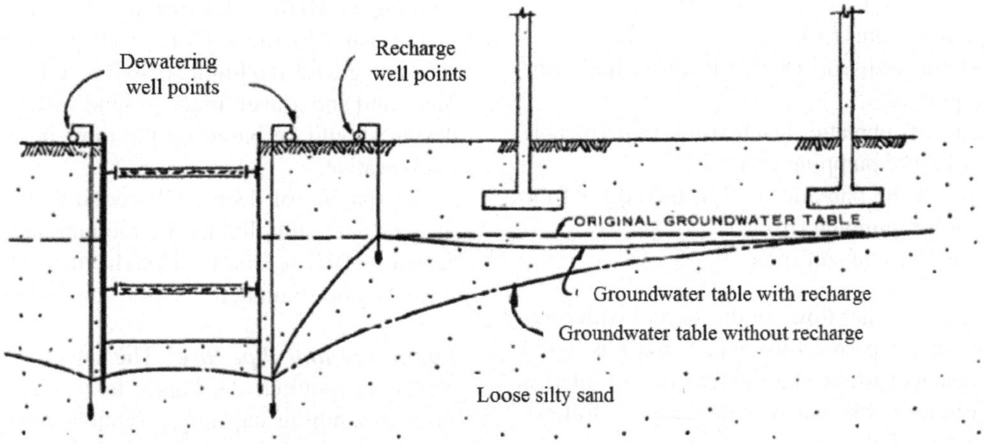

Figure 6.41 Drawdown curve with recharge water.

Table 6.9 Summary of Dewatering Techniques

Method	Application	Guideline
Sumps and ditches	Collect water entering an excavation or structure	Generally lower water level in stable gravel or well-graded sandy gravel by 1.5 m or less.
Conventional well point system	Dewater soils that can drained by gravity flow (gravel or sand)	Most commonly used. Drawdown limited to about 5 m per stage. May use multiple staged well points for large drawdown.
Vacuum well point system	Dewater soils with low permeability (silts)	Accelerate water flow. Vacuum effect limited to 5 m.
Jet-eductor well point system	Dewater soils that can be drained by gravity flow. Usually for large, deep excavations where large flow is required	Lower water table as much as 30 m from top of excavation. Particularly suitable for dewatering shafts and tunnels. Require two header pipes and two riser pipes or a pipe within a pipe.
Deep well system	Dewater soils that can be drained by gravity flow. Usually for large, deep excavations where large flow is required	Installed around periphery of excavation. Suitable for dewatering shafts and tunnels
Electroosmosis	Dewater soils that cannot be drained by gravity flow (silts, clayey silts and sands)	Direct electrical current increases hydraulic gradient causing flow.

Source: Modified from Department of the Army, 1983.

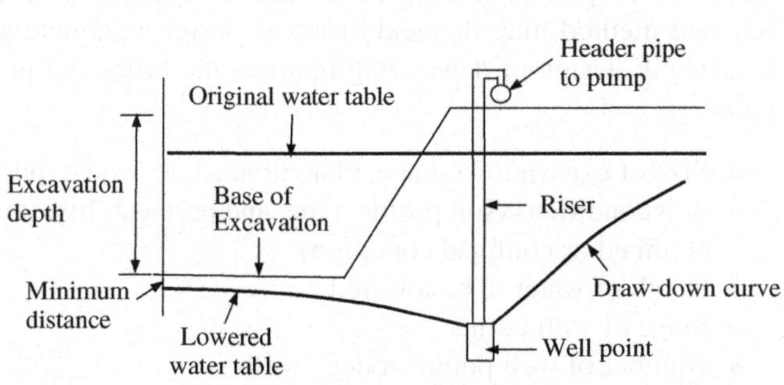

Figure 6.42 Depth of drawdown.

Single Well The rates of water flow from a single well in unconfined and confined permeable layers can be calculated using Equations (6.75) and (6.79).

Under the unconfined condition, there is a height of the free discharge surface, which can be estimated as follows:

$$h_s = \frac{C(h_{w0} - h_0)^2}{h_{w0}} \tag{6.80}$$

Ollos proposed a value of $C = 0.5$ (Hausmann, 1990). Sichardt (1928) suggested the following formula for the influence radius:

$$R_i = C'(h_{w0} - h_{w1})\sqrt{k} \tag{6.81}$$

where

R_i = influence radius (m)
C' = 3000 for wells or 1500–2000 for single-line well points
h_{w0} = height of phreatic level from impermeable layer before pumping (m)
h_{w1} = height of the phreatic level at the edge of the well after pumping (m)
k = permeability of soil (m/s)

Multiwells The rate of water flow (or discharge) with multiwells in an unconfined permeable layer (each has equal length and flow capacity) arranged around a point of interest (i.e., point P in Figure 6.43) can be calculated as follows (Forchheimer, 1930):

$$Q_w = \frac{\pi k (h_{w0}^2 - z^2)}{\ln R_i - (1/N_w)\ln(x_1 x_2 \cdots x_{Nw})} \tag{6.82}$$

where N_w is the number of wells.

When there is a circular arrangement of wells as shown in Figure 6.44, it can be simplified as follows:

$$Q_w = \frac{\pi k (h_{w0}^2 - z^2)}{\ln R_i - \ln a_w} \tag{6.83}$$

Well Penetration The solutions discussed so far are based on fully penetrating wells, that is, the wells reach the underlying impermeable layer. In reality, the wells may not reach an impermeable layer. In other words, they are partially penetrating. The rate of water flow, Q_w, calculated for a fully penetrating well, may be increased by 10–30% for a partially penetrating well (Hausmann, 1990).

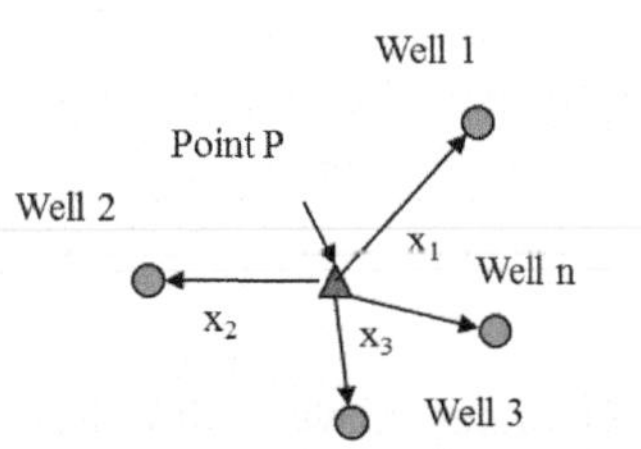

Figure 6.43 Multiwells in a random arrangement.

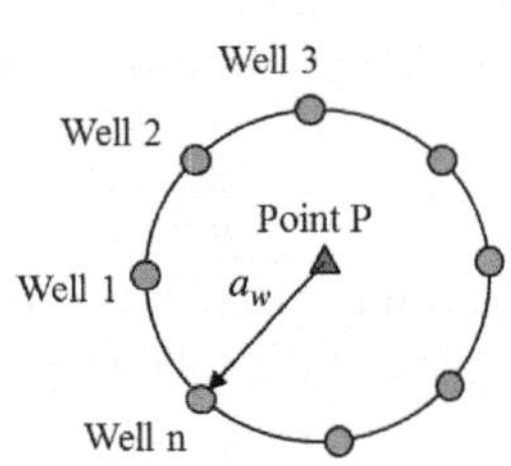

Figure 6.44 Multiwells in a circular arrangement.

Spacing of Wells The spacing of well points can be estimated using Figure 6.45 for well points in clean, uniform sand and gravel and Figure 6.46 for well points in clean, stratified sand and gravel. In the ground with stratified layers, the design should be based on the most permeable layer to be conservative.

The spacing of deep wells required equals the perimeter of the excavation divided by the number of wells required. Sichardt (1928) recommended the minimum spacing of deep wells be approximately 32 times the radius of the well.

Pump Size and Pipe Size The selection of pump size depends on required discharge, total dynamic head, suction lift, air-handling capacity, available power, cost, and system durability (Xanthakos et al., 1994). Typical diameters of deep well pumps range from 100 to 400 mm and their discharge capacities are from 0.3 to 11.0 m^3/min as shown in Table 6.10.

Header pipes are typically steel pipes with multiple inlets. The size of the header pipe should be selected to sufficiently transport water collected from different deep wells or well points.

Bottom Stability of Excavation Figure 6.47 shows an excavation with dewatering by well points. To maintain the bottom stability of excavation, the total overburden stress above the sand layer should be higher than the uplift pressure, that is,

$$\gamma z > \gamma_w(\Delta h_w + z) \tag{6.84}$$

where

γ = unit weight of the soil above the sand layer
z = thickness of the soil layer above the sand layer
Δh_w = original water head above the bottom of the excavation

6.5.4 Design Parameters and Procedure

Design Parameters Design of dewatering depends on a number of factors, which may determine whether sumps and ditches, well points, or deep wells should be selected. The selected method may demand different design parameters. In general, design of dewatering requires the following parameters:

- Size of excavation (shape, plan dimensions, and depth)
- Soil conditions (soil profile, type, and permeability; unconfined or confined condition)
- Depth of water to be lowered
- Stage of well points
- Number of well points or deep wells
- Spacing of well points or deep wells
- Pipe size
- Pump size

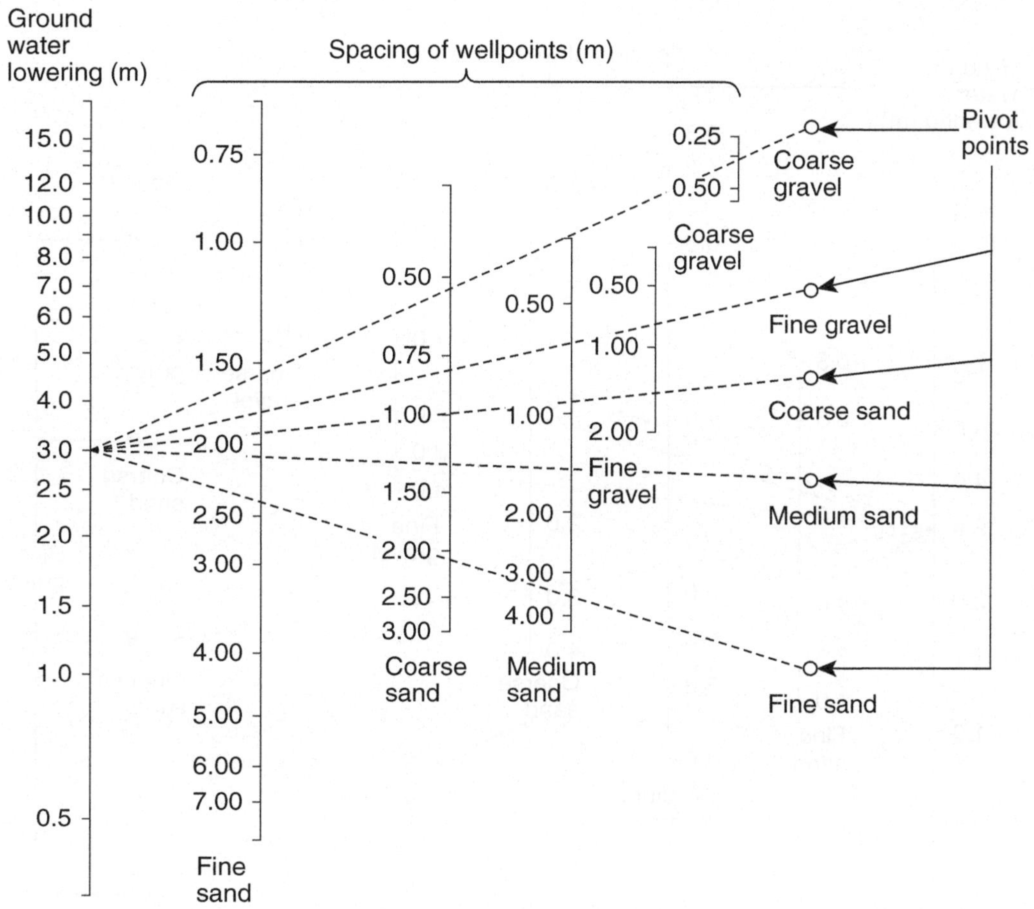

Figure 6.45 Design chart for well point spacing (unit: m) in clean, uniform sand and gravel (NAVFAC, 1982).

Table 6.10 Typical Deep Well Pumps

Pump Size (mm)	Minimum Well Size (mm)	Approximate Maximum Discharge Capacity (m^3/min)
102	127	0.3
143	152	0.6
152	203	1.7
203	254	2.3
254	305	4.5
305	356	6.8
356	406	9.1
406	457	11.4

Source: Mansur and Kaufman (1962).

Design Procedure Figure 6.48 shows a design layout of a rectangular excavation with deep wells as an example to illustrate the design procedure:

1. The excavation area is approximated to a circular area with a radius, r_0:

$$r_0 = \sqrt{\frac{LB}{\pi}} \tag{6.85}$$

2. Based on the required height of the water level at the center of the excavation after pumping, h_{w1}, the influence radius, R_i, can be calculated according to Equation (6.81).
3. Depending on whether it is an unconfined or confined condition, Equation (6.75) or (6.79) may be used to calculate the total required discharge, Q_w.

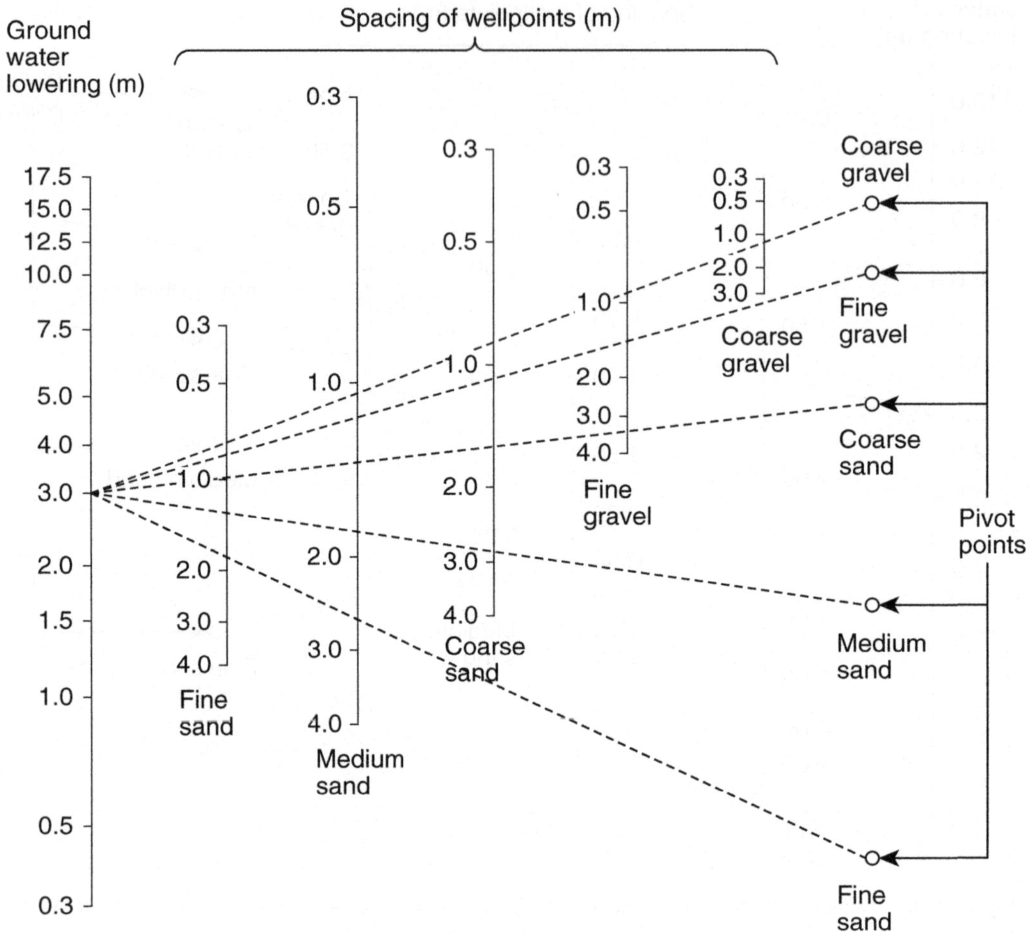

Figure 6.46 Design chart for well point spacing (unit: m) in clean, stratified sand and gravel (NAVFAC, 1982).

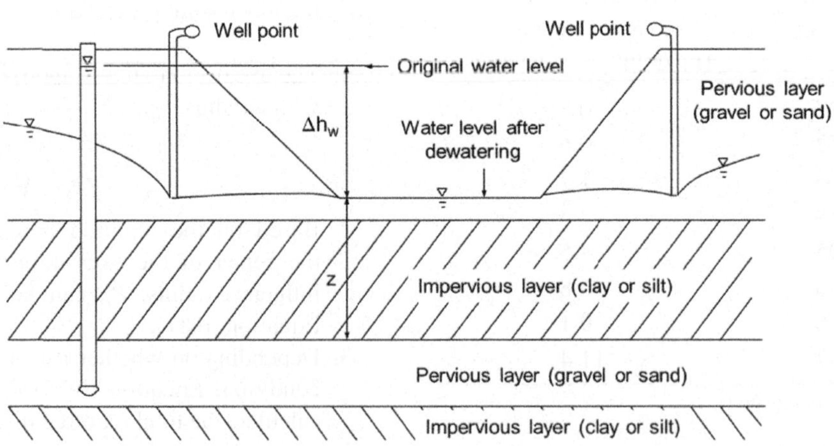

Figure 6.47 Bottom stability of excavation.

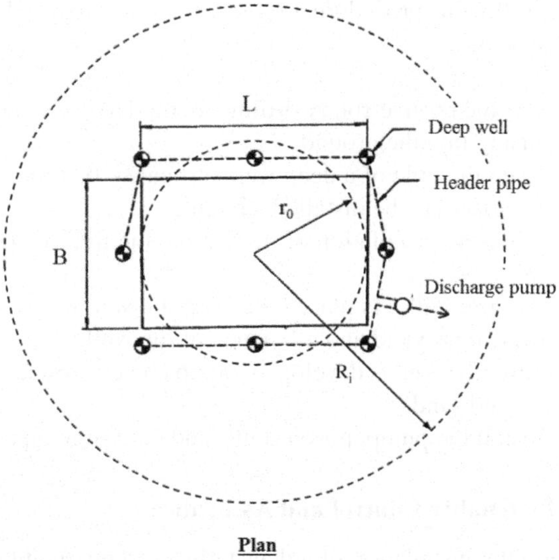

Figure 6.48 Design layout for deep wells (modified from Xanthakos et al., 1994).

4. Based on the discharge capacity of each well, Q_{w1}, the number of the well can be calculated as follows:

$$N_w \geq \frac{Q_w}{Q_{w1}}$$

5. Based on the number of wells, they are uniformly placed around the excavation as shown in Figure 6.48.
6. Based on the well layout and the total discharge, $Q_w = N_w Q_{w1}$, verify the water level after dewatering using Equation (6.82) or Equation (6.83).
7. Select a pump and the size of a header pipe.

6.5.5 Design Example

Problem

A job site requires an excavation of a rectangular area (220 m × 170 m) to a depth of 15 m, as shown in Example Figure 6.3. The existing groundwater table is at 5 m. Below the ground surface is 30-m-thick gravel with a permeability of 5.0 × 10^{-5} m/s, which is underlain by bedrock. The groundwater table should be lowered to 1.5 m below the bottom of the excavation. Deep wells are used to dewater the site. Calculate the total required discharge. If 200-mm-diameter deep wells are used, how many deep wells are required?

Solution

1. Calculate the equivalent radius of the excavation:

$$r_0 = \sqrt{\frac{LB}{\pi}} = \sqrt{\frac{220 \times 170}{3.14}} = 109 \text{ m}$$

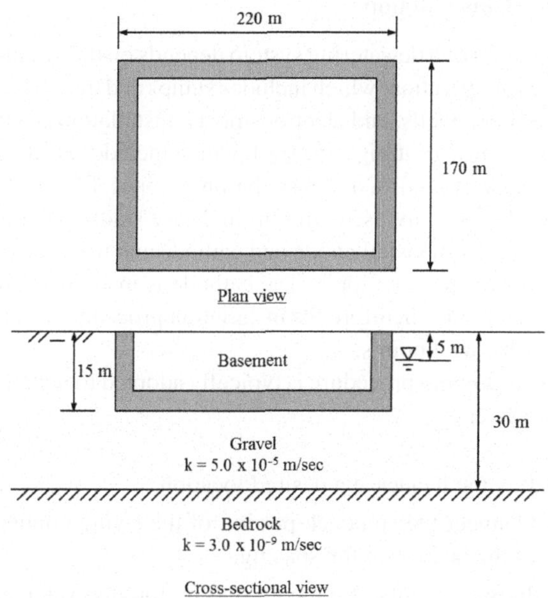

Example Figure 6.3

2. Calculate the influence radius: The height of the water level below the bottom of the excavation:

$$h_{w1} = 30 - 15 - 1.5 = 13.5 \text{ m}$$

The influence radius is

$$R_i = C'(h_{w0} - h_{w1})\sqrt{k}$$

$$= 3000 \times (25 - 13.5)\sqrt{0.00005} = 244 \text{ m}$$

3. Calculate the total required discharge: Since it is an unconfined condition, Equation (6.75) is used to calculate the total required discharge:

$$Q = \frac{\pi k(h_{w0}^2 - h_{w1}^2)}{\ln(R_i/r_0)}$$

$$= \frac{3.14 \times 5.0 \times 10^{-5} \times (25^2 - 13.5^2)}{\ln(244/109)}$$

$$= 0.0864 \text{ m}^3/\text{s}$$

4. Determine the number of deep wells: The discharge by each well is

$$Q_{w1} = \frac{\pi k(h_{w0}^2 - h_{w1}^2)}{\ln(R_i/r_0)}$$

$$= \frac{3.14 \times 5.0 \times 10^{-5} \times (25^2 - 13.5^2)}{\ln(244/0.10)} = 0.0089 \text{ m}^3/\text{s}$$

The number of wells is

$$N_w = \frac{Q_w}{Q_{w1}} = \frac{0.0864}{0.0089} = 9.7$$

Use 10 deep wells

6.5.6 Construction

Installation of a dewatering system depends on the system selected for operation, which includes sumps and trenches, well points, deep wells, and electroosmosis. Installation of sumps and trenches are straightforward, which includes excavation, placement of pumps, and installation of pipes. Electroosmosis has been rarely used with limited case studies. The main procedure is to install anode and cathode and the connection of wires for power supply. The cathode is made in the form of a well point; therefore, its installation procedure is similar to that for well points.

The following procedure is typically adopted to install well point systems:

- Lay the header at a desired location.
- Connect the stopcock portion of the swing connection to the header on the spacing.
- Jet well points into the ground by forcing water out of the tip of the well point under high pressure.
- If a well point is installed with a filter, a heavy steel casing should be installed first by jetting down the ground. The well point is placed inside the casing, the sand filter is tremied or poured in, and the casing is withdrawn.
- Connect the well point to the header and the pump.
- Start the pump by providing the vacuum and removing water flowing into the system.

The following procedure is typically adopted to install deep well systems:

- Use the reverse-rotary drilling method to drive and jet a casing into the ground.
- Keep the hole of a deep well vertical so that the screen and riser can be installed straight.
- After the installation of the screen, the filter is tremied in.
- After the filter is placed, develop the well to obtain the maximum yield and efficiency of the well.
- After the well is developed, pump and clear muddy water and sand.
- Install the pump, power units, and discharge pipes.

6.5.7 Quality Control and Assurance

During the installation of a dewatering system, quality control and assurance include:

- Locations and depths of pumps and wells
- Sizes of holes and wells
- Quality of filter material and installation
- Connections of all the pipes and with pumps

After the installation of the dewatering system, a full-scale pumping test should be performed to evaluate its performance for adequacy or need for any modification of the system.

PROBLEMS

6.1. Give five examples of possible geotechnical failures caused by water.

6.2. List five benefits of removing water from ground.

6.3. Explain the differences and relationships between drainage and dewatering.

6.4. What is filtration?

6.5. What are the three components of the water head according to Bernoulli's equation?

6.6. What is the key difference between total water head and pressure head?

6.7. In the following figure, the flow channel has two different soil specimens: Specimen 1 from points A to B and specimen 2 from points B to C. If the permeability of specimen 1 is twice that of specimen 2 (i.e., $k_1 = 2\,k_2$). Express the water head at Position 3 (h_{w3}) using the water heads h_{w1} and h_{w2}.

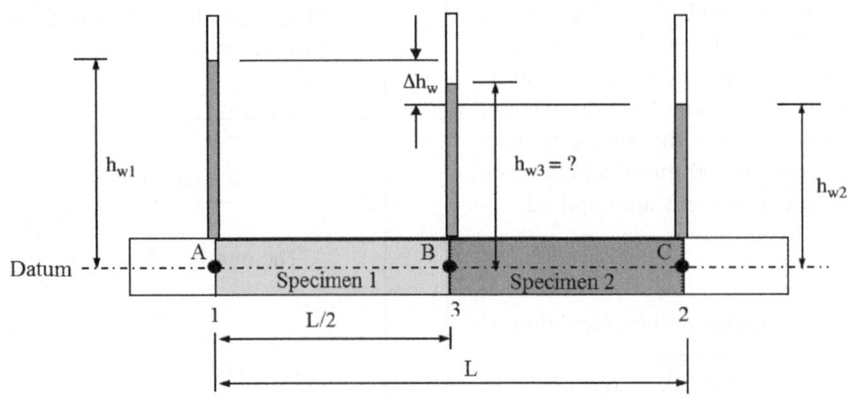

6.8. What are the basic requirements for a flow net?

6.9. If a flow net beneath a structure is drawn with eight equipotentials instead of four, will the flow rate be greater if there is no change in geometry or soil properties? Why or why not?

6.10. (1) Calculate the flow rate beneath the dam shown below and the uplift force per meter of dam. The soil permeability under the dam is $k = 7 \times 10^{-7}$ m/s and the 15-m-long sheet pile is 7 m behind the front of the dam. (2) Calculate the total, elevation, and pressure heads at points A, B, and C.

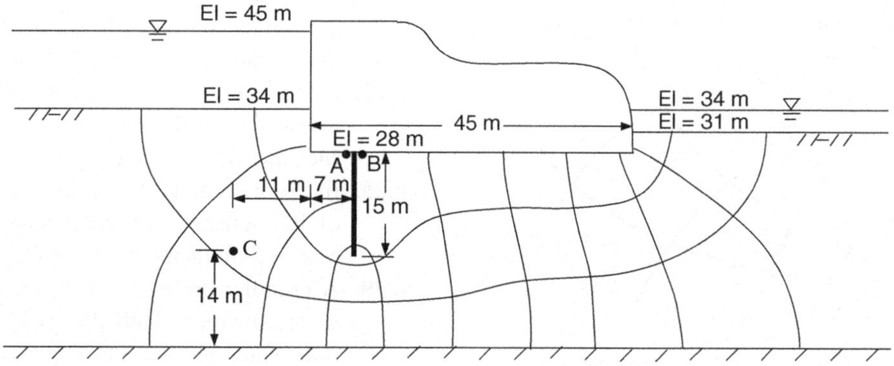

6.11. In the following figure, the inflow water pressure is 10 kPa. The soil sample is 0.5 m high and the total soil unit weight is 18 kN/m^3. The water above the soil sample is 0.3 m high. Calculate the effective vertical stress at the midheight of the sample. What will happen if the inflow water pressure is increased to 20 kPa?

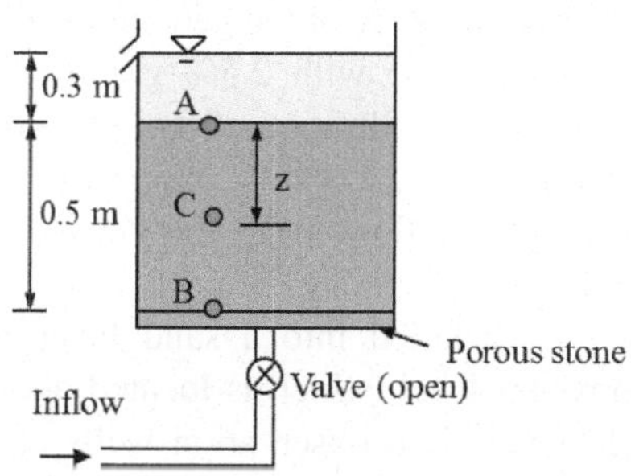

6.12. Explain the difference(s) between a geosynthetic filter and a geosynthetic separator.

6.13. A filter is used in a dam toe drain above a soil and the water flows from the soil to the filter. The soil permeability is 2×10^{-7} m/s. What is the minimum required permeability of the filter?

6.14. A geotextile is recommended as a filter for landfill drainage. The permeability of the soil is $k = 3 \times 10^{-6}$ cm/s, and the permeability of the geotextile filter is $k_f = 5 \times 10^{-6}$cm/s. Evaluate whether this geotextile meets the permeability requirement.

6.15. A nonwoven geotextile is used between the base course and the subgrade in a roadway. The soaked CBR value of the subgrade is 4%. Find the strength requirements of the geotextile based on the AASHTO M288.

6.16. Crushed stone with a maximum particle size of 50 mm is to be placed on a geotextile. Stone can be dropped from a height of 1.5 m during the construction. To ensure the survivability during the construction, what trapezoidal tear resistance and mass per unit area are required for this geotextile?

6.17. A granular filter with the gradation corresponding to the permeability of 15 m/day in Figure 6.15 is placed above a subgrade soil. Determine the filter particle size at 15% passing. If the particle sizes of the subgrade corresponding to 15 and 85% passing are 0.03 and 0.1 mm, respectively, check whether this granular filter is suitable for this project based on Terzaghi's filter criteria.

6.18. A geotextile filter is needed above a cohesionless soil. The soil has $D_r = 50\%$, a linear coefficient of uniformity $C'_u = 2.5$, $D'_{85} = 0.35$ mm. Determine the required O_{95} of the geotextile based on Giroud's filter retention criteria.

6.19. A nonwoven geotextile is selected to separate subgrade and base course in a roadway project. The subgrade soil is silty sand ($D_{85} = 0.5$ mm) with a uniformity coefficient (C_u) of 2.5 and has a CBR value of 3.0%. This geotextile is also used as a filter; therefore, it is necessary to meet the FHWA filter criteria. Please provide the specifications of this geotextile.

6.20. What is the hydraulic gradient of the water flow in a saturated slope with a slope angle of 18.6°?

6.21. What is the hydraulic gradient of the water flow in a drainage layer behind a vertical retaining wall facing?

6.22. A gravel contains 5% silt fines. The porosity of the gravel is 0.30. Estimate the effective porosity of this gravel.

6.23. A geosynthetic composite is used behind the retaining wall as shown below. The permeability of soil behind the retaining wall is $k = 2.3 \times 10^{-6}$ m/s. Determine the required transmissivity of the geosynthetic composite.

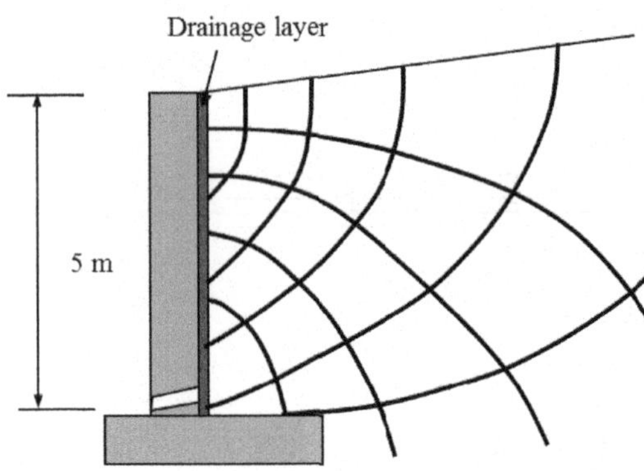

6.24. A geosynthetic filter is used to protect soil from the erosion by water flow in a reservoir project as shown below. Given $\Delta h_{w1} = 10$ m, $\Delta h_{w2} = 5$ m, the soil thickness h = 15 m, the soil permeability $k = 1.3 \times 10^{-6}$ m/s, the permittivity of the geosynthetic $\psi = 0.01 s^{-1}$. Find the factor of safety against the water flow through the geosynthetic layer.

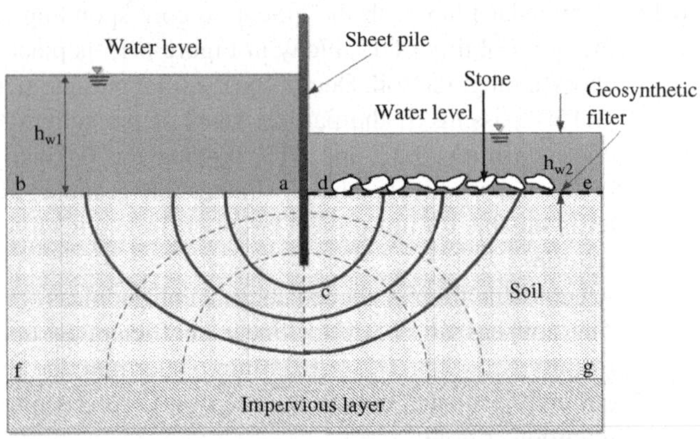

6.25. A highway is constructed by excavating down the ground with a drainage layer under the pavement. Given: $h_w = 10$ m, $h = 4$ m, $L_{R2} = 4$ m. The ground has a uniform soil with permeability $k = 6.5 \times 10^{-6}$ m/s. Calculate the total rate of the groundwater inflow.

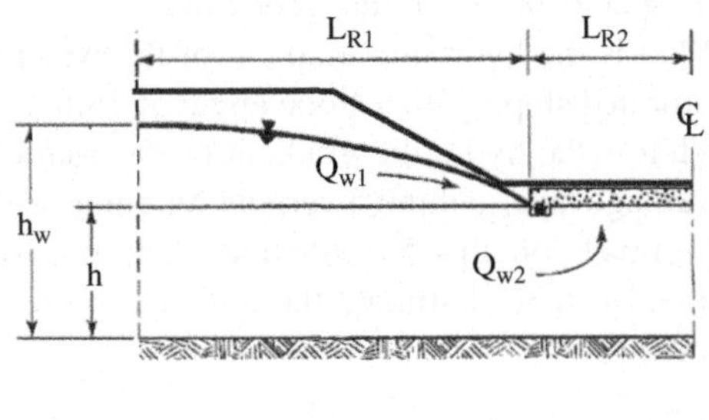

6.26. A two-lane asphalt pavement with a width of 7.2 m is constructed in an area with a maximum rainfall ratio of 35 mm/h. The pavement has a 4% transverse slope in one direction. A 300-mm-thick drainage base under the pavement is gravel with 5% silt fines and has a permeability of 0.014 m/s and porosity of 0.35. Calculate the total rate of water flow into the pavement section per unit length and the time for a 50% degree of drainage.

6.27. In Problem 6.26, calculate the maximum height of flow using the FHWA (1992), Giroud et al. (1992), and McEnroe (1993) methods.

6.28. In Problem 6.26, if smooth plastic pipes with spacing of 10 m are used as outlet drainage pipes, what is the required diameter of the pipe?

6.29. A granular drainage layer is installed beneath a 9-m-wide pavement with 8% grade to drain the infiltration water to the side slope. The effective porosity and the permeability of the granular layer are 35% and 2.5 m/day, respectively. Five days is the allowable time to have 50% drainage. Determine the required thickness of the drainage layer.

6.30. In Problem 6.29. If the granular drainage layer is replaced by a geocomposite material with a thickness of 5 mm and effective porosity of 60%, calculate the required transmissivity of the geocomposite layer.

6.31. A 9-m-wide slope with 2.5% grade in a landfill is subjected to an infiltration of 8 mm/h. Given: the soil permeability $k = 2.1 \times 10^{-7}$ m/s. Calculate the maximum height of flow using the Giroud et al. (1992) method.

6.32. A well is installed into a sand layer to lower the groundwater level, which is located above an impermeable layer. Two observation wells are installed at distances of 5 and 10 m from the center of the well to monitor the effectiveness of dewatering. Given: the radius of the well $r_0 = 0.5$ m, the sand permeability $k = 2.8 \times 10^{-5}$ m/s, water heads in two observation wells are 12 and 15 m, respectively, above the impermeable layer. Calculate the rate of water flow from the well.

6.33. A well is installed into a sand layer, which is located between two impermeable layers, to lower the groundwater level. Two observation wells are installed at distances of 5 and 10 m from the center of the well to monitor the effectiveness of dewatering. Given: the radius of the well $r_0 = 0.5$ m, the sand permeability $k = 2.8 \times 10^{-5}$ m/s, the thickness of the sand layer $h_{dr} = 20$ m, water heads in two observation wells above the lower impermeable layer are 12 and 15 m, respectively. Calculate the rate of water flow from the well.

6.34. If a pumping well is close to an existing building, dewatering may induce settlement of the building. What measures may be taken to minimize the settlement?

6.35. Single-line well points are used for dewatering at an excavation site, which consists of a sandy soil above impermeable bedrock. The groundwater before dewatering is 10 m above the bedrock and the soil permeability $k = 1.3 \times 10^{-5}$ m/s. The maximum drawdown depth is 4 m. Estimate the influence radius of this dewatering.

6.36. Multiwells are used for dewatering at an excavation site. The dimension of the excavation pit is 150 m × 200 m. The height of the phreatic level above the impermeable layer before pumping is 15 m. The height of the phreatic level at the center of the excavation pit after pumping is 6 m. The soil permeability $k = 2.3 \times 10^{-5}$ m/s. The wells have a diameter of 200 mm. Determine the number of wells required for this dewatering project.

6.37. Well points are used to lower a groundwater table by 5.0 m in uniform medium sand. What is the required spacing of well points according to the NAVFAC (1982) design chart?

6.38. In Problem 6.37, if the medium sand is stratified, what is the required spacing of well points?

6.39. A 10-m-thick clay layer is above a sand layer in an excavation project. The groundwater table before excavation and dewatering is 1.0 m below the ground surface. The excavation depth is 5 m from the ground surface. The groundwater table in the excavation pit is at the bottom of the pit. The saturated unit weight of the clay is 18.5 kN/m^3. Evaluate whether the bottom of the excavation is stable.

REFERENCES

AASHTO (1991). Report on Task Force 25, Joint Committee Report of AASHTO-AGCARTBA, American Association of State, Highway, and Transportation Officials, Washington, DC, Dec.

AASHTO (1993). *AASHTO Guide for Design of Pavement Structures*. AASHTO, Washington, DC.

AASHTO (2006). *Standard Specifications for Geotextiles—M 288, Standard Specifications for Transportation Materials and Methods of Sampling and Testing* (26th ed.). American Association of State Transportation and Highway Officials, Washington, DC.

American Society of Civil Engineers (ASCE) (1998). *Urban Subsurface Drainage*. ASCE Manuals and Reports on Engineering Practice No. 95. Washington, DC.

Barber, E. S., and Sawyer, C. L. (1952). "Highway subdrainage." *Proceedings,* Highway Research Board, 643–666.

Bell, R.R. and Hicks, R.G. (1980). Evaluation of Test Methods and Use Criteria for Geotechnical Fabrics in Highway Applications. Interim Report, FHWA/RD-80/021. FHWA, Washington, DC.

Casagrande, A. and Shannon, W. L. (1952). "Base course drainage for airport pavements." *Proceedings of the American Society of Civil Engineers*, 77, 792–814.

Cedergren, H. R., Arman, J. A., and O'Brien, K. H. (1972). *Development of Guidelines for the Design of Subsurface Drainage Systems for Highway Pavement Structural Sections*. Report No. FHWA-RD-72-30, Federal Highway Administration, Washington, DC.

Cedergren, H. R., Arman, J. A., and O'Brien, K. H. (1973). *Development of Guidelines for the Design of Subsurface Drainage Systems for Highway Pavement Structural Sections*. Report No. FHWA-RD-73-14, Federal Highway Administration, Washington, DC.

Chen, B.S. and Jensen, R.E. (2013). "Case studies of dewatering and foundation design: Retail warehouse in Taiwan." *Proceedings of the Seventh International Conference on Case Histories in Geotechnical Engineering*, April 20 to May 4, Chicago.

Christopher, B.C. and Zhao, A. (2001). *Design Manual for Roadway Geocomposite Underdrain Systems*. Internal Report, Christopher Consultants, Roswell, GA and Tenax Corporation, Baltimore, MD.

Department of the Army (1983). *Dewatering and Groundwater Control*. Army Technical Manual No. 5-818-5 (November 15). Available at http://armypubs.army.mil/eng/DR_pubs/dr_a/pdf/tm5_818_5.pdf.

FHWA (1992). *Demonstration Project 87: Drainage Pavement Systems, Participant Notebook*, Federal Highway Administration, Publication No. FHWA-SA-92-008, Washington, DC.

Forchheimer, P. (1930). *Hydraulik*. B. G. Teubner, Leipzig.

Giroud, J.P. (2010). "Development of criteria for geotextile and granular filters." *Proceedings of the 9th International Conference on Geosynthetics*, Brazil, 45–64.

Giroud, J.P., Badu-Tweneboah, K., and Bonaparte, R. (1992). "Rate of leakage through a composite liner due to geomembrane defects." *Geotext. Geomembr.*, 11(1): 1–28.

Giroud, J. P., Zhao, A. and Boneparte, R. (2000). "The myth of hydraulic transmissivity equivalency between geosynthetic and granular liquid collection layers." *Geosynthetics International* 7(4–6): 381–401.

Han, J. and Jiang, Y. (2013). "Use of geosynthetics for performance enhancement of earth structures in cold regions." *Sci. Cold Arid Regions*, 5(5): 517–529.

Henry, K.S. (1996). "Geotextile to mitigate frost effects in soils: A critical review." *Transport. Res. Rec.*, 1534: 5–11.

Holtz, R. D., Christopher, B. R., and Berg, R. R. (2008). *Geosynthetic Design and Construction Guidelines*. Publication No. FHWA NHI-07-092, National Highway Institute, Arlington, VA.

Huang, Y. H. (2004). *Pavement Analysis and Design* (2nd ed.). Pearson/Prentice Hall, Upper Saddle River, NJ.

Huasmann, M.R. (1990). *Engineering Principles of Ground Modification*. McGraw-Hill, New York.

Lawson, C.R. (1992). "Geotextile revetment filters." *Geotext. Geomembr.*, 11(4–6): 431–448.

Manning, R. (1891). "On the flow of water in open channels and pipes." *Transactions of the Institution of Civil Engineers of Ireland* 20, 161–207.

Mansur, C. I. and Kaufman, R. I. (1962). "Dewatering." Chapter 3 of Foundation Engineering, G.A. Leonards (ed.), McGraw-Hill, New York, 241–351.

McEnroe, B.M. (1993). "Maximum saturated depth over landfill liner." *J. Environ. Eng.*, 119(2): 262–270.

Moulton, L. K. (1980). *Highway Subdrainage Design*. Federal Highway Administration, Report No. FHWA-TS-80-224 (August). Available at http://isddc.dot.gov/OLPFiles/FHWA/009633.pdf.

Naval Facilities Engineering Command (NAVFAC) (1982). *Soil Mechanics Design Manual 7.1*, NAVFAC DM-7.1, May, Department of the Navy.

Sichardt, W. (1928). *Das Fassungsvermogen von Rohrbrunnen*. Springer, Berlin.

Steinberg, M. L. (1998). *Geomembranes and the Control of Expansive Soils*. McGraw-Hill Professional, New York.

Xanthakos, P.P., Abramson, L.W., and Bruce, D.A. (1994). *Ground Control and Improvement*. Wiley, New York.

CHAPTER 7

Preloading

7.1 INTRODUCTION

7.1.1 Basic Concept

Preloading is one of the traditionally but still commonly used ground improvement methods in practice. The basic concept of this technology is to reduce void ratio (i.e., compressibility) of geomaterial through consolidation (i.e., dissipation of excess pore water pressure) by applying loads on ground surface for a certain time period and then removing it for construction of a permanent structure. Figure 7.1 illustrates fill preloading followed by permanent structure construction. During the preloading, settlement develops with loading and time. When the fill is removed at the end of preloading, there is a rebound. Construction of the permanent structure induces new settlement due to the increase of the load. However, the induced settlement by the structure after preloading is expected to be much smaller than that without preloading. Figure 7.1 shows the continuous settlement without the removal of preloading. It is important in design to determine the time for the end of preloading. For soft clays, the consolidation may take longer time to complete due to their low permeability. If the time for preloading and construction of the structure exceeds the available time, vertical drains can be installed to shorten drainage distance thus accelerating the rate of consolidation and reducing the time for soil consolidation and settlement.

The American engineer, D. J. Moran, first proposed a sand drain technology for ground improvement of deep soft deposits in 1925. A U.S. patent was granted in 1926. Sand drains are formed by filling predrilled holes with highly permeable sand and arranged in regular patterns with certain spacing to shorten drainage distance of water in soil thus accelerating the rate of consolidation. The California Division of Highways, Materials and Research Department, conducted laboratory and field tests using this sand drain technology in 1933 (Rixner et al., 1986). In the late 1930s, W. Kjellman in Sweden developed the technology of prefabricated band-shaped vertical drains, each drain made of a cardboard core and a paper filter jacket. The paper filter jacket was later replaced with nonwoven geotextile. Nowadays prefabricated vertical drains (PVDs) have been widely used in the world for ground improvement.

Fill preloading requires the use of significant amount of fill. Transport and placement of fill take time and cost money. Sometimes, fill placement induces failure and large movement. To avoid these problems, another important development of the preloading technology is the idea of vacuum preloading, which was proposed by Kjellman (1952). Different from fill preloading, vacuum preloading utilizes atmospheric pressure on an air-tight system installed in the field by applying vacuum inside the system (Figure 7.2). A vacuum preloading system typically includes a vacuum pump, impervious membrane liner, sand drainage layer, water collection pipe, vertical drains, and a trench and clay revetment to seal the boundary and prevent leak. When the vacuum pressure is not sufficient, fill may be added on top of the system to increase the preloading pressure.

Therefore, the applied load in preloading can be fill, vacuum pressure, or a combination of fill and vacuum pressure. Water has also been filled into storage tanks to preloading soft foundations before they are used to store other liquid. Vacuum pressure will be removed after the completion of preloading while fill may be removed or remain as part of an earth structure. Permanent structures will be constructed after preloading and are expected to have small settlement. Soft clays typically have low shear strength, which may limit the magnitude of the applied load. Under such a condition, staged construction is needed to increase the allowable load by stages.

To increase the magnitude of the primary consolidation settlement during preloading and minimize the secondary settlement after preloading, surcharge loading can be used.

7.1.2 Suitability

Preloading is often cost effective to improve saturated, low strength, and highly compressible clays and silts when time is not a major concern. Vertical drains can be used to shorten the time for preloading if time is a major concern. Vertical drains are typically installed to a depth of 30 m (more than 60 m in some projects). Preloading is effective when the loading is higher than soil preconsolidation stress. Fill preloading is more suitable if fill material is inexpensive and readily available and/or part of the permanent structure (such as embankments). Vacuum preloading is more suitable for the areas where soils are too weak to support construction equipment and fill, and/or fill material is expensive and not readily available. The maximum pressure applied by vacuum preloading is limited by the atmosphere pressure. Under such

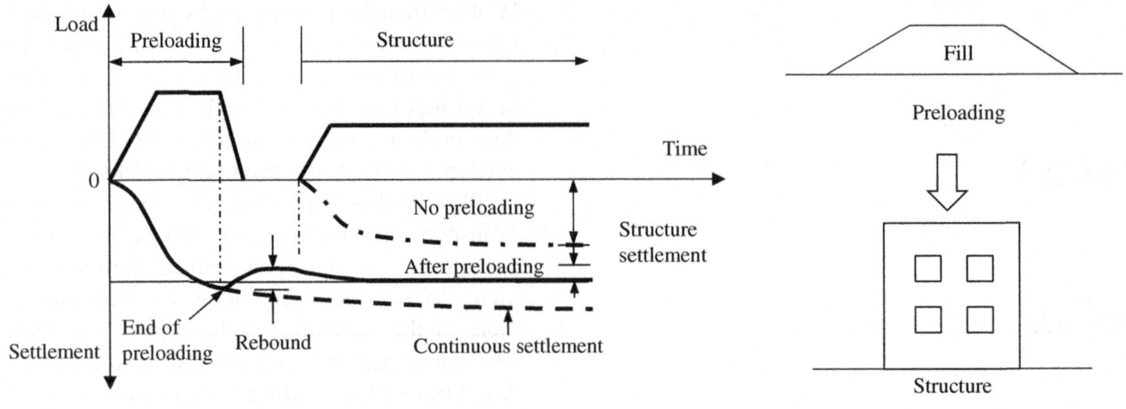

Figure 7.1 Fill preloading and application.

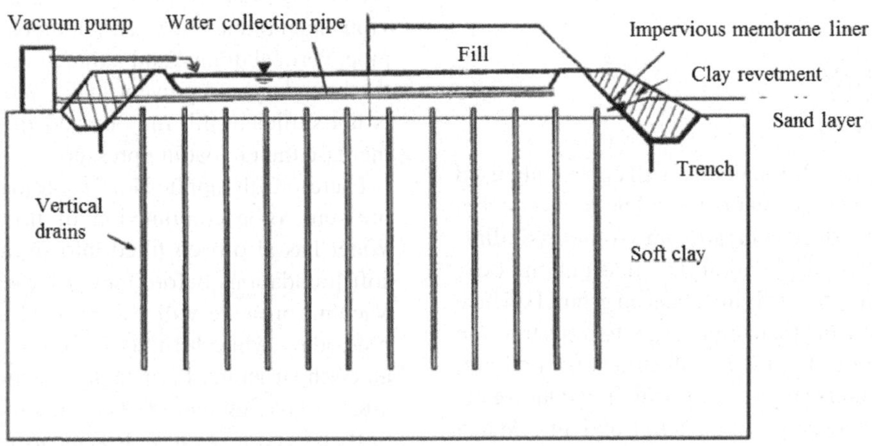

Figure 7.2 Vacuum preloading (Shang et al., 1998, reprinted with permission from NRC Research Press, National Research Council of Canada).

a situation, fill may be added on top of the vacuum system to increase the preloading pressure. When permeable layers exist close the ground surface, they may pose challenges for maintaining vacuum pressure.

7.1.3 Applications

Preloading has been commonly used and not limited to the following applications: (1) highways, (2) airports, (3) land reclamation, (4) storage tanks, and (5) buildings.

7.1.4 Advantages and Limitations

Preloading is a reliable ground improvement method and has a long track record of many successful case studies. It is often cost effective if fill material is readily available or part of a permanent structure. Fill preloading can be easily implemented in an open land with limited or no demand for specialty construction equipment, skilled workers, and quality control/assurance. The advantages of vacuum preloading are no fill material required, short construction period, and no need for heavy equipment. The main limitation of preloading is longer required time for consolidation, especially used for low permeable geomaterials. When vertical drains are used, they require high headspace for installation. This method may not work well within constrained areas because of material transportation and induced ground movement. The limitations of vacuum preloading are the magnitude of vacuum pressure limited by atmospheric pressure (theoretically 100 kPa) and potential cracks in the surrounding areas due to inward lateral soil movement, which will be discussed in next section.

7.2 PRINCIPLES

7.2.1 Precompression

The basic principle of preloading is to reduce the void ratio of geomaterial by consolidation, thus reducing

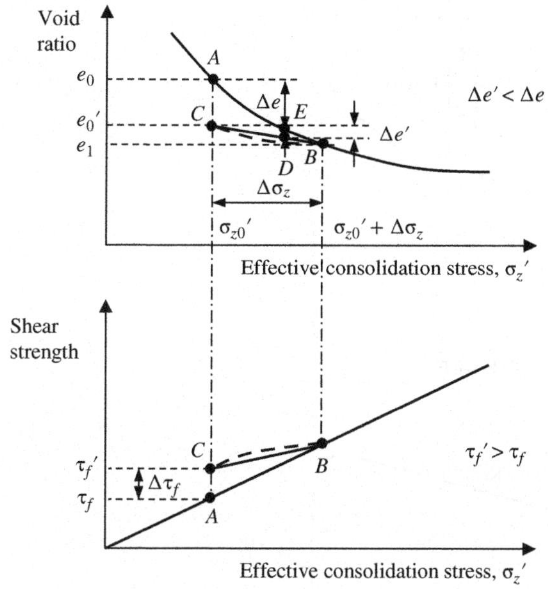

Figure 7.3 Principle of preloading (modified from Holtz and Kovacs, 1981).

the compressibility and increasing the strength of the geomaterial. Figure 7.3 illustrates this basic principle based on one-dimensional consolidation. The effective consolidation stress implies that the void ratio and shear strength of the geomaterial in Figure 7.3 are at the end of consolidation. The process of consolidation will be discussed later. When the consolidation stress on the geomaterial element increases from the initial overburden stress, σ'_{z0} (point *A*) to $\sigma'_{z0} + \Delta\sigma_z$ (point *B*) by preloading, the void ratio of the geomaterial decreases from e_0 to e_1 following the compression line. After the completion of the consolidation, the applied stress, $\Delta\sigma_z$, is removed, thus the consolidation stress decreases to σ'_{z0}. During this process, the geomaterial rebounds through the dashed line (a rebound line) from point *B* to *C* (at void ratio, e'_0). It is clear that e'_0 is less than e_0; therefore, the geomaterial has been densified. If a permanent structure is constructed afterward and the stress increases from point *C* to *D* (less than $\sigma'_{z0} + \Delta\sigma_z$), the induced void ratio reduction, $\Delta e'$, following the recompression line is much smaller than that induced by increasing the stress from point *A* to *E* (at the same stress as point *D*) following the compression line i.e. $\Delta e' << \Delta e''$. In other words, the geomaterial deformation after preloading is greatly reduced. It is desired that point *D* is between points *C* and *B* because the geomaterial deformation within the recompression line is small. It is undesired if point *D* is beyond point *B* because the geomaterial deformation on the compression line is large.

Due to the densification of the geomaterial after preloading, the shear strength of the geomaterial increases from point *A* to *C* before the reloading by the structure. The shear strength increase, also referred as strength gain, is important for the stability of fill surcharge during staged construction. The method for estimating the strength gain will be discussed in a later section.

7.2.2 Stress and Ground Movement

Fill preloading and vacuum preloading have three obvious differences: (1) fill preloading induces positive excess pore water pressure in geomaterial while vacuum preloading induces negative excess pore water pressure; (2) fill preloading induces unequal vertical and horizontal stresses, while vacuum preloading induces the same stresses (i.e., vacuum pressure) in all directions; and (3) fill preloading induces outward movement in the horizontal direction, while vacuum preloading induces inward movement in the horizontal directions. Figure 7.4 presents the stress paths of fill preloading and vacuum preloading. It is shown that the fill preloading results in a stress path *AB* toward the failure envelope; therefore, there is a possibility of failure during fill preloading. Geomaterial consolidation during a waiting period follows the stress path *BC*. However, vacuum preloading results in a horizontal stress path *AE*, which moves away from the failure envelope; therefore, no failure is possible during vacuum preloading. The stress path *AD* represents a K_0 consolidation (i.e., no horizontal deformation or stain, $\varepsilon_h = 0$). Figure 7.4 also presents the vertical and horizontal stresses in a triaxial condition from Points A, B, C, D, to E.

Both Figures 7.4 and 7.5 show that fill preloading induces outward horizontal movement (i.e., $\varepsilon_h < 0$) while vacuum preloading induces inward horizontal movement ($\varepsilon_h > 0$). The K_0 consolidation due to horizontal confinement has no horizontal movement (i.e., $\varepsilon_h = 0$). From odometer tests, Chai et al. (2005) found that when the vacuum pressure is higher than the stress requested to maintain the K_0 condition, there is inward horizontal movement, and the settlement by vacuum loading is less than that by fill loading. However, when the vacuum pressure is lower than the requested stress for the K_0 condition, their settlements are the same.

Figure 7.6 illustrates the variations of excess pore water pressure and effective stress profiles under fill and vacuum preloading. It is shown that at the moment of fill loading, the initial hydrostatic water pressure $u_0(z)$ is increased by the amount of additional fill stress $\Delta\sigma_z$ and the pore water pressure decreases with time to $u_t(z)$ at time *t* [i.e., the remaining positive excess pore water pressure $\Delta u(z) = u_t(z) - u_0(z)$]. As a result, the effective stress in geomaterial increases from the initial vertical overburden stress $\sigma'_{z0}(z)$ to $\sigma'_{z0}(z) + \Delta\sigma_z - \Delta u(z)$ at time *t*. On the other hand, at the moment of vacuum preloading, the initial hydrostatic water pressure $u_0(z)$ is reduced by the amount of vacuum pressure $-u_s$ and the pore water pressure increases with time to $u_t(z)$ at time *t* [i.e., the remaining negative excess pore water pressure $\Delta u(z) = u_t(z) - u_0(z)$]. As a result, the effective stress in soil increases from $\sigma'_{z0}(z)$ to $\sigma'_{z0}(z) + u_s - \Delta u(z)$ at time *t*. If the vacuum

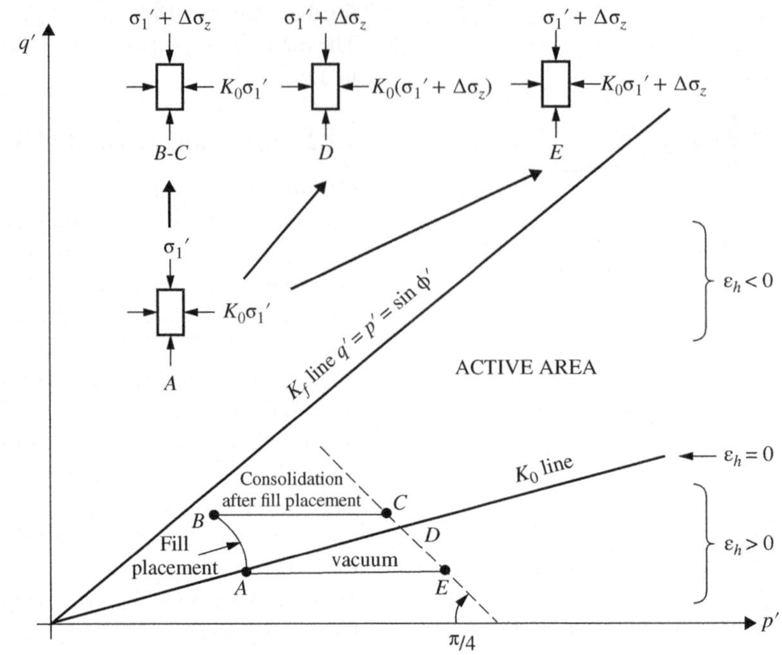

Figure 7.4 Stress paths of fill preloading and vacuum preloading (modified from Masse et al., 2001).

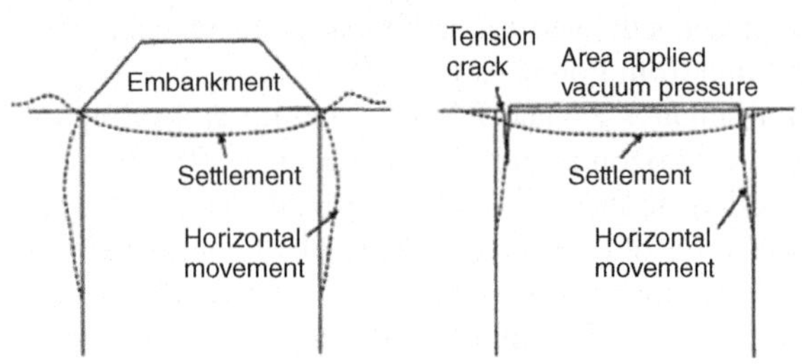

Figure 7.5 Ground movement by fill and vacuum preloading (after Chai et al., 2005).

pressure and the additional fill stress are equal and positive and negative excess pore water pressures dissipate at the same rate, the effective stress increases by fill and vacuum loadings are equivalent.

7.2.3 Consolidation Theory

General Problem To accelerate consolidation of soft soil, vertical drains can be installed in soft soil to provide another and short drainage path. Figure 7.7 illustrates a general axisymmetric problem that includes a vertical drain with higher permeability in the middle of a unit cell of soil cylinder. The unit cell has drainage surfaces on the top and bottom; therefore, the longest drainage distance in the vertical distance is h_{dr}. Due to the inclusion of the vertical drain, water can also flow radially toward the drain. After water reaches the drain, it will flow vertically inside the drain to the top or bottom surface.

Basic Equations Partial differential equation for axisymmetric water flow in soil as shown in Figure 7.6 can be expressed as follows:

$$c_r\left(\frac{1}{r}\frac{\partial u}{\partial r} + \frac{\partial^2 u}{\partial r^2}\right) + c_v\frac{\partial^2 u}{\partial z^2} = \frac{\partial \overline{u}}{\partial t} \tag{7.1}$$

where c_r = coefficient of consolidation in a horizontal (radial) direction
c_v = coefficient of consolidation in a vertical direction
r = radial distance
u = pore water pressure at a distance of r
u = average pore water pressure
t = time

The degree of consolidation, also referred as the rate of consolidation, is defined as

$$U = \frac{u_0 - u_t}{u_0} = 1 - \frac{u_t}{u_0} \tag{7.2}$$

where u_0 = initial excess pore water pressure
u_t = remaining excess pore water pressure at time t

The overall degree of consolidation of soil due to vertical and radial flow can be calculated as follows (Carillo, 1942):

$$U_{vr} = 1 - (1 - U_v)(1 - U_r) \tag{7.3}$$

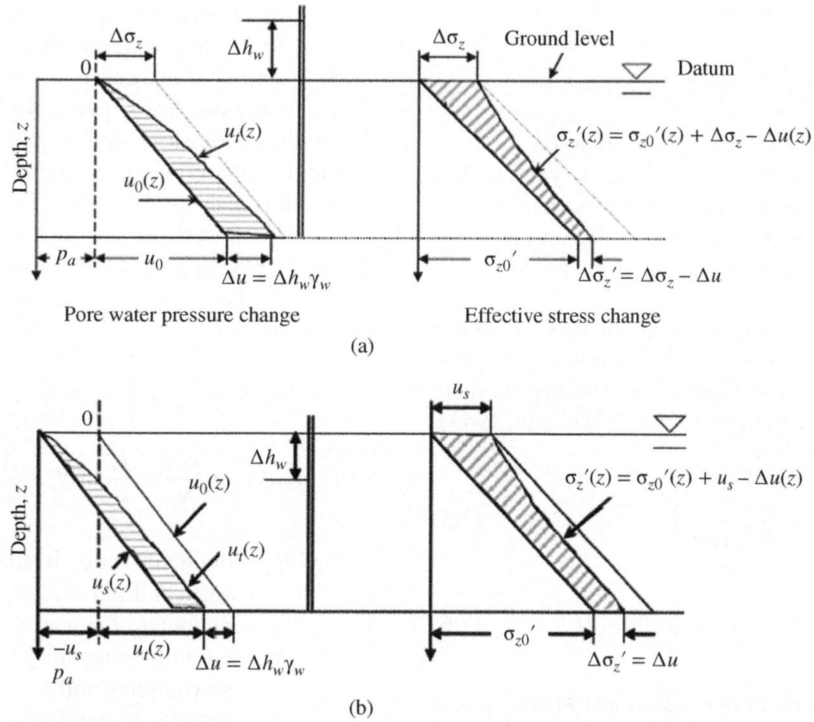

Figure 7.6 Variations of excess pore water pressure and effective stress profiles under (a) fill preloading and (b) vacuum preloading (after Chu and Yan, 2006).

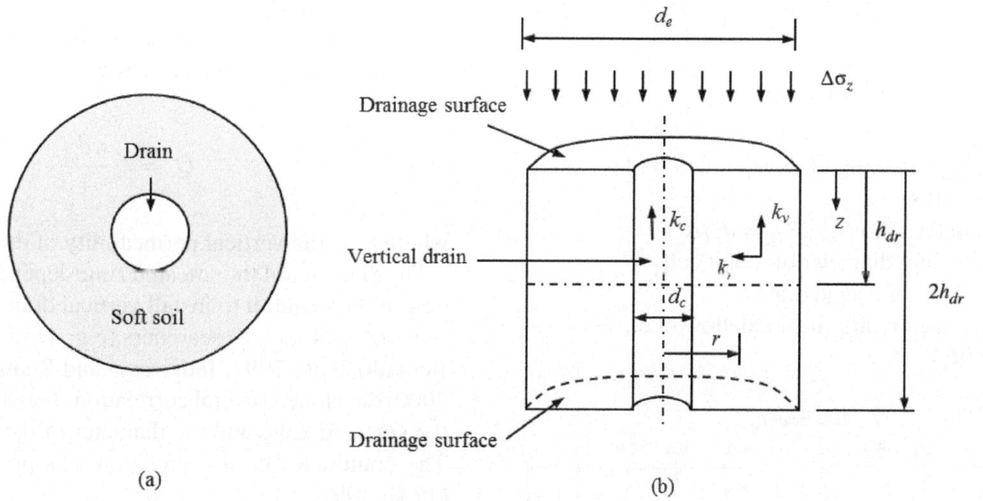

Figure 7.7 Axisymmetrical consolidation problem: (a) plan view and (b) cross-sectional view.

where U_{vr} = overall degree of consolidation
U_v = degree of consolidation in a vertical direction
U_r = degree of consolidation in a radial direction

Consolidation Due to Vertical Flow The degree of soil consolidation due to vertical flow can be calculated using the Terzaghi one-dimensional consolidation theory (Terzaghi, 1943):

$$U_v = 1 - \sum_{m=0}^{\infty} \frac{2}{M^2} e^{-M^2 T_v} \quad m = 0, 1, 2, 3, \ldots \tag{7.4}$$

where $M = \frac{(2m+1)\pi}{2}$

$T_v = \frac{c_v t}{h_{dr}^2}$, time factor

t = time

c_v = coefficient of vertical consolidation

h_{dr} = longest drainage distance due to vertical flow (if top and bottom drainage surfaces exist, half of the soil thickness between these two surfaces should be used)

The average degree of consolidation with the time factor under a uniform distribution of initial excess pore water pressure can also be presented in Figure 7.8. The approximate relationships for the curve in Figure 7.8 are (Terzaghi, 1943):

For U_v = 0–52.6%:

$$T_v = \frac{\pi}{4}\left(\frac{U_v}{100}\right)^2 \tag{7.5}$$

For U_v > 52.6%:

$$T_v = 1.781 - 0.933\log_{10}(100 - U_v) \tag{7.6}$$

Consolidation Due to Radial Flow Barron (1948) proposed a solution for an average degree of consolidation due to horizontal (radial) flow toward a free-draining sand drain:

$$U_r = 1 - \exp\left[-\frac{8}{F(N_D)}T_r\right],$$

$$F(N_D) = \frac{N_D^2}{N_D^2 - 1}\ln(N_D) - \frac{3N_D^2 - 1}{4N_D^2} \tag{7.7}$$

where U_r = average degree of consolidation due to radial flow

N_D = diameter ratio (i.e., $N_D = d_e/d_c$)

d_e = equivalent diameter of a unit cell

d_c = diameter of a sand drain

T_r = time factor due to radial flow (i.e., $T_r = c_r t/d_e^2$)

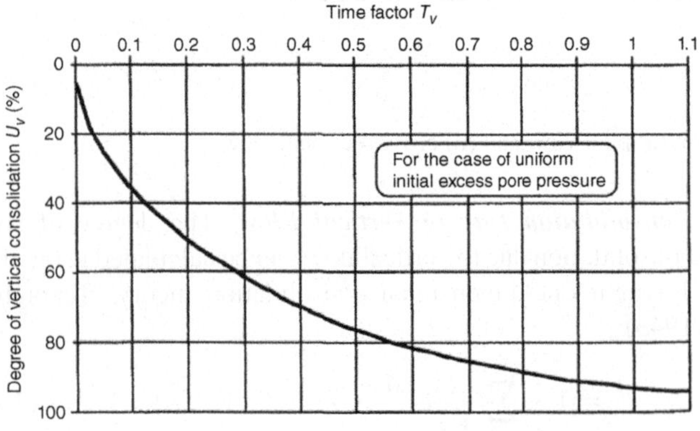

Figure 7.8 Average rate of consolidation due to vertical flow.

During installation of vertical drains, the surrounding soil around the drains is often disturbed. The disturbed zone with a certain thickness is often referred as the smeared zone, which has lower permeability than the original soil. In addition, vertical drains may have limited discharge capacities; therefore, water flow in the vertical drains may encounter resistance. This effect is referred to as well resistance. Based on the unit cell model in Figure 7.9, Hansbo (1981) proposed an average degree of consolidation of a sand drain due to radial flow considering smear and well resistance:

$$U_r = 1 - \exp\left[-\frac{8}{F_m(N_D)}T_r\right]$$

$$F_m(N_D) = \ln\frac{N_D}{N_s} + \frac{k_r}{k_s}\ln N_s - \frac{3}{4} + \pi z(2h_{dr} - z)\frac{k_r}{Q_c} \tag{7.8}$$

where N_s = diameter ratio of smeared zone to vertical drain (i.e., $N_s = d_s/d_c$)

d_s = diameter of smeared zone

k_r = radial permeability of undisturbed surrounding soil

k_s = radial permeability of smeared soil

z = depth in the ground at which the degree of consolidation is computed

h_{dr} = longest drainage distance due to vertical flow

Q_c = discharge capacity of a vertical drain, which can be expressed as follows:

$$Q_c = k_c\frac{\pi d_c^2}{4} \tag{7.9}$$

where k_c is the vertical permeability of drain well.

The diameter of the smeared zone depends on the shape and size of the mandrel to install vertical drains and the type and sensitivity of soil. Researchers (e.g., Holtz and Holm, 1973; Bergado et al., 1991; Indraratna and Redna, 1998; Bo et al., 2000) developed several correlations between the diameter of the smeared zone and the diameter of the drain or mandrel. The commonly used correlation was proposed by Hansbo (1981, 1997):

$$d_s = (1.5 \text{ to } 3.0)d_c \tag{7.10}$$

Installation of prefabricated vertical drains (PVDs) disturbs the soil surrounding the PVDs and reduces the radial permeability of the smeared zone. Researchers (e.g., Hansbo, 1981; Onoue, 1992; Indraratna and Redna, 1998; Hird and Moseley, 2000) found that the permeability of the smear zone can be estimated by

$$k_s = \frac{1}{\lambda}k_r \quad \text{typically } \lambda = 2\text{–}6 \tag{7.11}$$

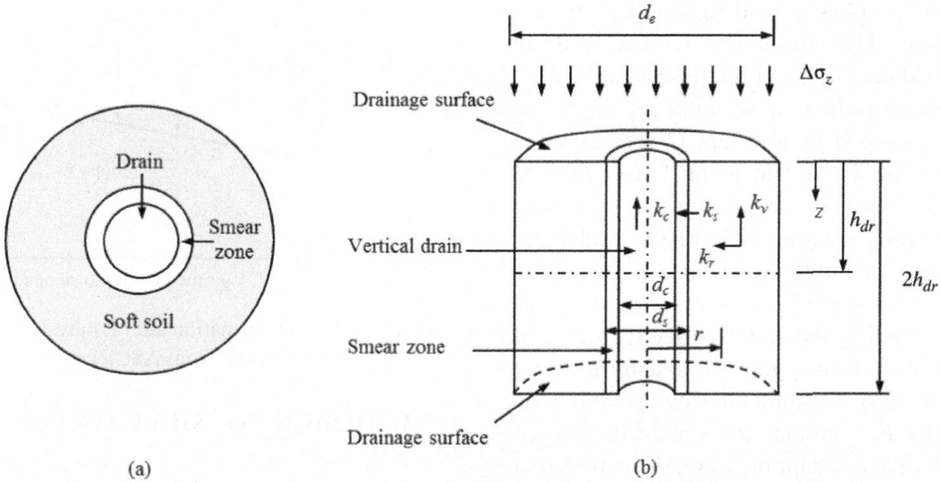

Figure 7.9 Axisymmetric unit-cell drainage model considering smear and well resistance.

Consolidation Due to Vacuum Pressure The average degree of consolidation of soil under vacuum preloading can be calculated as follows (Chu and Yan, 2005):

$$U = 1 - \frac{\int_0^{h_{\mathrm{dr}}} [u_t(z) - u_s(z)]\, dz}{\int_0^{h_{\mathrm{dr}}} [u_0(z) - u_s(z)]\, dz} \quad (7.12)$$

$$u_s(z) = \gamma_w z - u_s \quad (7.13)$$

where $u_0(z)$ = initial pore water pressure at depth z
$u_t(z)$ = pore water pressure at depth z and time t
$u_s(z)$ = vacuum pressure at depth z
u_s = vacuum pressure applied (typically 80 kPa)
If $u_s = 0$, the above equations can be used for fill preloading.

7.2.4 Vacuum and Fill Combined Preloading

The maximum vacuum pressure for preloading is limited by atmosphere pressure (theoretically 100 kPa). In practice, however, the achievable pressure typically ranges from 60 to 80 kPa (Bergado et al., 1998; Tang and Shang, 2000; Chu and Yan, 2005) due to loss of vacuum in the system and into the geomaterial. When the required pressure is higher than 80 kPa, additional fill can be added on the vacuum system thus creating a vacuum and fill combined preloading (Chu et al., 2000; Yan and Chu, 2005). Chai et al. (2006) pointed out that fill loading induces outward horizontal displacement while vacuum loading induces inward horizontal displacement. If they are used separately, both horizontal displacements are undesirable. However, when they are combined together, the horizontal displacement is minimized due to the opposite movements. Chai et al. (2013) proposed a method to estimate horizontal displacement under vacuum and fill combined loading. Readers are referred to Chai et al. (2013) for details.

Mohamedelhassan and Shang (2002) derived a solution for one-dimensional consolidation of vacuum and fill combined preloading. They considered the excess pore water pressure under combined loading is the sum of the excess pore water pressures under vacuum loading and fill loading separately. Based on the initial and boundary conditions, they obtained the average degree of consolidation under combined loading, which is the same as that by the Terzaghi one-dimensional consolidation solution. They further confirmed that the soil consolidation coefficient under vacuum, fill, or combined loading is nearly identical. Chai and Carter (2013) developed the solutions for consolidation of soil under combined vacuum pressure and fill loading considering vertical and radial flow. They also confirmed that the average degrees of consolidation under combined loading considering one-way vertical flow and radial flow are the same as those by the Terzaghi one-dimensional consolidation solution and the Hansbo's consolidation solution, respectively. However, under combined loading for two-way vertical flow, Chai and Carter (2013) suggested that the drainage distance should be the total thickness of the soft soil when the Terzaghi one-dimensional consolidation solution is used to calculate the degree of consolidation, because no vaccum pressure can be developed at the bottom boundary.

7.2.5 Surcharge Preloading

To shorten the time required for preloading and minimize postconstruction settlement, including the secondary settlement, surcharge preloading is often adopted in practice.

Surcharge preloading applies a load higher than what is needed for a final load. The extra load is removed after the completion of preloading. Nash (2001) illustrated the effect of surcharge on the reduction of secondary compression using the concept proposed by Bjerrum (1972) as shown in Figure 7.10. Assume the soil at point *A* has deposited for 1000 years, and it does not have any strain as an initial state. When a preloading pressure including a surcharge is applied on the ground, the initial state tends to move toward the instant compression state. Since the soil consolidates during and after the loading, the vertical effective stress and the strain increase. As a result, point *A* actually moves to point *B*. Assume primary consolidation completes in one year; therefore, point *B* represents the end of the primary consolidation. Under the constant pressure, the soil starts to have secondary compression under the same vertical stress. If the pressure is maintained up to 100 years, the secondary compression will be from point *B* to *D*. If the surcharge $\Delta\sigma_s$, is removed at point *C*, there is a slight rebound from point *C* to point *E*. After that, the soil will resume secondary compression under the final pressure. Up to 100 years, the secondary compression is from point *E* to *F*. Figure 7.10 clearly shows that the strain from point *E* to *F* is much smaller than that from point *C* to *D*. In other words, the secondary compression after surcharge is much reduced.

Stewart et al. (1994) and Alonso et al. (2000) attributed the reduction of secondary compression to the higher overconsolidation ratio (OCR) induced by removal of the surcharge after preloading.

Mesri et al. (1997) suggested that the deformation of soil after removal of surcharge consists of several components as shown in Figure 7.11. After the removal of the surcharge, there is a primary rebound (i.e., the dissipation of negative pore water pressure) up to t_{rp} followed by a secondary rebound up to t_s. Then a secondary compression occurs, but at a varying slope, C'_α at time *t*. If the time of interest, *t*, is known, a secant slope, C''_α, may be used for calculation.

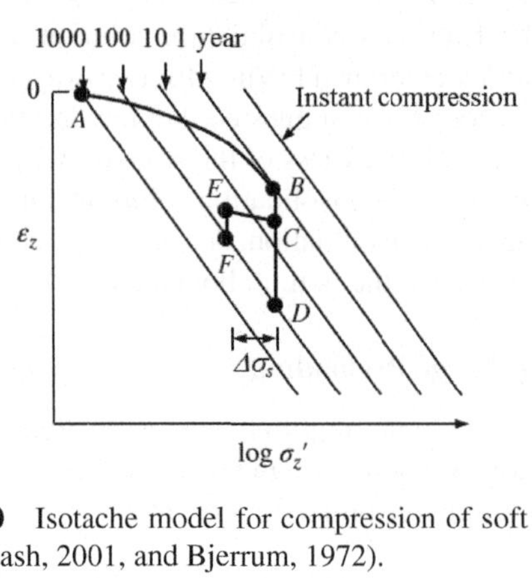

Figure 7.10 Isotache model for compression of soft clay (modified from Nash, 2001, and Bjerrum, 1972).

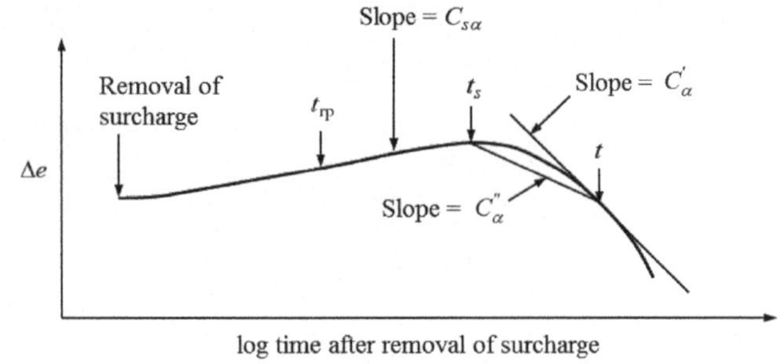

Figure 7.11 Deformation after removal of surcharge (Mesri et al., 1997, with permission from ASCE).

7.3 DESIGN CONSIDERATIONS

7.3.1 Vertical Drains

Types and Patterns Table 7.1 provides the types of vertical drains used in practice. Nowadays prefabricated band-shaped drains are most commonly used and PVDs are often referred to this type of drain.

In the field, vertical drains are often installed in a square or triangular pattern (Figure 7.12). The equivalent diameter of a unit cell for each vertical drain can be approximated based on the equivalent influence area of each vertical drain as follows:

$$d_e = 1.13s \quad \text{for a square pattern} \tag{7.14}$$

$$d_e = 1.06s \quad \text{for a triangular pattern} \tag{7.15}$$

where d_e = equivalent influence diameter
s = spacing between two adjacent vertical drains

Prefabricated Vertical Drains Equivalent diameter of a PVD based on perimeter equivalency and considering a correction factor is

$$d_c = \alpha_d \frac{2(b + t_g)}{\pi} \tag{7.16}$$

where d_c = equivalent diameter of a PVD
b = width of the PVD
t_g = thickness of the PVD
α_d = correction factor, typically 0.9–1.0

Federal Highway Administration *Prefabricated Vertical Drains Engineering Guidelines* (Rixner et al., 1986) suggested the following simplified formula:

$$d_c = \frac{b + t_g}{2} \tag{7.17}$$

The efficiency of PVD in discharging water depends on not only the discharge capacity of the PVD but also on the permeability of the surrounding soil and the length of the drain. Chu et al. (2004) suggested a PVD in a soil should have

Table 7.1 Types of Vertical Drains

Drain Type	Installation Method	Drain Diameter (m)	Typical Spacing (m)	Maximum Length (m)
Sand drain	Driven or vibratory closed-end mandrel (displacement type)	0.15–0.60	1.0–5.0	≤30
	Hollow-stem continuous flight auger (small displacement)	0.30–0.50	2.0–5.0	≤35
	Jetted (nondisplacement)	0.20–0.30	2.0–5.0	≤30
Prefabricated sand drain	Driven or vibratory closed-end mandrel; flight auger; rotary wash boring (displacement or nondisplacement)	0.06–0.15	1.0–4.0	≤30
Prefabricated band-shaped drain ("wick drain" or "strip drain")	Driven or vibratory closed-end mandrel (displacement or small displacement)	0.05–0.10 (equivalent diameter)	1.0–3.5	≤60

Source: Modified from Holtz et al. (1991).

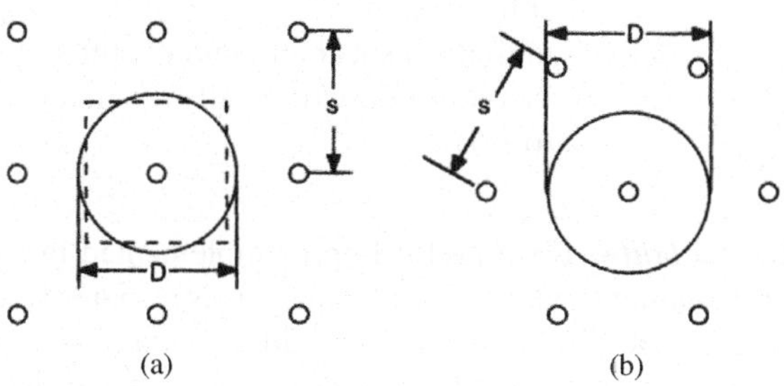

Figure 7.12 Vertical drain patterns: (a) square and (b) triangular patterns.

the following required discharge capacity to be considered as a free drain without any well resistance:

$$Q_c \geq 7.85\text{FS} \cdot k_r h_{\text{dr}}^2 \tag{7.18}$$

where Q_c = required discharge capacity of a PVD
FS = factor of safety (typically 4–6)
k_r = radial permeability of soil
h_{dr} = maximum drainage distance of PVD

In order to ensure the long-term performance of PVDs, geotextile filters wrapped around drainage cores also need to meet the soil retention, drainage, and clogging resistance requirements (Chu et al., 2004). The soil retention criteria for a geotextile filter are (Carroll, 1983):

$$O_{95} \leq (2\text{–}3)D_{85} \tag{7.19}$$

and

$$O_{50} \leq (10\text{–}12)D_{50} \tag{7.20}$$

where O_{95} = apparent opening size of geotextile filter
D_{85} = soil particle size of 85% finer
O_{50} = mean opening size of geotextile filter
D_{50} = soil particle size of 50% finer

The drainage criterion for a geotextile filter is

$$k_c \geq 10k_r \tag{7.21}$$

where k_c = permeability of the PVD.

The clogging resistance requirements for a geotextile filter (Wang and Chen, 1996) include

$$n_g \geq 30\% \tag{7.22}$$

$$O_{95} \geq 3D_{15} \tag{7.23}$$

$$O_{15} \geq (2\text{–}3)D_{10} \tag{7.24}$$

where n_g = porosity of filter
D_{15} = soil particle size of 15% finer
O_{15} = opening size of filter of 15% finer
D_{10} = soil particle size of 15% finer

PVDs should have enough tensile strength to survive during their installation. Bo et al. (2003) suggested that the tensile strengths of the core, filter, entire drain, and spliced drain should be higher than 1 kN at a tensile strain of 10% under both dry and wet conditions.

Equivalent Permeability Chai et al. (2001) proposed a simplified method to estimate the degree of consolidation of vertical drain-improved foundations. In their method, an equivalent vertical permeability can be estimated considering the existence of vertical drains as shown in Figure 7.13:

$$k_{\text{ve}} = \left(1 + \frac{2.5h_{\text{dr}}^2}{F_m\left(N_D\right) d_e^2} \frac{k_r}{k_v}\right) k_v \tag{7.25}$$

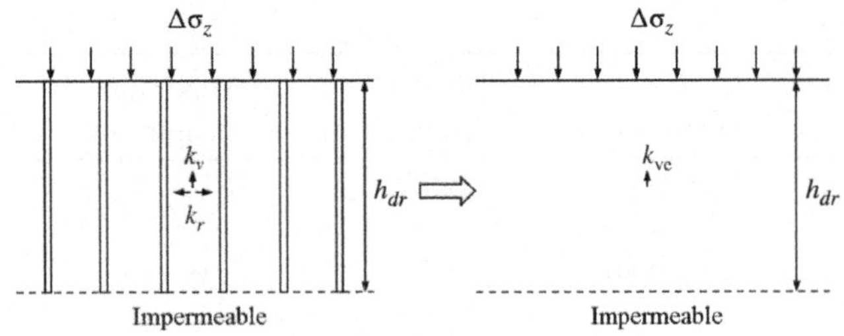

Figure 7.13 Equivalent permeability.

where
k_{ve} = equivalent vertical permeability
h_{dr} = maximum drainage distance of vertical drains
$F_m(N_D)$ = same as the one proposed by Hansbo (1981)

The equivalent vertical permeability is used in the Terzaghi one-dimensional consolidation solution to calculate the average degree of consolidation of vertical drain-improved foundations.

Optimum Penetration under Vacuum Preloading In the field, vacuum pressure often decreases with an increase of depth due to loss of vacuum. Chai et al. (2006) developed a solution for determining the optimum penetration depth of PVDs based on a simplified distribution of vacuum pressure with depth as shown in Figure 7.14:

$$h_{opt} = \left(\frac{k_{v1} - \sqrt{k_{v1}k_{v2}}}{k_{v1} - k_{v2}} \right) h \tag{7.26}$$

where
h_{opt} = optimum penetration depth
h = thickness of soft soil
k_{v1} = equivalent vertical permeability, which can be determined using Equation (7.25)
k_{v2} = vertical permeability of the soil below the PVD-improved zone

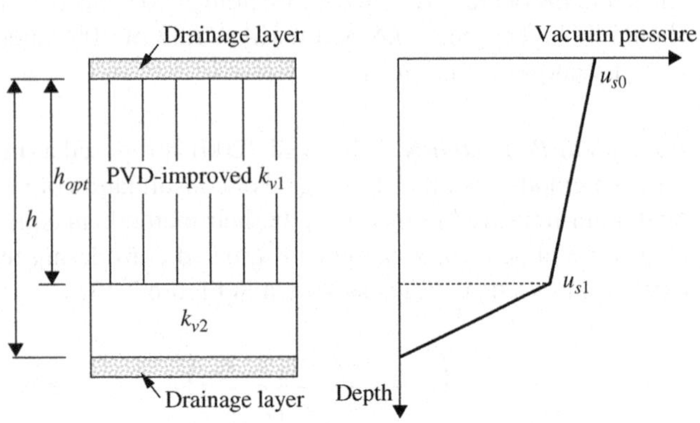

Figure 7.14 Optimum penetration depth of PVDs in vacuum preloading (after Chai et al., 2006).

An iterative procedure is needed to determine the optimum penetration depth of PVDs.

7.3.2 Preloading

Allowable Load Preloading has been mostly used to improve soft soils, which are often normally or underconsolidated. These soils, when subjected to fill loading, are more critical under an undrained condition than under a drained condition. Therefore, the ultimate bearing capacity of the soft foundation under an undrained condition is often used to estimate the allowable pressure for preloading:

$$p = \frac{q_{ult}}{FS} = \frac{N_c c_u}{FS} \tag{7.27}$$

where
p = allowable pressure for preloading
q_{ult} = ultimate bearing capacity
N_c = bearing capacity factor (5.14 for uniform soft soil)
c_u = undrained shear strength of soft soil
FS = factor of safety (typically ranging from 1.1 to 1.5)

Slope Stability If fill preloading is implemented in a form of embankments, slope stability analysis is also necessary to ensure the stability of embankments in addition to the bearing capacity requirement. Slope stability analysis can be conducted using limit equilibrium methods, such as Bishop's method based on circular failure and/or Spencer's method based on wedge failure, or numerical methods. If slope stability becomes a problem, geosynthetic reinforcement or other ground improvement method can be used to stabilize the embankment. The design of geosynthetic-reinforced embankments is discussed in Chapter 10.

Time-Dependent Loading In practice, loads are often not applied instantaneously, instead, they increase gradually. Figure 7.15 shows a typical loading pattern and its induced settlement. In this loading, the load increases linearly and then remains constant for a certain period. The consolidation solutions presented earlier are based on instantaneous loading. Taylor (1948) suggested a method to approximate the linearly increasing load to an instantaneous uniform load starting from the middle of the loading period (i.e., the dashed lines in Figure 7.15). This method can be used to calculate the degrees of consolidation under vertical and radial flow as discussed in Equations (7.4) and (7.8).

Olson (1977) developed the solution for one-dimensional consolidation under the linearly increasing load based on the Terzaghi consolidation theory as follows:

When $t \le t_1$,

$$\overline{u} = \sum_{m=0}^{\infty} \frac{2p}{M^3 T_{v1}} \sin \frac{Mz}{h_{dr}} (1 - e^{-M^2 T_v}) \tag{7.28}$$

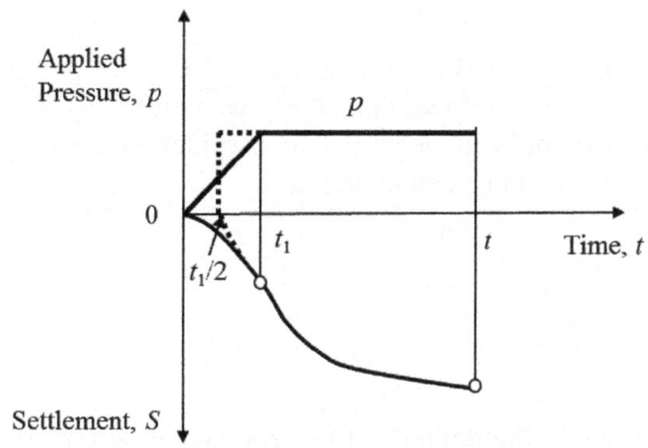

Figure 7.15 Time-dependent loading.

$$U_v = \frac{T_v}{T_{v1}}\left[1 - \frac{2}{T_v}\sum_{m=0}^{\infty}\frac{1}{M^4}\left(1 - e^{-M^2T_v}\right)\right] \quad (7.29)$$

where $\bar{u}$ is the average excess pore water pressure at depth z; $T_{v1} = c_v t_1/h_{dr}^2$; and other symbols are as defined earlier.

When $t > t_1$,

$$\bar{u} = \sum_{m=0}^{\infty}\frac{2p}{M^3T_{v1}}(e^{M^2T_{v1}} - 1)\sin\frac{Mz}{h_{dr}}(1 - e^{-M^2T_v}) \quad (7.30)$$

$$U_v = 1 - \frac{2}{T_{v1}}\sum_{m=0}^{\infty}\frac{1}{M^4}(e^{M^2T_{v1}} - 1)e^{-M^2T_v} \quad (7.31)$$

The equivalent permeability in Equation (7.25) proposed by Chai et al. (2001) can be used in Equation (7.29) or (7.31) to calculate the degree of consolidation of vertical drain-improved foundations. The solution developed by Zhu and Yin (2004) for the consolidation of soil with vertical and radial drainage under ramp loading considering smear effects can also be used for this purpose.

Time for Consolidation The time required for a desired degree of consolidation (say 80%) in one stage can be estimated using the consolidation theories discussed earlier. For an instantaneous load, if no vertical drain is included, Equation (7.4) can be used to estimate the time. For a linearly increasing load followed by a constant load, Taylor (1948) and Olson (1977)'s approaches can be used. If vertical drains are included, the Hansbo solution or the equivalent permeability method by Chai et al. (2001) is needed for this calculation. When vertical drains are used, the portion of consolidation contributed by vertical flow is typically small and a consolidation rate of 5–10% may be assumed for U_v as a starting point to estimate the required consolidation degree by radial flow, U_r, using Equation (7.3).

Strength Gain The undrained shear strength of a saturated cohesive soil can be estimated by (Ladd, 1991):

$$\frac{c_u}{\sigma_z'} = \chi_f(\text{OCR})^{\wedge} \quad (7.32)$$

where
- c_u = undrained shear strength of saturated cohesive soil
- χ_f = 0.22 ± 0.03 for homogeneous sedimentary clays (above A-line) or $\chi_f = 0.25 \pm 0.05$ for silts and organic clays (below A-line)
- OCR = overconsolidation ratio
- $\wedge$ = $0.88\,(1 - C_r/C_c)$
- C_r = recompression index
- C_c = compression index
- σ_z' = effective consolidation stress

The strength gain of the soil after consolidation can be estimated by

$$\Delta c_u = \chi_f(\text{OCR})^{\wedge}\Delta\sigma_z' = \chi_f(\text{OCR})^{\wedge}U_t\Delta\sigma_z \quad (7.33)$$

where
- Δc_u = strength gain
- $\Delta\sigma_z'$ = additional effective vertical consolidation stress induced by preloading
- $\Delta\sigma_z$ = additional total vertical consolidation stress induced by preloading
- U_t = degree of consolidation at time t (calculated from the previous step)

Since soft soils are often approximately normally consolidated (i.e., OCR = 1.0), $\chi_f = 0.25$ is commonly used. When the width of the load area is much larger than (typically at least three times) the thickness of the soft soil, Equation (7.33) can be simplified into

$$\Delta c_u = 0.25U_t p \quad (7.34)$$

where p = applied uniform pressure for preloading.

The total undrained shear strength after preloading is

$$c_{u1} = c_u + \Delta c_u \quad (7.35)$$

Accumulated Degree of Consolidation During a staged construction, each load may not be held until the end of the primary consolidation due to time constraint. Under such a condition, some excess pore water pressure still remains in soil from the previous loading(s) when the next load is applied. As a result, the excess pore water pressure accumulates from the previous loading period. A three-stage construction shown in Figure 7.16 is selected as an example to illustrate the procedure on how to calculate the accumulated degree of consolidation. In this illustration, only the consolidation due to vertical flow is considered using the approximate method suggested by Taylor (1948). However, this procedure can

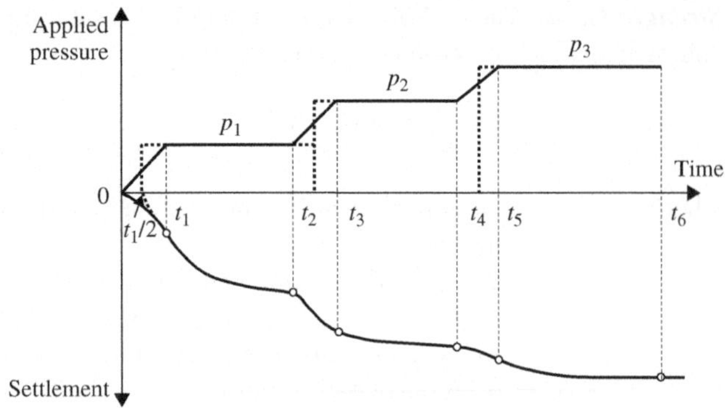

Figure 7.16 Applied pressure–settlement–time curve.

also be used for the consolidation due to radial flow. A similar procedure can be developed based on the method proposed by Olson (1977) using the equivalent permeability derived by Chai et al. (2001). To simplify the calculations, the incremental loading can be approximated to an instantaneous uniform loading with half of the duration, which starts at the middle of the actual loading.

For stage 1 at time t_1, the degree of consolidation, $U_{t1(p1)}$, under the loading, p_1, can be calculated by the time factor

$$T_{v1(p1)} = \frac{c_v t}{h_{dr}^2} = \frac{c_v(t_1 - t_1/2)}{h_{dr}^2} = \frac{c_v t_1}{2h_{dr}^2} \tag{7.36}$$

The excess pore water pressure at t_1 is

$$u_{t1(p1)} = u_{0(p1)}(1 - U_{t1(p1)}) \tag{7.37}$$

where $u_{0(p1)}$ is the initial excess pore water pressure induced by the instantaneous loading, p_1.

At time t_2, the degree of consolidation, $U_{t2(p1)}$, can be calculated by the time factor

$$T_{v2(p1)} = \frac{c_v t}{h_{dr}^2} = \frac{c_v(t_2 - t_1/2)}{h_{dr}^2} = \frac{c_v(2t_2 - t_1)}{2h_{dr}^2} \tag{7.38}$$

The excess pore water pressure at t_2 before the next stage of loading is

$$u_{t2(p1)} = u_{0(p1)}(1 - U_{t2(p1)}) \tag{7.39}$$

For stage 2 at time t_3, there are two loadings, p_1 and p_2, which have different time durations. The degrees of the consolidation under p_1 and $p_2 - p_1$, $U_{t3(p1)}$ and $U_{t3(p2-p1)}$ can be calculated by the following time factors, respectively,

$$T_{v3(p1)} = \frac{c_v t}{h_{dr}^2} = \frac{c_v(t_3 - t_1/2)}{h_{dr}^2} = \frac{c_v(2t_3 - t_1)}{2h_{dr}^2} \tag{7.40}$$

$$T_{v3(p2-p1)} = \frac{c_v t}{h_{dr}^2} = \frac{c_v(t_3 - t_2)}{2h_{dr}^2} \tag{7.41}$$

The excess pore water pressures under p_1 and $p_2 - p_1$ at t_3, respectively, are

$$u_{t3(p1)} = u_{0(p1)}(1 - U_{t3(p1)}) \tag{7.42}$$

$$u_{t3(p2-p1)} = u_{0(p2-p1)}(1 - U_{t3(p2-p1)}) \tag{7.43}$$

where $u_{0(p2-p1)}$ is the initial excess pore water pressure induced by the additional instantaneous loading, $p_2 - p_1$.

The accumulated degree of consolidation under p_1 and $p_2 - p_1$ at t_3 can be calculated as

$$U_{t3(p2)} = 1 - \frac{u_{t3(p1)} + u_{t3(p2-p1)}}{u_{0(p1)} + u_{0(p2-p1)}} = \frac{u_{0(p1)}U_{t3(p1)} + u_{0(p2-p1)}U_{t3(p2-p1)}}{u_{0(p1)} + u_{0(p2-p1)}} \tag{7.44}$$

At time t_4, the degrees of the consolidation under p_1 and $p_2 - p_1$ before the third stage of loading, $U_{t4(p1)}$ and $U_{t4(p2\text{-}p1)}$ can be calculated by the following time factors, respectively,

$$T_{v4(p1)} = \frac{c_v t}{h_{dr}^2} = \frac{c_v(t_4 - t_1/2)}{h_{dr}^2} = \frac{c_v(2t_4 - t_1)}{2h_{dr}^2} \tag{7.45}$$

$$T_{v4(p2-p1)} = \frac{c_v t}{h_{dr}^2} = \frac{c_v(2t_4 - t_2 - t_3)}{2h_{dr}^2} \tag{7.46}$$

The excess pore water pressures under p_1 and $p_2 - p_1$ at t_4, respectively, are

$$u_{t4(p1)} = u_{0(p1)}(1 - U_{t4(p1)}) \tag{7.47}$$

$$u_{t4(p2-p1)} = u_{0(p2-p1)}(1 - U_{t4(p2-p1)}) \tag{7.48}$$

The accumulated degree of consolidation under p_1 and $p_2 - p_1$ at t_4 can be calculated as

$$U_{t4(p2)} = 1 - \frac{u_{t4(p1)} + u_{t4(p2-p1)}}{u_{0(p1)} + u_{0(p2-p1)}} = \frac{u_{0(p1)}U_{t4(p1)} + u_{0(p2-p1)}U_{t4(p2-p1)}}{u_{0(p1)} + u_{0(p2-p1)}} \tag{7.49}$$

The accumulated degree of consolidation for stage 3 or a later stage can be calculated following the same procedure as discussed above.

The preceding procedure assumes that the excess pore water pressures induced by earlier and later pressures and the rates of their dissipations with time are independent and start from the time of their pressures applied separately. Chai et al. (2013) proposed a simplified method for calculating the degree of consolidation considering the combined effect of two pressures. In their method, the degree of consolidation, U_i, under the earlier pressure, p_i, at time, t_i, is first converted to that under the later instantaneous pressure, p_j, at time, t_i, when it is applied; i.e., $U_j = U_i\, p_i/p_j$. Based on the converted U_j, the equivalent time, t_j, can be calculated using the consolidation theory (Terzaghi's one-dimensional consolidation theory for vertical flow only or Hansbo's consolidation theory for radial flow only). The degree of consolidation under the pressure, p_j, at additional time, Δt, can be calculated using the total time, $t_j + \Delta t$. A preliminary study shows that these two methods yield similar results.

Settlement Calculation The methods presented in Chapter 2 can be used to calculate the immediate and final primary consolidation settlements after each stage of loading. The primary consolidation settlement at a certain time of preloading can be calculated by

$$S_{ct} = U_t S_c \tag{7.50}$$

where S_{ct} = settlement at time t
S_c = final primary consolidation settlement at the end of primary consolidation

As shown in Figure 7.16, the settlement corresponding to each time is calculated based on the above formula.

The secondary settlement after the end of primary consolidation can be calculated using the equation presented in Chapter 2.

7.3.3 Surcharge Effect

Time for Surcharge Removal Figure 7.17 shows the variations of excess pore water pressure and settlement during surcharge preloading and removal. The effective surcharge ratio is defined as (Mesri et al., 1997)

$$R_s = \frac{\sigma'_{zs}}{\sigma'_{zf}} - 1 \tag{7.51}$$

where σ'_{zs} = effective vertical surcharge stress in soil before removal of surcharge
σ'_{zf} = final vertical stress in soil after removal of surcharge

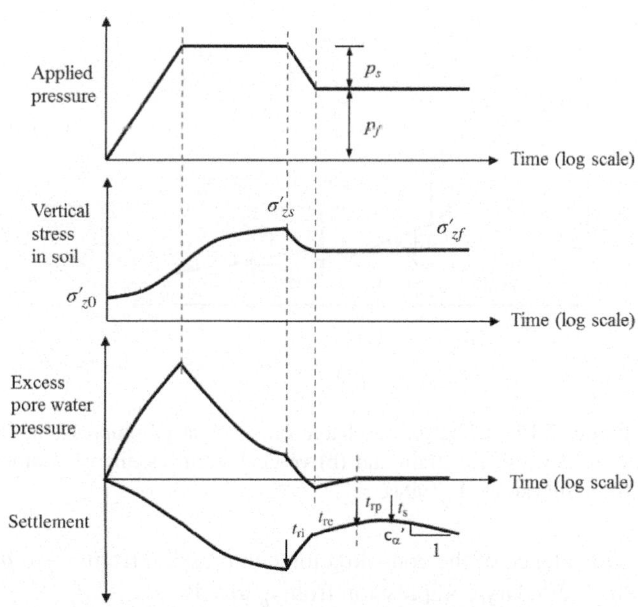

Figure 7.17 Vertical stress, excess pore water pressure, and settlement in soil during surcharge preloading and removal.

To ensure the benefit of surcharge, σ'_{zs} must be higher than σ'_{zf}.

To prevent the continued primary settlement after the removal of the surcharge, the achieved degree of consolidation should be

$$U_{tr} \geq \frac{p_f}{p_f + p_s} \tag{7.52}$$

where p_f = vertical applied pressure after removal of surcharge and p_s = vertical applied surcharge to be removed

This condition implies that the remaining excess pore water pressure at the beginning of surcharge removal, t_{ri}, is equal or less than the reduced stress induced by the removal of the surcharge pressure. Under such a condition, the removal of the surcharge pressure reduces the excess pore water pressure to zero or induces negative excess pore water pressure in soil at the end of surcharge removal, t_{re}. Therefore, no primary settlement occurs. If there is negative excess pore water pressure, there is a primary rebound (i.e., the dissipation of negative pore water pressure) from t_{re} up to t_{rp} followed by a secondary rebound up to t_s. After the secondary rebound, a secondary compression will start. Based on the required degree of consolidation in Equation (7.52), the required time for preloading can be determined following the procedure discussed in the previous section.

Rebound The primary rebound of the soil during and after the removal of the surcharge from t_{ri} to t_{rp} can be estimated by the following equation:

$$S_r = \frac{C_r h}{1 + e_0} \log \frac{\sigma'_{zf}}{\sigma'_{zs}} \tag{7.53}$$

where C_r = rebound index of soil
h = initial soil thickness
e_0 = initial void ratio of soil

Since σ'_{zs} is higher than σ'_{zf}, S_r is negative (i.e., rebound)

The time for end of primary rebound. t_{rp}, and the rate of rebound (also the rate of dissipation of positive and/or negative pore water pressure) from t_{ri} to t_{rp} can be estimated using the consolidation theories with a higher coefficient of consolidation due to a lower coefficient of volumetric compressibility. This primary rebound is followed by a secondary rebound, which is typically small and can be neglected for a conservative design in practice.

Secondary Compression There are two methods commonly used in practice to estimate the secondary compression after surcharge removal: (1) the Mesri et al. method and (2) the Ladd method.

1. The Mesri et al. Method Mesri et al. (1997) suggested secondary compression starts at time t_s and can be estimated as follows:

$$S_s = \frac{C'_\alpha h}{1+e_0} \log \frac{t}{t_s} \tag{7.54}$$

where C'_α = secondary compression index after surcharge
t_s = restart time for secondary settlement (calculated from the end of surcharge removal, t_{re})
t = time of interest after surcharge removal (calculated from t_{re})

Mesri et al. (2001) suggested the following empirical correlations between t_s/t_{rp} and R_s:

$$\frac{t_s}{t_{rp}} = 100R_s^{1.7} \quad \text{for inorganic and organic soft clays and silts} \tag{7.54}$$

$$\frac{t_s}{t_{rp}} = 10R_s \quad \text{for peat deposits} \tag{7.55}$$

where t_{rp} is the time for primary rebound, which can be estimated based on the method proposed by Mesri et al. (1978). Field observations show that most primary rebound after the removal of surcharge is completed within one month.

The secondary compression index after surcharge, C'_α, can be calculated based on the ratio of C'_α to C_α (secondary compression index before surcharge). The secondary compression index before surcharge can be measured or estimated based on the ratio of C_α to C_c as discussed in Chapter 2. Mesri et al. (1997, 2001) reported that the ratio of C'_α to C_α depends on the surcharge ratio R_s and the time ratio of t to t_s. The coefficient of secondary compression after surcharge, C'_α, can be estimated using Figure 7.18.

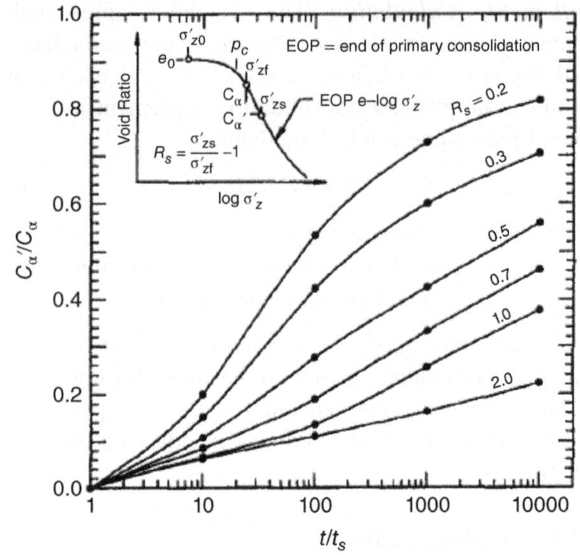

Figure 7.18 Coefficient of secondary compression after surcharge (Mesri et al., 1997, with permission from ASCE).

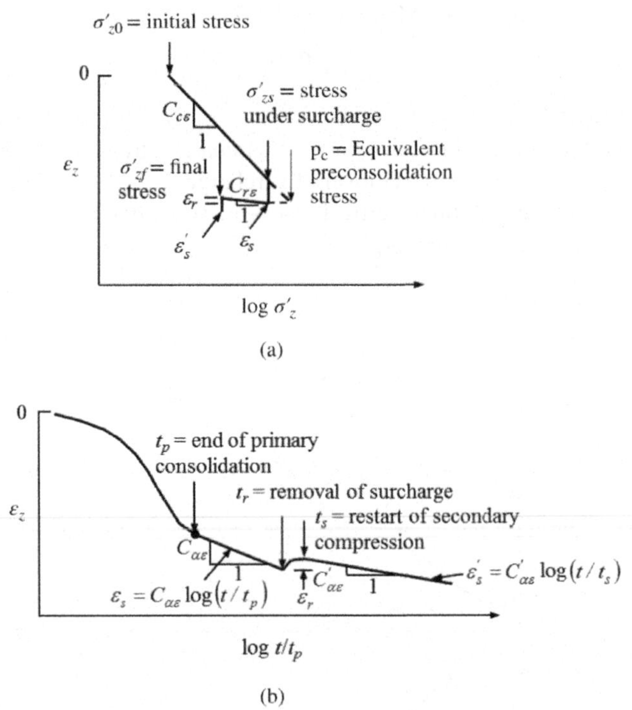

Figure 7.19 Effect of surcharge on secondary compression: (a) vertical stress vs. strain and (b) vertical strain vs. time (modified from Stewart et al., 1994).

2. The Ladd Method Ladd (1971) presented a method considering the benefit of surcharge on the reduction of soil secondary compression. Stewart et al. (1994) described the preloading and the effect of the surcharge on the secondary compression after the removal of surcharge based on Ladd (1971) in Figure 7.19. In Figure 7.19(a), when the vertical effective stress increases from the initial stress, σ'_{z0}, to the stress under surcharge, σ'_{zs}, the soil compression follows the compression line with a slope of $C_{c\varepsilon}$ (volumetric compression index of soil). Please note the volumetric compression index has a relationship with the compression index as follows: $C_{c\varepsilon} = C_c/(1+e_0)$, where e_0 is the initial void ratio of the soil. After the completion of the primary consolidation, the secondary settlement starts and continues to the point when the surcharge is removed. Stewart et al. (1994) equalized the secondary compression to the consolidation compression induced by the additional stress on the compression line so that an equivalent preconsolidation stress, p_c, is determined.The equivalence of the consolidation compression from σ'_{zs} to p_c the secondary compression from t_p to t_r is

$$C_{c\varepsilon} \log \frac{p_c}{\sigma'_{zs}} = C_{\alpha\varepsilon} \log \frac{t_r}{t_p} \tag{7.56}$$

$C_{c\varepsilon}$ = volumetric secondary compression index of soil.

Hence,

$$p_c = \sigma'_{zs}\left(\frac{t_r}{t_p}\right)^{C_{\alpha\varepsilon}/C_{c\varepsilon}} \tag{7.57}$$

After the removal of the surcharge, there is a slight rebound and then the secondary compression resumes under the final stress, σ'_{zf}. Stewart et al. (1994) defined an adjusted amount of surcharge (AAOS) as follows:

$$\text{AAOS} = \frac{p_c - \sigma'_{zf}}{\sigma'_{zf}} \tag{7.58}$$

where σ'_{zf} = final stress
p_c = equivalent preconsolidation stress

The preceding primary consolidation and secondary compression process is also illustrated in Figure 7.19(b) in a time scale. The first portion of the curve is the primary consolidation. After the end of the primary consolidation, t_p, the secondary compression occurs and increases with a slope, $C_{\alpha\varepsilon}$, in a logarithm scale. When the surcharge is removed at time t_r, there is a slight rebound, ε_r. The secondary compression restarts at t_s and increases at a reduced slope, $C'_{\alpha\varepsilon}$.

Ladd (1971) developed a relationship between the ratio of $C'_{\alpha\varepsilon}$ to $C_{\alpha\varepsilon}$ (normally consolidated) and AAOS, as shown in Figure 7.20. Conroy et al. (2010) updated this relationship as shown in Figure 7.21.

Ladd (1971) also developed a relationship between t_s/t_r and AAOS, as shown in Figure 7.20. It should be pointed out

REFERENCE (Location)	SOIL	$C_{\alpha\varepsilon}/C_{c\varepsilon}$	t_r/t_p	SYM
Koutsoftas et al. (1987)(Hong Kong)	Upper Marine CH PI ~ 55%	0.027 ± 0.001	≤ 1.0	△ ▽
Simons (1965) (Fornebu, Norway)	Marine CH PI ~ 30%	Est 0.04	1.0 2.0	○ ⊗
Haley & Aldrich (1987) (Hartford, Conn.)	Varved Clay W_N = 64±5%	0.02– 0.025	1.0– 1.8	⬡
GZA (1988) (Sommerville, MO)	Organic Silt W_l = 95% PI ~ 54%	0.046 ± 0.003	3.5 ± 0.5	□
MRCE (1989) (Syracuse, NY) (This project)	Varved Clay W_N = 46%	0.041 ± 0.003	21 ± 5	●

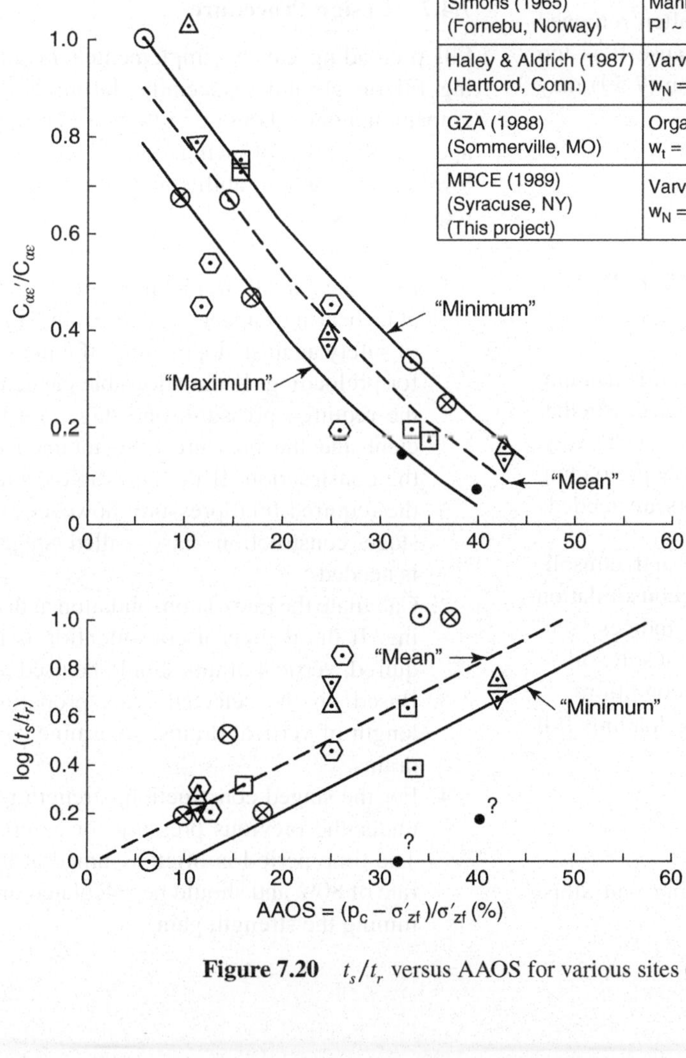

Figure 7.20 t_s/t_r versus AAOS for various sites (Ladd, 1971).

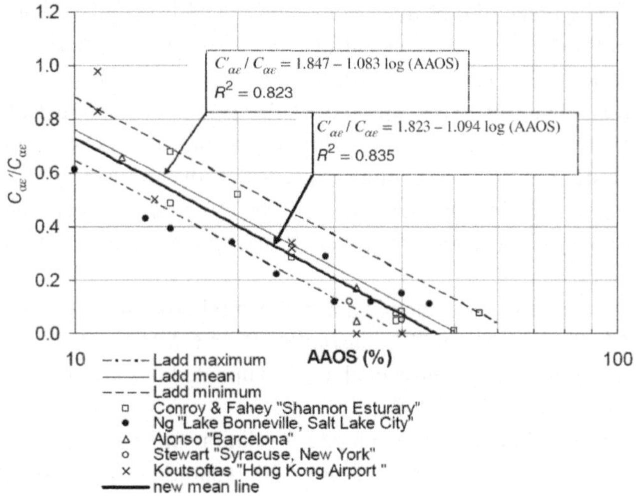

Figure 7.21 $C'_{\alpha\varepsilon}/C_{\alpha\varepsilon}$ versus AAOS for various sites (Conroy et al., 2010).

that the minimum line in Figures 7.20 and 7.21 corresponds to t_r/t_p close to 1.0.

In practice, the surcharge in preloading is often removed before the end of primary consolidation because it is too long to wait. Under such a condition, Equation (7.59) may be expressed as

$$p_c = \sigma'_{zs} \tag{7.59}$$

Under such a condition, AAOS = R_s.

7.4 DESIGN PARAMETERS AND PROCEDURES

7.4.1 Design Parameters

Depending on which method of preloading (i.e., fill, vacuum, or fill and vacuum combined preloading) is selected, whether vertical drains are used, and whether surcharge is necessary, some required design parameters are different. For preloading without vertical drains, the following parameters are needed:

- Undrained shear strength, permeability and consolidation coefficient, initial void ratio, preconsolidation stress, compression and recompression indices, secondary compression index, and thickness of soft soil
- Depth of groundwater table and drainage condition
- Size of preloading area and intensity of loading (fill pressure or vacuum pressure)
- Rate of loading
- Available time for preloading
- Final structure load
- Required factors of safety against bearing and slope failures
- Tolerable postconstruction settlement
- Service life

If preloading without vertical drains does not meet time and/or performance requirements, vertical drains are needed to accelerate consolidation. Additional parameters are needed for designing vertical drains as follows:

- Type, dimension, length, drainage capacity of vertical drains
- Pattern and spacing of vertical drains
- Size and permeability of smear zone

To accelerate preloading and reduce secondary compression, surcharge may be used. To design surcharge preloading, the following additional parameters are needed:

- Magnitude of surcharge
- Time for maintaining surcharge
- Time for removing surcharge

7.4.2 Design Procedure

Fill preloading can be implemented in the field by placing fill on ground (especially during the construction of embankments or dams) or filling water (especially in storage tanks). The following design procedure may be followed to ensure the stability and meet time and serviceability requirements:

1. Estimate an allowable pressure based on the allowable bearing capacity of the foundation and a factor of safety against slope failure if embankments are used for preloading. If the allowable capacity is higher than the required pressure, one-stage construction is sufficient and the pressure is maintained until the end of the construction. If the allowable capacity is less than the required final pressure; however, two- or multiple-stage construction (also called staged construction) is needed.
2. Calculate the rate of consolidation at the end of preloading. If the degree of consolidation is less than the required, vertical drains should be used and designed.
3. Based on the selected type, properties, pattern, and length of vertical drains, determine spacing of vertical drains.
4. For the staged construction, strength gain is calculated under the previous pressure for a certain time period. The time period is often assumed at the consolidation rate of 80% and should be calculated first prior to determining the strength gain.

5. Estimate a new allowable pressure based on the increased allowable bearing capacity. If the new allowable capacity is higher than the required pressure, the remaining pressure in addition to the previous pressure is applied and maintained until the end of the construction with a settlement calculation. Otherwise, repeat steps 2, 3, and 4 until all the required loads are applied.
6. Calculate the accumulated degree of consolidation based on the accumulated excess pore water pressure from each stage of loading and the dissipation of excess pore water pressure during preloading.
7. Calculate the settlement of the foundation for each corresponding stage.
8. Calculate the postconstruction settlement including secondary compression. If surcharge preloading is used, the reduced secondary consolidation index can be determined using the Ladd method or the Mesri et al. method.

7.5 DESIGN EXAMPLE

Design Example 7.1: Staged Construction

An unreinforced highway embankment is to be constructed on soft soil as shown in Example Figure 7.1. In order to ensure the stability of the embankment during the construction, staged construction may be necessary. Provide your design with a construction schedule, an expected settlement–time curve, and a postconstruction settlement. The construction should be completed within one year and the degree of consolidation at the end of the construction should reach at least 80%. The design life of this highway embankment is 100 years. PVDs may be used to shorten the construction period and accelerate the consolidation. The PVDs to be selected have the cross-sectional dimensions of 100 and 4 mm. The allowable discharge capacity of the drain is 0.000109 m^3/s. Assume there is no smear effect during the installation of PVDs. The required factor of safety during construction is 1.3. Provide the design of PVDs as well.

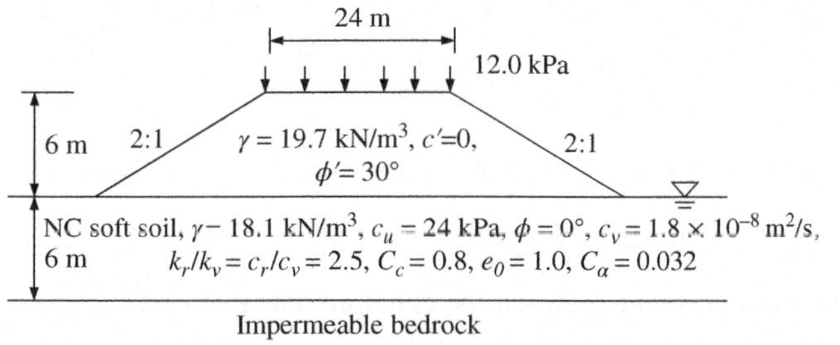

Example Figure 7.1 Design cross section and parameters.

Solution

Considering it is an unreinforced embankment, the effect of the limited depth to the embankment width ratio is not conservative. The bearing capacity factor, $N_c = 5.14$.

Stage 1

The permissible height of the fill for the first stage:

$$H_1 \leq \frac{N_c c_u}{\text{FS} \cdot \gamma} = \frac{5.14 \times 24}{1.3 \times 19.7} = 4.8 \text{ m}$$

Construct the first stage of the embankment up to 4.5 m at a construction rate of 0.3 m/week. This construction can be completed in 15 weeks, i.e., $t_1 = 15$ weeks $= 15 \times 7 = 105$ days. Assume the PVDs are arranged in an equilateral triangular pattern with spacing of 1.0 m and penetrate into the 6-m-thick soft soil. The equivalent influence diameter, d_e, is

$$d_e = 1.06s = 1.06 \times 1.0 = 1.06 \text{ m}$$

The total primary settlement in the first stage (the vertical overburden stress is calculated at the mid-depth of the soft soil and the width of the embankment is more than three times the thickness of the soft soil) is

$$S_{c1} = \frac{C_c h}{1+e_0} \log \frac{\sigma'_{z0} + \Delta\sigma_z}{\sigma'_{z0}} = \frac{0.8 \times 6}{1+1.0} \log \frac{(18.1-9.8) \times 3 + 4.5 \times 19.7}{(18.1-9.8) \times 3} = 1.58 \text{ m}$$

This settlement is about 35% the height of the fill to be constructed.

The coefficient of consolidation due to radial flow is

$$c_r = 2.5 \times c_v = 4.5 \times 10^{-8}\ \text{m}^2/\text{s}$$

The time factors for vertical and radial flow are

$$T_v = \frac{c_v t}{h_{\text{dr}}^2} = \frac{1.8 \times 10^{-8} \times (105/2) \times 24 \times 60 \times 60}{6^2} = 0.00227$$

$$T_r = \frac{c_r t}{d_e^2} = \frac{4.5 \times 10^{-8} \times (105/2) \times 24 \times 60 \times 60}{1.06^2} = 0.185$$

The degree of the consolidation due to vertical flow at t_1 is

$$U_v = \sqrt{\frac{4T_v}{\pi}} = 0.054 = 5.4\%$$

The equivalent diameter of PVDs is

$$d_c = \frac{b + t_g}{2} = \frac{100 + 4}{2} = 52\ \text{mm} = 0.052\ \text{m}$$

Estimation of k_v:

$$C_c = \frac{\Delta e}{\log \sigma'_{z2} - \log \sigma'_{z1}} \quad \text{and} \quad \Delta e = C_c(\log \sigma'_{z2} - \log \sigma'_{z1})$$

$$a_v = \frac{\Delta e}{\sigma'_{z2} - \sigma'_{z1}} = \frac{C_c(\log \sigma'_{z2} - \log \sigma'_{z1})}{\sigma'_{z2} - \sigma'_{z1}} = \frac{0.8 \times (\log 200 - \log 100)}{200 - 100} = 0.0024/\text{kPa}$$

$$m_v = \frac{a_v}{1 + e_0} = \frac{0.0024}{1 + 1.0} = 0.0012/\text{kPa}$$

$$k_v = c_v \gamma_w m_v = 1.8 \times 10^{-8} \times 9.8 \times 0.0012 = 2.12 \times 10^{-10}\ \text{m/s}$$

$$k_r = 2.5 k_v = 5.29 \times 10^{-10}\ \text{m/s}$$

The degree of consolidation due to radial flow can be calculated as follows:

$$N_D = \frac{d_e}{d_c} = \frac{1.06}{0.052} = 20$$

$$F_m(N_D) = \ln \frac{N_D}{N_s} + \frac{k_r}{k_s} \ln N_s - \frac{3}{4} + \pi z(2h_{\text{dr}} - z)\frac{k_r}{Q_c}$$

$$= \ln \frac{20}{1} + 0 - \frac{3}{4} + 3.14 \times \frac{6}{2}\left(2 \times 6 - \frac{6}{2}\right)\frac{5.29 \times 10^{-10}}{0.000109}$$

$$= 2.25$$

$$U_r = 1 - \exp\left[-\frac{8}{F_m(N_D)} T_r\right] = 1 - \exp\left(-\frac{8}{2.25} \times 0.185\right) = 0.482 = 48.2\%$$

The overall degree of consolidation is

$$U_{vr} = 1 - (1 - U_v)(1 - U_r) = 1 - (1 - 0.054) \times (1 - 0.482) = 0.510 = 51.0\%$$

The settlement at t_1 is

$$S_{t1} = U_{vr(t_1)} S_{c1} = 0.510 \times 1.58 = 0.81\ \text{m}$$

In order to reach at least 80% consolidation at the end of the first stage (assume $U_v = 10\%$), the required degree of consolidation due to radial flow is

$$U_r = 1 - \frac{1 - U_{vr}}{1 - U_v} = 1 - \frac{1 - 0.8}{1 - 0.10} = 0.778$$

The required time factor for the radial flow is

$$T_r = -1/8F_m(N_D)\ln(1 - U_r) = -1/8 \times 2.25 \times \ln(1 - 0.778) = 0.423$$

The required time is

$$t_2 - \frac{1}{2}t_1 = \frac{T_r d_e^2}{c_r} = \frac{0.423 \times 1.06^2}{4.5 \times 10^{-8}} = 1.04 \times 10^7\ \text{s} = 120\ \text{days}$$

$$t_2 = 120 + \frac{105}{2} = 172\ \text{days}$$

$$T_{v(t2)} = \frac{c_v t}{h_{\text{dr}}^2} = \frac{1.8 \times 10^{-8} \times 120 \times 24 \times 60 \times 60}{6^2} = 0.0052$$

$$U_v = \sqrt{\frac{4T_v}{\pi}} = 0.081 = 8.1\%$$

therefore, the assumed degree of consolidation at 10% is higher than the calculated. Adjustment of U_{vr} is needed:

$$U_{vr} = 1 - (1 - 0.081) \times (1 - 0.778) = 0.796$$

The settlement at t_2 is

$$S_{t2} = U_{vr(t_2)} S_{c1} = 0.796 \times 1.58 = 1.26\ \text{m}$$

The initial excess pore water pressure is

$$u_{01} = \gamma H_1 = 19.7 \times 4.5 = 89.8\ \text{kPa}$$

The excess pore water pressure at t_2 is

$$u_{t2} = u_{01} - U_{vr(t2)} u_{01} = 89.8 - 0.796 \times 89.8 = 18.3\ \text{kPa}$$

The strength gain due to consolidation can be calculated as follows:

$$\Delta c_u = 0.25 U_t\, \Delta\sigma_z = 0.25 \times 0.796 \times 19.7 \times 4.5 = 17.9\ \text{kPa}$$

Stage 2

The total permissible height of the fill for the second stage is:

$$H_2 \le \frac{N_c(c_u + \Delta c_u)}{\text{FS} \cdot \gamma} = \frac{5.14 \times (24 + 17.9)}{1.3 \times 19.7} = 8.8\ \text{m}$$

Considering the settlement of the embankment during and after construction, the actual height of fill to be added should be greater than that in the design section. Assume the total height of the fill to be added is 8 m, which is within the total permissible height for the second stage. The expected total primary settlement is

$$S_{c2} = \frac{C_c h}{1 + e_0} \log \frac{\sigma'_{z0} + \Delta\sigma_z}{\sigma'_{z0}} = \frac{0.8 \times 6}{1 + 1.0} \log \frac{(18.1 - 9.8) \times 3 + 8 \times 19.7}{(18.1 - 9.8) \times 3} = 2.08\ \text{m}$$

Therefore, the final height of the embankment after the primary settlement is close to 6 m.

The height of the fill to be added in the second stage is 8 − 4.5 = 3.5 m, which can be completed in 12 weeks if the same rate of construction is adopted. Therefore, the time right after placing the fill for the second stage, t_3, is

$$t_3 = t_2 + 12 \times 7 = 172 + 84 = 256\ \text{days}$$

The time for calculating the degree of consolidation under the first stage loading is $t_3 - t_1/2 = 256 - 105/2 = 204$ days.

$$T_{v(p1,t3)} = \frac{c_v(t_3 - t_1/2)}{h_{\mathrm{dr}}^2} = \frac{1.8\times10^{-8}\times204\times24\times60\times60}{6^2} = 0.0088$$

$$T_{r(p1,t3)} = \frac{c_r(t_3 - t_1/2)}{d_e^2} = \frac{4.5\times10^{-8}\times204\times24\times60\times60}{1.06^2} = 0.720$$

$$U_{v(p1,t3)} = \sqrt{\frac{4T_v}{\pi}} = 0.088 = 8.8\%$$

$$U_{r(p1,t3)} = 1 - \exp\left[-\frac{8}{F_m(N_D)}T_r\right] = 1 - \exp\left(-\frac{8}{2.25}\times0.720\right) = 0.923 = 92.3\%$$

The overall consolidation degree for the first stage loading at t_3 is

$$U_{vr(p1,t3)} = 1-(1-U_v)(1-U_r) = 1-(1-0.088)\times(1-0.923) = 0.931 = 93.1\%$$

The remaining excess pore water pressure induced by the first stage loading is

$$u_{t3(p1)} = u_{01}(1 - U_{vr(p1,t3)}) = 89.8\times(1-0.931) = 6.2\ \mathrm{kPa}$$

The time factors for the second stage loading are:

$$T_{v(p2-p1,t3)} = \frac{c_v(t_3 - t_2)}{2h_{\mathrm{dr}}^2} = \frac{1.8\times10^{-8}\times84\times24\times60\times60}{2\times6^2} = 0.0018$$

$$T_{r(p2-p1,t3)} = \frac{c_r(t_3 - t_2)}{2d_e^2} = \frac{4.5\times10^{-8}\times84\times24\times60\times60}{2\times1.06^2} = 0.148$$

$$U_{v(p2-p1,t3)} = \sqrt{\frac{4T_v}{\pi}} = 0.048 = 4.8\%$$

$$U_{r(p2-p1,t3)} = 1 - \exp\left[-\frac{8}{F_m(N_D)}T_r\right] = 1 - \exp\left(-\frac{8}{2.25}\times0.148\right) = 0.409 = 40.9\%$$

The overall consolidation degree for the second stage loading at t_3 is

$$U_{vr(p2-p1,t3)} = 1-(1-U_v)(1-U_r) = 1-(1-0.048)\times(1-0.409) = 0.437 = 43.7\%$$

The initial excess pore water pressure induced by the second stage loading is

$$u_{02} = 3.6\times19.7 = 71.9\ \mathrm{kPa}$$

The remaining excess pore water pressure induced by the second stage loading is

$$u_{t3(p2-p1)} = u_{02}(1 - U_{vr(p2-p1,t3)}) = 71.9\times(1-0.437) = 40.4\ \mathrm{kPa}$$

The total excess pore water pressure at t_3 is

$$u_{t3(p2)} = u_{t3(p1)} + u_{t3(p2-p1)} = 6.2 + 40.4 = 46.6\ \mathrm{kPa}$$

The overall degree of consolidation is

$$U_{vr(t3)} = 1 - \frac{u_{t3(p2)}}{u_{01}+u_{02}} = 1 - \frac{46.6}{89.8+40.4} = 0.712 = 71.2\%$$

The primary settlement at t_3 is

$$S_{t3} = U_{vr(t3)}S_{c2} = 0.712\times2.08 = 1.49\ \mathrm{m}$$

The time factors at the end of the year for the first stage loading are:

$$T_{v(p1,t4)} = \frac{c_v(2t_4 - t_1)}{2h_{\text{dr}}^2} = \frac{1.8 \times 10^{-8} \times (2 \times 365 - 105) \times 24 \times 60 \times 60}{2 \times 6^2} = 0.0135$$

$$T_{r(p1,t4)} = \frac{c_r(2t_4 - t_1)}{2d_e^2} = \frac{4.5 \times 10^{-8} \times (2 \times 365 - 105) \times 24 \times 60 \times 60}{2 \times 1.06^2} = 1.102$$

$$U_{v(p1,t4)} = \sqrt{\frac{4T_v}{\pi}} = 0.131 = 13.1\%$$

$$U_{r(p1,t4)} = 1 - \exp\left[-\frac{8}{F_m(N_D)}T_r\right] = 1 - \exp\left(-\frac{8}{2.25} \times 1.102\right) = 0.980 = 98.0\%$$

The overall consolidation degree for the first stage loading at t_3 is

$$U_{vr(p1,t4)} = 1 - (1 - U_v)(1 - U_r) = 1 - (1 - 0.131) \times (1 - 0.980) = 0.983 = 98.3\%$$

The remaining excess pore water pressure induced by the first stage loading is

$$u_{t4(p1)} = u_{01}(1 - U_{\text{vr}(p1,t4)}) = 89.8 \times (1 - 0.983) = 1.5 \text{ kPa}$$

The time factors at the end of the year for the second stage loading are:

$$T_{v(p2-p1,t4)} = \frac{c_v(2t_4 - t_3 - t_2)}{2h_{\text{dr}}^2} = \frac{1.8 \times 10^{-8} \times (2 \times 365 - 256 - 172) \times 24 \times 60 \times 60}{2 \times 6^2} = 0.0065$$

$$T_{r(p2-p1,t4)} = \frac{c_r(2t_4 - t_3 - t_2)}{2d_e^2} = \frac{4.5 \times 10^{-8} \times (2 \times 365 - 256 - 172) \times 24 \times 60 \times 60}{2 \times 1.06^2}$$

$$= 0.532$$

$$U_{v(p2-p1,t4)} = \sqrt{\frac{4T_v}{\pi}} = 0.091 = 9.1\%$$

$$U_{r(p2-p1,t4)} = 1 - \exp\left[-\frac{8}{F_m(N_D)}T_r\right] = 1 - \exp\left(-\frac{8}{2.25} \times 0.532\right) = 0.849 = 84.9\%$$

The overall consolidation degree for the second stage loading at t_3 is

$$U_{vr(p2-p1,t4)} = 1 - (1 - U_v)(1 - U_r) = 1 - (1 - 0.091) \times (1 - 0.849) = 0.863 = 86.3\%$$

The remaining excess pore water pressure induced by the second stage loading is

$$u_{t4(p2-p1)} = u_{02}(1 - U_{vr(p2,t4)}) = 71.9 \times (1 - 0.863) = 9.9 \text{ kPa}$$

The total excess pore water pressure at t_3 is

$$u_{t4(p2)} = u_{t4(p1)} + u_{t4(p2-p1)} = 1.5 + 9.9 = 11.4 \text{ kPa}$$

The overall degree of consolidation is

$$U_{vr(t4)} = 1 - \frac{u_{t3(p2)}}{u_{01} + u_{02}} = 1 - \frac{11.4}{89.8 + 71.9} = 0.929 = 92.9\%$$

The primary settlement at t_4 is

$$S_{t4} = U_{vr(t4)} S_{c2} = 0.929 \times 2.08 = 1.94 \text{ m}$$

The remaining settlement after t_4 is

$$S_{\text{rm}} = S_{c2} - S_{t4} = 2.08 - 1.94 = 0.14 \text{ m}$$

The time required to complete 99% consolidation is

$$t_p = -\frac{F_m(N_D)\ln(1-U_r)d_e^2}{8c_r} = -\frac{2.25\times\ln(1-0.99)\times 1.06^2}{8\times 4.5\times 10^{-8}} = 31732500s = 1.006\ \text{yr}$$

In addition, traffic loading would induce settlement in the soft soil and the embankment fill. Considering the embankment fill is well compacted, the settlement in the embankment fill is negligible. The settlement in the soft soil induced by the equivalent traffic loading of 12.0 kPa is

$$S_{\text{tr}} = \frac{C_c h}{1+e_0}\log\frac{\sigma'_{z0}+\Delta\sigma_z+\Delta\sigma_{\text{tr}}}{\sigma'_{z0}} - \sigma_{c2} = \frac{0.8\times 6}{1+1.0}\log\frac{(18.1-9.8)\times 3+8\times 19.7+12.0}{(18.1-9.8)\times 3} - 2.09$$

$$= 0.07\ \text{m}$$

The secondary settlement in 100 years (assume no change in C_α after the staged construction) is

$$S_s = \frac{C_\alpha h}{1+e_0}\log\left(\frac{t}{t_p}\right) = \frac{0.032\times 6}{1+1.0}\log\left(\frac{100}{1.006}\right) = 0.19\ \text{m}$$

Therefore, the total postconstruction settlement is

$$S_{\text{pc}} = S_{\text{rm}} + S_{\text{tr}} + S_s = 0.14 + 0.07 + 0.19 = 0.40\ \text{m}$$

The fill height–settlement–time relationships are shown in Example Figure 7.2.

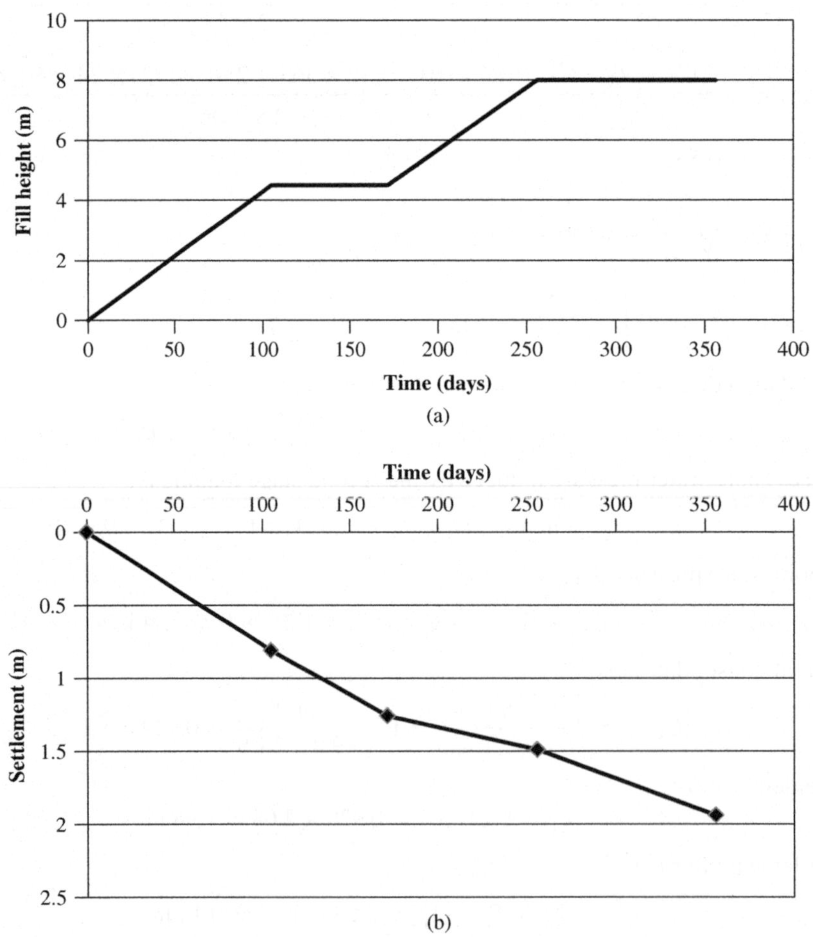

Example Figure 7.2 Design cross section and parameters: (a) fill height vs. time and (b) settlement vs. time.

In this example, the total settlement is larger than 2.0 m. Part of the embankment will settle below the groundwater table. Due to the buoyant force of the soil below the groundwater table, the actual pressure by the embankment is less than 8 × 19.7 = 158 kPa. In addition, due to the large compression of the soil, the soil consolidation coefficient may change, thus affecting the rate of consolidation. Therefore, in reality, the total settlement and the postconstruction settlement may be different. To accurately calculate these settlements, iterations are necessary.

Design Example 7.2: Surcharge Preloading

In Design Example 7.1, an additional 3-m fill is added for 6 months to have surcharge preloading. The recompression index of the soft soil is 0.16. What is the postconstruction settlement after removal of the surcharge?

Solution

First, it is necessary to check whether the soil strength at the end of the second stage is high enough to support the surcharge pressure. The soil strength gain at the end of the second stage from the initial strength is

$$\Delta c_u = 0.25 U_t \Delta\sigma_z = 0.25 \times 0.929 \times 19.7 \times 8 = 36.6 \text{ kPa}$$

The total permissible height of the fill for the surcharge stage is:

$$H_2 \leq \frac{N_c(c_u + \Delta c_u)}{\text{FS} \cdot \gamma} = \frac{5.14 \times (24 + 36.6)}{1.3 \times 19.7} = 12.2 \text{ m}$$

Therefore, an additional 3 m fill is OK. The surcharge pressure is

$$p_s = 3 \times 19.7 = 59.1 \text{ kPa}$$

Based on the construction rate, it takes 10 weeks to place additional 3 m fill. Therefore, the time right after placing the fill for the surcharge stage, t_4, is

$$t_5 = t_4 + 10 \times 7 = 365 + 70 = 435 \text{ days}$$

The surcharge preloading will end at

$$t_6 = t_4 + 6 \times 30 = 365 + 180 = 545 \text{ days}$$

The expected total primary settlement under the surcharge loading is

$$S_{c3} = \frac{C_c h}{1 + e_0} \log \frac{\sigma'_{z0} + \Delta\sigma_z}{\sigma'_{z0}}$$

$$= \frac{0.8 \times 6}{1 + 1.0} \log \frac{(18.1 - 9.8) \times 3 + 11 \times 19.7}{(18.1 - 9.8) \times 3}$$

$$= 2.37 \text{ m}$$

The time factors at the end of the surcharge loading for the first stage loading are:

$$T_{v(p1,t6)} = \frac{c_v(2t_6 - t_1)}{2h_{\text{dr}}^2}$$

$$= \frac{1.8 \times 10^{-8} \times (2 \times 545 - 105) \times 24 \times 60 \times 60}{2 \times 6^2}$$

$$= 0.0213$$

$$T_{r(p1,t6)} = \frac{c_r(2t_6 - t_1)}{2d_e^2}$$

$$= \frac{4.5 \times 10^{-8} \times (2 \times 545 - 105) \times 24 \times 60 \times 60}{2 \times 1.06^2}$$

$$= 1.704$$

$$U_{v(p1,t6)} = \sqrt{\frac{4T_v}{\pi}} = 0.165 = 16.5\%$$

$$U_{r(p1,t6)} = 1 - \exp\left[-\frac{8}{F_m(N_D)} T_r\right]$$

$$= 1 - \exp\left(-\frac{8}{2.25} \times 1.704\right) = 0.998 = 99.8\%$$

The overall consolidation degree for the first stage loading at t_6 is

$$U_{vr(p1,t6)} = 1 - (1 - U_v)(1 - U_r)$$

$$= 1 - (1 - 0.165) \times (1 - 0.998)$$

$$= 0.998 = 99.8\%$$

The remaining excess pore water pressure at the end of the surcharge induced by the first stage loading is

$$u_{t6(p1)} = u_{01}(1 - U_{vr(p1,t6)})$$

$$= 89.8 \times (1 - 0.998) = 0.2 \text{ kPa}$$

The time factors at the end of the year for the second stage loading are:

$$T_{v(p2-p1,t6)} = \frac{c_v(2t_6 - t_3 - t_2)}{2h_{dr}^2}$$

$$= \frac{1.8 \times 10^{-8} \times (2 \times 545 - 256 - 172) \times 24 \times 60 \times 60}{2 \times 6^2}$$

$$= 0.0143$$

$$T_{r(p2-p1,t6)} = \frac{c_r(2t_6 - t_3 - t_2)}{2d_e^2}$$

$$= \frac{4.5 \times 10^{-8} \times (2 \times 545 - 256 - 172) \times 24 \times 60 \times 60}{2 \times 1.06^2}$$

$$= 1.145$$

$$U_{v(p2-p1,t6)} = \sqrt{\frac{4T_v}{\pi}} = 0.135 = 13.5\%$$

$$U_{r(p2-p1,t6)} = 1 - \exp\left[-\frac{8}{F_m(N_D)}T_r\right]$$

$$= 1 - \exp\left(-\frac{8}{2.25} \times 1.145\right)$$

$$= 0.983 = 98.3\%$$

The overall consolidation degree for the second stage loading at t_6 is

$$U_{vr(p2-p1,t6)} = 1 - (1 - U_v)(1 - U_r)$$

$$= 1 - (1 - 0.135) \times (1 - 0.983)$$

$$= 0.985 = 98.5\%$$

The remaining excess pore water pressure induced by the second stage loading is

$$u_{t6(p2-p1)} = u_{02}(1 - U_{vr(p2,t6)})$$

$$= 71.9 \times (1 - 0.985) = 1.1 \text{ kPa}$$

The time factors at the end of the year for the surcharge loading (can also be considered as the third stage) are:

$$T_{v(p3-p2,t6)} = \frac{c_v(2t_6 - t_5 - t_4)}{2h_{dr}^2}$$

$$= \frac{1.8 \times 10^{-8} \times (2 \times 545 - 435 - 365) \times 24 \times 60 \times 60}{2 \times 6^2}$$

$$= 0.0063$$

$$T_{r(p3-p2,t6)} = \frac{c_r(2t_6 - t_5 - t_4)}{2d_e^2}$$

$$= \frac{4.5 \times 10^{-8} \times (2 \times 545 - 435 - 365) \times 24 \times 60 \times 60}{2 \times 1.06^2}$$

$$= 0.502$$

$$U_{v(p3-p2,t6)} = \sqrt{\frac{4T_v}{\pi}} = 0.089 = 8.9\%$$

$$U_{r(p3-p2,t6)} = 1 - \exp\left[-\frac{8}{F_m(N_D)}T_r\right]$$

$$= 1 - \exp\left(-\frac{8}{2.25} \times 0.502\right)$$

$$= 0.832 = 83.2\%$$

The overall consolidation degree for the surcharge stage loading at t_6 is

$$U_{vr(p3-p2,t6)} = 1 - (1 - U_v)(1 - U_r)$$

$$= 1 - (1 - 0.089) \times (1 - 0.832)$$

$$= 0.847 = 84.7\%$$

The remaining excess pore water pressure induced by the surcharge stage loading is

$$u_{t6(p3-p2)} = u_{03}(1 - U_{vr(p3,t6)})$$

$$= 59.1 \times (1 - 0.847) = 9.0 \text{ kPa}$$

The total excess pore water pressure at t_6 is

$$u_{t6(p3)} = u_{t6(p1)} + u_{t6(p2-p1)} + u_{t6(p3-p2)}$$

$$= 0.2 + 1.1 + 9.0 = 10.3 \text{ kPa}$$

The overall degree of consolidation at the end of the surcharge stage is

$$U_{vr(t6)} = 1 - \frac{u_{t6(p3)}}{u_{01} + u_{02} + u_{03}}$$

$$= 1 - \frac{10.3}{89.8 + 71.9 + 59.1}$$

$$= 0.953 = 95.3\%$$

The primary settlement at t_6 is

$$S_{t6} = U_{vr(t6)}S_{c3} = 0.953 \times 2.37 = 2.26 \text{ m}$$

Considering the primary settlement and potential secondary compression, remove 2.5 m surcharge fill, that is, $p_f = (11.0 - 2.5) \times 19.7 = 167$ kPa and $p_s = 2.5 \times 19.7 = 49$ kPa

The required degree of consolidation before removal of surcharge is

$$U_{\text{tr}} = \frac{p_f}{p_f + p_s} = \frac{167}{167 + 49}$$

$$= 0.773 = 77.3\% < U_{\text{vr}(t6)} \text{ (OK)}$$

The effective vertical stress in the middle of the soft soil at the end of surcharge loading is

$$\sigma'_{\text{zs}} = \sigma'_{z0} + U_{vr(t6)}(p_f + p_s)$$

$$= 25 + 0.953 \times (167 + 49) = 231 \text{ kPa}$$

Use the Ladd's method for the calculations of rebound and secondary compression. The final effective vertical stress in the middle of the soft soil after removal of the surcharge is

$$\sigma'_{\text{zf}} = \sigma'_{z0} + p_f = 25 + 167 = 192 \text{ kPa}$$

The amount of rebound can be calculated as follows:

$$S_{\text{rb}} = \frac{C_r}{1 + e_0} h \log \frac{\sigma'_{\text{zf}}}{\sigma'_{\text{zs}}}$$

$$= \frac{0.16}{1 + 1.0} \times 6 \times \log \frac{192}{231} = -0.04 \text{ m}$$

$$\text{AAOS} = R_s = \frac{\sigma'_{\text{zs}}}{\sigma'_{\text{zf}}} - 1 = \frac{231}{192} - 1 = 0.20 = 20\%$$

From Figure 7.20 (the minimum bound is used because t_r/t_p is close to 1.0), $\log(t_s/t_r) = 0.2$. Since $t_r = t_6 = 545$ days, $t_s = 10^{0.2} \times 545 = 864$ days.

From Figure 7.21 (also the minimum bound is used), $C'_\alpha/C_\alpha = 0.57$. Therefore, $C'_\alpha = 0.57 \times 0.032 = 0.018$. The secondary compression is

$$S_s = \frac{C'_\alpha}{1 + e_0} h \log \frac{t}{t_s}$$

$$= \frac{0.018}{1 + 1.0} \times 6 \times \log \frac{100 \times 365}{864} = 0.09 \text{ m}$$

In addition, traffic loading would induce settlement in the soft soil. The settlement induced by the equivalent traffic loading of 12.0 kPa along the recompression line is

$$S_{\text{tr}} = \frac{C_r h}{1 + e_0} \log \frac{\sigma'_{\text{zf}} + \Delta\sigma_{\text{tr}}}{\sigma'_{\text{zf}}} = \frac{0.16 \times 6}{1 + 1.0} \log \frac{192 + 12.0}{192}$$

$$= 0.01 \text{ m}$$

Therefore, the total postconstruction settlement after removal of the surcharge is

$$S_{\text{pc}} = S_{\text{rb}} + S_{\text{tr}} + S_s = -0.04 + 0.01 + 0.09 = 0.06 \text{ m}$$

This settlement is much smaller than that without surcharge preloading (i.e., 0.40 m).

7.6 CONSTRUCTION

7.6.1 Vertical Drains

Table 7.1 lists the installation methods for vertical drains. Sand drains are installed by backfilling sand into holes predrilled by a driven or vibratory closed-end mandrel, a hollow-stem continuous flight auger, or water jetting. Prefabricated sand drains are placed into holes formed by a driven or vibratory closed-end mandrel, flight auger, or rotary wash boring. PVDs are commonly installed by a driven or vibratory closed-end mandrel. Figure 7.22 shows the typical equipment for PVD installation, which includes an installation rig and a drain delivery system.

There are three types of rigs: static rig, static rig with water balancing system, and vibratory rig. The selection of a rig should consider the following factors: (1) soil type, (2) bearing capacity of ground, (3) depth of installation, and (4) productivity of rig. Static rig, static rig with water balancing system, and vibratory rig are suitable for normal ground conditions, very soft soil, and firm to stiff soil, respectively (Bo et al., 2003). A steel mandrel, typically of a rhombic or rectangular shape carries and protects PVDs from damage when driven into the ground.

A typical installation sequence starts with a PVD material in rolls threaded through the mandrel and attachment of an anchor at the bottom of the mandrel. The anchor can be a 13-mm-diameter cable, rebar, or steel anchor plate, which is used to anchor the PVD in the ground. The mandrel with the rig is positioned to a desired location, inserted into the ground, pushed to a required depth, and then withdrawn to the ground surface. The PVD with a certain length is cut off and left in a drainage layer. The rig is moved to the next location and the same sequence is repeated to install the next PVD.

7.6.2 Drainage Layer

To discharge water from drains, a drainage layer with a thickness of 300–500 mm is often placed on ground surface. The drainage layer is also used to support construction equipment when placed over soft soils. Under this condition, extra material of at least 300 mm may be added to accommodate aggregate loss due to contamination or a geotextile separator is used to maintain the integrity of the drainage layer. The drainage layer should be a good drainage material with less than 3% clays, such as gravel or sand, which has permeability not less than 10^{-5} m/s.

Strip drains may be placed horizontally and connected to PVDs instead of a drainage layer to discharge water. Typical strip drains are approximate 25 mm thick and 150 or 300 mm wide (FHWA, 2004).

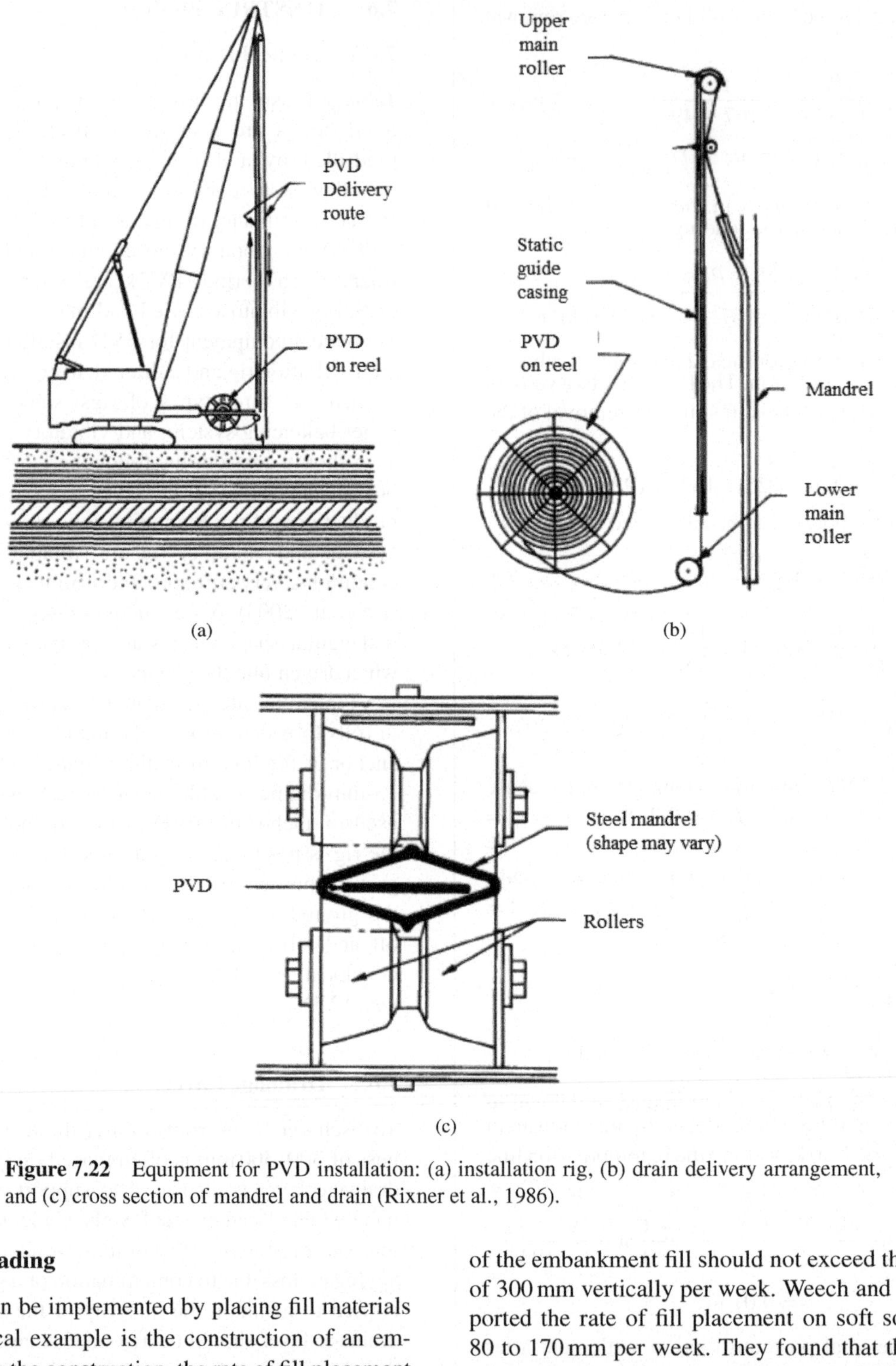

Figure 7.22 Equipment for PVD installation: (a) installation rig, (b) drain delivery arrangement, and (c) cross section of mandrel and drain (Rixner et al., 1986).

7.6.3 Fill Preloading

Fill preloading can be implemented by placing fill materials in layers. A typical example is the construction of an embankment. During the construction, the rate of fill placement and the height of the embankment are two important controlling factors. They should be properly controlled to ensure the stability of fill preloading. The rate of fill placement depends on the properties of soft soil. Typically, the placement of the embankment fill should not exceed the maximum rate of 300 mm vertically per week. Weech and Lister (2009) reported the rate of fill placement on soft soils ranged from 80 to 170 mm per week. They found that the installation of vertical drains allowed a faster rate of fill placement. The accumulated excess pore water pressure should be less than 50–60% of the applied stress. The height of fill in each stage should not be larger than that calculated based on the allowable bearing capacity as discussed in Section 7.3.2.

7.6.4 Vacuum Preloading

Vacuum preloading has two different systems (Figure 7.23): (1) with membrane and (2) without membrane (also referred as "membraneless system"). Nowadays, vacuum preloading mostly includes PVDs, which serve as drains and distribute vacuum pressure.

In the membrane vacuum preloading system as shown in Figure 7.23(a), there is a drainage layer under the membrane liner in which PVD heads and water collection pipes are included. This drainage layer is also used as a construction platform to support installation rigs for PVDs. To prevent leakage of the membrane, it is necessarily anchored into trenches around the perimeter (typically at least 1.0 m deep). When a permeable soil exists on the ground surface, the membrane should be anchored in the trenches below the bottom of this soil layer. Above the membrane, a water pond or fill material is placed to prevent its damage during the construction. Vacuum pumps are connected to the water collection pipes. Shang et al. (1998) reported that a single vacuum pump with a capacity of 7.5 MW could cover an area of 1000 m^2. If the area for vacuum preloading is large, multiple pumps may be used or the area to be treated may be divided into different preloading cells. Figure 7.24 shows a photo of multiple vacuum pumps on different preloading cells in one construction site. The photo clearly shows the layout of water collection pipes under the membrane and the trenches around the perimeter.

In the membraneless vacuum preloading system, the water collection pipes are connected to individual PVDs by connectors as shown in Figure 7.23(b) and Figure 7.25 and vacuum is distributed through these PVDs.

7.7 QUALITY CONTROL AND ASSURANCE

Proper quality control and assurance are important parts of the preloading technology, which contribute to the successful field performance.

Figure 7.24 A vacuum preloading site with multiple vacuum pumps.

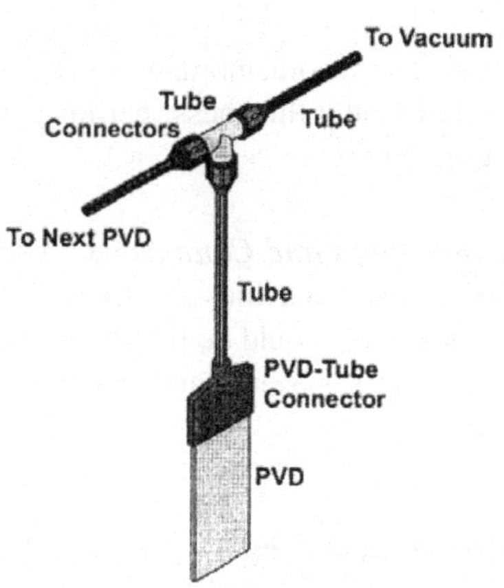

Figure 7.25 Connections between pipes and PVDs in a membraneless vacuum preloading system (Seah, 2006).

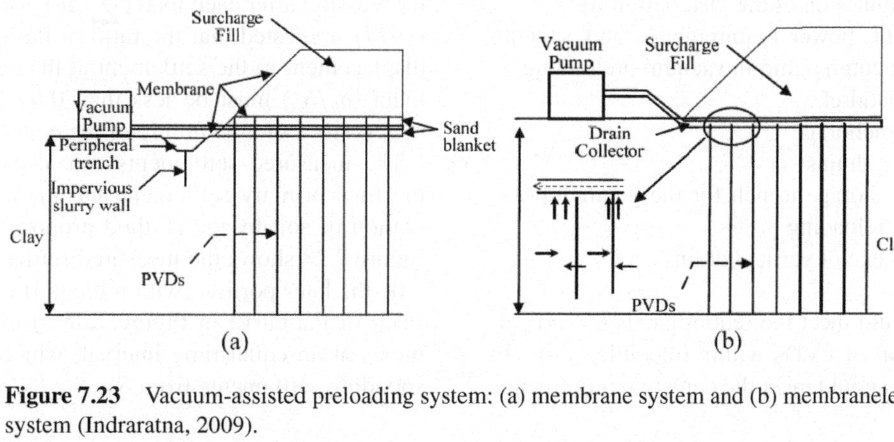

Figure 7.23 Vacuum-assisted preloading system: (a) membrane system and (b) membraneless system (Indraratna, 2009).

7.7.1 Materials

The type, dimension, and properties of the materials used in preloading should be evaluated prior to construction.

Vertical Drains If sand drains are used, the gradation and discharge capacity of sand should be obtained. If PVDs are used, they should be evaluated for filtration and drainage requirements. The discharge capacity of PVDs should be determined at normal stresses corresponding to the field horizontal stresses on PVDs at different depths. Hydraulic gradient, $i \leq 0.5$, should be used to measure the discharge capacity (Wang and Chen, 1996). PVDs including splicing drains should have sufficient tensile strengths.

Drainage Layer The sand used in the drainage layer should meet the filtration and drainage requirements. If the drainage layer is also used as a construction platform, it should meet the shear strength requirement.

Geomembrane The geomembrane used in vacuum preloading should meet the thickness, permittivity, and tensile strength requirements.

Water Collection Pipes and Connectors The pipes should have enough drainage capacities and meet filtration requirements. The connectors should be tightly connected to PVDs (in the membraneless system) and water collection pipes without any leakage.

7.7.2 Construction Details

Contractors should submit at least but not be limited to the following construction details for approval prior to construction:

- Size, type, weight, maximum pushing force, vibratory energy, and configuration of the installation rig
- Size, type, weight, power requirements, and vacuum pressure of the vacuum pump in vacuum preloading
- Dimensions of mandrel
- Details of drain anchorage
- Details of splicing drains
- Details of the anchorage trench for the geomembrane liner in vacuum preloading
- Installation sequence of vertical drains

The installation should meet the requirements of verticality, location, and depth of PVDs within tolerable limits. In fill preloading, the fill should meet the density requirements.

7.7.3 Field Monitoring

It is important to set up proper instrumentation for field monitoring during preloading. Typical instrumentation as shown in Figure 7.26 includes settlement plates for the measurement of embankment settlements, extensometer points for soft soil compressions, inclinometers for soft soil horizontal movements, piezometers for pore water pressures in the soft soil, groundwater observation well for the groundwater level, alignment stakes for ground surface movements, and survey points for embankment and ground movements. The instrumentation serves the following purposes: (1) to ensure the stability of embankment, (2) to assess the performance of preloading, (3) to determine the time for next stage loading, and (4) to provide field data to back-calculate and verify soil parameters used in the design.

Ground movement is an important piece of information to evaluate whether the soft foundation or embankment has a stability problem. The following are the warning signs for possible failure (Ye et al., 1994):

- Whether the crest and/or slope of the embankment develops cracks
- Whether the middle portion of the embankment has rapidly increasing settlement
- Whether the horizontal movement at the toe of the embankment accelerates
- Whether heave occurs at the toe of the embankment
- Whether the horizontal movement and heave at the toe and the pore water pressure continue increasing.

To prevent possible failure, the loading rate has to be properly controlled. Figure 7.27 shows three measured curves that can be used to control the loading rate: (1) horizontal displacement at the toe (δ_h) versus settlement at the center of the embankment (S_o), (2) applied pressure (p) versus horizontal displacement rate at the toe of the embankment (δ_h), and (3) applied pressure (p) versus accumulated excess pore water pressure after each load ($\sum \Delta u$). Matsuo and Kawamura (1977) suggested that the ratio of the maximum horizontal displacement to the settlement at the center of the embankment (δ_h/S_o) must be less than $0.694S_o^{-1.08}$ (δ_h and S_o in meters) to avoid the failure of the embankment.

The measured settlements can also be used to estimate the final primary settlement and the coefficient of consolidation of soil by the method proposed by Asaoka (1978). Figure 7.28 shows the measured settlement–time curve. Divide the later portion (within the portion under the constant load) of the curve in Figure 7.28(a) into a number of segments at an equal time interval, which result in the corresponding settlements from S_0, S_1, S_2, … , S_{n-1}, to S_n. The

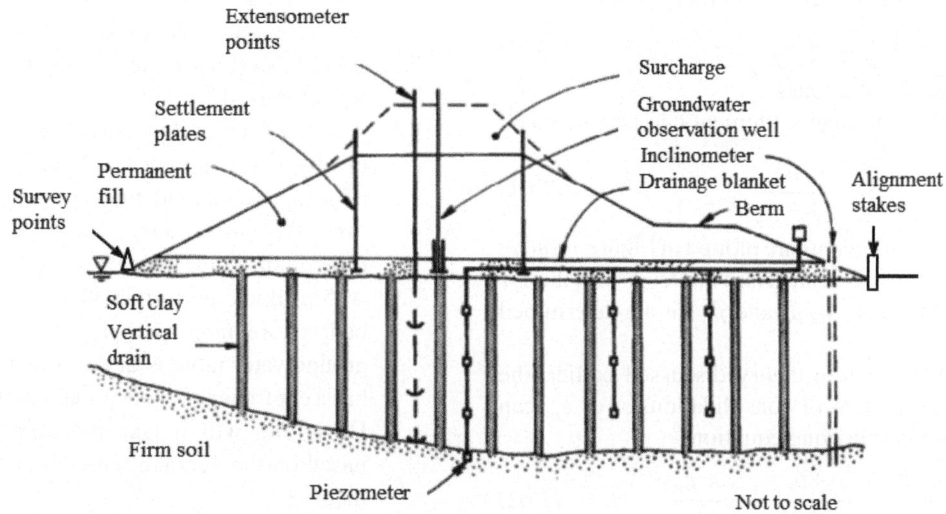

Figure 7.26 Instrumentation for field monitoring during preloading (Rixner et al., 1986).

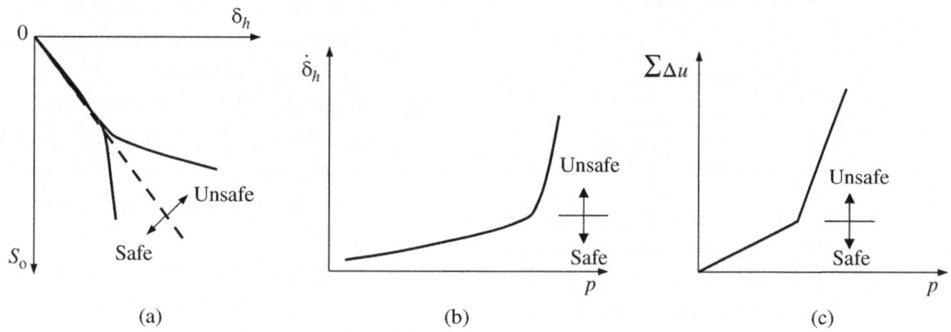

Figure 7.27 Measured curves for the control of loading rate.

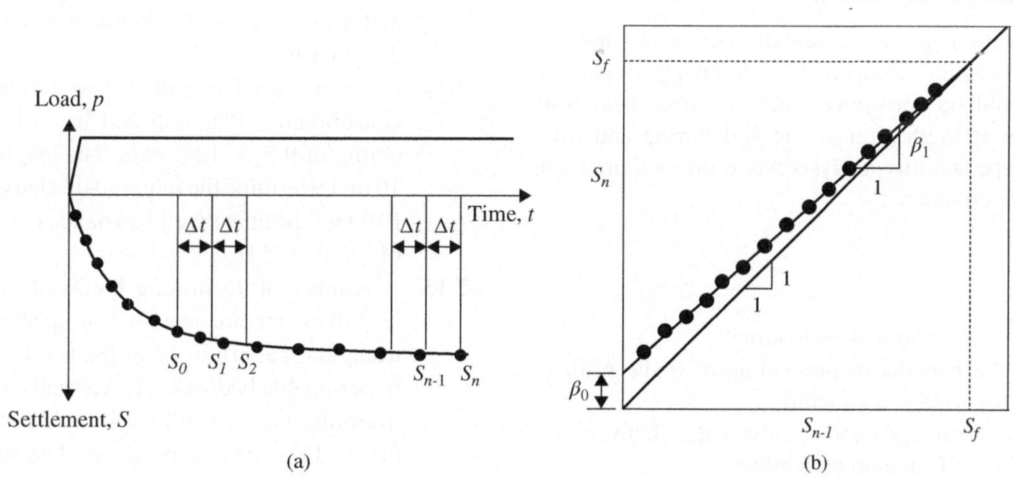

Figure 7.28 Asaoka's method for final settlement: (a) settlement–time curve and (b) settlement S_{n-1} vs. S_n.

settlements can be expressed as a first-order approximation:

$$S_n = \beta_0 + \beta_1 S_{n-1} \tag{7.60}$$

where β_0 and β_1 are two constants.

Set $S_n = S_{n-1} = S_f$, the final settlement can be calculated as follows:

$$S_f = \frac{\beta_0}{1 - \beta_1} \tag{7.61}$$

Graphically, these settlements are plotted in Figure 7.28(b). The data point line can be extended and intercept with the 1:1 line at the final settlement, S_f. β_0 and β_1 can be determined in Figure 7.28(b).

Based on the consolidation theory discussed earlier, the vertical or radial coefficient of consolidation, c_v or c_r, can be calculated from the following equation:

$$-\frac{\ln \beta_1}{\Delta t} = \frac{8c_r}{d_e^2 F(N_D)} + \frac{\pi^2 c_v}{4h_{\text{dr}}^2} \tag{7.62}$$

If there is no vertical drain

$$c_v = -\frac{4h_{\text{dr}}^2 \ln \beta_1}{\pi^2 \Delta t} \tag{7.63}$$

When

$$\frac{8c_r}{d_e^2 F(N_D)} >> \frac{\pi^2 c_v}{4h_{\text{dr}}^2}$$

$$c_r \approx -\frac{d_e^2 F(N_D) \ln \beta_1}{8\Delta t} \tag{7.64}$$

Pore water pressure in soil during preloading should be monitored to control the loading rate and estimate the rate of consolidation. Vacuum pressure should also be monitored during vacuum preloading to determine the actual vacuum pressure achieved in the field and detect any leakage.

7.7.4 Performance Evaluation

In addition to settlements, horizontal displacements, and rate of pore water pressure dissipation, additional performance evaluation should be performed, such as vane shear tests to evaluate the strength gain of the soil during and after preloading. Slope stability analysis based on measured soil strength may be conducted.

PROBLEMS

7.1. What is the principle of preloading?

7.2. Explain the benefits of preloading of a site prior to building a permanent structure.

7.3. Compare advantages and disadvantages between fill preloading and vacuum preloading.

7.4. Can vacuum preloading generate an applied pressure of 150 kPa? Why?

7.5. When fill preloading is combined with vacuum preloading, will the contributions from these two methods cancel out each other? Why?

7.6. What is a smear effect? Explain how the smear effect may affect the rate of consolidation.

7.7. What is well resistance? How can the well resistance be minimized?

7.8. A 10-m-thick uniform clay layer underlain by impermeable bedrock under preloading is expected to a total primary consolidation settlement of 200 mm. If the first 100-mm settlement takes 90 days, estimate the time required for the next 50-mm settlement.

7.9. A 5-m-thick uniform soft clay above impermeable bedrock requires preloading for improvement. The groundwater table is at the ground surface. This clay has a coefficient of consolidation of 1.5×10^{-7} m^2/s. How long will it take to reach 90% consolidation based on the Terzaghi one-dimensional consolidation theory?

7.10. In Problem 7.9, if 300-mm-diameter and 5-m-long free-draining sand drains with spacing of 1.5 m in a square pattern are used to accelerate the rate of consolidation, what is the average degree of consolidation due to horizontal flow in one year according to the Barron (1948) solution?

7.11. In Problem 7.10, if there is a smear zone induced by the installation of free-draining sand drains, the diameter of the smear zone is 1.5 times the diameter of the sand drain, and the permeability of the smear zone is half of the natural soil, what is the average degree of consolidation due to horizontal flow in one year according to the Hansbo (1981) solution?

7.12. Can vertical drains reduce secondary consolidation? Why?

7.13. A commercial PVD product has a width of 100 mm and a thickness of 6.25 mm. Estimate the equivalent diameter of the PVD.

7.14. PVDs are used in soft soil to accelerate the rate of consolidation. The soft soil has a horizontal permeability of 9.5×10^{-8} m/s. The length of the PVDs is 10 m. Determine the required discharge capacity of the PVD to eliminate well resistance considering a factor of safety of 5.0.

7.15. A number of 15-m-long PVDs of 75 mm in equivalent diameter are installed at spacing of 1.3 m in a triangular pattern in 15-m-thick soft soil underlain by impermeable bedrock. The vertical and horizontal permeability values of this soil are 2.5×10^{-9} m/s and 6.5×10^{-9} m/s, respectively. The smear zone size is 2.5 times the PVD equivalent diameter and its permeability is one third the vertical permeability of the natural soil. The allowable discharge capacity of the PVD is 0.00025 m^3/s. Calculate the equivalent vertical permeability of the PVD-improved soft foundation.

7.16. A storage tank with a diameter of 30 m (self-weight = 2000 tons) is to be constructed on a normally consolidated soft clay with a thickness of 7.5 m underlain by a dense sand layer. This soft clay has an undrained shear strength of 25 kPa, effective friction angle of 24°, saturated unit weight of 17.5 kN/m^3, and coefficient of consolidation and permeability equal to 2.7×10^{-8} m^2/s and 6.0×10^{-10} m/s. The groundwater table is near the ground surface. To minimize the service settlement, the tank is filled with water up to 6 m in one day to preload the soft soil for 6 months. Prefabricated vertical drains (PVDs) are proposed to accelerate the consolidation. The PVDs to be selected have the cross-sectional dimensions of 100 and 5 mm and penetrate the soft soil into the sand layer. The allowable discharge capacity of the drain is 0.00081 m^3/s under 50 kPa normal stress and 0.00014 m^3/s under 150 kPa. Assume there is no smear effect during the installation of PVDs. Determine the factor of safety against bearing failure at the end of water filling and the required spacing of PVDs (in a square pattern) to ensure 80% consolidation by the end of the 6-month preloading.

7.17. Vacuum preloading with PVDs is used to improve 20-m-thick soft soil. The equivalent vertical permeability of the PVD-improved soil is three times that before improvement. Determine the optimum penetration depth of PVDs for vacuum preloading.

7.18. What is staged construction? Under what situation(s) is staged construction needed?

7.19. Fill preloading is required to improve a soft foundation. The undrained shear strength of the foundation soil is c_u = 25kPa. The unit weight of fill material for preloading is γ_f = 18 kN/m^3. The required factor of safety against bearing failure is FS =1.3. Determine the maximum fill height under the current situation.

7.20. A large area of 2-m-high fill is placed on a normally consolidated homogeneous sedimentary clay for preloading. The unit weight of the fill is 19.2 kN/m^3. The undrained shear strength of the natural soil is 20 kPa. Estimate the undrained shear strengths of the soil when the degree of consolidation reaches 80%.

7.21. What are the main purposes for surcharge preloading?

7.22. A unreinforced highway embankment is to be constructed on normally consolidated soft soil underlain by permeable dense sand (assume no deformation in this layer), as shown in the following figure. In order to ensure the stability of the embankment during the construction, staged construction is necessary. In addition, the construction should be completed within 16 months and the degree of consolidation at the end of the construction should reach at least 80%. The traffic load will be applied after 16 months. The design life of this highway embankment is 75 years. PVDs will be placed in a triangular pattern to the top of the dense sand to shorten the construction period and accelerate the consolidation. The PVDs to be selected have the cross-sectional dimensions of 100 and 6 mm. The allowable discharge capacity of the drain is 0.00035 m^3/s. The average diameter of the smear zone around each PVD is 2.5 times the equivalent diameter of the PVD. The permeability of the smear zone is 46% that of the vertical permeability of the natural soil. The required factor of safety during construction is 1.3. Determine the spacing of PVDs, develop a construction schedule with an expected load–settlement–time curve, and calculate the postconstruction settlement.

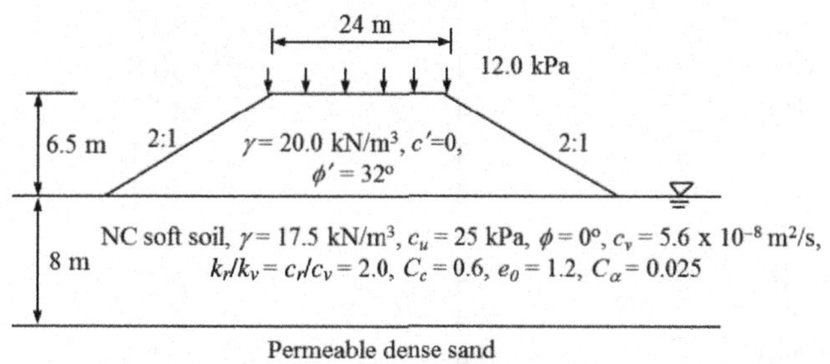

7.23. To minimize secondary settlement of an embankment on soft soil, surcharge preloading is adopted in this project as shown in the following figure. The construction including the removal of the surcharge should be completed within one year. At the time of surcharge removal, the degree of consolidation should reach at least 80%. PVDs will be placed in a triangular pattern to the top of the bedrock to shorten the construction period and accelerate the consolidation. The PVDs to be selected have the cross-sectional dimensions of 100 and 4 mm. The allowable discharge capacity of the drain is 0.00025 m^3/s. The average diameter of the smear zone around each PVD is 2.5 times the equivalent diameter of the PVD. The permeability of the smear zone is 60% that of the vertical permeability of the natural soil. The required factor of safety during construction is 1.3. Determine the spacing of PVDs required to meet the one-year construction time and estimate the secondary settlement after the removal of the surcharge in 75 years as compared with the one without surcharge preloading.

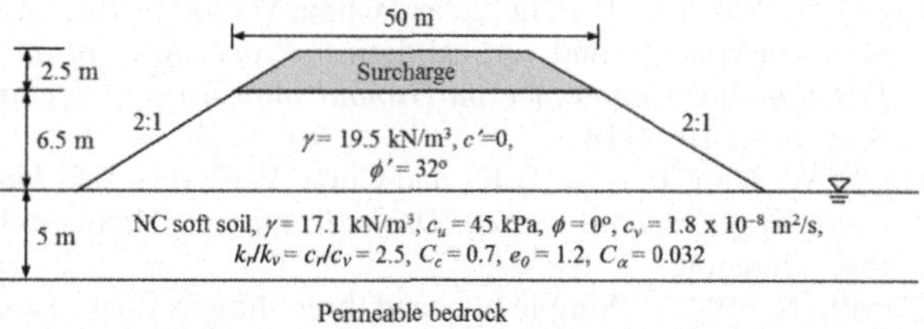

7.24. A 4-m-high fill is placed on a 10-m-thick soft clay deposit underlain by impermeable stiff clay. PVDs with cross-sectional dimensions of 125 mm wide and 5 mm thick were installed to the top of the stiff clay in a square pattern with a spacing of 2.0 m. The measured settlements with time are provided in the following table. Use Asaoka's method to: (1) estimate the final settlement and (2) calculate the coefficient of consolidation due to horizontal flow (ignore any consolidation due to vertical flow and assume free-draining PVDs and no smear effect).

Time (days)	Fill Height (m)	Settlement (mm)
0	1	0
25	2	10.2
50	3	20.4
75	4	30.8
100	4	40.2
125	4	45.1
150	4	49.1
175	4	52.0
200	4	54.1
225	4	56.0
250	4	57.0
275	4	57.8
300	4	58.5

REFERENCES

Alonso, E.E., Gens, A., and Lloret, A. (2000). "Precompression design fro secondary settlement reduction." *Geotechnique*, 50(6): 645–656.

Asaoka, A. (1978). "Observational procedure of settlement prediction." *Soils Foundations*, 18(4): 87–101.

Barron, R.A. (1948). "Consolidation of fine-grained soils by drain wells." *Transaction*, 113: 718–742.

Bergado, D.T., Assakami, H., Alfaro, M., and Balasubramaniam, A.S. (1991). "Smear effects of vertical drains on soft Bangkok clay." *J. Geotech. Eng.*, 117(10): 1509–1529.

Bergado, D.T., Chai, J.-C., Miura, N., and Balasubramaniam, A.S. (1998). "PVD improvement of soft Bangkok clay with combined vacuum and reduced sand embankment preloading." *Geotech. Eng. J.*, 29(1): 95–121.

Bjerrum, L. (1972). "Embankments on soft ground. State of the art report." *Proceedings of ASCE Specialty Conference on Performance of Earth and Earth-supported Structures*, Vol. 1. Purdue University, 1–54.

Bo, M.W., Bawajee, R., Chu, J., and Choa, V. (2000). "Investigation of smear zone around vertical drain." *Proceedings of the Third International Conference on Ground Improvement Techniques*, Singapore, 109–114.

Bo, M.W., Chu, J., Low, B.K., and Choa, V. (2003). *Soil Improvement: Prefabricated Vertical Drain Technique*. Thomson Learning, Singapore.

Carrillo N. (1942). "Simple two- and three-dimensional cases in the theory of consolidation of soils." *J. Math. Phys.*, 21: 1–5.

Carroll, R.G. (1983). "Geotextile filter criteria." *Transport. Res. Rec.*, No. 916, 46–53.

Chai, J.C. and Carter, J.P. (2013). "Consolidation theory for combined vacuum pressure and surcharge loading." *Proceedings of the 18th International Conference on Soil Mechanics and Geotechnical Engineering,* Paris 2013.

Chai, J., Ong, C.Y., Carter, J.P., and Bergado, D.T. (2013). "Lateral displacement under combined vacuum pressure and embankment loading." *Géotechnique*, 63(10), 842–856.

Chai, J.C., Carter, J.P., and Hayashi, S. (2006). "Vacuum consolidation and its combination with embankment loading." *Canad. Geotech. J.*, 43(10): 985–996.

Chai, J.-C., Hayashi, S., and Carter, J. P. (2005). "Characteristics of vacuum consolidation." *Proceedings of the 16th International Conference on Soil Mechanics and Geotechnical Engineering*. Osaka, Rotterdam, Millpress.

Chai, J.-C., Shen, S.-L., Miura, N., and Bergado, D.T. (2001). "Simplified method of modeling PVD-improved subsoil." *J. Geotechn. Geoenviron. Eng.*, 127(11): 965–972.

Chu, J., Yan, S. W., and Yang, H. (2000). "Soil improvement by vacuum preloading method for an oil storage station". *Geotechnique*, 50(6), 625–632.

Chu, J., Bo, M.W., and Choa, V. (2004). "Practical considerations for using vertical drains in soil improvement projects." *Geotext. Geomembr.*, 22: 101–117.

Chu, J. and Yan, S. W. (2005). "Application of the vacuum preloading method in land reclamation and soil improvement projects." Chapter 3, In *Ground Improvement—Case Histories*, B. Indraratna and J. Chu, eds., Elsevier, 91–118.

Chu, J. and Yan, S.W. (2005) "Estimation of degree of consolidation for vacuum preloading projects." *Int. J. Geomech.*, 5(2): 158–165.

Conroy, T., Fahey, D., Buggy, F., and Long, M. (2010). Control of secondary creep in soft alluvium soil using surcharge loading." BCRI Conference, Cork, 247–254.

FHWA (2004). *Ground Improvement Methods*. FHWA NHI-04-001. FHWA, Washington, DC.

Hansbo, S. (1981). "Consolidation of fine-grained soils by prefabricated drains." *Proceedings of the 10th International Conference on Soil Mechanics and Foundations Engineering*, Stockholm, Sweden, 3, 667–682.

Hansbo, S. (1997). "Aspects of vertical drain design: Darcian or non-Darcian flow." *Geotechnique*, 47: 983–992.

Hird, C.C. and Moseley, V.J. (2000). "Model study of seepage in smear zones around vertical drains in layered soil." *Geotechnique*, 50(1): 89–97.

Holtz, R.D., Lancellotta, R., Jamiolkowski, M.B., and Pedroni, R. (1991). *Prefabricated Vertical Drains: Design and Performance*, Butterworth-Heinemann. Oxford, UK.

Holtz, R.D. and Holm, G. (1973). "Excavation and sampling around some drains at Ska-Edeby, Sweden." *Proceedings of the Nordic Geotechnical Meeting*, Trondheim, Oslo, 79–85.

Holtz, R.D. and Kovacs, W.D. (1981). *An Introduction to Geotechnical Engineering*. Prentice Hall, Englewood Cliffs, NJ.

Indraratna, B. (2009). "Recent advances in the application of vertical drains and vacuum preloading in soft soil stabilization." EH Davis Memorial Lecture—Australian Geomechanics Society.

Indraratna, B. and Redna, I.W. (1998). "Laboratory determination of smear zone due to vertical drain installation." *J. Geotech. Eng.*, 124(2): 180–184.

Kjellman, W. (1952). "Consolidation of clayey soils by atmospheric pressure." *Proc. of a Conf. on Soil Stabilization*, Massachusetts Institute of Technology, Boston, 258–263.

Ladd, C.C. (1971). *Settlement Analysis for Cohesive Soils*. Research Report R71-2, Soils Publication 272, Massachusetts Institute of Technology, Cambridge, MA.

Masse, F., Spaulding, C.A., Wong, P.I.C., and Varaksin, S. (2001). "Vacuum consolidation: A review of 12 years of successful development." *Geo-Odyssey*, ASCE Geo-Institute, Virginia Tech, Blacksburg, VA, June 6–13.

Matsuo, M. and Kawamura, K. (1977). "Diagram for construction control of embankment on soft ground." *Soils and Foundations* 17(3), 37–52.

Mesri, G., Ajlouni, M.A., Feng, T.W., and Lo, D.O.K. (2001). "Surcharging of soft ground to reduce secondary settlement." In *Soft Soil Engineering*, C.F. Lee, C.K. Lau, C.W.W. Ng, A.K.L. Kwong, P.L.R. Pang, J.-H. Yin, and Z.Q. Yue (eds.). A.A. Balkema, London, UK.

Mesri, G., Stark, T.D., Ajlouni, M.A., and Chen, C.S. (1997). "Secondary compression of peat with or without surcharging." *J. Geotech. Geoenviron. Eng.*, 123(5): 411–421.

Mesri, G., Ullrich, C.R., and Choi, Y.K. (1978). "The rate of swelling of overconsolidated clays subjected to unloading." *Geotechnique*, 28(3): 281–307.

Mohamedelhassan, E. and Shang, J.Q. (2002). "Vacuum and surcharge combined one-dimensional consolidation of clay soils." *Canad. Geotech. Eng.*, 39: 1126–1138.

Nash, D. (2001). "Precompression design for secondary settlement reduction—discussion." *Geotechnique*, 51(9): 822–826.

Olson, R.E. (1977). "Consolidation under time-dependent loading." *J. Geotech. Eng. Div.*, 103(GT1): 55–59.

Onoue, A. (1992). "Precompression and vertical drain designs." Workshop on Applied Ground Improvement Techniques, Southeast Asian Geotechnical Society, AIT, 1–78.

Rixner, J.J., Kraemer, S.R., and Smith, A.D. (1986). *Prefabricated Vertical Drain, Vol. 1, Engineering Guidelines*. Report No. FHWA/RD-86/168, Federal Highway Administration, Washington, DC.

Seah, T.H. (2006). "Design and construction of ground improvement works at Suvarnabhumi Airport." *Geotechn. Eng. J. Southeast Asian Geotechn. Soc.*, 37: 171–188.

Shang, J.Q., Tang, M. and Miao, Z. (1998). "Vacuum preloading consolidation of reclaimed land: a field study." *Canad. Geotech. J.*, 35: 740–747.

Stewart, J.P., Lacy, H.S., and Ladd, C.C. (1994). "Settlement of large mat on deep compressible soil." In *Vertical and Horizontal Deformations of Foundations and Embankments*, Geotechnical Special Publication No. 40, A.T. Yeung and Felio (eds.), 1, 842–859, ASCE Press, Reston, VA.

Tang, M. and Shang, J.Q. (2000). "Vacuum preloading consolidation of Yaogiang Airport runway." *Geotechnique*, 50(6): 613–623.

Taylor, D.W. (1948). *Fundamentals of Soil Mechanics*. Wiley, New York.

Terzaghi, K. (1943). *Theoretical Soil Mechanics*. Wiley, New York.

Wang, T.-R. and Chen, W.-H. (1996). "Development in application of prefabricated drains in treatment of soft soils." *Proceedings of the Third Symposium on Weak Ground Improvement Using PVD*, China, 13–40.

Weech, C.N. and Lister, D.R. (2009). "Highway embankment construction over soft soils in the lower mainland of British Columbia." Presented in the Session "Slope and Embankment Engineering for Changing Environments" at the 2009 Annual Conference of the Transportation Association of Canada, Vancouver, British Columbia.

Yan, S. W. and Chu, J. (2005). "Soil improvement for a storage year using the combined vacuum and fill preloading method." *Canad. Geotechn. J.* 42(4), 1094–1104.

Ye, S.L., Han, J., and Ye, G.B. (1994). *Ground Improvement and Underpinning Technologies*, 2nd ed. Beijing, China Building Industry Press.

Zhu, G. and Yin, J.H. (2004). "Consolidation analysis of soil with vertical and horizontal drainage under ramp loading considering smear effects." *Geotext. Geomembr.*, 22: 63–74.

CHAPTER 8

Deep Mixing and Grouting

8.1 INTRODUCTION

Chemical agents (also referred to as binders) can be introduced into the ground and combined with existing geomaterials (referred to as soils or rocks in this chapter) to form hardened solid materials (i.e., improved geomaterials), which have higher strength and stiffness. The chemical agents can be lime, cement, silicate-based gel, and chemical solution. Two general methods are available to introduce and combine hardening agents with soils: mixing and grouting. The mixing method utilizes mechanical mixers or augers, while the grouting method utilizes pipes with high-pressure grouts. Mixing can occur near the surface (mainly for improving subgrade and base course) or at depths (to form columns or walls). The method for mixing hardening agents with soils at depths as shown in Figure 8.1(a) is often referred to as deep mixing. The details on surface mixing for roads and airfields can be found in the *Military Soils Engineering Manual* (FM5-410) (U.S. Department of the Army, 1997) or other documents and therefore will not be further discussed in this book.

Chemical binders can also be introduced to the ground by grouting. A typical grouting process involves the installation of a pipe into the ground and pumping of chemical agents as grout into the ground by high pressure as shown in Figure 8.1(b). Depending on type, composition, and consistency of the grout, soil type, and pumping pressure, grout may fill existing cavities in the soil as fill grouting, may penetrate into the soil through voids between soil particles as permeation grouting, may induce cracks in the soil and then penetrate into the cracks as hydrofracture grouting, and may displace and/or densify the soil as compaction grouting. When grouting is used to compensate ground loss due to construction activities, excavation, and tunneling, it is often referred to as compensation grouting. Jet grouting is a special ground improvement technique between conventional grouting and deep mixing. Jet grouting utilizes grout (sometimes with water and/or air) to erode the soil at depths and then mix the eroded soil with grout to form hardened columns or walls, as shown in Figure 8.1(c). Due to the hybrid nature of this technique, it is considered to be one of the deep mixing methods by some authors (e.g., Kitazume and Terashi, 2013) but considered to be one of the grouting methods by others (e.g., Warner, 2004). In this book, jet grouting is included in the section of grouting but its design procedure mostly follows that for deep mixing.

8.2 DEEP MIXING

8.2.1 Introduction

Basic Concept The deep mixing (DM) method mixes in situ soil with a hardening agent (cement, lime, slag, or other binders) at depths by augers. Deep mixing can be accomplished by a wet or dry method. A wet method uses the binder in a slurry form as shown in Figure 8.2(a), while a dry method uses the binder in a powder form in Figure 8.2(b)

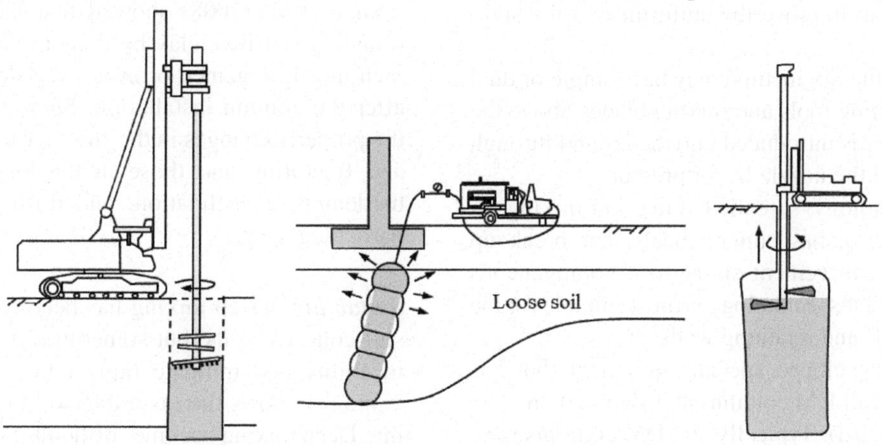

Figure 8.1 Deep mixing and grouting: (a) deep mixing, (b) grouting, and (c) jet grouting.

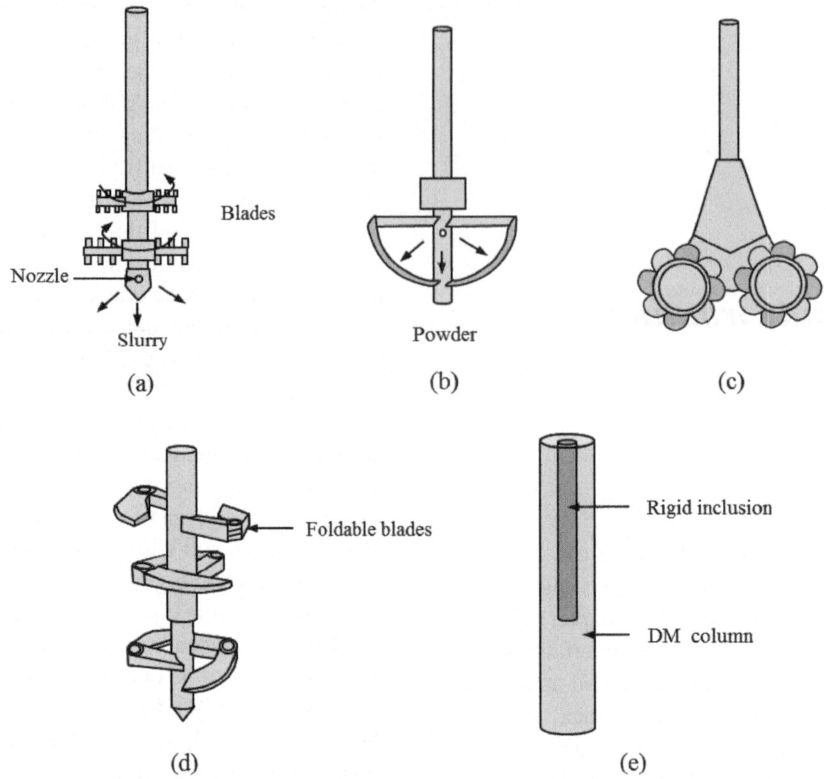

Figure 8.2 Different types of deep mixing techniques: (a) wet method, (b) dry method, (c) cutter soil mixing, (d) T-shape deep mixing, and (e) stiffened DM column.

The equipment for the wet method may have one to eight rotary hollow shafts with cutting tools and mixing blades above the tip. The binder slurry is introduced into the ground through each hollow shaft and exits from the nozzle while the shaft penetrates into the soil or is withdrawn. Some equipment has mixing blades rotating in opposite directions (i.e., double mixing) to improve the uniformity of the soil–binder mixture.

The equipment for the dry method may have single or dual rotary shafts with cutting tools and mixing blades above the tip. The binder powder is introduced into the ground through each hollow shaft and the nozzle by air pressure.

The new DM technology, so-called cutter soil mixing, as shown in Figure 8.2(c), has cutter wheels that break up the soil and mix it with cement slurry to a homogeneous soil–cement mixture. This technology is most suitable for the construction of cut-off and retaining walls.

Liu et al. (2012) developed special equipment that has foldable augers to install DM columns at different diameters as shown in Figure 8.2(d). Typically, the DM columns have a larger diameter near the top and a regular diameter for the rest of the columns. Due to the shape of the columns like the letter T, they are called the T-shaped DM columns.

Deep mixed columns can be stiffened by rigid inclusions (such as concrete piles, spun piles) in the center of the columns as shown in Figure 8.2(e) to increase the vertical and horizontal load capacities and the stiffness of the DM columns (Jamsawang et al., 2008; Bhandari and Han, 2009). This type of column is also called a composite column.

Shen et al. (2008) showed that the strength of the surrounding sensitive clay by deep mixing first decreased and then mostly regained or even exceeded its original strength after the column installation. Shen et al. (2008) attributed the property changes in the short term to the soil disturbance and fracturing and those in the long term to thixotropic hardening, consolidation, and diffusion of ions from the hardening agent.

Suitability Deep mixing has been mostly used to improve soft cohesive soils, but sometimes it is used to reduce permeability and mitigate liquefaction of cohesionless soils. Table 8.1 shows the favorable soil properties for deep mixing. Deep mixing becomes difficult if the ground is very stiff, very dense, and contains boulders or other obstructions. Typically, deep mixing requires unrestricted site access and overhead clearance due to large equipment used in most projects.

Table 8.1 Favorable Soil Properties for Deep Mixing

Property	Favorable Soil Chemistry
pH	Should be greater than 5
Natural water content	Should be less than 200% (dry method) and less than 60% (wet method)
Organic content	Should be less than 6% (wet method)
Loss on ignition	Should be less than 10%
Humus content[a]	Should be less than 1.0%
Electrical conductivity	Should be greater than 0.04 mS/mm

[a]Determined by loss on ignition after ashing for 1 h at 800°C.
Source: Elias et al. (2006).

Deep mixing can reach a depth of up to 70 m in marine work and 30 m for land operations.

Applications Columns have been used for many applications in soft soils: (1) support of superstructures, including buildings, walls, embankments, and the like, (2) waterfront and marine applications including quay walls, wharf structures, and breakwaters, (3) stabilization of slopes, (4) lateral support, (5) containment of water and pollutant, (6) liquefaction mitigation, and (7) vibration reduction. Deep mixed columns have also been used for roadway widening to support new embankments (Han et al., 2007) and mitigation of ground heave due to expansive soil (Puppala et al., 2008). In these applications, DM columns are used to increase bearing capacity, reduce settlement, enhance slope stability, provide lateral support, contain water and pollutant movement, mitigate liquefaction, and reduce vibration. DM columns are typically arranged in four different patterns as shown in Figure 8.3. Individual columns are used when an area replacement ratio is relatively low, for example, less than 50%. Individual columns have been mostly used to increase bearing capacity and reduce settlement. Block pattern is used to carry significant vertical and/or horizontal loads when a high area replacement ratio is needed, for example, more than 50%. Block pattern has been mostly used to improve the stability of large marine structures. This pattern has also been used to contain waste materials to prevent the leaching of hazardous chemicals. Panel or wall pattern has been commonly used as a retaining wall for lateral support, a seepage wall to cut off seepage, a curtain wall to contain waste materials, or a wall perpendicular to the centerline of the embankment to increase the stability. Grid pattern is between the wall pattern and the block pattern. It can be used for the applications suitable for wall and block patterns. A unique application of the grid pattern is to mitigate liquefaction of sandy soils. The liquefiable soils are contained inside the cells of grids.

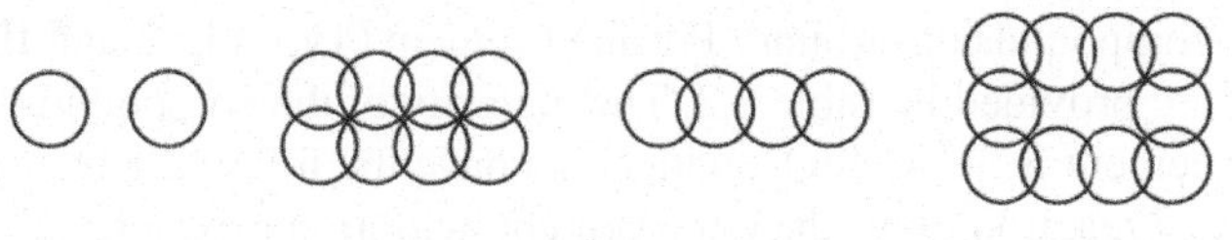

Figure 8.3 Patterns of columns: (a) individual column, (b) block, (c) wall, and (d) grid.

In recent years, columns have been largely used to support embankments over soft foundations. This application is considered as one of the most important applications of the column technologies. Deep mixed columns have also been increasingly combined with other technologies, such as geosynthetic reinforcement, PVDs, and rigid piles. Geosynthetic reinforcement above columns acts as a bridging layer to transfer the embankment load to the columns and reduce the differential settlement between the columns. The common applications of geosynthetic-reinforced column-supported embankments are: (1) bridge approach, (2) roadway widening, (3) subgrade improvement, and (4) support of storage tanks (Han and Gabr, 2002). The design of geosynthetic-reinforced column-supported embankments is discussed in Chapter 10. Concrete slabs have also been used above columns instead of geosynthetic reinforcement to support the embankment over soft foundations (e.g., Zheng et al., 2011).

When embankments are constructed over thick soft soils, the soft soils often do not have sufficient bearing capacities to support embankments. Under such a condition, columns can be used to increase the bearing capacity of the soft foundation. Unless the soft soils within the influence depth are fully reinforced by columns (often requiring significantly long columns, which are costly), the soft soils below the reinforced zone at a deeper depth still deform at a slow rate. To reduce the length of columns and accelerate the rate of consolidation of the soft soils below the reinforced zone, Xu et al. (2006) proposed the use of DM columns to improve soft soils at a shallower depth and PVDs to accelerate the rate of consolidation at a deeper depth. The concept of this combined technology is shown in Figure 8.4(a). A similar method was proposed by Liu et al. (2008), as shown in Figure 8.4(b) by combining dry jet mixed columns with PVDs; however, its primary purpose is to use PVDs to accelerate the dissipation of excess pore water pressure induced by dry jet mixing installation, thus minimizing the installation disturbance. The secondary purpose of this combined technology is to accelerate the consolidation rate of the soft soil between the columns.

Short DM columns can also be used between long rigid piles or long columns to increase the bearing capacity of soft soils at a shallower depth, thus reducing the loads carried by

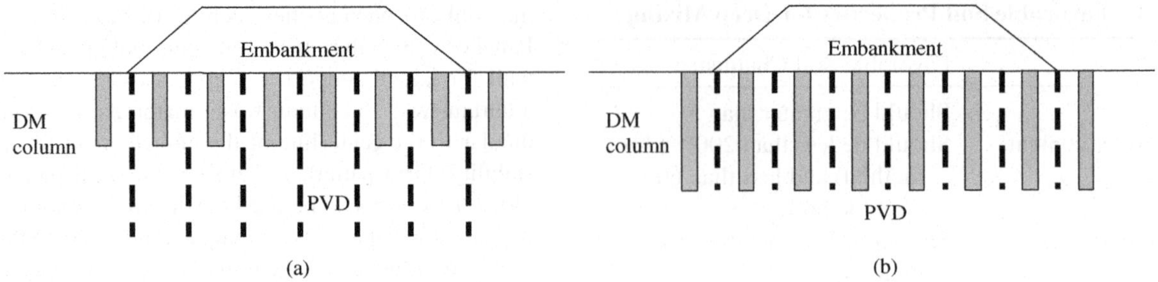

Figure 8.4 Deep mixed column–PVD combined method: (a) short columns and long PVDs and (b) equal length columns and PVDs (Han, 2012).

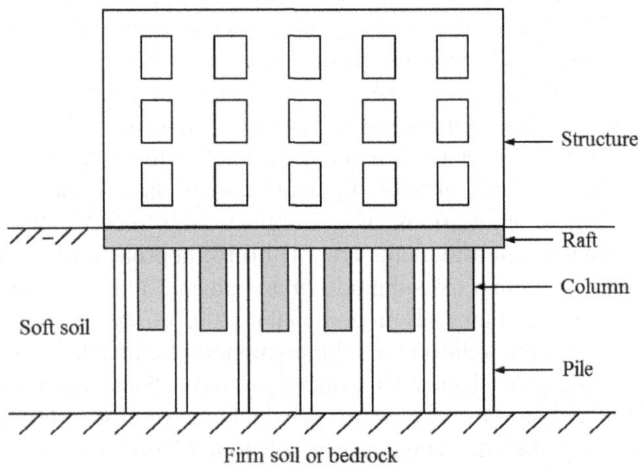

Figure 8.5 Column–pile combined method.

long rigid piles or long columns so that the spacing of the long rigid piles or long columns can be enlarged to create a more economic solution (Huang and Li, 2009; Zheng et al., 2009b). Figure 8.5 illustrates the concept of this combined method.

Advantages and Limitations The deep mixing method has the following advantages:

- Applicable for most soil types
- Installed at great depths
- Relatively fast installation
- Low noise and vibration level
- Formation of a DM wall for earth retaining and water barrier at the same location and time
- Less spoil soil, especially for the dry method

However, the deep mixing method may have the following limitations:

- Relatively high mobilization cost
- High variability in column quality
- Lack of standardized quality control methods

8.2.2 Principles

Chemical Reaction Lime stabilization involves hydration of binder, ion exchange reaction, and formation of pozzolanic reaction products. When quicklime (CaO) is mixed with a moist soil, it absorbs the moisture in the soil and has the following chemical reaction:

$$CaO + H_2O = Ca(OH)_2 + \text{heat} \tag{8.1}$$

This chemical reaction generates heat and reduces the moisture in the soil. The reduction of the moisture in the soil increases the strength of the soil. This chemical process is also called the hydration of binder. The hydrated lime dissolves into the water thus increasing the concentrations of the calcium ion, Ca^{2+} and the hydroxyl ion, OH^-. The calcium ion Ca^{2+} exchanges with cations on the surface of the clay particles. The ion exchange reaction between the binder and the soil can change the physical properties of the soil, such as the reduction of soil plasticity. Under a high concentration of OH^- (i.e., a high pH environment), silica and aluminum in the clay minerals dissolve into water and have reaction with calcium to form calcium-silicate and calcium-aluminate gels, which are stable and do not dissolve into water as long as the high pH environment is maintained. This reaction process is called the pozzolanic reaction. The products produced from this process are called the pozzolanic reaction products. These products turn the soil into a hardened solid with high strength and stiffness.

In addition to hydration of binder, ion exchange reaction, and formation of pozzolanic reaction products, the chemical stabilization of soils by cement has a reaction process to form cement hydration products. Cement has several compounds, which can have chemical reactions with water. The main compounds of ordinary Portland cement (Types I, II, and III) are provided in Table 8.2. The majority of the compounds in cement is $3CaO{\cdot}SiO_2$, which can have the following chemical reaction to produce the cement hydration product:

$$2(3CaO \cdot SiO_2) + 6H_2O = 3CaO \cdot 2SiO_2 \cdot 3H_2O + 3Ca(OH)_2 \tag{8.2}$$

Table 8.2 Main Components of Ordinary Portland Cement

Compound	Chemical Formula	Common Formula	Usual Range by Weight (%)
Tricalcium silicate	$3CaO \cdot SiO_2$	C_3S	40–65
Dicalcium silicate	$2CaO \cdot SiO_2$	C_2S	15–35
Tricalcium aluminate	$3CaO \cdot Al_2O_3$	C_3A	5–12
Tetracalcium Aluminoferrite	$4CaO \cdot Al_2O_3 \cdot Fe_2O_3$	C_4AF	6–10

Source: Modified from Mamlouk and Zaniewski (2011).

The cement hydration product, $3CaO \cdot 2SiO_2 \cdot 3H_2O$, is often expressed as $C_3S_2H_3$ or C-S-H. During this hydration, calcium hydroxide, $Ca(OH)_2$, is released, which contributes to the pozzolanic reaction as lime stabilization. The formation of hydration products over time leads to:

- Stiffening (loss of workability)
- Setting (solidification)
- Hardening (strength gain)

The formation of cement hydration and pozzolanic reaction products change the mechanical properties of the soil–cement mixture by increasing its strength and stiffness.

In addition to lime and cement, slag and fly ash have been used as binders for deep mixing. Figure 8.6 shows the chemical reactions of clay, cement, slag, and water and the reaction products. Figure 8.7 illustrates the chemical reactions and products formed in soil by different binder types.

The chemical reactions shown in Figures 8.6 and 8.7 happen within the DM columns. Ion exchanges may happen between the DM columns and the surrounding soil. Because cement contains a large amount of Fe_2O_3, CaO, and MgO ingredients, cations such as ferric Fe^{3+}, calcium Ca^{2+}, and magnesium Mg^{2+} are dissolved in the pore water of DM columns after the installation as a result of cement hydration. Under this condition, the concentration of cations in the pore water of DM columns is higher than that in the pore water of the surrounding soils. Due to this concentration gradient, multivalent cations diffused from the DM columns to the surrounding clay. Because the diffusion of cations is gradual, no significant change in cation concentrations occurs right after the installation. After 1 month, however, the mean concentration of Fe^{3+} increased approximately 10 times as compared with the original value (Shen et al., 2008). At the same time, the mean concentration of Ca^{2+} in the surrounding soil increased significantly. Similar changes happen to the concentrations of two monovalent cations of Na^+ and K^+. These chemical effects are limited to a transitional zone in the surrounding clay around the DM columns. In the transitional zone, the physical, chemical, and mechanical properties change with curing time. However, the effect on the property change in the surrounding soil is not considered in most designs in practice.

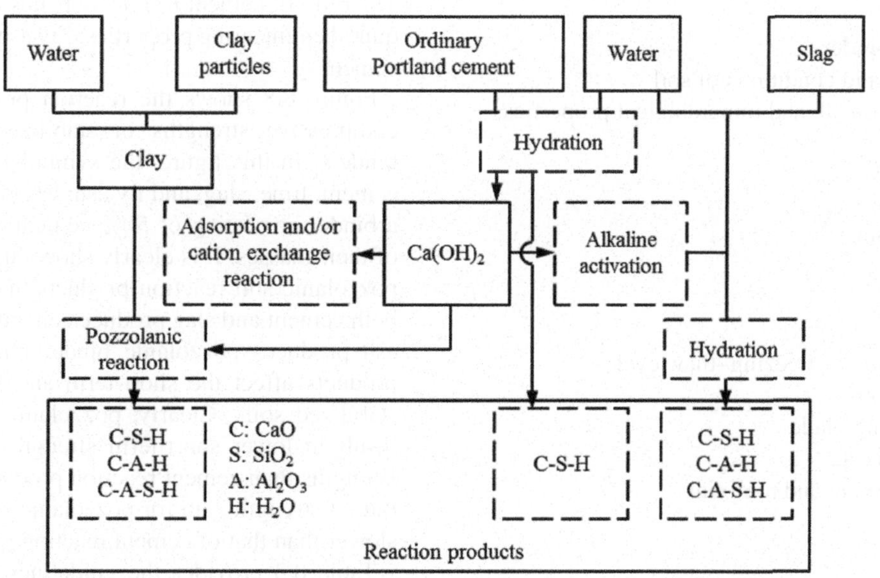

Figure 8.6 Chemical reactions of clay, cement, slag, and water (Saitoh et al., 1985, with permission from Taylor & Francis).

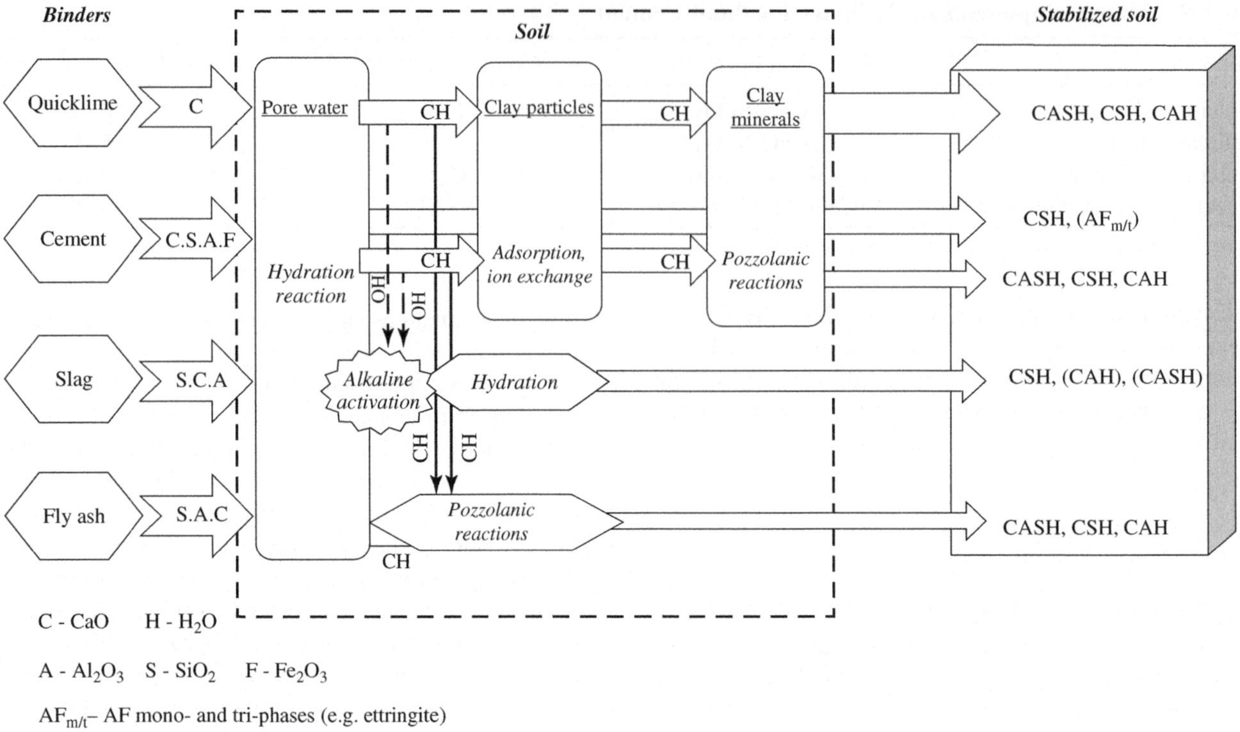

Figure 8.7 Principal chemical reactions and subsequent products formed in soil by different binder types (after Ahnberg and Johansson, 2005).

Properties of Stabilized Soils The properties of soil–binder mixtures depend on the following factors (Terashi, 2003):

1. Characteristics of hardening binder
   - Type of binder
   - Quality of binder
   - Additives
   - Quantity of binder
2. Characteristics and conditions of soil
   - Physical, chemical, and mineralogical properties
   - Organic content
   - pH of pore water
   - Water content
3. Curing conditions
   - Temperature
   - Curing time
   - Humidity
   - Wetting–drying or freezing–thaw cycles
4. Mixing conditions
   - Type of mixing blade
   - Degree of mixing
   - Timing of mixing and remixing
   - Mixing water

Binder Type Different binders have been used for deep mixing. The four commonly used types of lime for soil stabilization include calcium oxide (i.e., quicklime), calcium hydroxide (slaked or hydrated lime), wet hydroxide, and lime-based special lime (Kitazume and Terashi, 2013). Ordinary Portland cement and blast furnace slag cement are two commonly used cement types. Lime–cement mixtures have been increasingly used, which combine lime with cement. The ratio of cement to lime depends on applications and required engineering properties. Fly ash is sometimes added to cement.

Figure 8.8 shows the reaction products and unconfined compressive strengths of stabilized soils with different binders. In this figure, the symbols c, l, s, and f stand for cement, lime, slag, and fly ash; "cs (50:50)" is referred to as a binder consisting of 50% cement and 50% slag in terms of their masses. It is clearly shown that lime produces more pozzolanic soil reaction products than other binders, while both cement and slag produce cement reaction products. Fly ash produces pozzolanic binder reaction products. These products affect the short-term and long-term strengths of stabilized soils. Clearly, pozzolanic soil reaction products result in lower short-term strengths but higher long-term strengths than cement reaction products. In other words, the rate of strength gain for pozzolanic soil reaction products is slower than that of cement reaction products.

Table 8.3 provides the guidelines for the use of binders for different soil types. Lime-based binders are generally not suitable for peat; instead, cement-based binders should be used.

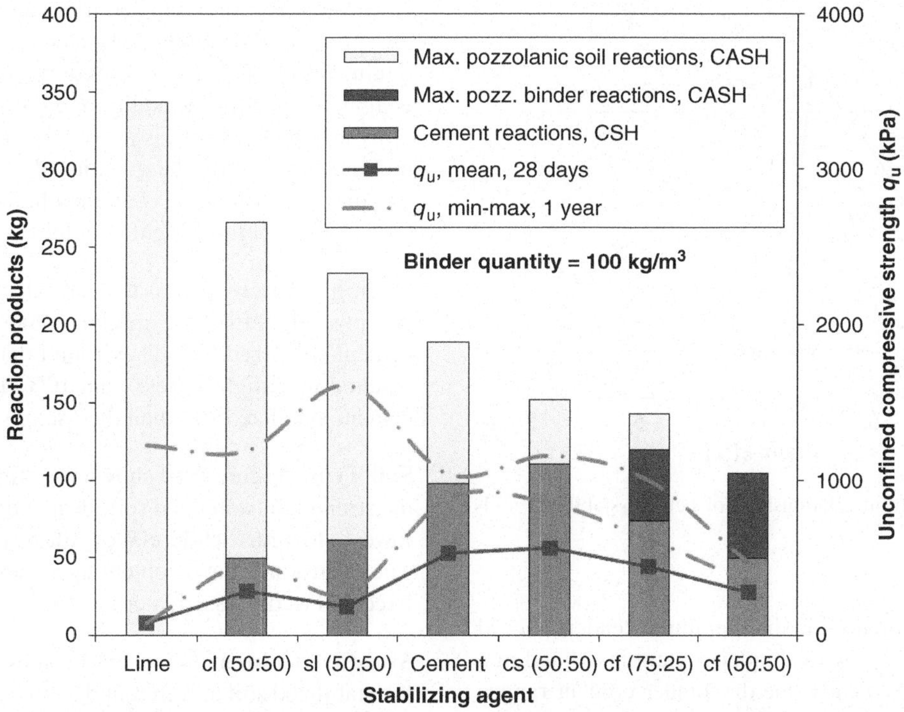

Figure 8.8 Production of reaction products in stabilized soils with different binder types (after Ahnberg and Johansson, 2005).

Table 8.3 Suitability of Binder for Soil Type[a]

Soil type Binder	Silt Organic Content = 0–2%	Clay Organic Content = 0–2%	Organic Soils (e.g., Gyttja, Organic Clay) Organic Content = 2–30%	Peat Organic Content = 50–100%
Cement	B	C	C	B
Cement + gypsum	C	C	B	B
Cement + furnace slag	B	B	B	A
Lime + cement	B	B	C	D
Lime + gypsum	B	B	B	D
Lime + slag	C	C	C	D
Lime + gypsum + slag	B	B	B	D
Lime + gypsum + cement	B	B	B	D
Lime	D	D	B	D

[a]A, very good binder in many cases; B, good in many cases; C, good in some cases; and D not suitable.
Souce: EuroSoilStab (2001).

Binder Content The binder content can be defined in three ways:

$$a_c = \frac{\text{Mass of dry binder}}{\text{Mass of dry soil}} \times 100\% \quad (8.2)$$

$$a_{cv} = \frac{\text{Mass of dry binder}}{\text{Volume of treated soil}} \quad (8.3)$$

$$a_{cw} = \frac{\text{Mass of dry binder}}{\text{Mass of natural soil}} \times 100\% \quad (8.4)$$

where a_c is dimensionless, has been commonly used in scientific research, and is referred to as dry binder content; a_{cv} is dimensional and convenient for estimating the amount of binder used in installation; therefore, it is more commonly used in practice; a_{cv} is referred to as volumetric binder content; a_{cw} is dimensionless, dependent on the moisture content of the soil, and convenient for estimating the amount of binder to be mixed with natural soil for lab testing. Also a_{cw} is referred to as the moist binder content. All these binder

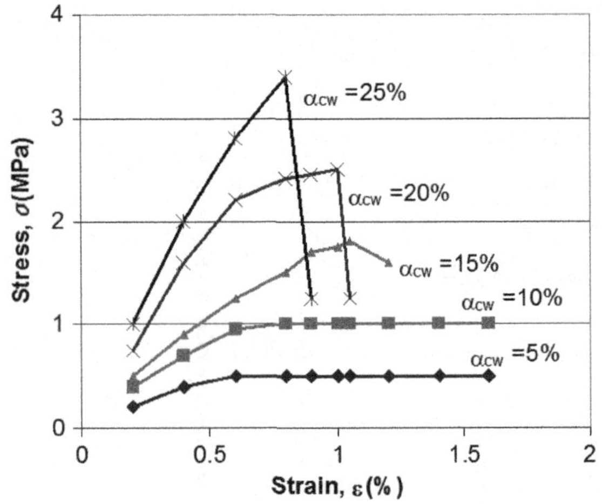

Figure 8.9 Stress–strain relationships of cement-stabilized soils (Ye et al., 1994).

content definitions are used in the literature. Readers should pay close attention to how the binder content is defined in a specific reference. Typically, the dry binder content ranges from 10 to 30% while the volumetric binder content ranges from 150 to 300 kg/m^3.

Figure 8.9 shows the typical stress–strain relationships of cement-stabilized soils obtained from unconfined compression tests. The figure shows that the cement-stabilized soil samples at higher cement contents have brittle behavior while those at lower cement contents have ductile behavior. The increase of the cement content increases the unconfined compressive strength of the sample.

Additive Several additives have been used in cement-stabilized soils for different purposes. For example, lignin $CaCO_3$ has been used for the purpose of water reduction. However, test results showed it has an insignificant effect on the strength of the mixture (Han et al., 2002). The amount of lignin $CaCO_3$ added in the mixture is typically 0.2% of the weight of cement. The addition of gypsum or 3-aminoethnol or fly ash into mixtures is proved to increase the strengths of mixtures and reduce the quantity of cement needed. Typical amount of gypsum and 3-aminoethnol used in mixtures are 2 and 0.05% of the weight of cement, respectively.

Mixing Water Kawasaki et al. (1978) investigated the possible effect of mixing water on the strength of cement-stabilized soil. They found that the use of tap water resulted in a slightly (less than 10%) higher strength of the cement-stabilized soil than the use of seawater.

Soil Type Figure 8.10 shows the effect of the soil type on the strength of stabilized soil. Peat and organic soils have the lowest strengths while clayey silt has the highest strength. The inclusion of sand content increases the strength of stabilized soil as discussed later.

Moisture Content Figure 8.11 shows that the strength of the stabilized soil at different dry binder content, a_c depends on the total moisture content, w_t which includes the moisture in the soil and the moisture introduced by the slurry binder. It is the same for both ordinary Portland cement and blast furnace cement that the increase of the total moisture content reduces the strength of the stabilized soil. Therefore, the dry method has the advantage by not introducing extra moisture to the soil.

Sand Content Figure 8.12 shows the effect of sand content on the unconfined compressive strengths of quicklime or cement-stabilized soils. Both results show that there was an optimum sand content and the samples with 60% sand reached the maximum unconfined compressive strengths.

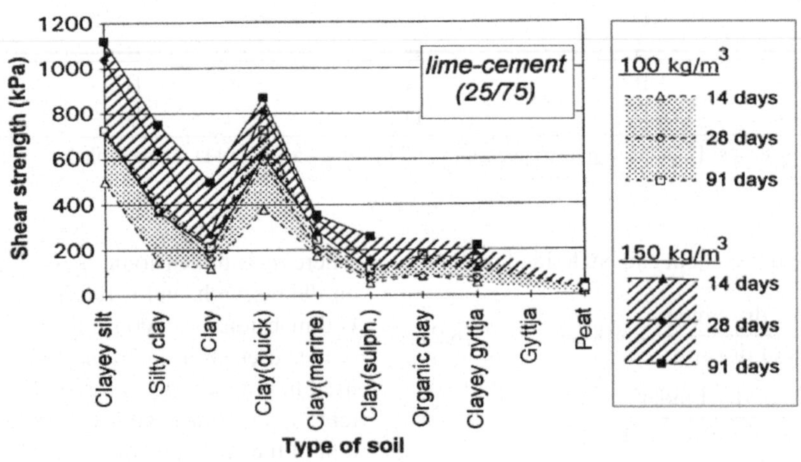

Figure 8.10 Effect of soil type (Holm, 2005).

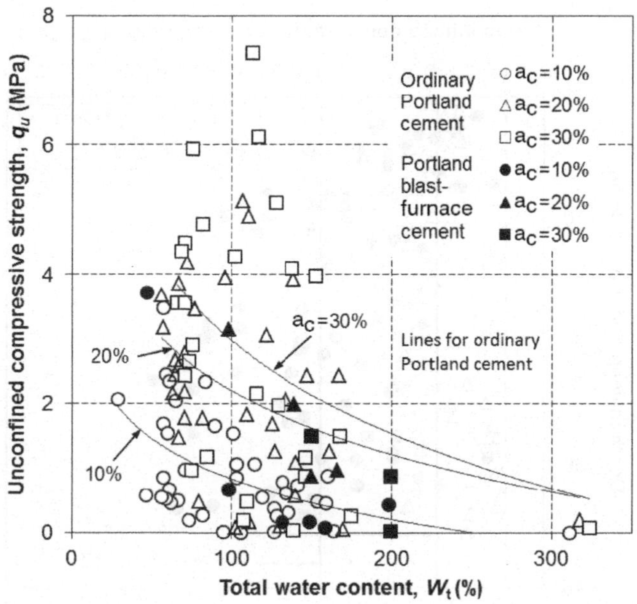

Figure 8.11 Effect of total moisture content (after Babasaki et al., 1996).

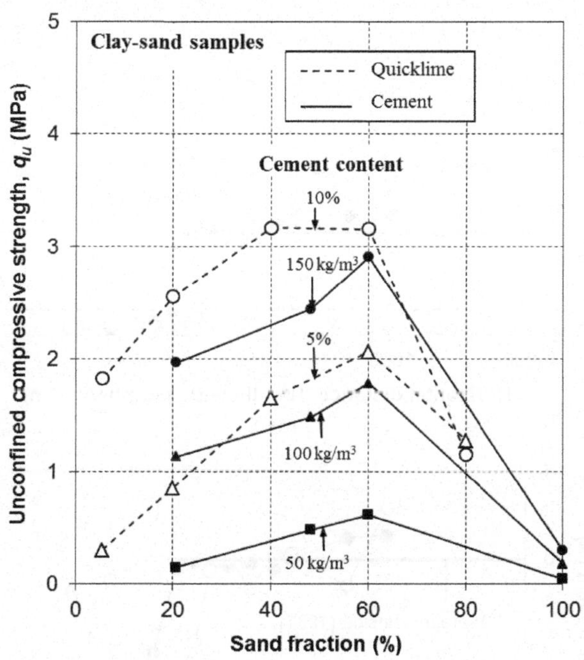

Figure 8.12 Effect of sand content (data for quicklime from Tersashi et al., 1977; data for cement from Niina et al., 1977).

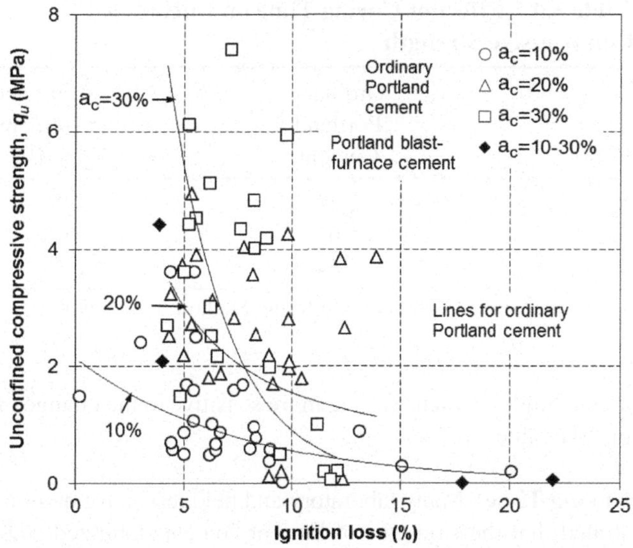

Figure 8.13 Effect of ignition loss (%) (modified from Babasaki et al., 1996).

Organic Content The effect of organic content can be evaluated in terms of the ignition loss. Figure 8.13 shows that the unconfined compressive strength of the stabilized soil decreases with the increase of the ignition loss. When the ignition loss is less than 15%, a higher binder content, a_c, may be used to achieve a higher strength. When the ignition loss is more than 15%, both ordinary Portland cement and Portland blast-furnace cement become less effective.

pH Value The hydration of lime or cement results in a high pH environment, which is needed for pozzolanic reactions. A low pH value soil (pH < 5) is not favorable for the hydration process of lime or cement and causes pozzolanic reaction products to be unstable. As a result, the unconfined compressive strength of the stabilized soil decreases with a decrease of the pH value as shown in Figure 8.14. The samples with

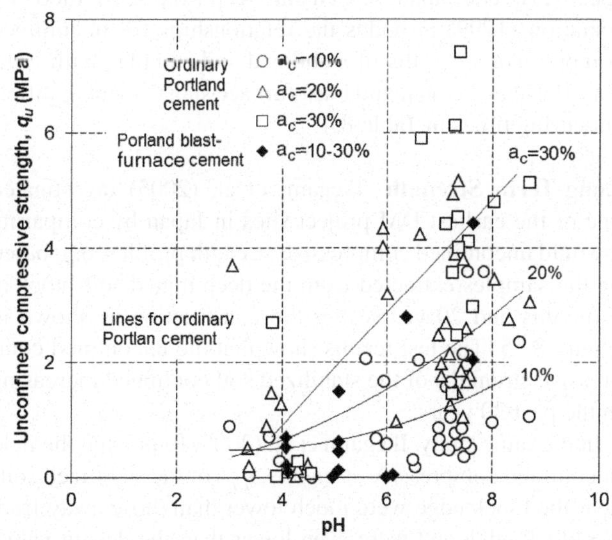

Figure 8.14 Effect of pH value of natural soil (Babasaki et al., 1996).

Table 8.4 Effect of Curing Time on Unconfined Compressive Strength

Ratio	Ordinary Portland Cement	Blast Furnace Slag Cement Type B
q_{u28}/q_{u7}	1.49	1.56
q_{u91}/q_{u7}	1.97	1.95
q_{u91}/q_{u28}	1.44	1.20

Source: Japanese Cement Deep Mixing Method Association (1999).

a high binder content, a_c, are more sensitive to the change of the pH value.

Curing Time Many laboratory and field tests have demonstrated that the strengths of lime or cement-stabilized soils increase with the curing time due to hydration, ion exchange between cement and clay, hardening, and solidification. The Chinese Ground Improvement Technical Code (China Academy of Building Research, 2000) suggests that the design strength should be based on the unconfined compressive strength of a cement-stabilized soil sample cured for 90 days. Due to the long test period criterion, which causes difficulties in design and construction, researchers have sought for alternatives by establishing correlation between quick test results and the 90-day test results. The explanation part of the Soil Improvement Technical Code (China Academy of Building Research, 2000) indicates that the unconfined compressive strengths of cement-stabilized soil samples cured for 7 and 30 days are, respectively, about 30–50% and 60–75% of the samples cured for 90 days, respectively. The Japanese Cement Deep Mixing Method Association (1999) provides the relationships for unconfined compressive strengths of stabilized soil samples with ordinary Portland cement and blast furnace slag cement at different curing times in Table 8.4.

Long-Term Strength Ikegami et al. (2005) investigated one of the earliest DM project sites in Japan by comparing the field unconfined compressive strength profiles, q_{uf}, based on the samples extracted from the deep mixed soil block at 3 months and 20 years after the construction as shown in Figure 8.15. The test results show that the unconfined compressive strengths of the stabilized soil continued increasing in the past 20 years.

In the same study, Ikegami et al. (2005) found that the field unconfined compressive strengths, q_{uf}, of the stabilized soils near the block edge were much lower than those away from the block edge and were even lower than the design value, q_{ud}, as shown in Figure 8.16. In addition, the Ca content near the edge was much lower than that away from the edge. There

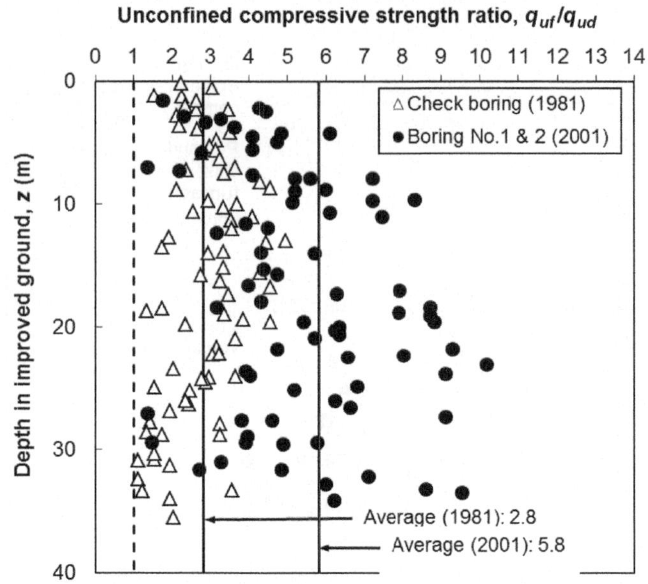

Figure 8.15 Unconfined compressive strengths of stabilized soil samples extracted from the field at 3 months and 20 years after the construction (modified from Ikegami et al., 2005).

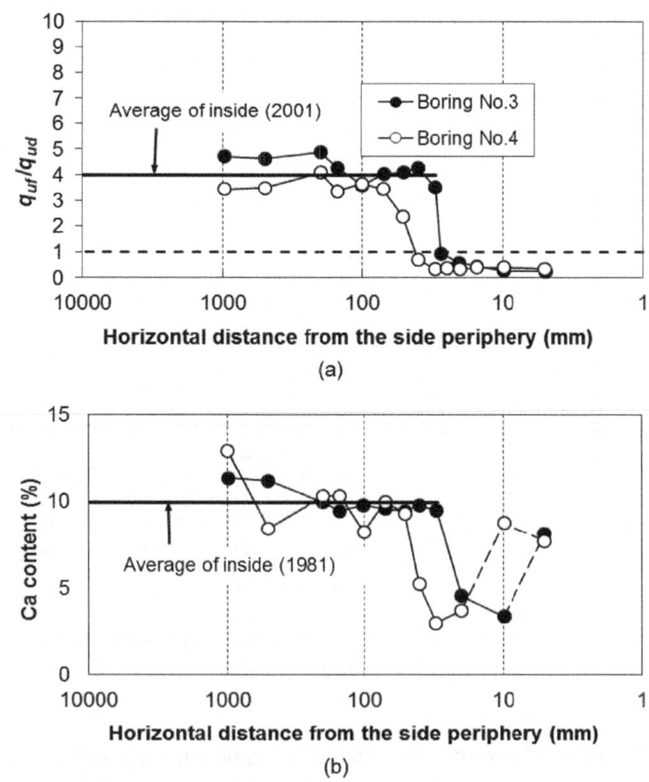

Figure 8.16 (a) Strength and (b) Ca content distributions near the periphery of in situ stabilized soil block (modified from Ikegami et al., 2005).

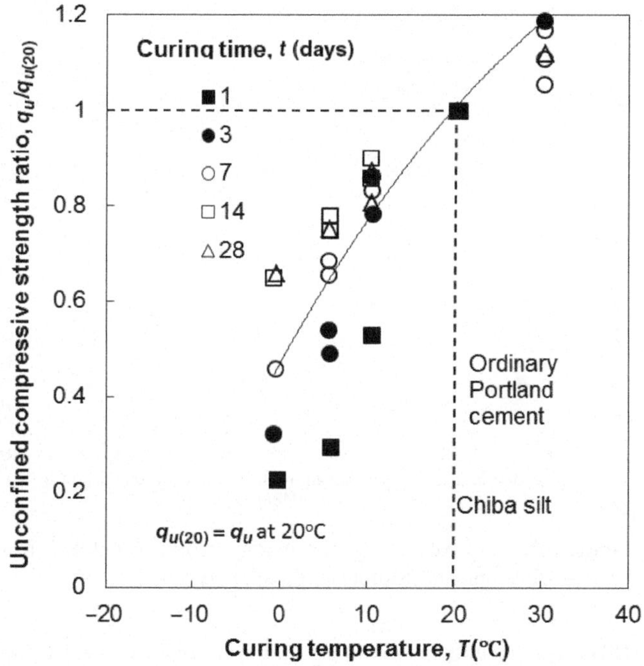

Figure 8.17 Effect of curing temperature (modified from Enami et al., 1985).

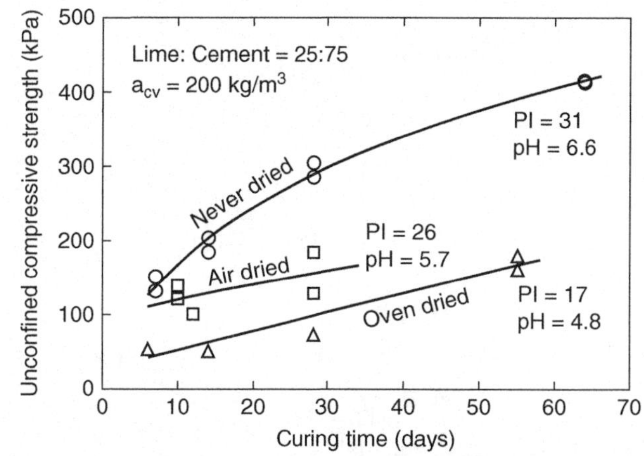

Figure 8.18 Effect of soil drying on the strength of stabilized soil (after Jacobson et al., 2005).

is an indication that the Ca ion leached out during the past 20 years. Fortunately, the area with the degraded strength is very limited.

Curing Temperature Figure 8.17 clearly shows that the curing temperature has a significant effect on the strength of the cement-stabilized silt. Increasing the curing temperature accelerates the early-age hydration reactions, which result in the initial increase in the strength of the mixture. A room temperature of 20°C is commonly used for preparation of samples in laboratory. The strength ratio, $q_u/q_{u(20)}$ shows that the samples prepared at lower temperature would result in lower strength. Ground temperature is typically lower than 20°C; therefore, cement-stabilized soil in the field typically has a lower strength than that in laboratory in addition to other factors.

Soil Dry and Rehydration Jacobson et al. (2005) conducted laboratory tests to evaluate the effect of drying and rehydrating the soil with 6% organic content on the strengths of the lime–cement stabilized soil samples. They found the strength of the mixture, when the air-dried or oven-dried soil was used, was much lower than that without any drying process as shown in Figure 8.18. Jacobson et al. (2005) attributed this strength decrease to the reason that the drying and rehydrating process made the organics more soluble and more dispersed throughout the mixture, which interfered with cementation.

Mixing Energy Shen et al. (2004) investigated the effect of mixing energy on the strengths of cement-stabilized soils in the laboratory using a model mixer. The energy consumption depends on the type of chemical agent (cement, lime, or none), the form of the agent (slurry or powder, or none), the content of the agent, the strength of soil, the installation speed, the rotation rate, and the mixing sequence. Figure 8.19 shows the relationship between the unconfined compressive strength ratio and the mixing energy. The strength ratio is defined as the strength (q_u) at a specific energy divided by the average strength (q_{ua}) at the final energy at the corresponding 7 or 28 day curing time. It is shown that an increase of the mixing energy increased the strength of the cement-stabilized soil and this increase became less

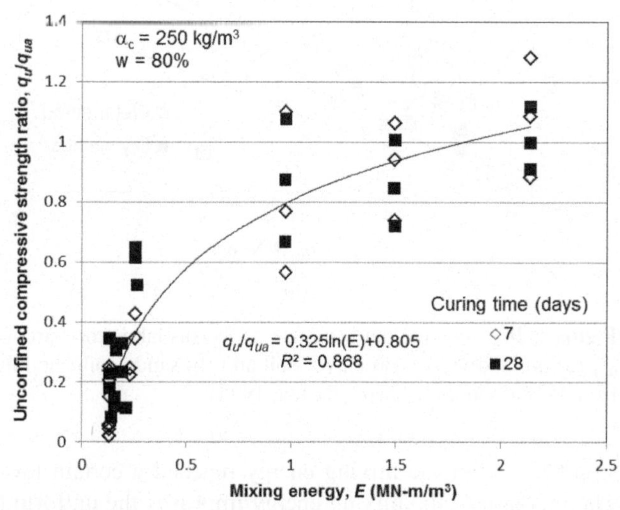

Figure 8.19 Effect of mixing energy on the strength of the stabilized soil (modified from Shen et al., 2004).

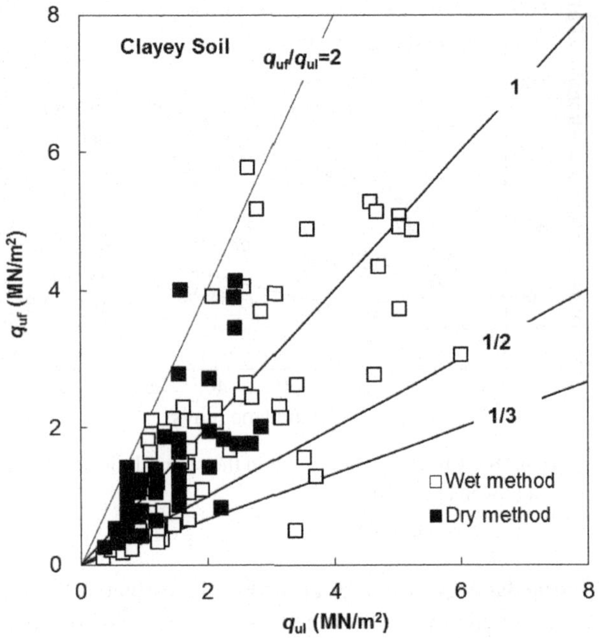

(a)

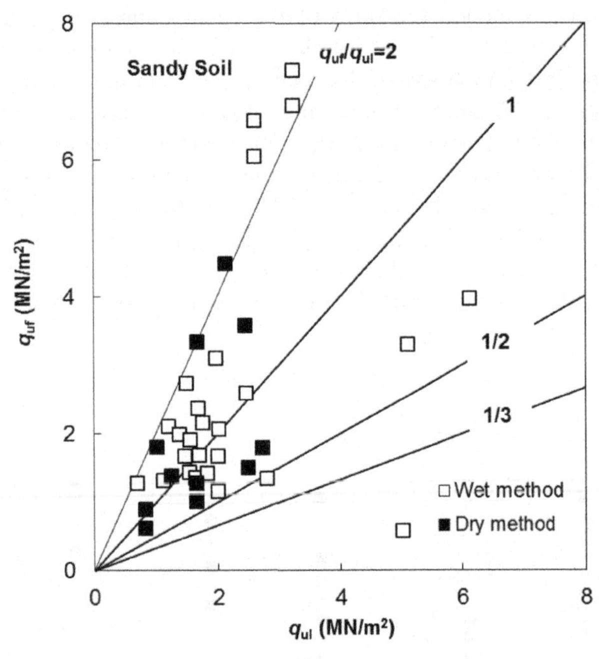

(b)

Figure 8.20 Average field strength q_{uf} versus laboratory strength q_{ul} for on-land works: (a) clayey soil and (b) sandy soil (modified from Public Works Research Center, 1999).

significant after the mixing energy reached a certain level. The increase of the mixing energy improves the uniformity of the mixture. The same results were obtained by Nishida and Miura (1996).

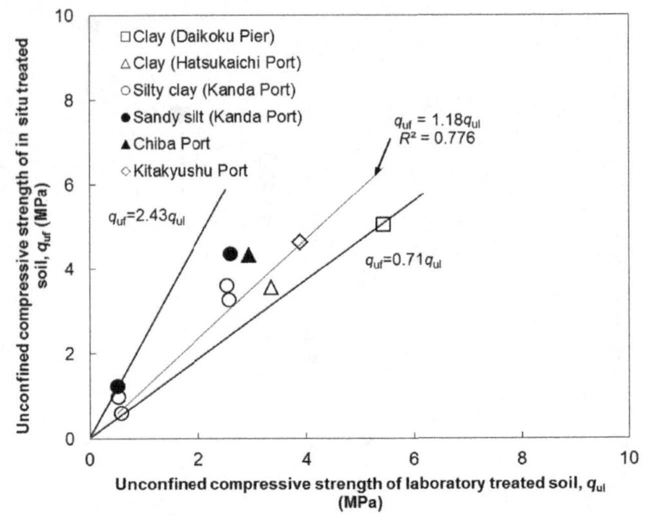

Figure 8.21 q_{uf} versus q_{ul} for marine works (modified from Coastal Development Institute of Technology, 2002).

Effective Strength Ahnberg and Johansson (2005) found that the mean effective cohesion of a stabilized soil is approximately $0.23q_u$ (q_u = unconfined compressive strength of the stabilized soil) and the mean effective friction angle is 31° to 33°.

Field Strength Due to possible differences in uniformity, curing conditions, and sample sizes, the properties of stabilized soils in the field may be different from those obtained from laboratory. Figures 8.20 and 8.21 show the comparisons of unconfined compressive strengths (q_{uf}) of stabilized soil samples obtained from the field with those (q_{ul}) based on the samples prepared in laboratory. The ratios of q_{uf}/q_{ul} for the stabilized soils used for on-land works range from $\frac{1}{3}$ to 2, but approximately half of the data are below the 1 : 1 line. However, the ratios of q_{uf}/q_{ul} for the stabilized soils used for marine works are mostly greater than 1.0. The higher q_{uf}/q_{ul} ratios for marine works in the field may be attributed to the fact that better uniformity of the mixtures is achieved by using large mixing equipment for large quantity of mixtures. The wet and dry methods of deep mixing resulted in similar relationships between the field strengths and the laboratory strengths of the stabilized soils.

Research was also done by the Building Center of Japan (1997) to investigate the scale effect on the strength of stabilized soils. Large column samples in diameters of 1.0–1.2 m and heights of 1.5–2.4 m extracted from the field were tested under compression for unconfined compressive strengths, $q_{u,c}$. Small samples with a diameter of 67 mm and a height of 130 mm were cored from the same site and tested for unconfined compressive strengths, q_u. Figure 8.22 shows the comparison of the unconfined compressive strengths from

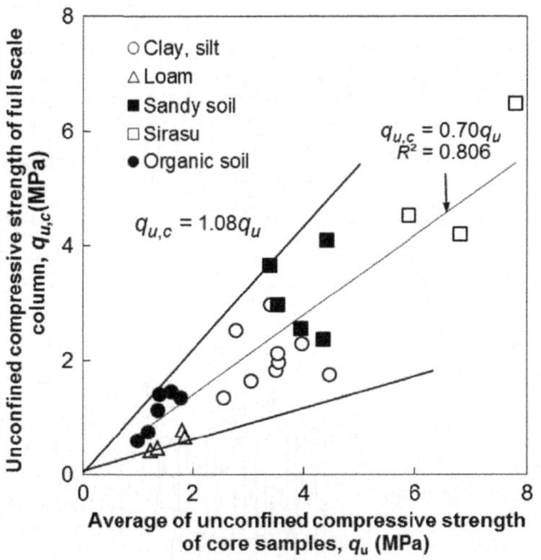

Figure 8.22 Full-scale column strengths versus unconfined compressive strengths of core samples (modified from Building Center of Japan, 1997).

full-scale column tests and small cored sample tests. It is shown that the full-scale column strengths, $q_{u,c}$ are approximately 70% of the cored sample strengths, q_u.

Variability Due to the variability in soil, binder, installation, and curing conditions, the strengths of the stabilized soils have variability. The Coastal Development Institute of Technology (1999) reported that the coefficients of variation (COV) for the strengths of the stabilized soils using the wet installation method for marine construction range from 0.2 to 0.5 with a value of 0.3 in most cases. Figure 8.23 shows the

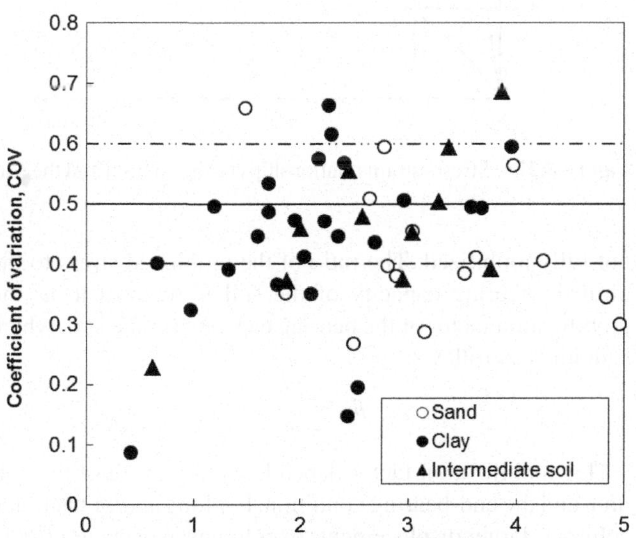

Figure 8.23 Variability of strength data for on-land works (modified from Matsuo, 2002).

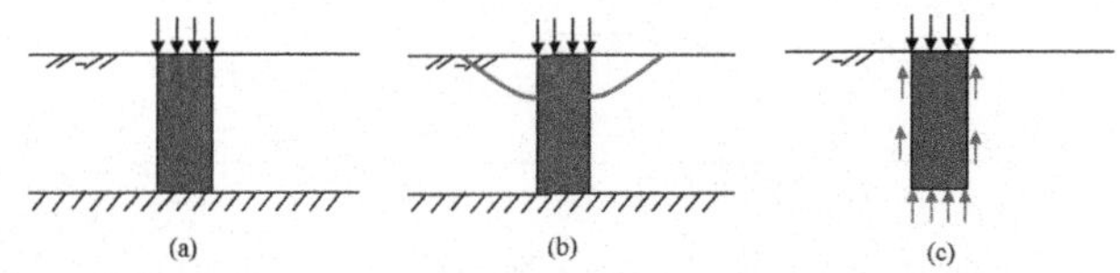

Figure 8.24 Failure modes of DM columns under vertical loads: (a) crushing failure, (b) shear failure, and (c) punching failure.

variability of the strength data for on-land works by Matsuo (2002). Therefore, the COV values for on-land works range from 0.4 to 0.6. Larsson (2005) reported the COV values of compression test results from 17 studies by researchers in different countries range from 0.15 to 0.80.

Possible Failure Modes Deep mixed columns have been installed in the field in a form of individual columns, blocks, walls, or grids. When individual columns are subjected to a vertical compressive load, they may have crushing, shear, and punching failures as illustrated in Figure 8.24. Columns may crush when the applied load is higher than the capacity of the columns. This failure more likely happens to DM columns because they are relatively brittle. The shear failure may happen to DM columns at a low binder content. The punching failure may happen to short DM columns without an end-bearing layer.

When columns are under embankment loading (i.e., a combination of vertical and horizontal loads), they may have the following possible failure modes (Figure 8.25): (1) sliding, (2) collapse (rotational), (3) bending, (4) circular shear, (5) horizontal shear, and (6) combined failure. The actual failure mode of the column depends on its strength, rigidity, length and diameter, location and spacing, end-bearing condition, the strength and stiffness of soft soil, and the height and slope angle of the embankment. Under certain conditions, there is a combined failure mode as shown in Figure 8.25(f), which includes bending and tensile failures.

When the DM columns are installed in a block, wall, or grid pattern, they can act together as a rigid gravity wall. As a result, they can have the same failure modes as those for a gravity wall as shown in Figure 8.26, which include sliding, overturning, bearing failure, and rotational failure. The methods for the gravity wall can be adopted for these analyses.

Stress Transfer Due to the modulus difference between DM columns and the surrounding soil, higher stresses develop on DM columns than those on the surrounding soil when the composite soil foundation is subjected to applied loads. This stress difference is often described by a stress concentration ratio, which is defined as the ratio of the stress on the column to that on the surrounding soil. The concept of

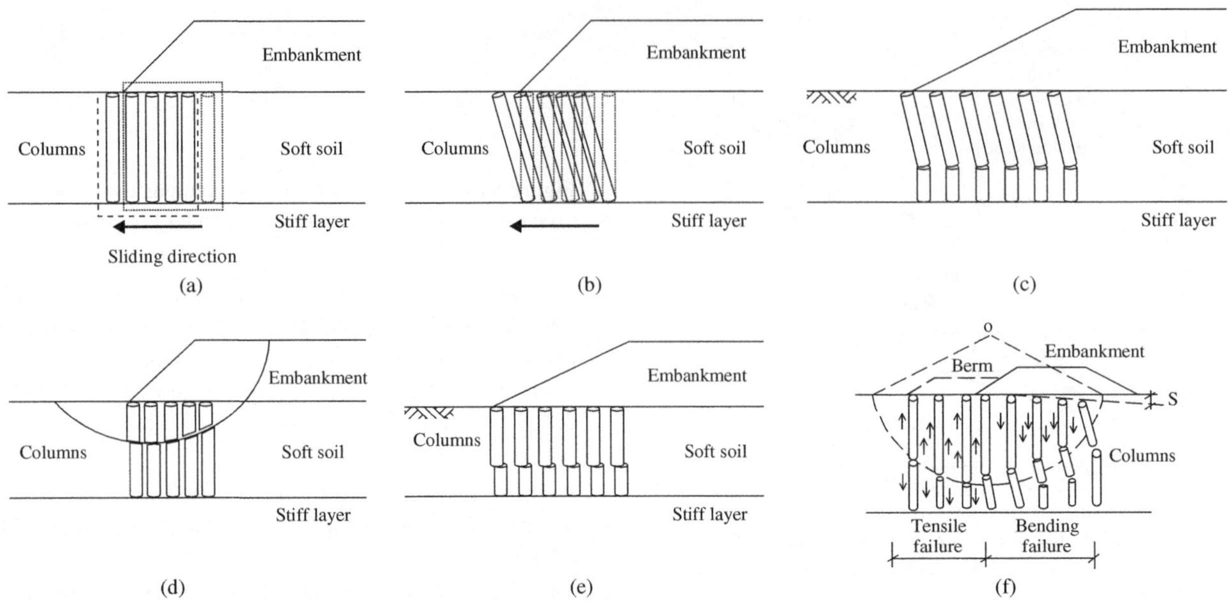

Figure 8.25 Possible failure modes of columns under embankments: (a) sliding, (b) collapse (rotational), (c) bending, (d) circular shear, (e) horizontal shear, and (f) combined (modified from Kitazume, 2008; Broms, 1999).

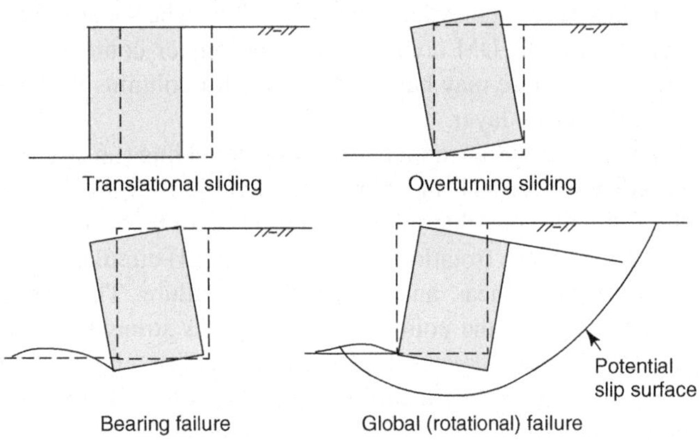

Figure 8.26 Gravity wall failures.

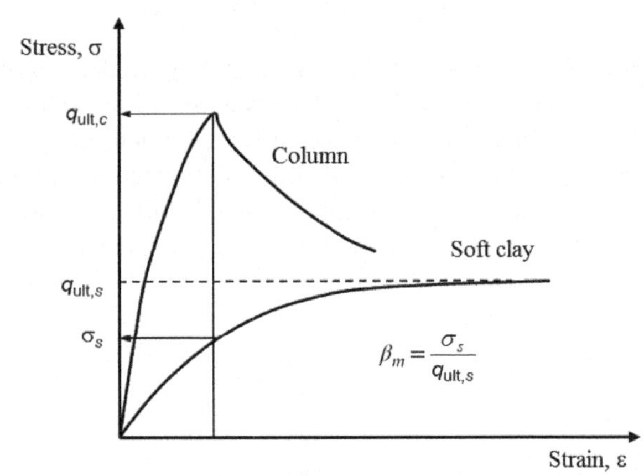

Figure 8.27 Stress–strain relationships of the column and the soil.

the stress concentration for the DM column-reinforced composite foundation is similar to that for the columns installed by the deep replacement method as discussed in Chapter 5. This concept has been confirmed by many full-scale plate loading tests and under embankments in the field or numerical analyses, for example, Ye et al. (1991), Lin and Wong (1999), Huang et al. (2009), and Jiang et al. (2013). Field data show that the stress concentration ratios for DM column-reinforced composite soil foundations mostly range from 5.0 to 10.0 under design loads.

Figure 8.27 shows the stress–strain relationships for the column and the soil. At an equal strain, more and more stresses are transferred to the column with the increase of the strain until the column reaches the ultimate strength. At the ultimate strength of the column, the soil strength is not fully mobilized. The ratio of the mobilized stress to the ultimate bearing capacity of the soil is referred to as the mobilization factor of the bearing capacity for the soil, which is defined as follows:

$$\beta_m = \frac{\sigma_s}{q_{\text{ult},s}}$$

This mobilization factor depends on the length of the column and the end-bearing condition. For long and/or frictional columns, larger displacements of columns may occur prior to the yielding of the column. The larger displacements allow the soil to mobilize its stress close to the ultimate bearing capacity of the soil thus resulting in a higher mobilization factor. However, for short and/or end-bearing columns, limited displacements of columns may occur prior to the yielding of the column. As a result, a lower soil stress is mobilized with a lower mobilization factor.

8.2.3 Design Considerations

Typical Properties of Stabilized Soil The typical properties of stabilized soils are provided in Table 8.5 based on the wet method of deep mixing and Table 8.6 based on the dry method of deep mixing. The effects of different factors on the unconfined compressive strengths of stabilized soils have been discussed above.

Table 8.5 Typical Properties of Stabilized Soils (Wet Method)

Property	Typical Range
Unconfined compressive strength, q_u	Up to 1.2 MPa for organic and very plastic clays, sludges 0.4–1.5 MPa for soft clays 0.7–2.5 MPa for medium/hard clays 1.0–3.0 MPa for silts 1.5–5.0 MPa for fine-medium sands
Permeability, k_c	10^{-6}–10^{-9} m/s
Young's modulus (E_{50}) (secant modulus at 50% q_u)	(50–1000) q_u for lab samples (Japan) (50–150) q_u for lab samples (US) (100–300) q_u (typical range)
Tensile strength	8–15% of q_u
Poisson's ratio	0.20–0.45 (typically 0.26)

Source: Modified from Elias et al. (2006).

Table 8.6 Typical Properties of Lime–Cement Stabilized Soils (Dry Method)

Property	Typical Range
Undrained shear strength	(10–50) c_u of soil (0.15–1.0 MPa)
Young's modulus	(50–200) c_u of lime–cement column (50–200) q_u of cement-treated soil
Strain at failure	<2%
Permeability (lime–cement)	About the same as for in situ soil
Permeability (lime)	10–100 × in situ soil permeability

Source: Modified from Elias et al. (2006).

Han et al. (2002) investigated the effect of cement content on the permeability of the stabilized soil and found that an increase of the cement content reduced the permeability of the stabilized soil. However, Chew et al. (2004) found that the permeability of the stabilized soil increased with an increase of the cement content. Åhnberg (2003) developed the empirical relationships between the permeability ratio of the homogeneous stabilized soil to the natural soil with the moisture content and the strength of the stabilized soils as follows:

$$\frac{k_c}{k} \approx 0.043 \exp\left(\frac{6w}{w_0} - 0.004\, q_u\right) \tag{8.5}$$

where k_c = permeability of the stabilized soil
k = permeability of natural soil
w = moisture content of the stabilized soil
w_0 = moisture content of the natural soil
q_u = unconfined compressive soil of the stabilized soil (unit: kPa)

Figure 8.28 shows that the permeability ratio increases with the moisture content but decreases with the strength of the stabilized soil. The permeability ratio can be more or less than 1.0. In other words, the permeability of the stabilized soil can be more or less than that of the natural soil. In addition, the higher strength samples of stabilized soil have lower permeability.

Young's modulus of a stabilized soil is defined as a secant modulus from the origin to 50% unconfined compressive strength (i.e., E_{50}) as shown in Figure 8.29. E_{50} has been correlated to unconfined compressive strength, q_u as follows:

$$E_{50} = \alpha_E q_u \tag{8.6}$$

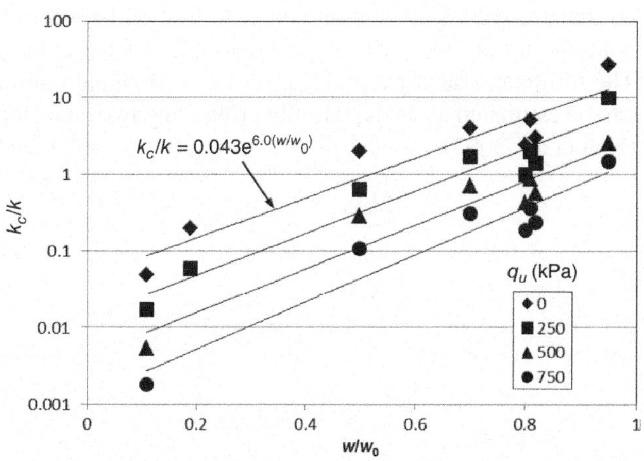

Figure 8.28 Permeability of stabilized soil (redrawn from data in Åhnberg, 2003).

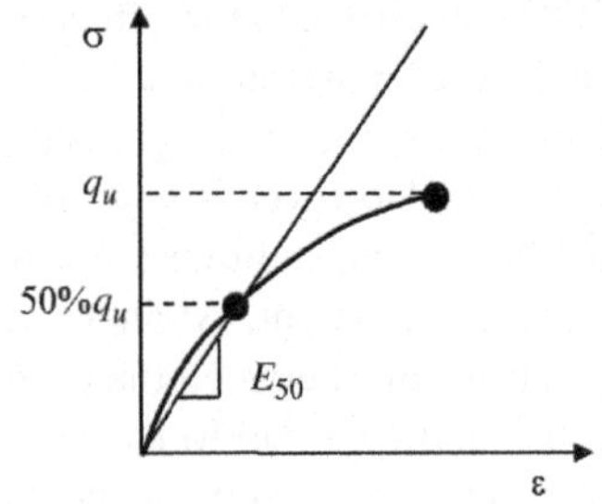

Figure 8.29 Young's modulus, E_{50}.

where α_E = constant, which varies in a wide range from 50 to 1000.
α_E = 100 has been commonly used and is also mostly conservative.

Typical Shapes and Dimensions of Columns The equipment used for the dry method is typically lighter than that for the wet method using rotary equipment. Deep mixing can install four different shapes of columns as shown in Figure 8.30. The columns installed by the dry method are typically circular and about 0.6–0.9 m in diameter, while the columns installed by the wet method can be of different shape and up to about 2.5 m in equivalent diameter. Cutter soil mixing installs panels, which are typically 2.4–2.8 m long and 0.55–1.2 m wide. For the individual column pattern, the area replacement ratio typically ranges from 0.2 to 0.5. Depths of treatment are typically less than 20 m for the dry method and 30 m for the wet method. The water to binder ratio for the wet method typically ranges from 0.6 to 1.0.

Bearing Capacity DM columns, as semirigid columns, have the behavior between flexible and rigid columns. The ultimate load capacity of an individual DM column depends on the strength of the column, the side friction between the column and the soil, and the toe resistance of the column. The ultimate load capacity ($Q_{\text{ult},c}$) of an individual column can be estimated as the lesser of the following two capacities (Han et al., 2002):

$$Q_{\text{ult},c} = q_{u,c} A_c \tag{8.7}$$

$$q_{u,c} = \alpha_1 \alpha_2 q_{\text{ul}} \tag{8.8}$$

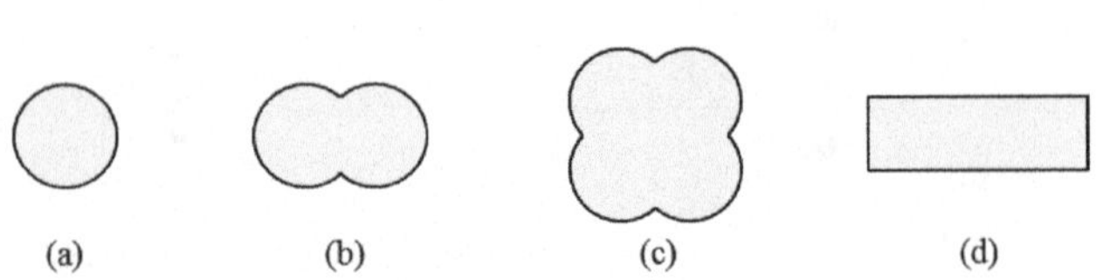

Figure 8.30 Shapes of deep mixed columns or walls: (a) circular, (b) dual, (c) quadral, and (d) panel.

$$Q_{\text{ult},c} = f_s U_c L_c + q_{\text{tm}} A_c \tag{8.9}$$

$$q_{\text{tm}} = \lambda_E q_t \tag{8.10}$$

where $q_{u,c}$ = field unconfined compressive strength of the column
q_{ul} = laboratory unconfined compressive strength of the stabilized soil sample
α_1 = laboratory to field strength correction factor (typically 0.3–1.0 for on-land works or 1.0 for marine works)
α_2 = small cored sample to full-scale column correction factor (typically 0.70)
A_c = cross-sectional area of columns
f_s = average skin friction between the column and the surrounding soil
U_c = perimeter of the column
L_c = column length
λ_E = mobilization factor of the end-bearing (typically ranging from 0.4 to 0.6)
q_{tm} = modified end-bearing capacity of the column toe
q_t = end-bearing capacity, estimated based on pile toe bearing capacity

The average skin friction is suggested in the Chinese Ground Improvement Technical Code (China Academy of Building Research, 2000) based on the soil type as follows: type I—very soft clay (OC > 5%, $e_0 > 1.5$), f_s = 5–8 kPa; type II—soft clay (OC > 5%, $1.0 < e_0 \leq 1.5$), f_s = 8–12 kPa; type III—clay (OC < 5%, $e_0 \leq 1.0$), f_s = 12–15 kPa, where OC is the organic content and e_0 is the in situ void ratio.

The Building Center of Japan (1997) provides the following guidelines for the skin friction and the end-bearing capacity:

For clayey soils:

$$f_s = c_u \tag{8.11}$$

$$q_{\text{tm}} = 6c_{\text{ut}} \tag{8.12}$$

For sandy soils:

$$f_s = \frac{10N}{3} \text{ (kPa)} \tag{8.13}$$

$$q_{\text{tm}} = 75N_t \text{ (kPa)} \tag{8.14}$$

where c_u = average cohesion of the soil along the column shaft (kPa)
c_{ut} = cohesion of the soil below the column toe (kPa)
N = average SPT N value of the soil along the column shaft
N_t = SPT N value below the column toe

The ultimate bearing capacity of a DM column composite foundation (q_{ult}) can be calculated as follows (Han et al. 2002):

$$q_{ult} = a_s \frac{Q_{ult,c}}{A_c} + \beta_m (1 - a_s) q_{ult,s} \tag{8.15}$$

where a_s = area replacement ratio of the column to the soil
β_m = mobilization factor of the bearing capacity of the soil (typically varying from 0.5 to 1.0 for frictional columns or 0.1 to 0.4 for end-bearing columns)
$q_{ult,s}$ = ultimate bearing capacity of the surrounding soil

A factor of safety of 2.0–3.0 may be used to calculate the allowable bearing capacity.

Settlement Due to the characteristics of the DM columns as semirigid columns, the methods for flexible or rigid columns may be used to estimate the settlement of the DM column-reinforced soft foundations, depending on the column rigidity. Three methods for settlement calculations are presented below.

Stress Reduction Method The same stress reduction method as that used for granular columns can be used for the end-bearing DM columns as follows:

$$S' = \frac{1}{1 + a_s(n-1)} S \tag{8.16}$$

where S = settlement of natural ground
a_s = area replacement ratio of columns
n = stress concentration ratio.

The stress concentration ratio, n, can be estimated using the design chart in Figure 8.31. This design chart was developed

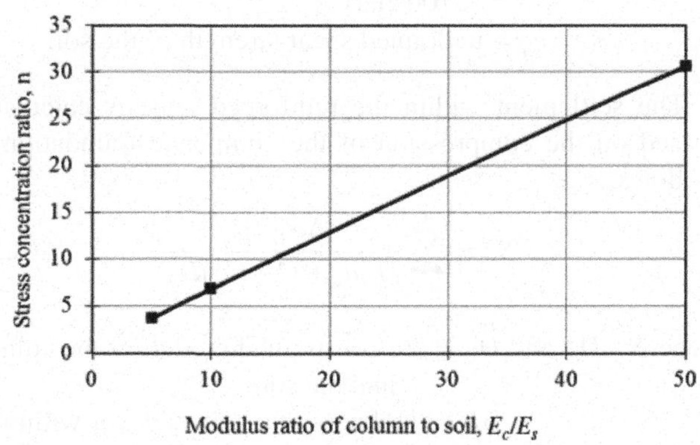

Figure 8.31 Stress concentration ratio versus modulus ratio of column to soil (Jiang et al., 2013). With permission of Springer Science and Business Media.

based on the assumption that both the column and the soil are in an elastic state, the loading plate or footing is rigid, and the column is end-bearing. Jiang et al. (2013) showed that the stress reduction method calculated the settlement of the DM column-reinforced foundation close to the numerical result with less than 10% error. Kitzsume and Terashi (2013) indicated that the modulus ratio of column to soil, E_c/E_s, typically ranges from 10 to 20 in many cases.

In reality, the column may yield under loading; therefore, the stress concentration ratio should be limited by the strength of the column. Jiang et al. (2014) developed a method to determine the limited stress concentration ratio considering the unconfined compressive strength of the column and the elastic behavior of the soil. Below is a summary of this development.

The maximum vertical stress the column can carry (ignoring the low effective overburden stress near the ground surface) is:

$$\Delta\sigma_{cmax} = q_u + K\Delta\sigma_s \tag{8.17}$$

where q_u = unconfined compressive strength of the column
K = coefficient of lateral earth pressure, which is between the coefficient at rest and the coefficient of passive earth pressure
$\Delta\sigma_s$ = vertical stress on the soil

Based on the force equilibrium of a unit cell under an equal strain condition, the vertical stress on the soil can be obtained as follows:

$$\Delta\sigma_s = \frac{\Delta\sigma_z - a_s q_u}{a_s K + (1 - a_s)} \tag{8.18}$$

where $\Delta\sigma_z$ is the average vertical stress on the unit cell.

Therefore, the maximum stress concentration ratio can be calculated as follows:

$$n_{max} = \frac{\Delta\sigma_{c\,max}}{\Delta\sigma_s} == \frac{q_u[a_s K + (1 - a_s)]}{\Delta\sigma_z - a_s q_u} + K \tag{8.19}$$

In the preceding equation, the applied average vertical stress must be higher than $a_s q_u$, otherwise, no column will yield. The elastic modulus of the DM column is assumed to be $100q_u$ and Figure 8.31 can be used to estimate the stress concentration ratio. If the column yields, the stress concentration ratio, n, determined from Figure 8.31, should be limited by the following equation:

$$n \le n_{max} \tag{8.20}$$

Based on the analysis, a K value of 0.6–0.7 results in a close comparison of the computed stress concentration ratio with the numerical result.

The stress concentration ratio discussed above is based on rigid loading. If DM columns are used to support embankments (not rigid loading), the measured stress concentration ratio typically ranges from 2.0 to 6.0, which is lower than that under a rigid loading. The calculated settlement using Equation (8.16) is the average settlement of the embankment.

Piled-Raft Method Han et al. (2009) verified that the method proposed by Horikoshi and Randolph (1999) for piled rafts can be used to calculate the settlement of DM column-reinforced soft foundations. The details of this method are presented in Chapter 5. This method is applicable to end-bearing or floating DM column-reinforced soft foundations under circular or square footings. An example will be presented later in this chapter to demonstrate how this method can be used to calculate the settlement of a DM column-reinforced soft foundation.

Column Penetration Method When columns do not reach a stiff bearing layer, they are often referred to as floating columns. The settlement of a floating column-reinforced soft foundation consists of the compression with the reinforced zone and that below the reinforced zone. Due to the stiffness difference between the columns and the soil, the columns under concentrated loads may have some penetration into the soil. Chai et al. (2010) and Pongsivasathit et al. (2013) proposed a method to calculate the settlement of the DM column-reinforced foundation underlain by a soft soil, in which the penetration of the columns was considered by treating the lower portion of the reinforced zone as an "unreinforced" layer as shown in Figure 8.32. Chai et al. (2010) and Pongsivasathit et al. (2013) obtained a simplified formula to estimate the thickness of the "unreinforced layer", which depends on the area replacement ratio, the improvement depth ratio (i.e., the column depth divided by the thickness of the soft soil), and the pressure-strength ratio as follows:

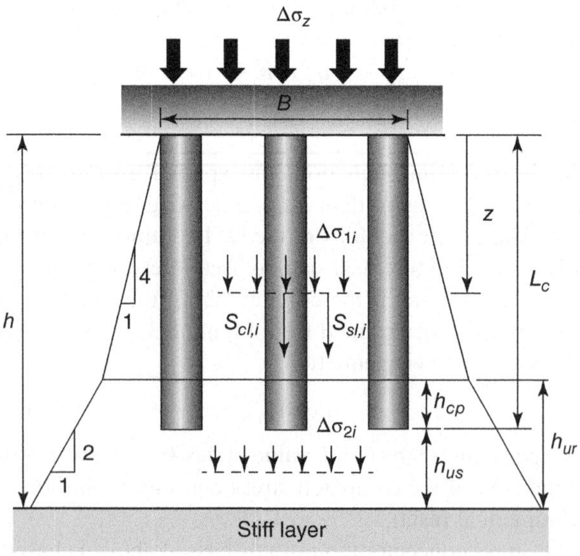

Figure 8.32 Column penetration method (modified from Chai et al., 2010; with permission from ICE Publishing).

$$h_{cp} = L_c f(a_s) g(\beta_L) h(\gamma_p) \tag{8.21}$$

where h_{cp} = thickness of the column penetration zone (treated as an "unreinforced" layer)
L_c = length of the column
$f(a_s)$ = influence factor of the area replacement ratio
$g(\beta_L)$ = influence factor of the improvement depth
$h(\gamma_p)$ = influence factor of the applied pressure to soil strength ratio

When a rigid slab is above the columns,

$$f(a_s) = \begin{cases} 0.75 - 2.5a_s, & a_s \geq 0.20 \\ 0.4 - 1.0a_s, & 0.20 < a_s \leq 0.45 \\ 0, & a_s > 0.45 \end{cases} \tag{8.22}$$

The improvement depth ratio and the pressure-strength ratio are:

$$g(\beta_L) = \begin{cases} 1.62 - 1.6\beta_L, & 0.2 \lesssim \beta_L \lesssim 0.7 \\ 0.5, & 0.7 \leq \beta_L \leq 0.9 \end{cases} \tag{8.23}$$

$$\beta_L = \frac{L_c}{h} \tag{8.24}$$

$$h(\gamma_p) = 0.27 \ln(\gamma_p) - 0.41 \tag{8.25}$$

$$\gamma_p = \frac{\Delta\sigma_z \, p_a^{\,1.5}}{c_u^{2.5}} \tag{8.26}$$

where $\Delta\sigma_z$ = average applied pressure
p_a = atmosphere pressure (approximately 100 kPa)
c_u = undrained shear strength of the soil

The settlement within the reinforced zone is calculated based on the compression of the composite foundation as follows:

$$S_1 = \sum_{i=1}^{n} \frac{\Delta\sigma_{1i} h_{1i}}{D_{ci} a_s + (1 - a_s) D_{si}} \tag{8.27}$$

where D_{ci} and D_{si} = are constrained moduli of the column and the soil
h_{1i} = thickness of a soil sublayer within the reinforced zone
$\Delta\sigma_{1i}$ = additional stress within the reinforced zone induced by the applied pressure, $\Delta\sigma_z$

The stress distribution method for pile groups included in AASHTO (2006) can be used to estimate this additional stress with a 4(H):1(V) stress distribution.

The settlement below the reinforced zone (including the penetration zone) can also be calculated using the stress distribution method for pile groups:

$$S_2 = \sum_{i=1}^{m} h_{2i} \frac{C_{ci}}{1 + e_{0i}} \log\left(1 + \frac{\Delta\sigma_{2i}}{\sigma'_{z0i}}\right) \tag{8.28}$$

where C_{ci} = compression index of the soil sublayer
e_{0i} = initial void ratio of the soil sublayer
h_{2i} = thickness of the soil sublayer within the unreinforced zone
σ'_{z0i} = initial effective overburden stress in the sublayer
$\Delta\sigma_{2i}$ = additional stress within the unreinforced zone estimated with a 2(H):1(V) stress distribution

The total settlement of the foundation is the sum of the settlements within the reinforced zone and below the reinforced zone. i.e.,

$$S' = S_1 + S_2 \tag{8.29}$$

Consolidation Lorenzo and Bergado (2003) derived an analytical solution to predict the consolidation rate of DM column-reinforced foundations based on a unit cell concept. In their model, it is assumed that the DM column has a higher permeability than the surrounding soil, and all the water draining out from the surrounding soil enters the DM column and then drains out vertically through the column. One-dimensional Terzaghi's solution was used to calculate the consolidation degree of the DM column. Lorenzo and Bergado (2003) did not model lateral drainage as Han and Ye (2001, 2002) did for the consolidation of stone column foundations. Miao et al. (2008) modeled the DM column-reinforced foundation as a composite foundation with a higher equivalent modulus than the untreated soil and analyzed the DM column-reinforced foundation over soft soil as a double layer system. They assumed that the DM columns are impervious. Similar to Miao et al. (2008), Chai and Pongsivasathit (2009) treated the DM column foundation as a composite foundation with a higher equivalent modulus as shown in Figure 8.33. However, Chai and Pongsivasathit (2009) proposed an equivalent permeability of the composite foundation, the same as the formula for vertical drains developed by Chai et al. (2001). Equivalent coefficient of volumetric compressibility and permeability of the composite foundation can be calculated as follows:

$$m_{v1} = \frac{1}{a_s D_c + (1 - a_s) D_s} \tag{8.30}$$

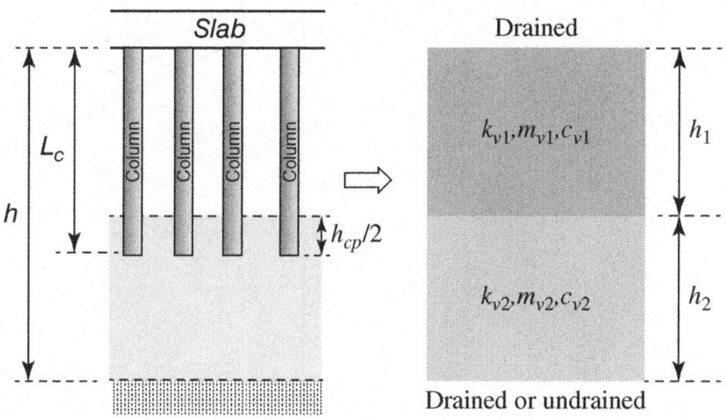

Figure 8.33 Consolidation of column-reinforced soft foundation over soft soil (Chai and Pongsivasathit, 2009).

$$k_{v1} = \left(1 + \frac{1.5 h_1^2}{F_m(N_D) d_e^2} \frac{k_r}{k_v}\right) k_v \tag{8.31}$$

$$F_m(N_D) = \ln \frac{N_D}{N_s} + \frac{k_r}{k_s} \ln(N_s) - \frac{3}{4} + \frac{8 h_1^2 k_r}{3 d_c^2 k_c} \tag{8.32}$$

where m_{v1} = coefficient of volumetric compressibility of the equivalent layer
D_c, D_s = constrained moduli of the column and the soil
h_1 = thickness of the equivalent layer
d_e = diameter of the unit cell
d_c = diameter of the column
k_r, k_v = permeability of the soil in the radial and vertical directions
k_s = permeability of the smear zone
k_c = permeability of the column
N_D = diameter ratio of the unit cell to the column
N_s = diameter ratio of the smear zone to the column

With the equivalent parameters, Zhu and Yin's (1999) closed-form solution for consolidation of two-layered soils can be used to calculate the degree of consolidation. Chai and Pongsivasathit (2009) pointed out that treating half of the penetration zone (i.e. $h_{cp}/2$) as part of the "unreinforced" layer using this simplified method resulted in a better comparison with the numerically computed degree of consolidation.

Huang et al. (2009) found that the consolidation of the DM column-reinforced foundation under the embankment could be accelerated by the columns with higher stiffness than the soft clay even though the columns had the same permeability as the soft clay. This finding was further confirmed by Jiang et al. (2013) in their numerical study. Figure 8.34 shows that

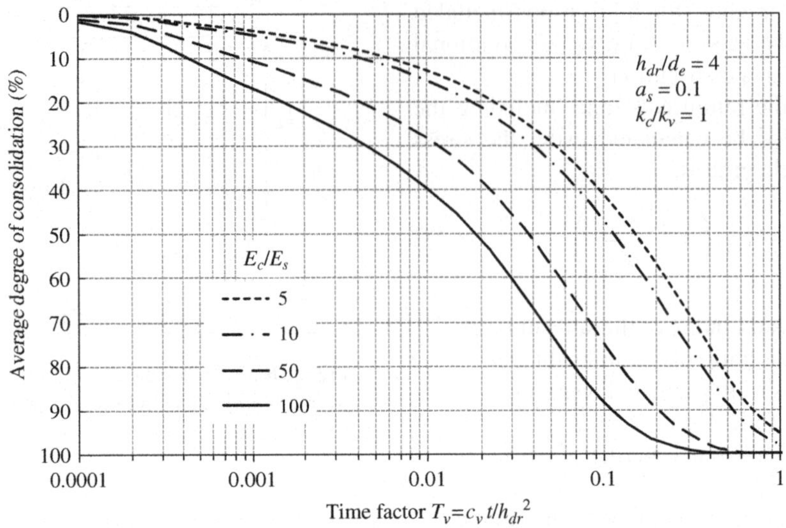

Figure 8.34 Effect of modulus ratio on rate of consolidation (Jiang et al., 2013, with kind permission from Springer Science and Business Media).

the degree of consolidation increased with the modulus ratio of the elastic column to the soil when the column permeability was the same as the soil permeability.

Stability of Column-Supported Embankments Kitazume et al. (2000) investigated the failure modes of DM columns subjected to a vertical load or a combination of vertical and lateral loads in centrifuge tests. They found that the bearing capacities of DM column-reinforced foundations under a combination of vertical and lateral loads were lower than those under vertical loads only. The DM columns under the combined loads ruptured at a lower column strength, but rotated at a higher column strength. Han et al. (2005) and Filz and Navin (2006) found numerically that DM columns under embankments might fail due to shear, bending, or rotation, depending on the strength of the columns. Rigid inclusions can be installed into chemically stabilized soil columns to increase the shear and bending resistance of columns. Kitazume (2008) proposed two external failure modes (sliding and collapse) and three internal failure modes (circular shear, horizontal shear, and bending) and developed the simplified methods to estimate the embankment pressures for collapse, horizontal shear, and bending failures. However, it is challenging to develop a theoretical solution for the combined failure mode. A numerical method with a strength reduction approach may be used to estimate the factor of safety for this combined failure mode. Limit equilibrium methods with composite undrained shear strengths have also been used for the stability analysis in practice as shown in Figure 8.35. The composite undrained shear strength of the DM column-reinforced soft foundation

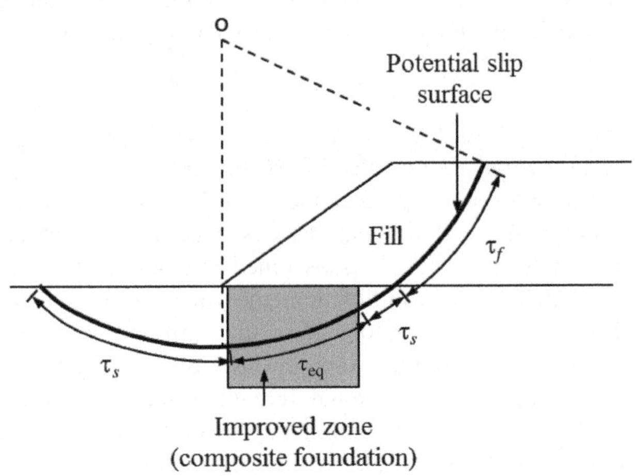

Figure 8.35 Stability of column-supported embankments.

is estimated as follows:

$$c_{u,\text{eq}} = a_s c_{u,c} + (1 - a_s) c_{u,s} \tag{8.33}$$

where $c_{u,\text{eq}}$ = undrained shear strength of the composite foundation
$c_{u,c}$ = undrained shear strength of the column
$c_{u,s}$ = undrained shear strength of the soil

Han et al. (2005) showed that the limit equilibrium method with a circular slip surface (e.g., Bishop's simplified method) overestimated the factor of safety of the embankment over a DM column-reinforced foundation as compared with that determined by the numerical method using the strength reduction approach as shown in Figure 8.36. The reason

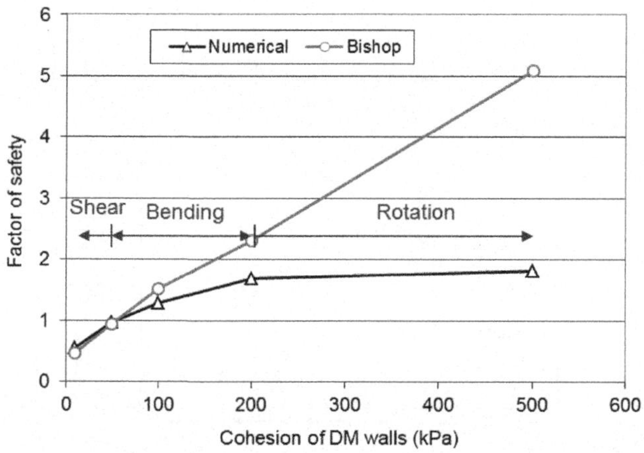

Figure 8.36 Limit equilibrium and numerical analyses of DM wall-supported embankments (Han et al., 2005, 2010).

for the overestimation by the limit equilibrium method is because it cannot model the bending and rotational failures of the columns.

According to the Swedish practice, the undrained shear strength of the column should be limited to within the range of 100–150 kPa. This limitation can minimize the difference in the calculated factor of safety between the limit equilibrium method and the numerical method.

Design for Excavation Deep mixed columns or walls have been commonly used to retain soils during excavation. DM walls are also served as curtain walls to minimize water from entering the excavation pit; therefore, no dewatering is necessary. For this application, cement binder is often used. DM walls are typically designed as gravity walls as shown in Figure 8.37 with the following minimum factors of safety:

- Factor of safety against sliding = 1.5
- Factor of safety against overturning = 1.5
- Factor of safety against base heave = 1.5
- Factor of safety against seepage = 1.5
- Factor of safety against global stability = 1.3

The typical dimensions of the DM wall shown in Figure 8.37 are based on the successful applications of DM walls in soft soils with high groundwater tables and excavation depths limited to 6 m in China (Han et al., 2002). These dimensions may vary when the site conditions and excavation depths are different. The following analyses should be performed to ensure the stability of the DM wall in an actual application.

Internal Sliding The DM wall should be designed with sufficient factors of safety against internal sliding and internal overturning at the elevation of the excavation base as shown in Figure 8.38. Depending on tolerable deformation of the wall, the lateral earth pressure behind the wall should be assumed to be active if a large deformation is tolerable or at rest if a small deformation is tolerable. The factor of safety against the internal sliding can be calculated as follows:

$$\mathrm{FS_{is}} = \frac{T_r}{P} = \frac{2T_r}{K\gamma H^2} \tag{8.34}$$

$$\mathrm{FS_{is}} = \frac{q_{\mathrm{uf}}B}{K\gamma H^2} \quad \text{for an undrained condition} \tag{8.35}$$

$$\mathrm{FS_{is}} = \frac{2(c'B + W\tan\phi')}{K\gamma H^2} \quad \text{for a drained condition} \tag{8.36}$$

where T_r = shear resistance by the DM wall
K = coefficient of lateral earth pressure of the soil
γ = unit weight of the soil
q_{uf} = field unconfined compressive strength of the stabilized soil
H = depth of excavation

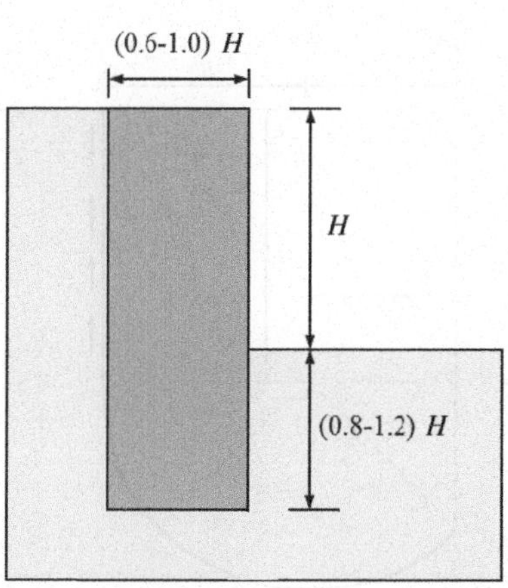

Figure 8.37 Typical cross section of DM wall for excavation.

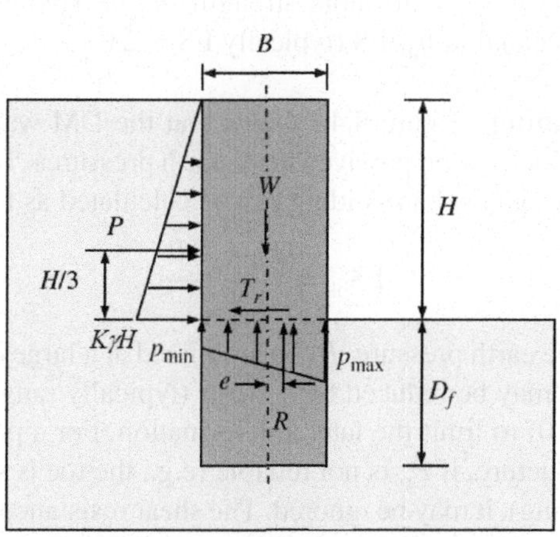

Figure 8.38 Internal stability analysis.

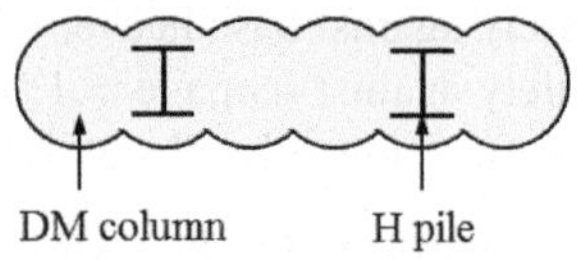

Figure 8.39 H-pile reinforced DM wall.

B = width of the wall
c' = effective cohesion of the wall
ϕ' = effective friction angle of the wall

The water pressure should be included if the groundwater table is above the excavation base.

If the factor of safety against internal sliding is not sufficient, the width of the wall can be increased with extra row(s) of columns and/or reinforcement (such as H pile) can be installed inside the DM wall as shown in Figure 8.39.

Internal Overturning The overturning stability can be evaluated by a factor of safety, an eccentricity, and yielding of the DM wall. The factor of safety against internal overturning is

$$\mathrm{FS_{io}} = \frac{3WB}{2PH} = \frac{3WB}{K\gamma H^3} \tag{8.37}$$

The eccentricity of the wall at the elevation of the excavation base is

$$e = \frac{PH}{3W} = \frac{K\gamma H^3}{6W} \tag{8.38}$$

The eccentricity, e, should be less than $\left(\frac{1}{6}\right) B$.

The maximum stress at the edge of the wall as calculated below should be less than the allowable strength of the stabilized soil:

$$p_{\max} = \frac{W}{B}\left(1 + \frac{6e}{B}\right) \le q_a \tag{8.39}$$

where q_a is the allowable strength of the stabilized soil, which is equal to q_{uf}/FS (typically FS = 2.0).

Base Sliding Figure 8.40 shows that the DM wall is subjected to active and passive lateral earth pressures. The factor of safety against base sliding can be calculated as follows:

$$\mathrm{FS_{bs}} = \frac{P_p + \tau_f B}{P_a} \tag{8.40}$$

Passive earth pressure, P_p, is mobilized at a large deformation; P_p may be reduced by a factor (typically ranging from 1.5 to 2.0) to limit the lateral deformation. For a permanent earth structure, if P_p is not reliable (e.g., the toe is subjected to scouring), it may be ignored. The shear resistance between the block and the soil, τ_f, can be estimated by $(\frac{2}{3}–1)$ times the shear strength of the soil.

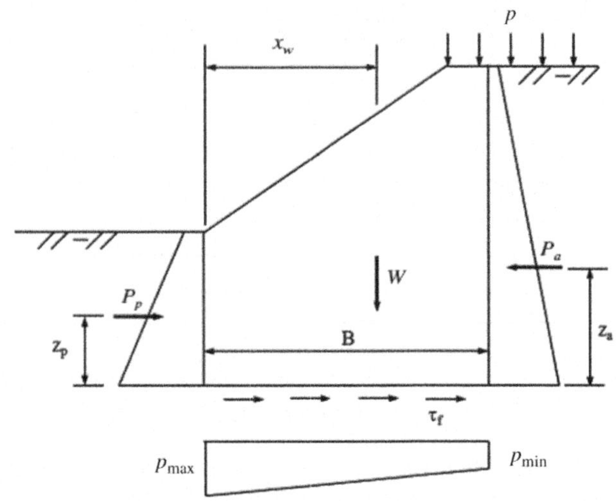

Figure 8.40 External stability analysis.

External Overturning As shown in Figure 8.40, the factor of safety against external overturning can be calculated as follows:

$$\mathrm{FS_{eo}} = \frac{P_p z_p + W x_w}{P_a z_a} \tag{8.41}$$

Same as that for the base sliding, the passive earth pressure may be reduced depending on tolerable deformation or ignored if not reliable.

Base Heave The DM wall should be designed against base heave in the excavation pit. Based on the potential failure mode as shown in Figure 8.41, Terzaghi (1943) proposed the following formula for the base stability:

$$\mathrm{FS_{bh}} = \frac{N_c c_u}{\gamma H - c_u(H + D_f)/B' - 2c_u D_f/B} \tag{8.42}$$

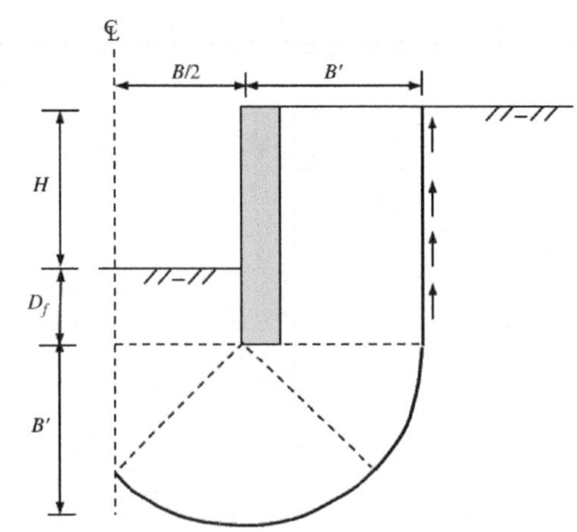

Figure 8.41 Base stability with wall embedment.

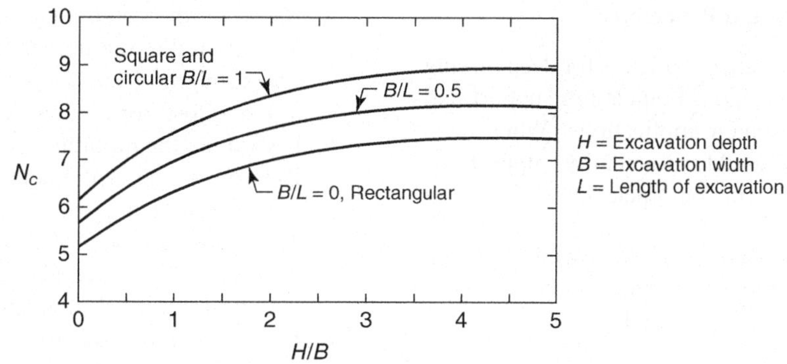

Figure 8.42 Bearing capacity factor for base stability (Terzaghi et al., 1996).

where
c_u = undrained shear strength of the soil
γ = unit weight of the soil
H = depth of excavation
D_f = embedment depth of wall
B' = depth of failure surface ($B' = B/\sqrt{2}$ in uniform soil or limited depth to a firm soil if $B' < B/\sqrt{2}$)
N_c = bearing capacity factor, which can be determined based on Figure 8.42

and other dimensions are shown in Figure 8.41.

When the embedment depth is equal to zero, the above equation can be simplified into

$$\text{FS}_{\text{bh}} = \frac{N_c c_u}{\gamma H - c_u(H/B')} \tag{8.43}$$

This formula has been commonly used for the design of ground anchors and soil nailed walls considering the base stability, which will be discussed further in Chapter 9.

Seepage Due to the head difference between inside and outside the excavation pit, water seepage is expected. Flow net can be drawn as shown in Figure 8.43 and can be used to estimate the amount of water seepage into the excavation pit. In addition, the maximum hydraulic gradient can be estimated. Based on the maximum hydraulic gradient, the factor of safety against seepage can be calculated as follows:

$$\text{FS}_{\text{sp}} = \frac{i_{\text{cr}}}{i_{\text{max}}} = \frac{\gamma'}{i_{\text{max}}\gamma_w} \tag{8.44}$$

where
i_{cr} = critical hydraulic gradient
i_{max} = maximum hydraulic gradient
γ' = effective unit weight of the soil
γ_w = unit weight of water

Global Stability A slope stability method can be used for a global stability analysis. It is important to point out that the global stability analysis should consider the seepage forces if there is a water head difference between inside and outside the excavation pit, as shown in Figure 8.43. The methods for slope stability analysis are discussed in Chapter 2.

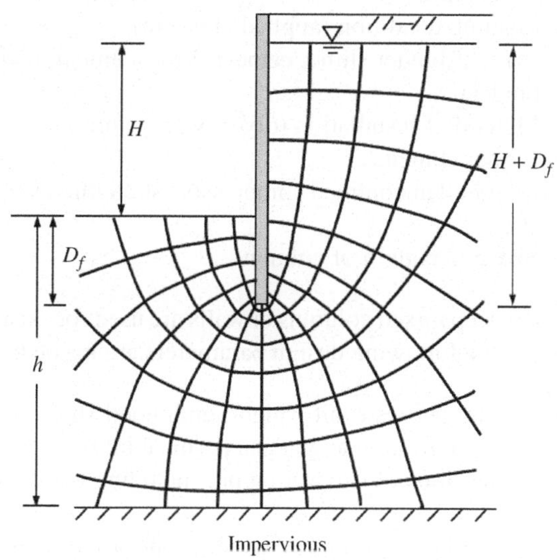

Figure 8.43 Flow net analysis.

Liquefaction Mitigation The deep mixing method has been used to form shear wall grids in ground to mitigate liquefaction. The DM wall itself is not liquefiable due to the cementation by binders. The shear wall grid also provides more shear resistance as a composite to cyclic shear stress induced by earthquake or reduce cyclic shear stress applied to the liquefiable soil inside the grid. A design method similar to that for the columns installed by the deep replacement method was developed by Nguyen et al. (2013). Details of the design method can be found in this publication. Kitazume and Terashi (2013) suggested that the grid cells formed by deep mixed walls should have the cell width less than 80% the cell depth to be effective in mitigating soil liquefaction.

8.2.4 Design Parameters and Procedure

Design Parameters The design parameters for deep mixing depend on the type of application. Foundation support and earth retaining are two common applications. When deep mixed columns or walls are used for foundation support, the design may include the following parameters:

- Soil type, natural moisture content, organic content, groundwater table, permeability or coefficient of consolidation and soil strength and modulus
- Depth of improvement
- Project requirements (allowable bearing capacity, tolerable settlement, factor of safety against slope failure for embankments)
- Loading condition (applied pressure)
- Type of binder (lime, cement, lime–cement, and other binder)
- Method of installation (dry or wet method)
- Binder content
- Required unconfined compressive strength of stabilized soil
- Size and pattern of columns.

When deep mixed columns or walls are used for excavation support, the following design parameters are needed:

- Soil properties (unit weight, undrained shear strength, effective friction angle, and permeability)
- Dimensions of excavation pit, including depth, length, and width
- Elevations of groundwater tables inside and outside the excavation pit
- Width of deep mixed wall
- Embedment depth of deep mixed wall
- Strength of stabilized soil
- Required factors of safety against sliding, overturning, base instability, seepage, and global instability

Design Procedure The design procedure for deep mixing also depends on the type of application. For foundation support, the following design procedure may be followed:

1. Based on the geotechnical conditions, select the type of method (dry or wet method) and type of binder for deep mixing.
2. Based on the geometry of the superstructure and the distribution and magnitude of loads, select the pattern of deep mixing (individual columns, blocks, walls, and grids).
3. Based on the required ultimate bearing capacity of the foundation and the soil strength, determine the required column capacity if the area replacement ratio is assumed or the area replacement ratio if the column strength is assumed.
4. Based on the column capacity requirement, determine the required minimum laboratory unconfined compressive strength of the stabilized soil considering the field to laboratory conversion factor and the scale factor.
5. Determine the minimum column length.
6. Determine the settlement of the column-reinforced foundation using the stress reduction method, the piled-raft method, or the column penetration method.
7. Determine the degree of consolidation of the column-reinforced foundation.
8. If columns are used to support embankments, stability analysis should be conducted. A numerical method is preferred, but a simplified limit equilibrium method may be used by limiting the column strength.
9. Iterations may be needed if at least one of the calculations does not meet the project requirement.

For earth retaining for excavation, the following design procedure may be used:

1. Based on the depth of excavation, a preliminary design of the deep mixed wall can be selected based on a typical cross section.
2. The deep mixed wall should be checked against the factor of safety of internal sliding. If not sufficient, the width of the wall should be increased and/or reinforcement (such as H piles) should be added.
3. The deep mixed wall should have sufficient stability against internal overturning by satisfying the factor of safety, eccentricity, and yielding strength requirements. If not sufficient, the width of the wall should be increased, reinforcement (such as H piles) should be added, and/or the strength of the stabilized soil should be increased.
4. In addition to internal stability, external stability against base sliding, external overturning, base heave, seepage, and global stability should be evaluated with sufficient factors of safety. The embedment depth of the wall is one of the important factors for this evaluation.

8.2.5 Design Example

Problem

A concrete raft with a dimension of 10 m × 10 m × 0.5 m thick is rested on a deep mixed column-supported foundation with an embedment depth of 0.5 m, as shown in Example Figure 8.1. Also, 4 × 4 rows of columns with a diameter of 0.8m and a length of 10 m are installed in uniform soft clay with spacing of 2.2 m in a square pattern. The soft clay is more than 20 m thick and has a saturated unit weight of 18 kN/m^3 and a drained modulus of 5 MPa and the ground water table is at 0.5 m below the surface. The columns have an unconfined compressive strength of 1 MPa and a saturated unit weight of 20 kN/m^3. The Poisson's ratios of the soft clay and the columns are 0.3 and 0.2, respectively. Estimate the settlement of the raft under a vertical load of 15 MN.

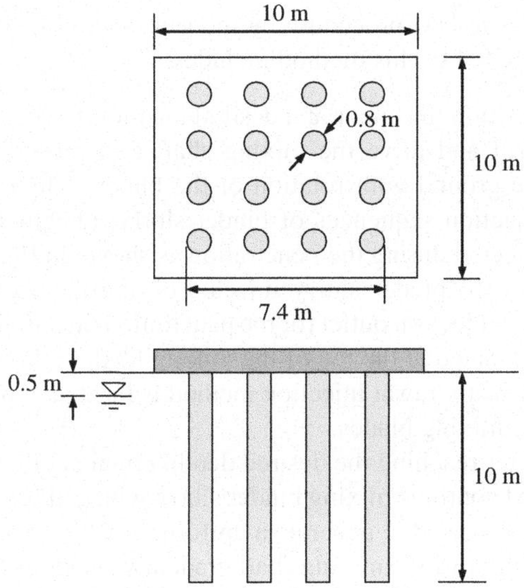

Example Figure 8.1

Solution

The piled-raft method developed by Horikoshi and Randolph (1999) and presented in Chapter 5 is selected for the calculation of the settlement of the DM column-reinforced soft foundations.

R_f factor is

$$R_f = \sqrt{\frac{N_{cl}s}{L_c}} = \sqrt{\frac{16 \times 2.2}{10}} = 1.88 \ < 2$$

therefore, the equivalent pier approach is suitable.

Cross-section area of group columns (i.e., equivalent pier) is

$$A_g = 7.4 \times 7.4 = 54.76 \text{ m}^2$$

Radius of raft is

$$r_r = \sqrt{\frac{A_r}{\pi}} = \sqrt{\frac{10 \times 10}{3.14}} = 5.64 \text{ m}$$

Radius of equivalent pier is

$$r_{eq} = \sqrt{\frac{A_g}{\pi}} = \sqrt{\frac{54.76}{3.14}} = 4.18 \text{ m}$$

Length of equivalent pier is

$$L_{eq} = L_c = 10 \text{ m}$$

Total cross-section area of columns is

$$A_{tc} = \sum A_{ci} = 16\pi r_0^2 = 16 \times 3.14 \times 0.4^2 = 8.04 \text{ m}^2$$

Elastic modulus of columns is

$$E_c = E_{50} = 100 q_u = 100 \text{ MPa}$$

Modulus of equivalent pier is

$$E_{eq} = E_s + (E_c - E_s)\frac{A_{tc}}{A_g} = 5 + (100 - 5)\frac{8.04}{54.76}$$
$$= 18.95 \text{ MPa}$$

Shear modulus of soil is

$$G_s = G_L = \frac{E_s}{2(1 + \nu_s)} = \frac{5}{2 \times (1 + 0.3)} = 1.92 \text{ MPa}$$

For a uniform soil,

$$\xi = \frac{G_L}{G_b} = 1 \text{ and } \rho = \frac{G_{avg}}{G_L} = 1$$

For a uniform pier,

$$\eta = \frac{r_{eq,b}}{r_{eq}} = 1$$

The pier length to radius ratio, $L_{eq}/r_{eq} = 10/4.18 = 2.4 < 5$, therefore, $C = 5$.

Maximum influence radius of the equivalent pier is

$$r_{m,p} = C r_{eq} + \{0.25 + \xi[2.5\rho(1 - \nu_s) - 0.25]\} L_{eq}$$
$$= 5 \times 4.18 + \{0.25 + 1 \times [2.5 \times 1 \times (1 - 0.3) - 0.25]\} \times 10$$
$$= 38.4 \text{ m}$$

$$\zeta = \ln\left(\frac{r_{m,p}}{r_{eq}}\right) = \ln\left(\frac{38.4}{4.18}\right) = 2.22$$

$$\lambda = \frac{E_{eq}}{G_L} = \frac{18.95}{1.92} = 9.87$$

$$\mu_L L_{eq} = \sqrt{\frac{2}{\zeta\lambda}} \cdot \frac{L_{eq}}{r_{eq}} = \sqrt{\frac{2}{2.22 \times 9.87}} \times \frac{10}{4.18} = 0.72$$

$$\tanh(\mu_L L_{eq}) = \tanh(0.72) = 0.619$$

Stiffness of the equivalent pier is

$$K_{eq} = G_L r_{eq} \left[\frac{\frac{4\eta}{(1 - \nu_s)\xi} + \rho \cdot \frac{2\pi}{\zeta} \cdot \frac{\tanh(\mu_L L_{eq})}{\mu_L L_{eq}} \cdot \frac{L_{eq}}{r_{eq}}}{1 + \frac{1}{\pi\lambda} \cdot \frac{4\eta}{(1-\nu_s)\xi} \cdot \frac{\tanh(\mu_L L_{eq})}{\mu_L L_{eq}} \cdot \frac{L_{eq}}{r_{eq}}} \right]$$

$$= 1.92 \times 4.18$$

$$\times \left[\frac{\frac{4\times 1}{(1-0.3)\times 1} + 1\times \frac{2\times 3.14}{2.22}\times \frac{\tanh(0.72)}{0.72}\times \frac{10}{4.18}}{1+\frac{1}{3.14\times 9.87}\times \frac{4\times 1}{(1-0.3)\times 1}\cdot \frac{\tanh(0.72)}{0.72}\times \frac{10}{4.18}} \right]$$

$$= 67.02 \text{ MN/m}$$

Stiffness of raft on soil (assuming a rigid raft) is

$$K_r = \frac{P_r}{S} = \frac{E_s A_r}{(1-v_s^2)B_r I_s} = \frac{5\times 100}{(1-0.3^2)\times 10\times 0.88}$$

$$= 62.34 \text{ MN/m}$$

The column length to radius ratio, $L_c/r_c = 10/0.4 = 25 > 5$, therefore, $C = 0$.

$$r_m = \{0.25 + \xi[2.5\rho(1-v_s) - 0.25]\}L_c$$

$$= \{0.25 + 1\times[2.5\times 1\times(1-0.3) - 0.25]\}\times 10$$

$$= 17.5 \text{ m}$$

Pier–raft interaction factor is

$$\alpha_{\text{rp}} = 1 - \frac{\ln\left(\frac{r_r}{r_{\text{eq}}}\right)}{\ln\left(\frac{r_m}{r_{\text{eq}}}\right)} = 1 - \frac{\ln\left(\frac{5.64}{4.18}\right)}{\ln\left(\frac{17.5}{4.18}\right)} = 0.790$$

Overall stiffness of the pier raft is

$$K_{\text{pr}} = \frac{P_{\text{eq}} + P_r}{S_{\text{pr}}} = \frac{K_{\text{eq}} + K_r(1-2\alpha_{\text{rp}})}{1-\left(\frac{K_r}{K_{\text{eq}}}\right)\alpha_{\text{rp}}^2}$$

$$= \frac{67.20 + 62.34\times(1-2\times 0.790)}{1-\left(\frac{62.34}{67.20}\right)\times 0.790^2}$$

$$= 73.57 \text{ MN/m}$$

Settlement of the pier raft is

$$S' = S_{\text{pr}} = \frac{P}{K_{\text{pr}}} = \frac{15}{73.57} = 0.204\text{ m} = 204\text{ mm}$$

8.2.6 Construction

Deep mixed columns can be installed by a wet or dry method. Different binders can be used for the installation. Table 8.7 provides the typical parameters for deep mixing in different countries.

The blade rotation number is defined as follows:

$$N_{\text{br}} = N_{\text{mb}}\left(\frac{R_p\alpha_{\text{cvi}}}{v_p\alpha_{\text{cv}}} + \frac{R_w}{v_w}\right) \tag{8.45}$$

where N_{mb} = total number of mixing blades
v_p = penetration rate (m/min)
v_w = withdrawal rate (m/min)
R_p = blade rotation rate during penetration
R_w = blade rotation rate during withdrawal

α_{cvi} = binder injection rate during penetration (kg/m^3)
α_{cv} = total volumetric binder content (kg/m^3)

The basic DM procedure by the wet method is illustrated in Figure 8.44. This method includes:

1. Position the auger at a desired location.
2. Drill and drive the mixing shaft at a preset rate into the ground with rotation of the blades. There are two injection sequences of binder slurry: (a) binder slurry injected during the penetration as shown in Figure 8.44 and (b) binder slurry injected during the withdrawal. The injection outlet for the penetration injection method is located at the toe of the mixing blades while that for the withdrawal injection method is located at the top of the mixing blades.
3. After reaching the desired depth, remain at the position and continue mixing binder slurry with soil for a certain time period to generate a uniform mix.
4. Withdraw the mixing shaft gradually at a preset rate and continue mixing the binder slurry with soil (for some machines, the direction of the mixing blade rotation is reversed during the withdrawal).
5. Complete the installation of the mixed column until the mixing shaft reaches the desired elevation (mostly at the ground surface but sometimes at a preset depth).

The spoil soil should be excavated and removed during or after the installation of each column to minimize the interference with the next installation.

The basic DM procedure by the dry method is illustrated in Figure 8.45. This includes:

1. Position the auger at a desired location.
2. Drill and drive the mixing shaft at a preset rate into the ground with rotation of the blades. There are two injection sequences of binder powder: (a) binder powder injected during the penetration and (b) binder powder injected during the withdrawal as shown in Figure 8.45. The injection outlet for the penetration injection method is located at the toe of the mixing blades while that for the withdrawal injection method is located at the top of the mixing blades.
3. Withdraw the mixing shaft gradually at a preset rate, inject binder powder, and mix binder powder with soil (for some machines, the direction of the mixing blade rotation is reversed during the withdrawal). Pressured air

Table 8.7 Typical Parameters for Deep Mixing

Region	Japan		Europe	United States
Reference	Kitazume and Terashi (2013)		Holm (2005)	Elias et al. (2006)
Method	Dry method	Wet method		
Penetration rate (m/min)	1.0–2.0	1.0	2–6	1.0
Withdrawal rate (m/min)	0.7–0.9	0.7–1.0	1.5–6	
Blade rotation rate—penetration (rpm)	24–32	20	80–200	20–45
Blade rotation rate—withdrawal (rpm)	48–64	40	10–30	
Blade rotation number (/min)	274–284	350–360	100–500	350 for clays; 400 to 450 for peaty soils
Binder content (kg/m^3)			100–250	150–300

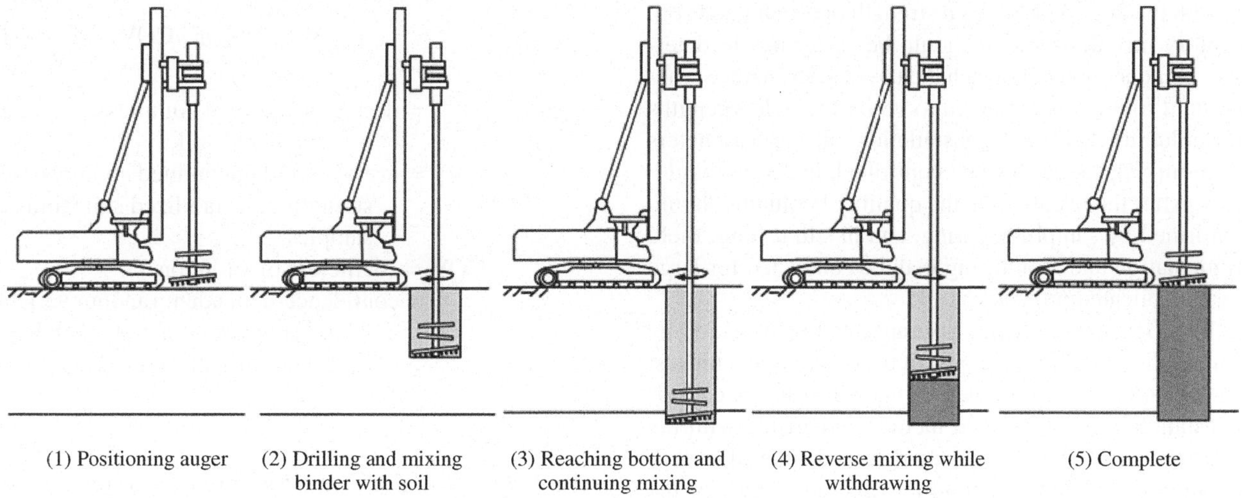

Figure 8.44 Installation procedure of wet deep mixing.

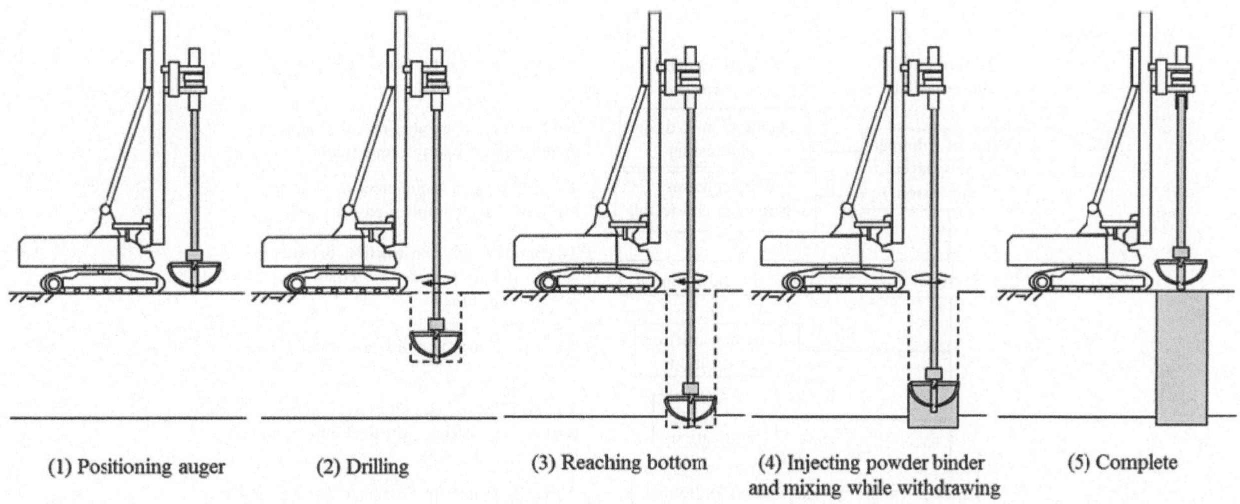

Figure 8.45 Installation procedure of dry deep mixing.

up to 450 kPa is used to inject powder binder at a desired rate (typically 4 m^3/min).

4. Complete the installation of the mixed column until the mixing shaft reaches the desired elevation (mostly at the ground surface but sometimes at a preset depth).

The dry method generates less spoil soil as compared with the wet method. The spoil soil should be cleaned after the installation of each column to minimize the interference with the next installation.

8.2.7 Quality Control and Assurance

The typical quality control (QC) and quality assurance (QA) procedure for deep mixing is presented in Figure 8.46. It starts with the design or targeted strength of a stabilized soil. Laboratory mix design tests should be performed to determine the binder content to reach the targeted strength. Sometimes, field trial is implemented to verify the field strengths of test columns. During the installation, all the parameters for deep mixing should be well controlled, including binder content. After the installation, the quality of columns should be confirmed by sampling, coring, and in situ testing. Field instrumentation and monitoring may be conducted for large and/or complicated projects.

Quality control and quality assurance can be achieved by evaluating the installation records of the columns and the test results of laboratory and field verification tests. Each column installation record should include a chart-log with the construction information, such as date and time of execution, length of column, penetration/withdrawal rates of the mixing shaft, blade rotation rate, pressure and flow rate of pumped slurry or injected powder, and total slurry or powder consumption per column. The following verification tests are commonly used for deep mixing:

1. Specimens of stabilized soils obtained from fresh columns with the wet grab method are tested for unconfined compressive strengths.
2. Core sampling is utilized to exhume stabilized soil samples from field columns after being hardened, and then the samples are tested in the laboratory for the unconfined compressive strengths. The triple tube core barrel is commonly used. It is recommended that large samples (greater than 76 mm in diameter and 150 mm long) should be obtained. Cored samples can be used to examine the uniformity of the mixture visually. The Coastal Development Institute of Technology in Japan (2002) provides a guideline for evaluating the unconfined compressive strength of a stabilized soil from core samples based on the coefficient of variation as follows:

$$q_{ud} \leq q_{uf}(1 - m \cdot \text{COV}) \tag{8.46}$$

where q_{ud} = design unconfined compressive strength of a stabilized soil
q_{uf} = average field unconfined compressive strength of a stabilized soil from core samples
COV = coefficient of variation
m = confidence-dependent random variable, which is provided in Table 8.8 based on the number of samples

Table 8.8 Random variable versus number of samples

Number of samples	1	2	3	4–6	7–8	≥9
m	1.8	1.7	1.6	1.5	1.4	1.3

Source: Futaki and Tamura (2002).

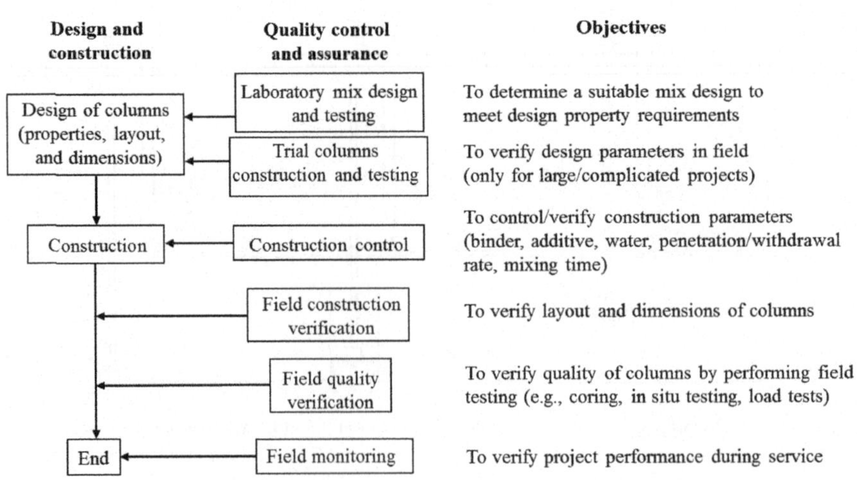

Figure 8.46 Typical QC/QA procedure for deep mixing (modified from Larsson, 2005).

3. Several in situ testing methods have been adopted for evaluating the quality of deep mixed columns in the field, which include the cone penetration test, the dynamic penetration test, the modified cone penetration test, the pullout resistance test, the pressuremeter test, the field column compression test, the plate loading test, the down-hole test, the cross-hole seismic geophysics, and the wave velocity test.

The dynamic penetration test (with hammer mass of 63.5 kg) and the cone penetration test are used for examining the uniformity of deep mixed columns with depth. These tests are mostly performed within a 7-day curing period. Empirical relationships are used for estimating unconfined compressive strengths of cement-stabilized soils (Ye et al., 1994):

$$q_u = 12.5N \tag{8.47}$$

$$q_u = \frac{1}{10} q_c \tag{8.48}$$

where q_u = unconfined compressive strength of stabilized soil (kPa)
N = number of blow counts by the dynamic penetration test
q_c = cone tip resistance (kPa)

The modified cone penetration test (also called the wing shear test) and the pullout resistance test, as shown in Figure 8.47, have been commonly used in Europe. In both tests, the devices record the penetration resistance when the two wings cut through the column. The profile of the resistance can be used to evaluate the uniformity and strength of the columns.

The field compression test is used to evaluate the strength of the column near the ground surface after the column head is exposed.

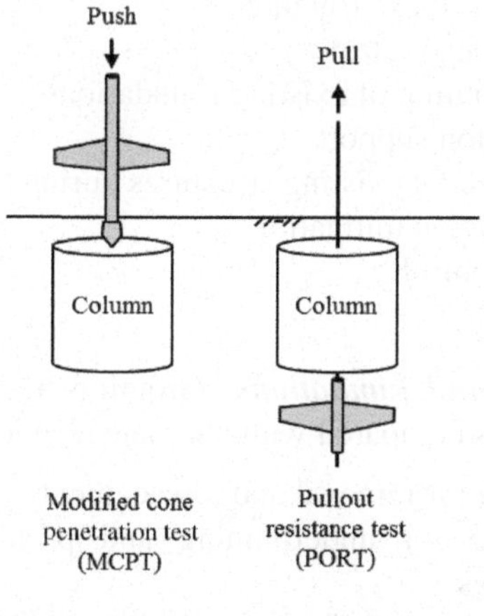

Figure 8.47 Modified cone penetration test and pullout resistance test.

Plate loading tests are also used to determine the ultimate bearing capacity of the single column and the composite foundation with column(s).

8.3 GROUTING

8.3.1 Introduction

Basic Concept Figure 8.48 shows typical grouting components and setup for a grouting project. The grouting components include a grout mixer, an agitator for storage and agitation, a grout pump, control and recording equipment, and a grout pipe. The grout mixer is to mix binder, water, and other necessary materials into uniform grout. The agitator continues agitation of the grout to prevent sedimentation and segregation. The pump generates pressure to inject the grout into the ground. The control and recording equipment controls, maintains, and records the required pump pressure and the rate of injected grout. The grout tube has packers, opening, or nozzle to allow the grout to enter the ground. The grout pipe is inserted into a predrilled hole or self-drills into the ground. In addition to the pump, the control valve, and the pressure gauge for the grout, the jet grouting technology may require a water pump and an air compressor to inject water and air, and the control valves and pressure gauges for water and air, depending on the option for the jet grouting system.

Grouting can be used in rock and soil. Grouting has five common types: (1) permeation grouting, (2) compaction grouting, (3) hydrofracture grouting, (4) compensation grouting, and (5) jet grouting, as shown in Figure 8.49.

Permeation grouting is to fill voids in soils by grout. Void or cavity filling is a special case of permeation grouting. Compaction grouting is to densify loose soils by injecting stiff and low mobility grout to displace soil particles. Hydrofracture grouting is to inject stiff grout under high pressure (up to 4 MPa) to fracture the soil mass and force the grout into the fractures. Compensation grouting is to compensate ground loss due to construction activities, such as excavation and tunneling. Jet grouting utilizes grout (sometimes with water and/or air) to erode the soil at depths and then mix the eroded soil with grout to form hardened columns or walls. The jet grouting technology has single, double, and triple systems as shown in Figure 8.50, which use different media (grout, air, and/or water) to erode the soil. These systems are used in different soil conditions and produce different sizes of columns or walls.

The main objectives of grouting are:

- Densification to prevent or arrest settlement and mitigate liquefaction
- Soil solidification to increase cohesion of granular soils
- Reduction of permeability and water control

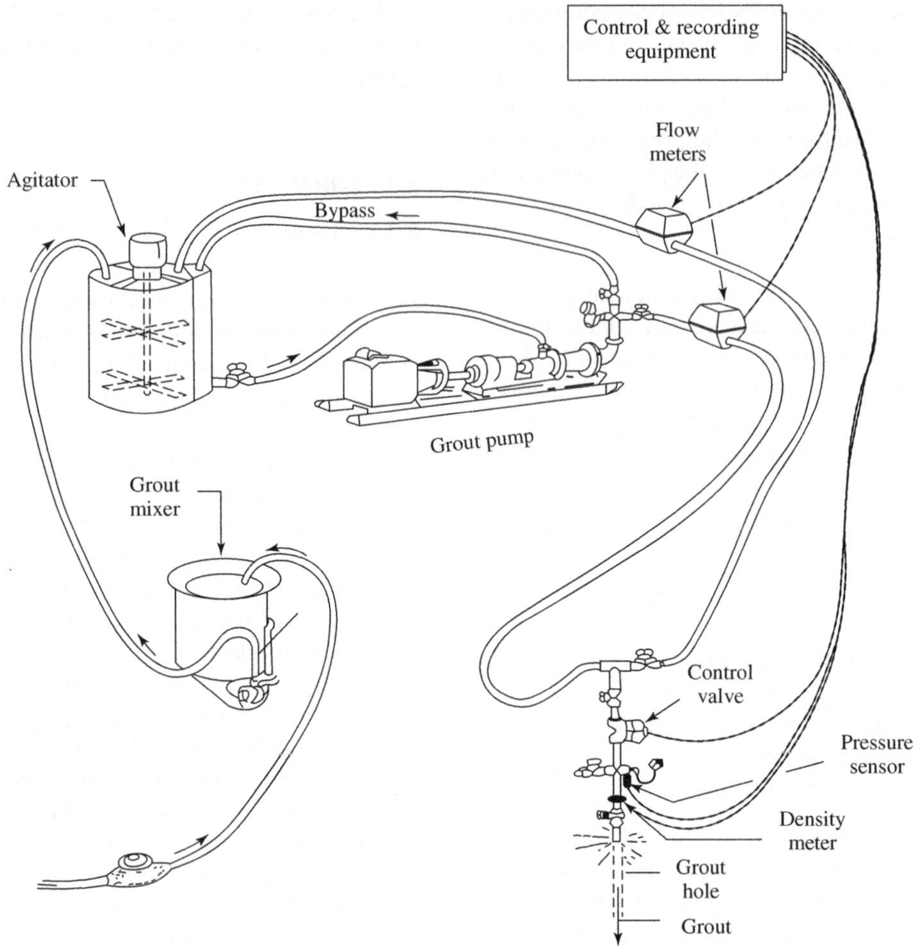

Figure 8.48 Typical grouting components and setup (Houlsby, 1990).

- Stabilization and reduction of expansion of clay soils
- Compensation for lost ground and filling large voids
- Additional support for existing structures

Suitability Figure 8.51 shows the suitability of different grouting methods. The permeation grouting is suitable for cohesionless soil (such as gravel and sand). The permeation grouting includes cement slurry grouting and chemical grouting. The cement slurry grouting is more suitable for gravel while the chemical grouting is more suitable for sand. Compaction grouting is mostly used for sand, but sometimes also used for silt and clay if dissipation of excess pore water pressure is permitted. The hydrofracture grouting is suitable for sand, silt, and clay and often used for compensation grouting purposes. The jet grouting method is suitable for all kinds of soil types.

Permeation grouting, compaction grouting, and hydrofracture grouting have also been used for decomposed rock and fissured rock.

Applications Grouting has been used for the following applications (Elias et al., 2006):

- Densification of granular soils
- Raising settled structures
- Settlement control
- Underpinning of existing foundations
- Excavation support
- Protection of existing structures during tunneling
- Liquefaction mitigation
- Water control

Advantages and Limitations Grouting has the following advantages as compared with alternate technologies:

- No need for removal and replacement
- Effective for underpinning and protecting existing structures
- Easy to access and operate within constrained space
- Low mobilization cost

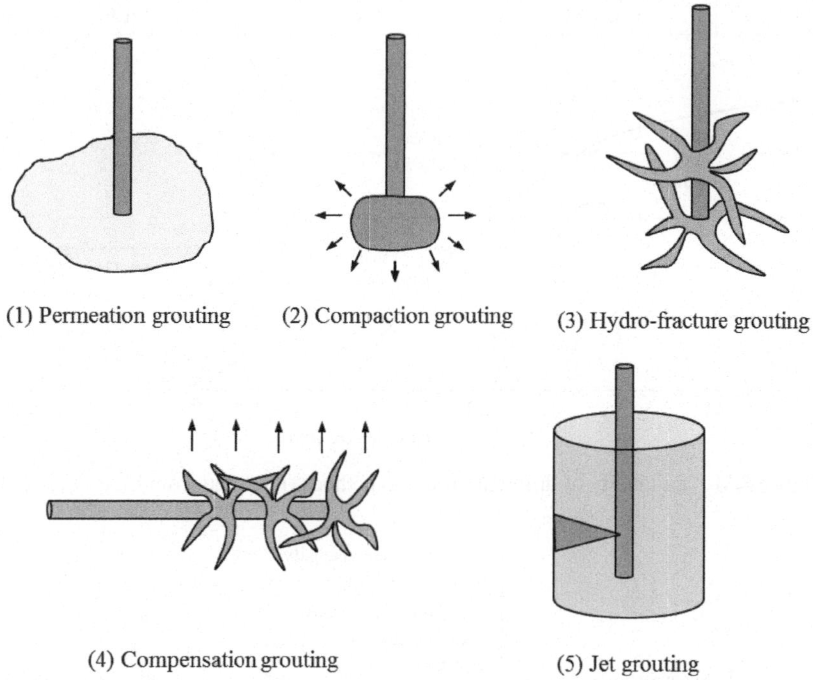

Figure 8.49 Types of grouting.

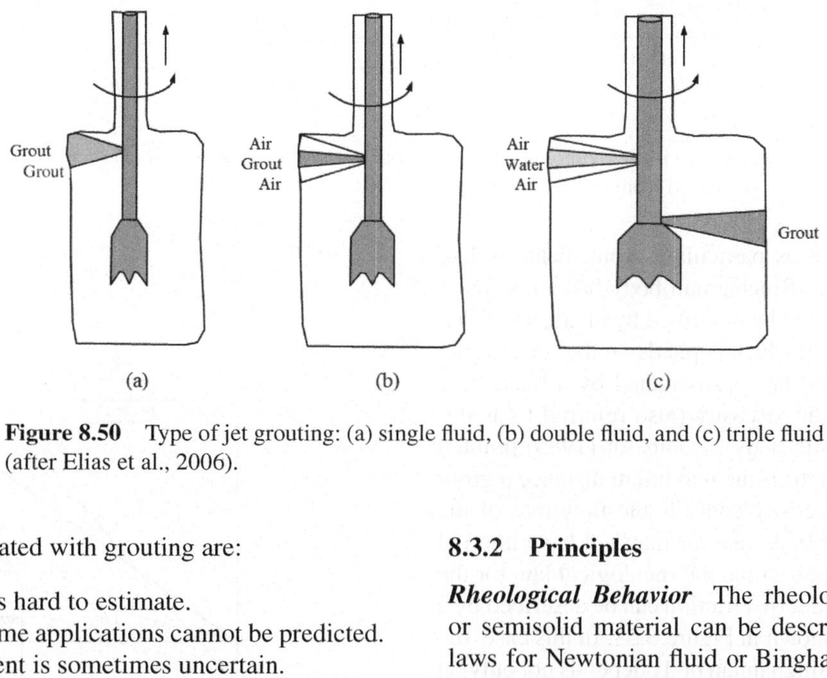

Figure 8.50 Type of jet grouting: (a) single fluid, (b) double fluid, and (c) triple fluid (after Elias et al., 2006).

The limitations associated with grouting are:

- Quantity of grout is hard to estimate.
- Effectiveness of some applications cannot be predicted.
- Area of improvement is sometimes uncertain.
- Grouting may cause ground movement and distresses to existing structures.
- Certain chemical grouts may contain toxicity and have adverse impact to groundwater and underground environment.
- Specialty contractors are required for the operation.

8.3.2 Principles

Rheological Behavior The rheological behavior of fluid or semisolid material can be described by the rheological laws for Newtonian fluid or Binghamian body as shown in Figure 8.52. In Figure 8.52, dv is the shear velocity change and dx is the segment of the specimen, and $\dot{\gamma} = \Delta v/\Delta x$ is the shear strain rate. The Newtonian fluid has a linear relationship between the shear stress, τ, and the shear strain rate, $\dot{\gamma}$. The slope of this linear line is the dynamic viscosity. Water is an example of Newtonian fluid. However, most

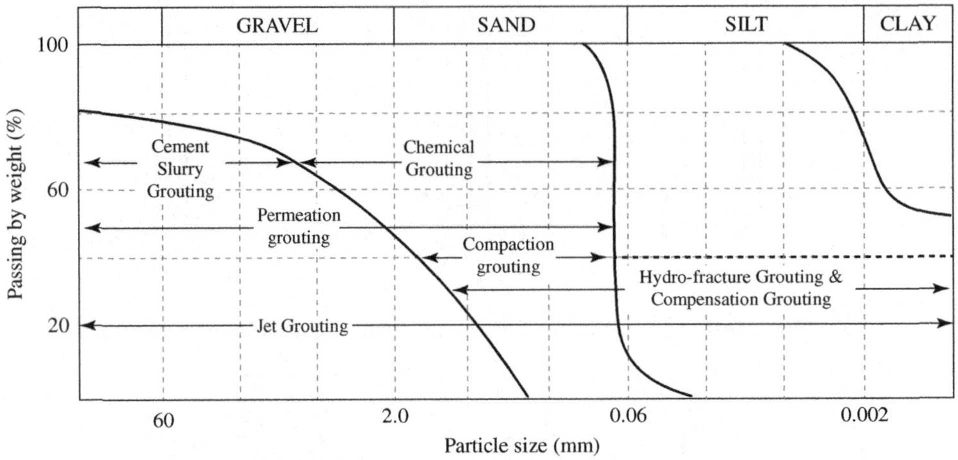

Figure 8.51 Suitability of different grouting methods (modified from Elias et al., 2006).

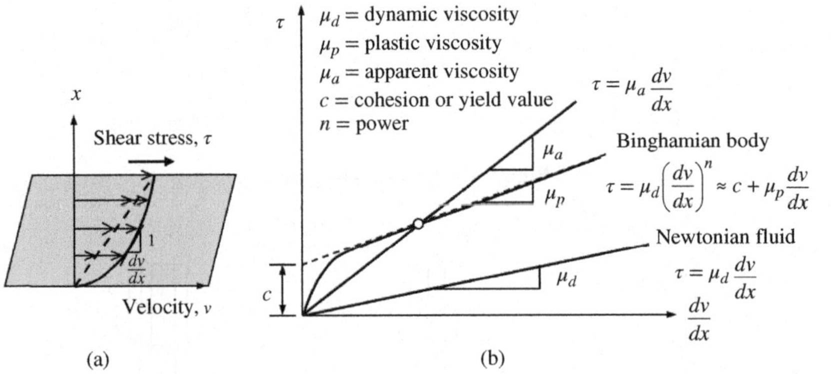

Figure 8.52 Basic rheological laws: (a) shear stress vs. velocity and (b) rheological laws (modified from De Paoli et al., 1992).

semisolid material, such as particulate grout, behaves like a Binghamian body. The Binghamian body has a nonlinear rheological behavior. It can be described by an apparent viscosity (i.e., a secant line), which depends on the shear strain rate. Alternatively, it can be approximated by a linear line with cohesion and plastic viscosity (also referred to as dynamic viscosity of plastic body). Lombardi (1985) pointed out that the cohesion controls the maximum distance a grout can reach while the viscosity controls the flow rate of the grout. The Binghamian body may or may not have internal friction. If it has internal friction, the rheological law for the Binghamian body with internal friction can be described by a rheological surface as shown in Figure 8.53. In this case, the shear resistance of the Binghamian body depends not only on the cohesion and the viscosity but also on the σ on its body. The pressure-induced resistance can be high if there is an internal friction angle ϕ of the Binghamian body under a high pressure. Under such a situation, the flow of the Binghamian body becomes difficult or impossible.

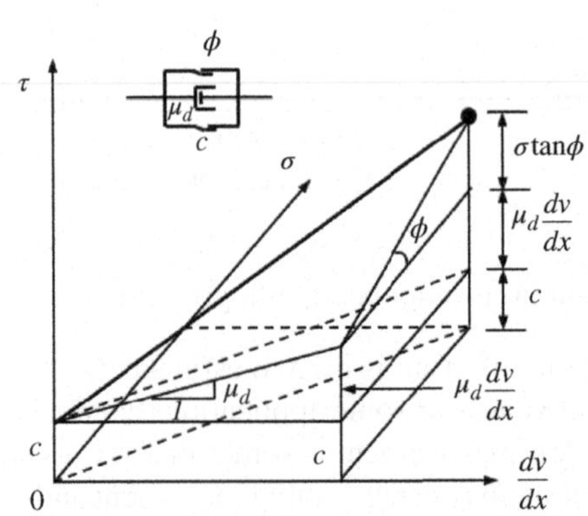

$$\tau = c + \mu_d \frac{dv}{dx} + \sigma \tan\phi$$

Figure 8.53 Rheological behavior of a Binghamian body (modified from Lombardi, 1985).

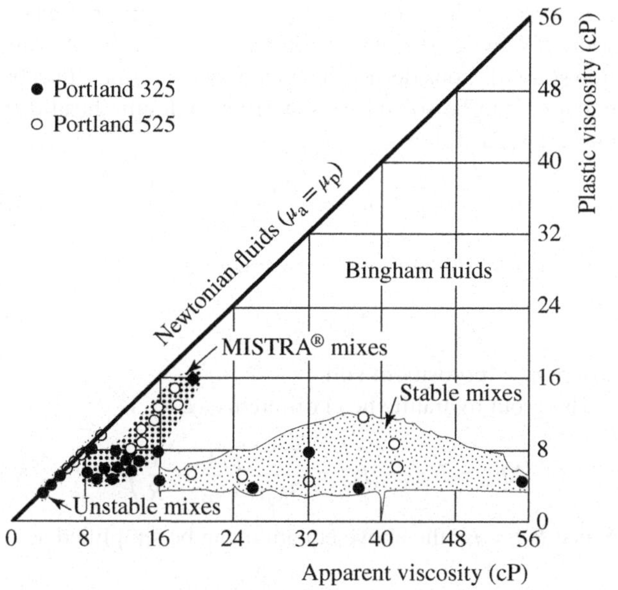

Figure 8.54 Apparent viscosity vs. plastic viscosity (De Paoli et al., 1992 with permission from ASCE).

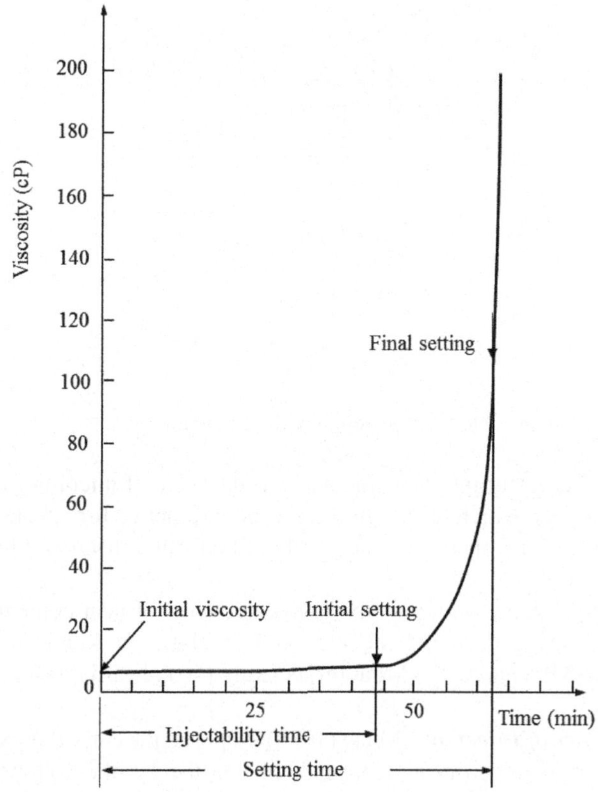

Figure 8.55 Viscosity–time curve for sodium silicate ground (modified from AFTES, 1991).

Figure 8.54 shows the behavior of different mixes with Portland cement in terms of apparent viscosity versus plastic viscosity. When the apparent viscosity, μ_a, is equal to the plastic viscosity, μ_p, the mix behaves as a Newtonian fluid, which is not stable. In an unstable mix, water can separate from a chemical agent (such as cement) due to pressure or self-weight. This phenomenon is also called "bleeding." Stable mixes are desired for grouting and deep mixing to get uniform stabilization of the soil. Figure 8.54 shows that stable mixes typically have lower plastic viscosity. The trademark mixes, MISTRA™, maintain the mixes stable by the use of bentonite.

Setting is a process for grout to harden. There are initial and final set times. The initial set happens when the grout starts hardening while the final set is at the end of hardening. Figure 8.55 shows the change of the viscosity of the grout with time. It is shown that the initial viscosity of the grout is low and increases gradually until the initial set time. After the initial set time, the viscosity increases exponentially. The injection time should be controlled before the initial set. Cementitious grout typically takes several to 24 hours to set, while chemical grout sets in minutes to hours, depending on the chemical agent (e.g., polyurethanes react instantly with water). Set time is also referred to as the gel time for chemical grouts.

Grout Materials The selection of a grout material should consider its mobility, penetrability, cohesion, bleed potential, setting time, and solubility.

Grout materials include four categories (Bruce et al., 1997):

1. Particulate (suspension or cementitious)
2. Collodial solutions
3. Pure solutions
4. Other grout materials

Particulate chemical agents, such as cement, lime, or a combination of cement and lime, have been commonly used in a slurry form like grout or in a powder form for deep mixing. Slag and fly ash have sometimes been used as well. Particulate grout is often a mixture of water and one or several particulate materials, such as cement, fly ash, clay, and/or sand. Littlejohn (1982) pointed out that the water to solid ratio is the key parameter for the properties and basic characteristics of particulate grout, such as stability, fluidity, rheology, strength, and durability. Table 8.9 shows the typical grout materials and their relative characteristics including viscosity, water : binder ratio, toxicity, strength, and relative cost. cP stands for centipoise, which is a dynamic viscosity unit. 1 cP = 0.001 N.s/m^2. Collodial solutions are silicate based while pure solutions are resins.

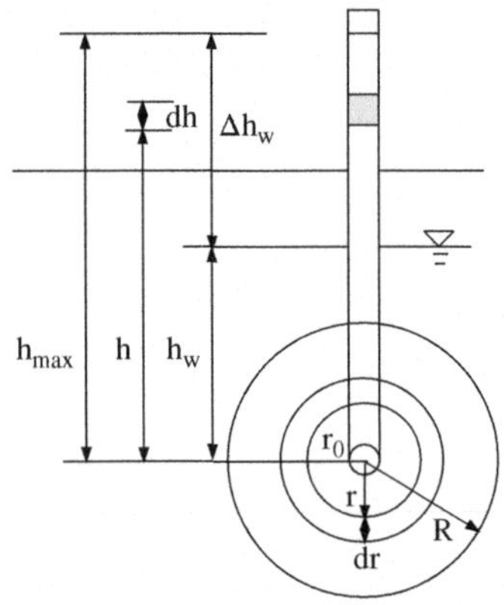

Figure 8.56 Model of injection with spherical flow.

Other grout materials include a wide range of uncommon materials, which have only been used in practice for special applications; therefore, they will not be further discussed in this book.

Typical unconfined compressive strengths of neat cement in sandy gravel and silicates and acrylates in sands are 1–10 MPa and 0.3–3 MPa, respectively (Woodward, 2005).

Theory of Injection Maag (1938) proposed the early theory of grout injection into ground. In this theory, the following assumptions are made: (1) the ground is uniform and isotropic, (2) the grout is Newtonian fluid, (3) the grout is injected from the bottom of the pipe, and (4) the grout to penetrate to a radius, R, by spherical flow (Figure 8.56).

According to Darcy's law, the quantity (volume) of grout injection, V_g is

$$V_g = k_g iAt = 4\pi r^2 k_g t \frac{-dh}{dr} \tag{8.49}$$

where k_g = permeability of grout in soil
i = hydraulic gradient of grout
A = spherical area at a distance, r, from the injection source
t = time
h = hydraulic head from the grout point
h_{max} = maximum hydraulic head including groundwater and grout pressure heads

$$-dh = \frac{V_g}{4\pi r^2 k_g t} dr \tag{8.50}$$

The above differential equation can be solved as follows:

$$h = \frac{V_g}{4\pi k_g t} \cdot \frac{1}{r} + C \tag{8.51}$$

where C = a constant.

The boundary conditions are (1) when $r = r_0$, (radius of the injection source) $h = h_{max}$ (maximum hydraulic head) and (2) when $r = R$, (radius of the influence from the injection source) $h = h_w$ (height of the water level from the injection source). Considering the boundary conditions, the above equation can be solved for the grout hydraulic head difference, Δh_w, as follows:

$$\Delta h_w = h_{max} - h_w = \frac{V_g}{4\pi k_g t}\left(\frac{1}{r_0} - \frac{1}{R}\right) \tag{8.52}$$

The quantity of the grout in the sphere is

$$V_g = \frac{4}{3}\pi R^3 n_p \tag{8.53}$$

where n_p = porosity of soil

The grout hydraulic head difference, Δh_w, is

$$\Delta h_w = h_{max} - h_w = \frac{R^3 n_p}{3k_g t}\left(\frac{1}{r_0} - \frac{1}{R}\right) \tag{8.54}$$

Since $R >> r_0$, the above equation can be simplified as

$$\Delta h_w = \frac{R^3 n_p}{3k_g t r_0} \tag{8.55}$$

$$t = \frac{R^3 n_p}{3k_g \Delta h_w r_0} \tag{8.56}$$

$$R = \sqrt[3]{\frac{3k_g \Delta h_w r_0 t}{n_p}} \tag{8.57}$$

If the shape of grout flow is cylindrical, the following equations can be obtained:

$$t = \frac{n_p R^3 \ln(R/r_0)}{2k_g \Delta h_w} \tag{8.58}$$

$$R = \sqrt[3]{\frac{2k_g \Delta h_w t}{n_p \ln(R/r_0)}} \tag{8.59}$$

The permeability of grout in soil depends on the viscosity of the grout and can be estimated based on the following relationship:

$$k_g = k\frac{\mu_w}{\mu_g} = \frac{k}{\beta_g} \tag{8.60}$$

where k = permeability of soil;
μ_g = viscosity of grout
μ_w = viscosity of water
β_g = grout to water viscosity ratio

Figure 8.57 shows the viscosity values of different grouts used in practice. Water has a viscosity of 1 cP at 20°C.

The permeability values of grouts with different diameters of mix grains are shown in Figure 8.58.

Table 8.9 Characteristics of Grout Materials

Description	Viscosity (cP) (water : binder ratio)	Toxicity	Strength	Relative Cost
Particulate grout				
Type I cement	High (50 cPs) (2:1)	Low	High	Low
Type III cement	Medium (15 cPs) (2:1)	Low	High	Low
Ultra-fine cement	Low (8 cPs) (2:1)	Low	High	Medium
Colloidal solution				
Silicates	Low (>6 cPs)	Low	Medium	Low
Solution grout				
Lignosulfites	Medium (>8 cPs)	High	Low	Medium
Polyurethane	High (>400 cPs)	High	High	High
Acrylamides	Low (1.2 cPs)	High	Low	Medium
Acrylates	Low (1.5 cPs)	Low	Low	Medium

Source: Modified from Elias et al. (2006).

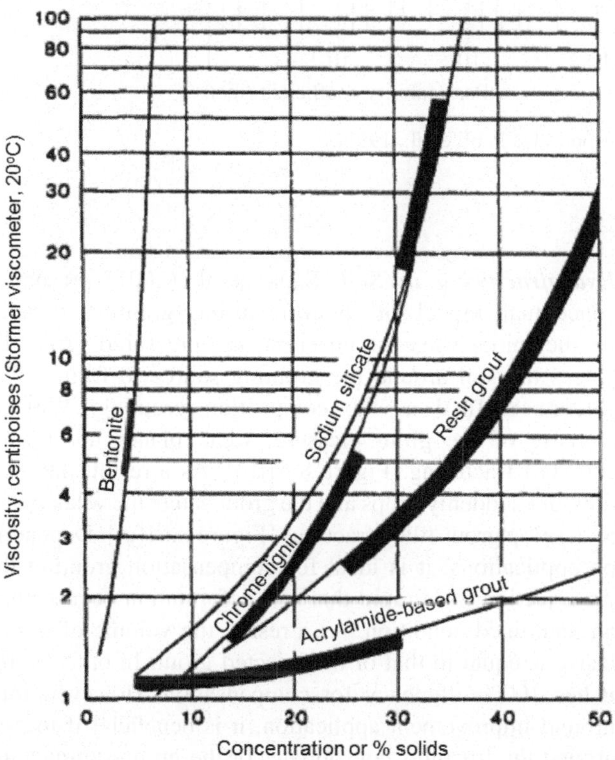

Figure 8.57 Viscosities of various grouts. Heavy lies indicate the concentrations normally used for field work (Karol, 1983, with permission from Taylor & Francis).

Compaction Grouting Compaction grouting injects stiff and low mobility grout to apply pressure on the surrounding soil and displace soil particles. A grout bulb is formed around the grouting source. If the grout bulb is simplified as a sphere, the pressure within the bulb is equal in all directions but decreases in the soil rapidly from the edge the spherical bulb to zero at a certain distance as shown in Figure 8.59.

Al-Alusi (1997) illustrated the soil unit weight increase by compaction grouting below.

Under the applied pressure, the soil within the influence range deforms and accumulates a volumetric strain, ε_v (Figure 8.60). The volumetric strain can be calculated based on the volume of grout, V_g, and the initial soil volume within the influence range, V_i, as follows:

$$\varepsilon_v = \frac{V_g}{V_i} \tag{8.61}$$

Assume the soil bulk modulus is K_{sb}, which can be expressed as

$$K_{sb} = \frac{p_g}{\varepsilon_v} \tag{8.62}$$

where p_g is grout pressure.

Hence

$$\frac{V_g}{V_i} = \frac{p_g}{K_{sb}} \tag{8.63}$$

The increase of the soil unit weight can be expressed as

$$\Delta\gamma = \frac{W_s}{V_i - V_g} - \gamma = \frac{\gamma}{1 - V_g/V_i} - \gamma = \frac{\gamma}{1 - p_g/K_{sb}} - \gamma \tag{8.64}$$

where W_s is the weight of the soil mass within the influence range and γ = initial unit weight of the soil.

The percent of unit weight increase can be expressed as follows:

$$\frac{\Delta\gamma}{\gamma} = \frac{1}{1 - p_g/K_{sb}} - 1 = \frac{p_g/K_{sb}}{1 - p_g/K_{sb}} \approx \frac{p_g}{K_{sb}} \tag{8.65}$$

The approximation in the above equation is made because $p_g/K_{sb} << 1$. The above equation shows that the increase of the unit weight of the soil is proportional to the grout pressure but reciprocal to the bulk modulus of the soil.

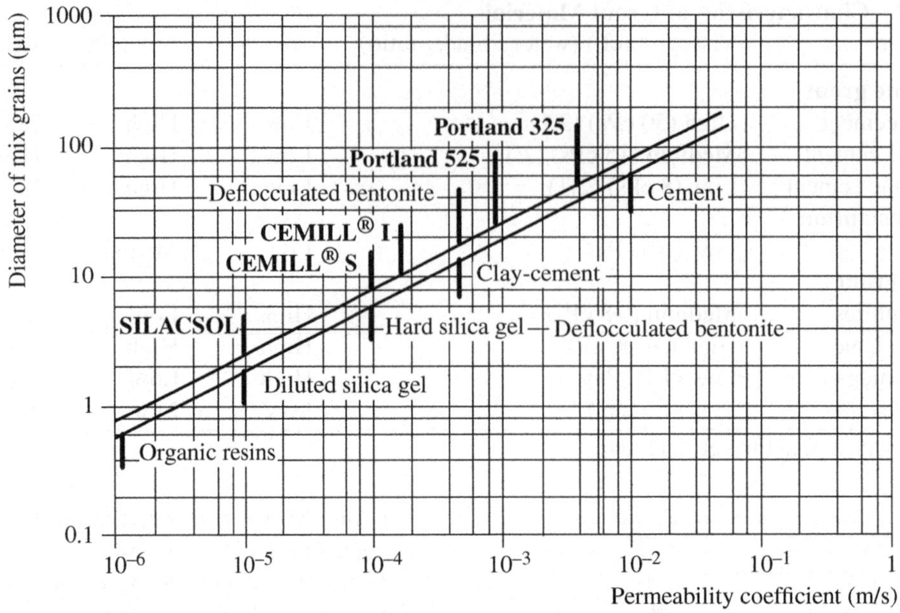

Figure 8.58 Permeability limits of grouts (De Paoli et al., 1992).

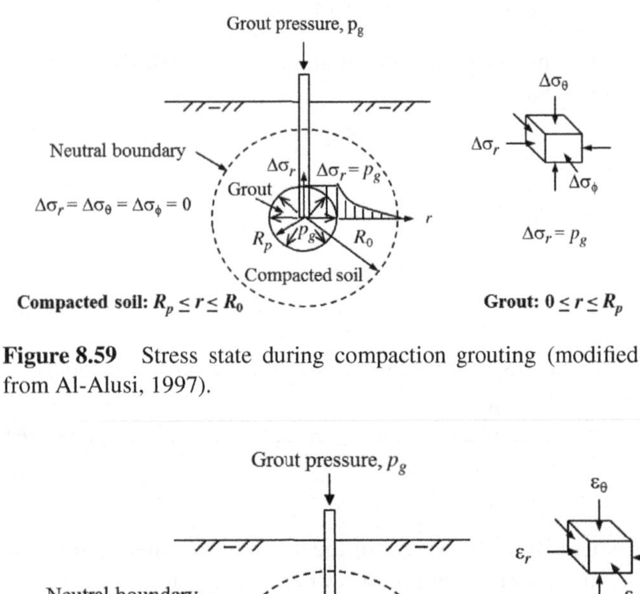

Figure 8.59 Stress state during compaction grouting (modified from Al-Alusi, 1997).

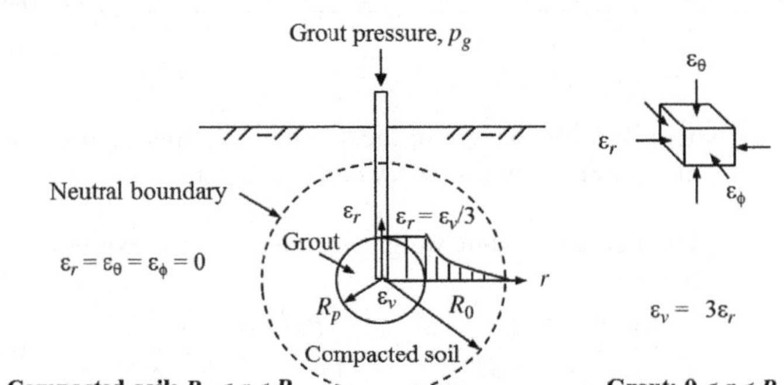

Figure 8.60 Strain state during compaction grouting (after Al-Alusi, 1997).

Hydrofracturing in Soil Soga et al. (2001) presents a conceptual model of hydrofracturing grouting in clay. At the initial stage of injection, the pressured grout displaces the soil around the grouting source to form a bulb [Figure 8.61(a)] as the compaction grouting. With an increase of the grout pressure, weak planes are formed and start fracturing [Figure 8.61(b)]. As a result, the grout pressure suddenly drops and the grout enters the weak planes to develop grout-filled fractures [Figure 8.61(c)]. Depending on applications, it is ideal for compensation grouting that the injection is so rapid that soil deformation occurs under an undrained condition. As a result, the volume of ground heave is equal to that of the injected grout. In other words, it has 100% efficiency for compensation. However, for a ground improvement application, it is beneficial if the soil around the fractures consolidates or the grout migrates into the voids around the fractures [Figure 8.61(d)]. In addition to soil consolidation and/or grout migration, the pressured grout may cause soil movement in an undesired direction and/or location and escape through extended fractures. Therefore, the compensation efficiency decreases with time.

Hydrofracturing can also happen in sandy soil if the grout is stiff and has low mobility. The formation of the fracture in sandy soil can be modeled using the Mohr circles as shown in Figure 8.62 and is discussed below.

Prior to grouting, the soil is at the at-rest state as the right Mohr circle. Considering the pore water pressure in the

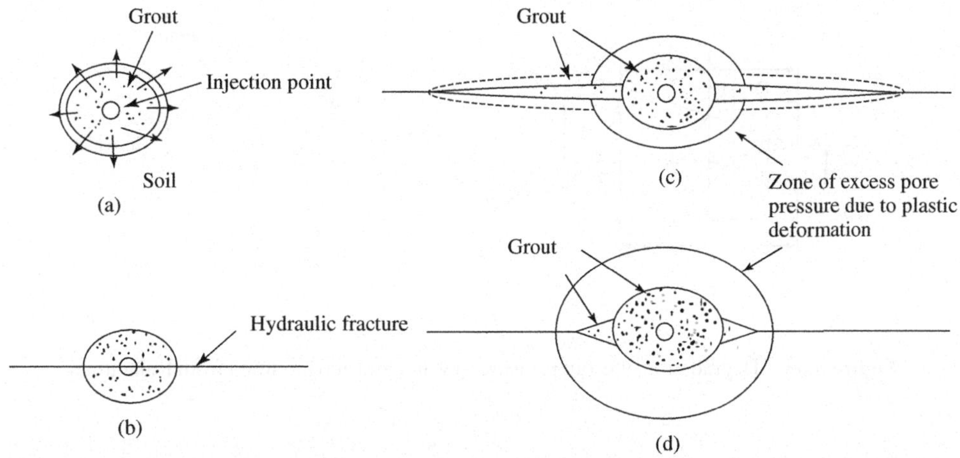

Figure 8.61 Conceptual modeling of hydrofracturing grouting in clay: (a) initial stage of grout injection, (b) initiation of fracture, (c) penetration of grout into fracture, low viscosity grout, (d) penetration of grout into fracture, high viscosity grout (Soga et al, 2001, with permission from ICE Publishing).

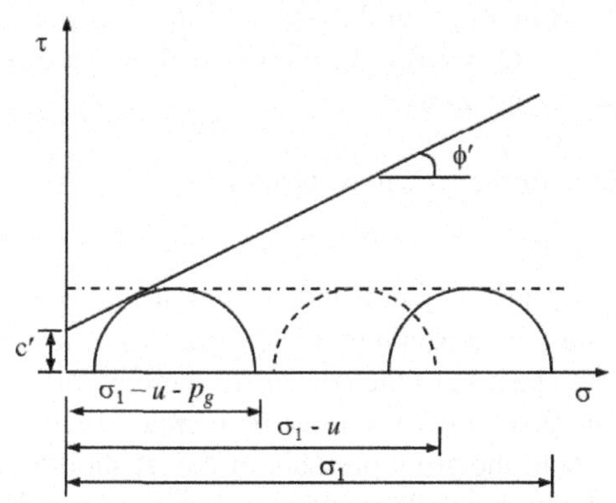

Figure 8.62 Failure mechanism of hydrofracturing of soil (modified from Herndon and Lenahan, 1976).

ground, the soil is at the state as the middle dashed Mohr circle. Due to the grouting pressure, p_g, the major and minor principal stresses in the soil are both reduced by p_g (σ_1' and σ_3' are the principal stresses in the vertical and circumferential directions, respectively). As a result, the Mohr circle moves toward the left. With an increase of p_g, the Mohr circle touches the Mohr–Coulomb failure envelope as expressed below:

$$\frac{\sigma_1' + \sigma_3'}{2} \sin\phi' = \frac{\sigma_1' - \sigma_3'}{2} - c' \cos\phi' \quad (8.66)$$

The grout pressure to fracture the soil can be solved as follows:

$$p_g = \frac{(\gamma_g h_{gp} - u)(1+K)}{2} - \frac{(\gamma_g h_{gp} - u)(1-K)}{2\sin\phi'} + \frac{c'}{\tan\phi'} \quad (8.67)$$

where p_g = grout pressure
γ_g = unit weight of the grout
h_{gp} = height of the grout in the pipe to the location of the grout injected
u = pore water pressure
K = principal stress ratio (commonly assumed to be the lateral earth pressure coefficient at rest, i.e., K_o)

Mechanisms of Jet Flow The mechanisms of jet flow during jet grouting are illustrated in Figure 8.63. Figure 8.63(a) shows the jet moves in air from the nozzle to a distance, x. The jet movement consists of three zones: initial flow (zone A), turbulent flow (zone B), and discontinuous flow (zone C). Right after the jet leaves the nozzle, it enters the initial laminar flow zone, which has uniform flow and uniform pressure. The uniform portion becomes smaller with the jet's further travel and ends at the distance, x_c. Within the initial flow zone, there are two subzones: zone D and zone E. The jet in zone D has uniform velocity and uniform pressure. Zone E is a transition zone from the initial flow to the turbulent flow. The energy loss within zone A is minor so that the jet pressure almost remains the same as shown in Figure 8.63(b). The length of zone A is an important parameter for jet grouting, which contributes to the erodibility of the soil. Starting from x_c, the jet has nonuniform velocity and nonuniform pressure; therefore, it enters the turbulent flow zone. The energy loss in the turbulent flow zone increases rapidly with the jet's travel distance. As a result, the jet velocity and pressure decrease significantly. At the end of the turbulent flow zone, the jet starts to turn into steam, which is a discontinuous flow and

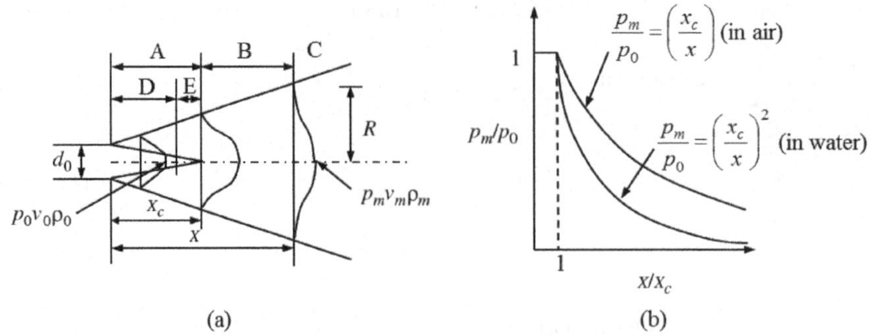

Figure 8.63 Degradation of grout pressure in water and air (modified from Rajaratnam, 1976).

has very low energy. Therefore, the effective jet travel distance is the total length of zone A and zone B. This distance dominates the range of the grout mixed with the soil. The larger effective jet travel distance results in a larger diameter jet-grouted column.

When the jet travels in air and the travel distance is beyond x_c, the jet pressure decreases with the distance as follows:

$$\frac{p_m}{p_0} = \frac{x_c}{x} \tag{8.68}$$

where p_0 = pressure of the jet emitted from the nozzle
p_m = pressure of the jet at a distance, x, from the nozzle

When the jet travels in water and the travel distance is beyond x_c, the jet pressure decreases with the distance as follows:

$$\frac{p_m}{p_0} = \left(\frac{x_c}{x}\right)^2 \tag{8.69}$$

The preceding equations show that the jet pressure decreases much faster in water than in air.

Experimental data shows that the distance of the initial flow zone depends on the size of the nozzle and the medium in which the jet travels (Ye et al., 1994), that is,

$$x_c = (75\text{–}100)d_0 \quad \text{in air} \tag{8.70}$$

$$x_c = (6\text{–}6.5)d_0 \quad \text{in water} \tag{8.71}$$

where d_0 is the diameter of the nozzle.

Miki and Nakanishi (1984) demonstrated in experimental tests that water jet traveling in air has the longest travel distance with the least pressure reduction, followed by the water jet enveloped by air jet traveling in water, and then the water jet traveling in water.

Formation of Jet-Grouted Columns The erosion of soil by jet grouting mainly relies on the impact force produced by grout jet. The force of the jet in air can be expressed as

$$P = \rho_g Q_g v \tag{8.72}$$

where ρ_g = density of the grout
v = average velocity of the jet
Q_g = rate of grout flow, which can be expressed as $Q_g = vA_0$ (A_0 = cross sectional area of the nozzle)

The force of the jet can be written as

$$P = \rho_g A_0 v^2 \tag{8.73}$$

The preceding equation indicates that the momentum of the jet is largely dependent on the velocity of the jet and also depends on the size of the nozzle. To increase the velocity of the jet, the grout pressure must be increased. To effectively erode the soil, the grout pressure of the jet should be at least 20 MPa. Coomber (1985) suggested that at least 40 MPa jet pressure is needed to erode cohesive soils.

During jet grouting, the jet at high pressure erodes the soil up to the effective travel distance. In the single jet system, the grout erodes the soil. In the double jet system, the grout with the compressed air erodes the soil. In the triple jet system, the water with the compressed air erodes the soil and the grout fills in the space left by the eroded soil. Since the water with compressed air can travel farther than the grout with air, which can travel farther than the grout, the triple jet system produces the largest column diameter, followed by the double jet system and then the single jet system. The eroded soil is mixed with the grout. Due to the pressure change, fine particles tend to migrate to the center and coarse particles stay at a distance from the center. At the end of the jet effective travel distance, the impact force is not high enough to erode the soil; however, it applies pressure to compress the soil. In the soil with high permeability, such as sand, the grout may permeate into the soil to a certain distance. Figure 8.64 shows the structure of the jet-grouted column in sand and clay. The column in the clay does not have the

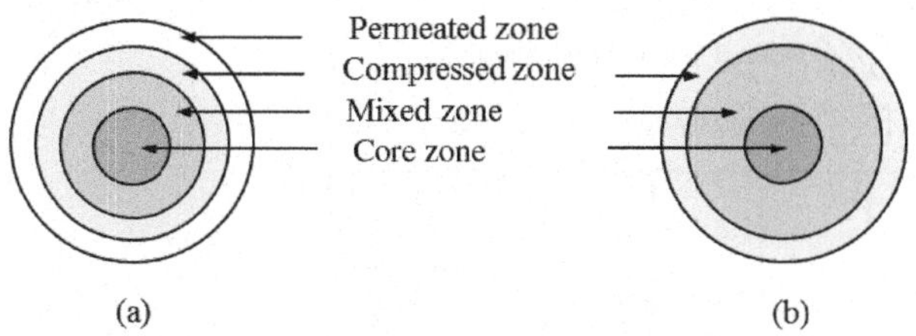

Figure 8.64 Structure of jet routed column: (a) sand and (b) clay (after Ye et al., 1994).

permeated zone. The core of the column is typically softer than the mixed zone due to the existence of fine particles. In the clay, there is no permeated zone because of its low permeability.

The process for the hardening of the grouted column is similar to that of the deep mixed column as discussed earlier; therefore, it will not be repeated here.

8.3.3 Design Considerations

Permeation Grouting

Groutability To ensure the groutability of grout into soil or rock, the following two parameters were proposed by Mitchell and Katti (1981) based on Terzaghi's filter criteria:

$$N_{gs} = \frac{(D_{15})_{soil}}{(D_{85})_{grout}} \quad \text{(for soil)} \tag{8.74}$$

$$N_{gr} = \frac{t_f}{(D_{95})_{grout}} \quad \text{(for rock)} \tag{8.75}$$

where N_{gs} = groutability of soil
N_{gr} = groutability of rock
$(D_{15})_{soil}$ = soil particle size corresponding to 15% passing
$(D_{85})_{grout}$ = grout particle size corresponding to 85% passing
t_f = width of fissure in rock

Cement grout is groutable in soil when $N_{gs} > 11$ but consistently groutable when $N_{gs} > 24$. When a clay–cement grout is used in soil, it is groutable when $N_{gs} > 5$. In rock, cement grout is groutable when $N_{gr} > 2$ but consistently groutable when $N_{gr} > 5$. Typical applications of grouts in granular soils are shown in Figure 8.65.

Required Hydraulic Head Considering a Newtonian fluid in a uniform isotropic soil, the required grout hydraulic head difference for grout to penetrate to a radius, R, by spherical flow can be estimated by the following formula (Raffle and Greenwood, 1961):

$$\Delta h_w = \frac{Q_g}{4\pi k}\left[\beta_g\left(\frac{1}{r_0} + \frac{1}{R}\right) + \frac{1}{R}\right] \tag{8.76}$$

where k = permeability of soil
β_g = grout to water viscosity ratio
r_0 = radius of spherical injection source
R = radius of penetration in the ground
Q_g = rate of grout injection

For a cylindrical injection source, r_0 in the above equation can be estimated as $0.5\sqrt{Ld}$, where L and d are the length and diameter of the injection source

The required grout pressure can be calculated as follows:

$$p_g = \gamma_w(h_w + \Delta h_w) - \gamma_g h_{gp} \tag{8.77}$$

where γ_g = unit weight of grout
γ_w = unit weight of water
h_{gp} = height of the grout to the injection point
h_w = height of the groundwater table to the injection point

Time to Penetration The time for a Newtonian fluid to penetrate to a radius, R, by spherical flow is (Raffle and Greenwood, 1961):

$$t = \frac{n_p r^2}{k\Delta h_w}\left[\frac{\beta_g}{3}\left(\frac{R^3}{r_0^3} - 1\right) - \frac{\beta_g - 1}{2}\left(\frac{R^2}{r_0^2} - 1\right)\right] \tag{8.78}$$

where n_p = porosity of the soil.

It is recommended that permeation grouting be used in gravel or sandy soils with less than 15% silts and clays. Otherwise, it may take too long to inject grout at permissible pressure.

Effect of Bingham Fluid A minimum hydraulic gradient is needed to overcome the yield strength if a Bingham fluid is used (Raffle and Greenwood, 1961):

$$i_{min} = \frac{4\tau_s}{d\gamma_g} \tag{8.79}$$

where τ_s = yield strength of the Bingham fluid
d = effective diameter of the pore in the soil
γ_g = unit weight of the grout

The yield strengths of typical chemical grout (Newtonian fluid), 5% bentonite suspension (Bingham fluid), and 0.67 water/cement ratio grout are 0, 2.3, and 4.0 N/m^2.

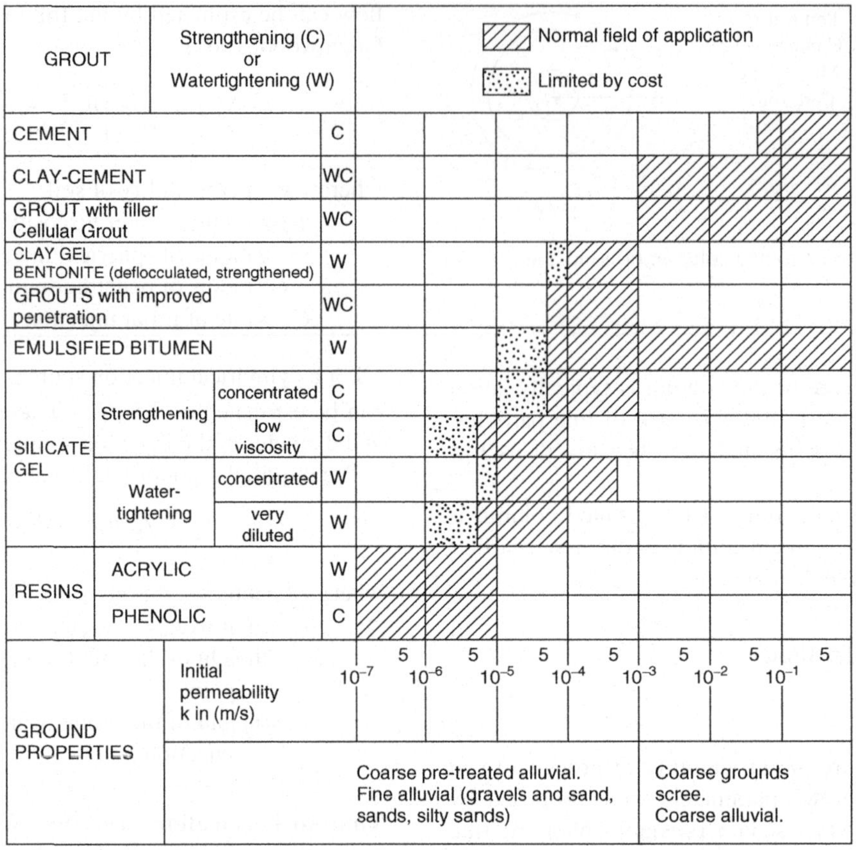

Figure 8.65 Typical applications of grouts in granular soils (AFTES, 1991).

The effective diameter of pore in the soil can be estimated based on the Kozeny equation:

$$d = 2\sqrt{\frac{8\mu_w k}{\gamma_w n_p}} \tag{8.80}$$

where d = pore diameter (mm)
μ_w = water viscosity (cP)
k = soil permeability (m/s)
γ_w = unit weight of water (kN/m^3)
n_p = porosity of the soil

Table 8.10 shows approximate hydraulic gradients to maintain the flow of a Bingham fluid in soil at different permeability.

Allowable Grout Pressure When the grout is a Bingham fluid, it requires much higher grout pressure. The injection pressure should be limited to avoid ground heave. The allowable injection pressure can be estimated as follows (Chinese Ground Improvement Manual Committee, 1988):

$$p_{ga} = 100(\alpha_p p + C_{gs}\beta_{gm}\lambda_{sc}z) \tag{8.81}$$

where p_{ga} = allowable grout pressure (kPa)
α_p = surcharge factor (typically 1–3)
p = surface surcharge (kPa)
C_{gs} = grouting sequence factor (1.0 for the primary grouting, 1.25 for the secondary grouting, and 1.5 for the tertiary grouting)
β_{gm} = grouting method factor (0.8 for downstage and 0.6 for upstage) λ_{sc} = soil characteristic factor, ranging from 0.5–1.5 (a lower value for loose and permeable soil but a higher value for dense and low permeable soil)
z = depth of the injection point from the ground surface (m)

The estimated allowable grout pressure should be verified in the field by performing a grout trial test. The allowable grout pressure can be determined from the test result as shown in Figure 8.66. When the grout pressure is higher than the allowable pressure, the grout volume rapidly increases due to ground heave.

Table 8.10 Hydraulic Gradient to Maintain Flow of Bingham Fluid

Soil Permeability (m/s)	Yield Strength (Pa)	Hydraulic Gradient
10^{-2}	1	1.2
	10	12
	100	120
	1000	1200
10^{-3}	1	4
	10	40
	100	400
	1000	—
10^{-4}	1	12
	10	120
	100	1200
	1000	—
10^{-5}	1	4
	10	40
	100	400
	1000	—

Source: Lttlejohn (1985).

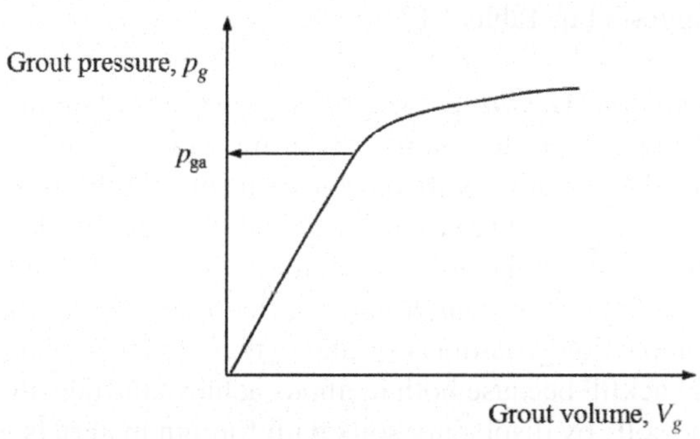

Figure 8.66 Grout pressure vs. grout volume.

The limited penetration radius can be estimated as follows (Xanthakos et al., 1994):

$$R_L = \frac{\gamma_g h_L d}{4\tau_s} + r_0 \tag{8.82}$$

where τ_s = yield strength of the Bingham fluid
d = effective diameter of the pore in the soil
γ_g = unit weight of the grout; and

h_L is the limited grout head at the injection point, which can be estimated as follows:

$$h_L = \frac{p_{ga} + h_g\gamma_g - h_w\gamma_w}{\gamma_g} \tag{8.83}$$

Penetration Radius AFTES (1991) suggested the following formula to estimate the penetration radius, R:

$$R = \sqrt{\frac{V_g}{n_p \pi L_g}} \tag{8.84}$$

where V_g = injected grout volume per pass
n_p = soil porosity
L_g = thickness of injected section or pass length

Ye et al. (1994) pointed out that the actual injected grout volume per pass may be less than that calculated using the above equation because (1) grout may not fill all the voids and (2) the water in the soil has occupied some of the voids. Therefore, the following formula may be used to estimate the injected grout volume:

$$V_g = C_{vc}\pi n_p R^2 L_g \tag{8.85}$$

where C_{vc} is the volume correction factor, which is provided in Table 8.11.

Research shows that the actual penetration radius of a grout depends on the following factors (Xanthakos et al., 1994):

- Geotechnical conditions (such as soil permeability or degree of fissuring in rock)
- Injection pressure
- Injected volume
- Characteristics of grout (viscosity and setting time)
- Desired efficiency

Grout Hole Pattern and Spacing The grout hole pattern and spacing for permeation grouting should be designed based on the grout penetration radius and the complete coverage of the grouted area within the improved depth. Depending on applications, grout holes may be arranged in one row or multiple rows.

For one row of grout holes as shown in Figure 8.67, the effective thickness of the grouted area can be calculated as follows (Ye et al., 1994):

$$b = 2\sqrt{R^2 - \left[(2-R) + \frac{R-(s-R)}{2}\right]^2} = 2\sqrt{R^2 - \frac{s^2}{4}} \tag{8.86}$$

Table 8.11 Grout Volume Correction Factor

Soil Type	Correction Factor, C_{vc}
Clay, silt, and fine sand	0.3–0.5
Medium to coarse sand	0.5–0.7
Gravel	0.7–1.0
Collapsible loess	0.5–0.8

Source: Ye et al. (1994).

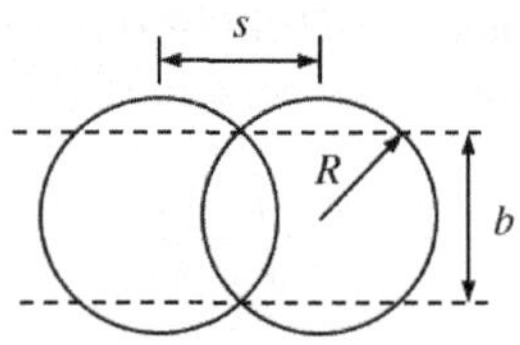

Figure 8.67 Layout of one row of grout holes.

If the required effective thickness is b, the spacing of grout holes can be calculated as follows (Ye et al., 1994):

$$s = 2\sqrt{R^2 - \frac{b^2}{4}} \tag{8.87}$$

For two rows of grout holes, two patterns as shown in Figure 8.68 may be used. The triangular pattern includes all the primary holes while the square pattern requires the primary holes and the secondary holes to cover all the improved area. For the triangular pattern, the horizontal spacing and the vertical spacing are $1.8R$ and $1.5R$, respectively. For the square pattern, the horizontal spacing and the vertical spacing are both $2R$ and there is a secondary hole in the middle of every four primary holes. The typical grout hole spacing is suggested in Table 8.12.

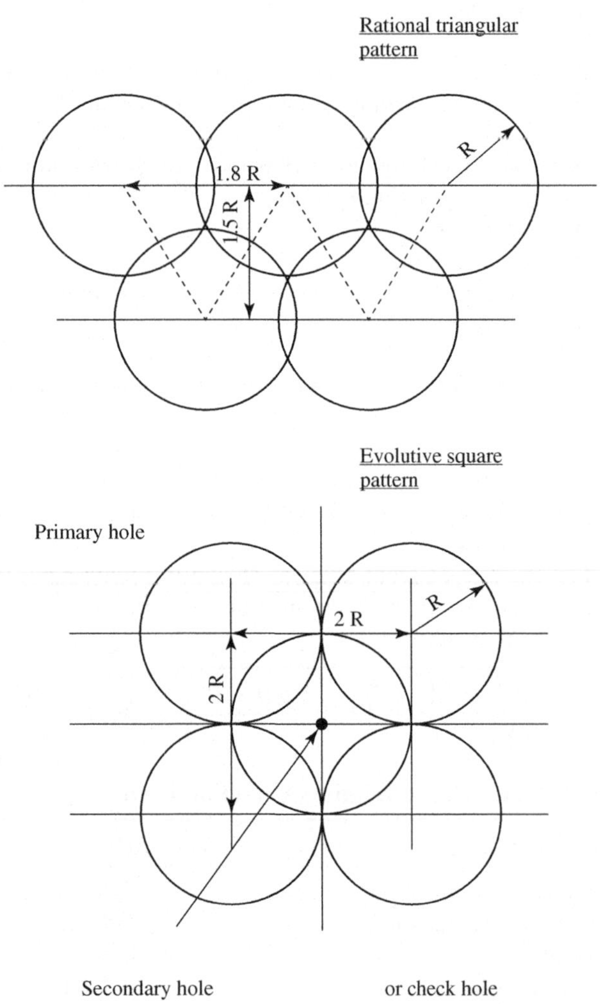

Figure 8.68 Layout of two rows of grout holes (AFTES, 1991).

Table 8.12 Typical Grout Hole Spacing

Medium to Be Grouted	Description	Grout Hole Spacing (m)
Soil, depth < 25 m	Fine sand	0.8–1.3
	Sand, sand and gravel	1.0–2.0
	Gravel	2–4
	Sand and gravel ($k_h > k_v$)	Watertight ground 3–5
Rock, depth < 25 m	Fine cracks	1–3
	Open cracks	2–4
Structure	Backing behind the vault	2–3
Cavity	Filling of large void	3–15

Source: AFTES (1991).

Compaction Grouting The basic principles of compaction grouting are (1) to improve bearing capacity and density of soil by injecting stiff mortar to form a bulb around the injection pipe and (2) to displace and compact weak soil by the formation of the bulb. In addition, the mortar should be a coherent mass not to enter pores or fractures. The mechanism of compaction grouting is similar to that of vibro-compaction with backfill because both methods achieve the densification of the soils by displacing soils with foreign materials.

Grout Hole Spacing The spacing of the grout holes can be determined based on the required density or void ratio for applications, such as liquefaction mitigation or increase of bearing capacity after compaction grouting. The following formula may be used to estimate the grout hole spacing:

$$s = \sqrt{\frac{\pi(1+e_0)d_c^2 L_c}{4(e_0 - e_1)L_c + 4(1+e_0)S}} \tag{8.88}$$

where e_0 = initial void ratio of soil
e_1 = void ratio of soil after compaction grouting
d_c = average diameter of bulb
L_c = length of grouted column
S = total surface heave

Alternatively, L_c can be the thickness of injected section or pass length and S is the surface heave per pass.

The grout hole spacing used in practice ranges from 0.9 to 4.5 m, but mostly from 1.5 to 2 m. Tight spacing with low grout pressure is used near the ground surface with low overburden stress, while large spacing with high grout pressure is used when there is sufficient overburden stress (e.g., 4.5 m spacing at an overburden stress of 9 m thick soil).

Compaction grouting is done in stages. The vertical interval of injections is typically 0.3–0.9 m.

The replacement ratio of the compaction grout volume to the treated volume typically ranges from 5 to 15%. Another general rule is that the spacing of grout holes is three to five times the diameter of the grouted bulb.

Overburden Stress To ensure the densification of soils during compaction grout, it is important to maintain sufficient overburden stress. It is a typical requirement that an overburden stress of 70 kPa is needed to achieve the maximum densification of the soils. Limited densification can be achieved with lower overburden stress. The overburden stress can come from overburden soils, surcharge loads, and/or foundation loads.

Injection Pressure The pressure at the point of injection usually ranges from 700 to 3000 kPa. The refusal pressure on most granular soil projects ranges from 3000 to 4000 kPa. It is common that the refusal pressure cannot be reached due to the surface heave. For most compaction grouting projects, ground heave is not desired. Ground heave is an indication that the grout pressure has exceeded the confining stress and fracturing occurs. Typically, the total ground heave is limited to be less than 13 mm except for releveling projects.

Injection Rate The injection rate of compaction grouting depends on the permeability of the soil. For a low permeable soil with a low overburden stress, a slow injection rate (0.01 to 0.02 m^3/min) should be used. For a free-draining soil with an intermediate overburden stress, a medium injection rate (0.02 –0.10 m^3/min) may be used. For a loose soil under a high overburden stress, a fast injection rate (0.10 –0.35 m^3/min) may be used.

Grout Material The main purpose of compaction grouting is to densify the surrounding soil. Therefore, it is not necessary for the grout material to have a high strength. Sand or sandy soil has been commonly used as the main component of the grout material with Type I or II Portland cement. The grout material typically has a slump at 25–75 mm.

Hydrofracture Grouting The primary objective of hydrofracture grouting is to apply high pressure in the ground to induce an interwoven network of grout-filled fractures, which provide reinforcement to the soils. Hydrofracture grouting may have some densification effect on the soils through the soil consolidation. Cementitious suspension grouts are most commonly used. Chemical grouts are sometimes used in fine-grained soils. Two different concepts of hydrofracture grouting are illustrated in Figure 8.69. When a stable cementitious grout is used, it fractures the soils under a high grout pressure and compact the soils above, below, and between thick grout lenses. The grouting

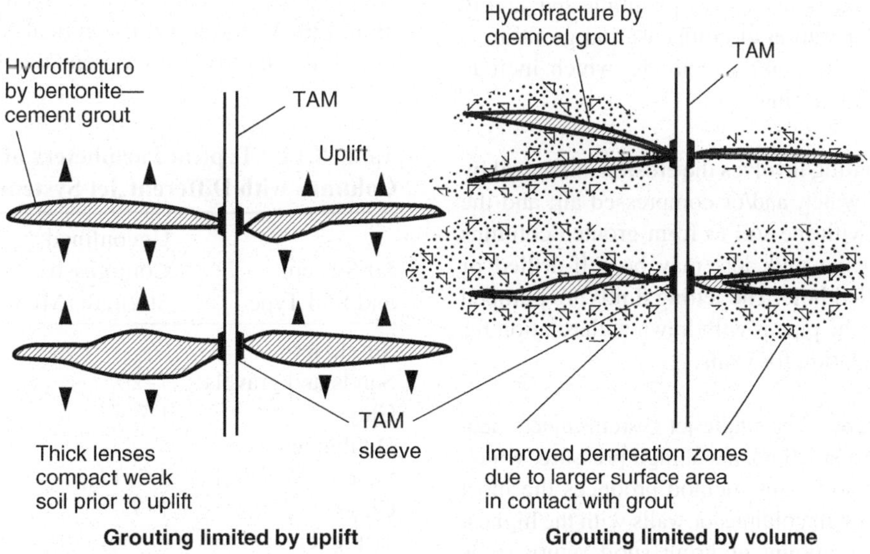

Figure 8.69 Concepts for hydrofracture grouting (Woodward, 2005, with permission from Taylor & Francis).

for this method is often limited by the surface heave. When chemical grout is used, it fractures the soils under a high grout pressure and permeates into the surrounding soils due to larger surface area in contact with grout. The grouting for this method is limited by grout volume. To induce fractures in the soils, the grout pressure should be higher than the allowable grout pressure discussed earlier. Sleeve ports, such as the Tube-a-Manchette (TAM), are commonly used to control the directions of grout flow. However, Warner (2004) pointed out that "Unfortunately, it is virtually impossible to control the direction, interconnection, or extent of fractures." Therefore, caution should be exercised during the hydrofracture operation.

Hydrofracture grouting is also used to compensate ground loss during construction activities, such as tunneling and excavation. This application is considered as compensation grouting.

Compensation Grouting Compensation grouting is designed to protect existing structures from potential damage as a result of ground movement or ground loss from adjacent or underground excavation. It is a general concept that a sufficient volume of grout is injected to the ground around or below existing structures to ensure the structures remain at their existing levels while excavation takes place below or nearby. Compensation grouting is typically conducted prior to as well as during the excavation to compensate for the change in stresses and ground loss before they influence the structure. Controlled hydrofracture grouting, compaction grouting, and a combination of one of these methods with permeation grouting have been used in practice. The prediction of ground movement or loss, which is beyond the scope of this book, is useful for planning and laying out grout holes. The operation of compensation grouting is always controlled by observation methods, which include instrumentation and monitoring.

Jet Grouting Jet grouting involves the erosion of the soil by cement grout, jets of water, and/or compressed air, and the mixture of the grout with the soil to form grouted columns or walls. The grout pipe with jets may be self-drilled or inserted into a predrilled hole. The jetting fluids are pumped at high pressure while the pipe is withdrawn with rotation for columns or without rotation for walls.

Selection of Jet System The single jet system injects neat cement grout through a small nozzle at high pressure, which is mixed with in situ soil. This method produces the most homogeneous soil–cement columns or walls with the highest strength and the least amount of grout spoil return. It is the simplest system among the three systems and has more choices of qualified specialty contractors.

Table 8.13 Technical and Economic Assessments

Soil Type	Technical Capability	Economic Preference
Gravel	Single, double, triple	Single
Clean sand	Single, double, triple	Double
Silty sand	Single, double, triple	Triple, double
Silt	Triple, double	Triple, double
Clay	Triple	Triple

Source: Elias et al. (2006).

The double jet system injects neat cement grout at a lower pressure, which is aided by a cone of compressed air. The air reduces the friction loss and allows the grout to travel a farther distance to produce a greater column diameter. However, the presence of the air reduces the strength of the column and produces more spoil return than the single jet.

The triple system injects water at high pressure, which is aided by a cone of compressed air. This process produces an air lifting effect, which erodes the soil. The grout is injected at a lower pressure through a separate nozzle below the water and air nozzles to fill the void created by the air lifting process.

Elias et al. (2006) provide a guideline for the selection of the jet system in terms of technical and economic assessments, as shown in Table 8.13.

Typical Parameters Table 8.14 provides the typical parameters of jet-grouted columns with different jet systems for a preliminary design.

Figure 8.70 shows that the diameter of the jet-grouted column depends on the rate of withdrawal, the density of the soil, and the pressure of water. Shen et al. (2013) and Flora et al. (2013) developed theoretical solutions for the prediction of the diameter of jet grouted columns.

Table 8.14 Typical Parameters of Jet-Grouted Columns with Different Jet Systems

Jet System and Soil Type	Unconfined Compressive Strength (MPa)	Column Diameter (m)
Single jet		
Sands and gravels	7–20	0.6–0.9
Clays	1.7–7	0.6–0.9
Double jet		
Sands and gravels	3.5–14	0.9–1.8
Clays	1–7	0.9–1.5
Triple jet		
Sands and gravels	3.5–10	1.5–2.5
Clays	1–5	0.9–1.8

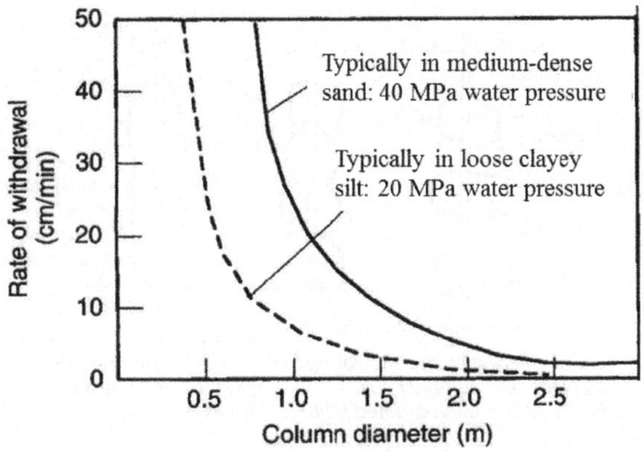

Figure 8.70 Effect of rate of withdrawal on column diameter (Woodward, 2005, with permission from Taylor & Francis).

Design for Foundation Support The design of jet-grouted columns is the same as that for deep mixed columns.

Design for Excavation Support The design of jet-grouted columns or walls for excavation support is the same as that for deep mixed columns.

8.3.4 Design Parameters and Procedure

The design parameters for permeation grouting include:

- Soil type, density, and D_{15}, or rock type, width of fissure
- Depth to be improved
- Groundwater table
- Type and properties of grout including grout D_{85} for soil or D_{95} for rock, grout unit weight, and grout viscosity
- Grout penetration radius and injection rate
- Grout pipe head height
- Yield strength if Bingham fluid is used
- Grout operation procedure (downstage or upstage)

The design procedure for permeation grouting includes:

1. Select grout material.
2. Check groutability.
3. Evaluate grout penetration radius and grout properties.
4. Calculate required grout head.
5. Determine required grout pressure.
6. Effect of Bingham fluid.
7. Verify against allowable grout pressure.

The design parameters and procedure for compaction grouting are similar to those for vibro-compaction.

No step-by-step design procedure is available for hydrofracture grouting and compensation grouting. Field trial or observational method should be used for these two methods.

The design parameters and procedure for jet grouting are similar to those for deep mixing.

8.3.5 Design Example

Problem

A site consists of a loose medium sand ($D_{15} = 0.1$ mm), which is located at a depth from 4.3 to 5.3 m. The groundwater is at the depth of 2 m. The permeability of this sand is 0.001 m/s. This loose sand has liquefaction potential; therefore, it should be improved. Permeation grouting is selected for this improvement. Design the permeation grouting.

Solution

1. Selection of grout material: Considering the soil to be grouted is a medium sand, chemical grout is suitable. Based on the toxicity and cost information, silicate solution is selected.
2. Groutability: The diluted silicate gel has a diameter ranging from 1.2 to 2 μm. Use $(D_{85})_{\text{grout}} = 2\,\mu\text{m}$ for the calculation as follows:
$$N_{gs} = \frac{(D_{15})_{\text{soil}}}{(D_{85})_{\text{grout}}} = \frac{0.1}{0.002} = 50 > 24$$
(consistently groutable)
3. Grout penetration radius and grout properties: Considering 1-m-thick loose sand, assume the radius of grout penetration, $R = 0.3$ m. The radius of the injection source, $r_0 = 0.015$ m. The rate of grout injection, $Q_g = 1\ \text{m}^3/\text{h}$. The unit weight of the grout is 11 kN/m^3. The viscosity of silicate solution, $\mu_g = 6$ cP. The grout to water viscosity ratio is
$$\beta_g = \frac{\mu_g}{\mu_w} = \frac{6}{1} = 6$$
4. Required grout hydraulic head: The required grout hydraulic head using the Raffle and Greenwood formula is
$$\Delta h_w = \frac{Q_g}{4\pi k}\left[\beta_g\left(\frac{1}{r_0} + \frac{1}{R}\right) + \frac{1}{R}\right] = \frac{1/(60 \times 60)}{4 \times 3.14 \times 0.001}$$
$$\times \left[6 \times \left(\frac{1}{0.015} + \frac{1}{0.3}\right) + \frac{1}{0.3}\right] = 9.4\ \text{m}$$
5. Required grout pressure: Assume the head of the grout pipe is 1 m above the ground. The injection point is at a depth of 5.0 m; therefore, $h_{gp} = 5.0 +$

1.0 = 6.0 m. The height of the groundwater to the injection point is $h_w = 3$ m. The required grout pressure can be estimated by the following formula:

$$p_g = \gamma_w(h_w + \Delta h_w) - \gamma_g h_{gp}$$
$$= 9.81 \times (3 + 9.4) - 11 \times 6 = 55.6 \text{ kPa}$$

6. Effect of Bingham fluid: Since the silicate solution is a Newtonian fluid, there is no yield strength. The effect of Bingham fluid does not exist.
7. Allowable grout pressure: It is necessary to check whether the required grout pressure exceeds the allowable grout pressure. Assume downstage grouting with primary holes. There is no surcharge on the site. The allowable grout pressure is

$$p_{ga} = 100(\alpha_p p + C_{gs}\beta_{gm}\lambda_{sc}z)$$
$$= 100 \times (2 \times 0 + 1 \times 0.8 \times 0.5 \times 5.0)$$
$$= 200 \text{ kPa} > p_g \qquad \text{(OK)}$$

8.3.6 Construction

Permeation Grouting The following procedure may be adopted for permeation grouting:

- Prepare grout in a batch before grouting or form grout by supplying grout components from several tanks during grouting.
- Drill a hole in soil to a desired depth.
- To form a grouted bulb, insert an injection pipe, pump the grout into the pipe and pressure it out from the bottom of the pipe.
- To form a grouted wall, place a sleeve port pipe (a steel or PVC pipe with holes or ports in intervals) first and then insert an injection pipe with two packers inside the sleeve port pipe. Pump grout and force it exit from one port at a time in the desired direction.
- After the preset grout pressure and/or volume is reached, raise or lower the injection pipe to the next interval depending on the installation (downstage or upstage) method until the whole required zone is grouted.

Compaction Grouting Compaction grouting is always done in stages, which can start from the top down (downstage) or from the bottom up (upstage) as shown in Figure 8.71. The upstage grouting includes (Brown, 2001):

- Drilling a hole to a desired depth.
- Placing a casing to the bottom of the hole.

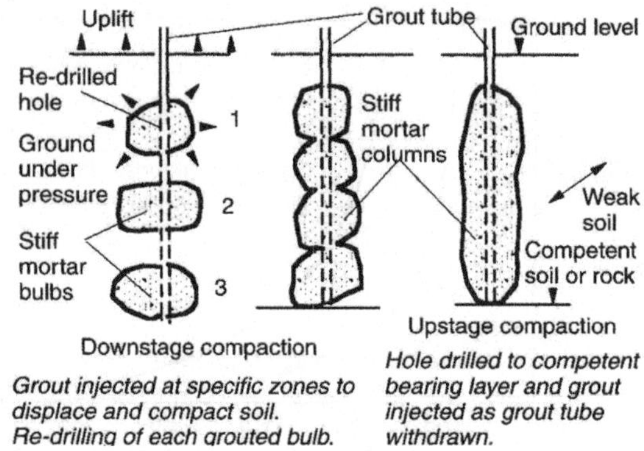

Figure 8.71 Compaction grouting procedure (Woodward, 2005, with permission from Taylor & Francis).

- Injecting the grout until a refusal is reached. The refusal can be the limited ground heave, the preset maximum pumping pressure, or the predetermined amount of grout.
- Raising the casing at a preset interval for each stage.
- Resuming injection of grout until reaching the refusal.
- Repeating the above steps until reaching the top zone.

The downstage grouting includes (Brown 2001):

- Drill a large hole (typically 75 mm in diameter) to the top of the zone to be densified (at least 1.2 m below the ground surface).
- Insert a casing (typically 50 mm in inner diameter) into the hole and fill the annular space outside the casing with rapid setting grout.
- Drill through the casing and deepen the hole in order of approximately 1–2 m as the stage length for the first stage.
- Inject grout until reaching the refusal pressure.
- After the previously injected grout is set, repeat the above steps until reaching the bottom of the zone to be injected.

To achieve better densification of soils, grouting is preferred to start from the perimeter to the center of an area to be improved.

Hydrofracture Grouting and Compensation Grouting The general construction procedure for hydrofracture grouting and compensation grouting are similar to those for the compaction grouting. However, it is important for hydrofracture grouting to carefully monitor ground movement and control the grout pressure. Compensation grouting should be strictly operated based on the timely monitoring of the ground movement and the displacements of existing structures to prevent any possible damages.

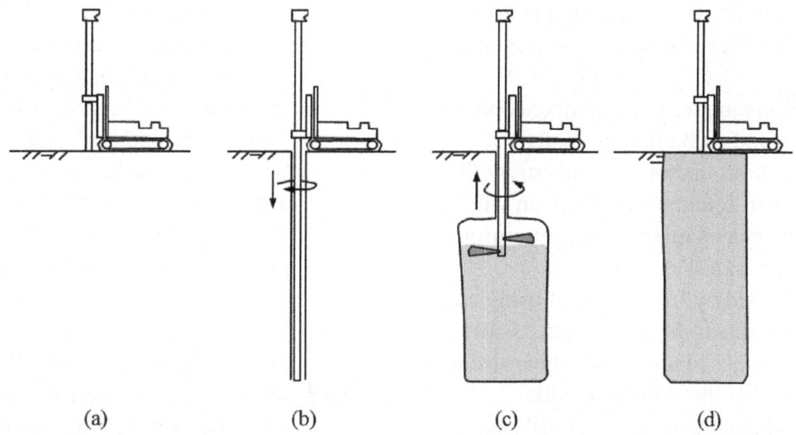

Figure 8.72 Jet grouting procedure: (a) positioning, (b) drilling, (c) jet grouting while withdrawing, and (d) complete.

Jet Grouting The procedure for jet grouting operation typically includes the following steps (Figure 8.72):

1. Position the drilling machine at a desired location for injection.
2. Drill into the ground to the desired depth. The tolerance of the drilling location should be less than 50 mm.
3. Insert the jet grouting pipe to the predrilled grout hole. Some machine installs the jet grouting pipe directly with an auger at the tip.
4. Once the pipe reaches the desired depth, start the jet grouting process by withdrawing the pipe and injecting grout until reaching a desired elevation. The operation should follow the specifications for injection pressure, injection rate, injection time, pipe withdrawal rate, and pipe rotation rate.
5. Clean up the injection pipe and other associated tools.
6. Move to the next jet grouting location.

8.3.7 Quality Control and Assurance

The general procedure for quality control and assurance are similar to that for deep mixing. Quality control for grouting may include the following steps:

- Evaluate the grouting record including the grout volume, injection pressure, rate of injection, time of injection.
- Inspect the integrity and uniformity of the grouted column or wall.
- Verify the dimensions of the grouted column or wall.
- Evaluate the strength of the grouted column.

Quality assurance includes the following field testing

- Core samples of grouted columns or walls to verify the strength and stiffness of the stabilized soil
- SPT and CPT
- Single column and composite foundation loading tests
- Cross-hole geophysical testing

PROBLEMS

8.1. How does deep mixing improve weak soils? How many types of deep mixing exist in practice?

8.2. What are the main differences for the installation of deep mixed columns by a wet method and the dry method?

8.3. What are the favorable soils for DM methods?

8.4. What are the typical patterns of deep mixed columns? Give an example of the favorable application for each pattern.

8.5. What is pozzolanic reaction and what are its products?

8.6. What is the difference between the composite column and the combined column technology?

8.7. Cement powder with a weight of 150 kg is mixed with 1.0 m^3 moist soil that has a unit weight of 18.1 kN/m^3 and a moisture content of 60%. Calculate the binder contents using three different definitions.

8.8. If the unconfined compressive strength of cement-stabilized soil with ordinary Portland cement at 7 day curing time is 1.2 MPa, estimate the unconfined compressive strengths of the same stabilized soil at 28 and 91 days.

8.9. List six main influence factors on the strength of soil–cement samples.

8.10. Explain why an increase of mix energy can increase the strength of the cement stabilized soil.

8.11. What are the main reasons why stabilized soils in the field often have lower strengths than those prepared in the laboratory?

8.12. What are typical failure modes of deep mixed columns under vertical loads and embankment loads?

8.13. Define the mobilization factor of soil strength in column-reinforced soft foundations. Explain why the mobilization factor for deep mixed column-reinforced soft foundations is often less than 1.0.

8.14. Given a soil is stabilized by lime–cement using a dry method and its undrained shear strength is 40 kPa, estimate the typical ranges of undrained shear strength and Young's modulus of the stabilized soil.

8.15. A natural soil has moisture content $w_0 = 40\%$ and permeability $k = 1.0 \times 10^{-7}$ m/s. The stabilized soil has moisture content $w = 35\%$ and unconfined compressive strength $q_u = 1.2$ MPa, what is the permeability of the stabilized soil?

8.16. A square rigid footing is to be constructed on soft clay on land that has an undrained shear strength of 15 kPa and is improved by deep mixed columns with an area replacement ratio of 30%. The diameter and length of deep mixed columns are 1 and 10 m, respectively. They are frictional columns. The field unconfined compressive strength of the column is 750 kPa. The average skin resistance between column and soil is 10 kPa. Calculate the ultimate bearing capacities of the deep mixed column and the deep mixed column-reinforced composite foundation.

8.17. A 3 m × 3 m wide rigid footing on a 10-m thick soft soil subjected to a column load of 1800 kN settles 200 mm. Deep mixed columns with an area replacement ratio of 0.15 and an unconfined compressive strength of 900 kPa are used to improve the soft soil (fully penetrated by columns). Poisson's ratio of soil is 0.3. Compute the maximum stress concentration ratio and the settlement of the soft foundation improved by DM columns.

8.18. A concrete raft with a dimension of 12 m × 12 m × 0.5 m thick is rested on a deep mixed column-reinforced soft foundation without embedment as shown in Problem Figure 3.18. The 4 × 4 rows of columns with a diameter of 1 m and a length of 14 m are installed in uniform soft clay with spacing of 3 m in a square pattern. The soft clay underlain by a firm soil layer is 25 m thick and has a saturated unit weight of 18 kN/m^3 and a drained modulus of 5 MPa and the groundwater table is at 2 m below the surface. The firm soil has a drained elastic modulus of 100 MPa. The columns have an unconfined compressive strength of 1 MPa and a saturated unit weight of 20 kNm^3. The Poisson ratios of the soft clay and the columns are 0.3 and 0.2, respectively. Estimate the settlement of the raft under a vertical load of 20 MN.

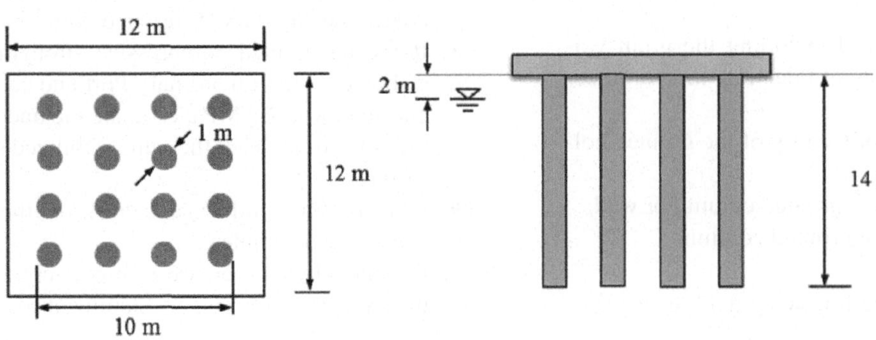

8.19. A deep mixed wall as shown below is used to retain soil during excavation (a square excavation pit of 20 × 20 × 6 m deep). The deep mixed wall has a unit weight of 19.5 kN/m^3 and the unconfined compressive strength of 1.2 MPa. The effective cohesion and effective friction angle of the wall is 250 kPa and 30°, respectively. Other dimensions and parameters are provided in the following figure. Evaluate the stability against the internal sliding and overturning and base heave.

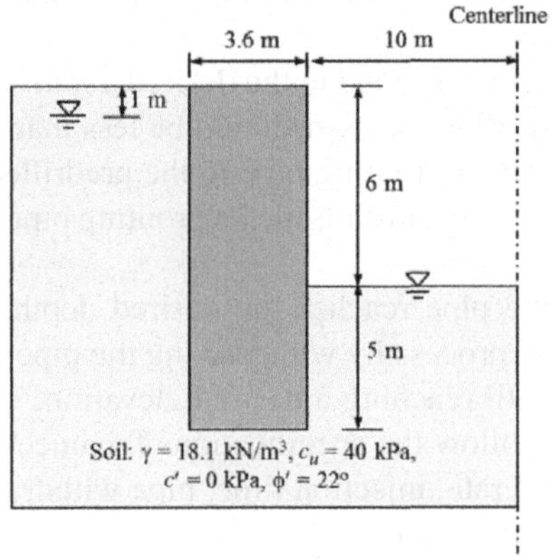

8.20. Five cored samples from deep mixed columns in the field are taken to the laboratory for unconfined compressive tests. The test results are provided below:

Sample Number	1	2	3	4	5
q_u **(MPa)**	1.1	0.8	1.0	0.9	1.2

The design unconfined compressive strength is 0.9 MPa. Evaluate whether the field column strength meets the design requirement.

8.21. What are five common types of grouting methods? What are favorable conditions for each method?

8.22. Can compaction grouting be used for fine-grained soils? Why?

8.23. What is the phenomenon of "bleeding" in grouting?

8.24. Describe how a Binghamian body behaves differently from a Newtonian fluid.

8.25. Compaction grouting with a grout pressure of 20 MPa injects stiff grout in the ground to densify cohesionless soil at 5 m below the ground surface. The soil bulk modulus at that depth is 200 MPa. What is percent of unit weight increase within the compacted zone?

8.26. A single pipe of jet grouting has a 3-mm-diameter nozzle. The grout pressure of the jet emitted from the nozzle is 30 MPa, calculate the pressures of the jet at 100 mm from the nozzle if the jet travels in air and in water, respectively.

8.27. A soil has a particle size corresponding to 15% passing, $D_{15} = 1$ mm. Determine the maximum grout particle size corresponding to 85% passing, D_{85}, if the soil is consistently groutable for cement grout.

8.28. Given the radius of grout penetration, $R = 0.2$ m, the radius of the injection source, $r_0 = 0.016$ m. The rate of grout injection, $Q_g = 1.0\,\text{m}^3/\text{h}$. The unit weight of the grout is 12 kN/m^3. The permeability of the sand to be improved is 0.001 m/s. The silicate solution is used for permeation grouting and its viscosity, $\mu_g = 5$ cP. Calculate the required grout hydraulic head.

8.29. Given the required grout hydraulic head is 10 m, the head of the grout pipe is 1.5 m above the ground. The injection point is at a depth of 4 m. The groundwater is 2 m above the injection point. Calculate the required grout pressure.

8.30. During downstage grouting with a primary hole, a surcharge of 100 kPa is applied on the site. Calculate the allowable grout pressure when the depth of the injection point is 6 m.

8.31. A Bingham fluid has a yield strength, $\tau_s = 5$ Pa. The soil has permeability, $k = 2 \times 10^{-5}$ m/s and porosity, $n_p = 35\%$. The unit weight of the grout is 12 kN/m^3. Calculate the minimum pressure gradient needed to overcome the yield strength.

8.32. A site consists of a fine sand ($D_{15} = 0.05$ mm), which is located at a depth from 3 to 5 m. The groundwater table is at the depth of 1 m. The permeability of this sand is 0.0005 m/s. This fine sand has liquefaction potential; therefore, it should be improved. Permeation grouting is selected for this improvement. Design the permeation grouting.

8.33. A site consists of uniform medium sand with 5% fine content ($D_{50} = 1.4$ mm). The thickness of this sand layer is 12 m from the ground surface. The minimal and maximum void ratio values are 0.45 and 0.98, respectively. The groundwater table is at 1.5 m from the ground surface. The unit weights of the sand above and below the groundwater table are 18 kN/m^3 and 19 kN/m^3, respectively. The SPT N_{60} value at a depth of 6.0 m is 5. This site is located in a seismically active area. The potential earthquake magnitude can be 7.0. The peak ground acceleration can reach as high as 0.3g. Compaction grouting is proposed to densify this possible liquefiable soil (please verify whether the sand is liquefiable first) to prevent potential liquefaction. Assume the average diameter of grout bulbs can reach 0.9 m and the heave of the ground is limited to be less than 25 mm during the installation. Determine the spacing of compaction grouting in a square pattern.

REFERENCES

AFTES (1991). "Recommendations on grouting for underground works." *Tunnel. Underground Space Technol.*, 6(4): 383–461.

Adler, W.F. (1979). "The mechanics of liquid impact." In *Treatise on Material Science and Technology*, vol. 16, C.M. Preece (ed.), New York, Academic Press, 127–183.

Ahnberg, H. (2003). "Measured permeabilities in stabilized Swedish soils." Proceedings of the 3rd International Conference on Grouting and Ground Treatment, *Geotechnical Special Publication* (120):1, 622–623.

Ahnberg, H., and Johansson, S.E. (2005). "Increase in strength with time in soils stabilized with different types of binder in relation to the type and amount of reaction products." International Conference on Deep Mixing Best Practice and Recent Advances, Deep Mixing, May 23–25, Stockholm, Sweden, 195–202.

Al-Alusi, H.R. (1997). "Compaction grouting: From practice to theory." *Grouting: Compaction, Remediation and Testing*, ASCE, GSP No.66, 43–53.

Babasaki, R., Terashi, M., Maekawa, A., Kawamura, M., and Fukazawa, E. (1996). "Factors influencing the strength of improved soil." *Japanese Geotechnical Society Technical Committee Report, Proc. of the 2nd International Conference on Ground Improvement Geosystems*, vol. 2, 913–918.

Bhandari, A. and Han, J. (2009). "Evaluation of high-capacity composite spun piles." *J. Transport. Res. Board*, 2116: 53–61

Broms, B.B. (1999). "Can Lime/Cement Columns Be Used in Singapore and Southeast Asia?" 3rd GRC Lecture, Nov. 19, Singapore, Nanyang Technological Univ. and NTU-PWD Geotechnical Research Centre.

Brown, R.W. (2001). *Practical Foundation Engineering Handbook* (2nd ed.). McGraw-Hill, New York.

Bruce, D.A., Littlejohn, G.S., and Naudts, A.M.C. (1997). "Grouting materials for ground treatment: A practitioner's guide." In *Grouting: Compaction, Remediation and Testing*, C. Vipulanandan (ed.). ASCE Geotechnical Special Publication (GSP) No. 66, 306–334.

Building Center of Japan (1997). *Design and Quality Control Guideline for Building Foundation Treatment*. Building Center of Japan, Tokyo (in Japanese).

Chai, J.C., Miura, N., Kirekawa, T., and Hino, T. (2010). "Settlement prediction for soft ground improved by columns." *Ground Improvement* 163(G12), 109–119.

Chai, J.C., Shen, S.L., Miura, N., and Bergado, D.T. (2001). "Simple method of modeling PVD improved subsoil." *Journal of Geotechnical and Geoenvironmental Engineering* 127(11), 965–972.

Chew, S.H., Kamruzzaman, A.H.M., and Lee, F.H. (2004). "Physicochemical and engineering behavior of cement treated clays." *J. Geotech. Geoenviron. Eng.*, 130(7): 696–706.

China Academy of Building Research (2000). *Ground Improvement Technical Code.* Beijing, The China Planning Press.

Chinese Ground Improvement Manual Committee (1988). *Ground Improvement Manual.* Beijing, China Building Industry Press.

Coastal Development Institute of Technology (1999). *Technical Manual for Deep Mixing with Reference to Marine Works*, Tokyo (in Japanese).

Coastal Development Institute of Technology (2002). *The Deep Mixing Method–Principle, Design and Construction.* Leiden, Netherlands, CRC Press.

Coomber, D.B. (1985). "Groundwater control by jet grouting." *Proc. 21st Reg. Conf. on Eng. Group of Geological Society*, September 15–19, Sheffield, England, 485–498.

De Paoli, B., Bosco, B., Granata, R., and Bruce, D.A. (1992). "Fundamental observations on cement based grouts (1): Traditional materials." *Proc. ASCE Conf. on Grouting, Soil Improvement and Geosynthetics*, New Orleans, 1, 474–485.

Elias, V., Welsh, J., Warren, J., Lukas, R., Collin, J.G., and Berg, R.R. (2006). *Ground Improvement Methods.* FHWA NHI-06-020, Federal Highway Administration, Washington, DC.

Enami, A., Yoshida, M., Hibino, S., Takahashi, M., and Akitani, K. (1985). "In situ measurement of temperature in soil cement columns and influence of curing temperature on unconfined compressive strength of soil cement." *Proc. of the 20th Annual Conference of the Japanese Society of Soil Mechanics and Foundation Engineering*, 1737–1740 (in Japanese).

EuroSoilStab (2001). *Development of Design and Construction Methods to Stabilise Soft Organic Soils.* Design guide soft soil stabilization. CT97-0351, EC Project No. BE 96-3177, Industrial & Materials Technologies Programme (BriteEuRam III), European Commission.

Filz, G.M., and Navin, M.P. (2006). *Stability of Column-supported Embankments.* Virginia Transportation Research Council, Charlottesville, VA.

Flora, A., Modon, G., Lirer, S., and Croce, P. (2013). "The diameter of single, double and triple fluid jet grouting columns: Prediction method and field trial results." *Geotechnique* 63(11), 934–994.

Futaki, M. and Tamura M. (2002). "The quality control in deep mixing method for the building foundation ground in Japan." *Proc. Tokyo Workshop 2002 on Deep Mixing*, Tokyo, 139–149.

Han, J. (2012). "Recent advances in column technologies to improve soft soils." In *Proceedings of International Conference on Ground Improvement and Ground Control*, B. Indraratna, C. Rujikiatkamjorn, and J. Vinod (eds.). Wollongong, Australia, 30 October to 2 November, Vol. 1, Research Publishing, 99–113.

Han, J., Chen, J.F., Hong, Z.S., and Shen, S.L. (2010). "Mitigation of levee failure using deep mixed columns and geosynthetics." *GeoMech. GeoEng.: Int. J.*, 5(1): 49–55.

Han, J. and Gabr, M.A. (2002). "Numerical analysis of geosynthetic-reinforced and pile-supported earth platforms over soft soil." *J. of Geotech. and Geoenviron. Eng.* 128(1), 44–53.

Han, J., Oztoprak, S., Parsons, R.L., and Huang, J. (2007). "Numerical analysis of foundation columns to support widening of embankments." *Comput. Geotech.*, 34(6): 435–448.

Han, J., Parsons, R.J., Sheth, A.R., and Huang, J. (2005). "Factors of safety against deep-seated failure of embankments over deep mixed columns." Deep Mixing 2005 Conference, Sweden, Vol. 1.2, May 23–25, 231–236.

Han, J., Yang, X.M., Chen, J.F., and Porbaha, A. (2009). "Settlement calculation of deep mixed foundations." *Proceedings of International Symposium on Deep Mixing and Admixture Stabilization*, Okinawa, Japan, May 19–21.

Han, J., Zhou, H. T., and Ye, F. (2002). "State of practice review of deep soil mixing techniques in China." *J. Transport. Res. Board*, No. 1808, Soil Mechanics 2002, Washington, DC Transportation Research Board of the National Academies, 49–57.

Herndon, J. and Lenahan, T. (1976). *Grouting in Soils.* Vol. 1: A State-of-the-art Report, FHWA-RD-76-26, FHWA, Washington DC.

Holm, G. (2005). "Deep mixing properties and applications." Powerpoint presentation, STAR Conference, Cambridge, UK, April 12.

Horikoshi, K. and Randolph, M.F. (1999). "Estimation of overall settlement of piled rafts." *Soils and Foundations* 39(2), 59–68.

Houlsby, A.C. (1990). "Construction and design of cement grouting." In *A Guide to Grouting in Rock Foundations.* Wiley, New York.

Huang, J., Han. J., and Oztoprak, S. (2009). "Coupled mechanical and hydraulic modeling of geosynthetic-reinforced column-supported embankments." *J. Geotech. Geoenviron. Eng.*, 135(8): 1011–1021.

Ikegami, M., Ichiba. T., Ohishi, K. and Terashi, M. (2005). "Long-term properties of cement treated soil 20 years after construction." *Proc 16th ICSMGE*, Tokyo.

Jacobson, J.R., Filz, G.M., and Mitchell, J.K. (2005). "Factors affecting strength of lime-cement columns based on a laboratory study of three organic soils." *Proceedings of Deep Mixing* 2005, 87–94.

Jamsawang, P., Bergado, D.T., Bhandari, A., and Voottipruex, P. (2008). "Investigation and simulation of behavior of stiffened deep cement mixing (SDCM) piles." *Int. J. Geotech. Eng.*, 2(3): 229–246.

Japanese Cement Deep Mixing Method Association (1999). *Cement Deep Mixing Method (CDM), Design and Construction Manual* (in Japanese).

Jiang, Y., Han, J., and Zheng, G. (2013). "Numerical analysis of consolidation of soft soils fully-penetrated by deep-mixed columns." *KSCE J. Civil Eng.*, 17(1): 96–105.

Jiang, Y., Han, J., and Zheng, G. (2014). "Influence of column plasticity on degree of consolidation of soft foundations improved by deep mixed columns." *Geomech. Eng.* 6(2), 179–194.

Karol, R.H. (1983). *Chemical Grouting.* Marcel Dekker, New York.

Kawasaki, T., Niina, A., Saitoh, S., and Babasaki, R. (1978). *Studies on Engineering Characteristics of Cement-based Stabilized Soil.* Takenaka Technical Research Report, vol. 19, 144–165 (in Japanese).

Kitazume, M. (2008). *Stability of Group Column Type DM Improved Ground under Embankment Loading Behavior of Sheet Pile Quay Wall.* Report of the Port and Airport Research Institute, 47(1).

Kitazume, M., Okano, K., and Miyajima, S. (2000). "Centrifuge model tests on failure envelope of column type mixing method improved ground." *Soils and Foundations* 40(4), 43–55.

Kitazume, M. and Terashi, M. (2013). *The Deep Mixing Method*. CRC Press, Boca Raton, FL.

Larsson, S. (2005). "State of practice report — Execution, monitoring and quality control." *Deep Mixing*, 2005: 732–786.

Lin, K.Q. and Wong, I.H. (1999). "Use of deep cement mixing to reduce settlements at bridge approaches." *J. Geotech. Geoenviron. Eng.*, 125(4): 309–320.

Littlejohn, G.S. (1982). "Design of cement based grouts." *Proc. ASCE Conf. on Grouting in Geotechnical Engineering* (February 10–12), New Orleans, LA, 35–48.

Littlejohn, G.S. (1985). "Chemical grouting." *Ground Engineering* 18(2), 13–16; (3), 23–28; (4), 29–34.

Liu, S.-Y., Du, Y.-J., Li, Y.-L., and Puppala, A.J. (2012). "Field investigations on performance of T-shaped Deep Mixed (TDM) soil-cement columns supported embankment over soft ground." *J. Geotech. Geoenviron. Eng.*, 138(6), 718–727.

Liu, S.Y., Han, J., Zhang, D.W., and Hong, Z.S. (2008). "A new DJM-PVD combined method for soft ground improvement." *Geosynth. Int. J.*, 15(1): 43–54.

Lombardi, G. (1985). "The role of cohesion in cement grouting of rock." *15th ICOLD*, Lausanne, Q58, R13.

Lorenzo, G.A., and Bergado, D.T. (2003). "New consolidation equation for soil-cement piles improved ground." *Canadian Geotechnical Journal* 40(2), 265–275.

Maag, E. (1938). "Ueber die Vergestigung und Dichtung das Baugrundes (Injektionen)." Erdbaukers der ETH.

Mamlouk, M.S. and Zaniewski, J.P. (2011). *Materials for Civil and Construction Engineers*, 3rd ed. Prentice Hall, Upper Saddle River, NJ.

Matsuo, O. (2002). "Determination of design parameters for deep mixing." Proc. Tokyo Workshop 2002 on Deep Mixing. Port and Airport Research Institute, 75–79.

Miki, G. and Nakanishi, W. (1984). "Technical progress of the jet grouting method and its newest type." *Proc. Int. Conf. on In Situ Soil and Rock Reinforcement*, Paris, Oct. 9–11, 195–200.

Mitchell, J.K. and Katti R.K. (1981). "Soil Improvement—State of the Art Report." *Proc. 10th ICSMFE*, Stockholm, 4, 509–565.

Nguyen, T.V., Rayamajhi, S.M., and Boulanger, R.W. (2013). "Design of DSM grids for liquefaction remediation." *J. Geotech. Geoenviron. Eng.*, 139(11), 1923–1933.

Niina, A., Saitoh, S., Babasaki, R., Tsutsumi, I., and Kawasaki, T. (1977). "Study on DMM using cement hardening agent: Part I." In *Proc. of the 12th Japanese Society Conference on Soil Mechanics and Foundation Engineering*, 1325–1328 (in Japanese).

Nishida, K. and Miura, N. (1996). "Study of the mixing energy on the deep mixing improvement of sensitive clay by DJM." (in Japanese). *J. Construct. Manage. Eng.*, JSCE, VI-27(516), 165–172.

Pongsivasathit, S., Chai, J.C., and Ding, W.Q. (2013). "Consolidation settlement of floating-column-improved soft clayey deposit." *Ground Improvement* 166(GI1), 44–58.

Public Works Research Center (1999). Technical Manual on Deep Mixing Method for On-land Works. (In Japanese).

Puppala, A.J., Madhyannapu, R.S., Nazarian, S., Yuan, D., and Hoyos, L. (2008). *Deep Soil Mixing Technology for Mitigation of Pavement Roughness*. FHWA/TX-08/0-5179-1, FHWA, Washington, DC.

Raffle, J.F. and Greenwood, D.A. (1961). "The relationship between the rheological characteristics of grouts and their capacity to permeate soils."In *Proc. 5th Int. Conf. on Soil Mechanics and Foundation Engineering*, vol. 2, 789–793.

Rajaratnam, N. (1976). *Turbulent Jets*. Elsevier, Amsterdam, Netherlands.

Saitoh, S., Suzuki, Y., and Shirai, K. (1985). "Hardening of soil improved by the deep mixing method." *Proc. of the 11th Int. Conf. on Soil Mechanics and Foundation Engineering*, 4, 2475–2478.

Shen, S.L., Han, J., and Du, Y.J. (2008). "Deep mixing induced property changes in sensitive marine clays." *J. Geotech. Geoenviron. Eng.*, 134(6): 845–854.

Shen, S.L., Han, J., and Miura, N. (2004). "Laboratory evaluation of mixing energy consumption and its influence on soil-cement strength." *J. Transport. Res. Board*, No. 1868, 23–30.

Shen, S.-L., Wang, Z.-F., Yang, J., and Ho, C.-E. (2013). "Generalized approach for prediction of jet grout column diameter." *Journal of Geotech. Geoenviron. Eng.* (139)12: 2060–2069.

Soga, K., Komiya, K., Akagi, H., Jafari, M.R., and Bolton, M.D. (2001). "Soil consolidation associated with grouting during shield tunnelling in soft clayey ground." *Géotechnique* 51(10): 835–846.

Terashi, M. (2003). "The State of practice in deep mixing methods." *Grouting and Ground Treatment, Grouting and Ground Treatment*, ASCE GSP 120, L. F. Johnsen, D. A. Bruce, and M. J. Byle (eds.), Reston, VA, 25–49.

Terashi, M., Okumura, T., and Mitsumoto, T. (1977). "Fundamental properties of lime-treated soils." *Report of the Port and Harbor Research Institute* 16(1), 3–28 (in Japanese).

Terzaghi, K. (1943). *Theoretical Soil Mechanics*. Wiley, New York.

Terzaghi, K., Peck, R.B., and Mesri, G. (1996). *Soil Mechanics in Engineering Practice*, 3rd ed. Wiley, New York.

U.S. Department of the Army (1997). *Military Soils Engineering: Field Manual C1- FM 5-410*, Washington DC.

Warner, J. (2004). *Practical Handbook of Grouting Soil, Rock, and Structures*. Wiley, Hoboken, NJ.

Woodward, J. (2005). *Introduction to Geotechnical Processes*, CRC Press, Boca Raton, FL.

Xanthakos, P.P., Abramson, L.W., and Bruce, D.A. (1994). *Ground Control and Improvement*. John Wiley & Sons, Inc., New York.

Xu, C., Ye, G.-B. Jiang, Z.-S., and Zhou, Q.-Z. (2006). Research on mechanism of combined improvement of soft soils based on field monitoring. *Chinese J. of Geotech. Eng.* 28(7), 918–921 (in Chinese).

Ye, S. L., Cai, W. M., and Han, J. (1992). "Soil improvement by deep mixing piles and stone columns." *Proc. of US-China Workshop on Cooperative Research in Geotechnical Engineering*, Shanghai, P.R. China, 57–73.

Ye, S.L., Han, J., and Ye, G.B. (1994). *Ground Improvement and Underpinning Technologies*, 2nd ed., Beijing China Building Industry Press.

Zheng, G., Jiang, Y., and Han, J. (2011). Performance of cement-fly ash-gravel pile-supported high-speed railway embankments over soft marine clay. *J. of Marine Georesources & Geotech.* 29(2), 145–161.

Zheng, J.-J., Liu, Y., and Xu, Z.-J. (2009b). Reliability-based design applied to multi-column composite foundation. Advances in Ground Improvement, ASCE GSP No. 188, 83–91.

Zhu, G. and Yin, J.H. (1999). "Consolidation of double soil layers under depth-dependent ramp load." *Geotechnique* 49(3), 415–421.

CHAPTER 9

In Situ Ground Reinforcement

9.1 INTRODUCTION

In situ ground reinforcement is a technique to stabilize existing unstable ground due to the change of geotechnical conditions by nature and/or human activities. For example, intensive precipitation may destabilize existing slopes and induce landslides due to increased soil weight, reduced soil strength, and water seepage. Scour of a slope toe in a river may destabilize the river bank. Excavation in the ground for wall or foundation construction (i.e., a cut situation) induces unbalanced forces and results in ground movement and even failure if not properly designed and protected. Underground tunneling may also induce ground movement and even collapse. Ground anchors, soil nails, micropiles, and slope stabilizing piles have been used as in situ ground reinforcement techniques to mitigate the preceding problems, as shown in Figure 9.1.

Ground anchors are also called tiebacks and rely on long prestressed steel tendons bonded in a stable mass at a greater depth and distance to provide tensile resistance to the unstable mass near the wall or slope surface. The tensile force induced by the prestressing of the steel tendons provides additional normal stresses to a critical slip surface so that the shear strength along the critical slip surface is increased thus resulting in a higher factor of safety against sliding. Soil nailing relies on shorter but closely spaced, passive structural inclusions to stabilize existing unstable ground. The tension in soil nails are mobilized during the soil movement. Therefore, soil movement is necessary for soil nails to be effective, while the tension in the ground anchors is mobilized mostly by the prestressing and the limited soil movement adds more tension to the anchors. Another major difference between ground anchors and soil nails is the bonded length between the steel tendon and the surrounding soil. Ground anchors have the bond lengths in the stable mass while soil nails have the bonded lengths in the stable and unstable masses. Both ground anchors and soil nails provide tensile resistance to the unstable ground. Micropiles, especially reticulated micropiles, rely on the axial and lateral capacities of the piles to provide bending and/or rotational resistance to the unstable soil behind the micropile wall due to lateral earth pressure. The design and construction of micropiles can be found in Sabatini et al. (2005) and will not be further discussed in this book. Slope stabilizing piles, depending on their stiffness and shear strength, may serve as shear keys (especially for granular columns, which were discussed in Chapter 5) or laterally loaded piles (especially for rigid piles or columns) to resist slope movement. When rigid piles are used to stabilize slopes, they are often designed as laterally loaded piles as in a deep foundation solution (e.g., Ito and Matsui, 1975); therefore, they will not be further discussed in this book.

9.2 GROUND ANCHORS

9.2.1 Introduction

Basic Concept Ground anchors are cement-grouted prestressed tendons installed in in situ soil or rock by transmitting applied tensile loads into ground to stabilize earth retaining structures or to provide uplift resistance to structures. Ground anchors are also referred to as "tiebacks" in practice. Figure 9.2 shows the basic components of a typical ground anchor, which includes three parts:

- Anchorage set, which consists of an anchor head, a bearing plate, and a trumpet
- Unbonded prestressing steel tendon
- Bonded steel tendon with grout

The anchorage component is to transmit the prestressing force from the prestressing steel to the ground surface or the supported structure. The unbonded steel is prestressed and can have elastic elongation and transfer the resistance from the bond length to a structure. A smooth plastic sleeve as a bond breaker is placed over the steel tendon to separate the prestressing steel from the surrounding grout. The bonded steel with grout can provide a tensile load into the ground; therefore, the bond length should be behind a critical slip surface. The term "tendon" refers to the prestressing steel strands or bars. Sheaths are smooth or corrugated pipes or tubes to protect the prestressing steel in the unbonded length from corrosion. Centralizers are used to ensure the steel strands or bars in the center so that there is enough grout around them. Spacers are used to separate the strands or bars so that they are properly bonded with grout. Portland cement is commonly used as a hardening agent for the grout.

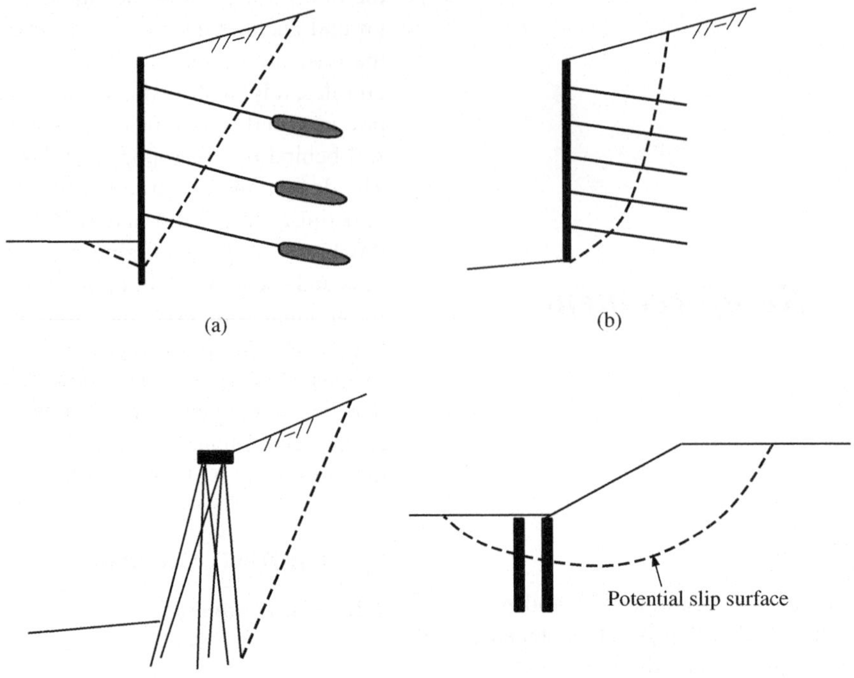

Figure 9.1 Types of in situ ground reinforcement: (a) ground anchor, (b) soil nailing, (c) micropiles, and (d) slope stabilizing piles.

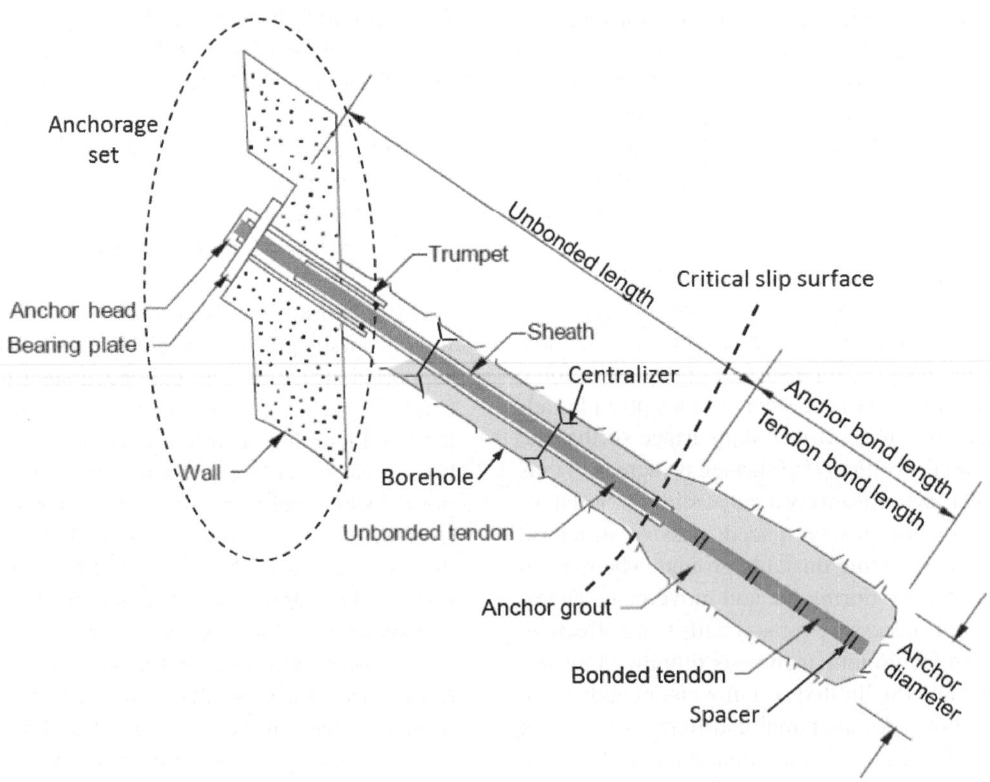

Figure 9.2 Components of a ground anchor (modified from Sabatini et al., 1999).

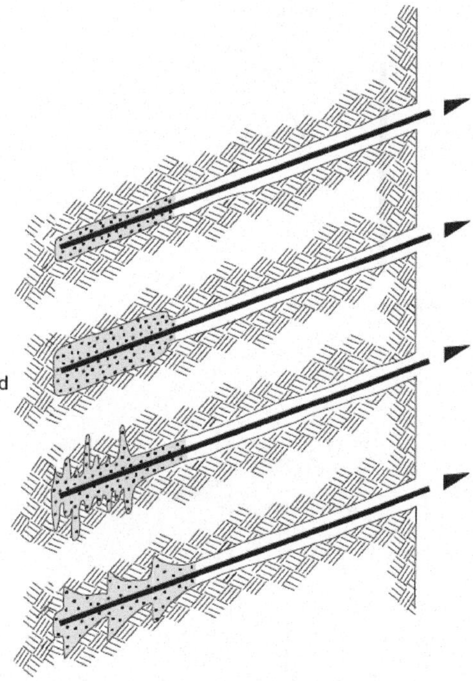

Figure 9.3 Types of ground anchors (Sabatini et al., 1999).

Most of ground anchors are subjected to tension; however, some ground anchors are designed to carry compressive loads. For those anchors, part or a complete length of the steel tendon is bonded with grout. An end plate or the bonded portion with grout provides the reaction for compressive loads. Since compression anchors are not that commonly used as tension anchors, they will not be further discussed in this chapter.

There are four types of ground anchors commonly used in practice as shown in Figure 9.3:

- Straight shaft gravity-grouted ground anchors
- Straight shaft pressure-grouted ground anchors
- Postgrouted ground anchors
- Under-reamed anchors

The details of the installation procedure for each type of ground anchor may be different and are discussed in Sabatini et al. (1999); however, the general procedure is the same, which includes the following:

- Drill a hole.
- Insert a steel tendon.
- Grout the hole within the bond length.
- Install the anchorage assembly.
- Prestress the steel tendon.

Suitability Ground anchors are suitable for a variety of geotechnical conditions. They can be used in situ soils, rocks, or other geomaterials. Different techniques may be used to install ground anchors in different geomaterials, mostly related to drilling and stability of the hole. Ground anchors may experience excessive creep deformations when they are installed in organic soil or soils with high plasticity. Caution should be exercised when ground anchors are used under such conditions.

Applications Ground anchors have been used permanently or temporarily in anchored systems. Permanent ground anchors are typically designed for a service life of 75–100 years. Temporary anchored systems are mostly used for earth support before permanent structures are installed. The service life of temporary anchored systems depends on project needs, but commonly ranges from 18 to 36 months.

Ground anchors are commonly used to provide lateral earth support during excavations, stabilize unstable slopes, and provide uplift resistance to foundations below a groundwater table. In addition, ground anchors can provide resistance to overturning, sliding, and earthquake loadings. Typical ground anchored systems include:

- Flexible anchored walls [Figure 9.4(a)]
- Ground anchor-stabilized slopes [Figure 9.4(b)]
- Structures by tie-down anchors [Figures 9.4(c) and (d)].

For excavations, ground anchors are mostly used with soldier beams and lagging walls but also used with continuous walls, such as sheet-pile walls.

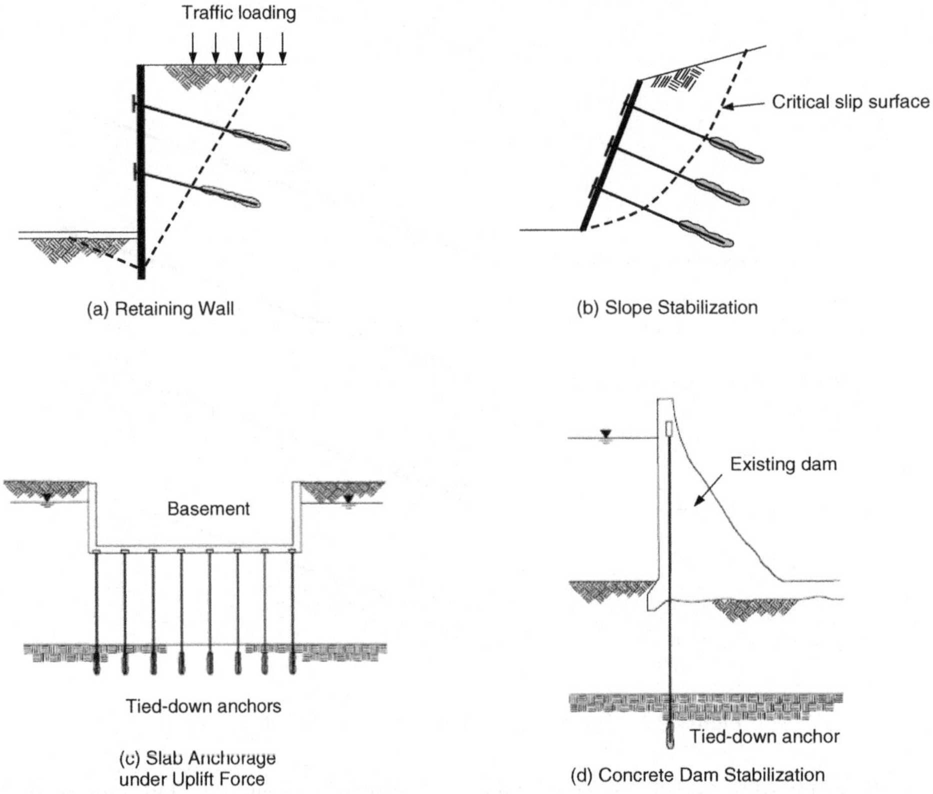

Figure 9.4 Applications of ground anchors and anchored systems: (a) highway retaining wall, (b) slope stabilization, (c) uplift slab, and (d) concrete dam stabilization (modified from Sabatini et al., 1999).

Advantages and Limitations Ground anchored systems have the following advantages over conventional earth retaining systems, such as gravity walls (Sabatini et al., 1999):

- No obstruction for workspace during excavations
- Tolerable to large lateral earth pressure without a significant increase of wall cross sections
- Used as temporary excavation support as well as a permanent structure
- No need for select backfill
- No need for deep foundation support
- Fast construction
- Less surface right-of-way issue

However, ground anchors may be limited by objects in the ground (such as underground tunnels and utility lines) and the ground right-of-way space. They may not work well in soft soils due to large and creep deformations. Difficulties may arise in constructing watertight connections at the anchor-structural slab interface below the groundwater table.

9.2.2 Principles

Load Transfer The contributions of ground anchors used in anchored wall systems during excavations are different from those of ground anchors used to stabilize unstable slopes and landslides.

When soil is excavated from one side of the wall, it creates an unbalanced force. The unbalanced force comes from lateral earth pressure on one side but no pressure on another side. The unbalanced force tends to induce soil movement. The soil movement is restrained by the wall facing element, such as sheet piles or lagging with soldier piles, which are fixed by prestressed steel tendons. This restraint becomes effective when the wall facing element is connected to the prestressed steel tendons by the anchorage assembly of the ground anchor, which provides tensile resistance to the wall facing element. At the same time, the unbonded steel tendon is subjected to a tensile force. With an increase of the unsupported wall height, the required tensile force in the prestressed steel tendon increases from T_1, T_2, to T_3 as show in Figure 9.5, and it extends from the top of the bond length to the end of the bond length. At the same time, the bond

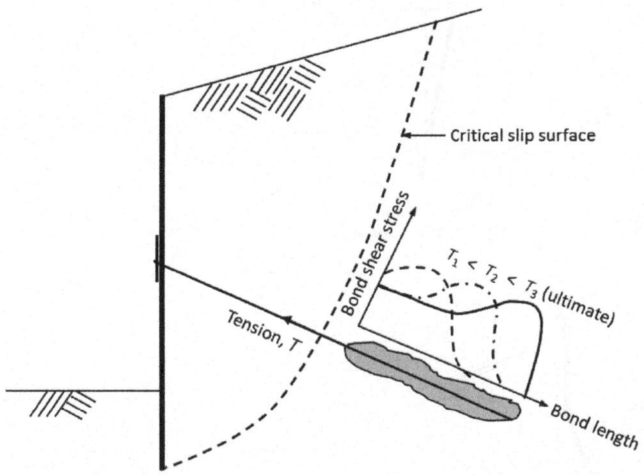

Figure 9.5 Load transfer in the anchor under tension.

shear strength between the grout and the surrounding soil within the bond length is mobilized. The bond shear strength is first fully mobilized near the top of the bond length and may decrease to a residual strength due to stress softening. As a result, more load is transferred to the next portion of the bond length. This load transfer process can continue toward the end of the bond length with an increase of the prestress applied. When the shear strength at the end of the bond length is full mobilized, the anchor approaches failure and no more load can be carried by the anchor.

When prestressed ground anchors are used to stabilize unstable slopes and landslides, they can provide thrusts to the potential slip surface and increase the normal stress on the potential slip surface. As a result, the shear strength is increased and the stability of the slopes is enhanced. Limit equilibrium methods for slope stability are used for this analysis.

Possible Failure Modes Figure 9.6 shows the possible failure modes of an anchored wall system, which include internal, facing, and external failure modes. Internal failures happen when the anchor ruptures or is pulled out. When the steel tendon is too weak, it will rupture as shown in Figure 9.6(a). The pullout of the steel tendon may occur at the interface between the grout and the surrounding soil or rock (Figure 9.6[b]) or between the steel tendon and the grout (Figure 9.6[c]) due to short bond length and/or low bond strength.

The facing failures may happen: (1) when the wall facing (such as sheet piles) does not have sufficient bending strength and/or stiffness (Figure 9.6[d]), (2) the wall facing does not have sufficient embedment depth so that there is a passive failure (Figure 9.6[e]), (3) the unanchored wall facing portion on the top fails (Figure 9.6[f]), and (4) the wall facing has an axial bearing or penetration failure (Figure 9.6[g]).

The external failures may happen as the anchored mass acts as a rigid body. The possible external failures include overturning (Figure 9.6[h]), sliding (Figure 9.6[i]), and global or deep-seated rotational failure (Figure 9.6[j]).

Internal Stability of Anchored Wall In a typical anchored wall, there are active, passive, and stable zones as shown in Figure 9.7. The active zone is right behind the wall facing but in front of the stable zone, while the passive zone is right in front of the embedded wall facing. The active zone and the stable zone are divided by a critical failure surface. The unbonded length of the anchor is mostly located within the active zone. The entire bond length of the anchor should be within the stable zone to provide tensile resistance to the active zone. The passive zone provides toe resistance to the wall. When there is sufficient toe resistance, the potential failure surface develops from the base of the excavation. When the toe resistance is low (i.e., there is a potential passive failure), however, the potential failure surface extends down to the effective embedment depth (D_a, to be determined based on laterally loaded piles) or the whole embedment depth and toward the bond length. The bond length should be beyond the potential failure surface due to insufficient toe resistance. Typically, anchors are inclined from the horizontal direction below.

Lateral Earth Pressure Terzaghi and Peck (1967) and Peck (1969) developed the apparent earth pressure diagrams as shown in Figure 9.8, based on the back-calculation from field measurements of strut loads in internally braced excavations. In this development, they assumed the following conditions:

- Large excavation with an excavation depth greater than 6 m
- Large wall movement to fully mobilize soil shear strength
- Groundwater table below the base of the excavation and no water pressure considered
- Homogeneous soil under a drained condition for sands and an undrained condition for clays
- The loading diagrams corresponding to the wall portion above the base of the excavation

When the walls are constructed in sands, the lateral earth pressure is uniform and the maximum lateral pressure is $0.65K_a\gamma H$, where K_a is the coefficient of lateral active earth pressure, γ is the unit weight of soil, and H is the excavation depth. When the walls are constructed in clays, the lateral earth pressure has a trapezoidal distribution. However, the distributions in the stiff-hard fissured clays and soft to medium clays are slightly different. The main difference exists at the bottom portion of the wall. The stability number defined below is used to divide the soil type between the

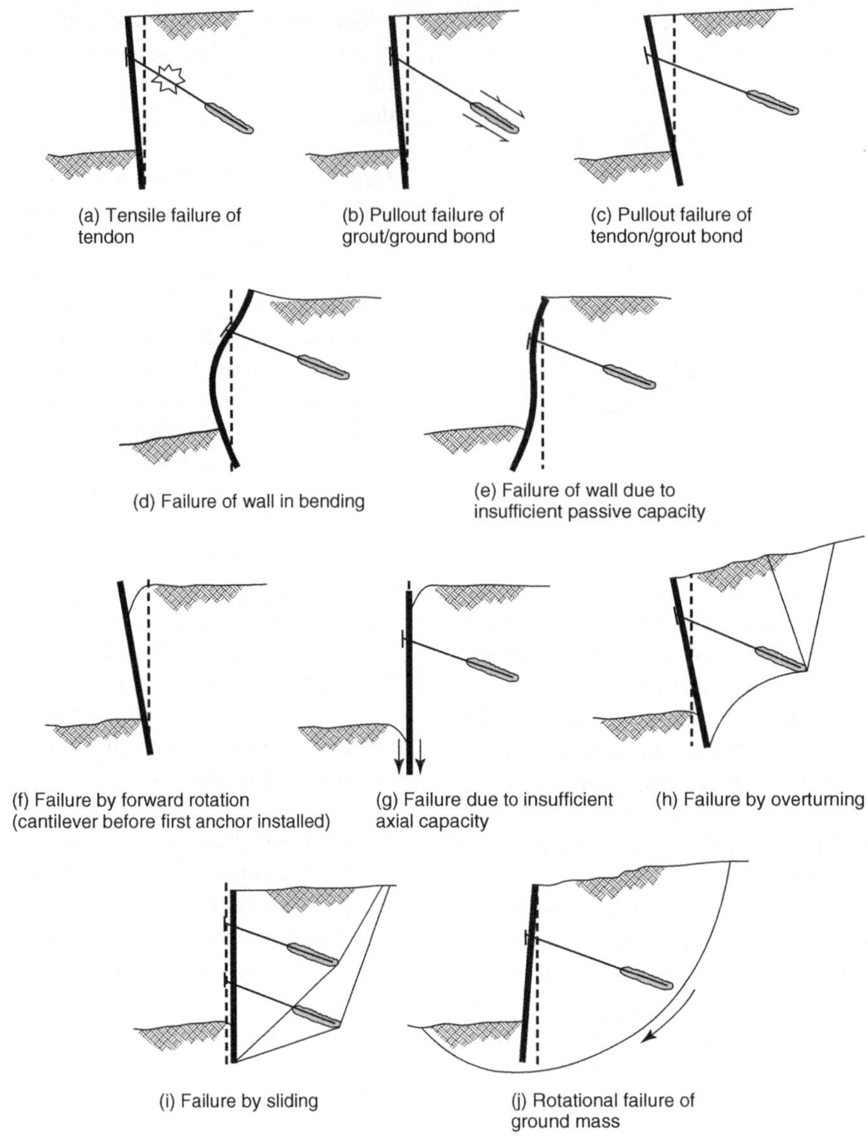

Figure 9.6 Possible failure modes (Sabatini et al., 1999).

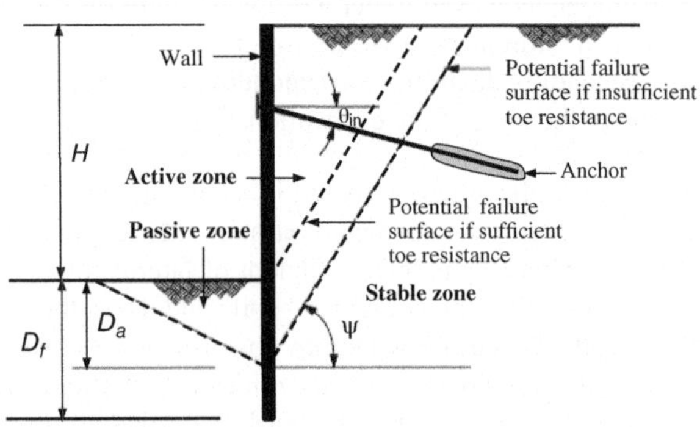

Figure 9.7 Internal stability of an anchored wall.

stiff-hard fissured clay and the soft to medium clay:

$$N_{sn} = \frac{\gamma H}{c_u} \tag{9.1}$$

where c_u = undrained shear strength of clay.

When $S_{sn} < 4$, the clay is considered a stiff-hard fissured clay. The maximum lateral pressure in the stiff-hard fissured clay ranges from $0.2\gamma H$ to $0.4\gamma H$. However, $N_{sn} \geq 4$, the clay is considered as a soft to medium clay. When the wall is in the soft to medium clay, the coefficient of lateral earth pressure

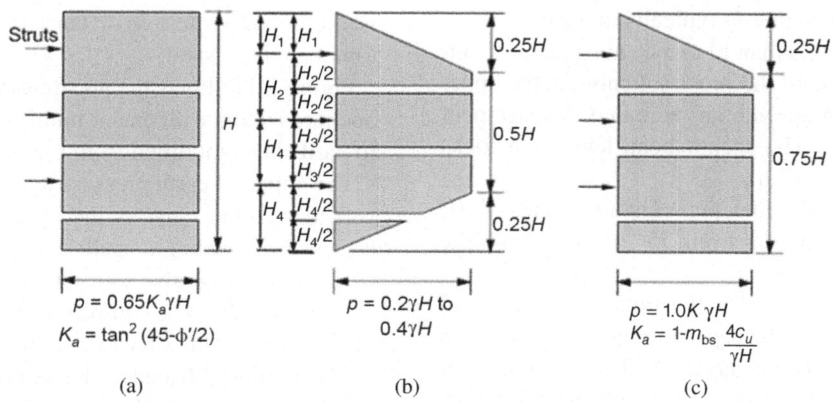

Figure 9.8 Terzaghi and Peck apparent pressure envelopes: (a) sand, (b) stiff-hard fissured clay, and (c) soft to medium clay (Terzaghi and Peck, 1967).

is equal to

$$K_a = 1 - m_{bs}\frac{4c_u}{\gamma H} \tag{9.2}$$

where m_{bs} is the empirical factor, which depends on the base stability in the soft clay. Peck (1969) suggested that when $N_{sn} > 6$, $m_{bs} = 0.4$ and, otherwise, $m_{bs} = 1.0$.

Each strut load is equal to the tributary area of the apparent earth pressure as shown in Figure 9.8.

9.2.3 Design Considerations

Typical Design Parameters Figure 9.9 shows the typical design cross section of an anchored wall system. This design section is based on the condition that the embedded vertical facing element has sufficient toe resistance against passive failure.

Diameter of drill holes for anchors is typically smaller than 150 mm. When the stability of holes becomes an issue, hollow-stem augers are used to install anchors with a typical hole diameter of approximately 300 mm.

The total lengths of most anchors range from 9 to 18 m. AASHTO (2012) suggests that the unbonded length should be at least 4.5 m. The bonded length should start beyond the critical failure surface at a distance of χ = greater of 1.5 m or $H/5$ (H = design height) as shown in Figure 9.9. The overburden cover above the bond length should be at least 4.5 m.

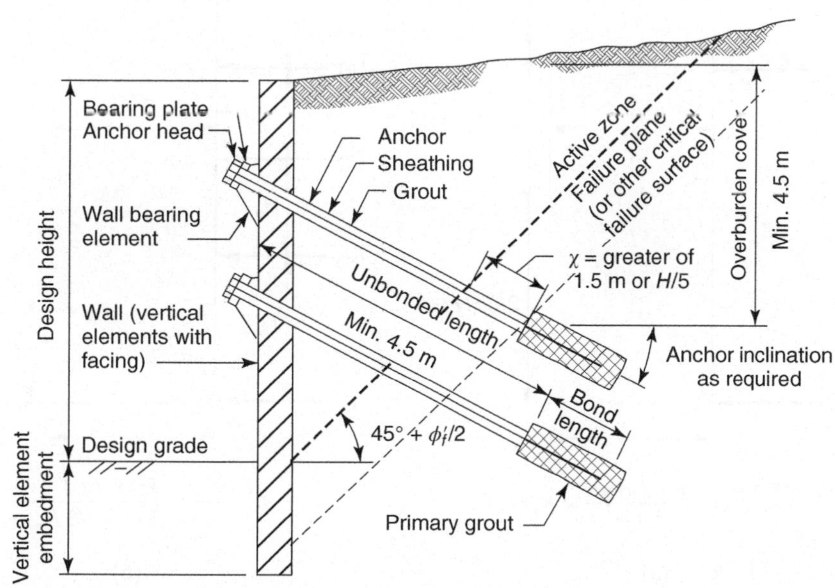

Figure 9.9 Typical design cross section of an anchored wall system (modified from AASHTO, 2012).

The anchor bond lengths in soils typically range from 4.5 to 12.0 m. Longer bond length than 12 m may not gain much additional capacity because of the stress softening in the upper portion of the bond length with an increased displacement as shown in Figure 9.5. The anchor bond lengths in rocks typically range from 3 to 10 m.

Ground anchors are installed with inclination angles of 10° to 45° but commonly ranging from 15° to 30° below the horizontal direction.

AASHTO (2012) suggests that the minimum horizontal spacing (i.e., direction into the paper in Figure 9.9) between adjacent anchors should be the larger of $3d$ (d = diameter of the bonded zone) and 1.5 m. The vertical spacing of anchors is typically 1.8 to 3.0 m.

The typical design load of a ground anchor is between 260 and 1160 kN (Sabatini et al., 1999).

Apparent Earth Pressure Diagram Sabatini et al. (1999) proposed the apparent earth pressure diagrams modified from Figure 9.8(a) for anchored walls in sands as shown in Figure 9.10 for one or multiple levels of anchors. Due to the displacements of the wall, the toe support of the wall, and the overburden stresses with depth, the earth pressure distributions behind the walls are trapezoidal. These distributions are simplified based on the measured earth pressures in the field. Please note that the lateral earth pressure, p, in Figure 9.10 is different from that in Figure 9.8. However, the total loads in Figure 9.8(a) and Figure 9.10 are the same, that is, $0.65K_a\gamma H^2$ where $K_a = \tan^2(45°\text{-}\phi/2)$, ϕ = effective frictional angle of sand.

Figure 9.11 shows the apparent earth pressure profiles for anchored walls with one or multiple levels of anchors in stiff to hard clays, modified from Figure 9.8(b).

The apparent earth pressure diagram for anchored walls in soft to medium clays is the same as that in Figure 9.8(c). The use of anchored walls in soft to medium clays ($N_{sn} > 4$) depends on the soil condition below the base of the excavation. When a competent layer exists below the base of the excavation, a reasonable toe support to the sheet pile wall can be established so that temporary or permanent walls can be constructed. However, if no competent layer exists or a competent layer exists at a greater depth, walls may develop excessive deformations due to minimum toe support. Therefore, this condition is not suitable for permanent walls. The apparent earth pressure diagram shown in Figure 9.8(c) can be used to estimate the loads of anchors in soft to medium clays for excavations. Sabatini et al. (1999) indicated that the m_{bs} factor proposed by Peck (1969) may not be conservative because the potential bearing failure below the base of the excavation was not considered in the previous development. Therefore, Sabatini et al. (1999) suggested (1) when $4 < N_{sn} \leq 5.14$, $K_a = 0.22$ and (2) when $N_{sn} > 5.14$, the equation developed by Henkel (1971) should be used as follows:

$$K_a = 1 - \frac{4c_u}{\gamma H} + 2.83\frac{D_a}{H}\left(1 - \frac{5.14c_{ub}}{\gamma H}\right) \tag{9.3}$$

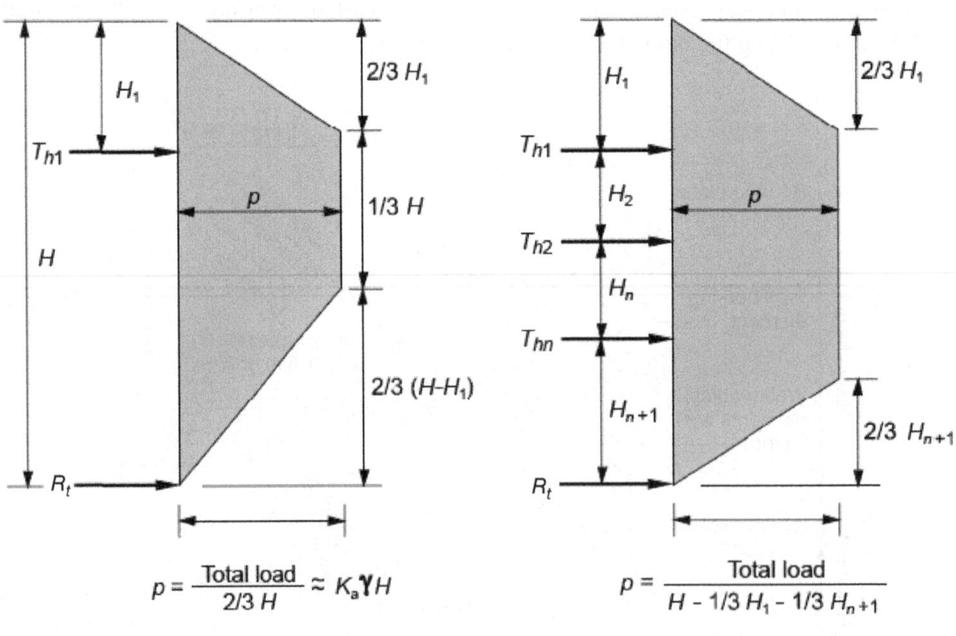

Figure 9.10 Apparent earth pressure diagrams for anchored walls in sands: (a) one level and (b) multiple levels (Sabatini et al., 1999).

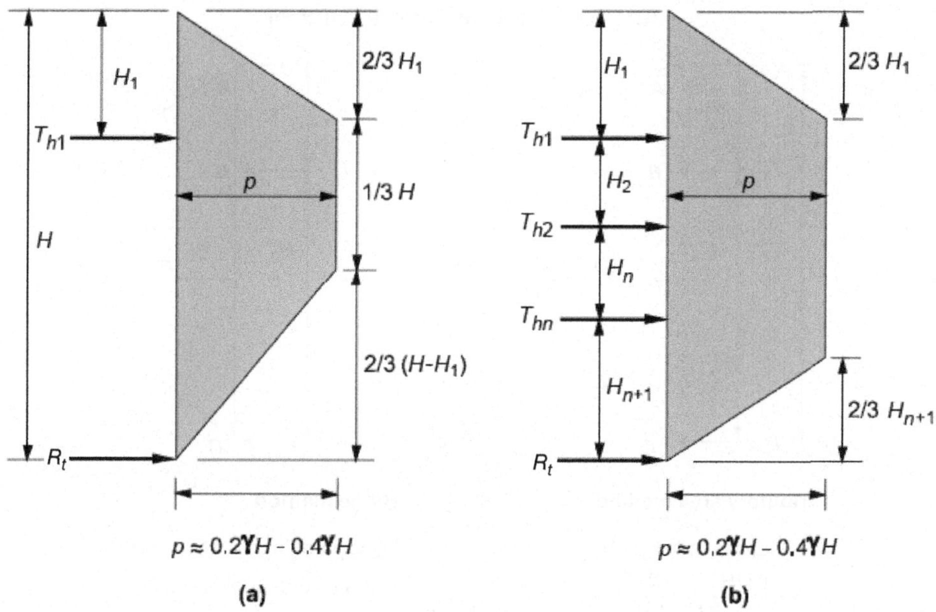

Figure 9.11 Apparent earth pressure envelopes for anchored walls in stiff to hard clays: (a) one level and (b) multiple levels (Sabatini et al., 1999).

where D_a = depth of the failure surface below the base of the excavation
c_{ub} = undrained shear strength of the soil below the base of the excavation

The earth pressure due to large-area uniform surface surcharge can be calculated as follows:

$$\Delta\sigma_h = K_a p_s \qquad (9.4)$$

where p_s is the vertical surface surcharge. Traffic loading is typically assumed to be a uniform pressure with a magnitude of 12 kPa.

Anchor Load The anchor loads can be calculated by a tributary area method or a hinge method as shown in Figure 9.12 for one level of anchors and Figure 9.13 for multiple levels of anchors. The tributary area method is easier for hand calculations; therefore, it is more commonly used in practice.

Since anchors are installed with spacing horizontally, the total horizontal anchor load carried by one anchor at a certain level can be calculated as follows:

$$T_h = T_{hi} s_h \qquad (9.5)$$

where T_{hi} = horizontal anchor load at the ith level per lineal wall length
s_h = the horizontal spacing between anchors.

Considering most anchors are inclined, the total anchor load is

$$T = \frac{T_h}{\cos\theta_{in}} \qquad (9.6)$$

where θ_{in} is the inclination angle of the anchor below the horizontal direction.

The vertical component of the total anchor load is

$$T_v = T\sin\theta_{in} \qquad (9.7)$$

The total vertical component of the anchor load should be used to check against the bearing capacity of the facing element.

Potential Critical Failure Surface The potential critical failure surface is often assumed to start from the corner of the excavation base at an inclination angle of $45° + \phi'/2$ from the horizontal, as shown in Figure 9.9, where ϕ' = effective friction angle of soil. However, when the embedded wall below the excavation does not have sufficient toe resistance, the critical potential failure surface can extend to the depth close or at the base of the wall as shown in Figure 9.7. These failure planes are often referred to as the Rankine failure planes.

Unbonded Anchor Length Sabatini et al. (1999) suggested that the unbonded length of ground anchors in rock and soil should be at least 4.5 m for strand tendons and 3 m for bar tendons. These minimum values can minimize large reductions in anchor loads due to potential seating losses

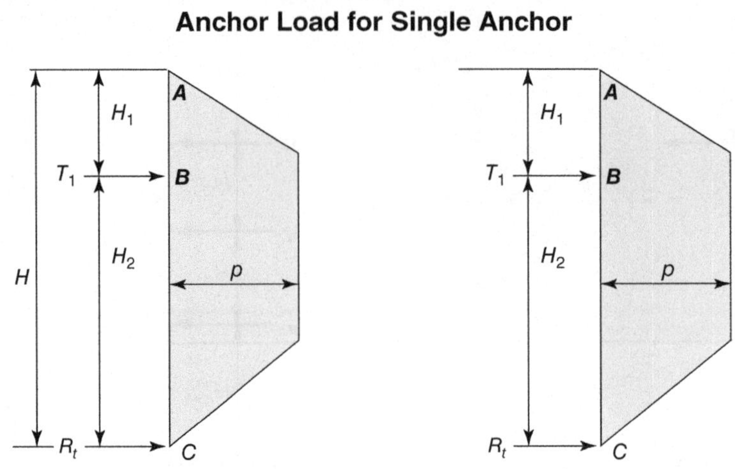

Tributary area method

T_1 = load over length $H_1 + H_2/2$

$$= \frac{(23H^2 - 10HH_1)}{54(H - H_1)}p$$

R_t = load over length $H_2/2$

$$= \frac{2}{3}Hp - T_1$$

Hinge method

T_1 calculated from $\Sigma M_C = 0$

R_t = total earth pressure − T_1

Figure 9.12 Calculation of anchor load with a single anchor (modified from Sabatini et al., 1999).

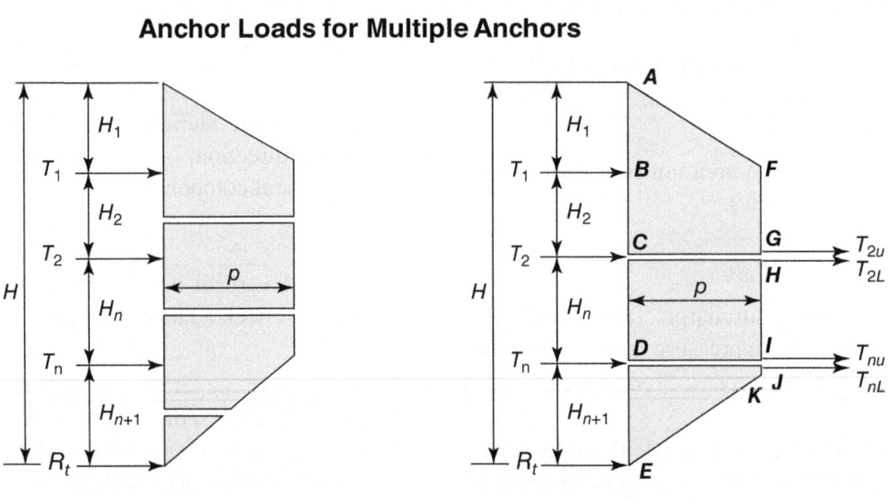

Tributary area method

T_1 = load over length $H_1 + H_2/2$

$$= \left(\frac{2H_1}{3} + \frac{H_2}{2}\right) \cdot p$$

T_2 = load over length $H_2/2 + H_n/2$

$$= \left(\frac{H_2}{2} + \frac{H_n}{2}\right) \cdot p$$

T_n = load over length $H_n/2 + H_{n+1}/2$

$$= \left(\frac{H_n}{2} + \frac{23H_{n+1}}{48}\right) \cdot p$$

R_t = load over length $H_2/2$

$$= \left(\frac{3}{16}H_{n+1}\right) \cdot p$$

Hinge method

T_1 calculated from $\Sigma M_C = 0$

T_{2u} = total earth pressure ($ABCGF$) − T_1

T_{2L} calculated from $\Sigma M_D = 0$

T_{nu} = total earth pressure ($CDIH$) − T_{2L}

T_{nL} calculated from $\Sigma M_E = 0$

R_t = total earth pressure - $T_1 - T_2 - T_n$

$T_2 = T_{2u} + T_{2L}$

$T_n = T_{nu} + T_{nL}$

Figure 9.13 Calculation of anchor loads with multiple anchors (modified from Sabatini et al., 1999).

during load transfer to the structure after anchor load testing. In addition to these minimum requirements, the unbonded length should be extended at least $H/5$ (H = the wall height), or 1.5 m behind the critical potential failure surface.

Anchor Bond Capacity Anchor bond capacity (or anchor pullout capacity) depends on several factors such as:

- Method of drilling including quality of drill hole cleaning and period of time that the drill hole is left open
- Diameter of the drill hole
- Method and pressure used in grouting
- Anchor bond length

For a straight shaft anchor in soil or rock, the anchor bond capacity can be estimated by the following equation:

$$Q = \pi d_{\mathrm{DH}} \tau_a L_b \qquad (9.8)$$

where d_{DH} = diameter of anchor drill hole
τ_a = anchor bond stress
L_b = anchor bond length

The ultimate unit bond stresses for anchors in cohesive soils, cohesionless soils, and rock for a preliminary design are provided in Tables 9.1, 9.2, and 9.3, respectively.

A factor of safety of 2.0 is typically used to calculate the allowable load capacity of anchors in soils while a factor of 3.0 is used for anchors in rocks due to possible discontinuities in the rock mass.

For a final design, preproduction tests may be needed, such as pullout tests and/or extended creep tests, to confirm or establish the relationship between anchor length and anchor load capacity.

Tensile Strength of Steel Tendon To avoid the rupture of the steel tendon, it should be designed with a sufficient factor of safety. Sabatini et al. (1999) recommended that the tendon load should not exceed 60% of the specified minimum tensile strength (SMTS) for final design and 80% of SMTS for temporary loading conditions (e.g., during loading test). The lock-off load after the loading test should not exceed 70% of SMTS. The properties of prestressing steel

Table 9.1 Ultimate Unit Bond Stresses for Anchors in Cohesive Soils

Anchor/Soil Type (Grout Pressure)	Soil Stiffness or Unconfined Compressive Strength (kPa)	Ultimate Unit Bond Stress, τ_a (kPa)
Gravity grouted anchors (<350 kPa)		
Silt-clay mixtures	Stiff to very stiff, 100–400	30–75
Pressure grouted anchors (350–2800 kPa)		
High plasticity clay	Stiff, 100–250	30–100
	Very stiff, 250–400	75 –180
Medium plasticity clay	Stiff, 100–250	100–260
	Very stiff, 250–400	145–365
Medium plasticity sandy silt	Very stiff, 250–400	290–395

Source: AASHTO (2012).

Table 9.2 Ultimate Unit Bond Stresses for Anchors in Cohesionless Soils

Anchor/Soil Type (Grout Pressure)	Soil Compactness or SPT $(N_1)_{60}$	Ultimate Unit Bond Stress, τ_a (kPa)
Gravity-grouted anchors (<350 kPa)		
Sand or sand-gravel mixtures	Medium dense to dense, 11–50	75–145
Pressure-grouted anchors (<350 to 2800 kPa)		
Fine to medium sand	Medium dense to dense, 11–30	85–395
	Medium dense, 11–30	115–700
Medium to coarse sand with gravel	Dense to very dense, 30–50	260–1000
Silty sands	-	175–425
Sandy gravel	Medium dense to dense, 11–40	220–1450
	Dense to very dense, 40 to ≥ 50	290–1450
Glacial till	Dense, 31–50	315–550

Source: AASHTO (2012).

Table 9.3 Ultimate Unit Bond Stresses for Anchors in Rock

Rock Type	Ultimate Unit Bond Stress, τ_a (kPa)
Granite or basalt	1800–3250
Dolomitic limestone	1450–2200
Soft limestone	1050–1450
Slates and hard shales	850–1450
Sandstones	850–1800
Weathered sandstones	750–850
Soft shales	210–850

Source: AASHTO (2012).

bars and 15-mm-diameter prestressing steel strands are provided in Tables 9.4 and 9.5.

Tendon and Trumpet Opening Sizes The trumpet opening size depends on the type and size of the tendon for each anchor and the class of corrosion protection. Table 9.6 provides the suggested minimum trumpet opening size for each corresponding tendon size.

Corrosion of steel in soil with the presence of water is always a concern due to the electrochemical reaction. There are two classes of protection recommended by the Post-Tensioning Institute (1996) as shown in Table 9.7.

The Post-Tensioning Institute (1996) requires class I corrosion protection if ground anchors are used for permanent applications with an unknown or known aggressive condition or with a nonaggressive condition for critical structures. Class II corrosion protection is recommended for temporary applications with an unknown or known aggressive condition or permanent applications with a nonaggressive condition for noncritical structures.

Table 9.5 Properties of 15-mm Diameter Grade 1860 Prestressing Steel Strands

Number of 15-mm Diameter Strands	Cross-Section Area (mm^2)	Ultimate Strength (kN)
1	140	261
3	420	782
4	560	1043
5	700	1304
7	980	1825
9	1260	2346
12	1680	3128
15	2100	3911
19	2660	4953

Source: ASTM A416-12, *Standard Specification for Steel Strand, Uncoated Seven-Wire for Prestressed Concrete*. Reprinted with permission from ASTM International.

Table 9.6 Suggested Trumpet Opening Size for Corresponding Tendon Size

	Minimum Trumpet Opening Size (mm)	
Tendon Type and Size	Class II Corrosion Protection	Class I Corrosion Protection
Number of 15-mm diameter strands		
4	102	150
7	115	165
9	127	178
11	140	191
13	153	203
17	165	216
Bar diameter (mm)		
16	64	89
32	70	95
36	76	102

Source: Sabatini et al. (1999).

Table 9.4 Properties of Prestressing Steel Bars

Steel Grade (MPa)	Nominal Diameter (mm)	Ultimate Stress, f_{pu} (MPa)	Nominal Cross Section Area, A_{ps} (mm^2)	Ultimate Strength, $A_{ps} f_{pu}$ (kN)
	26	1035	548	568
1035	32	1035	806	835
	36	1035	1019	1055
	45	1035	1716	1779
	64	1035	3348	3461
	26	1104	548	605
1104	32	1104	806	890
	36	1104	1019	1125

Source: ASTM A722-12, *Standard Specification for Uncoated High-Strength Steel Bars for Prestressing Concrete*. Reprinted with permission from ASTM International.

Table 9.7 Corrosion Protection Requirements

Class	Requirements for Protection		
	Anchorage Assembly	Unbonded Length	Bond Length
I (Encapsulated tendon)	1. Trumpet 2. Cover if exposed	1. Encapsulate tendons composed of individual grease-filled extruded strand sheaths with a common smooth sheath 2. Encapsulate tendons composed of individual grease-filled strand sheaths with grout-filled smooth sheath 3. Use smooth bond breaker over grout-filled bar sheath	1. Grout-filled encapsulation or 2. Fusion-bonded epoxy
II (grout protected tendon)	1. Trumpet 2. Cover if exposed	1. Grease- filled sheath, or 2. Heat shrink sleeve	Grout

Source: Post-Tensioning Institute (1996).

Design of Wall Facing Structure In addition to the design of ground anchors, the wall facing structure, such as soldier beams and lagging, sheet piles, or wales and permanent facing, should be designed. The design should consider the bending moment of the structural elements, depth of embedment of the wall facing below the base of the excavation, axial capacity of the wall facing, and lateral capacity of the embedded portion of the wall facing. Details on the design of the wall facing structure can be found in Sabatini et al. (1999).

Compound and Global Slope Stability A factor of safety of 1.2–1.3 is typically used for the stability analysis for slope and landslide stabilization by ground anchors. However, when anchored walls are used to support critical or deformation-sensitive structures, they should be designed for a higher factor of safety (e.g., 1.5). The slope stability analysis is typically done by software, which is developed based on limit equilibrium methods. Planar and/or circular slip surfaces may be considered. Basic concepts and common methods for slope stability analysis based on limit equilibrium methods are discussed in Chapter 2.

Compound slope failure is referred to as the case with a slip surface partially outside and within the reinforced zone by anchors or below/above the toe of the wall facing. Slip surfaces (1) and (3) shown in Figure 9.14 are those for compound slope failures. For the case with the slip surface (1), anchor A does not provide any contribution to the slope stability but part of the bonded length of anchor B providess the tensile resistance to the slope stability. For the case with the slip surface (3), neither anchor A nor anchor B provides any contribution to the slope stability; however, the

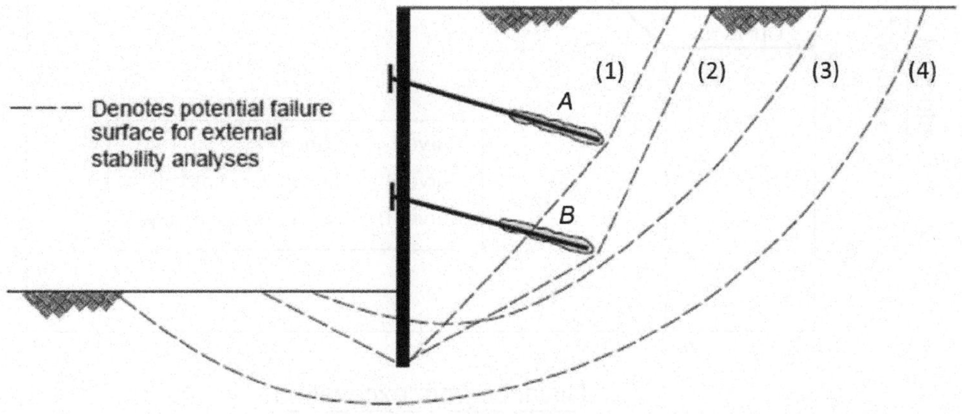

Figure 9.14 Failure surfaces for compound and global stability analysis (after Sabatini et al., 1999).

embedment of the wall facing provides passive resistance to the slope stability. The passive resistance can be estimated using the methods proposed by Broms (1965) or Wang and Reese (1993) for laterally loaded piles if solder piles and lagging are used or using the method for an embedded retaining wall if a continuous wall facing is used.

Global or deep-seated slope failure is referred to as the failure beyond the reinforced zone by anchors and/or below the toe of the wall facing, for example, slip surfaces (2) and (4), in which no anchor or wall facing provides any direct contribution to the slope stability along these slip surfaces. It should be pointed out that even though some slip surfaces do not pass the anchors or the wall facing, they still contribute to the stability by pushing the slip surfaces back and deeper into the ground.

Base Stability When an excavation occurs in soft soil, the analysis for possible bearing capacity failure due to the base of the excavation should be performed. The same approach as presented for base stability of the excavation pit with the deep mixed or jet-grouted walls in Section 8.2.3 of Chapter 8 can be used.

Creep Deformation of Anchor Ground anchors may develop excessive deformations under sustained loading because of creep of soils. The following soils may have high creep potentials (Sabatini et al., 1999):

- Organic soils
- Clayey soils with an average liquidity index (LI) greater than 0.2
- Clayey soils with an average liquid limit (LL) greater than 50
- Clayey soils with an average plasticity index (PI) greater than 20

In these soils, conservative anchor design loads and working bond stress values should be used for permanent anchors.

Wall and Ground Movements Walls and ground move during and after excavations. The magnitude of the movement depends on several factors:

- Soil type
- Excavation depth
- Existence of toe and/or top slope
- Rigidity of facing elements
- Distance from wall facing
- Time of installation and location of the first anchor
- Design and installation methods of ground anchors
- Creep deformation of ground anchors
- Water
- Surface surcharge including traffic loading
- Seismic loading
- Design factor of safety

Anchored walls constructed in sands and stiff clays typically have an average lateral wall movement of 0.2%*H* (*H* = wall height) and a maximum lateral wall movement of 0.5%*H* (Sabatini et al., 1999). The typical settlement profiles behind the wall facing are presented in Figure 9.15. These profiles are approximated from collected field data. Soft to

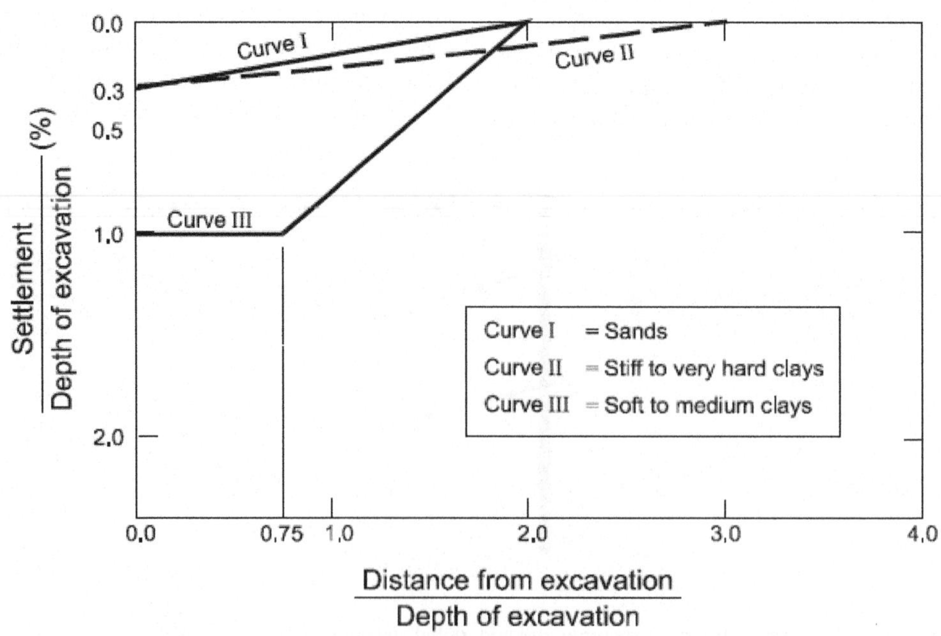

Figure 9.15 Settlement profiles behind anchored walls (Sabatini et al., 1999).

medium clays result in the largest settlement while stiff to very hard clays have the largest influence distance. When there are nearby existing structures, it is important to limit the settlement to a tolerable value.

9.2.4 Design Parameters and Procedure

Design Parameters The design parameters for ground anchors may include the following parameters (mainly for anchored walls):

- Type of application (temporary or permanent, critical or noncritical)
- Project requirements (tolerable settlement, factor of safety against slope failure)
- Construction constraints
- Geometry of project (such as depth of excavation)
- Type of wall facing
- Site subsurface conditions (type and properties of geomaterial, groundwater table, aggressive or nonaggressive for corrosion)
- Loading conditions (surcharge, water pressure or seismic)
- Number of ground anchor levels
- Method of installation of anchors
- Length and inclination of anchors
- Depth of the upper level of ground anchor
- Bond length

Design Procedure The detailed design procedure for an anchored wall is provided by Sabatini et al. (1999) as follows:

1. Establish project requirements including type of project (temporary and/or permanent), project geometry, external loading (water, surcharge or seismic), performance criteria, and construction constraints (right-of-way limitations, nearby structures, and existing utility lines).
2. Evaluate site subsurface conditions and relevant properties of in situ geomaterial.
3. Establish design requirements, including factors of safety and level of corrosion protection.
4. Based on type of geomaterial, select lateral earth pressure distribution behind the wall. Add water pressure and surcharge for total lateral pressure calculation if they exist.
5. Calculate horizontal ground anchor loads by adjusting vertical anchor locations to achieve the optimum wall bending moment distribution.
6. Determine required anchor inclination based on construction constraints and geotechnical conditions.
7. Calculate a vertical force component and a force along the anchor from each horizontal anchor load.
8. Evaluate horizontal spacing of anchors based on wall type and experience. Calculate individual anchor loads.
9. Select the type of ground anchors.
10. Evaluate the embedment depth and cross section of the wall by calculating vertical and lateral capacities of the wall below excavation base
11. Calculate factors of safety for internal and external stability of the anchored system and check them against design requirements.
12. Estimate maximum lateral wall movements and ground surface settlements. Revise design if necessary.

If any of the calculated values in Steps 10, 11, and 12 does not meet the design requirement, adjust design parameters of anchors and/or the wall and repeat the above design procedure.

9.2.5 Design Example

An excavation is needed for the construction of a high-rise building in uniform medium dense sand with an SPT N value of 20. The groundwater table is at 20 m below the ground surface. The depth of the excavation is 9 m. The unit weight and friction angle of the sand are 18 kN/m^3 and 34°, respectively. A temporary anchored wall as shown in Example Figure 9.1 is proposed to retain the soil during the excavation. The horizontal spacing of the anchors at the same elevation is 2 m. Sheet piles are used as the wall facing and have a sufficient embedment depth for toe resistance. Determine the bond and total lengths of the upper and lower anchors.

Solution

1. The angle for the potential critical failure plane from the horizontal is

$$\psi = 45^\circ + \frac{\phi}{2} = 45^\circ + \frac{34^\circ}{2} = 62^\circ$$

2. The bond length of the anchors should start from the location of greater of 1.5 m or $H/5$ (i.e., 1.8 m) beyond the critical failure plane; therefore, $\chi = 1.8$ m.

   The horizontal distance for the upper anchor from the back of the wall facing to the possible start point of the bond length, x_1, is

$$x_1 = \frac{H - H_1}{\tan\psi + \tan\theta_{in}} + \frac{\chi}{\sin(\theta_{in} + \psi)} \cos\theta_{in}$$
$$= \frac{9 - 3}{\tan 62^\circ + \tan 15^\circ} + \frac{1.8}{\sin(15^\circ + 62^\circ)} \cos 15^\circ$$
$$= 4.6 \text{ m}$$

Considering the minimum overburden depth $z_{min} = 4.5$ m, the horizontal distance for the upper anchor from the back of the wall facing to the possible start point of the bond length, x_1, is

$$x_1' = \frac{z_{min} - H_1}{\tan\theta_{in}} = \frac{4.5-3}{\tan 15^\circ} = 5.6 \text{ m}$$

Since $x'_1 > x_1$, 5.6 m is used to calculate the unbonded length as follows:

$$L_{ub1} = \frac{x_1'}{\cos\theta_{in}} = \frac{5.6}{\cos 15^\circ} = 5.8 \text{ m}$$

Since the overburden depth of the lower anchor is obviously greater than $z_{min} = 4.5$ m, the horizontal distance for the lower anchor from the back of the wall facing to the possible start point of the bond length, x_2, is

$$x_2 = \frac{H - H_1 - H_2}{\tan\psi + \tan\theta_{in}} + \frac{\chi}{\sin(\theta_{in}+\psi)}\cos\theta_{in}$$
$$= \frac{9-3-3}{\tan 62^\circ + \tan 15^\circ} + \frac{1.8}{\sin(15^\circ + 62^\circ)}\cos 15^\circ$$
$$= 3.2 \text{ m}$$

$$L_{ub2} = \frac{x_2}{\cos\theta_{in}} = \frac{3.2}{\cos 15^\circ} = 3.4 \text{ m}$$

3. The apparent earth pressure diagram is used to estimate the anchor loads. Since the anchored wall is to be constructed in sand, the diagram for sand is used and plotted in Example Figure 9.1.

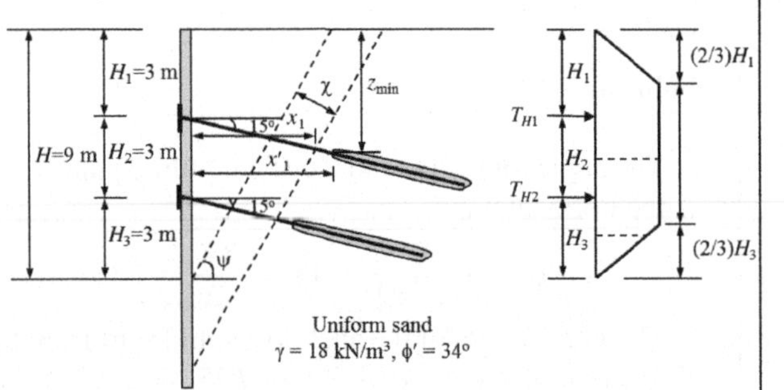

Example Figure 9.1

The coefficient of the active lateral earth pressure is

$$K_a = \tan^2\left(45^\circ - \frac{\phi}{2}\right) = \tan^2\left(45^\circ - \frac{34^\circ}{2}\right)$$
$$= 0.283$$

The total lateral thrust is

$$P_a = 0.65 K_a \gamma H^2 = 0.65 \times 0.283 \times 18 \times 9^2$$
$$= 268 \text{ kN/m}$$

The maximum lateral earth pressure is

$$p = \frac{P_a}{H - H_1/3 - H_3/3} = \frac{268}{9 - 3/3 - 3/3}$$
$$= 38.3 \text{ kPa}$$

4. The horizontal anchor loads for the two anchors can be calculated using the tributary area method as follows:

$$T_{H1} = \left(\frac{2}{3}H_1 + \frac{H_2}{2}\right)p = \left(\frac{2}{3}\times 3 + \frac{3}{2}\right)\times 38.3$$
$$= 134 \text{ kN/m}$$

$$T_{H2} = \left(\frac{H_2}{2} + \frac{23}{48}H_3\right)p = \left(\frac{3}{2} + \frac{23}{48}\times 3\right)\times 38.3$$
$$= 112 \text{ kN/m}$$

Considering the horizontal spacing of the anchors at 2.0 m and the inclination of the anchors is 15°, the anchor design loads for the upper and lower anchors are

$$T_{d(H1)} = \frac{T_{H1}s_h}{\cos\theta_{in}} = \frac{134 \times 2.0}{\cos(15^\circ)} = 277 \text{ kN}$$

$$T_{d(H2)} = \frac{T_{H2}s_h}{\cos\theta_{in}} = \frac{112 \times 2.0}{\cos(15^\circ)} = 232 \text{ kN}$$

5. Since the site aggressive condition is unknown and the anchored wall is temporary, class II corrosion protection is selected. A 26-mm diameter, grade 1035 prestressing steel bar with 60% of the SMTS equal to 341 kN is suitable for the anchor. The minimum trumpet opening size of 64 mm, corresponding to a 26-mm-diameter bar, based on class II corrosion protection can be selected.
6. Considering the medium sand has an SPT N value of 20, an ultimate unit bond stress, $\tau_a = 400$ kPa, is selected. For a straight shaft anchor with a diameter of drill holes of 64 mm (same as the trumpet opening) in sand, the bond length can be estimated by the following equation:

$$L_{b1} = \frac{T_{d(H1)}}{\pi d_{DH}\tau_a} = \frac{277}{3.14 \times 0.064 \times 400} = 3.4 \text{ m}$$

$$L_{b2} = \frac{T_{d(H2)}}{\pi d_{DH}\tau_a} = \frac{232}{3.14 \times 0.064 \times 400} = 2.9 \text{ m}$$

Therefore, the total lengths for the upper and lower anchors are

$$L_1 = L_{\text{ub1}} + L_{b1} = 5.8 + 3.4 = 9.2 \text{ m}$$

$$L_2 = L_{\text{ub2}} + L_{b2} = 3.4 + 2.9 = 6.3 \text{ m}$$

9.2.6 Construction

The installation of ground anchors typically includes the following steps:

1. Drilling. Rotary, percussion, rotary/percussive, or auger drilling can be used to drill holes in soil or rock for ground anchors. The method of drilling should be selected to minimize the disturbance to soil or rock and nearby existing structures.
2. Inserting tendon. The tendon should be inserted into the drill hole to the desired depth and inclination angle.
3. Installing trumpet and anchorage. Depending on the method for corrosion protection, a specific material may be used around the unbonded lengths of the tendons, for example, corrosion inhibiting grease or grout. The bearing plate or the anchor head to be used for load testing should not contact with the corrosion protection material.
4. Grouting anchor. The grout is injected from the toe of the drill hole before or after inserting the tendon. The top of the grout should not be in contact with the bottom of the trumpet.
5. Testing and stressing the prestressed ground anchor. Each anchor should be subjected to stress for testing and evaluation. This is also part of quality assurance.
6. Lock-off. After the load testing, the anchor should be locked off by reducing the load to the lock-off load and transferring the load to the anchorage assembly. Typically, the lock-off load is 80% of the prestressed anchor service load.
7. Cutting tendon protrusions. Extra tendon beyond the anchorage assembly should be cut by a saw.

9.2.7 Quality Control and Assurance

All the materials to be used for the anchors should be inspected and verified prior to installation, including

- Pretressing steel strand or bar
- Anchorage and trumpet
- Spacers
- Centralizers
- Portland cement
- Bearing plates
- Corrosion protection materials

The European Standard for Execution of Special Geotechnical Work—Ground Anchors (British Standards Institution, 2000) allows the initial drilling alignment to be deviated within $\pm 2^{\circ}$ from the desired axis of the borehole and the drilling to be deviated within 1/30 of the anchor length.

During the grouting operation, the following information should be recorded:

- Type of mixer
- Water/cement ratio
- Types of additives (if any)
- Grout pressure
- Type of cement
- Strength test samples (if any)
- Volume of first and second stage grout

Grout strength should be measured and confirmed by cube samples before any load test is performed. Grout typically should have its required strength of 20–30 MPa in 7–10 days.

For anchored walls, each anchor should be tested for its load capacity and load–deformation behavior after installation. AASHTO (2012) requires all production anchors should be subjected to load testing and stressing. The test load is typically 125–150% of the design load of the anchor. At the end of each load testing, the anchor should be locked off to remove any slack in the anchored wall system to minimize postconstruction displacements. The acceptance or rejection of ground anchors is relied on the results of three different tests: (1) performance test, (2) proof test, and (3) extended creep test. Proof tests are most commonly adopted in the field.

A performance test involves several load cycles with increasing magnitudes until the test load. After each load cycle, elastic and residual deformations of the anchor are recorded. At the test load, the load is held for 10 min to evaluate the time-dependent deformation (i.e., creep) of the ground anchor. This test is used to verify anchor capacity and establish load–deformation behavior.

A proof test applies loads in several increments until the test load, which is held for 10 min before unloaded to the initial alignment load. This test is used to examine whether a test anchor is acceptable for applications.

An extended creep test is to evaluate creep deformations of an anchor under different load intensities for certain durations. Each test typically lasts 8 h. This test is done for cohesive soil with PI > 20 or LL > 50.

9.3 SOIL NAILING

9.3.1 Introduction

Basic Concept Soil nailing is a technique to install closely spaced, passive structural inclusions to stabilize existing unstable ground due to the change of geotechnical conditions by nature and/or human activities. Common natural causes are precipitation and/or erosion, while the common human activity is partial removal of the ground for project needs. The basic procedure for installing a soil nail consists of drilling a hole in the ground, placing a steel bar in the hole, and grouting the hole. Soil nails can be installed on existing or cut slopes and walls during excavation. Figure 9.16 shows a typical cross section of soil nailing, which includes multiple soil nails, temporary and/or permanent facing, and drainage system. There are different types of nails in practice, which include

- Grouted nails
- Self-drilling nails
- Jet grouted nails
- Helical nails
- Driven nails
- Shoot-in nails

The most commonly used soil nails are the grouted nails; therefore, this book mainly focuses on the grouted nails. Some of the above nails are installed for temporary applications but others for permanent applications. Most of nails are installed in inclination angles of 10°–20° below the horizontal. Different from ground anchors, soil nails are not prestressed when they are installed. Tension develops during the service due to ground movements. Sometimes, nails are installed vertically or perpendicularly to critical slip surfaces, especially for slope stabilization. This application is similar to that by micropiles and will not be further discussed in this book. Typical temporary facing is installed by steel mesh with shotcreting (i.e., spraying concrete). Cast-in-place, prefabricated concrete facing, or other type of facing may be installed as the permanent wall facing in front of the temporary wall facing.

Figure 9.17 shows the details of facing connection between soil nails, temporary facing (shotcrete), and permanent facing. A bearing plate and washers are used to fix the soil nail on the temporary facing. Steel reinforcements are connected with the bearing plate by studded heads before the permanent concrete facing is cast in place on the temporary facing.

Suitability Soil nailing is suitable for vertical or near-vertical excavations in both soils and weathered rocks. It is also suitable for stabilizing steep unstable terrain of soils or weathered rocks. The favorable geomaterials for soil nailing installation include (Lazarte et al., 2003):

- Stiff to hard fine-grained soils
- Dense to very dense granular soils with some apparent cohesion
- Weathered rock with no weak plane
- Glacial soils
- Ground that can stand unsupported on a vertical or sloped cut of 1–2 m for 1–2 two days.

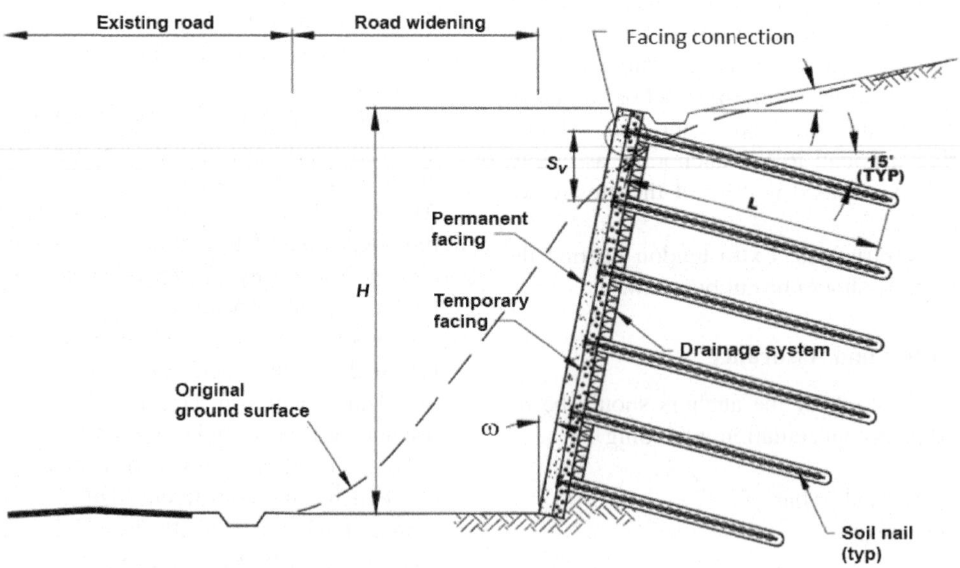

Figure 9.16 Typical cross section of soil nailing (Lazarte et al., 2003).

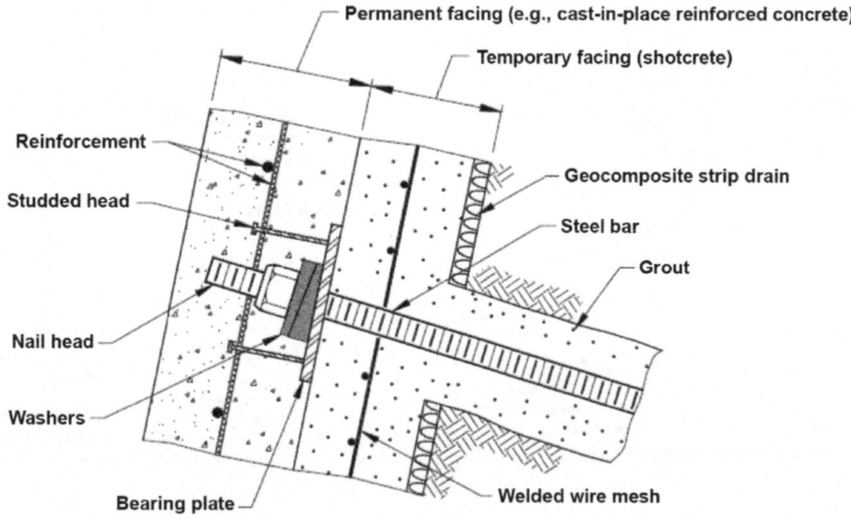

Figure 9.17 Details of facing connection (Lazarte et al., 2003).

The unfavorable geomaterial conditions for soil nailing are (Lazarte et al., 2003):

- Dry, poorly graded cohesionless soils
- Soils with high groundwater
- Soils with large boulders or cobbles
- Soft to very soft fine-grained soils
- Organic soils
- Highly corrosive soil (e.g., cinder or slag) or groundwater
- Weathered rock with unfavorable weak planes and karst
- Loess.

Applications Soil nailing has been used for different applications:

- Vertical or near-vertical excavations
- End slope removal to widen existing bridge abutments
- Tunnel portals
- Repair or stabilization of existing earth retaining structures
- Repair or stabilization of existing natural slopes.

Figure 9.18 shows the examples of three common soil nailing applications.

Advantages and Limitations Soil nailing has the following advantages as compared with alternate technologies (Lazarte et al, 2003; Elias et al., 2004):

- Less right-of-way requirements than ground anchors
- Less concrete usage than conventional reinforced concrete retaining walls
- No need for deep foundations or structural elements below the base of excavations
- Rapid installation with less material than ground anchors
- Less disruptive to traffic and less environmental impact
- More accessible to a job site and low overhead requirements due to the use of light construction equipment
- Higher redundancy because of the larger number of nails than ground anchors
- More flexible to deformation than conventional rigid structures
- Easy incorporation of temporary wall facing into permanent wall facing
- More cost effective than conventional rigid retaining walls.

The possible limitations of soil nailing are (Lazarte et al, 2003; Elias et al., 2004):

- Not appropriate for applications with strict deformation requirements
- Restriction by nearby existing utilities and structures
- Requirement for permanent underground easements
- Difficult to install below the groundwater table
- Low nail capacity and large creep deformation in high-plasticity cohesive soils

9.3.2 Principle

Failure Modes Similar to anchored walls, soil-nailed walls may have internal, external, and facing failures as shown in Figures 9.19, 9.20, and 9.21.

The internal failures as shown in Figure 9.19 include nail–soil pullout, bar–grout pullout, nail tensile failure, and bending and/or shear failure through the reinforced mass.

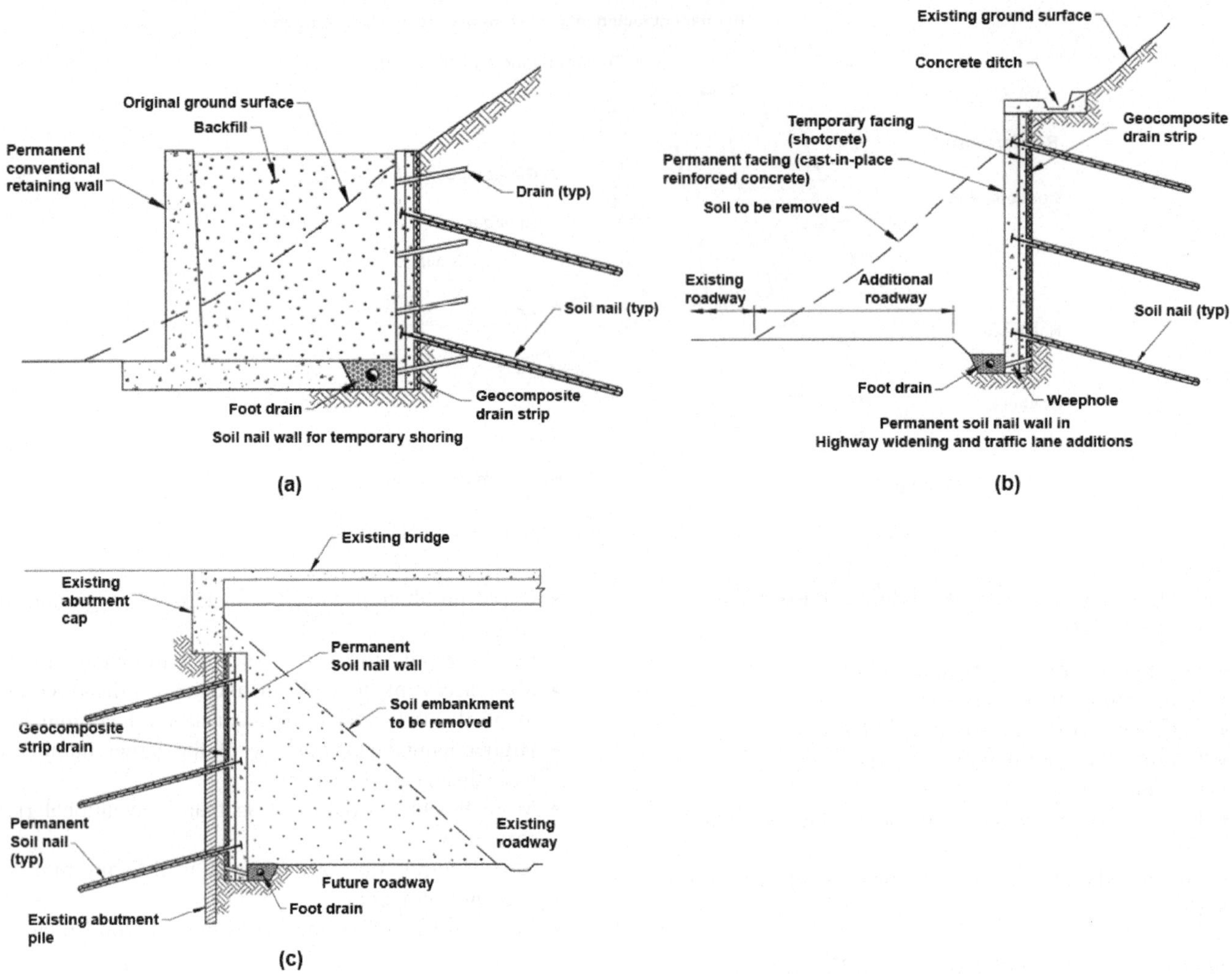

Figure 9.18 Examples of soil nailing: (a) temporary shoring, (b) roadway widening, and (c) end slope removal (Byrne et al., 1998).

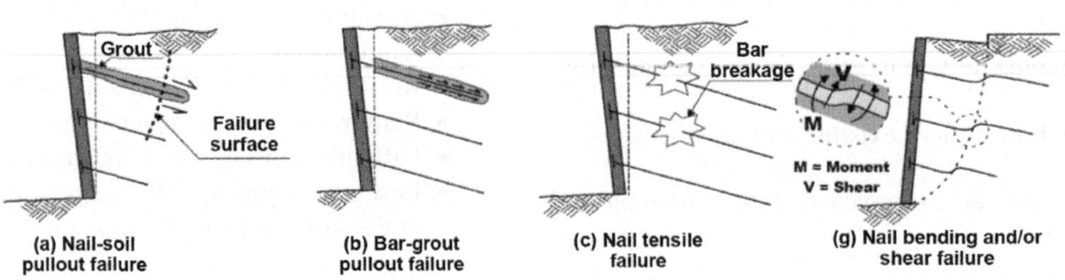

Figure 9.19 Internal failure modes of soil nailed walls (Lazarte et al., 2003).

The external failures as shown in Figure 9.20 include compound failure, sliding failure, and global failure.

The facing failures as shown in Figure 9.21 include facing flexure failure, facing punching shear failure, and facing headed-stub connection failure.

Load Transfer The deformation and load transfer of a soil-nailed wall is illustrated in Figure 9.22. On a level ground, the excavation is first progressed to a depth corresponding to excavation phase 1. Due to the excavation, there is an unsupported face with unbalanced forces, which induce

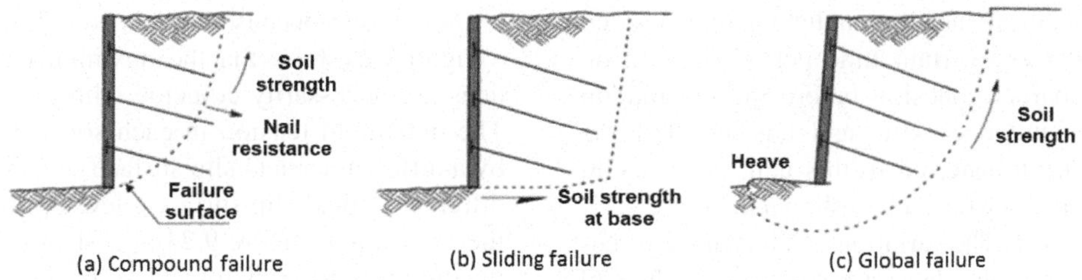

Figure 9.20 External failure modes of soil nailed walls (Lazarte et al., 2003).

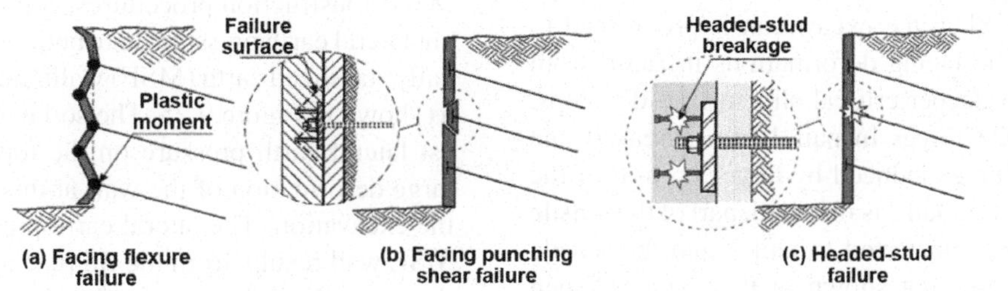

Figure 9.21 Facing failure modes of soil nailed walls (Lazarte et al., 2003).

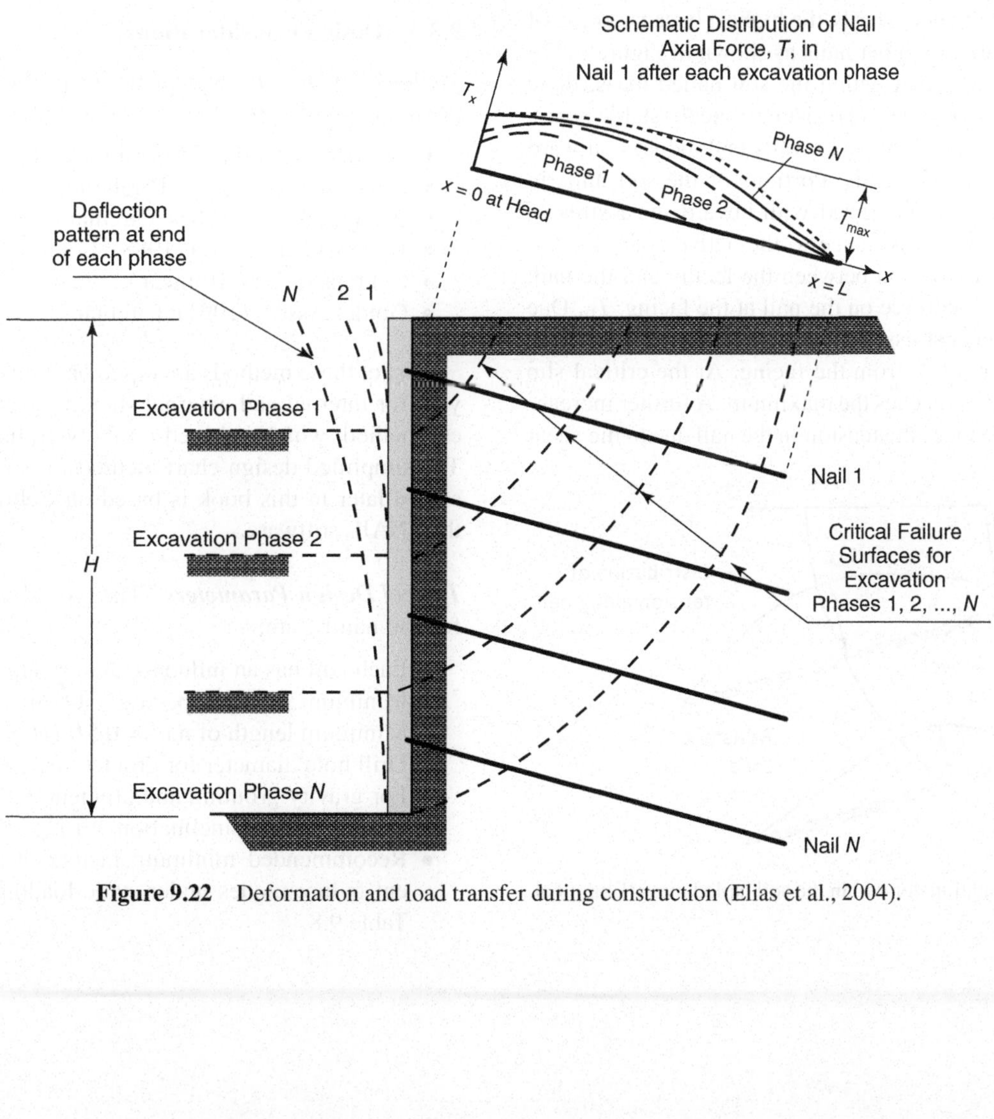

Figure 9.22 Deformation and load transfer during construction (Elias et al., 2004).

facing movement and a critical slip surface. This is a basic requirement for soil nailing installation that the unsupported facing should stand for a certain time period, relying on soil cohesion or apparent cohesion before the installation of the first nail and shotcrete. After nail 1 is installed, the unbalanced forces are transferred to this nail. As a result, tensile forces develop along the nail, as shown in Figure 9.22, corresponding to phase 1. The variations of the tensile forces along the nail result from the shear stresses between the nail and the surrounding soil. The installation of steel mesh and shotcrete helps more loads transferred to the nail through the facing. This is why the tensile force at the nail head is not equal to zero. When the excavation is progressed to excavation phase 2, the facing deformations increase from line 1 to line 2 and a deeper critical slip surface develops. As a result, the tensile forces in nail 1 are increased due to more unbalanced forces induced by larger portion of the unsupported facing. After nail 2 is installed, part of the tensile loads in nail 1 will be transferred to nail 2 and the tensile loads in nail 1 decrease (not shown in the figure). When the excavation process continues to excavation phase N, the tensile loads in nail 1 will continue changing. It should be pointed out that the tensile loads in nail 1 at the end of the construction are not necessarily the highest because some of the loads are shared by other nails as shown in Figure 9.22.

Figure 9.23 shows that within the soil nailed mass, there are two zones: active zone and resistant zone (or stable zone). The active zone tends to move laterally and induces outward shear stresses along the nails. Portions of the soil nails in the resistant zone are mobilized with inward shear stresses to provide the tensile resistance to the active zone. Due to the mechanical connection between the facing and the nail, there is a connection force on the nail at the facing, T_0. Due to the outward shear stresses, the tension in the nail increases with the distance away from the facing. At the critical slip surface, the tension reaches the maximum. A further increase of the distance reduces the tension in the nail due to the shear

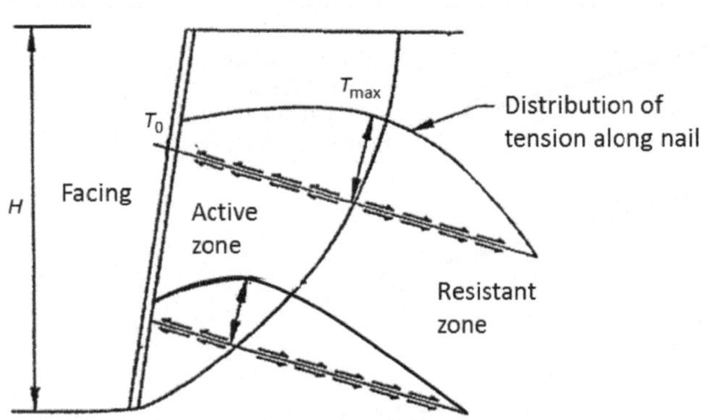

Figure 9.23 Load transfer within the soil nailed mass (Elias et al., 2004).

stresses in the opposite direction. At the end of the nail, the tension is zero because there is no pullout resistance.

Figure 9.24 shows that the maximum tension line in the nails does not necessarily coincide with the critical slip surface. The maximum tension in each soil nail may be controlled by a different critical slip surface. As shown in Figure 9.22, different critical slip surfaces develop during the process of the excavation. Figure 9.24 also shows that the locus of the maximum nail axial force is typically within (0.3–0.4) H behind the wall facing where H is the wall height.

Lateral Earth Pressure Distribution Due to the differences in the construction procedures, soil nailed walls have different lateral earth pressure distributions from those in mechanically stabilized earth (MSE) walls and ground anchored walls as shown in Figure 9.25. The soil nailed walls have the highest lateral earth pressure on the top of the wall due to the large deformation of the wall at this location resulting from the excavation. The lateral earth pressure distribution of the MSE wall results from the fill placement from the bottom to the top and will be discussed in Chapter 10. The distribution of the lateral earth pressure in an anchored wall is discussed in the previous section.

9.3.3 Design Considerations

Methods for Design Several design methods have been proposed and used in practice to design soil nailing:

- German method (1979): Bilinear slip surface
- Davis method (1981): Parabolic
- French method (1983): Circular or of any shape
- Kinematical method (1988): Log spiral
- Caltrans (1991): Bilinear
- Golder Assoc. (1991): Circular

Most of these methods are based on limit equilibrium analyses for internal and external stability except the kinematical method, which is based on the working stress analysis. The simplified design chart method for soil nails to be presented later in this book is based on Caltrans (1991) using the SNAIL software.

Typical Design Parameters The typical design parameters for soil nailing are:

- Each nail has an influence area $\leq 4\ m^2$
- Minimum soil nail spacing = 1.0 m
- Minimum length of nail = 0.5H (H = height of wall)
- Drill hole diameter for grouted nails = 100–200 mm
- For gravity grouting and efficient nail tensile capacity, a minimum nail inclination = 15°
- Recommended minimum factors of safety for different failure modes under static loading are provided in Table 9.8.

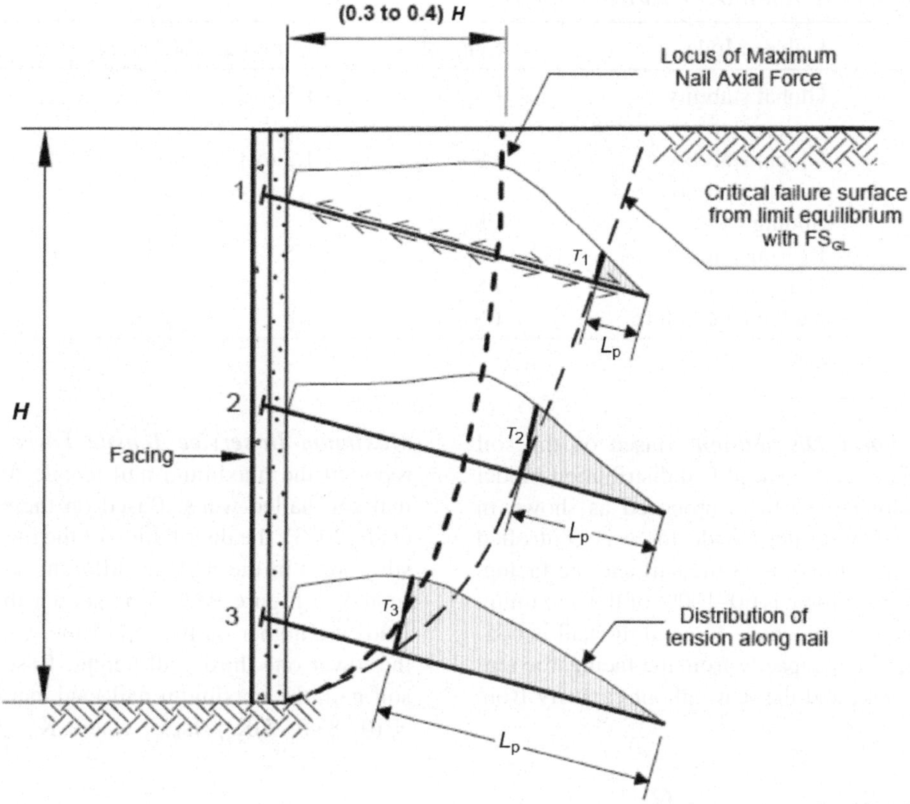

Figure 9.24 Locations of maximum tensile forces and critical failure surface (Byrne et al., 1998).

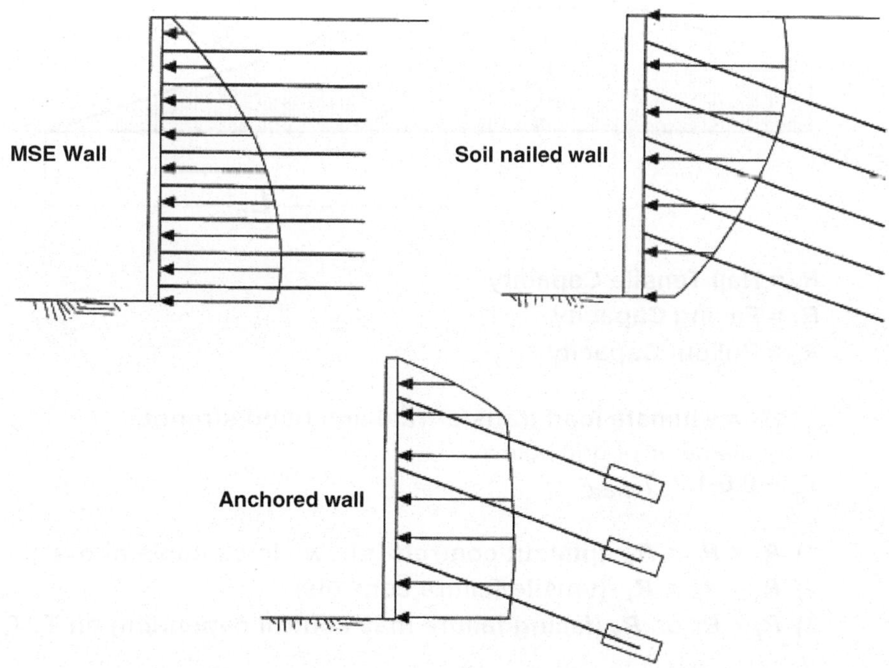

Figure 9.25 Lateral earth pressure distributions (Byrne et al., 1998).

Table 9.8 Recommended Minimum Factors of Safety

Type of Analysis	Failure Mode	Symbol	Temporary Structure	Permanent Structure
External stability	Global stability (long term)	FS_G	1.35	1.5
	Global stability (excavation)	FS_G	1.2 to 1.3	1.2 to 1.3
	Sliding	FS_{SL}	1.3	1.5
	Bearing capacity	FS_{BC}	2.5	3.0
Internal stability	Nail pullout	FS_P	2.0	2.0
	Nail tensile failure	FS_T	1.8	1.8

Source: after Lazarte et al. (2003).

Simplified Tensile Force Distribution Based on the soil nail behavior and field data, a simplified distribution model for tensile forces along the nail is proposed as shown in Figure 9.26. At the facing, the tensile force is controlled by the connection strength between the nail and the facing. Typically, the connection force is 60–100% of the maximum tensile force. The maximum tensile force in the nail is controlled by the front pullout capacity from the facing, the tensile strength of the nail, and the rear pullout capacity from the resistant zone.

Maximum In-service Tensile Force Byrne et al. (1998) reported the maximum nail tensile forces measured in actual soil nailed walls. Based on the suggestion by Lazarte et al. (2003), the distribution of the maximum in-service tensile forces in the nails at different nail head depths is presented in Figure 9.27. It is shown that the maximum nail loads are higher on the top of the wall and decrease within the lower one-third wall height. Based on the depth of the soil nail, the maximum nail load can be calculated as $T_{max} = (0.38{\sim}0.75)K_a \gamma H s_h s_v$, where K_a = coefficient of active

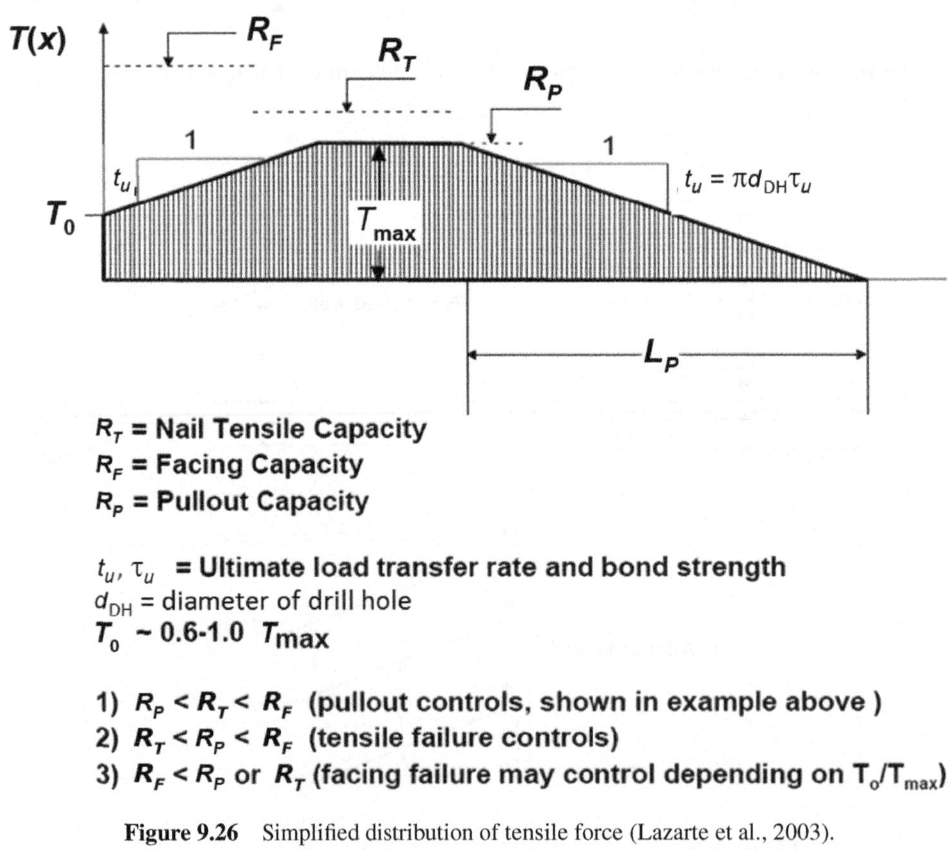

Figure 9.26 Simplified distribution of tensile force (Lazarte et al., 2003).

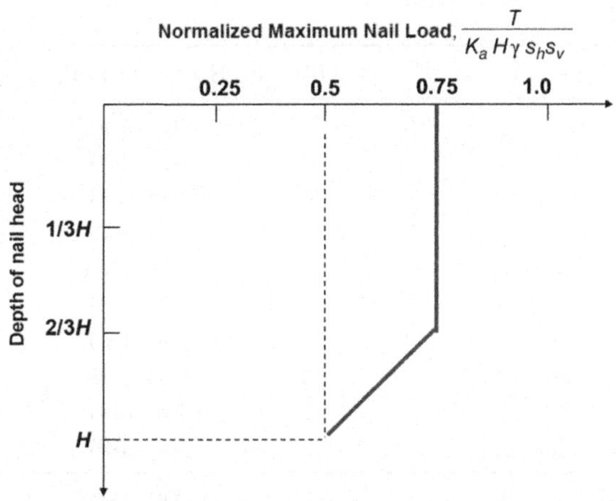

Figure 9.27 Approximation for maximum in-service tensile force (drawn from the suggestion in Lazarte et al. (2003)).

lateral earth pressure, γ = unit weight of soil, H = height of the wall, s_h = horizontal spacing of nails, and s_v = vertical spacing of nails. This calculated maximum nail load corresponds to the long-term in-service soil nail force instead of that at the end-of-construction or at failure.

Ultimate Pullout Capacity The front pullout capacity of the nail can be calculated as follows:

$$R_F = T_0 + \pi\tau_u d_{\rm DH} L_a \tag{9.9}$$

where T_0 = connection strength between the nail and the facing
τ_u = ultimate bond strength between the nail and the soil or rock
d_{DH} = average or effective diameter of the drill hole
L_a = length of the nail in the active zone.

The rear pullout capacity of the nail can be calculated as follows:

$$T_{\rm po} = \pi T_u d_{\rm DH} L_p \tag{9.10}$$

where L_p is the nail length in the resistant zone.

The ultimate bond strength is provided in Table 9.9.

To prevent the pullout of the nail from the resistant zone, $T_{po} \geq FS_{po}T_{max}$, where FS_{po} = factor of safety against pullout (typically 2.0). The required nail length in the resistant zone can be calculated based on this requirement. To prevent the pullout of the nail from the active zone and facing, $R_F \geq FS_{po}T_{max}$. This requirement demands sufficient connection strength between the nail and the facing. The total nail length is the sum of the nail lengths in the active and resistant zones.

Design Charts for Nail Force and Length Design charts for preliminary determination of nail force and length are developed in Lazarte et al. (2003) using the SNAIL computer program and also shown in Figures 9.28–9.31. These design charts are developed based on the following conditions:

- Homogeneous ground conditions
- Nail inclination angle = 15° (below the horizontal direction)
- Global factor of safety, FS_G = 1.35 for minimum nail length and FS_G = 1.0 for maximum design nail force
- Nail diameter, d_{DH} = 100 mm
- Normalize cohesion, c^* = 0.02

The normalized cohesion, c^*, can be expressed as follows:

$$c^* = \frac{c}{\gamma H} \tag{9.11}$$

where c = cohesion of the soil
γ = unit weight of the soil
H = height of the wall

The normalized pullout resistance, μ_{po}, is expressed as follows:

$$\mu_{\rm po} = \frac{T_u d_{\rm DH}}{{\rm FS}_{po}\gamma s_h s_v} \tag{9.12}$$

where τ_u = ultimate bond strength between nail and soil or rock
s_h = horizontal spacing of nails
s_v = vertical spacing of nails
FS_{po} = factor of safety against pullout (typically 2.0)

Based on the normalized cohesion, c^*, and the normalized pullout resistance, μ_{po}, the minimum nail length to wall height ratio, L/H, and the normalized maximum design nail force, t_{max-s} for each nail, can be determined from Figures 9.28–9.31.

The maximum design nail force, T_{max-s} for each nail, can be determined as follows:

$$T_{\max-s} = t_{\max-s}\gamma H s_h s_v \tag{9.13}$$

It should be noted that the design charts from Figures 9.28–9.31 are developed based on specific conditions as listed above. When the project conditions are different from these conditions, the determined L/H ratio and t_{max-s} should be corrected as follows:

$$\left(\frac{L}{H}\right)_{\rm corrected} = C_{1L}C_{2L}C_{3L}\left(\frac{L}{H}\right)_{\rm Chart} \tag{9.14}$$

$$(t_{\max-s})_{\rm corrected} = C_{1F}C_{2F}(t_{\max-s})_{\rm Chart} \tag{9.15}$$

Table 9.9 Estimated Bond Strength of Soil Nails in Soil and Rock

Material	Construction Method	Soil/Rock Type	Ultimate Bond Strength, τ_u (kPa)
Rock	Rotary Drilled	Marl/limestone	300–400
		Phyllite	100–300
		Chalk	500–600
		Soft dolomite	400–600
		Fissured dolomite	600–1000
		Weathered sandstone	200–300
		Weathered shale	100–150
		Weathered schist	100–175
		Basalt	500–600
		Slate/Hard shale	300–400
Cohesionless soils	Rotary drilled	Sand/gravel	100–180
		Silty sand	100–150
		Silt	60–75
		Piedmont residual	40–120
		Fine colluvium	75–150
	Driven casing	Sand/gravel	
		Low overburden	190–240
		High overburden	280–430
		Dense moraine	380–480
		Colluvium	100–180
	Augered	Silty sand fill	20–40
		Silty fine sand	55–90
		Silty clayey sand	60–140
	Jet grouted	Sand	380
		Sand/gravel	700
Fine-Grained Soils	Rotary drilled	Silty clay	35–50
	Driven casing	Clayey silt	90–140
	Augered	Loess	25–75
		Soft clay	20–30
		Stiff clay	40–60
		Stiff clayey silt	40–100
		Calcareous sandy clay	90–140

Source: Elias and Juran (1991).

where C_{1L} = length correction factor for drill hole diameter
C_{2L} = length correction factor for soil cohesion
C_{3L} = length correction factor for global factor of safety
C_{1F} = force correction factor for drill hole diameter
C_{2F} = force correction factor for soil cohesion

Also C_{1L} and C_{1F} can be determined from Figure 9.32; C_{2L}, C_{2F}, and C_{3L} can be determined from the following relationships:

$$C_{2L} = -4.0c^* + 1.09 \geq 0.85 \quad (9.16)$$

$$C_{2F} = -4.0c^* + 1.09 \geq 0.85 \quad (9.17)$$

$$C_{3L} = 0.52FS + 0.30 \geq 1.0 \quad (9.18)$$

With the known maximum design nail load, the maximum design nail head load for each nail can be estimated by

$$T_0 = T_{\max-s}[0.6 + 0.2(s_{\max} - 1)] \quad (9.19)$$

where T_0 = maximum design nail load at the nail head
$s_{\max}$ = maximum soil nail spacing (maximum of s_v and s_h) (unit: m)

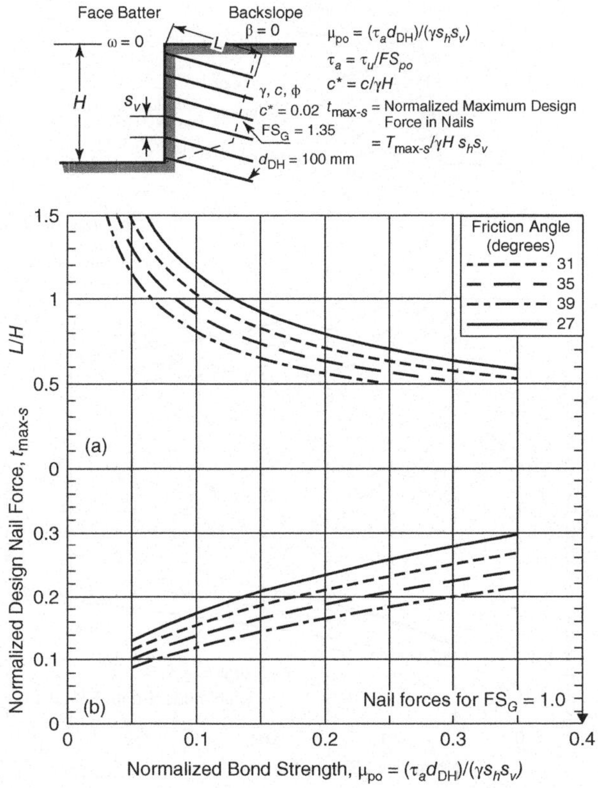

Figure 9.28 Design chart for minimum nail length and maximum design force, $t_{max\text{-}s}$ (batter = 0° and blackslope = 0°) (Lazarte et al., 2003).

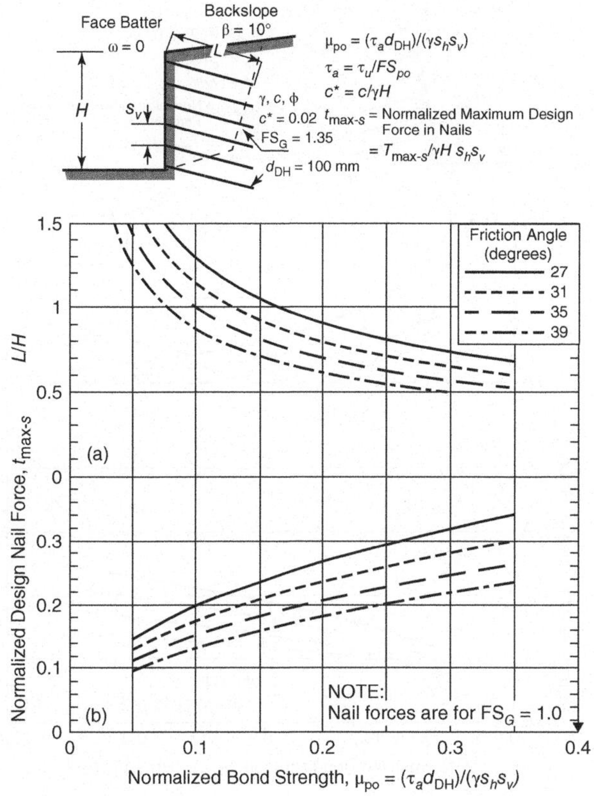

Figure 9.29 Design chart for minimum nail length and maximum design force, $t_{max\text{-}s}$ (batter = 0° and blackslope = 10°) (Lazarte et al., 2003).

The required cross-sectional area of the nail bar is

$$A_{nb} = \frac{T_{max\,-s} \mathrm{FS}_T}{f_y} \tag{9.20}$$

where A_{nb} = required cross-sectional area of the nail bar
f_y = steel yield strength
FS_T = factor of safety against rupture of the nail bar (typically 1.8 for static loading)

The properties of threaded bars are provided in Table 9.10.

Internal Sliding Analysis The internal sliding stability can be analyzed using the simplified wedge method shown in Figure 9.33.

The factor of safety against the internal sliding through the nails can be defined as follows:

$$\mathrm{F_{is}} = \frac{\sum \text{resisting forces}}{\sum \text{driving forces}} \tag{9.21}$$

The normal and tangent forces on the failure plane are

$$N_F = (W + Q_T)\cos\psi + T_{\mathrm{EQ}}\cos(\psi - \theta_{\mathrm{in}}) \tag{9.22}$$

$$S_F = (W+Q_T)\sin\psi - T_{\mathrm{EQ}}\sin(\psi-\theta_{\mathrm{in}}) = c_m L_F + N_F \tan\phi_m \tag{9.23}$$

$$c_m = \frac{c'}{\mathrm{FS_{is}}} \tag{9.24}$$

$$\tan\phi_m = \frac{\tan\phi'}{\mathrm{FS_{is}}} \tag{9.25}$$

where ω = wall face batter angle (from vertical)
β = slope angle
ϕ' = soil effective angle of internal friction
c' = soil effective cohesion
ψ = inclination of failure plane
θ_{in} = nail inclination
L_F = length of failure plane
W = weight of sliding mass
Q_T = surcharge load
T_{EQ} = equivalent nail force
N_F = normal force on failure surface
S_F = shear force on failure surface
$\mathrm{FS_{is}}$ = factor of safety against internal sliding

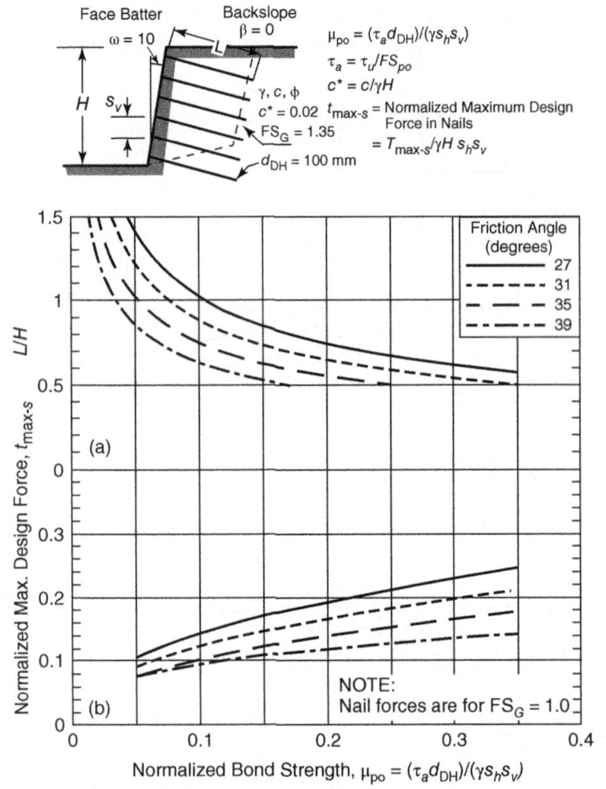

Figure 9.30 Design chart for minimum nail length and maximum design force, $t_{max\text{-}s}$ (batter = 10° and blackslope = 0°) (Lazarte et al., 2003).

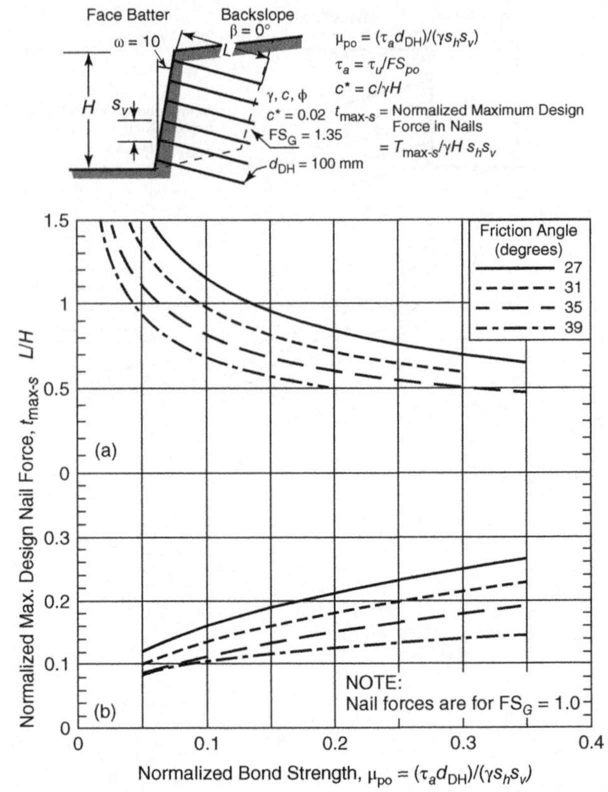

Figure 9.31 Design chart for minimum nail length and maximum design force, $t_{max\text{-}s}$ (batter = 10° and blackslope = 10°) (Lazarte et al., 2003).

At different slip plane angles, ψ, with the required factor of safety, the required equivalent nail forces can be solved in the above equations. The required maximum nail force will be used to design the nails.

Alternatively, based on the known total nail capacity, the factor of safety against the internal sliding can be solved from the above equations.

External Base Sliding Analysis The external base sliding analysis can be performed by treating the nailed mass as a rigid body subjected to external forces as shown in Figure 9.34. This analysis is similar to that for the MSE wall to be discussed in Chapter 10. The factor of safety against external sliding can be calculated as follows:

$$FS_{bs} = \frac{R_{bs}}{P_{ah}} \tag{9.26}$$

$$R_{bs} = c_b B_L + (W + Q_D + P_a \sin\beta)\tan\phi_b \tag{9.27}$$

$$P_{ah} = P_a \cos\beta \tag{9.28}$$

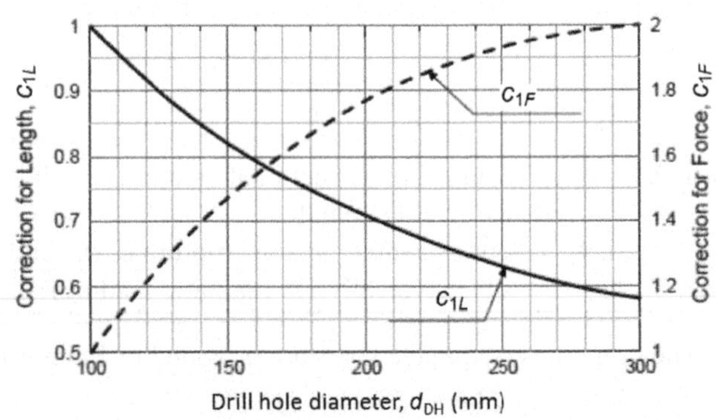

Figure 9.32 Correction factors for nail length and force due to drill hole diameter (Lazarte et al., 2003).

The active lateral earth thrust is

$$P_a = \frac{\gamma H_1^2}{2} K_a \tag{9.29}$$

$$H_1 = H + (B_L - H\tan\omega)\tan\beta_{eq} \tag{9.30}$$

where H = wall height
ΔH = slope rise up to bench (if present)

Table 9.10 Threaded Bar Properties (Grades 420 and 525 MPa)

Nominal Bar Designation (mm)	Cross-Sectional Area (mm^2)	Normal Unit Weight (kg/m)	Maximum Diameter with Threads (Mm)	ASTM Grade Metric	Yield Strength (MPa)	Maximum Axial Load (kN)
19	284	2.24	21.8	420/525	420/525	118/147
22	387	3.04	25.1	420/525	420/525	160/200
25	510	3.98	28.4	420/525	420/525	211/264
29	645	5.06	32.0	420/525	420/525	267/334
32	819	6.41	36.3	420/525	420/525	339/424
36	1006	7.91	40.9	420/525	420/525	417/520
43	1452	11.39	47.2	420/525	420/525	601/751

Source: ASTM A615-14, *Standard Specification for Deformed and Plain Carbon-Steel Bars for Concrete Reinforcement.* Reprinted with permission from ASTM International.

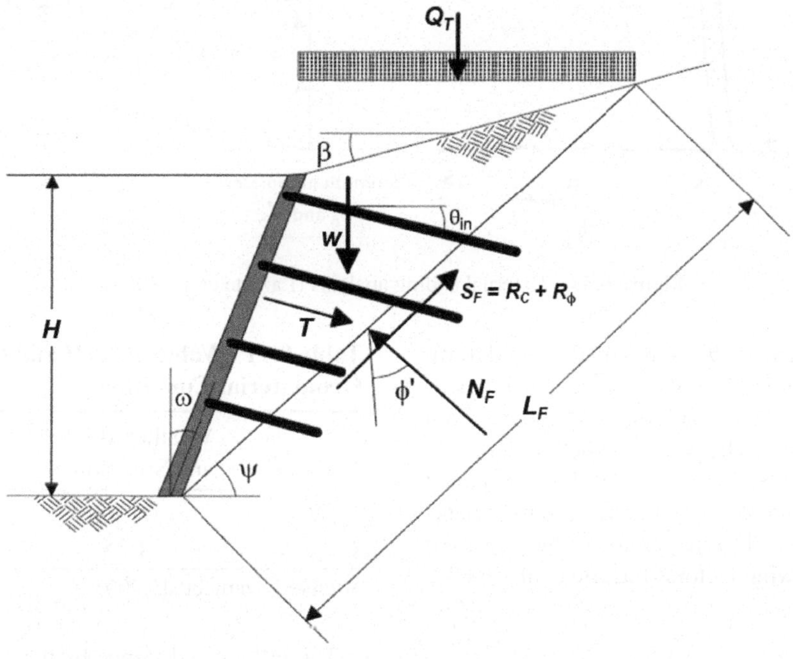

Figure 9.33 Simplified internal sliding analysis (Lazarte et al., 2003).

β = backslope angle;
β_{eq} = equivalent backslope angle [for broken slopes $\beta_{eq} = \tan^{-1}(0.5\Delta H/H)$, for infinite slopes $\beta_{eq} = \beta$]
ω = face batter angle
θ_f = inclination of wall face from horizontal (i.e., $\theta_f = \omega + 90°$)
c_b = soil cohesion along the base
B_L = length of the horizontal failure surface where c_b is effectively acting
W = weight of soil nail block
Q_D = permanent portion of total surcharge load Q_T
ϕ'_b = effective angle of internal friction of the base (remolded or residual values may be needed if significant movement takes place)
ϕ' = effective friction angle of soil behind soil nail block
δ = wall–soil interface friction angle [for a broken slope, $\delta = \beta_{eq}$, for infinite slope, $\delta = \beta$]
γ = total unit weight of soil mass
H_1 = effective height over which the earth pressure acts [$H_1 = H + (B_L - H \tan \omega) \tan \beta_{eq}$] and K_a = active earth pressure coefficient for soil behind the soil nail wall system

Base Stability When an excavation occurs in soft soil, the analysis for possible bearing capacity failure due to the base of the excavation should also be performed using the same approach as presented for base stability of the excavation pit with the deep mixed or jet-grouted walls in Section 8.2.3 of Chapter 8.

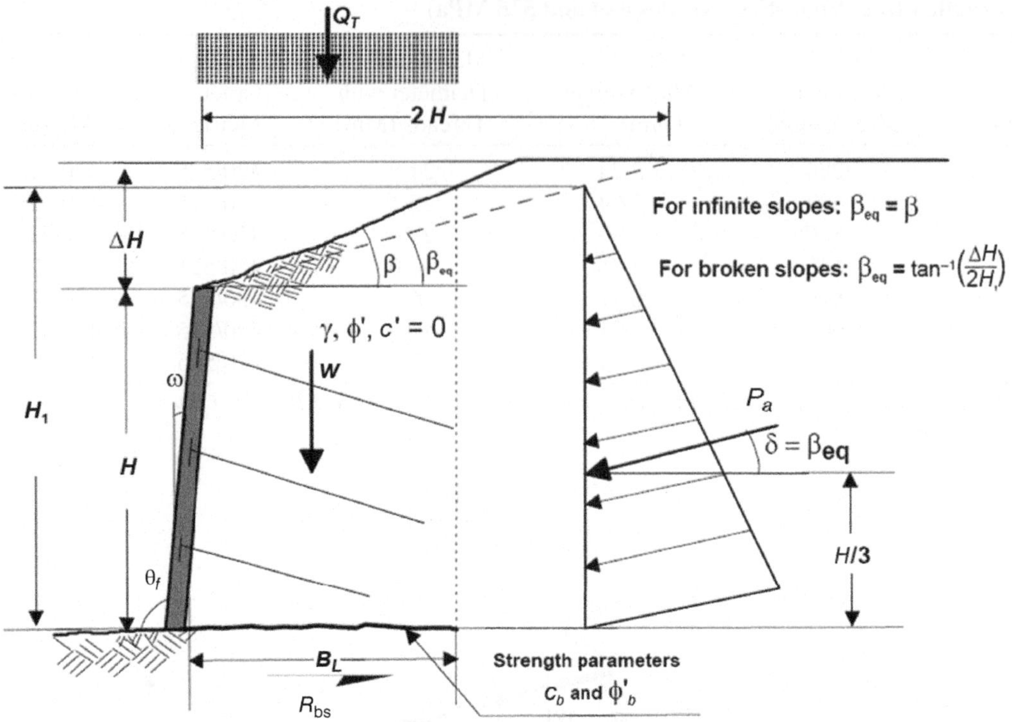

Figure 9.34 External sliding analysis (Lazarte et a.., 2003).

Deformation of Soil Nailed Wall Soil nailed walls deform during and after construction. Most of the movement happens during or right after excavation of the soil in front of the wall due to the unbalanced forces. The postconstruction deformation is related to stress relaxation and creep. For soil nailed walls, the maximum horizontal and vertical displacements occur at the top of the wall. The magnitude of the deformation depends on the following factors (Lazarte et al., 2003):

- Wall height
- Wall geometry
- Geomaterial type and properties
- Design factor of safety
- Nail length to wall height ratio
- Nail inclination angle
- Magnitude of surcharge

Based on empirical data with typical values of L/H = 0.7–1.0, no surcharge, and global factor of safety = 1.5, the maximum long-term horizontal and vertical wall displacements at the top of the wall can be estimated as follows:

$$\delta_h = \delta_v = \left(\frac{\delta_0}{H}\right) H \qquad (9.31)$$

where δ_h = horizontal displacement
δ_v = vertical displacement
δ_0/H = displacement to height ratio, which can be determined from Table 9.11

Table 9.11 Value of δ_0/H and C as Function of Geomaterial Condition

Variable	Weathered Rock and Stiff Soil	Sandy Soil	Fine-Grained Soil
δ_0/H	1/1000	1/500	1/333
C	1.25	0.8	0.7

Source: Lazarte et al. (2003).

The influence distance by the soil nailed wall as shown in Figure 9.35 can be estimated by the following equation:

$$\frac{x_{DEF}}{H} = C(1 - \sin\omega) \qquad (9.32)$$

where ω = wall batter angle
C = constant, provided in Table 9.11.

Drainage Considerations Drainage is one of the important parts of the soil nailed wall design, which includes surface drainage, wall facing drainage, internal drainage, and toe drainage, as shown in Figure 9.36. The objective of the drains is to reduce pore water pressure in the wall system. Details on drainage design can be found in Chapter 6.

Facing Design In addition to the above designs, the wall facing, including the connection between the nails and the

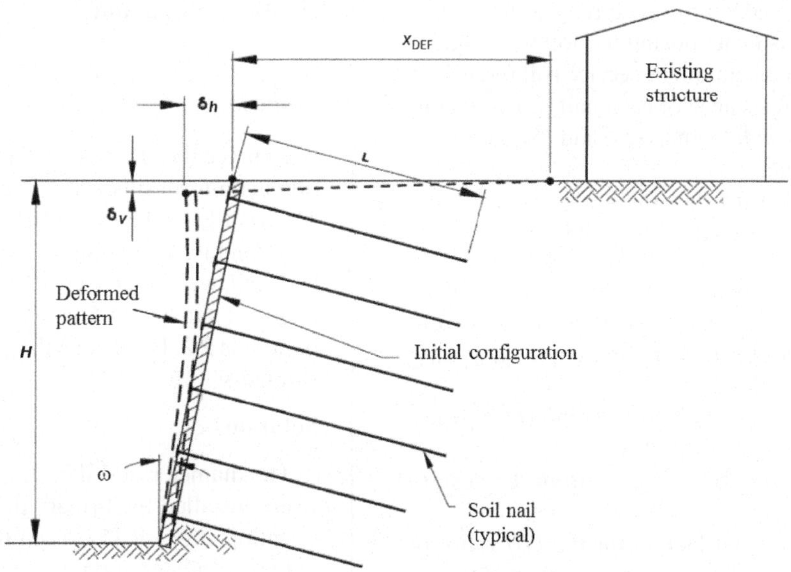

Figure 9.35 Deformations of soil nailed wall (Byrne et al., 1998).

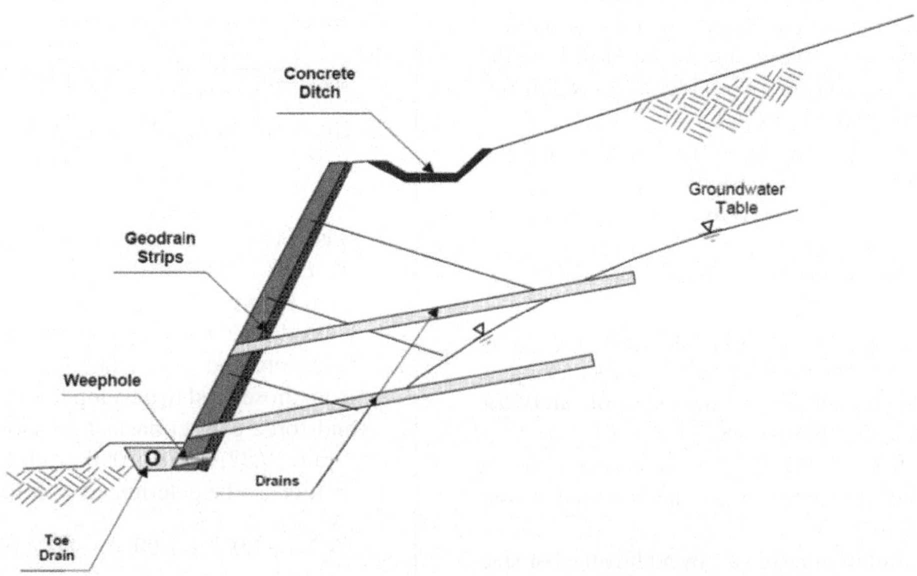

Figure 9.36 Typical drainage of soil nailed wall (Lazarte et al. 2003).

facing, should be designed. Most of the wall facing design is related to the structural (flexural and shear) capacities of the facing, which can be found in Lazarte et al. (2003).

9.3.4 Design Parameters and Procedure

Design Parameters Typical design parameters for soil nailed walls include:

- Type of application (temporary or permanent)
- Wall geometry (wall height, wall batter, backslope)
- Geomaterial type (e.g., soil or rock) and properties (cohesion, friction angle, unit weight)
- Groundwater table
- Drill hole diameter
- Soil nail layout (vertical and horizontal spacing, length, and inclination)
- Soil nail material type (e.g., steel bar grade)
- Nail installation method
- Ultimate bond strength between nail and geomaterial
- Distance to existing utilities and/or structures
- Required factors of safety

Design Procedure The maximum nail force and the minimum nail length can be estimated using the following three methods: (1) based on the maximum in-service nail force distribution and the pullout resistance of each nail, (2) based on the internal and external sliding analyses, and (3) based on the simplified design charts.

The preliminary design using the simplified design charts includes the following steps (Lazarte et al. 2003):

1. For a specific project application, calculate normalized pullout resistance (μ_{po}) based on the wall batter (ω), backslope (β), soil effective friction angle (φ'), and ultimate bond strength (τ_u).
2. Obtain the nail length to wall height ratio (*L*/*H*) from the design charts.
3. Obtain the normalized force ($t_{max\text{-}s}$) from the design charts.
4. Determine the correction factors for the *L*/*H* ratio to account for the drill hole diameter other than 100 mm (i.e., C_{1L}), the c^* value other than 0.02 (i.e., C_{2L}), and the global factor of safety other than 1.35 (i.e., C_{3L}).
5. Determine the correction factors for the normalized maximum nail force to account for the drill hole diameter other than 100 mm (4 in.) (i.e., C_{1F}) and the c^* value other than 0.02 (i.e., C_{2F}).
6. Apply the correction factors to the *L*/*H* ratio and the normalized force from the design charts.
7. Multiply the corrected *L*/*H* ratio by the wall height to obtain the soil nail length.
8. Calculate the maximum design load in the nail $T_{max\text{-}s}$ using the corrected value of $t_{max\text{-}s}$.
9. Calculate the required cross-sectional area (A_{nb}) of the nail bar based on the calculated nail axial force, $T_{max\text{-}s}$, the steel yield strength, f_y, and the factor of safety for the nail bar tensile strength, FS_T.
10. Select the first available bar size that has a cross-sectional area larger than the required cross-sectional area.
11. Calculate the minimum nail size by adding the bar size with two times the minimum grout cover thickness of 25 mm for two sides and verify that the minimum nail size is smaller or equal to the drill hole size.
12. If the length and/or nail diameter are not appropriate, select another nail spacing and/or drill hole diameter, recalculate the normalized pullout resistance, and start the process again.

Additional analyses may be performed for:

- External sliding
- Base stability
- Deformation
- Drainage

9.3.5 Design Example

Problem

A permanent soil nailed wall is proposed for an excavation in uniform silty sand. The excavation depth is 9 m. The wall has a batter of 10° without a top slope. The silty sand has cohesion of 2 kPa, friction angle of 32°, and unit weight of 18.5 kN/m^3. The required global factor of safety is 1.5 and the factor of safety against pullout is 2.0. Determine the layout, length, and size of soil nails.

Solution

1. The diameter of drill holes is 150 mm and soil nails are installed by a rotary drilled method at an inclination angle of 15°. The ultimate bond strength between nail and soil is 125 kPa. Assume soil nails are placed in spacing of 1.5 m in vertical and horizontal directions. The normalized pullout resistance, μ_{po}, can be calculated as follows:

$$\mu_{po} = \frac{\tau_u d_{DH}}{FS_{po}\gamma s_h s_v} = \frac{125 \times 0.15}{2.0 \times 18.5 \times 1.5 \times 1.5} = 0.225$$

2. The normalized cohesion, c^*, is

$$c^* = \frac{c}{\gamma H} = \frac{2.0}{18.5 \times 9} = 0.0120$$

3. Based on $\mu_{po} = 0.225$, $\phi = 32°$, $\omega = 10°$, and $\beta = 0°$, $L/H = 0.62$, and $t_{max\text{-}s} = 0.22$ are determined from Figure 9.30.
4. Considering the fact that the drill hole diameter, cohesion, and global factor of safety are different from those used to develop this design chart. Length and force correction factors should be used. From Figure 9.32, $C_{1L} = 0.83$ and $C_{1F} = 1.47$. Other factors can be determined as follows:

$$C_{2L} = -4.0c^* + 1.09 = -4.0 \times 0.0120 + 1.09 = 1.04$$

$$C_{3L} = 0.52\text{FS} + 0.30 = 0.52 \times 1.5 + 0.30 = 1.08$$

$$C_{2F} = -4.0c^* + 1.09 = -4.0 \times 0.0120 + 1.09 = 1.04$$

5. The corrected nail length to wall height ratio is

$$\left(\frac{L}{H}\right)_{\text{corrected}} = C_{1L}C_{2L}C_{3L}\left(\frac{L}{H}\right)_{\text{Chart}}$$
$$= 0.83 \times 1.04 \times 1.08 \times 0.62$$
$$= 0.58 > 0.5 \quad \text{(OK)}$$

The design length of soil nails is

$$L = 0.58 \times 9 = 5.2 \text{ m}$$

select 5.5 m for construction.

6. The corrected normalized nail force for each nail is

$$(t_{\max-s})_{\text{corrected}} = C_{1F}C_{2F}(t_{\max-s})_{\text{Chart}}$$
$$= 1.47 \times 1.04 \times 0.22 = 0.34$$

The maximum design nail force for each nail is

$$T_{\max-s} = t_{\max-s}\gamma H s_h s_v$$
$$= 0.34 \times 18.5 \times 9 \times 1.5 \times 1.5 = 126 \text{ kN}$$

The maximum design nail force at the nail head is

$$T_0 = T_{\max-s}[0.6 + 0.2(s_{\max} - 1)]$$
$$= 126 \times [0.6 + 0.2 \times (1.5 - 1)] = 114 \text{ kN}$$

7. The required cross-sectional area of the nail bar (f_y = 420 MPa and FS_T = 1.8) is

$$A_{nb} = \frac{T_{\max-s}FS_T}{f_y} = \frac{126 \times 1.8}{420{,}000}$$
$$= 0.000541 \text{ m}^2 = 541 \text{ mm}^2$$

8. Select 29-mm threaded bar. Considering the minimum grout cover thickness of 25 mm, the minimum size of the nail is

$$d = 29 + 2 \times 25 = 79 \text{ mm} < 150 \text{ mm} \quad \text{(OK)}$$

9.3.6 Construction

Soil nails can be installed by different methods. The installation procedure discussed below is focused on the most commonly used grouted nails. Figure 9.37 shows the basic procedure for soil nailing installation:

1. Excavate a small cut. Typically, the cut is limited to 1–2 m (mostly the initial cut is limited to 1.0–1.2 m), depending on the geotechnical conditions.
2. Drill a nail hole to a desired length. Nail holes can be drilled by a rotary, percussion, auger, or rotary/percussion drilling method. The most commonly used drill method in the United States is the open-hole installation using auger drilling (especially the hollow-stem auger). Casing may be used for some installation methods to maintain the stability of the hole, but it slows down the installation and increases the cost.
3. Install and grout the nail, including the installation of strip drains. Nail bars are inserted into predrilled, straight-shafted drill holes, which are filled with clean cement grout, mostly by gravity. Injection or regrout under pressure may be adopted sometimes to increase the bond strength, especially in poor or weak geomaterials. At the same time, geocomposite strip drains are installed with the geotextile side against the geomaterial to avoid contamination of the shotcrete.
4. Place temporary facing. Temporary facing (sometimes permanent facing) has welded wire mesh, which is installed before shotcrete is applied. Both dry and wet mixes have been used for the shotcrete, but wet mix is more commonly used. The plastic mix is sprayed on the mesh at a high velocity by compressed air. Bearing plates, hex nuts, and washers are installed to connect the nails with the facing. Precast concrete facing has been used in permanent applications to meet aesthetic, environmental, and durability needs.

Repeat the preceding procedure until the whole wall is constructed as shown in steps 5 and 6 in Figure 9.37.

9.3.7 Quality Control and Assurance

Quality control for soil nailing installation typically includes the following procedures:

- Verification of the quality of all the materials used
- Inspection of corrosion protection of nails
- Inspection of nail bars free of damage and required length
- Verification of the stability of excavated wall facing
- Verification of the size and length of drill holes
- Verification of nails installed at the desired inclination, spacing, and length
- Verification of sizes and locations of centralizers
- Measurement of amount of grout used in each hole
- Verification of shotcrete placed to the required thickness
- Verification of proper placement of welded wire mesh, bearing plates, and other connection parts
- Verification of proper installation of drains

Quality assurance during construction should ensure the following items (Lazarte et al. 2003):

- Construction completed in accordance with plans and specifications
- No excavation height exceeding an allowable value
- Not caved nail drill holes during nail installation
- Nail bars of the right size and type (i.e., steel grade, length, diameter)
- Appropriate corrosion protection systems
- Properly grouting, installation of facing rebar and mesh, and shotcrete
- Sufficient grout strength from grout cubes
- Sufficient shotcrete strength from cores
- Nail pullout capacity from field testing meeting the requirements
- Drainage properly installed

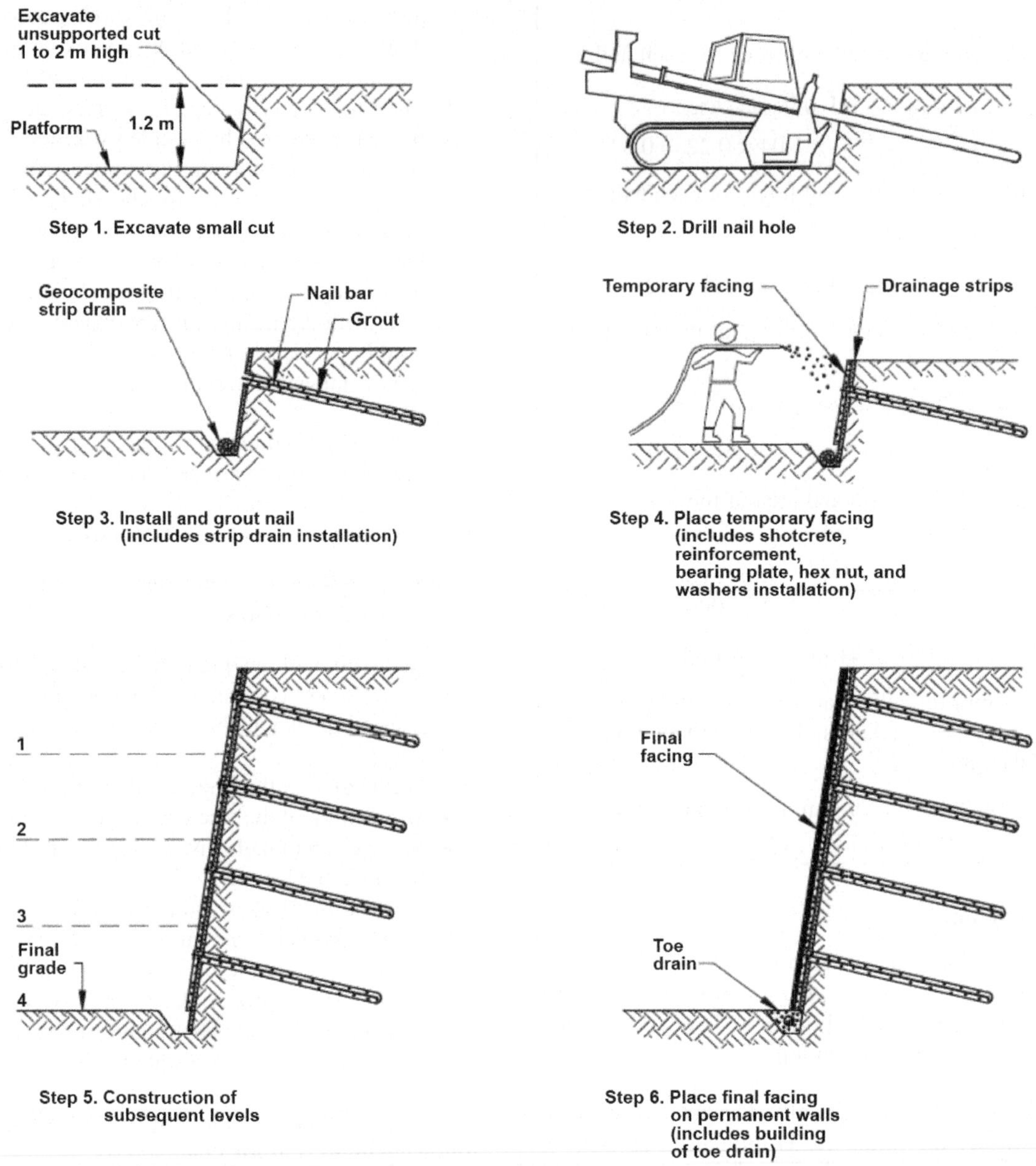

Figure 9.37 Installation sequence of soil nails (Lazarte et al. 2003).

PROBLEMS

9.1. Define in situ ground reinforcement and list four commonly used in situ ground reinforcement techniques.

9.2. What are the key differences between ground anchors and soil nails?

9.3. What are the key components of a ground anchor? Describe the functions of spacer, centralizer, and sheath in the ground anchor.

9.4. List five possible applications of ground anchors.

9.5. What are the advantages and limitations of ground anchors and anchored systems?

9.6. Describe the load transfer mechanism of a ground anchor under tension to stabilize a wall.

9.7. List possible failure modes of a ground anchor that may result from insufficient anchorage length/capacity. Explain why.

9.8. What measures or design considerations are needed to minimize prestressing steel bars and steel strands from corrosion in aggressive soil environment?

9.9. What is the major concern for ground anchors to be installed in organic soils or soils with high plasticity?

9.10. One level of ground anchors are installed at two-third height of a 3-m-high wall in sand during excavation.

The sand has a unit weight of 19.0 kN/m^3 and a friction angle of 34°. A competent soil exists at the base of the excavation and the groundwater table is at greater depth. Draw an apparent earth pressure diagram for the anchored wall and determine the horizontal anchor load per level using the tributary area method and the hinge method.

9.11. In Problem 9.10, if the excavation happens in a clay with a unit weight of 17.8 kN/m^3 and an undrained shear strength of 50 kPa, draw an apparent earth pressure diagram for the anchored wall and determine the horizontal anchor load per level using the tributary area method and the hinge method.

9.12. In Problem 9.10, if the anchored wall is subjected to a large area uniform traffic load of 12 kPa, what are the apparent earth pressure diagram and the horizontal anchor load per level?

9.13. Three levels of ground anchors are installed at equal spacing from the top to bottom of the 12-m-deep excavation in a clay with a unit weight of 18.2 kN/m^3 and an undrained shear strength of 50 kPa. The wall has an embedment of 3 m. Draw an apparent earth pressure diagram for the anchored wall and determine the horizontal anchor load at each level using the tributary area method and the hinge method.

9.14. One level of ground anchors are spaced at 2 m in the horizontal direction and installed at an inclination angle of 15° below the horizontal direction. The horizontal anchor load per level is 100 kN/m. Calculate the total anchor load on each anchor.

9.15. A straight shaft anchor is installed in medium dense fine to medium sand by pressure grouting. The diameter of the anchor drill hole is 60 mm. The anchor bond length is 1.5 m. Calculate the anchor bond capacity.

9.16. In Problem 9.15, if the same anchor is installed in stiff medium plasticity clay, what is the anchor bond capacity?

9.17. A 6-m-deep excavation is to be constructed in a medium to coarse sand with gravel [SPT $(N_1)_{60}$ = 30]. A site investigation at the site shows the soil unit weight and effective frictional angle of 20 kN/m^3 and 38°, respectively. The wall facing has sufficient toe resistance. (1) Determine the total load carried by the pressure-grouted anchors installed as shown in the following figure. Assume horizontal spacing of the anchors is 2 m. (2) Determine the sizes of the bar and trumpet opening. (A steel of grade 1035 is selected and the soil is not aggressive). (3) Determine the unbonded and bond lengths of the anchor. Since pressure grouting is used, care must be taken for the minimum overburden depth and critical failure surface requirements.

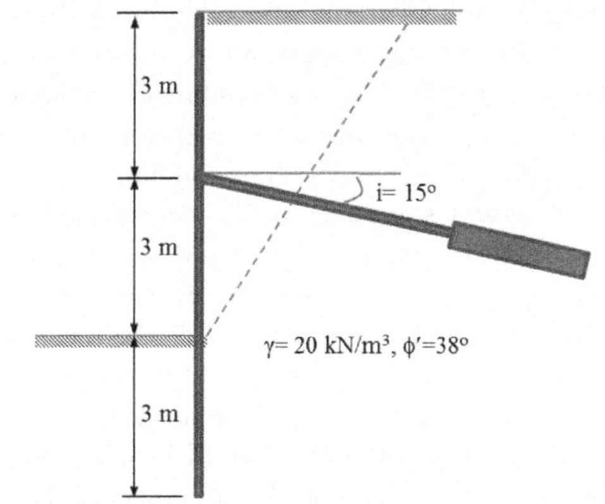

9.18. A 9-m-deep excavation is to be constructed in a medium plasticity sandy silt with an unconfined compressive strength of 400 kPa. A site investigation at the site shows the soil unit weight and effective frictional angle of 19.1 kN/m^3 and 25°, respectively. Two levels of ground anchors are needed: one at 6 m and another at 3 m from the bottom of the excavation. The wall embedment depth below the bottom of the excavation is 4.5 m, which provides sufficient toe resistance to the wall facing. (1) Determine the total loads carried by the pressure-grouted anchors. Assume horizontal spacing of the anchors is 2 m. (2) Determine the sizes of the bar and trumpet opening (A steel of grade 1035 is selected and the soil is environmentally aggressive.) (3) Determine the unbonded and bond lengths of the anchors. Since pressure grouting is used, care must be taken for the minimum overburden depth and critical failure surface requirements.

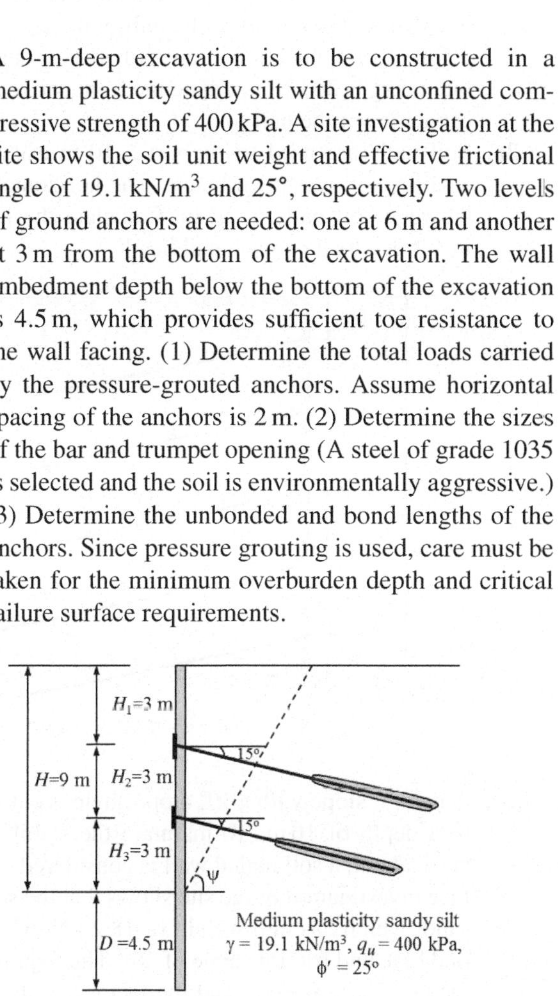

9.19. An 8.5-m-deep excavation with an anchored wall is to be constructed in stiff to hard clays at a distance of 25 m in front of an existing building. Estimate the induced settlement by the excavation at the closest distance of the existing building.

9.20. Define soil nailing technique for in situ ground reinforcement. What are the key components of soil nailing?

9.21. What are the favorable site conditions for soil nailing?

9.22. What are the major applications of soil nailing?

9.23. What are possible modes of failure of soil nailed walls?

9.24. Discuss the load transfer mechanism during the excavation and the installation of soil nails.

9.25. Explain why soil nailed walls, mechanically stabilized earth (MSE) walls, and ground anchored walls have different lateral earth pressure distributions.

9.26. Explain why in soil nailed walls, the locations of the maximum tensile forces are not necessarily the same as those for the critical failure surfaces.

9.27. A 8.0-m-long soil nail is installed in weathered shale by a rotary drill method. The diameter of the drill hole is 150 mm, the length of the nail in the active zone is 2.5 m. The connection strength between the nail and the facing is 30 kN. Estimate the front and rear pullout capacities of the nail.

9.28. Soil nails are used to support a 10-m excavation in a sandy soil as shown below. The horizontal spacing between soil nails is 2.5 m. Calculate the maximum in-service load on each nail head.

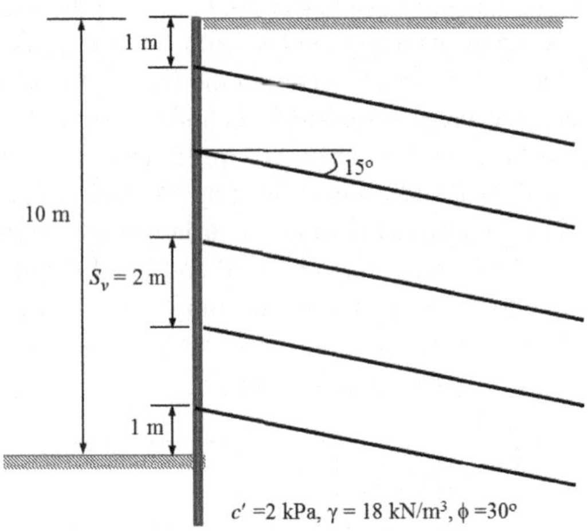

9.29. A natural slope with a 10° slope angle is cut vertically to a depth of 10 m. To maintain the stability of this vertical cut, a soil nailed wall is considered. A subsurface investigation of the site shows that the soil is silty sand, which has unit weight of 18.2 kN/m^3, cohesion of 3 kPa, and friction angle of 36°. The required global factor of safety is 1.5 and the factor of safety against pullout is 2.0. Design the layout, length, and diameter of soil nails required to retain the vertical cut. A rotary drilling is carried out to install soil nails.

9.30. Estimate the vertical and horizontal deflections at the top of soil nail wall if the depth of the excavation =10 m, the designed length of soil nails = 9 m, the wall batter angle = 10°, and the global factor of safety of 1.5. The soil nailing is constructed to retain fine-grained soil. An existing building is located at a distance of 6 m. Check whether the construction of the soil nail wall has any effect on the existing building.

REFERENCES

American Association of State Highway and Transportation Officials (AASHTO) (2012). *AASHTO LRFD Bridge Design Specifications*. Section 11: "Walls, Abutments, and Piers." Washington, DC, AASHTO.

ASTM International (2012). ASTM A722/A722M-12, *Standard Specification for Uncoated High-Strength Steel Bars for Prestressing Concrete*, West Conshohocken, PA, ASTM International.

ASTM International (2012). ASTM A416/A416M-12a, *Standard Specification for Steel Strand, Uncoated Seven-Wire for Prestressed Concrete*, West Conshohocken, PA, ASTM International.

ASTM International (2014). ASTM A615/A615M-14, *Standard Specification for Deformed and Plain Carbon-Steel Bars for Concrete Reinforcement*. West Conshohocken, PA, ASTM International.

British Standards Institution (2000). *Execution of special geotechnical work: Ground anchors*. BS EN 1537, London, UK.

Broms, B.B. (1965). "Design of laterally loaded piles." *J. Soil Mecha. Found. Div.*, (SM3: 79–99.

Byrne, R.J., Cotton, D., Porterfield, J., Wolschlag, C., and Ueblacker, G. (1998). *Manual for Design and Construction Monitoring of Soil Nail Walls*. Report FHWA-SA-96-69R, Federal Highway Administration, Washington, DC.

Elias, V. and Juran, I. (1991). *Soil Nailing for Stabilization of Highway Slopes and Excavations*. Publication FHWA-RD-89-198, Federal Highway Administration, Washington, DC.

Elias, V., Welsh, J., Warren, J., Lukas, R., Collin, J.G., and Berg, R.R. (2004). *Ground Improvement Methods*. FHWA-NHI-04-001, FHWA, Washington, DC.

Ito, T., and Matsui, T. (1975). "Methods to estimate lateral force acting on stabilizing piles." *Soils Found.*, 15(4), December: 43–59.

Lazarte, C.A., Elias, V., Espinoza, R.D., and Sabatini, P.J. (2003). *Geotechnical Engineering Circular No. 7 Soil Nail Walls*. Publication FHWA-IF-03-017, FHWA, Washington, DC.

Peck, R.B. (1969). "Deep excavations and tunneling in soft ground—State of art report." *Proceedings of the 7th International Conference on Soil Mechanics and Foundation Engineering*, Mexico City, Mexico, 225–290.

Post-Tensioning Institute (1996). *Recommendations for Prestressed Rock and Soil Anchors* (3rd ed.). Phoenix, AZ, Post-Tensioning Institute.

Sabatini, P.J., Pass, D.G., and Bachus, R.C. (1999). *Geotechnical Engineering Circular No. 4: Ground Anchors and Anchored Systems*. FHWA-IF-99-015, FHWA, Washington, DC.

Sabatini, P.J., Tanyu, B., Armour, T., Groneck, P., and Keeley, J. (2005) *Micropile Design and Construction*. Publication No. FHWA NHI-05-03, Federal Highway Administration, Washington, DC.

Terzaghi, K and Peck, R.B. (1967). *Soil Mechanics in Engineering Practice*. Wiley, New York.

Wang, S-T. and Reese, L.C. (1993). *COM624P, Laterally Loaded Pile Analysis Program for the Microcomputer, Version* 2.0. Report FHWA-SA-91-048, Federal Highway Administration, Washington, DC.

CHAPTER 10

Fill Reinforcement

10.1 INTRODUCTION

In Chapter 9, in situ ground reinforcement is used to stabilize the ground during excavation and under unstable natural conditions. Other types of reinforcement have also been used to stabilize fill for construction of slopes, embankments, walls, foundations, and roads. The reinforcement installed during the placement of fill is referred to as fill reinforcement in this book. Fill reinforcement has a long history. More than 1000 years ago (in the Han dynasty), Chinese people built earth retaining walls using local sand and weed for border security and paths to the Western world. Figure 10.1 shows the remains of the Han Great Wall. This wall is located at Dunhuang in Gansu Province, China, which is about 500 km west to the Jiayuguan Pass. The photo clearly shows that fill and reinforcement were placed in layers. The vertical spacing between reinforcement layers is 120–150 mm. The remains of the wall are 0.5 to over 2.0 m high. The modern

Figure 10.1 The Han Great Wall (photo courtesy of Xue).

fill reinforcement technology for retaining wall construction was pioneered by Henri Vidal, a French architect and engineer, in the early 1960s. More durable and stronger steel strip reinforcement was used instead of natural weed. In addition, wall facing was included in front of the wall for erosion protection.

Nowadays fill reinforcement includes geosynthetic reinforcement and metallic reinforcement. In fill reinforcement, fill and geosynthetic or metallic reinforcement are placed in layers. Soil is strong in compression but weak in tension. Reinforcements are factory-made polymeric or metallic materials that can carry tension. When reinforcements are included in soil, they make the reinforced soil strong in both compression and tension. Geosynthetics, including woven geotextile, geogrid, and geocell, are used to reinforce slopes, embankments, earth retaining walls, foundations, and roads. Metallic reinforcements, including steel strips and steel meshes, are mainly used for earth retaining walls. Chapter 2 discusses types, functions, and properties of geosynthetics. As the function of reinforcement, geosynthetic provides tension or confining stress to fill through the mechanisms of anchorage, tensioned membrane, and lateral restraint by friction, interlocking, or close confinement as shown in Figure 10.2. One or multiple mechanisms may exist in one application. For example, anchorage is the key mechanism for reinforced slopes and earth walls, while friction by geotextile, interlocking by geogrid, or closed confinement by geocell contributes to lateral restraint of granular fill in roads. Tensioned membrane effect may become important in roadway applications when differential settlement is large, such as embankments over voids. Geosynthetic reinforcement can also limit the development of slip surfaces in reinforced foundations and change the boundary conditions on the top of weak foundations, thus increasing bearing capacity.

This chapter presents the principles, design, construction, and quality control/assurance of fill reinforcement for reinforced slopes, embankments, earth retaining walls, foundations, and roads.

10.2 GEOSYNTHETIC-REINFORCED SLOPES

10.2.1 Introduction

Most design codes or manuals define slopes as having a face inclination angle less than 70° from the horizontal line while walls have an inclination angle equal or greater than 70°. This division is not theoretical but artificial. However, design of reinforced slopes mostly follows limit equilibrium methods while design of reinforced walls follows the lateral earth pressure method. The limit equilibrium and lateral earth pressure methods result in different requirements for tensile strength of geosynthetic reinforcement (Han and Leshchinsky, 2006). In this section, design and

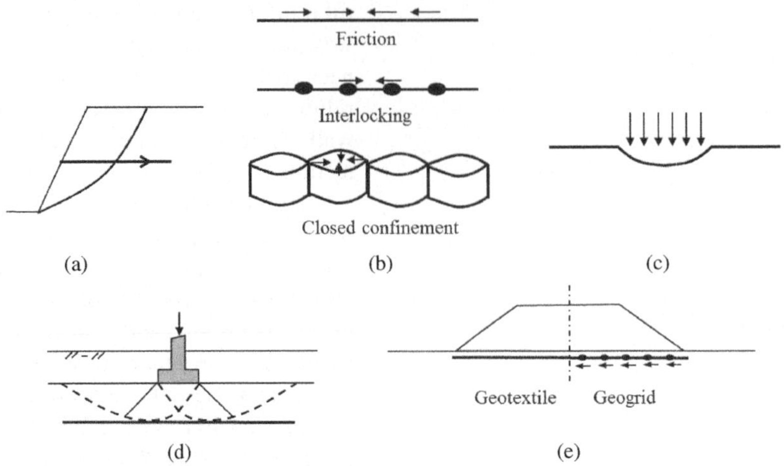

Figure 10.2 Mechanisms of geosynthetic reinforcement: (a) anchorage, (b) lateral restraint, (c) tensioned membrane, (d) limitation of slip surface, and (e) change of boundary.

construction of geosynthetic-reinforced slopes will be introduced. In a later section of this chapter, design and construction of geosynthetic-reinforced earth walls will be discussed together with metallic-reinforced earth walls.

Basic Concept Geosynthetic-reinforced slopes are formed by compacted fill and geosynthetics, which are placed in an alternating order. Geosynethtic reinforcement provides tensile resistance fill needs to maintain its stability. Geosynthetic-reinforced slopes can be built in different angles; however, they are more attractive when steep slopes are needed. Geosynthetic-reinforced slopes can have different face options including vegetated face, which can blend with the natural environment.

Suitability Most geosynthetic-reinforced slopes are constructed on strong and firm foundations. When weak foundations are encountered, they should be excavated and replaced, improved, or reinforced by extra geosynthetic reinforcements. No problematic geomaterial should be used as the fill material. Fill material should meet specific gradation and plasticity requirements

Applications Geosynthetic-reinforced slopes have been used when there is elevation difference. Common applications include embankments, levees, dams, riverbanks, and landfills. They have also been used to repair failed slopes.

Advantages and Limitations Geosynthetic-reinforced slopes have the following advantages:

- Can optimize space in terms of available budge.
- Have different facing options based on appearance, inclination, site conditions, and cost.
- Can have ecology-friendly vegetation.
- Are easy and speedy for construction.
- Does not require special labor and equipment.
- Allows the use of non-select fill.
- Are tolerant to different settlement.

Geosynthetic-reinforced slopes require large space as compared with walls. Surficial slope stability and erosion are two most common problems associated with geosynethetic-reinforced slopes.

10.2.2 Principles

A slope with cohesionless soil can be stable when its slope angle is less than the friction angle of the soil. However, when the slope angle is larger than the friction angle of the soil, a slip surface will develop and part of the slope will slide down. This slip surface divides the slope into an active zone and a stable zone. Geosynthetics have been successfully used to reinforce slopes and maintain their stability. The basic principle of geosynthetic-reinforced slopes is that geosynthetic reinforcement prevents the soil mass in the active zone from sliding down by providing necessary tensile resistance and resisting moment through its anchorage in the stable soil. Long and strong geosynthetic layers are often used as primary reinforcements while short and weak geosynthetic layers are sometimes used between primary reinforcements as secondary reinforcements. Geosynthetic-reinforced slopes have different types in practice, mostly based on the type of facing. During the design/analysis, all possible failure modes should be examined and prevented by the use of geosynthetic reinforcements or other means. Limit equilibrium method is the most commonly used method for the stability analysis of geosynthetic-reinforced slopes. Leshchinsky (1999) and

Duncan and Wright (2005) provided a comprehensive procedure for the design of steep slopes reinforced by geosynthetics using limit equilibrium methods. Numerical methods, based on the strength reduction concept have been increasingly adopted recently.

Type of Reinforced Slope Geosynthetic-reinforced slopes are mostly built with vegetated face and hard facing. The selection of slope facing is often dependent on slope angle and soil type as suggested by Collin (1996) in Table 10.1. Geosynthetic reinforcement can be wrapped or not wrapped at the slope face. The wrapped face enables the geosynthetic reinforcement to have more pullout resistance from the front and stabilize the soil exposed at the face. At a flatter slope angle (less than 45°), the facing provides the area for vegetation, which protects soil particles from being eroded. At a steeper slope angle (greater than 45°), vegetation growth is difficult, and, significant soil erosion likely happens; therefore, permanent erosion blanket or hard facing is used to protect soil particles from being eroded. Sometimes, steps are created for plant growth or special plants are directly planted into the steep slopes.

Failure Modes Geosynthetic-reinforced slopes have four possible failure modes: (1) surficial failure, (2) internal failure, (3) compound failure, and (4) global failure, as

Table 10.1 Options of Slope Facings

Slope Face Angle and Soil Type	Type of Facing			
	When Geosynthetic Is Not Wrapped at Face		When Geosynthetic Is Wrapped at Face	
	Vegetated Face[a]	Hard Facing[b]	Vegetated Face[a]	Hard Facing[b]
>45° (>1H:1V) All soil types	Not recommended	Gabions	Sod Permanent erosion blanket w/seed	Wire baskets Stone Shotcrete
35°–45° (~1.4H:1V to 1H:1V) Clean sands (SP)[c] Rounded gravel (GP)	Not recommended	Gabions Soil cement Sod	Permanent erosion blanket w/seed	Wire baskets Stone Shotcrete
35°–45° (~1.4H:1V to 1H:1V) Silts (ML) Sandy silts (ML) Silty sands (SM)	Bioreinforcement Drainage composite[d] Gabions	Soil-cement Stone veneer Sod	Permanent erosion blanket w/seed	Wire baskets Stone Shotcrete
35°–45° (~1.4H:1V to 1H:1V) Clayed sands (SC) Well-graded sands & gravels (SW & GW)	Temporary erosion blanket w/seed or sod Permanent erosion mat w/ seed or sod	Hard facing not needed	Geosynthetic wrap generally not needed	Geosynthetic wrap not needed
25°–35° (~2H:1V to 1.4H:1V) All soil types	Temporary erosion blanket w/seed or sod Permanent erosion mat w/ seed or sod	Hard facing not needed	Geosynthetic wrap generally not needed	Geosynthetic wrap not needed

[a]Vertical spacing of reinforcement (primary/secondary) shall be no greater than 400 mm with primary reinforcements spaced no greater than 800 mm when secondary reinforcements is used.
[b]Vertical spacing of primary reinforcement shall be no greater than 800 mm.
[c]Unified soil classification.
[d]Geosynthetic or natural horizontal drainage layers to intercept and drain the saturated soil at the face of the slope.
Source: Collin (1996) and Holtz et al. (2008).

shown in Figure 10.3. In all these modes, there are an active soil wedge and a stable soil mass. The active soil wedge is above the slip surface, while the stable soil mass is below the slip surface.

Surficial slope failure, mostly triggered by rainfall, is a common problem for slopes due to poor compaction, low overburden stress, loss of cohesion, saturation, and seepage force. The surficial failure typically happens within 1.2 m deep (Evans, 1972). Secondary weak and short secondary geosynthetic reinforcements can be placed between long and strong primary geosynthetic reinforcements to mitigate surficial slope instability.

Internal failure happens within the reinforced fill. The slip surface of internal failure can be circular or planar. Geosynthetic interface can be a potential slip plane. Internal failure may occur when geosynthetic reinforcements are too weak or too short for pullout resistance.

Compound failure happens when the slip surface partially cuts through the reinforced zone and the unreinforced zone. Such a failure mode can be prevented by extending the reinforcement length.

Global failure, also called deep-seated failure, is one of the major concerns when a reinforced slope is constructed on a weak foundation or a toe slope. This failure mode can be prevented by ground improvement of the weak foundation and geosynthetic reinforcement at the bottom of the slope. It is doable but not economic by extending the lengths of all the reinforcement layers in the slope to prevent the global failure.

Slope Stability Analysis Limit equilibrium methods have been commonly used to analyze the stability of the slope at each failure mode using a factor of safety. Bishop's simplified method (Bishop, 1955) and Spencer's method (Spencer, 1981) are two commonly used methods for slope stability analysis and were discussed in Chapter 2. Bishop's simplified method assumes a circular slip surface, while Spencer's method can be used for two-part or three-part wedge failure. Surficial and internal failures happen within the reinforced zone. The compound failure is partially through unreinforced and reinforced zones. Surficial slope stability is often analyzed by assuming an infinite slope because the thickness of the active soil is much smaller than its length. Global slope failure, has a critical slip surface beyond the reinforced soil mass. Therefore, the global slope stability analysis is the same as that for an unreinforced slope, which has been discussed in Chapter 2. In addition to circular slip failures, transitional failure should be checked because it may occur along a reinforcement layer or foundation interface. The methods for internal, surficial, compound, and translational stability analyses are discussed in the following sections.

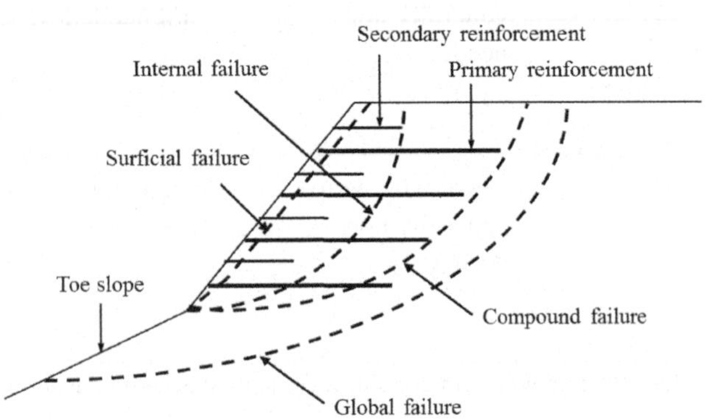

Figure 10.3 Possible failure modes of reinforced slope.

10.2.3 Design and Analysis

Performance Requirements Reinforced slopes should be designed to meet the following performance requirements:

1. Factors of safety against internal, surficial, compound, translational, and global failures under static loading should be greater than 1.3–1.5. Higher factors of safety should be used if a slope supports sensitive structures.
2. Factors of safety against internal, surficial, compound, and global failures under seismic loading should be greater than 1.1.
3. Postconstruction settlement should be based on project requirements.

Preliminary Design Design of geosynthetic-reinforced slopes requires the determination of geosynthetic layout (including length and spacing) and geosynthetic tensile strength based on the performance requirements. The determination of these parameters typically demands a trial-and-error process, which is tedious. The charts developed by Schmertmann et al. (1987), as shown in Figure 10.4, can estimate these parameters as a preliminary design. These charts were developed using limit equilibrium analyses of reinforced slopes with circular, wedge, and smooth noncircular slip surfaces based on the following conditions:

- Geosynthetic reinforcement
- Uniform, cohesionless reinforced and retained fills (i.e., $c = 0$)
- No pore water pressure within fills (i.e., $u = 0$)

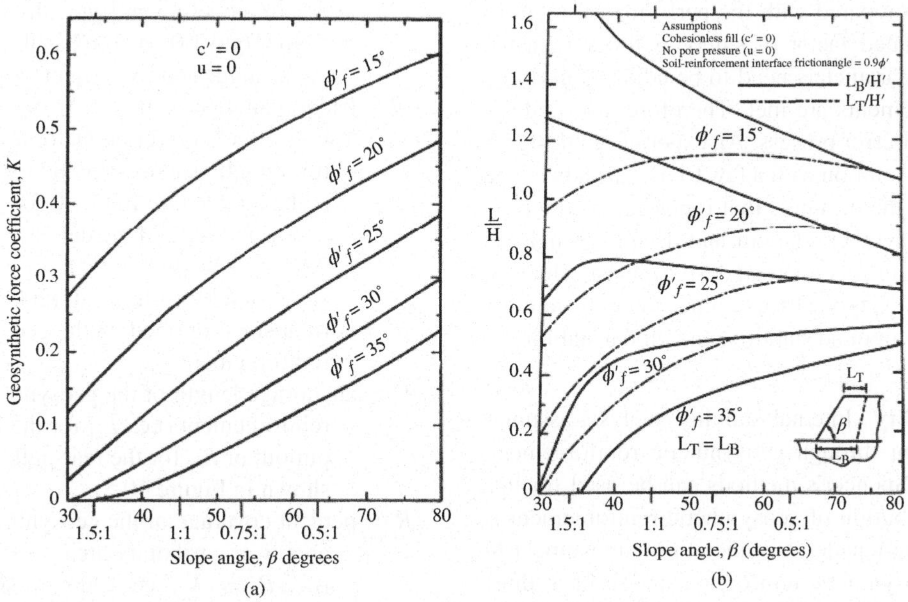

Figure 10.4 Design chart for geosynthetic-reinforced slopes: (a) force coefficient and (b) length ration (after Schmertmann et al., 1987, with permission from Industrial Fabrics Association International).

- Horizontal firm foundation (i.e., no slip surface into the foundation)
- No seismic force
- Uniform surcharge on the top of the slope with a magnitude less than $0.2\gamma_r H$ (γ_r is the unit weight of the reinforced fill and H is the height of the slope)
- Interface friction angle between geosynthetic and fill = $0.9\phi'_r$.

This preliminary design includes three design steps:

1. Use the following factored friction angle of the reinforced fill to determine the force coefficient K from Figure 10.4(a):

$$\phi'_f = \arctan\left(\frac{\tan\phi'_r}{FS}\right) \tag{10.1}$$

where ϕ'_f = factored effective friction angle of reinforced fill
ϕ'_r = effective friction angle of reinforced fill
FS = factor of safety

2. Calculate the total maximum required tensile strength of reinforcements as follows:

$$T_{\max} = 0.5K\gamma_r(H')^2 \tag{10.2}$$

where K = geosynthetic force coefficient from Figure 10.4(a)
γ_r = unit weight of reinforced fill
$H' = H + p/\gamma_r$
H = height of reinforced slope
p = uniform surcharge of slope

The distribution of the total maximum required tensile strength to reinforcements depends on the slope height. When $H < 6$ m, $T_{\max}$ is distributed uniformly to each reinforcement (i.e., $T_{\max}/N_R$ for each reinforcement, N_R = number of reinforcements). When $H > 6$ m, the slope can be divided into two zones or three zones in equal height. For two zones, $\frac{3}{4}\,T_{\max}$ should be distributed to the bottom zone, while $\frac{1}{4}\,T_{\max}$ should be distributed to the top zone. For three zones, $\frac{1}{2}\,T_{\max}$, $\frac{1}{3}\,T_{\max}$, and $\frac{1}{6}\,T_{\max}$ should be distributed to the bottom, middle, and top zones, respectively. The number of reinforcements in each zone can be calculated based on assumed spacing.

3. Determine the required top and bottom reinforcement lengths, L_T and L_B, from Figure 10.4(b). The reinforcement lengths between the top and bottom reinforcements can be interpolated linearly.

Slope Stability Analysis Different from the design, slope stability analysis is conducted based on known parameters to calculate the minimum factors of safety with different failure modes, which are compared with the performance requirements. If the calculated factor of safety is lower than the required value, the parameters need to be adjusted until the performance requirements are met. Therefore, the analysis process is a trial-and-error process. An experienced designer can quickly find the solution with a few trials. The advantage of the analysis over the design is that it provides more flexibility to adjust geosynthetic layout and design strength so that a more efficient and economical design can be achieved. Slope stability analysis typically examines internal stability, surficial stability, compound stability, and global stability.

1. Internal Stability Internal stability analysis assumes slip surfaces passing through geosynthetic reinforcement. Both Bishop's and Spencer's methods can be used for this purpose. The contribution of geosynthetic reinforcement is considered providing tensile resistance to the movement of the active soil. Geosynthetic reinforcement can have three different failure modes in soil, as shown in Figure 10.5: (1) tensile rupture, (2) pullout from stable soil (i.e., rear pullout), and (3) pullout from face (front pullout).

The allowable tensile strength of geosynthetic reinforcement can be determined as the ultimate tensile strength divided by the reduction factors due to creep, installation damage, and durability as discussed in Chapter 2. The pullout resistance of geosynthetic reinforcement can be estimated by the following equation:

$$T_{\text{po}} = 2F^{*}\alpha_{\text{se}}\sigma_z' L_a R_c \tag{10.3}$$

where F^* = pullout resistance (or friction-bearing-interaction) factor (commonly assumed as $F^* = \tan\delta = C_i \tan\phi$)

δ = interface friction angle between geosynthetic and soil

C_i = interaction coefficient [typically ranging from 0.6 to 1.0 and 0.67 as the default value suggested by Berg et al. (2009)]

ϕ = friction angle of soil

α_{se} = a scale effect correction factor to account for a nonlinear stress reduction over the embedded length [0.6–1.0 for geosynthetics and the default values of 0.6 for geotextile and 0.8 for geogrid as suggested by Holtz et al. (2008)]

σ_z' = normal stress applied on the geosynthetic reinforcement

L_a = anchorage length of the geosynthetic reinforcement (i.e., L_{af} for the front pullout or L_{ar} for the rear pullout as shown in Figure 10.5)

R_c = percent coverage of the geosynthetic area over a tributary area [i.e., $R_c = (A_g/A) \cdot 100\%$] as shown in Figure 10.6.

Percent coverage of geosynthetic reinforcement is sometimes used in practice to create a more economic design. The locations of geosynthetic layers below and above the current elevation should be offset to avoid creating large vertical

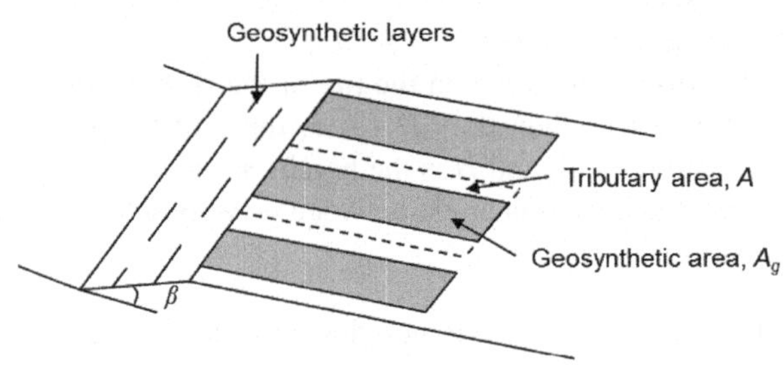

Figure 10.6 Percent coverage of geosynthetic reinforcement.

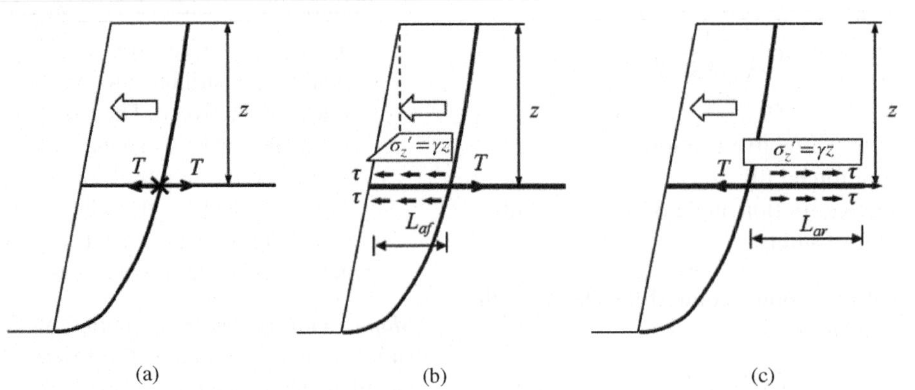

Figure 10.5 Possible failure modes of geosynthetic reinforcement in slope: (a) rupture, (b) front pullout, and (c) rear pullout (modified from Han and Leshchinsky, 2006).

unreinforced spacing as show in Figure 10.6. Due to the reduction of the cross section of geosynthetic reinforcement, the allowable tensile strength of the geosynthetic reinforcement should be multiplied by the percent coverage before used in design.

The contribution of geosynthetic reinforcement to the stability of reinforcement is to provide an anchorage force to the active soil mass through the stable soil. This anchorage force is often considered providing additional resisting moment to the active soil. Holtz et al. (2008) suggested adding this resisting moment provided by geosynthetic reinforcement to the resisting moment provided by soil in the calculation of factor of safety as follows:

$$\mathrm{FS}_r = \frac{M_r + M_g}{M_d} \tag{10.4}$$

where FS_r = factor of safety of reinforced slope
M_r = resisting moment provided by soil
M_g = resisting moment provided by geosynthetic reinforcement
M_d = driving moment due to soil weight and surcharge

When the reinforcement force for each layer is explicitly accounted for, Equation (10.4) can be rewritten as

$$\mathrm{FS}_r = \frac{M_r + \sum_{i=1}^{N} T_{\mathrm{ai}} y_{gi}}{M_d} = \mathrm{FS}_u + \frac{\sum_{i=1}^{N} T_{\mathrm{ai}} y_{gi}}{M_d} \tag{10.5}$$

where i = ith soil slice
N = total number of soil slices
T_{ai} = allowable tensile strength provided by the geosynthetic, which is the least of the rupture strength, and rear and front pullout capacities of the geosynthetic (when no geosynthetic intersects with the base of the slip surface, $T_{\mathrm{ai}} = 0$)
y_{gi} = vertical distance from the center of circle to the reinforcement (see Figure 10.7)
FS_u = factor of safety for unreinforced slope, which can be calculated following the procedure discussed in Chapter 2

The stability of a reinforced slope under seismic loading can be analyzed by considering a pseudoforce in each soil slice as shown in Figure 10.8:

$$\mathrm{FS}_r = \frac{M_r + \sum_{i=1}^{N} T_{\mathrm{ai}} y_{gi}}{M_d + \sum_{i=1}^{N} k_{\mathrm{hs}} W_i y_i} \tag{10.6}$$

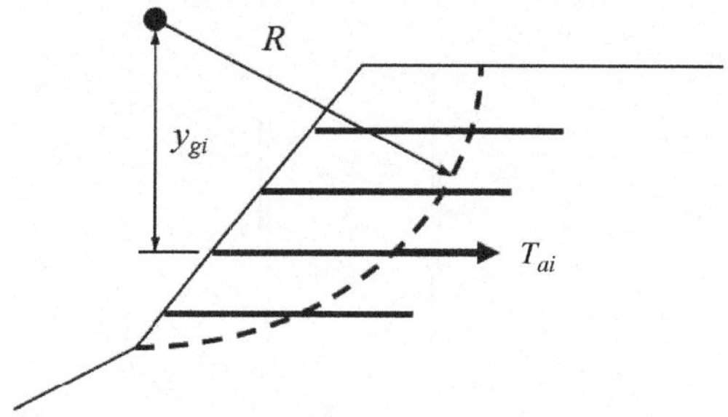

Figure 10.7 Notations used in circular slip surface stability analysis.

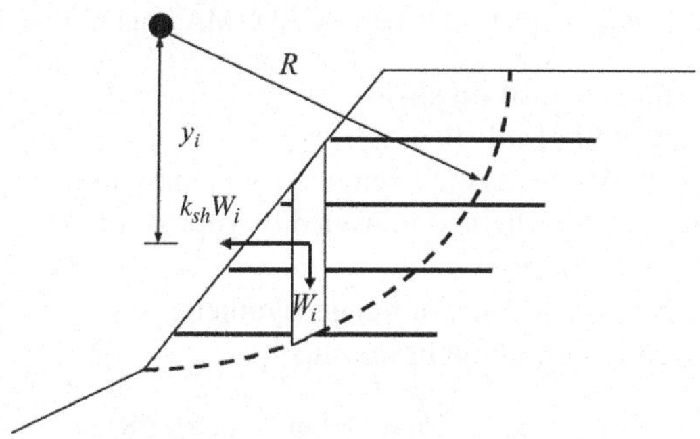

Figure 10.8 Consideration of horizontal seismic loading.

where y_i = vertical distance from the center of circle to the centroid of soil slice
k_{hs} = horizontal seismic coefficient
W_i = weight of fill slice

Considering the flexibility of reinforced slopes, the horizontal seismic coefficient can be taken as half of the seismic ground coefficient as suggested by AASHTO (2012) AASHTO LRFD Bridge Design Specifications (Section 11: Walls, Abutments, and Piers).

A more rigorous solution for a reinforced slope was obtained and is provided in the manual of the ReSSA software (ADAMA Engineering, Inc.), which follows the Bishop procedure of derivations for force and moment equilibrium (Figure 10.9) as follows:

$$\mathrm{FS}_r = \frac{\sum_{i=1}^{N} \{[c_i \Delta x_i + (W_i - T_{\mathrm{ai}} \sin\theta_{\mathrm{ri}}) \tan\phi_i]/m_{\mathrm{ai}}\}}{\sum_{i=1}^{N} [W_i \sin\alpha_i - T_{\mathrm{ai}} \cos(\alpha_i - \theta_{\mathrm{ri}})]} \tag{10.7}$$

where c_i = cohesion of fill
ϕ_i = friction angle of fill

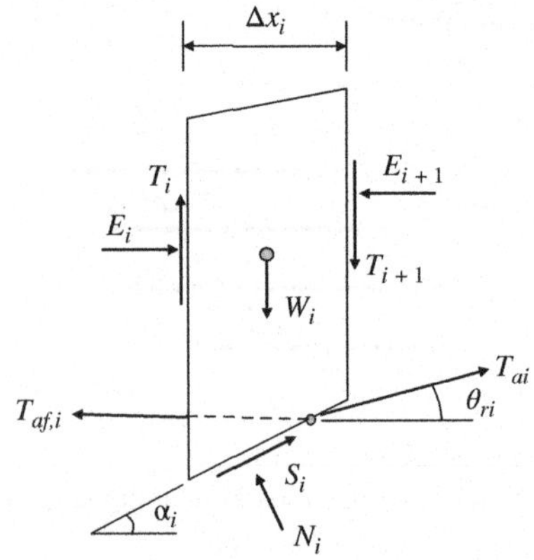

Figure 10.9 Force diagram in the modified Bishop's simplified method under static load (Courtesy ADAMA Engineering, Inc.).

Δx_i = thickness of fill slice
W_i = weight of fill slice
T_{ai} = allowable tensile strength provided by the geosynthetic (controlled by rupture or pullout)
θ_{ri} = rotational angle of the geosynthetic
α_i = base inclination of the slice

$$m_{\alpha i} = \cos \alpha_i + \tan \phi \sin \alpha_i / \text{FS}$$

The rotational angle of the geosynthetic from the horizontal direction may be considered in the design. The possible angle ranges from 0° (no rotation) to the inclination of the slip surface (full rotation).

2. Surficial Stability Due to potential shallow slip surfaces, as shown in Figure 10.10, the factor of safety of surficial slope stability is estimated based on the theory for an infinite slope (Collin, 1996):

$$\text{FS} = \frac{c'H + (\gamma_{\text{sat}} - \gamma_w)Hz\cos^2\beta \tan\phi' + T_a(\cos\beta \sin\beta + \sin^2\beta \tan\phi')}{\gamma_{\text{sat}} Hz \cos\beta \sin\beta} \tag{10.8}$$

where c' = effective cohesion of soil
ϕ' = effective friction angle of soil
H = slope height
γ_{sat} = saturated unit weight of soil
γ_w = unit weight of water
z = depth of slip surface
β = slope angle
T_a = total allowable tensile strength of geosynthetic layers intersecting with the slip surafce (controlled by rupture or pullout of geosynthetic)

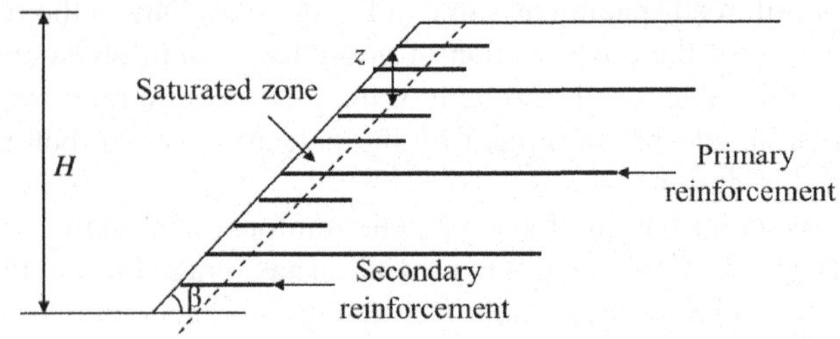

Figure 10.10 Secondary reinforcement layout.

When front pullout is a dominant failure mode, geosynthetic can be wrapped around at the slope face to increase its pullout capacity as shown in Figure 10.11.

3. Compound Stability The analysis of compound stability is similar to that of internal stability. Only the geosynthetic layers intersected by the slip surface should be considered in the stability analysis because they contribute to the stability of the slope. The formulas presented in the internal stability section can be used for this analysis.

4. Global Stability The global stability analysis examines slip surfaces beyond the reinforced fill and often becomes a concern when a slope is constructed on a weak foundation or a toe slope. This analysis has nothing to do with geosynthetic reinforcement; therefore, it is the same as that for an unreinforced slope.

Except for the surficial slope stability analysis, most of the above analyses are based on circular slip surfaces. Under certain conditions, such as the existence of a thin weak soil layer under the slope, translational failure may be more critical than circular slip failure. Translational stability analysis can utilize two-part and three-part wedge analyses, which are based on Spencer's method.

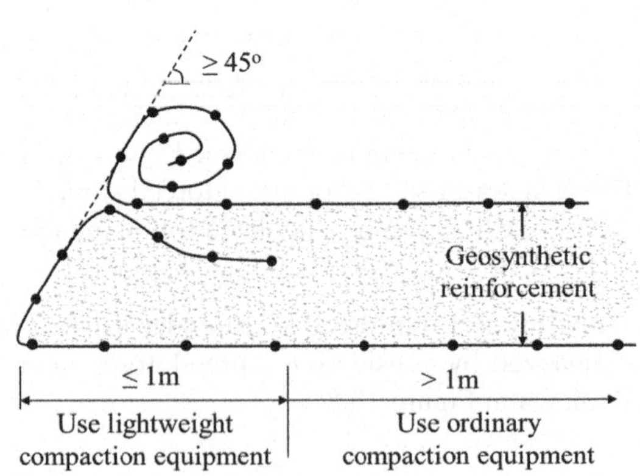

Figure 10.11 Wrapped-around facing.

Computer-Assisted Design A number of software available on the market can assist the analysis of reinforced slopes considering all the preceding failure modes. Most software is limit-equilibrium-based and some are based on limit analysis. Use of numerical software has been increased for slope stability analyses. With the rapid increase of computer speed, an analysis searching through thousands of slip surfaces using limit-equilibrium-based software may only take a few seconds to a few minutes. Numerical software may require more refined data and take longer than the software based on the limit equilibrium method to complete an analysis from a few minutes to a few hours, which are often acceptable for practical applications.

10.2.4 Design Parameters and Procedure

Design Parameters Design parameters for geosynthetic-reinforced slopes include the following:

- Geometry of slope (height, slope angle, reinforced fill zone, retained fill zone, top slope if any, and foundation soil including toe slope if any)
- Properties of reinforced fill (unit weight, cohesion, and friction angle)
- Properties of retained fill, top and/or toe slope, and foundation soil (unit weight, cohesion, and friction angle)
- Layout of geosynthetic reinforcements (length and spacing)
- Properties of geosynthetic reinforcements (allowable tensile strength and interaction coefficient with fill)
- Surcharge

The recommended gradation and plasticity for backfill used for reinforced slopes are provided in Table 10.2.

Design Procedure A preliminary design may be conducted first to understand the basic requirements for geosynthetic layout and tensile strengths. An analysis can be performed later to adjust the design parameters to achieve more efficient and economical design. The following analysis procedure is typically adopted in practice:

1. Run a stability analysis for the unreinforced slope, identify the critical slip surface, and compute the minimum factor of safety.
2. Place geosynthetic reinforcements in equal spacing and equal length. The lowermost reinforcement should be placed at the bottom of the slope if the critical slip surface is below the top of the foundation soil under the unreinforced slope. The lowermost reinforcement may be placed above the bottom of the slope if the critical slip surface is above the foundation soil. The reinforcement length should be at least 1 m beyond the critical slip surface toward the stable zone.
3. Assign the properties of geosynthetic reinforcements.
4. Run a stability analysis for the reinforced slope and identify the critical slip surface and compute the minimum factor of safety. If the computed minimum factor of safety is at the target value or slightly greater than the target value, the procedure is complete.
5. If the computed minimum factor of safety is higher or lower than the target value, adjust layout and properties of geosynthetic reinforcements and repeat steps 4 and 5 until the desired factor of safety is achieved.

The preceding procedure is only for the slope stability analysis on one failure mode. The same procedure can be used for the analysis on other failure modes. It is always a good idea to check all the failure modes. These analyses can be accomplished using design software.

Table 10.2 Recommended Backfill for Geosynthetic-Reinforced Slopes

Sieve Size (mm)	Percent Passing
20	75–100
4.75	20–100
0.425	0–60
0.075	0–50
Plasticity index (PI) ≤ 20	

Source: Elias et al. (2001).

Design Example 10.1

A geosynthetic-reinforced slope as shown in Example Figure 10.1 is to be built on a firm foundation. The reinforced fill has a unit weight of 19.6 kN/m^3 and a friction angle of 30° (assume the cohesion of this fill is equal to 0). Three geosynthetic layers are used. The geosynthetic layers have an ultimate tensile strength of 43.8 kN/m. The installation damage factor, the durability factor, and the creep factor of this product is 1.2, 1.2, and 2.6, respectively. The interaction coefficient of geosynthetic and fill is 0.8 and the scale effect correction factor is 0.8. The geosynthetic layers have 100% coverage. The center of a slip circle is located right above the toe of this slope and the radius of this circle is 3.6 m. Divide the soil mass above the slip surface into five slices and use the ordinary method to calculate the factor of the slope (1) without geosynthetics and (2) with geosynthetics.

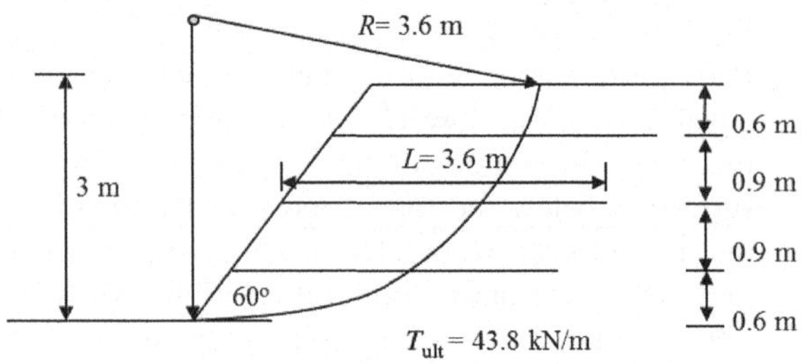

Example Figure 10.1 Geometry for reinforced slope.

Solution

For demonstration purposes, the ordinary method is used herein.

1. Slices. The soil mass above the slip surface is divided into five slices as shown in Example Figure 10.2 (not to scale).

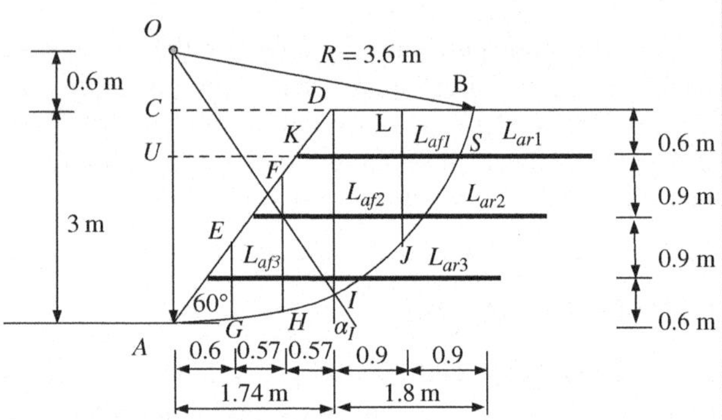

Example Figure 10.2 Slices for reinforced slope analysis.

The distance CD can be calculated as

$$CD = \frac{AC}{\tan 60^\circ} = \frac{3}{1.73} = 1.74\,\text{m}$$

The distance DB can be calculated as

$$DB = CB - CD = \sqrt{OB^2 - OC^2} - CD$$

$$= \sqrt{3.6^2 - 0.6^2} - 1.74 = 1.8\,\text{m}$$

The α angles at points G, H, I, J, and B can be calculated as follows:

$$\sin \alpha_G = \frac{0.6}{3.6} = 0.167 \qquad \alpha_G = 9.6^\circ$$

$$\sin \alpha_H = \frac{1.17}{3.6} = 0.325 \qquad \alpha_H = 19.0^\circ$$

$$\sin \alpha_I = \frac{1.74}{3.6} = 0.483 \qquad \alpha_I = 28.9^\circ$$

$$\sin \alpha_J = \frac{2.64}{3.6} = 0.733 \qquad \alpha_J = 47.1^\circ$$

$$\sin \alpha_B = \frac{3.54}{3.6} = 0.983 \qquad \alpha_B = 79.5^\circ$$

The α angles in the middle between two adjacent points A, G, H, I, J, and B can be calculated as follows:

$$\alpha_{AG} = \frac{\alpha_A + \alpha_G}{2} = \frac{0 + 9.6^\circ}{2} = 4.8^\circ$$

$$\alpha_{GH} = \frac{\alpha_G + \alpha_H}{2} = \frac{9.6^\circ + 19.0^\circ}{2} = 14.3^\circ$$

$$\alpha_{HI} = \frac{\alpha_H + \alpha_I}{2} = \frac{19.0^\circ + 28.9^\circ}{2} = 24.0^\circ$$

$$\alpha_{IJ} = \frac{\alpha_I + \alpha_J}{2} = \frac{28.9^\circ + 47.1^\circ}{2} = 37.6^\circ$$

$$\alpha_{JB} = \frac{\alpha_J + \alpha_B}{2} = \frac{47.1^\circ + 79.5^\circ}{2} = 63.3^\circ$$

The lengths of EG, FH, DI, and LJ can be calculated as

$$EG = 0.6 \tan 60^\circ - (3.6 - 3.6 \cos 9.6^\circ)$$

$$= 0.6 \times 1.73 - (3.6 - 3.6 \times 0.986) = 0.99\,\text{m}$$

$$FH = 1.17 \tan 60^\circ - (3.6 - 3.6 \cos 19^\circ)$$

$$= 1.17 \times 1.73 - (3.6 - 3.6 \times 0.946) = 1.83\,\text{m}$$

$$DI = 1.74 \tan 60^\circ - (3.6 - 3.6 \cos 28.9^\circ)$$

$$= 1.74 \times 1.73 - (3.6 - 3.6 \times 0.875) = 2.57\,\text{m}$$

$$LJ = 3 - (3.6 - 3.6 \cos 47.1^\circ)$$

$$= 3 - (3.6 - 3.6 \times 0.681) = 1.85\,\text{m}$$

The approximate areas of AEG, $EGHF$, $FHID$, $DILJ$, and LJB can be calculated as

$$AEG = 1/2 \times 0.6 \times EG = 0.296\,\text{m}^2$$

$$EGHF = 1/2 \times 0.57 \times (EG + FH)$$

$$= 1/2 \times 0.57 \times (0.99 + 1.83) = 0.804\,\text{m}^2$$

$$FHID = 1/2 \times 0.57 \times (FH + DI)$$

$$= 1/2 \times 0.57 \times (1.83 + 2.57) = 1.254\,\text{m}^2$$

$$DIJL = 1/2 \times 0.9 \times (DI + LJ)$$
$$= 1/2 \times 0.9 \times (2.57 + 1.85) = 1.989\ \text{m}^2$$
$$LJB = 1/2 \times 0.9 \times LJ$$
$$= 1/2 \times 0.9 \times 1.85 = 0.833\ \text{m}^2$$

2. Slice weights. The weights of *AEG*, *EGHF*, *FHID*, *DILJ*, and *LJB* can be calculated as

$$W_{AEG} = 0.296 \times 19.6 = 5.8\ \text{kN/m}$$
$$W_{EGHF} = 0.804 \times 19.6 = 15.8\ \text{kN/m}$$
$$W_{FHID} = 1.254 \times 19.6 = 24.6\ \text{kN/m}$$
$$W_{DIJL} = 1.989 \times 19.6 = 39.0\ \text{kN/m}$$
$$W_{LJB} = 0.833 \times 19.6 = 16.3\ \text{kN/m}$$

The normal force along the slip surface for each slice can be calculated as

$$N_{AG} = W_{AEG} \cos \alpha_{AG} = 5.8 \times \cos 4.8^\circ$$
$$= 5.8\ \text{kN/m}$$
$$N_{GH} = W_{EGHF} \cos \alpha_{GH} = 15.8 \times \cos 14.3^\circ$$
$$= 15.3\ \text{kN/m}$$
$$N_{HI} = W_{FHID} \cos \alpha_{HI} = 24.6 \times \cos 24^\circ$$
$$= 22.5\ \text{kN/m}$$
$$N_{IJ} = W_{DIJL} \cos \alpha_{IJ} = 39.0 \times \cos 37.6^\circ$$
$$= 30.9\ \text{kN/m}$$
$$N_{JB} = W_{LJB} \cos \alpha_{JB} = 16.3 \times \cos 63.3^\circ$$
$$= 7.3\ \text{kN/m}$$

The shear force along the slip surface for each slice can be calculated as (since $c = 0$)

$$T_{AG} = N_{AG} \tan \phi = 5.8 \tan 30^\circ = 3.3\ \text{kN/m}$$
$$T_{GH} = N_{GH} \tan \phi = 15.3 \tan 30^\circ = 8.8\ \text{kN/m}$$
$$T_{HI} = N_{HI} \tan \phi = 22.5 \tan 30^\circ = 13.0\ \text{kN/m}$$
$$T_{IJ} = N_{IJ} \tan \phi = 30.9 \tan 30^\circ = 17.8\ \text{kN/m}$$
$$T_{JB} = T_{JB} \tan \phi = 7.3 \tan 30^\circ = 4.2\ \text{kN/m}$$

3. Moments. The driving moment is the sum of the weight of each slice multiplying the arm distance from the center of each slice to the origin:

$$M_d = W_{AEG} \cdot x_1 + W_{EGHF} \cdot x_2 + W_{FHID} \cdot x_3$$
$$+ W_{DIJL} \cdot x_4 + W_{LJB} \cdot x_5$$
$$= 5.8 \times \frac{0.6}{2} + 15.8 \times \left(0.6 + \frac{0.57}{2}\right)$$
$$+ 24.6 \times \left(1.17 + \frac{0.57}{2}\right)$$
$$+ 39.0 \times \left(1.74 + \frac{0.9}{2}\right) + 16.3 \times \left(2.64 + \frac{0.9}{2}\right)$$
$$= 1.74 + 13.98 + 35.79 + 85.41$$
$$+ 50.37 = 187\ \text{kN} \cdot \text{m/m}$$

The resisting moment for the unreinforced case is the sum of the shear force of each slice multiplying the radius of the circle:

$$M_r = T_{AG} \cdot R + T_{GH} \cdot R + T_{HI} \cdot R + T_{IJ} \cdot R + T_{JB} \cdot R$$
$$= 3.3 \times 3.6 + 8.8 \times 3.6 + 13.0 \times 3.6$$
$$+ 17.8 \times 3.6 + 4.2 \times 3.6$$
$$= 170\ \text{kN} \cdot \text{m/m}$$

4. Factor of safety for the unreinforced slope. The factor of safety for the unreinforced case is

$$\text{FS}_u = \frac{M_r}{M_d} = \frac{170}{187} = 0.9$$

5. Reinforced slope. The design strength of geosynthetic layers is

$$T_a = \frac{T_{\text{ult}}}{\text{RF}_{\text{ID}}\text{RF}_D\text{RF}_{\text{CR}}} = \frac{43.8}{1.2 \times 1.2 \times 2.6}$$
$$= 11.7\ \text{kN/m}$$

The anchorage length of each geosynthetic layer can be calculated as follows:

$$L_{\text{ar1}} = \text{UK} + L - \text{US} = \frac{2.4}{\tan 60^\circ}$$
$$+ 3.6 - \sqrt{3.6^2 - 1.2^2} = 1.59\ \text{m}$$
$$L_{\text{af1}} = 3.6 - 1.59 = 2.01\ \text{m}$$
$$L_{\text{ar2}} = \frac{1.5}{\tan 60^\circ} + 3.6 - \sqrt{3.6^2 - 2.1^2} = 1.53\ \text{m},$$
$$L_{\text{af2}} = 3.6 - 1.53 = 2.07\ \text{m}$$
$$L_{\text{ar3}} = \frac{0.6}{\tan 60^\circ} + 3.6 - \sqrt{3.6^2 - 3^2} = 1.95\ \text{m},$$
$$L_{\text{af3}} = 3.6 - 1.95 = 1.65\ \text{m}$$

The allowable pullout capacity of each geosynthetic layer from the rear is (assume the factor of safety for pullout is 1.5)

$$T_{\text{po1}} = \frac{2\sigma_z' L_{\text{ar1}} \alpha_{\text{se}} C_i \tan \phi R_c}{\text{FS}_{\text{po}}}$$
$$= 2 \times 19.6 \times 0.6 \times 1.59 \times 0.8 \times 0.8$$
$$\times \tan 30^\circ \times 1.0/1.5 = 9.2\,\text{kN/m}$$

$$T_{\text{po2}} = \frac{2\sigma_z' L_{\text{ar2}} \alpha_{\text{se}} C_i \tan \phi R_c}{\text{FS}_{\text{po}}}$$
$$= 2 \times 19.6 \times 1.5 \times 1.53 \times 0.8 \times 0.8$$
$$\times \tan 30^\circ \times 1.0/1.5 = 22.2\,\text{kN/m}$$

$$T_{\text{po3}} = \frac{2\sigma_z' L_{\text{ar3}} \alpha_{\text{se}} C_i \tan \phi R_c}{\text{FS}_{\text{po}}}$$
$$= 2 \times 19.6 \times 2.4 \times 1.95 \times 0.8 \times 0.8$$
$$\times \tan 30^\circ \times 1.0/1.5 = 45.2\,\text{kN/m}$$

The allowable pullout capacity of each geosynthetic layer from the front is (assume the factor of safety for pullout is 1.5) as follows.

The weight of fill above the geosynthetic layer 1 in the active side and the pullout capacity of this geosynthetic layer are:

$$W_1 = \sigma_z' L_{\text{af1}} = \frac{1}{2}\left(2.01 + 2.01 - \frac{0.6}{\tan 60^\circ}\right)$$
$$\times 0.6 \times 19.6 = 21.6\,\text{kN/m}$$

$$T'_{\text{po1}} = \frac{2W_1 \alpha_{\text{se}} C_i \tan \phi R_c}{\text{FS}_{\text{po}}} = 2 \times 21.6 \times 0.8$$
$$\times 0.8 \times \tan 30^\circ \times 1.0/1.5 = 10.6\,\text{kN/m}$$

The weight of fill above the geosynthetic layer 2 in the active side and the pullout capacity of this geosynthetic layer are:

$$W_2 = \sigma_z' L_{\text{af2}} = \frac{1}{2}\left(2.07 + 2.07 - \frac{1.5}{\tan 60^\circ}\right)$$
$$\times 1.5 \times 19.6 = 48.1\,\text{kN/m}$$

$$T'_{\text{po2}} = \frac{2W_2 \alpha_{\text{se}} C_i \tan \phi R_c}{\text{FS}_{\text{po}}} = 2 \times 48.1 \times 0.8$$
$$\times 0.8 \times \tan 30^\circ \times 1.0/1.5 = 23.7\,\text{kN/m}$$

The weight of fill above the geosynthetic layer 3 in the active side and the pullout capacity of this geosynthetic layer are:

$$W_3 = \sigma_z' L_{\text{af3}} = \frac{1}{2}\left[1.65 + \max\left(1.65 - \frac{2.4}{\tan 60^\circ}, 0\right)\right]$$
$$\times 2.4 \times 19.6 = 45.0\,\text{kN/m}$$

$$T'_{\text{po3}} = \frac{2W_3 \alpha_{\text{se}} C_i \tan \phi R_c}{\text{FS}_{\text{po}}} = 2 \times 45 \times 0.8 \times 0.8$$
$$\times \tan 30^\circ \times 1.0/1.5 = 22.2\,\text{kN/m}$$

Since T_{po1} is less than T'_{po1} and T_a, the pullout from the rear will control. The design strength for geosynthetic layer 1 is $T_{a1} = 9.2$ kN/m.

Since T_{po2} and T'_{po2} are both greater than T_a, the rupture of geosynthetic will control. The design strength for geosynthetic layer 2 is $T_{a2} = 11.7$ kN/m.

Since T_{po3} and T'_{po3} are both greater than T_a, the rupture of geosynthetic will control. The design strength for geosynthetic layer 3 is $T_{a3} = 11.7$ kN/m.

The total resisting moment provided by the three layers of geosynthetics is

$$M_g = \sum T_{ai} y_{gi} = 9.2 \times 1.2 + 11.7 \times 2.1 + 11.7 \times 3$$
$$= 70.7\,\text{kN} \cdot \text{m/m}$$

The factor of safety for the reinforced case is

$$\text{FS}_r = \frac{M_r + M_g}{M_d} = \frac{170 + 70.7}{187} = 1.28$$

10.2.5 Construction

Construction of a reinforced slope is relatively easy, which mainly involves placement of geosynthetic and fill followed by compaction and installation of facing. The following construction sequence is commonly adopted in the field:

1. Prepare a site by removing large objects (e.g., trees), leveling ground, dewatering, filling existing ponds and local depressed area, and so forth.
2. Place the lowermost geosynthetic or the first lift of fill depending on design. If the geosynthetic is placed first, the principal strength direction of the reinforcement should be oriented toward the slope facing.
3. Place and compact fill on the reinforcement. A vibratory roller or plate compactor can be used to compact granular fill while a rubber-tired compactor can be used for cohesive fill. A lightweight compactor should be used

close to the slope face to avoid soil movement toward the face. The maximum lift thickness of granular fill should be limited to 300 mm, while the maximum lift thickness of cohesive fill should be limited to 200 mm. The fill should be compacted to at least 95% the maximum dry density determined by laboratory standard Proctor tests.

4. Construct slope face. An appropriate face should be selected based on the recommendations in Table 10.1. At a flatter slope (slope angle $< 45°$), no wrapped face is needed, but the vertical spacing of geosynthetics should be limited to the maximum of 0.4 m. The geosynthetic layers are extended to the slope face. At a steeper slope (slope angle $> 45°$), gabion stone can be placed at the face. Alternatively, wrapped face can be constructed as shown in Figure 10.11. Welded wire baskets are often used in the field to help install the face.

The general construction procedure for fill compaction discussed in Chapter 3 should be followed.

10.2.6 Quality Control and Assurance

Quality control and assurance typically include the following items:

- Gradation and plasticity index of backfill
- Allowable tensile strength of geosynthetic reinforcement
- Installation damage to geosynthetic
- Density of compacted fill
- Drainage system components and installation
- Vertical spacing of primary and secondary reinforcements
- Lengths and orientations of primary and secondary reinforcements including wrapped face return
- Slope angle, alignment, and top elevation
- Lateral movement and settlement of slope

Field test methods, such as the nuclear gauge method, the sand cone method, and the rubber balloon method, have been commonly used for quality assurance of the density of the compacted fill.

10.3 GEOSYNTHETIC-REINFORCED EMBANKMENTS

10.3.1 Introduction

Basic Concept Embankment is a man-made mound of earth fill to confine water or support roadways. Hydraulic dams, since their magnitudes are much larger than typical embankments and their requirements are very different, are not considered in this book. Embankments, especially when constructed on soft soils, may encounter bearing capacity, stability, and settlement problems. A variety of techniques are available to geotechnical engineers to address these problems, including preloading and/or staged construction, use of lightweight fill, overexcavation and replacement, geosynthetic reinforcement, and column technologies. The advantages and disadvantages of these techniques are discussed in Magnan (1994).

Among these techniques, geosynthetic reinforcement is one of the commonly used ones. Geosynthetic reinforcement is placed at the base of the embankment to increase bearing capacity, enhance stability, and reduce differential settlement. Therefore, it is also referred to as basal reinforcement. High-strength geotextile and uniaxial geogrid have been commonly used for this application.

Suitability Geosynthetic-reinforced embankments are suitable and cost effective when they are constructed over weak foundations without strict settlement and time constraints. Settlement can be significantly reduced when columns are used under the geosynthetic-reinforced embankment. This method will be discussed in the next section. Vertical drains can be used to shorten construction and consolidation time, which have been discussed in Chapter 7. A variety of fill materials, such as gravel, sand, and clay, can be used for embankment construction. Lightweight fill is used under some occasions when extremely soft soil is encountered and/or rapid construction is needed.

Applications Geosynthetic-reinforced embankments have been commonly used for levees or dikes, tailing dams, highways, and railways. They have also been used to widen existing embankments.

Advantages and Limitations The use of basal reinforcement can speed up the construction, avoid potential bearing failure and slope instability, increase the height of the embankment, and eliminate stabilizing berms. Construction of geosynthetic-reinforced embankments is relatively easy and does not require specialty contractors and equipment except earthmoving and compaction equipment.

However, geosynthetic reinforcement is not effective in reducing total settlement of the embankment. Construction takes time and often requires field monitoring and construction rate control to avoid possible bearing or slope failure.

10.3.2 Principles

Failure Modes Geosynthetic-reinforced embankments over soft soils may have the following failure modes (Figure 10.12): (1) local bearing failure due to lateral squeeze

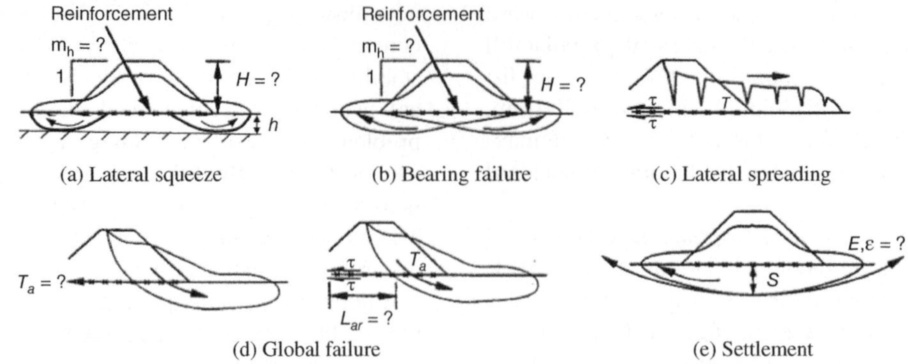

Figure 10.12 Potential failure modes of geosynthetic-reinforced embankments (modified from Fowler and Koerner, 1987, with permission from Industrial Fabrics Association International).

of a thin soft soil from the toe of the embankment, (2) general bearing failure, (3) lateral spreading above the geosynthetic, (4) global failure, and (5) excessive total and/or differential settlement. For lateral squeeze and bearing failure, the two key controlling factors are the embankment height, H, and the slope angle. For lateral spreading, the two key controlling factors are the interface shear resistance between geosynthetic and embankment fill or soft soil, τ, and the tensile strength of the geosynthetic, T. For the global failure, the geosynthetic reinforcement can fail due to rupture (depending on the tensile strength of the geosynthetic, T) or pullout (depending on the interface shear resistance between geosynthetic and embankment fill or soft soil, τ and the anchorage length, L), which is the same as what has been discussed in the reinforced slope design in the previous section. The elastic modulus, E, and the tolerable strain, ϵ, of the geosynthetic reinforcement may affect the settlement of the embankment (especially the differential settlement).

Basal Reinforcement Mechanism In addition to providing tensile resistance, Houlsby and Jewell (1988) attributed the geosynthetic to provide additional contribution by altering the interface characteristics between the embankment and the soft foundation. The construction of an embankment induces not only vertical stresses but also shear stresses at the interface between the embankment and the soft foundation. Without a geosynthetic layer, the shear stresses on the top of the soft foundation are outward as shown in Figure 10.13. The outward shear stresses reduce the ultimate bearing capacity of the soft foundation as compared with that under vertical stresses only without any shear stress as shown in Figure 10.14. When the shear stress on the top of the uniform soft foundation $\tau = 0$ (i.e., a smooth surface), the bearing capacity factor $N_c = 5.14$. When the outward shear stress $\tau = c_u$, the bearing capacity factor N_c is 2.57, which is half of that without the shear stress. However, Figure 10.13 also shows that the geosynthetic reinforcement resists the shear stresses induced by the lateral thrust by the embankment fill. At the same time, the geosynthetic reinforcement applies inward shear stresses to the soft foundation so that the ultimate bearing capacity of the soft foundation is increased. When the inward shear stress $\tau = c_u$, the top surface of the soft foundation is fully restrained by the reinforcement as a rough surface. The bearing capacity factor of a uniform soft foundation with a rough surface is $N_c = 5.71$ (Giroud and Han, 2004a).

10.3.3 Design Considerations

Performance Requirements The performance requirements tabulated in Table 10.3 are commonly used in practice (Holtz et al., 2008) to design geosynthetic-reinforced embankments.

Local Bearing Failure Local bearing failure (or lateral squeeze or extrusion failure) may occur when a thin soft soil existing above a firm soil and below the embankment toe as shown in Figure 10.15. When the soft soil has a uniform undrained shear strength, c_u, and is underlain by a rough firm soil, the factor of safety against the local bearing failure can be estimated as follows (Jewell, 1988):

$$\text{FS} = \frac{4c_u}{\gamma_e H + p} + \frac{(1+\alpha_{\text{sr}})c_u}{\gamma_e h \tan\beta} \tag{10.9}$$

where
c_u = undrained shear strength of soft soil
γ_e = unit weight of embankment soil
h = thickness of soft soil
H = height of embankment
p = surcharge on the crest of embankment
β = slope angle
α_{sr} = ratio of shear stress to undrained shear strength of soft soil (ranging from −1 to 1)

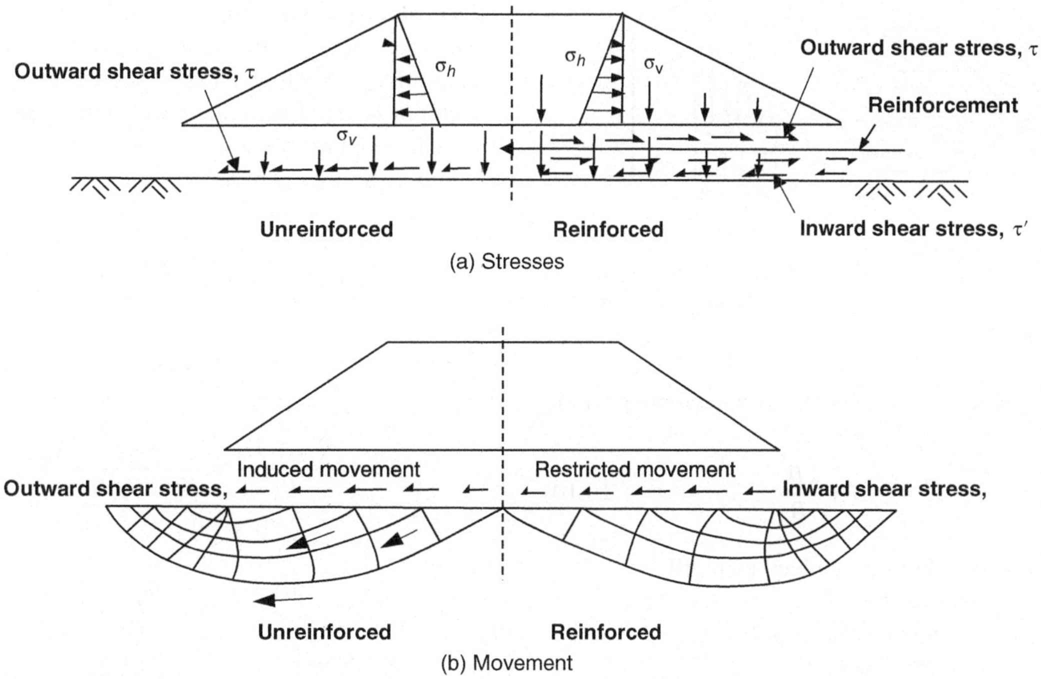

Figure 10.13 Basal reinforcement mechanism (modified from Houlsby and Jewell, 1988).

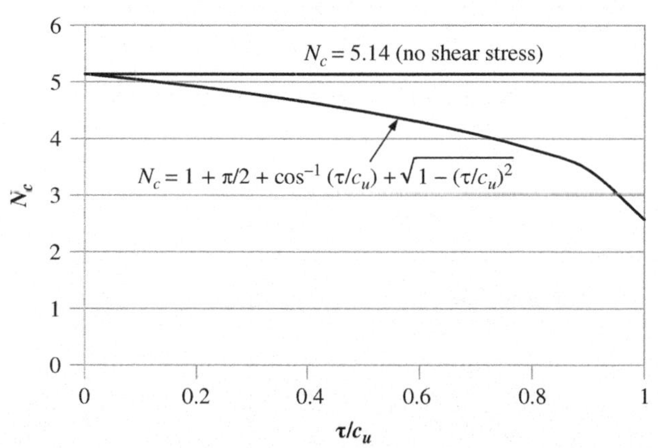

Figure 10.14 Effect of outward shear stress on bearing capacity factor (modified from Houlsby and Jewell, 1988).

Table 10.3 Performance Requirements for Geosynthetic-Reinforced Embankments

Failure Mode	Requirement
General bearing capacity	FS > 1.5–2.0
Local bearing failure	FS > 1.3–2.0
Global rotational failure	FS > 1.3 at end of construction and 1.5 in long term
Lateral spreading	FS > 1.5
Failure under seismic loading	FS > 1.1
Settlement and differential settlement	Project dependent

When $\alpha_{sr} = 1$, it represents a fully restrained upper surface with inward shear stresses (i.e., a rough surface). Under this condition, this formula is similar to that obtained by Silvestri (1983) in which a constant of 4.14 instead of 4 is used. When $\alpha_{sr} = 0$, it represents a free upper surface (i.e., a smooth surface). For an unreinforced embankment, $\alpha_{sr} = -1$ (i.e., outward shear stresses).

General Bearing Failure The general bearing failure may happen when the reinforced embankment behaves as an equivalent rigid footing. When the soft soil is deeper than 1.6 times the width of the footing, the bearing capacity factor is $N_c = 5.14$, which assumes a smooth interface between the footing and the soft soil. When the soft soil has a limited thickness (e.g., underlain by bedrock or a firm soil stratum), the bearing capacity factor, N_c, can be estimated based on the ratio of the average width of the embankment to the thickness of soft soil (Figure 10.16). The boundary conditions noted in the figure are referred to as the surface roughness of the firm stratum.

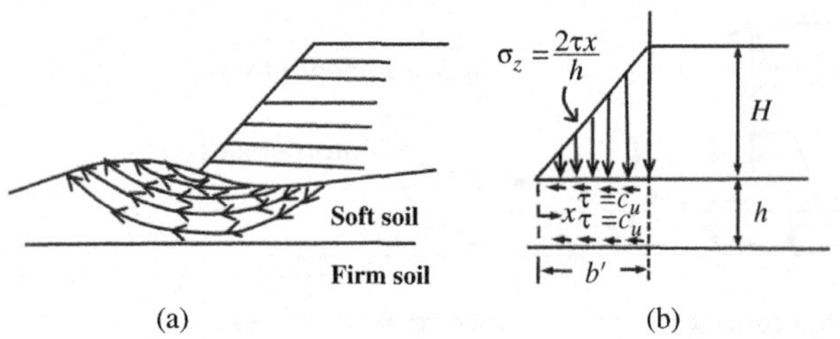

Figure 10.15 (a) Local bearing failure mode and (b) model for analysis (after Silvestri 1983, with permission from *Canadian Geotechnical Journal*).

The average applied pressure on the average footing width, p', is

$$p' = \gamma_e H + \frac{pB_t}{B} \tag{10.10}$$

where γ_e = unit weight of the embankment fill
H = embankment height
B = average embankment width
B_t = embankment top width
p = surcharge on the crest of the embankment

The factor of safety against the general bearing failure is

$$\text{FS} = \frac{N_c c_u}{p'} = \frac{N_c c_u B}{\gamma_e HB + pB_t} \tag{10.11}$$

Lateral Spreading Due to the lateral thrust from the active earth pressure, the soil wedge may slide above or below the geosynthetic reinforcement as shown in Figure 10.17. The factor of safety against these two lateral spreading modes can be calculated as follows:

Sliding above the reinforcement:

$$\text{FS} = \frac{T_{r1}}{P_a} = \frac{0.5\gamma_e H_b C_{i1} \tan\phi'}{K_a(0.5\gamma_e H^2 + pH)} \tag{10.12}$$

Sliding below the reinforcement:

$$\text{FS} = \frac{T_{r2} + T_a}{P_a} = \frac{bC_{i2}c_u + T_a}{K_a(0.5\gamma_e H^2 + pH)} \tag{10.13}$$

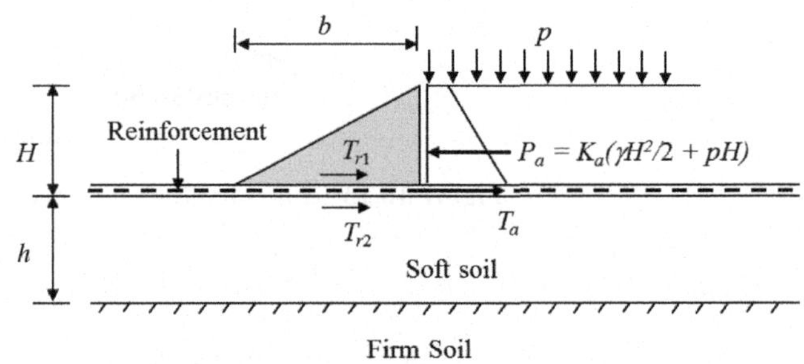

Figure 10.17 Lateral spreading analysis.

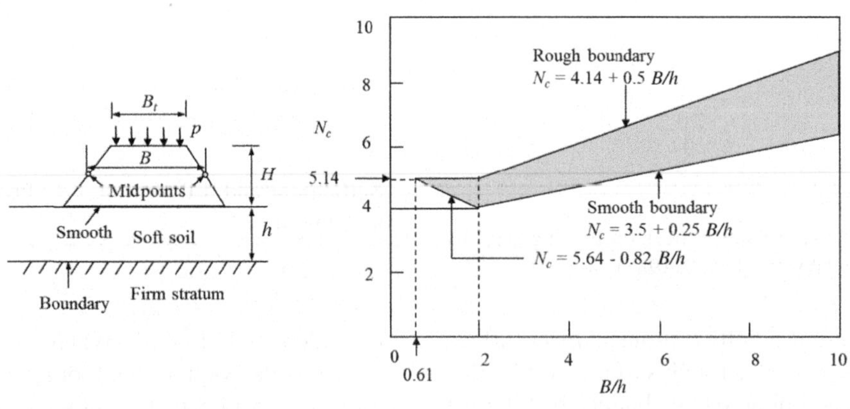

Figure 10.16 (a) Embankment dimensions and soil conditions and (b) bearing capacity factor considering limited thickness of soft soil (modified from Bonaparte et al. (1978)).

where T_{r1}, T_{r2} = shear resistance above and below the geosynthetic reinforcement, respectively
P_a = lateral active thrust from the embankment
γ_e = unit weight of the embankment fill
ϕ' = effective friction angle of the embankment fill
c_u = undrained shear strength of the soft soil
T_a = allowable tensile strength of the geosynthetic reinforcement
C_{i1}, C_{i2} = interaction coefficients between geosynthetic and embankment fill and between geosynthetic and soft soil, respectively
p = surcharge on the crest of embankment

Design of geosynthetic-reinforced embankments over soft soils should consider the strain compatibility among geosynthetic, embankment fill, and soft soil (especially sensitive soil). Elias et al. (2006) suggested the allowable tensile strength of the geosynthetic reinforcement should be selected in terms of a tolerable strain limit according to the quality of embankment fill and/or properties of soft soil. For a cohesionless fill, the tolerable strain limit of the geosynthetic is 5–10%. For a cohesive fill, the tolerable strain limit of the geosynthetic is 2%. When an embankment is constructed on peat, the tolerable strain limit of the geosynthetic is 2–10%.

Circular Slip Failure Circular slip failure likely happens when embankments are constructed on thick soft soils. The method for a circular slip analysis of an embankment over soft soil as shown in Figure 10.18 is similar to that for a global slope stability of a reinforced slope. Different from reinforced slopes, geosynthetic reinforcement is commonly placed at the base of the embankment. The rotation of geosynthetic reinforcement may be considered depending on properties of foundation soil and tolerable deformation. Holtz et al. (2008) have the following suggestion for the angle of geosynthetic rotation:

$\theta_{ri} = 0$ for brittle, strain-sensitive foundation soil (e.g., leached marine clay) or where a crust layer is considered in the analysis for increased support.

$\theta_{ri} = \alpha_i/2$ for soil depth to embankment width ratio $h/B \leq 0.4$ and moderate to highly compressible soil (e.g., soft clay, peat).

$\theta_{ri} = \alpha_i$ for $h/B > 0.4$ and highly compressible soil (e.g., soft clay, peat) and design geosynthetic elongation $\varepsilon_g \geq 10\%$ and large tolerable deformation.

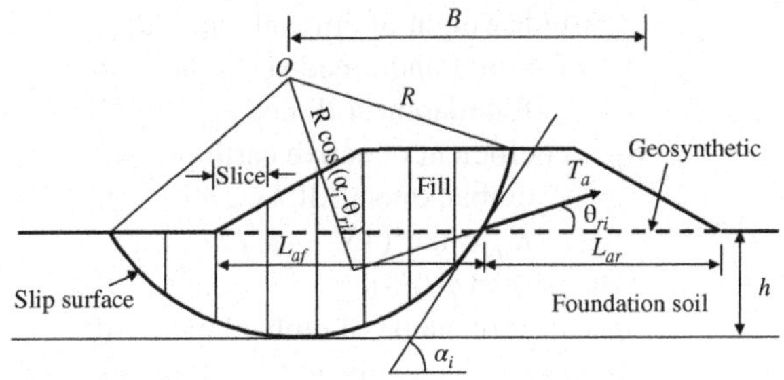

Figure 10.18 Circular slip analysis.

The tensile resistance of the geosynthetic reinforcement T_a is limited by its tensile strength or pullout capacity. The capacities of geosynthetic pullout from the front and the rear should be calculated and the lesser of these two capacities should be used in design.

Translational Failure Translational failure likely happens when a shallow weak foundation exists under an embankment. It can be analyzed using the two-part wedge and/or three-part wedge methods as discussed in Section 10.2.3.

Simplified Method Milligan and Busbridge (1983) developed a simplified method for the design of geosynthetic-reinforced embankments over weak foundations. This method considers deep-seated circular failure and shallow translational failure as shown in Figure 10.19. In Figure 10.19(a), the deep-seated circular failure surface develops in the soft foundation, and the Rankine failure plane develops in the embankment fill. In Milligan and Busbridge (1983), no surcharge on the embankment was considered. Their solutions can be easily expanded by considering the surcharge on the embankment. The total embankment thrust consists of the thrusts by the fill P_{a1} and the surcharge P_{a2}. Considering the foundation soil is uniform and undrained, the center of the failure circle is above the midpoint of the slope. Based on moment equilibrium, the following equation can be derived to calculate the required tensile strength of geosynthetic reinforcement:

$$\frac{T_f}{\gamma_e H^2} = \frac{1}{2Y}\left[2YZ - \frac{1}{12\tan\beta} + K_a\left(Y - \frac{1}{3}\right) + 2K_a\frac{p}{\gamma_e H}\left(Y - \frac{1}{2}\right) + Z^2 - \frac{2c_{uf}}{\gamma_e H}(Y+Z)^2\theta_1\right] \tag{10.14}$$

where T_f = factored tensile strength of geosynthetic reinforcement, i.e., T_g/FS
T_g = required tensile strength of geosynthetic reinforcement
FS = factor of safety

γ_e = unit weight of embankment fill
c_{uf} = factored undrained shear strength of foundation soil, i.e., $c_{uf} = c_u/FS$
K_a = coefficient of active earth pressure of embankment fill, i.e., $K_a = \tan^2(45° - \phi_f/2)$
$\phi_f = \text{arc}(\tan\phi/FS)$
ϕ = friction angle of embankment fill
p = surcharge on the crest of embankment
$Y = y/H$ and y = vertical distance from the center of circle to the geosynthetic reinforcement
$Z = h/H$ and h = thickness of weak foundation soil
$\tan\theta_1 = \sqrt{2Z/Y + (Z/Y)^2}$
H = height of embankment
β = slope angle

The factored tensile strength of geosynthetic reinforcement can be determined as the maximum value of T_f by varying the Y value. As shown in Equation (10.14), Y should be greater than $\frac{1}{3}$ without a surcharge or $\frac{1}{2}$ with a surcharge.

For the shallow translational failure in Figure 10.19(b), the slip surface is formed by the Rankine failure planes. Based on the force equilibrium, the required tensile strength of geosynthetic reinforcement can be derived as follows to calculate the required tensile strength of geosynthetic reinforcement:

$$\frac{T_f}{\gamma_e H^2} = \frac{K_a}{2} + K_a \frac{p}{\gamma_e H} + \frac{ph}{\gamma_e H^2} + Z - \frac{c_{uf}}{\gamma_e H}\left(4Z + \frac{1}{\tan\beta}\right) \quad (10.15)$$

The preceding equation does not involve the variable Y. The factored tensile strength of geosynthetic reinforcement can be calculated with known parameters. The final factored tensile strength of geosynthetic reinforcement should be the larger value of T_f from these two calculations. It should be noted that this simplified method is based on the assumption that the Rankine earth pressure in the embankment fill is fully mobilized. To satisfy this condition, the slip surface should not cross the centerline of the embankment. In other words, the average width of the embankment should be wide enough. Otherwise, the solution would result in a conservative design. Figure 10.20 shows the comparison of the calculated results by the simplified method and the limit equilibrium (LE) methods based on the Bishop approach and the Spencer three-part wedge method at the factor of safety equal to 1.0. The comparison shows that the simplified method overpredicts the required tensile strengths in geosynthetic reinforcement as compared with the LE methods. The analysis also shows that the wedge failure controls when $h/H \le 0.25$.

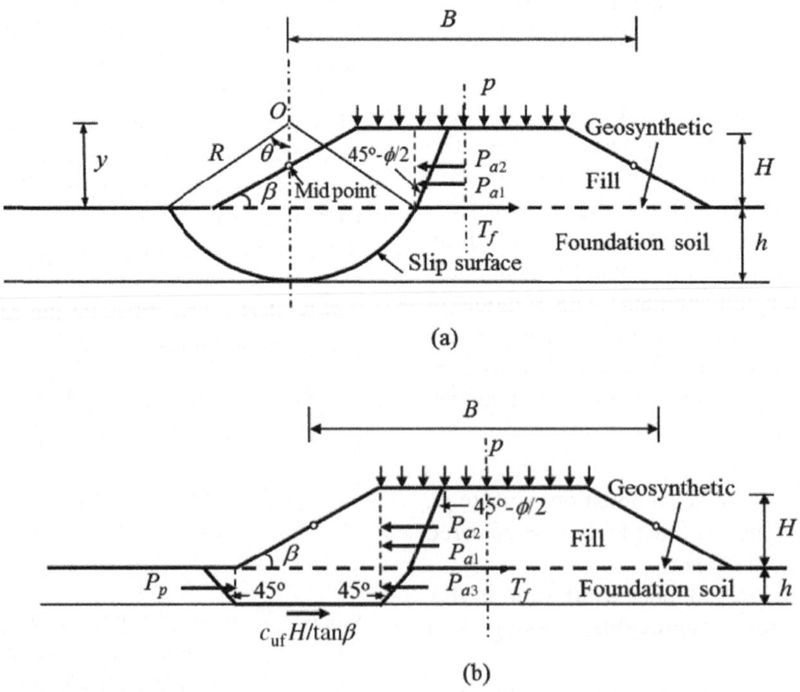

Figure 10.19 (a) Deep-seated circular failure and (b) shallow translational failure (modified from Milligan and Busbridge, 1983).

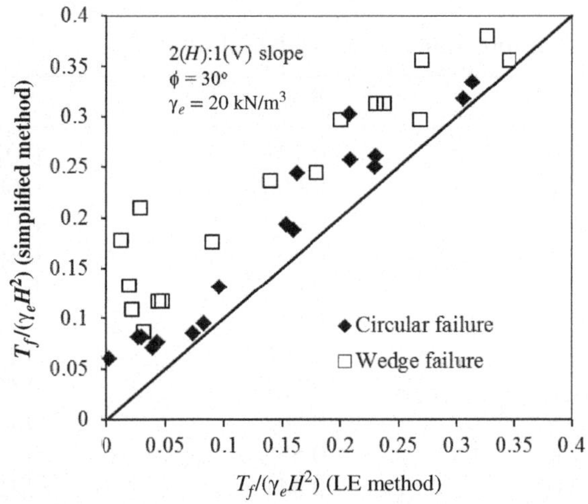

Figure 10.20 Comparison of the calculated results by simplified and LE methods.

Computer-Assisted Design A number of software available on the market can assist the analysis of reinforced embankments considering all the above failure modes. Most software is limit-equilibrium-based and some are based on limit analysis. Numerical software has been increasingly used for slope stability analyses. With the rapid increase of computer speeds, an analysis searching through thousands of slip surfaces using limit-equilibrium-based software may only take a few seconds to a few minutes. Numerical software takes longer to complete an analysis from a few minutes to a few hours, which are often acceptable for practical applications.

10.3.4 Design Parameters and Procedure

Design Parameters Design parameters for geosynthetic-reinforced embankments include the following:

- Geometry of embankment (height, side slope angle, foundation soil, and depth to firm soil or bedrock if any)
- Properties of embankment fill (unit weight, cohesion, and friction angle)
- Properties of foundation soil (unit weight, cohesion, and friction angle)
- Location and length of geosynthetic reinforcements
- Properties of geosynthetic reinforcements (allowable tensile strength and interaction coefficient with fill or foundation soil)
- Surcharge

Design Procedure The following procedure may be used for design/analysis of geosynthetic-reinforced embankments over soft foundations:

1. Calculate the factors of safety against local and general bearing failures for an unreinforced embankment.
2. Run a stability analysis for the unreinforced embankment, identify the critical slip surface, and compute the minimum factor of safety.
3. If at least one of the factors of safety is less than the required, geosynthetic reinforcement is needed. Place a full-length geosynthetic reinforcement at the base of the embankment (two or more reinforcement layers may be used if a single reinforcement layer is not enough). Assign the properties of geosynthetic reinforcements (including tensile strength and interaction coefficients with fill and foundation soil).
4. Calculate the factors of safety against local and general bearing failure, and lateral spreading for the reinforced embankment. Adjust the parameters until all the calculated factors of safety are equal or slightly greater than the required.
5. Run rotational and translational stability analyses for the reinforced embankment and identify the critical slip surface and compute the minimum factor of safety. If the computed minimum factor of safety is at the target value or slightly greater than the target value, the procedure is complete. If the computed minimum factor of safety is higher or lower than the target value, adjust layout and properties of geosynthetic reinforcements until the desired factor of safety is achieved.

For a small project or a preliminary design, the simplified method developed by Milligan and Busbridge (1983) may be used.

Design Example 10.2

A 4.5-m-high embankment is to be constructed on a 1.5-m-thick uniform soft soil with an undrained shear strength of 15 kPa (the groundwater table is near the ground surface). The embankment has 2 (horizontal): 1 (vertical) side slopes. The unit weight and the friction angle of the embankment fill are 20 kN/m^3 and 34°, respectively. The uniform soft soil is underlain by bedrock. The required factor of safety is 1.5. Determine the required tensile strength of the geosynthetic reinforcement using the simplified method.

Solution

The simplified method developed by Miligan and Busbridge (1983) was based on factored soil strength. The factored cohesion of the soft soil is

$$c_{\mathrm{uf}} = \frac{c_u}{\mathrm{FS}} = \frac{15}{1.5} = 10\,\mathrm{kPa}$$

The factored friction angle of the embankment fill is

$$\phi_f = \arctan\left[\frac{\tan(\phi)}{\mathrm{FS}}\right] = \arctan\left[\frac{\tan(34°)}{1.5}\right] = 24.2°$$

The active earth pressure coefficient of the embankment fill is

$$K_a = \tan^2\left(45^\circ - \frac{\phi_f}{2}\right)$$

$$= \tan^2\left(45^\circ - \frac{24.2^\circ}{2}\right) = 0.418$$

The ratio of soft soil depth to embankment height is

$$Z = \frac{h}{H} = \frac{1.5}{4.5} = 0.333$$

Since $Z \geq 0.25$, the circular failure mode controls.

$$\tan\theta_1 = \sqrt{2Z/Y + (Z/Y)^2}$$

$$= \sqrt{2 \times 0.333/Y + (0.333/Y)^2}$$

$$= \sqrt{0.667/Y + 0.109/Y^2}$$

Equation (10.14) can be used to calculate the required factored tensile strength of the geosynthetic reinforcement.

$$\frac{T_f}{\gamma_e H^2} = \frac{1}{2Y}\left[2YZ - \frac{1}{12\tan\beta} + K_a\left(Y - \frac{1}{3}\right) + 2K_a\frac{p}{\gamma_e H}\left(Y - \frac{1}{2}\right)\right] + Z^2 - \frac{2c_{\mathrm{uf}}}{\gamma_e H}(Y+Z)^2\theta_1$$

$$= \frac{1}{2Y}\left[\begin{array}{c} 2 \times 0.333Y - \dfrac{1}{12 \times 0.5} \\ + 0.418\,(Y - 0.33) + 2 \\ \times 0.418 \times 0 + 0.33^2 \\ -\dfrac{2 \times 10}{20 \times 4.5}(Y + 0.333)^2\theta_1 \end{array}\right]$$

$$= \frac{1}{2Y}[0.667Y - 0.167 + 0.418\,(Y - 0.33) + 0.945 - 0.222(Y + 0.333)^2\theta_1]$$

To solve the preceding equation, Y should be varied as shown in Example Table 10.4 and the maximum value of T_f can be found within the range of Y as $\frac{T_f}{\gamma_e H^2} = 0.125$.

Therefore, the required tensile strength of the geosynthetic reinforcement is

$$T_g = \mathrm{FS} \cdot T_f = \mathrm{FS} \cdot \frac{T_f}{\gamma_e H^2}\gamma_e H^2$$

$$= 1.5 \times 0.125 \times 20 \times 4.5^2 = 50.5\ \mathrm{kN/m}$$

Example Table 10.4 Calculation of T_f

Y	θ_1	T_f
1.0	0.723	0.076
1.1	0.696	0.089
1.2	0.672	0.099
1.3	0.650	0.107
1.4	0.631	0.112
1.5	0.613	0.117
1.6	0.596	0.120
1.7	0.581	0.122
1.8	0.567	0.124
1.9	0.553	0.124
2.0	0.541	0.125
2.1	0.530	0.125
2.2	0.519	0.124
2.3	0.509	0.123
2.4	0.499	0.122
2.5	0.490	0.121

10.3.5 Construction

The construction of geosynthetic-reinforced embankments typically involves the following steps: site preparation, placement of geosynthetic, placement of fill and compaction, and construction monitoring.

Site preparation can vary depending on site conditions. It may include removal of large objects (e.g., trees), leveling of ground, dewatering, filling of existing ponds and local depressed areas, and the like. Caution should be taken not to disturb the soil below the designed elevation, especially for sensitive soil. If the site is too weak for human activities and/or light equipment, a construction platform needs be constructed first. A low-strength sacrificial geosynthetic that is often helpful to stabilize the ground can be placed first and followed with a thin (100–150 mm) layer of soil.

Geosynthetic should be placed with the principal strength direction perpendicular to the centerline of the embankment. Unless geosynthetic seam or joint strength is used in the design, no seam or joint should be allowed in the direction parallel to the centerline of the embankment. After the placement of the geosynthetic, it should be manually pulled straight to remove wrinkles. The surface of geosynthetic should be visually inspected for possible defects, including holes, rips, and tears. Defects should be removed and the material replaced.

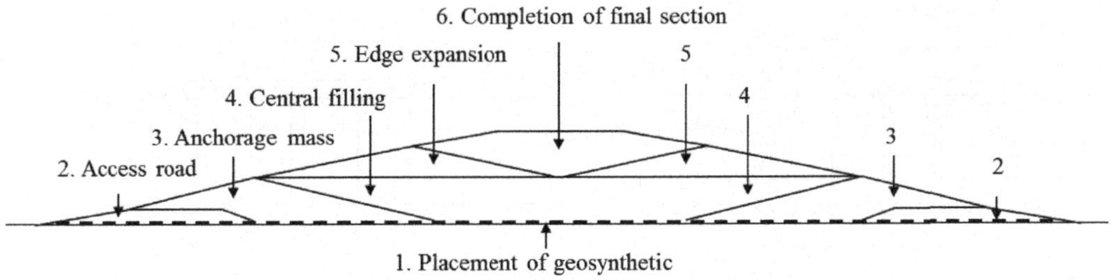

Figure 10.21 Sequence of construction of geosynthetic-reinforced embankment on extremely weak foundation (modified from Haliburton et al., 1977).

The sequence and rate of fill placement depend on the properties of soft foundations and the size and number of construction equipment. A typical rate of fill placement on soft foundations should not exceed 0.3 m vertically every week. To avoid mudwave on extremely soft foundations, a special sequence of construction from 1 to 6, as shown in Figure 10.21 is suggested by Haliburton et al. (1977), which involves the construction of two end dump access roads, placement of anchorage masses, central filling, edge expansion, and completion of final section.

Staged construction may be needed when high embankments are constructed on soft foundations. The basic idea is to build the full-height embankments in more than one stage. After the completion of each stage, there is a waiting period for soils to gain their strengths through consolidation before the next stage starts. The design procedure of staged construction is discussed in Chapter 7.

10.3.6 Quality Control and Assurance

In addition to the quality control and assurance of embankment fill and geosynthetic products, it is important to control the construction rate and have field monitoring during the construction, especially on sensitive or weak foundation soil. Depending on the size and importance of the embankment, different instrumentations can be adopted. Settlement and lateral movement of embankment toes are basic monitoring items. More comprehensive monitoring can include settlement plates or horizontal inclinometer tubes at the base of the embankment, piezometers and extensometers at different depths of the soft foundation, and vertical inclinometer tubes at the toes of the embankment. Criterion for each measurement to control the rate of fill placement or halt construction should be predetermined before the construction. It is a typical requirement that the accumulated excess pore water pressure should be less than 50–60% of the applied stress. The excess pore water pressure measurements can also estimate the degree of consolidation of the soft soil.

10.4 GEOSYNTHETIC-REINFORCED COLUMN-SUPPORTED EMBANKMENTS

10.4.1 Introduction

Basic Concept Many studies have shown that geosynthetic reinforcement can be used to reduce differential settlements; however, they have limited contributions in reducing total settlements of embankments over soft soils. When geosynthetic-reinforced embankments are connected to structures (such as bridges), large total settlements often create bumps at the interface between embankments and structures, which are not tolerable in terms of serviceability. In addition, geosynthetic-reinforced embankments often require staged construction to avoid failure during construction, which takes time. Vertical drains can be used to accelerate consolidation but required construction time and/or postconstruction settlement are often still the issues for projects. Column or pile-supported embankments have been increasingly used recently as an expedite construction technique. When piles are used, pile caps are often needed to reduce clear spacing between piles. Large columns installed by ground improvement techniques do not require caps. The column technologies by deep replacement and deep mixing or grouting are discussed in Chapters 5 and 8. The terms "column" and "pile" are both used in the literature. In this section, the term "pile" is used when there is a cap.

Column or pile-supported embankment systems have been used with or without geosynthetic reinforcement. A system without geosynthetic reinforcement is referred to herein as the conventional column-supported (CCS) or conventional pile-supported (CPS) embankment system while the system with geosynthetic reinforcement is referred to as the geosynthetic-reinforced column or pile-supported (GRCS or GRPS) embankment system (Figure 10.22). For a CPS embankment, inclined piles are commonly used near side slopes to carry the lateral thrust from the embankment. In addition, piles need to be closely spaced and/or have large pile caps in order to transfer surcharge loads through soil arching to

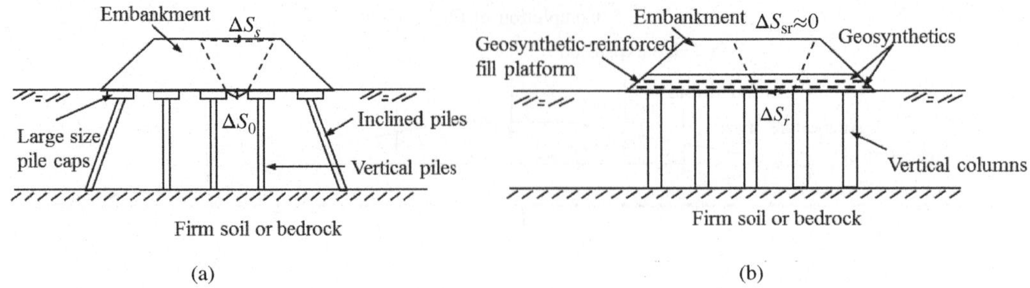

Figure 10.22 Column/pile-supported embankments: (a) CPS and (b) GRCS embankments.

the piles and minimize deflection of the soil between pile caps and the deflection being reflected to the embankment surface. In the GRCS embankment system, the geosynthetic reinforcement carries the lateral thrust from the embankment, creates a stiffened fill platform to enhance the load transfer from the soil to the columns, and reduces the differential settlement between columns. One single high-strength geosynthetic layer may be placed over the columns acting as a tensioned membrane or multiple layers of geosynthetics with adequate strengths may be placed within granular fill to form a load transfer platform. As a result, the GRCS embankment system does not require inclined columns, large caps, and close column spacing. Therefore, the GRCS embankment system creates a more cost-effective solution.

Suitability The GRCS embankments are suitable for accelerated construction of embankments over soft foundations, which are connected to stationary structures (such as bridge abutments) or existing embankments with strict total and differential settlement requirements. The suitable depth of improvement ranges from 5 to 20 m.

Applications The GRCS embankments have been mostly used for bridge approach and embankment widening. They have also been used to support pavements, storage tanks, and buildings.

Advantages and Limitations In the column or pile-supported embankment system, the columns or piles carry most of the loads from the embankment, and the soil is only subjected to small loads. The benefits associated with the use of column or pile-supported embankments are as follows: (1) allows construction of the embankment in a single stage without prolonged waiting time, (2) significantly reduces total and differential settlements, and (3) reduces or eliminates global stability concerns. The use of geosynthetic reinforcement eliminates inclined piles and reduces the size of pile caps.

The GRCS embankments are expensive as compared with other ground improvement methods. They often require columns to be seated on firm soil or bedrock to be more effective.

10.4.2 Principles

Load Transfer Mechanisms The interactions among column, foundation soil, embankment fill, and geosynthetic reinforcement as shown in Figure 10.23 can be described as follows. Under the influence of fill weight, W, the embankment fill mass between columns has a tendency to move downward, due to the presence of the soft foundation soil. This movement is restrained by shear resistance, τ, from the fill above the columns. The shear resistance reduces the pressure, p_{sl}, acting on the geosynthetic but increases the pressure applied onto the columns, p_{cl}. This load transfer mechanism was termed the soil arching effect by Terzaghi (1943).

Compared with the unreinforced case, the inclusion of geosynthetic reinforcement is expected to reduce the displacement of the embankment fill between the columns. The reduction of the displacement would reduce the shear stresses in the embankment so that the effect of soil arching in the embankment would be minimized. As a result, the load transferred by soil arching to the columns is reduced. At the same time, however, the load on the columns is increased by the vertical components of the tension force in the reinforcement. A single geosynthetic layer behaves as a tensioned

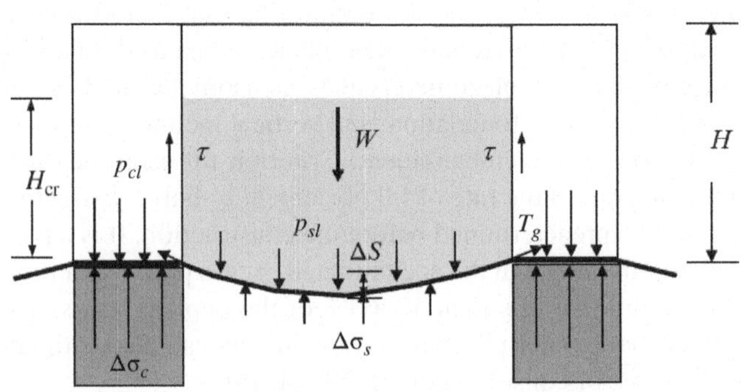

Figure 10.23 Load transfer mechanisms (modified from Han, 1999; Han and Gabr, 2002, with permission from ASCE).

membrane, while a multilayer system acts as a stiffened platform (or like a beam) due to the interlock of reinforcement with the surrounding soil. Underneath the geosynthetic reinforcement, stresses are concentrated on columns and less stresses are applied on soil between columns. Stress concentration ratio is defined as the average vertical stress on the columns to that on the soil, that is, $n = \Delta\sigma_c/\Delta\sigma_s$, which has been discussed in Chapter 5. Soil arching, tensioned membrane effect, and stress concentration also depend on the relative stiffness of the columns to the soil. A rigid column promotes the differential settlement between the columns and the soft soil so that there is more soil arching, tensioned membrane effect, and stress concentration. Han and Gabr (2002) have confirmed these phenomena in their numerical analysis. Therefore, the mechanisms of load transfer can be considered as a combination of soil arching, tensioned membrane effect, and stress concentration. The load transfer contributed by each mechanism depends on a number of factors, including the number of geosynthetic and the tensile stiffness of geosynthetics, the properties of embankment fill and foundation soils, and the moduli of columns and soil.

The interaction between the embankment and the column foundation is discussed in detail in Chapter 5. This section mainly addresses the design and construction of the load transfer platform.

Soil Arching Different soil arching models have been proposed or used by researchers, as shown in Figure 10.24, to analyze GRCS embankments: (a) vertical slip surfaces (Russell and Pierpoint, 2008; British Standard 8006, 2010; Chen et al., 2008); (b) semispherical dome (Hewlett and Randolph, 1988; Kempfert et al., 2004); and (c) triangular wedge (Carlsson, 1987; Miki, 1997; and Collin, 2003).

The vertical slip surfaces model was originally proposed by Terzaghi (1943) in which the shear resistance develops along the vertical slip surfaces as shown in Figure 10.24(a) so that the vertical stress between the stationary supports decreases. Russell and Pierpoint (1997) modified Terzaghi's arching theory by considering the three-dimensional shape of the settling soil mass in the embankment above the geosynthetic between columns. This modified method is also referred to as the adapted Terzaghi method. The British Standard BS8006 (2010) also assumed vertical slip surfaces; however, it considered the columns acting similar to buried rigid pipes in a conduit. The average vertical stress on the top of the columns is estimated using Marston's formula for positive projecting subsurface conduits. The degree of soil arching is commonly expressed by a soil arching ratio, ρ_{sa}, or stress reduction ratio (SRR) as follows:

$$\rho_{sa} = \text{SRR} = \frac{p_{sl}}{\gamma_e H + p} \tag{10.16}$$

The soil arching ratio, ρ_{sa}, ranges from 0 to 1; $\rho_{sa} = 0$ represents complete soil arching, while $\rho_{sa} = 1$ represents no soil arching.

Hewlett and Randolph (1988) proposed a semispherical dome model based on the mechanism observed in model tests in which they assumed the soil above the columns forms a semispherical soil arching as shown in Figure 10.24(b). This model considers possible failure of soil arching either at the crown of the arch or at the top of the column and ignores foundation support. Based on the limit equilibrium, a solution for the vertical stress between the columns was derived. Kempfert et al. (2004) also assumed semispherical domed arch between columns and derived a solution based on lower bound plasticity theory, pilot-scale tests, and numerical analysis. The Kempfert et al. (2004) method considers the foundation support underneath the geosynthetic reinforcement.

Carlsson (1987), Miki (1997), and Collin (2003) considered the formation of triangular soil wedges between columns, as shown in Figure 10.24(c). They assumed different θ angles for the soil wedge and the vertical stress between the columns induced by the weight of the soil wedge, which is carried by the geosynthetic reinforcement. All the embankment fill outside the wedge and surcharge are carried by the columns.

All the preceding models assume full mobilization of soil arching between rigid supports. Terzaghi (1936) and McNulty (1965) observed in experimental trapdoor tests that

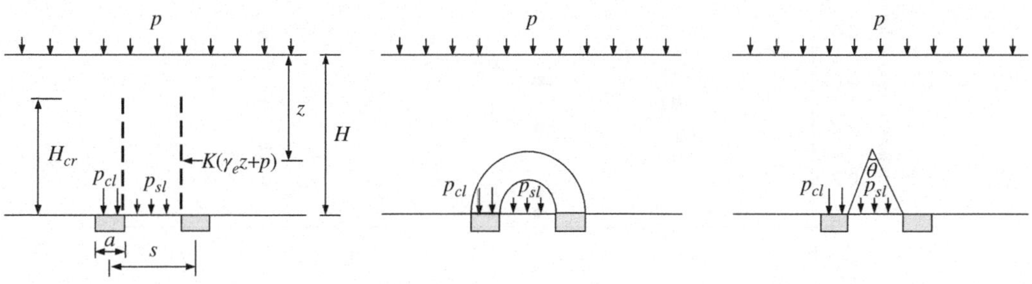

Figure 10.24 Soil arching models: (a) vertical slip surfaces, (b) semispherical dome, and (c) triangular wedge.

soil arching developed with an increase of the displacement of the trapdoor. This phenomenon was confirmed by Han and Gabr (2002) in their numerical analysis of GRCS embankments in which the soil arching was fully mobilized at approximately 6.5% differential settlement of the clear spacing of the columns at the elevation of column heads. Filz and Smith (2006) and McGuire and Filz (2008) examined these soil arching models and found that they resulted in very different vertical stresses between columns. Sloan et al. (2011) showed that the adapted Terzaghi model and the Hewlett and Randolph model had reasonable comparisons with field-measured vertical stresses. Therefore, these two methods will be presented in the design procedure.

Based on field test data and laboratory experimental tests, van Eekelen and Bezuijen (2012) found that the vertical stresses on the top of the reinforcement were not uniform and they could be better modeled by an inversely triangular distribution.

Critical Height To minimize the deflection at the top of the columns to be extended to the top of the embankment as shown in Figure 10.24(a), it is important to maintain the critical height lower than the embankment height. It is a general believe that the critical height depends on the clear spacing of columns. However, Filz et al. (2012) showed that this height also depends on the diameter of columns. Table 10.4 lists the suggested critical heights by different researchers. The criterion, $H_{cr} \geq 1.4\ (s - a)$, has been commonly used in practice.

Foundation Soil Resistance The foundation soil resistance has a significant effect on the tension development in the geosynthetic reinforcement. Figure 10.25 shows that consideration of the foundation soil resistance reduced the tension more than 70%. Han and Wayne (2000) showed that the measured stress concentration ratios of GRCS embankments were higher than those of CCS embankments; however, they

Table 10.4 Suggested Critical Height

Reference	Recommendation[a]	Note
Hewlett and Randolph (1988)	$H_{cr} \geq 1.0(s - a)$	Based on laboratory model test data and theoretical solution
Han and Gabr (2002)	$H_{cr} \geq 1.3(s - a)$	Numerical result
Chen et al. (2008a)	$H_{cr} \geq (1.4\text{–}1.6)(s - a)$	Based on laboratory model test data
Chen et al. (2010)	$H_{cr} \geq (1.1\text{–}1.5)(s - a)$	Based on field test data
British Standard BS8006 (2010)	$H_{cr} \geq 1.4(s - a)$	
Filz et al. (2012)	$H_{cr} \geq 1.15s' + 1.44d$	Based on laboratory model tests and field test data

[a] s = center-to-center column spacing; s' = distance from the edge of a column to the centroid of a column array; a = width of square column; and d = diameter of column.

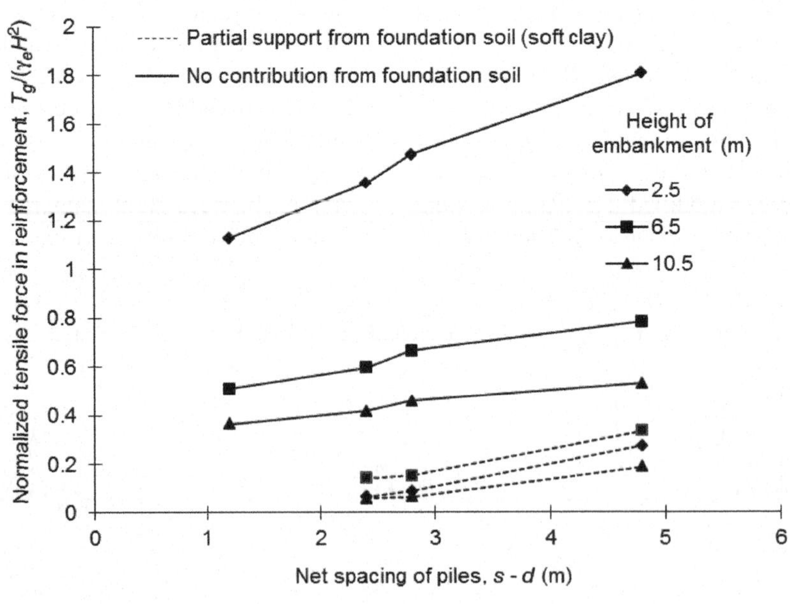

Figure 10.25 Effect of foundation soil resistance (modified from Jones et al., 1990).

were lower than those determined by plate loading tests as shown in Figure 10.26. This is because plate loading tests used rigid plates. Numerical results by Han and Gabr (2002) also show that the inclusion of geosynthetic reinforcement increased the stress concentration ratio by creating a stiffer fill platform. Therefore, it is conservative to use the stress concentration ratio determined by a plate loading test to estimate the foundation soil resistance if no void is expected under the platform. As discussed in Chapters 7 and 8, the stress concentration ratios for flexible, semirigid, and rigid columns under rigid loading are 2–5, 5–10, and 10–25, respectively. Alzamora et al. (2000) used a stress concentration ratio of 25 to estimate the foundation soil resistance in the design of geogrid-reinforced platform over jet-grouted columns in soft soil. Van Eekelen et al. (2011) suggested using a subgrade reaction modulus to account for foundation soil resistance. van Eekelen and Bezuijen (2012) also suggested that the foundation soil resistance should be accounted for within the entire reinforcement area instead of strips between columns.

Due to possible development of voids under the platform, no foundation soil resistance should be considered when the settlement between columns is caused by (Han, 1999):

- Consolidation of underconsolidated soil (OCR < 1.0)
- Liquefaction
- Lowering of groundwater
- Dissipation of pore water pressure not due to embankment loading
- Sinkholes due to Karstic collapse
- Soil dissolution
- Municipal solid waste settlement
- Soil loss due to adjacent excavation and tunneling.

When the foundation soil resistance is considered, the net vertical stress on the geosynthetic reinforcement is the difference between the vertical stress on the top of the reinforcement and the foundation soil resistance.

Tension in Reinforcement Once the geosynthetic reinforcement is subjected to the vertical stress from the embankment fill, it deforms and develops tension. The relationship between the deflection and the tension at certain vertical stress can be described by the tensioned membrane theory. Different tensioned membrane solutions have been proposed, for example, John (1987), Carlsson (1987), Giroud et al. (1990), and British Standard BS8006 (2010), which result in similar results. All these tensioned membrane solutions were developed based on a plane strain condition or a strip between two supports. To account for the three-dimensional effect as shown in Figure 10.27, Rogbeck et al. (1998) suggested a three-dimensional relative coverage area factor, f_{3D}, to convert the calculated tension under the plane strain

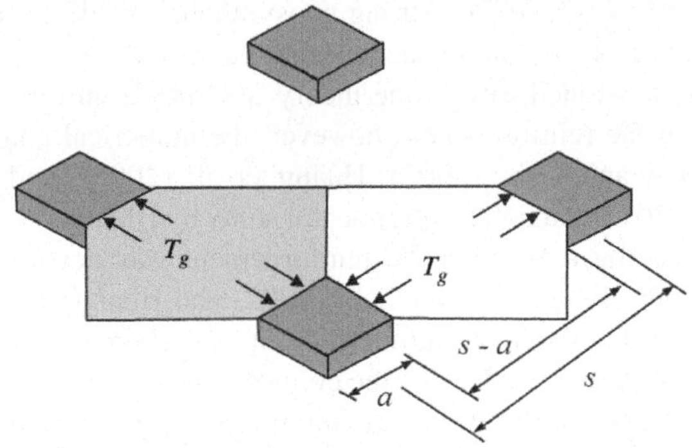

Figure 10.27 Three-dimensional relative coverage area (modified from Rogbeck et al., 1998).

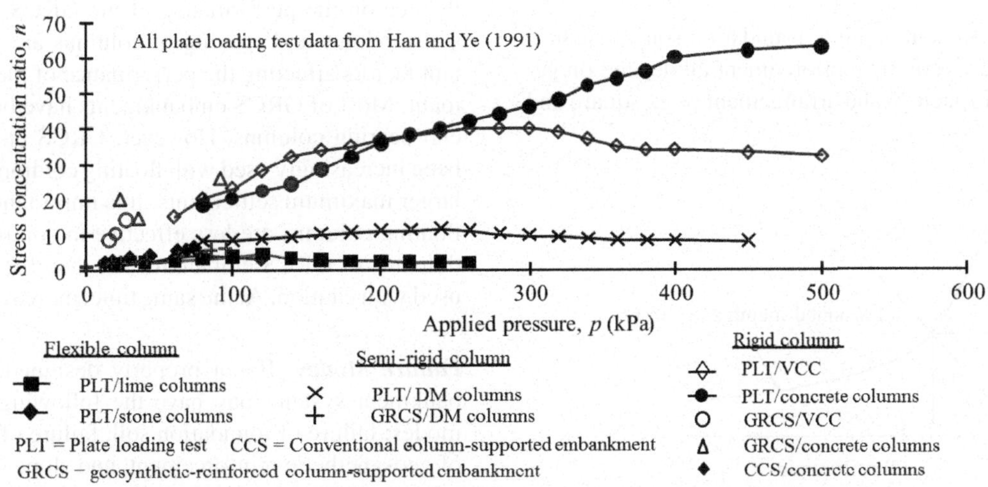

Figure 10.26 Field-measured stress concentration ratios (Han and Wayne, 2000).

condition to that under the three-dimensional condition as follows:

$$f_{3D} = 1 + \frac{s - a}{2a} \quad (10.17)$$

This factor was developed based on the diamond area as compared with the strip area between two pile caps.

According to the tensioned membrane theory, the tension in the reinforcement decreases with the increase of its strain or deflection as shown in Figure 10.28. However, the tension in the reinforcement is also proportional to the strain if the reinforcement has constant tensile stiffness, J. The intersection point between these two lines meets the tension–strain compatibility and results in the predicted tensile strain in the reinforcement. The British Standard BS8006 (2010) suggests an upper strain limit of 6% for load transfer in a high embankment (i.e., the critical height is below the embankment height) or 3% in a low embankment (i.e., the critical height extends to the surface) and 2% for creep in the service life. Collin (2004) suggested a 5% tensile strain for design. If the calculated tensile strain or tension exceeds the design requirement, a stiff or strong geosynthetic should be used. Alternatively, column spacing can be adjusted.

The tensioned membrane theory assumes a uniform tension in the reinforcement; however, the numerical analyses by Han and Gabr (2002), Huang et al. (2005), and Han et al. (2012b) and the experimental study by Bhandari (2010) showed the tension in the reinforcement was not uniform along the span. Han and Gabr (2002) and Bhandari (2010) showed that the maximum tension of a single reinforcement developed at the edges of the column top. However, Huang et al. (2005) showed the maximum tension of the upper reinforcement was located on the top of the columns, while the maximum tension of the lower reinforcement was located at the midspan between columns when three reinforcements were included in the load transfer platform.

Influence Factors Soil arching, foundation soil resistance, and tension in geosynthetic reinforcement all depend on geometry (size and spacing) and arrangement (e.g., square and triangular patterns) of columns, modulus ratio of column to soil, strengths of columns and soil, column end-bearing condition, embankment height, properties of platform fill and embankment fill, number, stiffness, and strength of geosynthetic reinforcement, type of loading (static or dynamic), and rate of construction. Han and Gabr (2002), Bhandari (2010), and Huang and Han (2010) and Han et al. (2012b) investigated most of these influence factors using finite-difference and discrete-element methods.

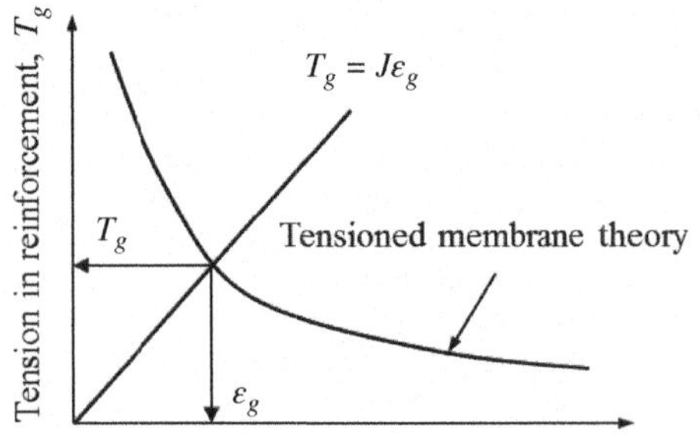

Figure 10.28 Tension–strain compatibility.

Han and Gabr (2002) showed that an increase of column modulus promotes differential settlement between columns and soil, soil arching, stress concentration on columns, reduces maximum settlement, and increases tension in geosynthetic reinforcement. Soil arching was fully mobilized at the column modulus of 1 GPa. When the column modulus was less than 100 MPa, geosynthetic reinforcement did not have any effect on soil arching. It is important to point out that geosynthetic reinforcement is less effective when flexible columns are used because less differential settlements occur between columns. Kempfert et al. (2004) suggested that the modulus ratio of column to soil be larger than 100 to ensure full mobilization of soil arching. The use of geosynthetic reinforcement and an increase of the reinforcement stiffness reduce maximum and differential settlements, minimize soil arching, promote stress concentration, and attract more tension in the reinforcement. An increase of embankment height increases the maximum settlement and the differential settlement at the top of columns but reduces the differential settlement on the top of the embankment, promotes soil arching and stress concentration, and increases the tension in the geosynthetic. Bhandari (2010) showed that cyclic loading minimizes soil arching and stress concentration more significantly in the unreinforced embankment than in the geosynthetic-reinforced embankment. Huang and Han (2010) found that the construction rate has a significant influence on the performance of the GRCS embankment but soil modulus and spacing of columns are two most important factors affecting the performance of the GRCS embankment. Most of GRCS embankments have been supported by end-bearing columns. However, GRCS embankments have been increasingly used with floating columns, which result in larger maximum settlements. It is important to point out that floating columns are less effective to control maximum settlements of GRCS embankments; therefore, they should be used with caution. At the same time, more research is needed.

Failure Modes If not properly designed, the GRCS embankment system may have the following possible failure modes: failure of foundation soil, failure of columns, failure of geosynethetic reinforcement, and slope instability.

When CCS embankments are constructed over soft soil, the soft soil between the columns may fail due to low bearing capacity. The inclusion of geosynthetic reinforcement above the columns reduces the load transmitted to the soft soil so that such a failure mode can be prevented.

The columns under the embankment may have the following possible failures modes:

- Column heads punching into or through embankment fill
- Compression failure of column heads or shafts
- End-bearing failure of columns
- Bending failure of columns
- Shear failure of columns

The inclusion of geosynthetic reinforcement increases the resistance against the punching of column heads into the fill and minimizes the chance of bending failure of columns by reducing lateral thrust from the embankment. However, it may require more load capacities of columns since more load is transferred onto the columns.

The geosynthetic reinforcement above column heads may fail due to rupture or pullout from the soil, especially when the reinforcement is near the edge of the embankment. The reinforcement can also experience excess elongation due to low modulus and/or creep deformation. The tension in a geosynthetic layer when acting as a tensioned membrane would be reduced as the deflection/elongation of the reinforcement increases (i.e., stress relaxation).

The embankment system may encounter the following possible slope instability situations:

- Lateral spreading along the geosynthetic due to the thrust from the embankment
- Local slope instability
- Slope instability outside the first row of columns
- Slope instability through columns
- Global slope instability below columns

10.4.3 Design Considerations

Percent Coverage Based on the performance investigation of conventional pile-supported embankments, Rathmayer (1975) recommended design criteria as shown in Figure 10.29. The required percent coverage of pile caps, defined as the percentage of the total area of pile caps to that of foundation footprint, depends on the quality of fill materials. For columnar systems, the percent coverage is equivalent to the area replacement ratio. The percent coverage of pile caps or column heads for 13 actual GRCS embankments is plotted in Figure 10.29 for comparison purposes (Han, 1999). As shown in Figure 10.29, the percentage with geosynthetic reinforcement is much lower than that suggested by Rathmayer (1975) for the CCS embankments. The percent coverage of the GRCS embankment systems is consistently less than 20%.

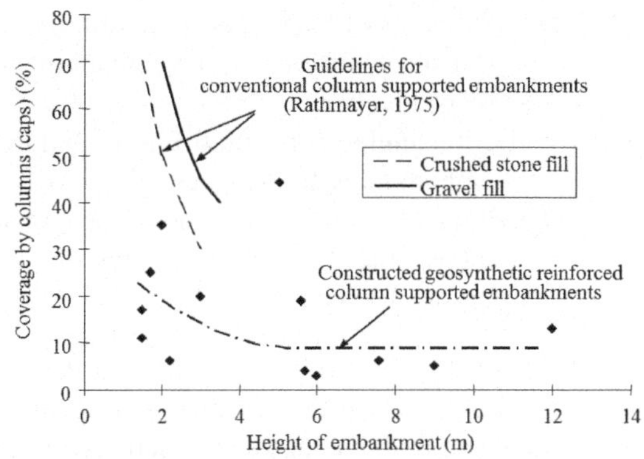

Figure 10.29 Percent coverage of columns in column-supported embankments (Han, 1999).

The reduction of percent coverage creates a more economical solution for embankment systems. The percent coverage for GRCS embankment systems mostly ranges from 5 to 30%.

Stress above Geosynthetic Reinforcement The vertical stress applied on the geosynthetic reinforcement between columns is reduced due to the soil arching effect. The vertical stress below the geosynthetic reinforcement is further reduced by the tensioned membrane effect. The vertical stress applied on the geosynthetic reinforcement is a key variable for computing the tension in geosynthetic reinforcement. The adapted Terzaghi method and the Hewlett and Randolph method are introduced below to calculate the vertical stress on the geosynthetic reinforcement. Most of research so far has been based on a square pattern of columns; therefore, the design based on this pattern is presented below. The design guidelines for GRCS embankments on columns in other patterns can be found in Sloan et al. (2011).

1. Adapted Terzaghi Method Sloan et al. (2011) expanded the adapted Terzaghi method by Russell and Pierpoint (1997) to a double layer system, which includes an embankment fill over a load transfer platform as shown in Figure 10.30. When the critical height, H_{cr}, is greater than the embankment height, H, the average vertical stress above the geosynthetic reinforcement can be calculated by

$$p_{sl} = \frac{\gamma_1}{\alpha_1}(1 - e^{-\alpha_1 H_1}) + \frac{\gamma_2}{\alpha_2} e^{-\alpha_1 H_1}(1 - e^{-\alpha_2 H_2}) + p e^{-\alpha_1 H_1 - \alpha_2 H_2} \quad (10.18)$$

$$\alpha_1 = \frac{4aK_1 \tan \phi_1}{s^2 - a^2}; \quad \alpha_2 = \frac{4aK_2 \tan \phi_2}{s^2 - a^2}$$

where γ_1 and γ_2 = unit weight of the platform fill and the embankment fill, respectively

K_1 and K_2 = coefficients of lateral earth pressure in the load transfer platform and the embankment fill, respectively

ϕ_1 and ϕ_2 = friction angles of the platform fill and the embankment fill, respectively

p = surface surcharge on the top of the embankment; other symbols are shown in Figure 10.30.

When the critical height, H_{cr}, is greater than the platform thickness, H_1, but less than the embankment height, H, the average vertical stress above the geosynthetic reinforcement can be calculated by

$$p_{sl} = \frac{\gamma_1}{\alpha_1}(1 - e^{-\alpha_1 H_1}) + \frac{\gamma_2}{\alpha_2} e^{-\alpha_1 H_1}[1 - e^{-\alpha_2 (H_{cr} - H_1)}] + [p + (H - H_{cr})\gamma_2] e^{-\alpha_1 H_1 - \alpha_2 (H_{cr} - H_1)} \tag{10.19}$$

When the critical height, H_{cr}, is within the platform ($H_{cr} < H_1$), the average vertical stress above the geosynthetic reinforcement can be calculated by

$$p_{sl} = \frac{\gamma_1}{\alpha_1}(1 - e^{-\alpha_1 H_{cr}}) + [p + (H_1 - H_{cr})\gamma_1 + H_2\gamma_2] e^{-\alpha_1 H_{cr}} \tag{10.20}$$

The platform thickness typically ranges from 0.5 to 1.0 m. Sloan et al. (2011) suggested that K_1 and K_2 should be 0.75 as the solution with this value had a reasonable match with field data.

2. The Hewlett and Randolph Method Hewlett and Randolph (1988) assumed the soil above the columns forms a semispherical soil arching as shown in Figure 10.24(b). The design considers possible failure of soil arching either at the crown of the arch or at the top of the column. The vertical stress above the reinforcement can be determined as the greater of the values in the following two equations:

At the crown,

$$p_{sl} = (\gamma_e H + p)\left\{\left(1 - \frac{a}{s}\right)^{2(K_p - 1)}\left[1 - \frac{s}{\sqrt{2}H}\left(\frac{2K_p - 2}{2K_p - 3}\right)\right] + \frac{s - a}{\sqrt{2}H}\left(\frac{2K_p - 2}{2K_p - 3}\right)\right\} \tag{10.21}$$

On the top of the column,

$$p_{sl} = \frac{\gamma_e H + p}{\left(\frac{2K_p}{K_p + 1}\right)\left[\left(1 - \frac{a}{s}\right)^{1 - K_p} - \left(1 - \frac{a}{s}\right)\left(1 + \frac{a}{s}K_p\right)\right] + \left(1 - \frac{a^2}{s^2}\right)} \tag{10.22}$$

where K_p is the coefficient of passive earth pressure of fill.

Hewlett and Randolph (1988) suggested that the thickness of well-compacted high-grade fill ($K_p > 3$) should not be less than the spacing of columns. When two layers of fill materials (such as the load transfer platform and the embankment fill) exist in the embankment, the weighted-average friction angle can be obtained from the friction angles of the platform fill and the embankment fill within the height above the column heads equal to the spacing of columns.

The design methods for soil arching presented above are based on full mobilization of soil arching between rigid supports, which likely happen when end-bearing rigid columns are used to support embankments over soft soil. When the columns have a modulus more than 100 times that of the soft soil, they can be considered as rigid columns or rigid supports. Limited soil arching between flexible columns (such as stone columns and sand columns) may be mobilized due to small differential settlement between columns. For semirigid columns (such as deep mixed columns and jet-grouted

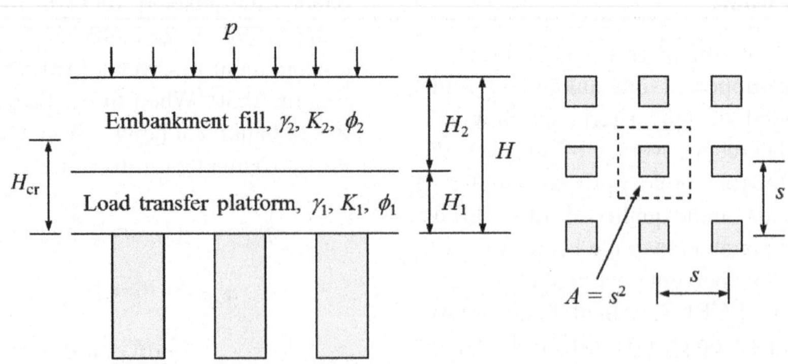

Figure 10.30 Cross-section and column layout for the adapted Terzaghi method: (a) cross section and (b) column layout (modified from Sloan et al., 2011).

columns), partial soil arching may develop. Currently, no design guideline is available for soil arching in flexible and semirigid column-supported embankments and therefore requires further research.

Strain and Tension in Geosynthetic Reinforcement Geosynthetic reinforcement under applied vertical stresses behaves as a tensioned membrane. A number of methods are available to estimate the strain and tension developed in the geosynthetic reinforcement. Two methods introduced below can be used to calculate the tensile strain in reinforcement:

1. Catenary Method The method presented in John (1987) included the calculation of the strain developed in the geosynthetic reinforcement:

$$\varepsilon_g = \frac{1}{2}\sqrt{1 + 16\frac{\Delta S_g^2}{(s-a)^2}} + \frac{s-a}{8\Delta S_g} \times \ln\left(\frac{4\Delta S_g}{s-a} + \sqrt{1 + \frac{16\Delta S_g^2}{(s-a)^2}}\right) - 1 \quad (10.23)$$

and

$$T_g = \frac{p_{\text{sl}} - \Delta\sigma_s}{4a}(s^2 - a^2)\sqrt{1 + \frac{(s-a)^2}{16\Delta S_g^2}} \quad (10.24)$$

where ε_g = strain developed in the geosynthetic reinforcement
ΔS_g = maximum deflection of the geosynthetic reinforcement
T_g = tension developed in the geosynthetic reinforcement
p_{sl} = average vertical stress on the geosynthetic reinforcement
$\Delta\sigma_s$ = average vertical stress (soil resistance) below the geosynthetic reinforcement

2. Parabolic Method Under a plane strain condition, the British Standard BS8006 (2010) suggested a formula to estimate the tension in geosynthetic reinforcement between columns based on a parabolic deflection shape. Considering the square pattern of square columns and the foundation soil resistance, the following formula can be obtained (Russell and Pierpoint, 1997):

$$T_g = \frac{(p_{\text{sl}} - \Delta\sigma_s)(s^2 - a^2)}{4a}\sqrt{1 + \frac{1}{6\varepsilon_g}} \quad (10.25)$$

Giroud (1995) suggested that the tensile strain of the reinforcement can be approximated with the midspan deflection of the reinforcement when the deflection to the clear span is small:

$$\varepsilon_g = \frac{8\Delta S_g^2}{3(s-a)^2} \quad (10.26)$$

This formula should be limited to a maximum midspan deflection of 300 mm. When Equation (10.26) is substituted into Equation (10.25), it yields the same equation as Equation (10.24).

When a geosynthetic reinforcement has a constant tensile stiffness, J, the tensile strain, $\varepsilon_g = T_g/J$, can be substituted into Equation (10.25), which leads to the formula (McGuire and Filz, 2008)

$$96T_g^3 - \frac{6(p_{\text{sl}} - \Delta\sigma_s)^2(s^2 - a^2)^2}{a^2}T_g - \frac{(p_{\text{sl}} - \Delta\sigma_s)^2(s^2 - a^2)^2}{a^2}J = 0$$

This formula satisfies the tension–strain compatibility requirement and the tension in the reinforcement can be solved.

Foundation Soil Resistance The average vertical stress on the geosynthetic reinforcement can be determined based on the methods discussed above. John (1987) assumed $\Delta\sigma_s = 0.15\gamma H$ for the foundation soil resistance. Alzamora et al. (2000) used a stress concentration ratio of 25 to estimate the foundation soil resistance considering the weight of the embankment fill:

$$\Delta\sigma_s = \frac{\gamma_e H}{1 + (a^2/s^2)(n-1)} \quad (10.27)$$

where n is the stress concentration ratio (assume 25 for rigid columns).

Traffic load should not be considered in the calculation of the foundation soil resistance because it is not a permanent load.

The foundation soil resistance should be less than the allowable bearing capacity to limit the differential settlement between the columns:

$$q_a = \frac{q_{\text{ult}}}{\text{FS}} = \frac{5.14c_u}{\text{FS}} \geq \Delta\sigma_s \quad (10.28)$$

where FS is the factor of safety (typically use 2.0) and c_u is the undrained shear strength of soft soil. If the preceding condition cannot be satisfied, the column spacing should be reduced.

As discussed earlier, the foundation soil resistance may be ignored to be conservative if voids are expected under the load transfer platform.

Lateral Spreading The method to evaluate the possible lateral spreading is the same as that for the geosynthetic-reinforced embankment presented in Section 10.3.2. The required tensile strength of the geosynthetic reinforcement due to lateral spreading needs be added to the required strength of the geosynthetic reinforcement under vertical stresses. Alternatively, additional geosynthetic reinforcement is added to accommodate this need.

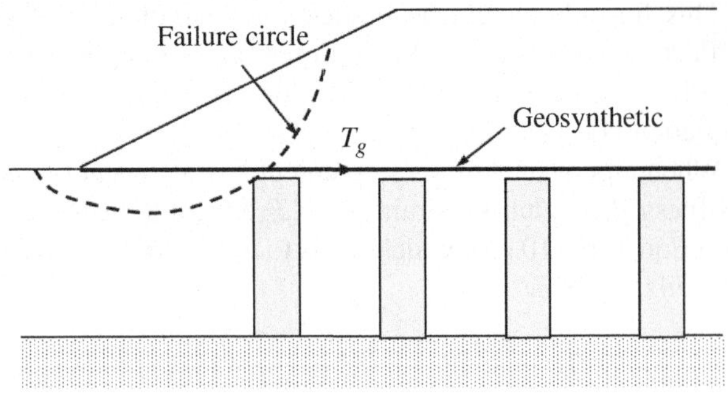

Figure 10.31 Stability of edge slope.

Edge Slope Stability The stability of an edge slope can be analyzed using typical limit equilibrium methods, such as the simplified Bishop method with a consideration of the contribution by geosynthetic reinforcement as shown in Figure 10.31. The tensile strength or pullout resistance (whichever is less) of geosynthetic reinforcement is included in the stability analysis as contributing a resisting moment to the system in addition to that by the soil.

Global Slope Stability The global slope stability of GRCS embankments can be analyzed in the same way as that discussed in previous chapters.

10.4.4 Design Parameters and Procedure

Design Parameters The following parameters are typically needed for the design of geosynthetic-reinforced column-supported embankments:

- Geometry of embankment (height, width, and slope angle)
- Properties of embankment fill (unit weight and friction angle)
- Thickness of load transfer platform
- Properties of load transfer platform (unit weight and friction angle)
- Size, spacing, length, and pattern of columns
- Strength, modulus, and end-bearing condition of columns
- Surcharge on embankment
- Tolerable total and differential settlements

Elias et al. (2006) suggested a granular fill to be used in the load transfer platform as listed in Table 10.5.

Table 10.5 Recommended Backfill for Load Transfer Platform

Sieve Size (mm)	Percent Passing
200	100
38	95–100
4.75	40–65
0.425	20–40
0.075	0–15
Plasticity index of soil particles passing No. 40 sieve, PI < 20	

Source: Elias et al. (2006).

Design Procedure The design procedure to select a geosynthetic reinforcement to carry vertical loads under the embankment is as follows:

1. Calculate the modulus ratio of column to soil. If the ratio is greater than 100, proceed with further calculations. If the ratio is less than 100, the following procedure may not be used or may be used with great caution.
2. Calculate the critical height.
3. Use the adapted Terzaghi method or the Hewlett and Randolph method to calculate the average vertical stress above the geosynthetic reinforcement.
4. Estimate the foundation soil resistance. Check the foundation soil resistance against the allowable bearing capacity.
5. Based on a tolerable maximum deflection of the geosynthetic reinforcement, use the catenary method or the parabolic method to calculate the strain in the geosynthetic reinforcement.
6. Calculate the tension in the geosynthetic reinforcement using Equation (10.24) or (10.25) to calculate the tension in the geosynthetic.
7. Select a geosynthetic product with the tensile strength at the required tensile strain higher than the calculated tension.

Design Example 10.3

A 5-m-high embankment will be constructed on a soft soil. To accelerate the construction and minimize post-construction settlement, the geosynthetic-reinforced column embankment system is considered in the design. Concrete columns with a diameter of 1 m are installed in a square pattern with spacing of 2.5 m. The elastic modulus of the concrete is 1 GPa. The undrained shear strength and the elastic modulus of the soft soil are 20 kPa and 4 MPa, respectively. Within the embankment, there is a 0.60-m-thick load transfer platform with geosynthetic reinforcement. The unit weights of the platform fill and the embankment fill are 20 and 18 kN/m^3, respectively. The friction angle of the platform fill and the embankment fill are 38° and 30°, respectively. The surcharge on top of the embankment is 13 kPa. The allowable maximum differential settlement between the columns at the base of the embankment is 0.15 m. Calculate the required tensile strength of the geosynthetic reinforcement and its corresponding tensile strain (using the adapted Terzaghi method).

Solution

1. The modulus ratio of columns to soil is $E_c/E_s = 1000/4 = 250 > 100$; therefore, the methods for fully mobilized soil arching can be used.

   The equivalent width of the column is

$$a = \sqrt{\frac{\pi d^2}{4}} = \sqrt{\frac{3.14 \times 1^2}{4}} = 0.89 \text{ m}$$

2. The critical height is $H_{cr} = 1.4(s - a) = 1.4 \times (2.5 - 0.89) = 2.26 < 5$ m (embankment height).
3. Since the critical height is greater than the platform fill thickness, but less than the embankment height, the following formula in the adapted Terzaghi method can be used to calculate the average vertical stress:

$$p_{sl} = \frac{\gamma_1}{\alpha_1}(1 - e^{-\alpha_1 H_1}) + \frac{\gamma_2}{\alpha_2} e^{-\alpha_1 H_1}[1 - e^{-\alpha_2(H_{cr} - H_1)}] + [p + (H - H_{cr})\gamma_2]e^{-\alpha_1 H_1 - \alpha_2(H_{cr} - H_1)}$$

   Assume $K_1 = K_2 = 0.75$.

$$\alpha_1 = \frac{4aK_1 \tan\phi_1}{s^2 - a^2} = \frac{4 \times 0.89 \times 0.75 \times \tan 38°}{2.5^2 - 0.89^2} = 0.38$$

$$\alpha_2 = \frac{4aK_2 \tan\phi_2}{s^2 - a^2} = \frac{4 \times 0.89 \times 0.75 \times \tan 30°}{2.5^2 - 0.89^2} = 0.28$$

$$p_{sl} = \frac{\gamma_1}{\alpha_1}(1 - e^{-\alpha_1 H_1}) + \frac{\gamma_2}{\alpha_2} e^{-\alpha_1 H_1}[1 - e^{-\alpha_2(H_{cr} - H_1)}] + [p + (H - H_{cr})\gamma_2]e^{-\alpha_1 H_1 - \alpha_2(H_{cr} - H_1)} = 56.8 \text{ kPa}$$

4. The foundation soil resistance (using $n = 25$) is

$$\Delta\sigma_s = \frac{\gamma_1 H_1 + \gamma_2 H_2}{1 + (a^2/s^2)(n - 1)} = 22.7 \text{ kPa}$$

   The ultimate bearing capacity of the soft soil is

$$q_{ult} = 5.14c_u = 5.14 \times 22.7 = 102.8 \text{ kPa}$$

   The allowable bearing capacity of the soft soil is

$$q_a = \frac{q_{ult}}{FS} = \frac{102.8}{2.0} = 51.4 \text{ kPa} > \Delta\sigma_s \text{ (OK)}$$

5. Use the Giroud method to calculate the tensile strain in the reinforcement:

$$\varepsilon_g = \frac{8\Delta S_g^2}{3(s - a)^2} = 2.3\%$$

6. The tension in the reinforcement can be calculated as follows:

$$T_g = \frac{(p_{sl} - \Delta\sigma_s)(s^2 - a^2)}{4a}\sqrt{1 + \frac{1}{6\varepsilon_g}} = 151 \text{ kN/m}$$

7. Select a geosynthetic product with a tensile strength at 2.3% strain higher than 151 kN/m.

10.4.5 Construction

The procedure for installing columns is the same as that described in Chapters 5 and 8, depending on which column technology is selected. The procedure for placing and compacting embankment fill is the same as that described in the previous section for geosynthetic-reinforced embankments. The following procedure is commonly adopted in practice to construct the load transfer platform:

- Prepare the surface of the foundation soil to the desired elevation by removing any deleterious objects and materials including extra column heads above the elevation and/or filling depressed areas including extension of column heads below the elevation.
- Place the granular fill in lifts (limit uncompacted lift thickness to 250 mm for a heavy compactor or 150 mm for a hand-operated compactor).
- Compact the granular fill to at least 95% maximum dry density determined using the modified Proctor method within ±2% of the optimum moisture content.
- Place geosynthetic reinforcement(s) at design elevation(s). The lowermost geosynthetic reinforcement should be placed at least 100 mm above column heads to avoid any puncture failure. When uniaxial geosynthetic reinforcements are used, they should be placed in orthotropic directions to have a uniform tensile strength in both directions. Geosynthetic reinforcement should be pretensioned manually to prevent sags and wrinkles.
- Continue with the placement and construction of the embankment fill.

10.4.6 Quality Control and Assurance

The quality control and assurance should include the following items:

- Examine location, elevation, and diameter of columns. Larger column spacing and/or smaller and uneven column heads will adversely affect the performance of the load transfer platform; therefore, they should be fixed by adding more columns and/or enlarging column heads before the placement of fill and geosynthetic.

- Verify quality of granular fill and geosynthetic reinforcement before placement.
- Verify the density and thickness of each fill lift after compaction.

It is recommended that geosynthetic-reinforced column-supported embankments should be monitored during and after construction, especially for large projects. Instrumentation can include settlements and vertical stresses on the top and between columns, strains in geosynthetics, and lateral movement of embankment toe and soil below the toe.

10.5 MECHANICALLY STABILIZED EARTH WALLS

10.5.1 Introduction

Basic Concept Mechanically stabilized earth (MSE) wall has been successfully used in practice for transportation and other applications since the 1960s. It consists of wall facing with more than 70° inclination angle, reinforced fill, retained soil, foundation soil, and drainage layer behind the wall facing as shown in Figure 10.32. Wall facing can be geosynthetic-wrapped around, stone-filled gabion baskets, modular blocks, and concrete panels. The reinforcement can be geosynthetic reinforcement (geogrid or woven geotextile) or metallic reinforcement (steel strip or steel mesh), which provides tensile resistance to minimize active fill movement and maintain the stability of the wall. The reinforcement and compacted fill are placed in an alternating manner. Each reinforcement is connected to wall face by a mechanical connector or friction. Geosynthetic reinforcement is also referred to as extensile reinforcement while metallic reinforcement is referred to as inextensible reinforcement. A drainage layer (geosynthetic or stone) behind the wall facing is often needed for drainage. When crushed stone or gravel is used as the drainage layer, it also eases the compaction behind the wall facing. A drainage layer is sometimes placed behind the reinforced fill to enhance external stability by reducing excess pore water pressure behind the reinforced fill. Under certain

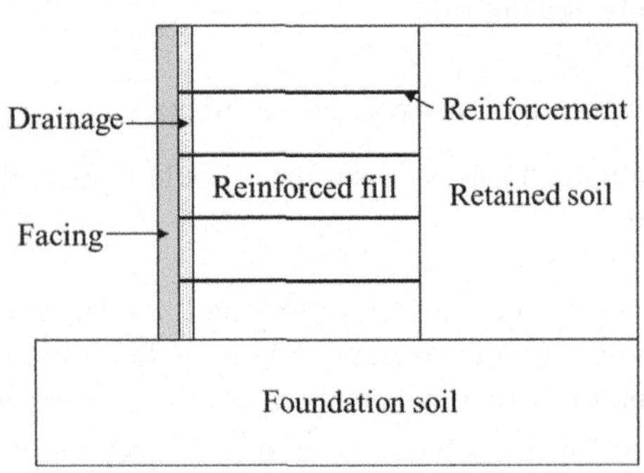

Figure 10.32 Components of mechanically stabilized earth wall.

conditions, there is an unreinforced or reinforced slope on the top or below the MSE wall.

Suitability The MSE walls are suitable when large elevation changes are needed. They can be built in less land space as compared with slopes. Free-draining granular fill with a plasticity index less than 6 is often required for the backfill material. Most MSE walls are constructed on unyielding foundations. If weak foundations exist, they should be removed and replaced, improved, or considered in the global stability analysis. MSE walls have performed well in seismic areas because of their flexibility.

Applications The MSE walls have been commonly used for bridge abutments, highways, railways, and commercial and residential areas with raised elevations for bridge approach or hill terrain. They can be temporary or permanent earth structures. They have also been used to repair failed slopes.

Advantages and Limitations The MSE walls are relatively easy to build and do not require specialty contractors and equipment (mostly earthwork equipment; however, panel walls may require cranes during the installation of large concrete panels). They require less right-of-way space as compared with slopes. They are cost effective and more tolerant to differential settlement as compared with gravity walls.

The MSE walls often require high-quality backfill materials. Durability of metallic and geosynthetic reinforcements in chemically active backfill may reduce service life. Poor drainage and global instability are two major reasons for some failed MSE walls.

10.5.2 Principles

Failure Modes The MSE walls may have the following modes of failure: (1) external, (2) internal, (3) local, (4) compound, and (5) global (or deep seated) as shown in Figure 10.33. The external failure includes sliding, overturning, and bearing failure. The internal failure may result from the rupture, pullout, and connection failure of reinforcement. The local failure including toppling and bulging may happen due to large unreinforced space. Compound and global failures are in the form of slope failure and their slip surfaces are partially or completely out of the reinforced zone. The above failure modes happen when the strength limits are exceeded. It is also considered unacceptable when deformations of walls exceed serviceability limits.

Potential Slip Surface and Tension in Reinforcement The reinforced fill is divided into active and stable zones by a slip surface as shown in Figure 10.34. The active zone tends to slide down under its self-weight, while the reinforcement

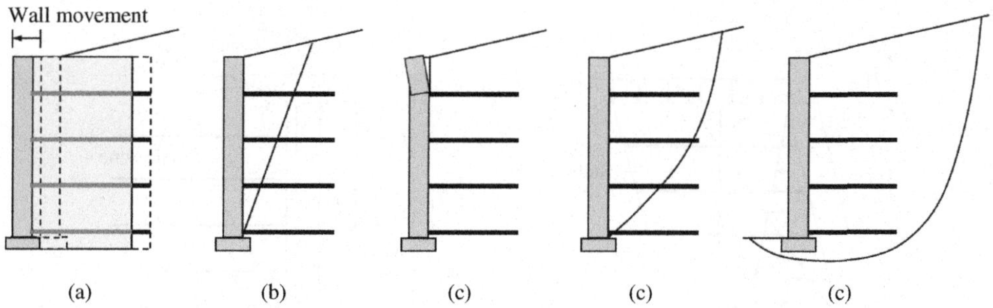

Figure 10.33 Failure modes of MSE walls: (a) external, (b) internal, (c) local, (d) compound, and (e) global.

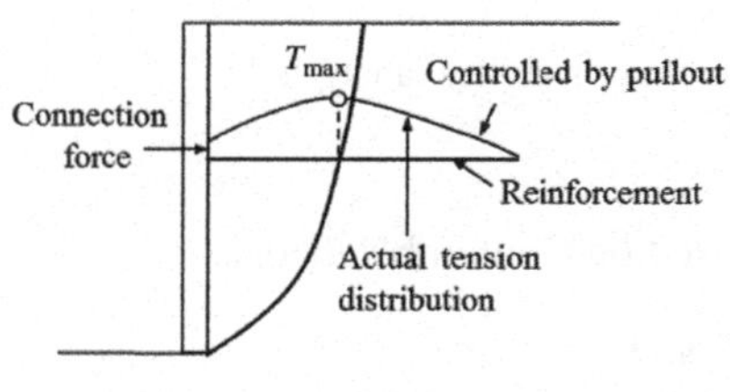

Figure 10.34 Potential slip surface and tension in reinforcement.

anchored in the stable zone provides tensile resistance to stabilize the active zone. The slip surface inside the reinforced fill is not necessarily planar and the location and shape of the slip surface depend on the soil strength, the batter of the wall facing, and the stiffness of reinforcement. Metallic reinforcement has higher tensile stiffness, which minimizes wall movement. As a result, the potential slip surface is close to the wall facing. On the other hand, geosynthetic reinforcement has lower tensile stiffness, which allows more wall movement. As a result, the potential slip surface extends more deeply into the reinforced fill. However, a planar or bi-planar slip surface is often assumed for design of geosynthetic-reinforced earth walls or metallic-reinforced earth walls, respectively, based on Coulomb or Rankine's theory for the internal analysis of an MSE wall to be discussed in later sections. Both experimental and theoretical results show that maximum tension in each reinforcement develops at the slip surface and the tension decreases toward the stable zone and the back of the wall face (Han and Leshchinsky, 2006). The rear end of a reinforcement is controlled by pullout capacity of the reinforcement, while the front end is controlled by the connection force between reinforcement and wall facing. In most design codes, however, the maximum tension extends to the back of the wall facing as an approximation, which is conservative.

Active Earth Pressure Theory The analysis of a geosynthetic-reinforced earth wall is mainly based on active earth pressure theory (i.e., Rankine or Coulomb's theory). Figures 10.35 and 10.36 shows the force diagrams for external and internal analyses. Active earth pressures develop behind the reinforced fill and within the reinforced fill (i.e., behind the wall face). Coulomb's active earth pressure theory can be used to describe the distributions of these lateral earth pressures.

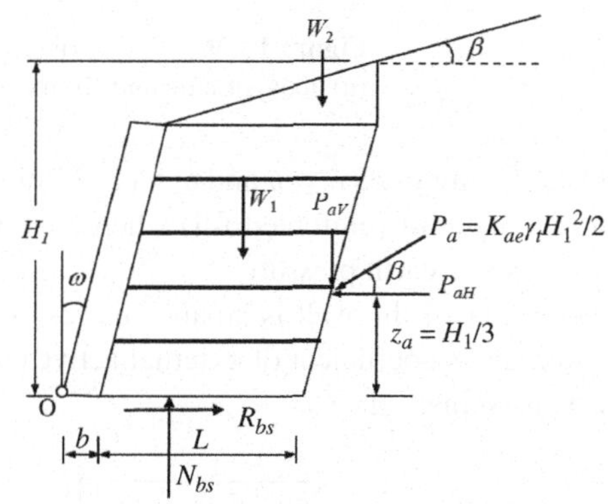

Figure 10.35 Force diagrams for external stability analysis.

For external stability analysis, the active thrust from the retained soil can be calculated based on the coefficient of external active earth pressure, the unit weight of the retained soil, and the height of the wall, including the top slope height. AASHTO (2012) assumes the inclination angle of the thrust is the same as the top slope angle. The coefficient of external earth pressure can be expressed based on Coulomb's solution as follows:

$$K_{ae} = \frac{\cos^2(\phi_t + \omega)}{\cos^2\omega\cos(\omega - \delta)\left[1 + \sqrt{\frac{\sin(\phi_t+\delta)\sin(\phi_t-\beta)}{\cos(\omega-\delta)\cos(\omega+\beta)}}\right]^2} \tag{10.29a}$$

where ω = batter of wall facing
β = top slope angle
ϕ_t = friction angle of the retained soil
δ = interface friction angle between reinforced fill and retained soil

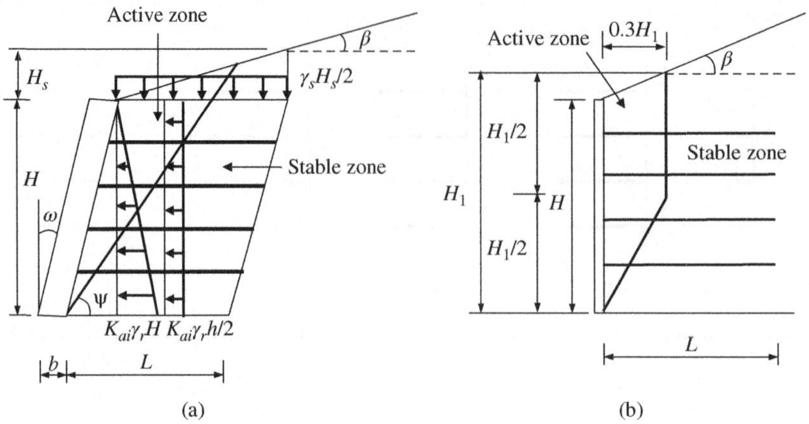

Figure 10.36 Force diagrams and slip planes for internal stability analysis: (a) geosynthetic and (b) metallic reinforcements.

AASHTO (2012) suggests $\delta = \beta$ and $\delta \leq (2/3)\min(\phi_r, \phi_t)$ (ϕ_r = friction angle of reinforced fill) when computing the coefficient of active earth pressure.

When the batter of the wall is small (i.e., less than 10°), however, Rankine's coefficient of external active earth pressure may be used instead:

$$K_{ae} = \cos\beta \left[\frac{\cos\beta - \sqrt{\cos^2\beta - \cos^2\phi_t}}{\cos\beta + \sqrt{\cos^2\beta - \cos^2\phi_t}}\right] \quad (10.29b)$$

The lateral component of the active thrust, P_{aH}, is the driving force for possible lateral sliding and overturning of the reinforced mass as a rigid body. The sliding resistance develops at the base of the reinforced fill. The soil weights, W_1 and W_2, and the vertical component of the active thrust, P_{aV}, are the stabilizing forces to resist the possible rotation induced by the lateral component of the active thrust, P_{aH}. The vertical applied load as a result of W_1, W_2, and P_{aV}, should be checked against the allowable bearing capacity of the foundation. It should be noted that different design manuals or guidelines may have different definitions of the width of the reinforced zone or the length of reinforcement. AASHTO (2012) defines the reinforced zone and the reinforcement length start from the back of the wall face. However, the National Concrete Masonry Association (2010) defines the reinforced zone and the reinforcement length start from the front of the wall face.

For internal stability analysis, the reinforced fill is divided into an active zone and a stable zone, as shown in Figure 10.36. Reinforcements anchored in the stable zone provide tensile resistance to the active zone. The potential failure plane, which divides the active and stable zones, can be determined by the inclination angle, ψ, in the following equation based on Coulomb's theory:

$$\tan(\psi - \phi_r) = \frac{-\tan(\phi_r - \beta) + \sqrt{\tan(\phi_r - \beta)[\tan(\phi_r - \beta) + \cot(\phi_r + \omega)][1 + \tan(\delta - \omega)\cot(\phi_r + \omega)]}}{1 + \tan(\delta - \omega)[\tan(\phi_r - \beta) + \cot(\phi_r + \omega)]} \quad (10.30)$$

When the wall batter, ω, is smaller than 10°, Rankine's theory can be used to determine this inclination angle, that is, $\psi = 45° + \phi_r/2$. When metallic reinforcements are used, the slip surface becomes vertical at the midheight of the wall.

The active lateral earth pressure applied on the back of the wall facing can be induced by the active earth pressure of the reinforced fill and the surcharge on the top of the wall. The coefficient of internal active earth pressure for the internal analysis can also be calculated using Equation (10.29). For simplification and approximation, AASHTO (2012) suggests an average vertical stress on the top of the wall acting as the surcharge to represent an actual slope. In addition, AASHTO (2012) assumes a smooth interface between the wall facing and the reinforced fill so that Equation (10.29) can be simplified into:

$$K_{ai} = \frac{\cos^2(\omega + \phi_r)}{\cos^3\omega[1 + \sin\phi_r/\cos\omega]^2} \quad (10.31)$$

When the wall batter is smaller than 10°, the coefficient of internal active earth pressure can be further simplified into:

$$K_{ai} = \tan^2\left(45° - \frac{\phi_r}{2}\right) \quad (10.32)$$

The variation of lateral earth pressure coefficient ratio depends on the depth below the top of the wall as shown in Figure 10.37. For metallic reinforcement, the lateral earth

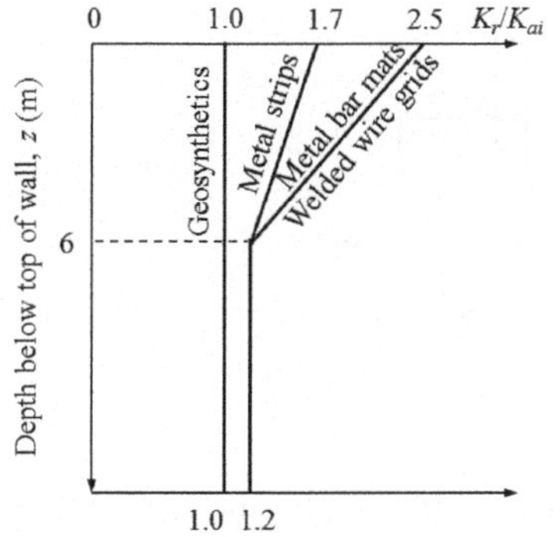

Figure 10.37 Variation of lateral earth pressure coefficient ratio with depth (Berg et al., 2009).

pressure coefficient decreases with depth down to 6 m and then becomes constant with depth. For geosynthetic reinforcements, the lateral earth pressure coefficient is constant with depth and equal to K_{ai}. The active lateral earth pressure on the back of the wall facing is carried by the reinforcements at different elevations. The tensile resistance provided by each reinforcement depends on the tensile rupture strength, pullout capacity, and connection strength with facing.

10.5.3 Design Considerations

Performance Requirements The performance requirements tabulated in Table 10.6 are commonly used in practice for the design of MSE walls (Holtz et al., 2008). Both British Standard BS8006 (2010) and AASHTO (2012) require that the reinforcement length should be at least 70% the wall height under a normal condition. However, the National Concrete Masonry Association (2010) suggests that the reinforcement length should be at least 60% the wall height. Research shows that the MSE walls may have large deformations if the reinforcement length is less than 60% the wall height.

Table 10.6 Performance Requirements for Geosynthetic-Reinforced Earth Walls

Location	Failure Mode	Requirement
External	Base sliding	FS > 1.5
	Bearing failure	FS > 2.0–2.5
	Overturning	FS > 2.0
	Geosynthetic rupture	FS > 1.5
Internal	Pullout	FS > 1.5 with 1.0 m embedment length
	Connection	FS > 1.5
	Deep-seated	FS > 1.3
Overall	Settlement and differential settlement	Project-dependent
Seismic	All failure modes	FS > 75% static FS

External Stability Analysis

1. Base Sliding In the analysis of base sliding, the reinforced fill is treated as a rigid block subjected to lateral load from the retained soil. The factor of safety against base sliding can be calculated as follows:

$$FS_{bs} = \frac{\text{Sliding resisting force}}{\text{Sliding driving force}} = \frac{c_b L + (W_1 + W_2 + P_{aV}) \tan \delta_b}{P_{aH}}$$

where c_b and δ_b are the interface cohesion and friction angle between wall base and foundation soil and other symbols are shown in Figure 10.38.

2. Overturning The factor of safety against overturning about the toe of the wall as shown in Figure 10.38 can be calculated as follows:

$$FS_{ot} = \frac{\text{Resisting moment}}{\text{Driving moment}} = \frac{W_1 x_1 + W_2 x_2 + P_{aV} L}{P_{aH} z_a} \tag{10.33}$$

where x_1 and x_2 are the distances from the wall face to the centroids of the triangular and rectangular blocks.

The overturning of the reinforced fill can also be controlled by limiting the eccentricity at the base of the reinforced zone to be less than (1/6) L on soil or (1/4) L on rock. The eccentricity, e, can be calculated in the same way as a footing subjected to a moment as follows:

$$e = \frac{L}{2} - \frac{W_1 x_1 + W_2 x_2 + P_{aV} L - P_{aH}(H_1/3)}{W_1 + W_2 + P_{aV}} \tag{10.34}$$

3. Bearing Failure Due to the driving moment by the thrust from the retained soil, the applied pressure at the base of the wall is not uniform. The resultant reaction force, N_{bs}, has an eccentricity, e, from the center of the wall base. For the convenience of calculating the factor of safety against bearing failure, a Meyerhof-type distribution method can be used to estimate the equivalent uniform applied pressure

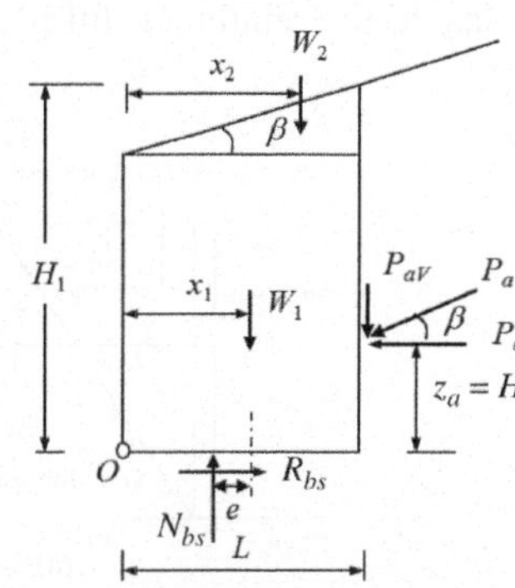

Figure 10.38 Force diagrams for external stability analysis.

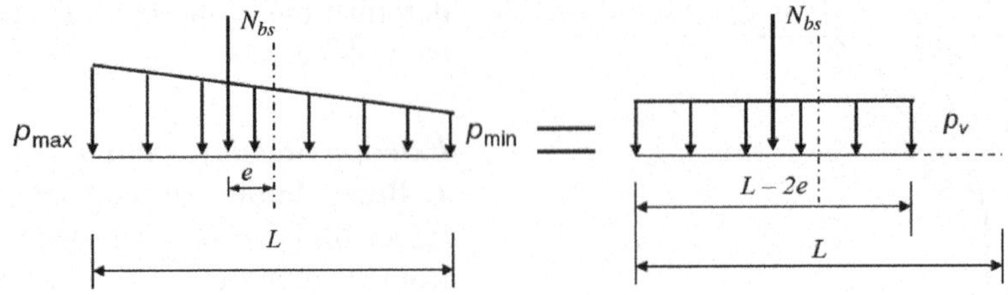

Figure 10.39 Applied pressure equivalency.

(Figure 10.39) as follows:

$$p_v = \frac{N_{bs}}{L-2e} = \frac{W_1 + W_2 + P_{aV}}{L-2e} \quad (10.35)$$

In the design of MSE walls, the embedment effect on the bearing capacity is often ignored to be conservative. The ultimate bearing capacity of the MSE wall foundation with the equivalent base width, $L-2e$, is

$$q_{ult} = c_f N_c + 0.5(L-2e)\gamma_f N_\gamma \quad (10.36)$$

Therefore, the factor of safety against bearing failure can be calculated as follows:

$$FS_{bc} = \frac{q_{ult}}{p_v} = \frac{c_f N_c + 0.5(L-2e)\gamma_f N_\gamma}{W_1 + W_2 + P_{aV}}(L-2e) \quad (10.37)$$

where N_c and N_γ are bearing capacity factors of the foundation.

Internal Stability Analysis The reinforced fill is divided into active and stable zones by a slip surface, as shown in Figure 10.40. The active zone tends to slide down, while the reinforcements anchored in the stable zone provide tensile resistance to stabilize the active zone. Both experimental and theoretical results show that maximum tension in each reinforcement develops at the slip surface and the tension decreases toward the stable zone and the back of the wall face (Han and Leshchinsky, 2006). The rear end of each reinforcement is controlled by the pullout capacity of the reinforcement while the front end is controlled by the connection force between reinforcement and wall facing. In most design codes, however, the maximum tension extends to the back of the wall facing as an approximation, which is conservative. The slip surface inside the reinforced fill is not necessarily planar. However, a planar slip surface is often assumed based on Coulombs or Rankine's theory for the internal analysis of an MSE wall.

In internal stability analysis, it is assumed that each reinforcement carries the lateral pressure within the tributary distance from the midpoint of the upper spacing to the midpoint of the lower spacing, as shown in Figure 10.41 as an example. The uppermost reinforcement carries all the lateral pressure above the reinforcement and the lateral pressure down to the midpoint of the lower space. Similarly, the lowermost reinforcement carries the lateral pressure from the midpoint of the upper space to all the lateral pressure below the reinforcement. The maximum tension in each reinforcement can be calculated using the following equation:

$$T_{i,\max} = \frac{1}{2}(p_{ai} + p_{a(i+1)})s_{vi}, \; i = 1, 2, \ldots, n \quad (10.38)$$

where n is the total number of reinforcement.

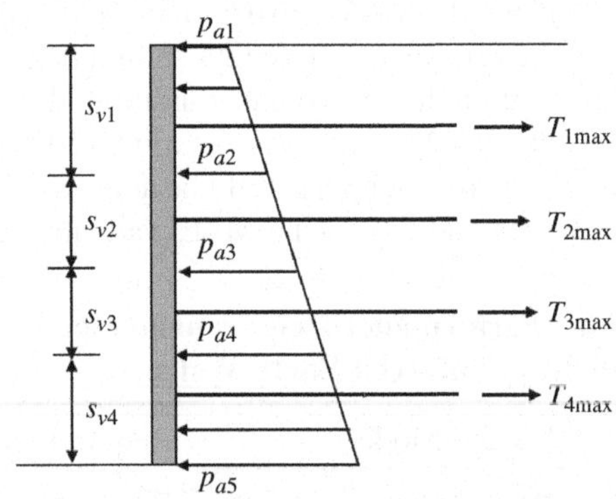

Figure 10.41 Calculation of maximum tension in reinforcement.

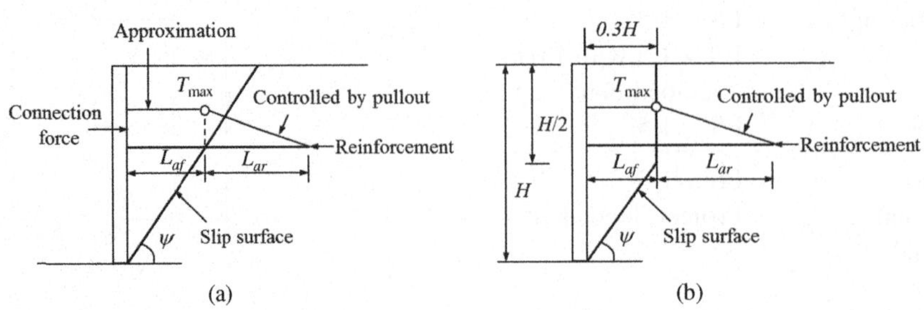

Figure 10.40 Simplified slip surface and tension distribution: (a) geosynthetic and (b) metallic reinforcements.

The reinforcement should be designed with a required factor of safety against possible rupture and pullout as follows:

$$FS_{rp} = \frac{T_a}{T_{max}} \quad (10.39a)$$

$$FS_{po} = \frac{T_{po}}{T_{max}} \quad (10.39b)$$

where FS_{rp}, FS_{po} = factor of safety against rupture and pullout, respectively
T_a = long-term allowable design strength of reinforcement after considering reduction factors of creep, installation, and durability for geosynthetic reinforcement or corrected cross-sectional area of metallic reinforcement due to corrosion
T_{po} = pullout capacity of reinforcement

The pullout capacity of reinforcement in fill can be calculated as follows:

$$T_{po} = 2F^*\alpha_{se}R_c\sigma_z L_a = 2C_i R_c \sigma_z L_a \quad (10.40)$$

where α_{se} = scale effect correction factor (default value = 1.0 for all steel reinforcements, 0.8 for geogrids, and 0.6 for geotextiles)
R_c = percent coverage of reinforcement
σ_z = vertical stress on reinforcement at depth z from top of wall
L_a = anchorage length in fill
F^* = pullout friction factor

The default values for F^* are provided in Figure 10.42, where ϕ_r is the friction angle of fill and C_u is uniformity of fill material ($C_u \geq 4.0$) and C_i is the interaction coefficient between reinforcement and fill (i.e., $C_i = F^*\alpha_{se}$).

It is also necessary to check the factor of safety against internal sliding along each reinforcement. This calculation is similar to the base sliding in the external stability analysis.

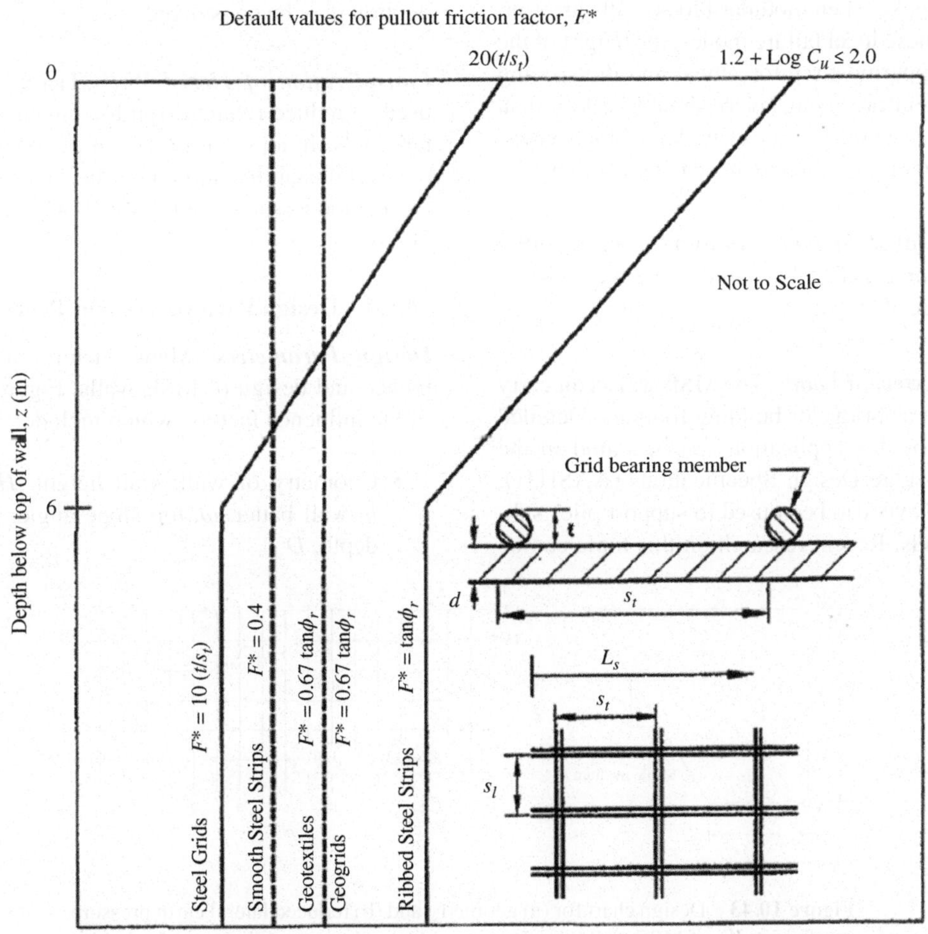

Figure 10.42 Default values for pullout friction factor, F^* (after Elias et al., 2001).

Instead of the interface cohesion and friction angle between wall base and foundation soil, the interface friction angle between reinforced fill and reinforcement should be used.

In order for reinforcement to mobilize its maximum tensile strength, the connection between reinforcement and wall facing should have enough connection strength. Under a normal condition, the force at the connection is lower than that at the failure plane. For a conservative design, the connection force is often assumed to be equal to the maximum tensile force, $T_{\max}$. Therefore, the factor of safety against the connection failure can be calculated as follows:

$$FS_{cn} = \frac{T_{cn}}{T_{\max}} \tag{10.41}$$

The allowable connection strength of reinforcement, T_{cn}, can be determined by connection tests. The modes of connection failure can be rupture and pullout. Reinforcement can be connected to wall facing mechanically or by friction.

Local Stability Analysis Possible local failures include toppling of the blocks above the uppermost reinforcement and bulging between blocks when modular block walls are constructed. To avoid these local failure modes, the height of the blocks above the uppermost reinforcement and the spacing between adjacent reinforcement layers should be less than two times the block depth. For most modular block walls, the allowable maximum reinforcement spacing is 0.6 m.

Compound and Global Stability Analyses The stability analyses of compound and global failures are the same as those for the reinforced slopes.

Consideration of External Load The MSE walls can carry external loads, such as bridge or building footings. Detailed design procedure for this application can be found in the ASSHTO LRFD Bridge Design Specifications (AASHTO, 2012). MSE walls have also been used to support piles subjected to lateral loads. Recent research on this topic can be found in publications by Pierson et al. (2009, 2011), and Huang et al. (2011, 2013).

Consideration of Earthquake Loading The MSE walls have performed well in past earthquake events due to their flexibility. Details on the design procedure for MSE walls under earthquake loading can be found in the ASSHTO LRFD Bridge Design Specifications (AASHTO, 2012).

Consideration of Stable Boundary An MSE wall is sometimes bounded by a stable boundary, such as bedrock and a soil nailed wall. When the stable boundary is close to the wall face or the reinforced fill, the internal or external active lateral earth pressure cannot be fully mobilized; therefore, the earth pressure is reduced. The reduced active lateral earth pressure coefficient, K_a', can be estimated using the design chart in Figure 10.43. When point A is at the back of the wall face, K_a' is the internal active lateral earth pressure coefficient. When point A is at the back of the reinforced fill, K_a' is the external active lateral earth pressure coefficient. When the base width to wall height ratio is greater than 0.2, no reduction may be considered.

Consideration of Tiered Wall Tiered MSE walls may be used to reduce required reinforcement tensile strengths for tall walls. In addition to design of individual wall stability, the overall stability should be checked using the slope stability method as shown in Figure 10.44 (Leshchinsky and Han, 2004).

10.5.4 Design Parameters and Procedure

Design Parameters Many factors influence the performance and design of MSE walls. Figure 10.45 shows most of the influence factors, which include:

- Geometry of wall: wall height, H, top slope height, h, wall batter, ω, top slope angle, β, wall embedment depth, D

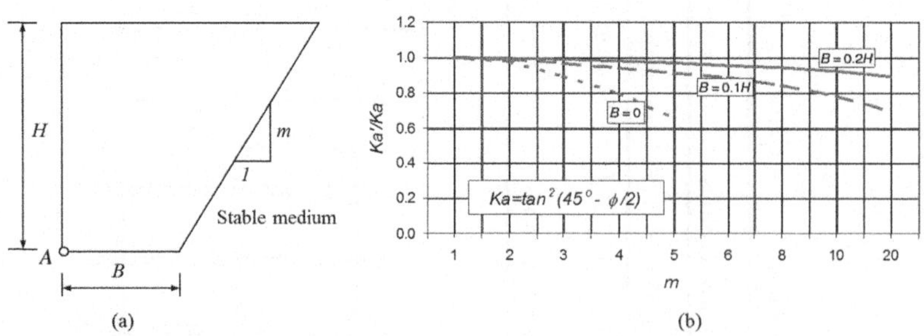

Figure 10.43 Design chart for (a) geometry and (b) reduced lateral earth pressure coefficient, K_a', of MSE wall bounded by stable medium (modified from Leshchinsky et al., 2004).

- Properties of reinforced fill: friction angle, ϕ_r, and unit weight, γ_r
- Properties of retained soil: friction angle, ϕ_t, and unit weight, γ_t
- Properties of foundation soil: cohesion, c_f friction angle, ϕ_f, and unit weight, γ_f
- Property of top slope soil: unit weight, γ_s
- Reinforcement: length, L, front anchorage length, L_{af}, and rear anchorage length, L_{ar}, tributary distance, s_{vi}, allowable tensile strength, T_a, pullout capacity, T_{po}, and allowable connection strength, T_{cn}
- Surcharge, p

Proper selection of reinforced fill is important for the performance of geosynthetic-reinforced earth walls. AASHTO (2012) requires the backfill for the reinforced zone should meet the gradation limits as provided in Table 10.7. Poor-quality backfill material is one of the reasons that causes failure of geosynthetic or metallic-reinforced earth walls. It is well understood that an increase of the plasticity index of backfill increases construction and creep deformations and results in poorer workability. An increase of fines in backfill results in poorer drainage and maintains higher excess pore water pressure, which may cause wall failure. No cohesion should be considered for reinforced fill and retained soil. A decrease of friction angle of backfill increases horizontal stress and reduces pullout capacity of geosynthetic or metallic reinforcement so that stronger and longer reinforcements are required. Backfill should also be checked for its pH value so that a proper geosynthetic type can be selected based on Table 10.8.

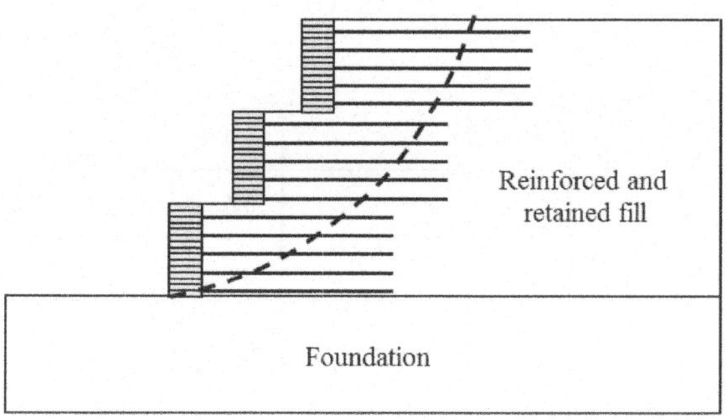

Figure 10.44 Overall stability of tiered wall (modified from Leshchinsky and Han, 2004).

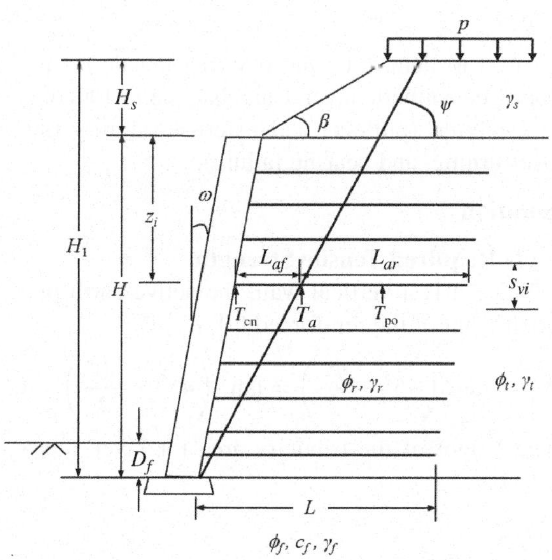

Figure 10.45 Influence factors for MSE wall.

Table 10.7 Required Gradation Limits for Mechanically Stabilized Earth Walls

Sieve Size	Percent Passing
102 mm	100
0.425 mm (No. 40)	0–60
0.075 mm (No. 200)	0–15
Plasticity Index (PI) shall not exceed 6	

Source: Elias et al. (2001).

Traffic loading is a dynamic load, which does not remain on the top of the wall all the time. To be conservative, the basic design principle is to consider the traffic loading in the calculation of driving force and/or driving moment but not to consider it in the calculation of resisting force and/or resisting moment as shown in Figure 10.46.

Design Procedure The following design procedure is commonly used in practice to design MSE walls:

1. Select reinforcement type (geosynthetic versus metallic reinforcement) and wall facing type.
2. Layout reinforcements based on the minimum length and spacing requirements.
3. Calculate coefficients of lateral earth pressure for external and internal analyses.
4. Calculate factors of safety against external failures including sliding, overturning, and bearing failure. If any calculated factor of safety does not meet the

Table 10.8 pH Limit of Backfill for Geosynthetic Reinforcement

Base Polymer	Criterion	Test Method
Polyester (PET)	$3 < pH < 9$	AASHTO T-289-91
Polypropylene (PP) and high-density polyethylene (HDPE)	$pH > 3$	AASHTO T-289-91

Source: Elias et al. (2001).

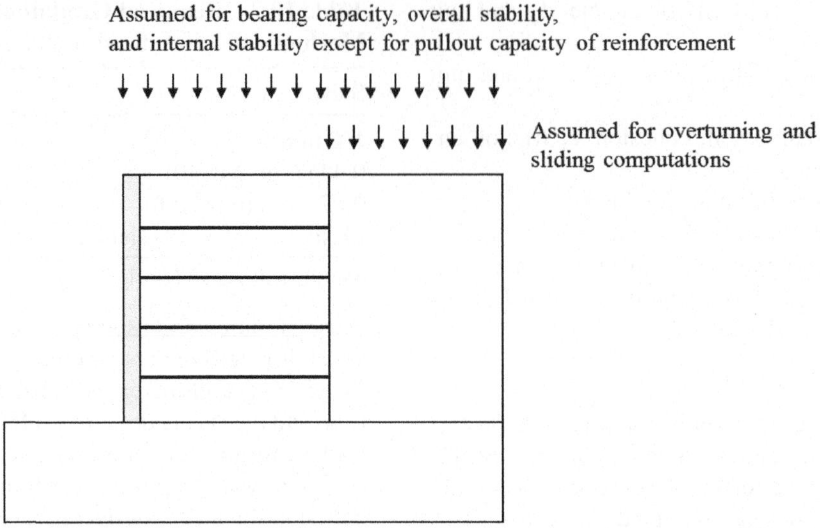

Figure 10.46 Consideration of traffic loading in analysis (Elias et al., 2001).

performance requirement, extend reinforcement length until all the requirements are met.

5. Calculate the maximum tension in each reinforcement and then select the grade of reinforcement to meet the strength requirement based on the desired factor of safety. Also, check the maximum tension against the pullout capacity. If the pullout capacity is not sufficient, extend the length of the reinforcement.
6. Check the maximum tension against the connection strength of reinforcement. If the connection strength is not sufficient, change connector or reinforcement type or reduce vertical spacing of reinforcements.
7. Calculate the factor of safety against compound failure. If the factor of safety is not sufficient, extend the reinforcement and/or increase tensile strength of reinforcements intersecting with the critical slip surface.
8. Calculate the factor of safety against global failure. If the factor of safety against the global failure is not sufficient, improve the foundation soil by basal reinforcement or other ground improvement techniques.

Design Example 10.4

A vertical geosynthetic-reinforced MSE wall is shown in Example Figure 10.3. The reinforced fill, retained fill, and foundation soil have the same properties as shown in the figure. The traffic loading on top of the wall is 15 kPa. The AASHTO (2012) design method

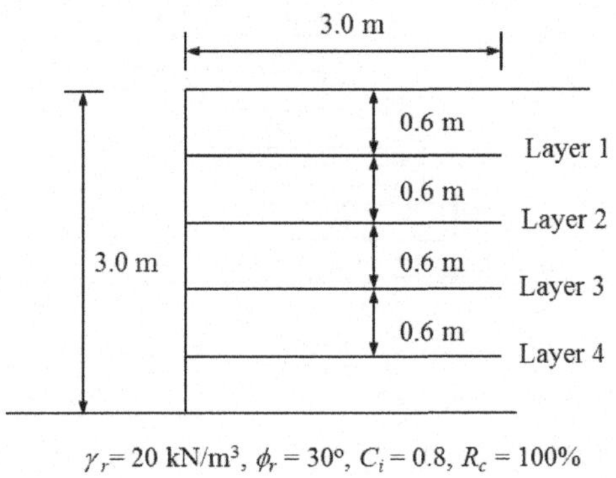

Example Figure 10.3 Geometry for MSE wall.

is used to determine the required tensile strength of each geosynthetic layer. Calculate the factors of safety for internal (pullout) and external stability (sliding, overturning, and bearing failure).

Solution

(1) Required Tensile Strength

Since it is a vertical wall, the active earth pressure coefficient of the reinforced fill, K_{ai}, is

$$K_{ai} = \tan^2\left(45° - \frac{\phi_r}{2}\right) = \tan^2\left(45° - \frac{30°}{2}\right) = 0.333$$

The height of the tributary area for layer 1 is $s_{v1} = 0.9$ m.

The required tensile strength for layer 1 is

$$T_{1\max} = 1/2K_{ai}\gamma_r s_{v1}^2 + K_{ai} p s_{v1}$$
$$= 1/2 \times 0.333 \times 20 \times 0.9^2 + 0.333 \times 15 \times 0.9$$
$$= 7.20\ \text{kN/m}$$

The height of the tributary area for layer 2 is $s_{v2} = 0.6$m.

The lateral earth pressure at the elevation of layer 2 ($z_2 = 1.2$ m) is

$$\sigma_h = K_{ai}\gamma_r z_2 + K_{ai} p$$
$$= 0.333 \times 20 \times 1.2 + 0.333 \times 15 = 13\ \text{kPa}$$

The required tensile strength for layer 2 is

$$T_{2\max} = \sigma_h s_{v2} = 13 \times 0.6 = 7.8\ \text{kN/m}$$

The height of the tributary area for layer 3 is $s_{v3} =$ 0.6 m.

The lateral earth pressure at the elevation of layer 3 ($z_3 = 1.8$ m) is

$$\sigma_h = K_{ai}\gamma_r z_3 + K_{ai} p = 0.333 \times 20 \times 1.8 + 0.333 \times 15$$
$$= 17\ \text{kPa}$$

The required tensile strength for layer 3 is

$$T_{3\max} = \sigma_h s_{v3} = 17 \times 0.6 = 10.2\ \text{kN/m}$$

The height of the tributary area for layer 4 is $s_{v4} =$ 0.9 m.

The lateral earth pressure at the middle of the tributary area for layer 4 ($z_4 = 2.55$ m) is

$$\sigma_h = K_{ai}\gamma_r z_4 + K_{ai} p = 0.333 \times 20 \times 2.55 + 0.333 \times 15$$
$$= 22\ \text{kPa}$$

The required tensile strength for layer 4 is

$$T_{4\max} = \sigma_h s_{v4} = 22 \times 0.9 = 19.8\ \text{kN/m}$$

(2) FS against Pullout

Since the top slope is flat, the critical failure plane within the reinforced zone is the Rankine failure plane, that is,

$$\psi = 45^\circ + \frac{\phi_r}{2} = 45^\circ + \frac{30^\circ}{2} = 60^\circ$$

The rear anchorage length of each layer (H_i is the height of the geosynthetic reinforcement, i, from the bottom of the wall) is

$$\text{Layer 1}: \quad L_{ar1} = L - \frac{H_1}{\tan\psi} = 3.0 - \frac{2.4}{\tan 60^\circ}$$
$$= 1.61\ \text{m}$$

$$\text{Layer 2}: \quad L_{ar2} = L - \frac{H_2}{\tan\psi} = 3.0 - \frac{1.8}{\tan 60^\circ} = 1.96\ \text{m}$$

$$\text{Layer 3}: \quad L_{ar3} = L - \frac{H_3}{\tan\psi} = 3.0 - \frac{1.2}{\tan 60^\circ} = 2.31\ \text{m}$$

$$\text{Layer 4}: \quad L_{ar4} = L - \frac{H_4}{\tan\psi} = 3.0 - \frac{0.6}{\tan 60^\circ} = 2.65\ \text{m}$$

The pullout capacity of each layer and the corresponding factor of safety are:

Layer 1:

$$T_{po1} = 2\gamma_r(H - H_1)L_{ar1}C_i R_c \tan\phi_r$$
$$= 2 \times 20 \times (3 - 2.4) \times 1.61 \times 0.8 \times 1 \times \tan 30^\circ$$
$$= 17.85\ \text{kN/m}$$
$$\text{FS}_{po1} = \frac{17.85}{7.20} = 2.48 > 1.5\ \text{(OK)}$$

Layer 2:

$$T_{po2} = 2\gamma_r(H - H_2)L_{ar2}C_i R_c \tan\phi_r$$
$$= 2 \times 20 \times (3 - 1.8) \times 1.96 \times 0.8 \times 1 \times \tan 30^\circ$$
$$= 43.45\ \text{kN/m}$$
$$\text{FS}_{po2} = \frac{43.45}{7.8} = 5.57 > 1.5\ \text{(OK)}$$

Layer 3:

$$T_{po3} = 2\gamma_r(H - H_3)L_{ar3}C_i R_c \tan\phi_r$$
$$= 2 \times 20 \times (3 - 1.2) \times 2.31 \times 0.8 \times 1 \times \tan 30^\circ$$
$$= 76.82\ \text{kN/m}$$
$$\text{FS}_{po3} = \frac{76.82}{10.2} = 7.53 > 1.5\ \text{(OK)}$$

Layer 4:

$$T_{po4} = 2\gamma_r(H - H_4)L_{ar4}C_i R_c \tan\phi_r$$
$$= 2 \times 20 \times (3 - 0.6) \times 2.65 \times 0.8 \times 1 \times \tan 30^\circ$$
$$= 117.5\ \text{kN/m}$$
$$\text{FS}_{po4} = \frac{117.5}{19.8} = 5.93 > 1.5\ \text{(OK)}$$

(3) FS against Base Sliding

The coefficient of lateral earth pressure for the external stability when $\beta = 0$ and $\phi_t = 30^\circ$ is

$$K_{ae} = \cos\beta \frac{\cos\beta - \sqrt{\cos^2\beta - \cos^2\phi_t}}{\cos\beta + \sqrt{\cos^2\beta - \cos^2\phi_t}}$$
$$= \cos(0)\frac{\cos(0) - \sqrt{\cos^2(0) - \cos^2(30)}}{\cos(0) + \sqrt{\cos^2(0) - \cos^2(30)}} = 0.333$$

The lateral forces due to soil and surcharge are

$$P_{aH1} = 1/2K_{ae}\gamma_t H^2 = 1/2 \times 0.333 \times 20 \times 3^2$$
$$= 30\,\text{kN/m}$$

$$P_{aH2} = K_{ae}pH = 0.333 \times 15 \times 3 = 15\,\text{kN/m}$$

The vertical force due to soil (surcharge is ignored for conservative design) is

$$W_1 = \gamma_r LH = 20 \times 3 \times 3 = 180\,\text{kN/m}$$

The factor of safety against sliding along the base is

$$FS_{bs} = \frac{W_1 \tan \delta_b}{P_{aH1} + P_{aH2}} = \frac{180 \times \tan 30^\circ}{30 + 15} = 2.31 > 1.5 \text{ (OK)}$$

(4) FS against Overturning

The driving moment due to soil pressure and surcharge is

$$M_d = P_{aH1} \times (1/3H) + P_{aH2} \times (1/2H)$$
$$= 30 \times (1/3 \times 3) + 15 \times (1/2 \times 3)$$
$$= 52.5\,\text{kN} \cdot \text{m/m}$$

The resisting moment by the weight of the reinforced soil (surcharge is ignored for conservative design) is

$$M_r = W_1 \times (1/2L) = 180 \times (1/2 \times 3) = 270\,\text{kN} \cdot \text{m/m}$$

The factor of safety against overturning is

$$FS_{ot} = \frac{M_r}{M_d} = \frac{270}{52.5} = 5.14 > 2.0 \quad \text{(OK)}$$

(5) FS against Bearing Failure

The resisting moment by the weight of the reinforced soil and surcharge is

$$M_r = W_1 \times (1/2L) + pL \times (1/2L)$$
$$= 180 \times (1/2 \times 3) + 15 \times 3 \times (1/2 \times 3)$$
$$= 337.5\,\text{kN} \cdot \text{m/m}$$

The eccentricity for bearing capacity calculation is

$$e = \frac{L}{2} - \frac{M_r - M_d}{W_1 + pL} = \frac{3}{2} - \frac{337.5 - 52.5}{180 + 15 \times 3} = 0.23\,\text{m}$$

The equivalent applied bearing pressure is

$$p_v = \frac{W_1 + pL}{L - 2e} = \frac{180 + 15 \times 3}{3 - 2 \times 0.23} = 88.6\,\text{kPa}$$

The ultimate bearing capacity is

$$q_{ult} = 0.5\gamma_f(L - 2e)N_\gamma$$
$$= 0.5 \times 20 \times (3 - 2 \times 0.23) \times 22.4$$
$$= 568.9\,\text{kPa}$$

The factor of safety against bearing failure is

$$FS_{bc} = \frac{q_{ult}}{p_v} = \frac{568.9}{88.6} = 6.42 > 2.5 \text{ (OK)}$$

10.5.5 Construction

The following construction sequence is commonly adopted in the field to construction MSE walls: site preparation, construction of leveling pads, erection of facing (panels, modular blocks, wire baskets, or gabions), placement and compaction of fill, and placement of reinforcement.

Site preparation can vary depending on site conditions. It may include removal of large objects (e.g., trees), leveling of ground to a desired elevation, removal of unsuitable materials (such as organic matters, vegetation, etc.). If foundation soil is too weak, other ground improvement measures may be taken.

Leveling pads (mostly unreinforced concrete but sometimes made of gravel) are casted or placed on the foundation soil to support facing panels, modular blocks, wire baskets, or gabions. Pads are typically 300 mm wide and 150 mm thick.

Installation of wall facing depends on the type of wall facing. The first row of facing panels needs to be braced to maintain their stability and alignment. Upper facing panels are placed on and connected to lower panels after reinforcement and fill are placed. Wire baskets have built-in bracing components. Modular blocks and gabions do not need any bracing.

When the compacted fill reaches the design level of a reinforcement, the reinforcement is placed on the top of the fill and connected to the facing. Different wall systems may have different connection mechanisms (mechanical or frictional). Reinforcement should be manually pulled straight to remove wrinkles before placement of next lift of fill.

Repeat the above procedure until the wall reaches the desired height.

10.5.6 Quality Control and Assurance

Quality control and assurance include the following items:

- Check the quality, gradation, pH value, and electrochemical properties of backfill material and the grades of reinforcements and corrosion protection (for metallic reinforcements only).
- Inspect precast panels or modular blocks for any damage and quality issues.
- Ensure the elevation and alignment of leveling pads.
- Ensure panels or blocks are erected or placed within the tolerance of vertical and horizontal alignments.

- Verify the density of compacted fill and its corresponding moisture content.
- Ensure the reinforcements are placed in a correct orientation.
- Inspect that the reinforcements are correctly connected to the panels or blocks.
- Ensure drainage systems behind the wall face and within the reinforced fill are properly installed.

10.6 GEOSYNTHETIC-REINFORCED FOUNDATIONS

10.6.1 Introduction

Basic Concept When problematic geomaterial exists near ground surface, one of the most economic and efficient ground improvement technologies is overexcavation and replacement. Chapter 4 discusses the design and construction for overexcavation and replacement. The ultimate bearing capacity of replaced foundations may be limited by the strength of the replacing fill and the thickness of the replaced zone. Reinforcements can be placed in layers (mostly two to four layers) within the replaced foundation to increase its ultimate bearing capacity by providing confinement and tensile resistance, distributing stresses to a wider area, and preventing punching failure.

Suitability This technology is most suitable for spread footings, such as isolated footings and continuous footings. Reinforcements, such as geogrid, woven geotextile, geocell, and metallic reinforcement, have been used in reinforced foundations. Metallic reinforcements (steel strips and meshes) have been used in the laboratory for research purposes; however, they are rarely used in practice. Aggregate, sand, clay, and fly ash have been used as the fill materials in geosynthetic-reinforced foundations (Chen, 2007). However, aggregate and sand have been mostly used in the field.

Applications Geosynthetic-reinforced foundations are mostly used to improve replaced fill under footings.

Advantages and Limitations The use of geosynthetic reinforcements can increase bearing capacity, reduce settlement, and allow the use of low-quality fill or a thinner replaced zone.

Numerous studies have been done in the past, most of which were laboratory model tests and numerical analyses. Few large plate loading tests and field tests were conducted for geosynthetic-reinforced foundations. For example, Adams and Collin (1997) and Demir et al. (2013) reported plate loading tests with plate sizes varying from 0.3 m × 0.3 m to 0.9 m × 0.9 m on geogrid-reinforced sand. Miura et al. (1995) reported a case study in which a geogrid-reinforced gravel foundation was used to support a 2.8-m-wide box culvert. Most experimental and numerical studies show that geosynthetic reinforcement is effective in increasing the ultimate bearing capacity but is less effective in reducing settlement.

10.6.2 Principles

Failure Modes Six potential failure modes of geosynthetic-reinforced foundations have been identified by researchers (Binquet and Lee, 1975; Huang and Menq, 1997; Wayne et al., 1998; Chen, 2007), as shown in Figure 10.47. The shallow failure as shown in Figure 10.47(a) happens when the distance from the base

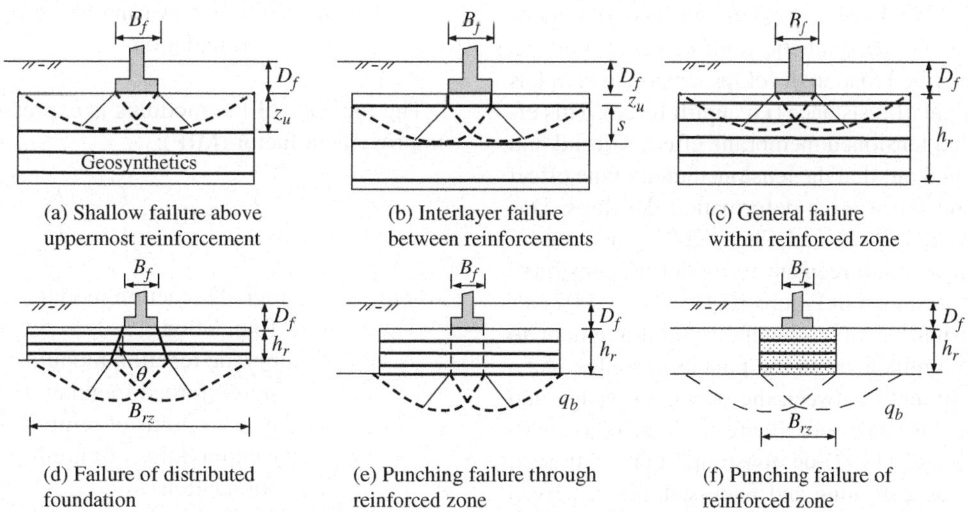

Figure 10.47 Potential failure modes of reinforced foundations (modified from Binquet and Lee, 1975; Huang and Menq, 1997, and Wayne et al., 1998; and Chen, 2007).

of the footing to the uppermost reinforcement is too large and/or the fill is too weak. Interlayer failure as shown in Figure 10.47(b) can develop when the spacing between reinforcements are too large. Within a thick reinforced foundation with multiple layers of reinforcement, a general failure, which involves one or more layers of reinforcements, may develop, as shown in Figure 10.47(c). Applied stress at the base of the footing distributes to the underlying soil. When the reinforced zone is thin and the underlying soil is weak, the distributed foundation may fail due to the low strength of the weak soil, as shown in Figure 10.47(d). When the reinforced foundation is thin but wide and underlain by weak soil, the footing may punch through the reinforced zone as shown in Figure 10.47(e). Figure 10.47(f) shows the punching of a reinforced zone into the weak soil due to the narrow reinforced zone and the existence of the underlying weak soil.

Effects of Geosynthetic Reinforcement The inclusion of geosynthetic reinforcement in replacing fill can have the following effects: (1) limited depth effect, (2) lateral restraint effect, (3) tensioned membrane effect, (4) lateral confinement effect, (5) deep-footing effect, and (6) wide-slab effect. A stable geosynthetic reinforcement (i.e., having sufficient tensile strength and pullout capacity) can serve as a boundary, limiting potential slip surfaces from going deeper. Mandel and Salencon (1972) demonstrated that limited depth of soil by a rigid base resulted in a higher bearing capacity. The uppermost reinforcement and close reinforcement spacing have this effect. Perkins and Ismeik (1997) and Giroud and Han (2013) indicated that geosynthetic reinforcement can provide lateral restraint to granular fill through interlocking with geogrid apertures, friction by geotextile, and closed confinement inside geocell pockets. Lateral restraint to granular fill adds apparent cohesion and increases modulus of the fill. When geosynthetic reinforcement deforms locally under an applied load, it develops tension, which has a vertical component of resistance (i.e., uplift force). This effect is the so-called tensioned membrane effect. Giroud and Han (2004a) pointed out that the tensioned membrane effect becomes important when large deformation develops. Das (1998), Wayne et al. (1998), and Chen (2007) suggested a complete rotation of reinforcement to model the tensioned membrane effect as shown in Figure 10.48. Chen (2007) attributed the contribution of geosynthetic reinforcement to providing lateral confinement to the punching wedge. As a result, the shear strength between the punching wedge and the surrounding soil is increased. Huang and Tatsuoka (1990) and Huang and Menq (1997) demonstrated that reinforced foundations have deep-footing and wide-slab effects. Even

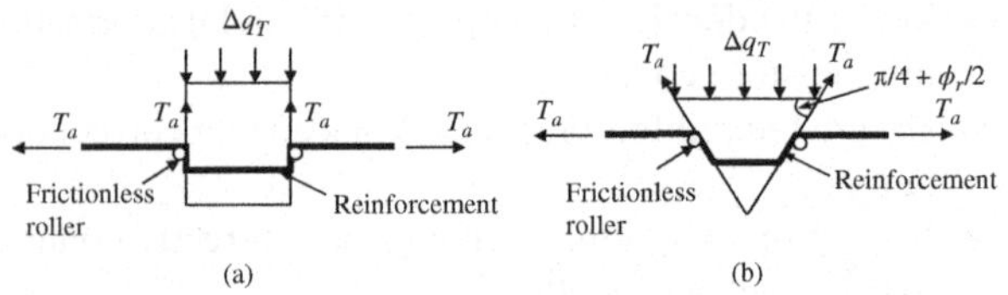

Figure 10.48 Complete rotation of geosynthetic reinforcement: (a) vertical punching and (b) triangular active wedges (modified from Das, 1998; Wayne et al. 1998; Chen, 2007).

if the reinforcements have the same length as the width of the footing, they can increase the bearing capacity by extending the footing to a deeper embedment depth. The inclusion of reinforcement helps distribute the applied load to a wide area as shown in Figure 10.47(d) so that the applied stress on the weak soil is reduced, the bearing capacity is increased, and the settlement is reduced. This effect has been confirmed by many experimental studies, for example, Gabr and Hart (1996), Han et al. (2013), and Qian et al. (2013).

When static plate loading tests are conducted on unreinforced and reinforced foundations, typical results (e.g., Guido et al., 1986; Adams and Collin 1997; Dong et al., 2010; Pokharel et al., 2010) are obtained as shown in Figure 10.49. Figure 10.49 shows that reinforcement increases bearing capacity and modulus but reduces settlement of the foundation. The increase of bearing capacity is often expressed as the bearing capacity ratio (BCR) as follows:

$$\mathrm{BCR} = \frac{q_{\mathrm{ult},r}}{q_{\mathrm{ult},u}} \tag{10.42}$$

where $q_{\mathrm{ult},u}$ = ultimate bearing capacity of unreinforced foundation
$q_{\mathrm{ult},r}$ = ultimate bearing capacity of reinforced foundation

The increase of the modulus is expressed as the modulus improvement factor (MIF):

$$\mathrm{MIF} = \frac{k_r}{k_u} = \frac{E_r}{E_u} \tag{10.43}$$

where k_r = subgrade reaction modulus of reinforced foundation
k_u = subgrade reaction modulus of unreinforced foundation
E_r = elastic modulus of reinforced foundation
E_u = elastic modulus of unreinforced foundation

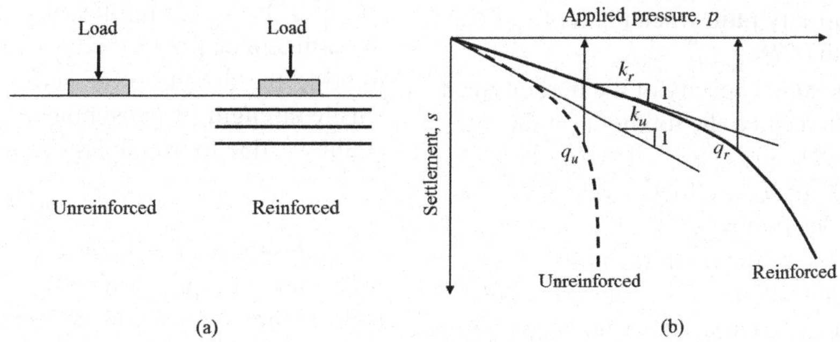

Figure 10.49 Plate loading tests on unreinforced and reinforced foundations: (a) plate loading test and (b) pressure vs. settlement.

Past studies show that BCR values mostly range from 1.0 to 3.0 while MIF values range from 1.0 to 2.0. The reason for the MIF to have a smaller value than the BCR value is that the MIF is determined at a smaller deformation at which reinforcement is less mobilized.

10.6.3 Design Considerations

Bearing Capacity against Shallow Failure To prevent the possible failure above the uppermost reinforcement, the reinforcement should be placed close to the base of the footing so that the potential slip surface is bounded by the reinforcement as shown in Figure 10.47(a). Binquet and Lee (1975) indicated that shallow failure likely happens when $z_u/B_f > \frac{2}{3}$. Mandel and Salencon (1972) developed a solution for a footing on sand bounded by a rigid base. Figure 10.50 shows that the BCR increases with an increase of sand friction angle, ϕ_r, but decreases toward 1.0 with an increase of the distance to the uppermost reinforcement, z_u. In this figure, B_f is the width of the footing. The ultimate bearing capacity of the footing against shallow failure, $q_{\text{ult},r}$, is

$$q_{\text{ult},r} = \text{BCR}_{\text{sf}} \cdot q_{\text{ult},u} \tag{10.44}$$

where BCR_{sf} = bearing capacity ratio due to shallow failure
$q_{\text{ult},u}$ = ultimate bearing capacity of unreinforced foundation

Test data from three laboratory studies generally match these theoretical curves. However, when z_u is small, overburden stress on the uppermost reinforcement is low so that pullout capacity of the reinforcement is limited. Under such a condition, the slip surface extends below the uppermost reinforcement. Therefore, the interlayer failure between reinforcements or general failure in the reinforced zone controls.

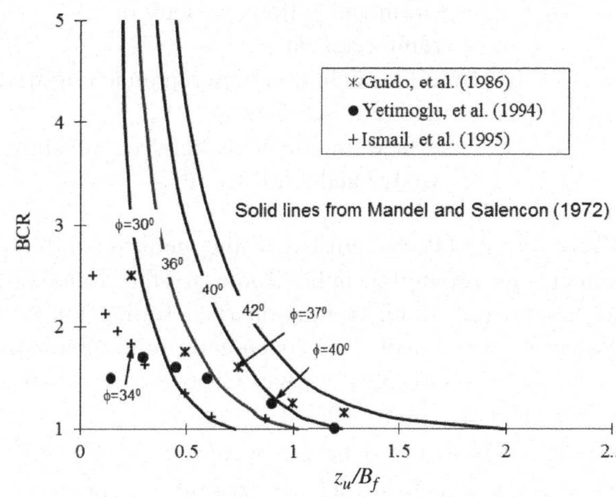

Figure 10.50 Bearing capacity ratio due to shallow failure above the uppermost reinforcement (after Wayne et al., 1998, with permission from ASCE).

Bearing Capacity against Interlayer Failure When the spacing between reinforcements is large, the failure can happen between the uppermost reinforcement and the next reinforcement as shown in Figure 10.47(b). Above the uppermost reinforcement, there is a punching failure. The formula for the ultimate bearing capacity of a footing due to punching failure has been discussed in Chapter 4. The ultimate bearing capacity against interlayer failure is contributed by the base resistance, the tensile effect of uppermost reinforcement, and the side resistance of the punching wedge:

$$q_{\text{ult},r} = \text{BCR} \cdot q_b + \frac{2C_1 c_a z_u}{B_f} + 2\gamma_r' z_u^2 \left(1 + \frac{2D_f}{z_u}\right) \frac{K_s \tan\phi_r}{B_f} + \frac{2C_1 T_a \tan\delta}{B_f} - \gamma_r' z_u \tag{10.45}$$

where BCR = bearing capacity ratio when the footing is at the depth of $D_f + z_u$

q_b = ultimate bearing capacity of the unreinforced foundation when the footing is at the depth of $D_f + z_u$

C_1 = constant (2 for square footing and 1 for continuous footing)

ϕ_r = friction angle of the reinforced fill

γ_r' = effective unit weight of the reinforced fill

K_s = coefficient of punching shear proposed by Meyerhof and Hanna (1978) (see Chapter 4)

T_a = allowable tensile strength of reinforcement (lesser of tensile strength at a specific strain and pullout capacity of reinforcement)

c_a = interface cohesion between punching wedge and reinforced fill

δ = interface friction angle between punching wedge and reinforce fill

Wayne et al. (1998) attributed the geosynthetic reinforcement to providing uplift force to the foundation. They assumed a complete rotation (i.e., setting $\tan\delta = 1$ in Equation (10.45)) of the geosynthetic reinforcement at the edges of the punching wedge. Chen (2007) attributed the geosynthetic reinforcement to providing additional confining stress to the punching wedge; $\delta = \left(\frac{2}{3}\right)\phi_r$ is a typical assumption. Test results show most measured tensile strains in geosynthetic reinforcement at failure were less than 3% (Chen, 2007) when a square plate size of 457 mm × 457 mm was used. It is expected that larger footings generate more deformation, which can result in higher strain in geosynthetic reinforcement. Without any field data, a tensile strain of 3% in the reinforcement may be used to calculate the ultimate bearing capacity of a reinforced foundation by limiting the overstress of geosynthetic reinforcement. The tensile strength of geosynthetic reinforcement at a specific strain, ε_g, due to overstress can be calculated by

$$T_g = J\varepsilon_g \tag{10.46}$$

where J is the tensile stiffness of geosynthetic reinforcement.

The pullout capacity of geosynthetic reinforcement on each side of the footing can be estimated using the following equation:

$$T_{\text{po}} = 2C_i[c_r + \gamma_r'(D_f + z)\tan\phi_r]L_a \quad \text{(for a continuous footing)} \tag{10.47}$$

$$T_{\text{po}} = \frac{2C_i[c_r + \gamma_r'(D_f + z)\tan\phi_r]A_a}{B_f} \quad \text{(for a square footing)} \tag{10.48}$$

where z, L_a, and A_a are defined in Figure 10.51; and A_a can be calculated as follows:

Square footing with uniaxial and biaxial reinforcements:

$$A_a = B_f\frac{B_{\text{rz}} - B_f}{2} \tag{10.49}$$

Square footing with uniform reinforcements:

$$A_a = \frac{B_{\text{rz}}^2 - B_f^2}{4} \tag{10.50}$$

As Dong et al. (2010) pointed out, biaxial geogrid is strong in the machine and cross-machine directions but weak in other directions, especially at 45° to the loading direction. Therefore, it is conservative to ignore the contributions of biaxial geogrid in the directions different from the machine and cross-machine directions. Dong et al. (2010) also found that triaxial geogrid has uniform tensile strength and stiffness

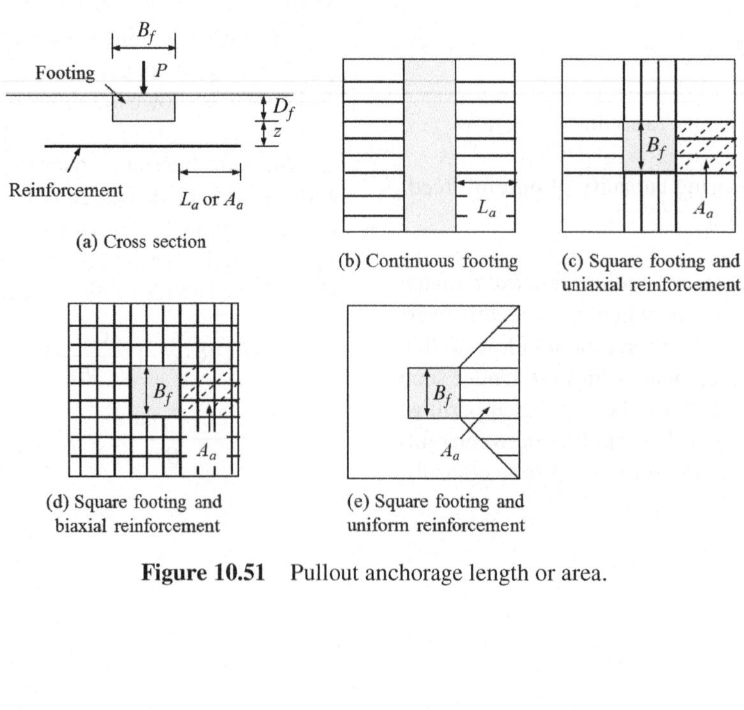

Figure 10.51 Pullout anchorage length or area.

in all the directions; therefore, it can be designed as a uniform reinforcement.

When the spacing is the same or smaller than the distance to the uppermost reinforcement, the ultimate bearing capacity against interlayer failure is always higher than that against the shallow failure above the uppermost reinforcement. Therefore, no analysis is needed if the ultimate bearing capacity against shallow failure is sufficient.

Bearing Capacity against General Failure within Reinforced Zone When a reinforced foundation is relatively thick and has multiple reinforcement layers, a general failure likely occurs within the reinforced zone as shown in Figure 10.47(c). Chen (2007) derived a theoretical solution for this failure mode. In this derivation, the contribution of geosynthetic reinforcement is to increase lateral confining stress between the active wedge and the passive wedge. The ultimate bearing capacity of the reinforced foundation due to the general failure within the reinforced zone can be expressed as follows:

$$q_{\text{ult},r} = q_{\text{ult},u} + \Delta q_T \tag{10.51}$$

where Δq_T is the additional bearing capacity due to tensile resistance provided by geosynthetic reinforcement, which can determined by the following formula:

$$\Delta q_T = \sum_{i=1}^{N_R} \frac{4T_{\text{ai}}[z_u + (i-1)s_v]r_T}{B_f^2} \tag{10.52}$$

where T_{ai} = allowable tensile strength of the ith layer of reinforcement due to overstress or pullout

$$r_T = \begin{cases} 1, \text{ for continuous footing} \\ 3\left[1 - 2\frac{z_u+(i-1)s_v}{B_f}\tan\left(\frac{\pi}{4} - \frac{\phi_r}{2}\right)\right], \text{ for square footing} \\ \quad \text{and } z_u + (i-1)s_v < \frac{B_f}{2}\tan\left(\frac{\pi}{4} + \frac{\phi_r}{2}\right) \\ 3\left(\frac{1}{2} - \frac{z_u+(i-1)s_v}{2h_f}\right), \text{ for square footing} \\ \quad \text{and } z_u + (i-1)s_v \geq \frac{B_f}{2}\tan\left(\frac{\pi}{4} + \frac{\phi_r}{2}\right) \end{cases}$$

where h_f is the depth of failure surface, that is,

$$h_f = \frac{B_f}{2\cos(\pi/4 + \phi_r/2)} e^{(\pi/4+\phi_r/2)\tan\phi_r}$$

It should be pointed out that geosynthetic reinforcements below the depth of the failure surface are not included in the calculation.

Under a large deformation, if the geosynthetic reinforcement is assumed to have a complete rotation as shown in Figure 10.48, the ultimate bearing capacity of the reinforced foundation can be calculated as follows:

$$\Delta q_T = \sum_{i=1}^{N_R} \frac{2C_2 T_{\text{ai}} \sin(\pi/4 + \phi_r/2)}{B_f} \tag{10.53}$$

where

$$C_2 = \begin{cases} 1, \text{ for square footing} \\ 2B_f - 4\left[z_u + (i-1)s_v\right]\tan\left(\frac{\pi}{4} - \frac{\phi_r}{2}\right), \\ \quad \text{for continuous footing} \end{cases}$$

Bearing Capacity against Distributed Foundation Failure Huang and Menq (1997) suggested reinforced foundations have deep-footing and wide-slab effects. These effects can be represented by distributed foundations as shown in Figure 10.47(d). The distributed foundation has a deeper and wider base. The similar method for the analysis of the distributed foundation through the replaced zone in Chapter 4 can be used for the reinforced foundation. The key is how to determine the distribution angle. An empirical relation was proposed by Huang and Menq (1997) based on laboratory tests on reinforced sandy foundations as follows:

$$\tan\theta = 0.680 - 2.071\frac{s_v}{B_f} + 0.743\text{CR} + 0.030\frac{B_{\text{rz}}}{B_f} + 0.076N_R \tag{10.54}$$

where θ = stress distribution angle from the reinforced zone to the underlying soil
CR = coverage ratio of the area of reinforcement strips or ribs over total reinforced area
N_R = number of reinforcements

The above relation is only valid for the following conditions:

$$\tan\theta > 0$$
$$0.25 < \frac{s_v}{B_f} < 0.5$$
$$0.02 \leq \text{CR} \leq 1.0$$
$$1 < B_{\text{rz}}/B_f \leq 10$$
$$1 \leq N_R \leq 5$$

Wayne et al. (1998) found that the distribution angle for unreinforced foundations is 26.7°, while the distribution angle for reinforced foundations is 45°.

Bearing Capacity against Punching Failure through Reinforced Zone The punching failure through the reinforced zone as shown in Figure 10.47(e) likely happens when the reinforced zone is thin and/or the underlying soil is weak. The ultimate bearing capacity against punching failure through the reinforced zone is contributed by the base resistance, the tensile effect of reinforcement, and the side resistance of the punching wedge:

$$q_{\text{ult},r} = q_b + \frac{2C_1c_ah_r}{B_f} + C_1\gamma_r'h_r^2\left(1+\frac{2D_f}{h_r}\right)\frac{K_s\tan\phi_r}{B_f} + \frac{2C_1T_a\tan\delta}{B_f} - \gamma_r'h_r \quad (10.55)$$

where q_b = ultimate bearing capacity of the foundation with a width of B_f at a depth of $D_f + h_r$
T_a = total allowable tensile strength of all reinforcement layers in one side of footing

Bearing Capacity against Punching Failure of Reinforced Zone The punching failure of the reinforced zone into the underlain soil as shown in Figure 10.47(f) likely happens when the reinforced zone is too narrow and the underlying soil is too weak. Since the reinforced zone acts as part of the rigid footing, geosynthetic reinforcement does not have any reinforcing effect. The same formula for the punching failure of a replaced zone into weak soil, as presented in Chapter 4, can be used to calculate the ultimate bearing capacity against the punching failure of the reinforced zone.

Critical Failure Mode Although six potential failure modes exist for a reinforced foundation, the actual failure of the foundation is always controlled by a critical mode. As such, the ultimate bearing capacity due to the critical failure mode is smaller than that due to any of the other modes of failure. The critical failure mode depends on the reinforcement configuration (distance of the reinforcement to the base of the footing, reinforced zone thickness, width and length of the reinforcement, and spacing of subsequent layers) and the properties of the reinforcement and soil, and the pullout interaction coefficient between soil and reinforcement. As mentioned previously, the failure above the uppermost reinforcement may be the critical mode when the distance to the uppermost reinforcement is too large. The punching failure may be the controlling mode for the case when the reinforced zone is thin and underlain by a layer of soft soil. The general failure may control when the reinforced foundation is thick and have multiple layers of reinforcements.

The design methods discussed above are based on planar reinforcements, such as geogrid, geotextile, and steel mesh. Geocell, a three-dimensional reinforcement, can be used to increase the bearing capacity as well. No systematic design method is available to calculate the bearing capacity of geocell-reinforced foundations over weak soil; therefore, more research is needed.

Settlement of Reinforced Foundation One of the benefits of geosynthetic in the reinforced foundation is to provide lateral restraint to granular fill so that its elastic modulus is increased. Based on laboratory and field plate loading tests, the elastic modulus of the reinforced foundation is higher than that of the unreinforced foundation. The following formula can be used to estimate the elastic modulus of the reinforced foundation, E_r:

$$E_r = \text{MIF} \cdot E_u \quad (10.56)$$

where MIF = modulus improvement factor
E_u = elastic modulus of unreinforced fill

The higher modulus of the reinforced fill reduces the settlement of the foundation. The method to calculate the settlement of the replaced zone in Chapter 4 can be used for the reinforced foundation by considering the modulus improvement factor, MIF.

10.6.4 Design Parameters and Procedure

Design Parameters The design parameters for geosynthetic-reinforced foundations include:

- Type of footing, square or continuous footing
- Width of footing, B_f
- Embedment depth of footing, D_f
- Distance from base of footing to uppermost reinforcement, z_u
- Spacing between reinforcements, s_v
- Number of reinforcements, N_R
- Tensile stiffness of reinforcement, J
- Allowable tensile strength of reinforcement at a specified strain, T_a
- Thickness of reinforced zone, h_r
- Width of reinforced zone, B_{rz}
- Friction angle and unit weight of reinforced fill, ϕ_r and γ_r
- Ultimate bearing capacity of underlying weak soil, q_b

Design Procedure The following procedure may be adopted to design a geosynthetic-reinforced foundation:

1. Follow the procedure for the overexcavation and replacement method to design a replaced zone without geosynthetic reinforcement.
2. Evaluate whether geosynthetic reinforcement is needed. If the designed replaced zone does not meet the performance requirements or there is a need to use low-quality fill and/or reduce the thickness of the replaced zone, geosynthetic reinforcement(s) may be used.
3. Layout geosynthetic reinforcement(s). Depending on the thickness of the replaced zone and the need for performance, one to four reinforcements may be placed. In soft soil, one reinforcement layer may be placed on the top of the soft soil to assist placement and compaction of fill. The spacing of reinforcements may be

selected in terms of lift thickness. The practical minimum spacing is 150 mm. The length and width of reinforcement(s) may be the same as those of the replaced zone at the base.

4. Calculate the ultimate bearing capacities and factors of safety in terms of different failure modes. If the calculated minimum factor of safety against critical bearing failure is less than the required, adjust the layout in step 3 and repeat the calculations until a satisfactory result is reached.
5. Calculate the settlement of the footing on the reinforced zone. If the calculated settlement is larger than the allowed value, adjust the layout in step 3 and repeat the calculations until a satisfactory result is reached.
6. Calculate the volumes of excavation and fill material and quantity of geosynthetic reinforcement.

Design Example 10.5

In the design example in Chapter 4, the thickness of the replaced zone is reduced to 700 mm. Three uniform geosynthetic layers are placed within the replaced zone as shown in Example Figure 10.4. The properties of geosynthetics are as follows: secant modulus at 3% is 500 kN/m and interaction coefficient with fill is 0.9. Calculate the ultimate bearing capacity of the reinforced foundation and the factor of safety at the applied pressure.

Solution

1. Ultimate bearing capacity and factor of safety without geosynthetic reinforcement: The ultimate bearing capacities of the 700-mm-thick replaced zone without geosynthetic reinforcement can be calculated following the same procedure presented in Chapter 4 as follows:

General failure: $q_{\text{ult}} = 2093$ kPa

Punching failure through the replaced zone: $q_{\text{ult}} = 426$ kPa (under short term) and 604 kPa (long term).

Failure of distributed foundation: $q_{\text{ult}} = 663$ kPa (under short term) and 1365 kPa (under long term).

Punching of the replaced zone into in situ soil: $q_{\text{ult}} = 879$ kPa (under short term) and 2387 kPa (under long term).

The minimum ultimate bearing capacity is $q_{\text{ult},u} = 426$ kPa due to punching failure of the footing through the replaced zone under the short term. Therefore, the minimum factor of safety is

$$\text{FS}_u = \frac{426}{170} = 2.51 < 3.0$$

(geosynthetic reinforcement is needed)

2. Ultimate bearing capacity and factor of safety with geosynthetic reinforcement: Since only the factor of safety against punching failure through the replaced zone under short term does not meet the requirement, the analysis of punching failure through the reinforced foundation is needed.

The anchorage area for a uniform geosynthetic layer is

$$A_a = 1/4(B_{\text{rz}}^2 - B_f^2) = 1/4 \times (2^2 - 1^2) = 0.75\ \text{m}^2$$

The pullout capacities for three reinforcements are

$$T_{\text{po1}} = \frac{2C_i[c_r + \gamma_r'(D_f + z)\tan\phi_r]A_a}{B}$$

$$= \frac{2 \times 0.9 \times [0 + (20 - 9.81) \times (1.0 + 0.2)\ \tan 38°] \times (0.75)}{1}$$

$$= 12.9\ \text{kN/m}$$

$$T_{\text{po2}} = \frac{2 \times 0.9 \times [0 + 10.19 \times (1.0 + 0.45) \tan 38°] \times (0.75)}{1}$$

$$= 15.1\ \text{kN/m}$$

$$T_{\text{po3}} = \frac{2 \times 0.9 \times [0 + 10.19 \times (1.0 + 0.7) \tan 38°] \times (0.75)}{1}$$

$$= 17.2\ \text{kN/m}$$

The tensile resistance due to overstress at the strain of 3% is

$$T_g = \varepsilon_g J = 0.03 \times 500 = 15.0\ \text{kN/m}$$

The design tensile strengths for three reinforcements are $T_{a1} = 12.9$ kN/m, $T_{a2} = 15.0$ kN/m, and $T_{a3} = 15.0$ kN/m.

The ultimate bearing capacity against punching failure through the reinforced zone is

$$q_{\text{ult},r} = q_b + \frac{2C_1 c_a h_r}{B_f} + C_1\gamma_r' h_r^2 \left(1 + \frac{2D_f}{h_r}\right)\frac{K_s \tan\phi_r}{B_f} + \frac{2C_1 T_a \tan\delta}{B_f} - \gamma_r' h_r$$

$$= 235 + \frac{2 \times 2 \times 0 \times 0.7}{1} + 2 \times (20 - 9.81) \times 0.7^2 \times \left(1 + \frac{2 \times 1}{0.7}\right) \times \frac{7 \times \tan 38°}{1} + \frac{2 \times 2 \times (12.9 + 15.0 + 15.0) \times \tan(0.67 \times 38)}{1} - (20 - 9.81) \times 0.7 = 520\ \text{kPa}$$

The factor of safety of the reinforced foundation is

$$FS_r = \frac{520}{170} = 3.1 > 3.0 \quad \text{(OK)}$$

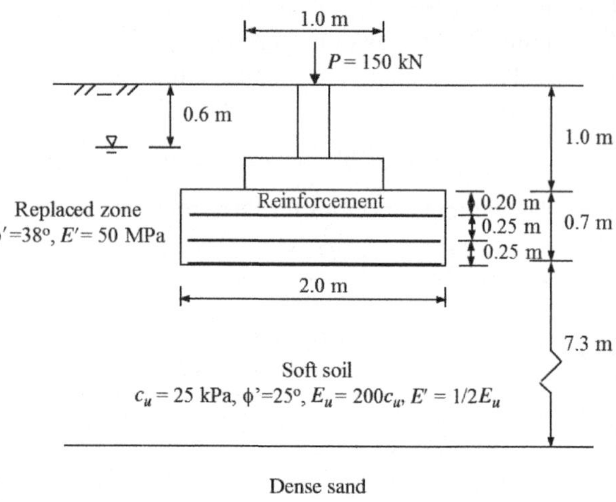

Example Figure 10.4 Cross section of reinforced foundation.

10.6.5 Construction

The construction procedure for geosynthetic-reinforced foundations is similar to that for overexcavation and replacement. The only difference is the placement of geosynthetic reinforcement, which should be placed on compacted fill. Typically, biaxial geogrid, triaxial geogrid, or woven geotextile is used for this application. Geocell has been used for this application as well. Under a continuous footing, the machine direction of uniaxial geogrid may be placed perpendicularly to the centerline of the footing. When uniaxial geogrids are placed under square footings, they should be oriented in perpendicular directions between two adjacent layers.

10.6.6 Quality Control and Assurance

In addition to the quality control and assurance of overexcavation and replacement, it is important to have appropriate quality control and assurance of geosynthetic products and placement. To verify designed bearing capacity and modulus of a reinforced foundation, static plate loading tests may be performed.

10.7 GEOSYNTHETIC-REINFORCED ROADS

10.7.1 Introduction

Basic Concept Geosynthetics (geotextile, geogrid, geocell, geocomposite, and geomembrane) have been increasingly used for the construction of roadways to reduce required base thickness and prolong roadway life. Their functions in the roadway applications include separation, filtration, drainage, reinforcement, and/or barrier. Chapter 6 discusses the use of geosynthetics for separation, filtration, drainage, and barrier. Woven geotextile and geogrid (planar reinforcements) and geocell (a three-dimensional reinforcement) can be placed at the interface between subgrade and base course or within the base course. They interact with subgrade and base course materials and provide lateral confinement to prevent bearing failure of subgrade and base courses, reduce rut depth, and minimize the chance of fatigue failure of pavement surfaces. The geosynthetic reinforcement at the interface can increase the bearing capacity of subgrade and slow down the deterioration of the base course. The geosynthetic reinforcement within the base course can increase base course stiffness. Geocell has been increasingly used with onsite or recycled geomaterials for roadway applications (Han and Thakur, 2014). Geosynthetic reinforcement used to increase subgrade bearing capacity is often referred to as subgrade improvement while geosynthetic reinforcement used to increase base course stiffness is often referred to as base reinforcement or stabilization.

Suitability Geotextile, geogrid, and geocell are suitable for subgrade of different California Bearing Ratio (CBR) values, as shown in Table 10.9. Problematic subgrade may include rounded sand, silty sand, silt, silty clay, clay, organic soil, and peat. Geotextile and geogrid typically require well-graded base course materials because they interact with particles by interlocking and friction. Low-quality geomaterials, such as poorly graded granular fill, sand, and recycled asphalt pavement (RAP) aggregate, can be used with geocell through closed confinement (Han et al., 2011; Han and Thakur, 2014). Subgrade improvement by geosynthetics has been used for low-strength subgrade of CBR lower than 3% while base reinforcement by geosynthetics has been used for low- to medium-strength subgrade of CBR up to 8% with noticeable benefits to different degrees (Berg et al., 2000).

Table 10.9 Use of Geosynthetics for Different Subgrade CBR Values

	CBR Value (%)		
Function	Unsoaked	Soaked	Geosynthetic Product
Separation	>8	>3	Nonwoven geotextile
Reinforcement	8–3	3–1	Woven geotextile, geogrid, or geocell
Reinforcement and separation	<3	<1	Woven geotextile, geogrid + nonwoven geotextile, and geocell + nonwoven geotextile

Source: Modified from Berg et al. (2000).

Applications Geosynthetics for roadway applications include unpaved roads, paved roads, and railroads. Unpaved roads (i.e., without a pavement surface) are commonly constructed as local, low-volume roads and temporary construction platforms. AASHTO (2008) defined low-volume roads as those expected to carry traffic loads for less than 100,000 ESALs (equivalent single-axle loads). Paved roads include flexible (asphalt) and rigid (concrete) pavements. However, geosynthetics have been used as reinforcement mostly for unpaved roads and flexible pavements. The use of geosynthetic reinforcement in rigid pavements is mostly related to asphalt overlays, which is part of pavement rehabilitation; therefore, this application will not be discussed in this chapter. There have been some research and applications of geosynthetics for railroads; however, no well-developed design methods are available for this application. Geocells, as three-dimensional geosynthetics, have been increasingly researched and used in the past few years. The recent advances in the use of geocells for unpaved and paved roads and railroads can be found in the publications, for example, Han et al. (2011a), Pokharel et al. (2010), Acharya et al. (2012), Yang et al. (2012), Han et al. (2013), Leshchinsky and Ling (2013), and Yang et al. (2013).

Advantages and Limitations The use of geosynthetic reinforcement can prolong roadway life by typically 2–6 times. Under the same service life, geosynthetic reinforcement can reduce required base thickness. Geosynthetics can be easily installed without any specialty worker and construction equipment. Geosynthetic-reinforced roads are often cost effective in life-cycle analysis. Geocell has been used with recycled asphalt pavement (RAP) aggregate, recycled concrete, and recycled ballast to create a more economic and sustainable solution (Han et al., 2011a; Han et al., 2012; Han and Thakur, 2014).

Different geosynthetic properties, such as tensile stiffness, interface characteristics, and aperture shape, size, thickness, junction strength, and junction stiffness of geogrid, affect the performance of geosynthetic reinforcement (Webster, 1992). No design method can quantify all these factors; therefore, calibration of a design method for a specific product is often needed. No well-accepted quality assurance method is available to evaluate the performance of geosynthetic-reinforced roads right after construction. Geosynthetic-reinforced roads may have higher initial cost especially when no or small base thickness reduction is allowed. Compaction of infill material in geocell pockets is sometimes difficult and challenging, especially with tall geocell. Height of geocell should be limited to not more than 300 mm to minimize compaction difficulty.

10.7.2 Principles

Roadway Structure Figure 10.52 shows the roadway structures for unpaved road, paved road, and railroad, which are layered systems. The unpaved road typically has two layers: base and subgrade. Paved roads and railroads have more layers. Ballast in the railroad is equivalent to base course in the paved road while subballast is equivalent to subbase. Therefore, the terms "base or base course" and "subbase" will be used instead of "ballast" and "subballast" in this discussion unless otherwise noted. Subbase may not be used for some projects. Railroads are actually one type of unpaved roads. The required strengths and moduli of layers decrease with the depth because less stress is transferred to the deeper depth by the above layer(s). Base and subbase can be different types of materials. However, geosynthetic reinforcement has mostly been used with granular materials because geosynthetic reinforcement can provide confinement needed by granular materials to become strong and stiff.

Geosynthetic reinforcement, including biaxial or triaxial geogrid, woven geotextile, and geocell, can be placed at different elevations of the roadway system. When subgrade is weak, the top of the subgrade is an ideal location for geosynthetic reinforcement. Geosynthetics can also be placed within bases and subbases. California Department of Transportation (2012) suggested that if the thickness of the base course is more than 450 mm, the first geogrid layer should be placed at the interface between base course and subgrade and the second geogrid layer should be added within the base course. To avoid the interference of tamping machines' operation, geosynthetic reinforcement should be placed in ballast at least 200 mm below the bottom of the tie (or sleeper) (Coleman, 1990).

Load Transfer Unpaved roads, paved roads, and railroads have channelized traffic, which can be quantified in terms of number of passes of vehicle or axle. An axle load of 80 kN is considered as a standard axle load in the 1993 AASHTO

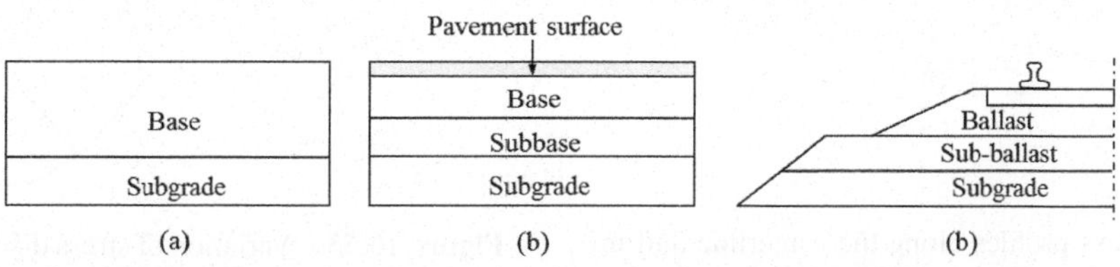

Figure 10.52 Roadway structures: (a) unpaved road, (b) paved road, and (c) railroad.

Pavement Design Guide. However, unpaved areas, such as construction sites and log yards, may not have channelized traffic. It is not clear how repeated loading on unpaved areas and their service life should be quantified. In practice, design methods for unpaved roads have been often used for unpaved areas (Giroud and Han, 2013).

The roadway structure, including the surface, base, and subbase, distributes the traffic load to the subgrade and the subgrade carries the load. The load transfer or stress distribution through the unreinforced roadway structure can be determined using the layered elastic theory (Burmister, 1958). Figure 10.53 shows the distributed stresses in unpaved and paved roads, which are used for design. These stresses induce vertical compressive strains within the layers or a tensile strain at the bottom of the surface layer. Figure 10.54 shows the vertical stresses along the centerline of the load and the interface between base and subgrade. Under the load, the base course with a higher modulus reduces the vertical stress on the subgrade under the load and results in more uniform distribution of the vertical stress at the interface than the base with an equal modulus to the subgrade. It should be noted that the vertical stress away from the center of the load under the base with the higher modulus is higher than that under the base with the equal modulus to maintain the force equilibrium.

Giroud and Noiray (1981) used a stress distribution angle method to estimate the stress at the interface between base and subgrade and assumed that the stress distribution angle is fixed at 31° for the design of geotextile-reinforced unpaved roads. Based on experimental data, Giroud and Han (2004a) proposed the stress distribution angle decreasing with the number of cycles as shown in Figure 10.55. The reduction of the stress distribution angle results from the deterioration of the base course. The lateral restraint of geosynthetic reinforcement can slow down the rate of deterioration of the base course.

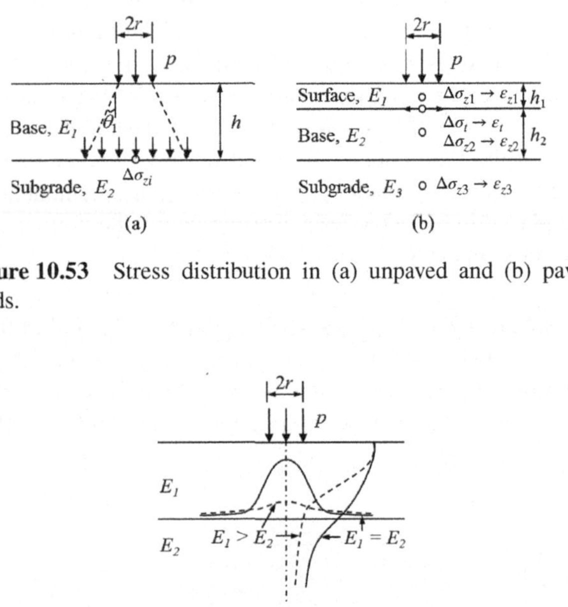

Figure 10.53 Stress distribution in (a) unpaved and (b) paved roads.

In the unpaved road design, the distributed stress on the top of the subgrade, $\Delta\sigma_{zi}$, should be lower than the bearing capacity of the subgrade. The vertical stresses in the surface, base, and subgrade, $\Delta\sigma_{z1}$, $\Delta\sigma_{z2}$, $\Delta\sigma_{z3}$, in Figure 10.53(b) can be used to calculate the vertical compressive strains in these layers, ε_{z1}, ε_{z2}, ε_{z3}, which are used to calculate the rut of the pavement.

The theory of elasticity shows that the high-modulus surface layer develops compressive stresses in the upper part of the layer but tensile stresses in the lower part. Since the granular base in the unpaved road cannot take any tensile stress, it deteriorates with repeated loading. The tensile stress at the base of the pavement surface, $\Delta\sigma_t$, in Figure 10.53(b) can be used to calculate the tensile strain at this location, ε_t, which is the key parameter for the fatigue life of the pavement.

The improvement of the unpaved road by geosynthetic reinforcement includes the improvement of stress distribution and subgrade bearing capacity; therefore, it is often referred to as subgrade improvement. The improvement of the paved road by geosynthetic reinforcement includes the improvement of stress distribution, base modulus, and structural capacity; therefore, it is referred to as base reinforcement. The geosynthetic reinforcement at the bottom of the base can provide lateral restraint to granular base through friction between woven geotextile and granular soil and confinement of granular soil by geogrid or geocell. Giroud and Han (2013) considered the confinement of granular soil through interlocking with geogrid apertures as an open confinement but the confinement of granular soil in geocell pockets as a closed confinement. All types of soils in geocell pockets are confined due to the three-dimensional configuration. However, granular soil can be effectively confined by interlocking with geogrid apertures only when the soil particles are strong and have appropriate sizes relative to the aperture size. Based on an experimental study, Brown et al. (2007) suggested that the aperture size, defined as the diameter of the largest sphere

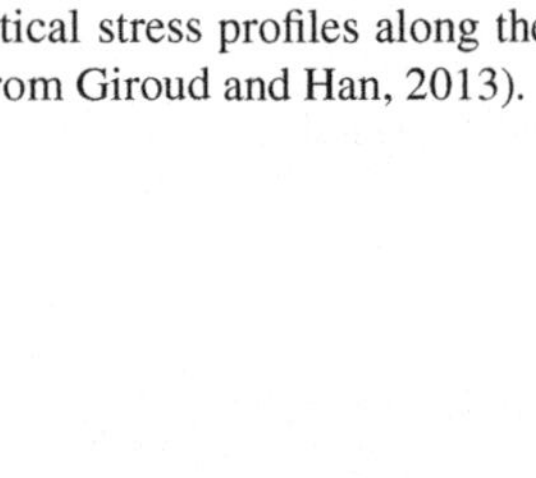

Figure 10.54 Vertical stress profiles along the centerline and interface (modified from Giroud and Han, 2013).

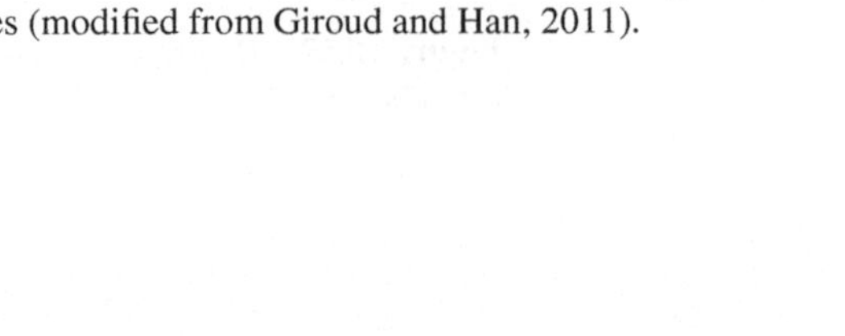

Figure 10.55 Variation of stress distribution angle with number of cycles (modified from Giroud and Han, 2011).

passing through the aperture, should be 1.2–1.6 times the nominal size of a uniform granular soil (i.e., less than 10% soil particles retained on the sieve of this size). In addition to lateral restraint, tensioned membrane effect is another mechanism for geosynthetic reinforcement in roadway applications. The tensioned membrane effect is valid when the traffic is channelized. However, Giroud and Han (2004a) indicated that the tensioned membrane effect becomes important only if a large deformation (i.e., rut for roadway applications) of 100 mm or more develops. From large cyclic plate-loading tests, Qian et al. (2013) concluded that the tensioned membrane effect was recognized when the permanent deformation was larger than 33% (i.e., one third) of the base thickness. Such large deformations are not tolerable for paved roads and railroads. Even for unpaved roads, ruts are periodically backfilled as part of road maintenance and the traffic may not be channelized at the same location. As a result, the tensioned membrane effect is minimized. Since the traffic in unpaved areas is not channelized the tensioned membrane effect is hard to develop and be quantified.

Bearing Capacity Figure 10.56 shows general rutting behavior of roads subjected to traffic and the basic design philosophy for unpaved and paved roads. When a road has an unstable base course or subgrade, rut increases with the number of traffic passes at an accelerated rate. Such behavior can be considered as the process of bearing failure. When a road has stable base course and subgrade, rut increases with the number of passes at a decreasing rate or eventually reaches an equilibrium state (i.e., only elastic deformation without any permanent deformation). Such behavior is often referred to as resilient behavior. Unpaved roads are often designed based on bearing failure, while paved roads are designed based on resilient behavior.

Bearing failure in unpaved roads is progressive, as shown in Figure 10.57. Soil movement develops under wheel loading, which includes vertical compression and lateral movement. Lateral movement often results in subgrade heave. Geosynthetic reinforcement provides vertical restraint to subgrade heave so as to maintain the stability of the road. Field measurements have shown that vertical restraint of

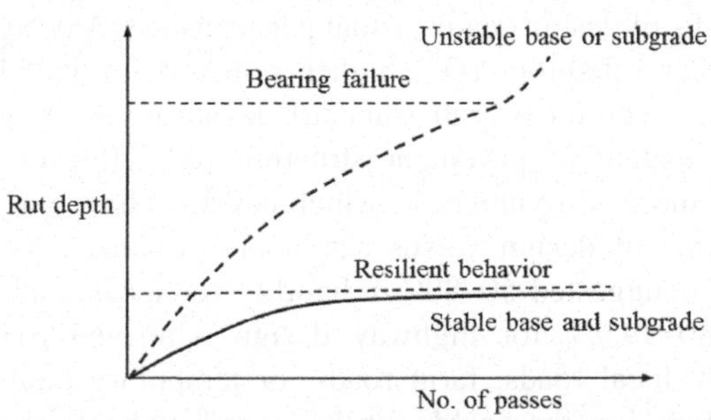

Figure 10.56 Design philosophy for unpaved and paved roads.

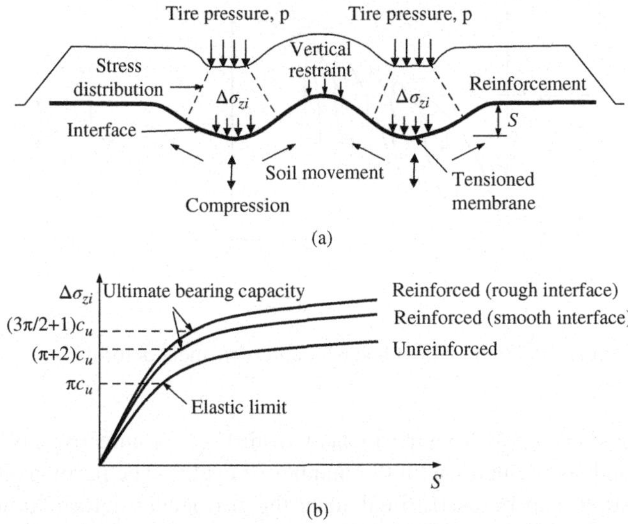

Figure 10.57 (a) Soil movement and vertical restraint and (b) bearing capacity factors (modified from Giroud and Noiray, 1981; Giroud and Han, 2004a; Giroud and Han, 2013).

subgrade can be achieved with small strains (e.g., 1%) in the geogrid (Giroud and Han, 2013). To limit large deformation of an unreinforced unpaved road, Giroud and Noiray (1981) suggested limiting the applied pressure on the top of subgrade lower than soil elastic limit, which is πc_u (i.e., $3.14c_u$, c_u is undrained shear strength of a saturated, cohesive soil). For a geotextile-reinforced unpaved road, Giroud and Noiray (1981) suggested the applied pressure can reach the ultimate bearing capacity of the subgrade due to the vertical restraint before the road develops excessive deformation. Considering the smooth interface between the woven geotextile and the subgrade, the ultimate bearing capacity of the subgrade is $(\pi + 2)c_u$ (i.e., $5.14c_u$). Considering a rough interface between the geogrid with interlocked aggregate and the subgrade, the ultimate bearing capacity of the subgrade is $(3\pi/2 + 1)c_u$ (i.e., $5.71c_u$) (Giroud and Han, 2004a). Therefore, the use of woven geotextile and geogrid results in bearing capacity increases by 64 and 82%, respectively. It should be pointed out that these bearing capacity factors (i.e., 3.14, 5.14, and 5.71) are for the two-dimensional situation that prevails in case of rutting (Giroud and Han, 2013). Both Giroud and Noiray (1981) and Giroud and Han (2004a) used the two-dimensional case, which is slightly conservative, as compared with an axisymmetric case.

Resilient Behavior and Permanent Deformation Figure 10.58 shows the vertical strains of a sample subjected to cyclic loading with a deviator stress from 0 to $\sigma_1 - \sigma_3$. The total strain of the sample under each loading cycle consists of elastic strain and plastic strain. The elastic strain, ε_r, is recoverable while the plastic strain

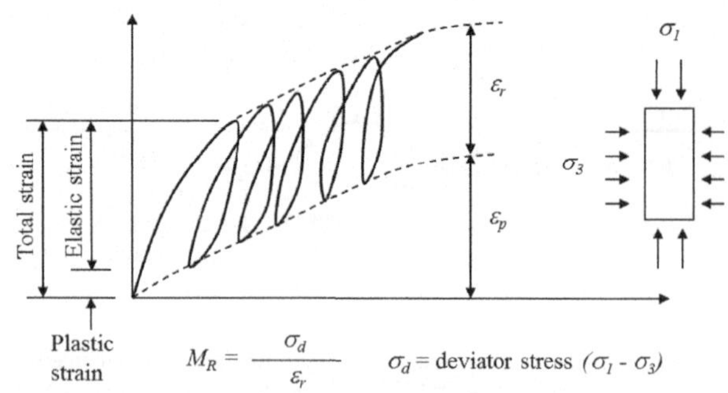

Figure 10.58 Resilient behavior and permanent deformation.

(also referred to as permanent strain), ε_p, is unrecoverable and accumulates with the number of cycles. The permanent strain can be used to calculate the permanent deformation of each pavement layer. The sum of the permanent deformations of all the layers is the rut on the road. The resilient behavior is often described by a resilient modulus, M_R, which is defined as the ratio of the deviator stress divided by the elastic strain. Resilient modulus is an elastic modulus but it is under cyclic loading. Cyclic triaxial tests and field tests [e.g., falling weight deflectometer (FWD), lightweight FWD, dynamic cone pentrometer (DCP)] can be used to determine or estimate the resilient modulus of each layer. Correlations are also available to estimate its value.

Cyclic triaxial tests show that geosynthetic reinforcement placed horizontally in the sample slightly increases the resilient modulus of granular material but significantly reduces its permanent deformation (Mengelt et al., 2000; Moghaddas-Nejad and Small, 2003; Perkins, et al. 2004; Nazzal, 2007; Wayne et al., 2011). This result has been confirmed by Yang and Han (2013) based on their theoretical solution.

The accumulated permanent strain can be estimated by the formula developed by Tseng and Lytton (1989):

$$\frac{\varepsilon_p}{\varepsilon_z} = \left(\frac{\varepsilon_0}{\varepsilon_r}\right) e^{-(\rho/N_a)^\beta} \tag{10.57}$$

where
ε_p = accumulated permanent strain in a layer
ε_z = average vertical resilient strain in a layer
$(\varepsilon_0/\varepsilon_r)$, ρ, β = material constants obtained from permanent deformation test
N_a = number of cycles (axles)

This model is also often referred to as a damage model. The average vertical resilient strain in each layer, ε_z, can be calculated based on the layered elastic theory under a static load as discussed earlier.

It should be pointed out that the layered elastic theory and the accumulated permanent strain model were based on the assumption that pavement layers are isotropic. Due to the inclusion of geosynthetic reinforcement, however, geosynthetic-reinforced layers are anisotropic. These approaches may not be exactly valid for geosynthetic-reinforced roads; therefore, further research is needed to resolve this issue. As an approximation, equivalent resilient modulus may be determined based on measured distributed stresses and permanent deformations (Giroud and Han, 2013).

Fatigue Failure Fatigue failure is considered when an asphalt surface exists in a paved road. Under a wheel load, a tensile strain, ε_t, develops at the bottom of the pavement surface. Similar to vertical compressive strains, the tensile strain accumulates with the number of cycles and the pavement surface has fatigue failure when the accumulated tensile strain reaches the limit. The lateral restraint of geosynthetic reinforcement to base course can increase the modulus of the base course and reduce the tensile strain in the pavement surface. However, limited research has been done to quantify the benefit of geosynthetic reinforcement in minimizing fatigue failure. Therefore, the design for geosynthetic reinforcement for this benefit will not be discussed.

Serviceability and Reliability Due to the variability of pavement structures (subgrade, subbase, base, and surface layers), traffic loading, and design methodologies, pavements have been designed based on reliability as discussed in the 1993 AASHTO *Design Guide*. Reliability is the probability for the actual road performance (or serviceability) to exceed or equal the design road performance. As schematically shown in Figure 10.59, the dots represent the actual individual performance with a statistical distribution, while the performance curve represents the average performance of the road. The road performance decreases from the initial serviceability (p_0) toward the terminal serviceability (p_t), at which point major rehabilitation or reconstruction is required. If the design curve matches the average performance curve, there is an equal chance of failure or success in terms of design versus actual performance. A design with a higher reliability (i.e., higher standard normal deviate, Z_R, at a certain overall standard deviation, s_0) requires a more expensive pavement structure (e.g., thicker and/or using more geosynthetics), which has less chance of failure in terms of design versus actual performance. AASHTO (1993) suggested 50–80% reliability for local road design and 80–99.9% for highway design. Unpaved roads are mostly local roads, farm roads, or temporary haul roads; therefore, it is reasonable for these roads to be designed at a reliability of 50%.

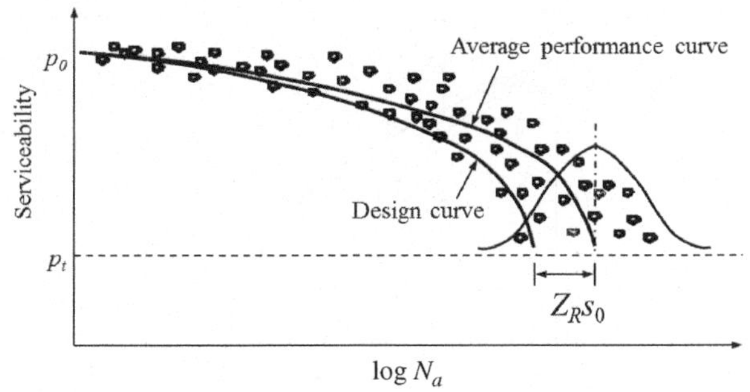

Figure 10.59 Schematic representation of design vs. performance (modified from AASHTO, 1993).

Mechanistic-Empirical Pavement Design The 1993 AASHTO *Pavement Design Guide* (AASHTO, 1993) was developed based on the AASHTO road test data using statistical methods. The AASHTO Mechanistic-Empirical Pavement Design Guide (MEPDG) (AASHTO, 2008) is a newly introduced pavement design method. It has comprehensive considerations of climate, traffic, materials, structure, pavement response, damage accumulation, and distress. Pavement response is based on the layered elastic theory. Damage accumulation is based on empirical models for different distresses. The linkage between the pavement response and the distress is through the damage model. The damage models in the AASHTO MEPDG software are calibrated with national data. They need to be calibrated with local data to improve the accuracy of these models. The current MEPDG software does not consider any benefit of geosynthetic reinforcement. In recent years, researchers (e.g., Perkins et al., 2004; Kown et al., 2007; Yang et al., 2013) have conducted studies to develop methods for quantifying the benefits of geosynthetic reinforcement in unpaved and paved roads in the framework of MEPDG. Details of these developments can be found in these publications and will not be presented herein.

10.7.3 Design Considerations for Unpaved Roads

Design of geosynthetic-reinforced unpaved roads has been based on the concepts of bearing failure and an increase of bearing capacity by the use of geosynthetics. A tensioned membrane effect is sometimes considered in the design; however, its effect is minimal unless an excessive rut depth is allowed. Several design methods (Steward et al., 1977; Giroud and Noiray, 1981; Tingle and Webster, 2003; Giroud and Han, 2004a, 2004b), have been developed so far and have been used for many projects. Most of these methods are empirical or semiempirical.

The design method proposed by Giroud and Han (2004a, 2004b) is recent, generic, and comprehensive and can be used for unreinforced and geotextile, geogrid, and geocell-reinforced unpaved roads; therefore, it is introduced in this section.

Quality of Base Course The Giroud and Han method assumed a high-quality base course, which is sufficient to support traffic loading during the service life. Based on Hammitt (1970), a base course with CBR of 10% should be able to sustain a wheel load of 45 kN with a tire pressure of 550 kPa for 1000 passes, while a base course with CBR of 15% should be able to sustain a wheel load of 45 kN with a tire pressure of 550 kPa for 10,000 passes. Therefore, a base course with a CBR value of 15% or higher is often needed for unpaved roads.

Stress Distribution In this method, a simplified stress distribution approach is used to estimate the additional vertical stress at the interface between the base and the subgrade based on Burmister's two-layer elastic solution (Burmister, 1958) for a circular loading area as follows:

$$\Delta\sigma_{zi} = \frac{r^2 p}{(r + h\tan\theta_1)^2} \tag{10.58}$$

$$\tan\theta_1 = \tan\theta_0\left[1 + 0.204\left(\frac{E_{bc}}{E_{sg}} - 1\right)\right] \tag{10.59}$$

where $\Delta\sigma_{zi}$ = additional vertical stress at the interface between base and subgrade
p = tire pressure
θ_1 = stress distribution angle in the base course
θ_0 = reference stress distribution angle for a uniform medium defined by $E_{bc} = E_{sg}$
E_{bc} = modulus of elasticity of base course
E_{sg} = modulus of elasticity of subgrade
r = radius of the loading area
h = thickness of the base course

The preceding equation shows that an increase of the modulus ratio of the upper layer to the lower layer increases the stress distribution angle, θ_1, so that the vertical stress at the interface decreases. Using empirical correlations for base course and subgrade in the 1993 AASHTO *Pavement Design Guide*, the modulus ratio can be expressed in terms of their CBR values. Based on the collected data for unreinforced bases over weak subgrade, Giroud and Han (2004a) found that the modulus ratio of base course to subgrade ranged from 1.0 to 5.0. Therefore, the following equation is obtained:

$$R_E = \frac{E_{bc}}{E_{sg}} = \frac{3.48\mathrm{CBR}_{bc}^{0.3}}{\mathrm{CBR}_{sg}} \le 5.0 \tag{10.60}$$

where R_E = modulus ratio of base to subgrade
CBR_{bc} = CBR of base course (%)
CBR_{sg} = CBR of subgrade (%)

Based on the laboratory test data obtained by Gabr (2001), Giroud and Han (2004a) proposed the following formula to account for the variation of the stress distribution angle with the number of loading cycles:

$$\frac{1}{\tan\theta} = \frac{1 + k_{ar}\log N_a}{\tan\theta_1} = \frac{1}{\tan\theta_1} + \lambda_{\text{ar}}\log N_a \qquad (10.61)$$

where θ = stress distribution angle at a specific number of loading cycles
k_{ar} = constant related to distribution angle reduction
λ_{ar} = distribution angle reduction rate
N_a = number of loading cycles

The decrease of the stress distribution angle can be explained as the deterioration of the base quality. The bearing capacity failure (assuming a 75-mm rut depth) happens when the stress distribution angle decreases to a certain value and the stress at the interface exceeds the bearing capacity of the subgrade. The inclusion of the geosynthetic reinforcement at the interface can slow down the deterioration rate of the base quality due to lateral restraint. The deterioration rate also depends on the thickness of the base course. A thick base course has a low deterioration rate because low stresses develop at the bottom of the base. Giroud and Han (2004a) correlated this deterioration rate to the aperture stability modulus for two specific biaxial geogrid products for geogrid-reinforced unpaved roads. Webster (1992) and Collin et al. (1996) found that the aperture stability modulus of a geogrid was shown to give good correlation with the measured performance of paved roads. Measurement of the aperture stability modulus is presented in a draft test method by Kinney (2000). The aperture stability modulus is measured by applying a torque on a central node of geogrid apertures rigidly fixed by a steel frame and measuring the corresponding rotation angle. The torque divided by the rotation angle is the aperture stability modulus, which has a unit of m-N/°.

Bearing Capacity Giroud and Noiray (1981) suggested that the bearing capacity factors, N_c, for the failure of subgrade in unreinforced and geotextile-reinforced roads be 3.14 and 5.14, respectively. Considering the rough interface between the geogrid with stones and the subgrade, the bearing capacity factor of 5.71 was suggested by Giroud and Han (2004a) for a geogrid-reinforced road. The bearing capacities of the subgrade are mobilized at a level of deformation that starts to become unacceptable to project requirements. A 75-mm rut is typically considered as the serviceability limit of failure for an unpaved road. At a small deformation, the bearing capacity is less mobilized. Giroud and Han (2004a) assumed that the mobilization factor is proportional to the rut depth until the serviceability limit, f_{sl} (i.e., 75 mm), as shown in Figure 10.60. In addition to the deformation, the mobilization factor depends on the size of the loading area (i.e., the radius of the tire contact area) and the base thickness.

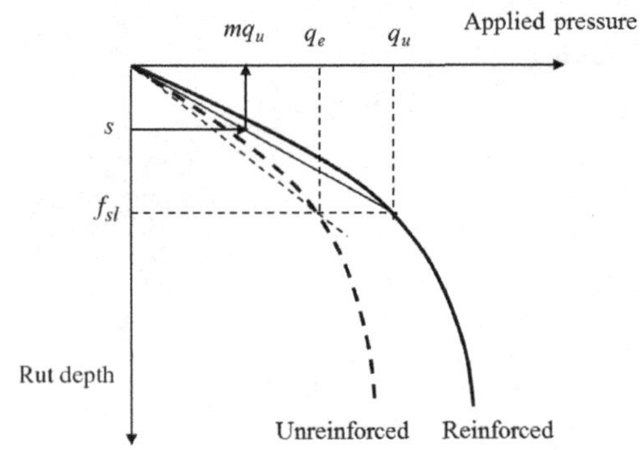

Figure 10.60 Bearing capacity mobilization factor.

Giroud and Han (2004a) proposed the following approximation for the bearing capacity mobilization factor, m_{bc}:

$$m_{\text{bc}} = \left(\frac{S}{f_{\text{sl}}}\right)\left\{1 - \xi\exp\left[-\zeta\left(\frac{r}{h}\right)^n\right]\right\} \qquad (10.62)$$

where S = rut depth of the unpaved road
f_{sl} = serviceability limit (i.e., 75 mm)
r = radius of the tire contact area
h = base thickness

The constants, ξ, ζ, and n are determined through the calibration process.

Applied Pressure versus Bearing Capacity Based on the relationship between the applied pressure on the top of the subgrade and the bearing capacity of the subgrade, the following equation can be obtained:

$$\Delta\sigma_{\text{zi}} = \frac{P}{\pi(r + h\tan\theta)^2} = m_{\text{bc}}N_c c_u \qquad (10.63)$$

where P = wheel load (= 1/2 axle load)
N_c = bearing capacity factor of subgrade
c_u = undrained shear strength of subgrade

Hence,

$$h = \frac{r}{\tan\theta}\left(\sqrt{\frac{P}{\pi r^2 m_{\text{bc}}N_c c_u}} - 1\right) \qquad (10.64)$$

Combining Equations (10.61), (10.62), and (10.64) yields

$$h = \frac{1 + k_{\text{ar}}\log N_a}{\tan\theta_0[1 + 0.204(E_{\text{bc}}/E_{\text{sg}} - 1)]}\left(\sqrt{\frac{P}{\pi r^2 m_{\text{bc}}N_c c_u}} - 1\right)r \qquad (10.65)$$

The undrained shear strength of the subgrade, c_u, can be determined by unconfined compressive strength test, unconsolidated undrained triaxial test, or vane shear test. It can also be estimated using the correlation with subgrade CBR as follows:

$$c_u = (2.90 \text{ to } 4.34)\text{CBR}_{\text{sg}} \qquad (10.66)$$

where c_u = undrained shear strength of subgrade (kPa)
CBR_{sg} = CBR of subgrade in percentage

The higher bound is used for a high-plasticity cohesive soil while the lower bound is used for a low-plasticity cohesive soil.

Calibration The distribution angle reduction rate, k_{ar}, in Equation (10.61) should be calibrated according to a specific geosynthetic product and thickness of base course using cyclic plate loading test data or field traifficking test data. Based on laboratory cyclic plate loading test data and field trafficking test data, Giroud and Han (2004a) obtained the following relationship:

$$\eta = \frac{\text{Field base course thickness}}{\text{Laboratory base course thickness}} = 0.689 \quad (10.67)$$

When cyclic plate loading test data are used for calibration, the η factor in Equation (10.67) should be used to get the field performance. In addition to unreinforced and geotextile-reinforced unpaved roads, Giroud and Han (2004b) calibrated the above method for two punched-drawn polypropylene biaxial geogrids as follows:

$$h = \frac{0.868 + (0.661 - 1.006J_a^2)(r/h)^{1.5}\log N_a}{1 + 0.204(E_{bc}/E_{sg} - 1)} \times \left(\sqrt{\frac{P}{\pi r^2 m_{bc} N_c c_u}} - 1\right) \quad (10.68)$$

where $m_{bc} = (S/f_{sl})\{1 - 0.9\exp[-(r/h)^2]\}$

J_a = aperture stability modulus of geogrid (m-N/degree)

For unreinforced unpaved roads, $J_a = 0$ and $N_c = 3.14$. For geotextile-reinforced unpaved roads, $J_a = 0$ and $N_c = 5.14$. For geogrid-reinforced unpaved roads, $J_a > 0$ and $N_c = 5.71$. The two punched-drawn biaxial geogrids, which were used for calibration of the design method, have the J_a values of 0.32 and 0.64 m-N/°. The modulus ratio, E_{bc}/E_{sg}, should be limited to be less than 5.0. Since the above equation contains h in the both sides of the equation, the required base thickness, h, can be solved by iterations.

Han and Pokharel (2015) calibrated the method for geocell-reinforced unpaved roads using cyclic plate loading test data and accelerated moving wheel test data as follows:

$$h = \frac{0.868 + 0.52(r/h)^{1.5}\log N_a}{1 + 0.204(E_{bc}/E_{sg} - 1)}\left(\sqrt{\frac{P}{\pi r^2 m_{bc} N_c c_u}} - 1\right) \quad (10.69)$$

Geocells used for the calibration were a novel polymeric alloy type. Due to geocell confinement, the modulus ratio for geocell-reinforced granular fill over subgrade can be estimated as follows:

$$R_E = \text{MIF}\frac{E_{bc}}{E_{sg}} = \text{MIF}\frac{3.48\text{CBR}_{bc}^{0.3}}{\text{CBR}_{sg}} \le 7.6 \quad (10.70)$$

where MIF = modulus improvement factor

Before the design of unpaved roads, it is necessary to check whether subgrade is sufficient to support traffic loading without a base course. The load capacity provided by the subgrade is

$$P_{h=0} = \left(\frac{S}{f_{sl}}\right)\pi r^2 N_c c_u \quad (10.71)$$

If the applied load is less than $P_{h=0}$, no base course is necessary. However, typically a thin base course of 100–150 mm is placed to protect the subgrade. If the applied load is higher than $P_{h=0}$, a base course is needed.

Proper Use Han and Giroud (2012) suggested the proper use of the above method. Below are key points:

- This method is generic but was calibrated for specific geosynthetic products.
- Large cyclic plate loading tests or field trafficking tests are required to calibrate this method.
- This method assumes saturated and fine-grained subgrade.
- Remolded strength should be used if subgrade is sensitive.
- Base course should have sufficient strength and stiffness.
- When geogrid is placed on soft subgrade, the base course should meet the filtration requirements.
- This method has 50% reliability if average subgrade strength is used.
- Base course thickness should be used to verify this method against field data.

The equation of the Giroud and Han method for unreinforced unpaved roads was calibrated against the average performance of unreinforced unpaved roads tested by Hammitt (1970); therefore, the design reliability is 50%. A design method with a higher reliability can be developed, but it will result in a more expensive design.

10.7.4 Design Parameters and Procedure for Unpaved Roads

Design Parameters The following design parameters are needed for geosynthetic-reinforced unpaved roads:

- Subgrade strength (undrained shear strength or CBR)
- Base quality (modulus or CBR)
- Wheel load and tire pressure

- Geosynthetic type and properties
- Allowable rut depth
- Number of passes

Design Procedure The following procedure can be followed to design a geosynthetic-reinforced unpaved road:

1. Based on subgrade strength, wheel load, and tire pressure, evaluate whether the subgrade is sufficient to support traffic. If it is not sufficient, a base course is needed.
2. Based on base course CBR, evaluate whether the base course has sufficient capacity to support traffic. If it is not sufficient, a better quality base course should be selected or the base course should be stabilized by extra layer of planar geosynthetic reinforcement or geocell, or chemical agents (cement or asphalt).
3. Through iterations, determine the required base course thickness based on the number of passes.

10.7.5 Design Considerations for Paved Roads

Design of geosynthetics for paved roads (commonly referred as base reinforcement) has been conducted by modifying the 1993 AASHTO *Pavement Design Guide* including the benefits of geosynthetics. Two common design parameters, traffic benefit ratio (TBR) and layer coefficient ratio (LCR), have been proposed to consider the benefits of geosynthetics. The purposes of geosynthetics used as base reinforcement are to reduce the required base thickness and prolong pavement life. In the past few years, great efforts have been made to develop a mechanistic-based design method for this application. However, no such method has been adopted by government agencies, or the geosynthetic industry; therefore, it will not be further discussed.

Equivalent Traffic In the 1993 AASHTO *Pavement Design Guide*, the number of 18-kip (80-kN) ESAL applications is used to define the pavement life in terms of the number of the equivalent traffics based on standard highway truck axle loads and can be estimated by

$$\log_{10}(W_{18}) = Z_R s_0 + 9.36\log_{10}(\mathrm{SN}+1) - 0.20 + \frac{\log_{10}\left[\frac{\Delta\mathrm{PSI}}{4.2-1.5}\right]}{0.40 + \frac{1094}{(\mathrm{SN}+1)^{5.19}}} + 2.32\log_{10}\left(\frac{M_R}{0.0069}\right) - 8.07 \quad (10.72)$$

where W_{18} = predicted number of 18-kip (80-kN) ESALs
Z_R = standard normal deviate (dimensionless)
s_0 = combined standard error of the traffic prediction and performance prediction (dimensionless), 0.45 commonly used
ΔPSI = difference between the initial present serviceability index, p_0, and the design terminal present serviceability index, p_t (dimensionless)
SN = structural number of pavement layer(s)
and M_R = resilient modulus of roadbed (base, subbase, or subgrade) (MPa)

This formula was originally developed based on English units. The constant, 0.0069, is used in this equation to convert M_R a metric unit of MPa to an English unit of psi.

The suggested levels of reliability, R_L, is provided in Table 10.10.

The standard normal deviate can be calculated in terms of the level of reliability as shown in Table 10.11.

Table 10.10 Suggested Level of Reliability, R_L

Functional Classification	Recommended Level of Reliability	
	Urban	Rural
Interstate and other freeways	85–99.9	80–99.9
Principal arterials	80–99	75–95
Collectors	80–95	75–95
Local	50–80	50–80

Source: AASHTO (1993).

Table 10.11 Standard Normal Deviate

Reliability, R_L (%)	Standard Normal Deviate, Z_R
50	0.000
60	−0.253
70	−0.524
75	−0.674
80	−0.841
85	−1.037
90	−1.282
92	−1.405
94	−1.555
96	−1.751
98	−2.054
99	−2.327

The serviceability index loss, ΔPSI, is defined as the difference between the initial present serviceability index and the terminal present serviceability index, that is,

$$\Delta \mathrm{PSI} = p_0 - p_t \tag{10.73}$$

where p_0 is the initial present serviceability index (use 4.2 for flexible pavements based on the AASHTO road test); and p_t is the terminal present serviceability index (use 2.5 or higher for major highways, 2.0 for highways with lesser traffic volume, and less than 2.0 for minor highways).

Resilient Modulus Based on Equation (10.72), the number of ESALs also depends on the resilient moduli of pavement structural layers (surface, base, subbase, and subgrade), which define the elastic response of each layer under repeated loading. The resilient modulus can be measured using a dynamic triaxial test apparatus. The resilient modulus of a fine-grained subgrade with a soaked CBR of 10% or less can be estimated using the following correlation (AASHTO, 1993):

$$M_R(\mathrm{MPa}) = 10.35\mathrm{CBR}_{\mathrm{sg}} \tag{10.74a}$$

NCHRP (2004) suggested the following formula for the resilient modulus of the subgrade:

$$M_R(\mathrm{MPa}) = 17.6\mathrm{CBR}_{\mathrm{sg}}^{0.64} \tag{10.74b}$$

The moduli of granular base and subbase can be estimated based on the correlations discussed in the next section.

Structural Number A structural number is defined in the 1993 AASHTO *Pavement Design Guide* to represent the capacity of the pavement structure in addition to the subgrade, which is expressed as

$$\mathrm{SN} = 0.04(a_1 h_1 + a_2 h_2 m_2 + a_3 h_3 m_3) \tag{10.75}$$

where a_1, a_2, a_3 = layer coefficients representative of surface, base, and subbase courses, respectively, which can be found in the 1993 AASHTO Guide

h_1, h_2, h_3 = actual thicknesses (mm) of surface, base, and subbase layers, respectively

m_2, m_3 = drainage coefficients for base and subbase layers, respectively

The constant, 0.394, was used to convert pavement layer thickness from a metric unit of mm to an English unit of inch

The layer coefficient of dense-graded asphalt concrete can be estimated based on the data provided in a table in the 1993 AASHTO *Pavement Design Guide* as follows:

$$a_1 = 0.398 \log_{10} E_{\mathrm{ac}} - 0.953 \tag{10.76}$$

Typical elastic modulus of asphalt concrete, E_{ac}, at 20°C ranges from 2100 to 3100 MPa.

The modulus and layer coefficient of granular base can be estimated by the following equations:

$$E_{\mathrm{bc}}(\mathrm{MPa}) = 29.4\mathrm{CBR}_{bc}^{0.4358} \tag{10.77}$$

$$a_2 = 0.249\log_{10}\left(\frac{E_{\mathrm{bc}}}{0.0069}\right) - 0.977 \tag{10.78}$$

where CBR_{bc} = the CBR value of the base course (in percent).

The modulus and layer coefficient of granular subbase can be estimated by the following equations:

$$E_{\mathrm{sb}}(\mathrm{MPa}) = 36.0\mathrm{CBR}_{sb}^{0.30} \tag{10.79}$$

where CBR_{sb} = the CBR value of the subbase course (in percent).

$$a_3 = 0.207\log_{10}\left(\frac{E_{\mathrm{sb}}}{0.0069}\right) - 0.754 \tag{10.80}$$

The drainage coefficient recommended by AASHTO (1993) is provided in Table 10.12.

Table 10.12 Recommended Drainage Coefficient

Quality of Drainage	Time for Water Removed within	% Time Pavement Exposed to Moisture Levels Approaching Saturation			
		<1%	1–5%	5–25%	>25%
Excellent	2 h	1.40–1.35	1.35–1.30	1.30–1.20	1.20
Good	1 day	1.35–1.25	1.25–1.15	1.15–1.00	1.00
Fair	1 week	1.25–1.15	1.15–1.05	1.00–0.80	0.80
Poor	1 month	1.15–1.05	1.05–0.80	0.80–0.60	0.60
Very poor	Water will not drain	1.05–0.95	0.95–0.75	0.75–0.40	0.40

Source: AASHTO (1993).

Table 10.13 Minimum Thickness of Asphalt Concrete and Granular Base[a]

Traffic (ESALs)	Asphalt Concrete (mm)	Granular Base (mm)
≤50,000	25 (or surface treatment)	100
50,001–150,000	50	100
150,001–500,000	63	100
500,001–2,000,000	75	150
2,000,001–7,000,000	87	150
>7,000,000	100	150

[a]Individual design agencies may modify the above information thickness for their own use.
Source: AASHTO (1993).

For practical and economic considerations, the 1993 AASHTO *Pavement Design Guide* recommended the minimum thickness requirements for the asphalt concrete and the aggregate base based on the level of traffic, as shown in Table 10.13.

Benefit of Geosynthetic Reinforcement Two empirical methods are available to consider the effects of the geosynthetic reinforcement for paved roads: (1) the traffic benefit ratio (TBR) method and (2) the layer coefficient ratio (LCR) method. Both methods modify the 1993 AASHTO *Pavement Design Guide*.

TBR is defined as the ratio of the number of cycles necessary to reach a given rut depth for a test section containing reinforcement divided by the number of cycles necessary to reach this same rut depth for an unreinforced section with the same section thickness and subgrade properties. Laboratory and field test results show that the TBR values for geotextiles, geogrids, and geocells range from 1.5 to 10, 1.5 to 70, and 10 to 12.5, respectively (Berg et al., 2000; Han et al., 2011b, 2012b). Typically, a TBR value of 2–6 has been used in the design. The TBR value can be used to calculate the extended pavement life as follows:

$$W_{18r} = \text{TBR} \cdot W_{18} \tag{10.81}$$

where W_{18r} = extended pavement life in ESALs with the use of geosynthetic reinforcement
W_{18} = pavement life in ESALs without the use of geosynthetic reinforcement, which can be determined from Equation (10.72)

Instead of prolonging the pavement life, the geosynthetic can also be used to reduce the required thickness of the base course. The following procedure was proposed by Holtz et al. (2008) using TBR to estimate the required base thickness:

- Based on the predicted ESALs (i.e., W_{18}), determine the required structural number SN_u for an unreinforced paved road using Equation (10.72).
- Considering the benefit of the geosynthetic, the same pavement section is expected to have an extended pavement life (i.e., $\text{TBR} \cdot W_{18}$). The expected pavement life has an equivalent structural number SN_r, which can be determined using Equation (10.72).
- The benefit of geosynthetic reinforcement can be considered as an increase of the structural number, that is $\Delta\text{SN} = \text{SN}_r - \text{SN}_u$.

The required thickness of the reinforced base can be calculated as

$$h_{2r} = \frac{(\text{SN}_u - \Delta\text{SN}) - 0.04a_1h_1}{0.04a_2m_2} \tag{10.82}$$

This approach assumes that the benefit of geosynthetic reinforcement in the increase of the structural number at the base thicknesses of h_2 and h_{2r} is the same.

Alternatively, the reduced thickness of the base course in the geosynthetic-reinforced section can be determined based on the following formula:

$$h_{2r} = \frac{\text{SN}_u - 0.04a_1h_1}{\text{LCR} \cdot (0.04a_2m_2)} \tag{10.83}$$

The LCR value can be determined based on box tests (Montanelli et al., 1997). Alternatively, this LCR value can be estimated based on MIF (Giroud and Han, 2013) as follows:

$$\text{LCR} = \frac{0.249\log_{10}(\text{MIF} \cdot E_{\text{bc}}/0.0069) - 0.977}{0.249\log_{10}(E_{\text{bc}}/0.0069) - 0.977} \tag{10.84}$$

The MIF value for geosynthetic reinforcement of granular base typically ranges from 1.0 to 2.0.

10.7.6 Design Parameters and Procedure for Paved Roads

Design Parameters The following design parameters are needed for geosynthetic-reinforced paved roads:

- Pavement structure, including type and thickness of asphalt surface, base course, subbase course, and subgrade
- Modulus or CBR values of pavement layers
- Drainage properties of pavement layers except asphalt surface
- Design traffic in terms of ESALs
- Level of reliability and standard deviation

- Initial and terminal present serviceability indices
- Benefit of geosynthetic reinforcement (MIF, LCR, or TBR)

Design Procedure The following procedure can be followed to design a geosynthetic-reinforced paved road:

1. Based on properties of pavement layers (modulus or CBR values), determine layer coefficients for all pavement layers.
2. Based on drainage properties of pavement layers, select drainage coefficients.
3. Based on design traffic, level of reliability, initial and terminal present serviceability indices, layer coefficients, and drainage coefficients, determine the structural number and required thickness for each pavement layer without geosynthetic reinforcement. Compare the calculated layer thickness against the minimum layer thickness requirement and make adjustments if necessary.
4. Based on the available benefit parameter of geosynthetic reinforcement (MIF, LCR, or TBR), calculate the extended pavement life if the base thickness is kept the same or determine the reduced base thickness if the pavement life is kept the same.

10.7.7 Design Examples

Example 10.6 Design of Unpaved Roads

An unpaved haul road needs to be designed for 1000 passes of trucks with single axles (the front axle is ignored due to light load). The axle load of the trucks used on this site is 72 kN and the tire pressure is 552 kPa. The clayey subgrade CBR is 1.5% and the base CBR is 20%. The allowable rut depth is 75 mm. Design this unpaved road section without geosynthetic, with a geotextile, and punch-drawn polypropylene geogrid ($J_a = 0.32$ m-N/°) using the Giroud and Han method.

Solution

Use the following relationship to estimate the undrained shear strength of subgrade:

$$c_u = 30\ \text{CBR}_{\text{sg}} = 30 \times 1.5 = 45\ \text{kPa}$$

Since the axle load is 72 kN, the wheel load is $P = 71/2 = 36$ kN.

The radius of the equivalent tire contact area is

$$r = \sqrt{\frac{P}{\pi p}} = \sqrt{\frac{36}{3.14 \times 552}} = 0.144\ \text{m}$$

Check the need for base course. The load capacity provided by the subgrade is

$$P_{h=0} = \left(\frac{S}{f_{\text{sl}}}\right) \pi r^2 N_c c_u$$

Input data: $S = 75$ mm, $f_{\text{sl}} = 75$ mm, $r = 0.144$ m, $N_c = 3.14$, and $c_u = 45$ kPa:

$$P_{h=0} = \left(\frac{S}{f_{\text{sl}}}\right) \pi r^2 N_c c_u = \left(\frac{75}{75}\right) \times 3.14 \times 0.144^2 \times 3.14 \times 45 = 9.2\ \text{kN} < 36\ \text{kN}$$

Therefore, a base course is needed.

Check the quality of base course. Since the base course has CBR = 20%, it is sufficient to sustain the traffic of the 36-kN wheel load and 552-kPa tire pressure.

The modulus ratio of base course to subgrade is

$$R_E = \min\left(\frac{3.48\text{CBR}_{\text{bc}}^{0.3}}{\text{CBR}_{\text{sg}}}, 5.0\right) = \min\left(\frac{3.48 \times 20^{0.3}}{1.5}, 5.0\right) = 5.0$$

The modulus ratio factor is

$$f_E = 1 + 0.204(R_E - 1) = 1 + 0.204 \times (5 - 1) = 1.816$$

(a) For the unreinforced case, assume $h = 0.33$ m. The bearing capacity mobilization factor, m, can be calculated as

$$m_{bc} = \left(\frac{S}{f_{sl}}\right)\left\{1 - 0.9\exp\left[-\left(\frac{r}{h}\right)^2\right]\right\} = \left(\frac{75}{75}\right)\left\{1 - 0.9\exp\left[-\left(\frac{0.144}{0.330}\right)^2\right]\right\} = 0.250$$

The base thickness can be calculated as follows:

$$h = \frac{0.868 + (0.661 - 1.006J_a^2)(r/h)^{1.5}\log N_a}{f_E}\left[\sqrt{\frac{P/(\pi r^2)}{m_{bc}N_c c_u}} - 1\right]r$$

$$= \frac{0.868 + (0.661 - 1.006 \times 0)(0.144/0.330)^{1.5}\log 1000}{1.816}\left[\sqrt{\frac{36/(3.14 \times 0.144^2)}{0.250 \times 3.14 \times 45}} - 1\right] \times 0.144$$

$$= 0.33\text{ m}$$

Therefore, the required base thickness for the unreinforced case is 0.33 m.

(b) For the reinforced case (geotextile), assume $h = 0.21$ m. The bearing capacity mobilization factor, m, can be calculated as

$$m_{bc} = \left(\frac{S}{f_{sl}}\right)\left\{1 - 0.9\exp\left[-\left(\frac{r}{h}\right)^2\right]\right\} = \left(\frac{75}{75}\right)\left\{1 - 0.9\exp\left[-1 \times \left(\frac{0.144}{0.21}\right)^2\right]\right\} = 0.427$$

The base thickness can be calculated as follows:

$$h = \frac{0.868 + (0.661 - 1.006J_a^2)\left(\frac{r}{h}\right)^{1.5}\log N_a}{f_E}\left[\sqrt{\frac{P/(\pi r^2)}{m_{bc}N_c c_u}} - 1\right]r$$

$$= \frac{0.868 + (0.661 - 1.006 \times 0^2)\left(\frac{0.144}{0.21}\right)^{1.5}\log 1000}{1.816}\left[\sqrt{\frac{36/(3.14 \times 0.144^2)}{0.427 \times 5.14 \times 45}} - 1\right] \times 0.144$$

$$= 0.21\text{ m}$$

Therefore, the required base thickness for the geotextile-reinforced case is 0.21 m.

(c) For the reinforced case (geogrid), assume $h = 0.18$ m. The bearing capacity mobilization factor, m_{bc}, can be calculated as

$$m_{bc} = \left(\frac{S}{f_{sl}}\right)\left\{1 - 0.9\exp\left[-\left(\frac{r}{h}\right)^2\right]\right\} = \left(\frac{75}{75}\right)\left\{1 - 0.9\exp\left[-1 \times \left(\frac{0.144}{0.18}\right)^2\right]\right\} = 0.513$$

The base thickness can be calculated as follows:

$$h = \frac{0.868 + (0.661 - 1.006J_a^2)\left(\frac{r}{h}\right)^{1.5}\log N_a}{f_E}\left[\sqrt{\frac{P/(\pi r^2)}{m_{bc}N_c c_u}} - 1\right]r$$

$$= \frac{0.868 + (0.661 - 1.006 \times 0.32^2)\left(\frac{0.144}{0.18}\right)^{1.5}\log 1000}{1.816}\left[\sqrt{\frac{36/(3.14 \times 0.144^2)}{0.513 \times 5.71 \times 45}} - 1\right] \times 0.144$$

$$= 0.17\text{ m}$$

Try $h = 0.17$ m:

$$m_{bc} = \left(\frac{S}{f_{sl}}\right)\left\{1 - 0.9\exp\left[-\left(\frac{r}{h}\right)^2\right]\right\} = \left(\frac{75}{75}\right)\left\{1 - 0.9\exp\left[-\left(\frac{0.144}{0.17}\right)^2\right]\right\} = 0.548$$

$$h = \frac{0.868 + (0.661 - 1.006J_a^2)\left(\frac{r}{h}\right)^{1.5} \log N_a}{f_E}\left[\sqrt{\frac{P/(\pi r^2)}{m_{bc}N_c c_u}} - 1\right]r$$

$$= \frac{0.868 + (0.661 - 1.006 \times 0.32^2)\left(\frac{0.144}{0.17}\right)^{1.5} \log 1000}{1.816}\left[\sqrt{\frac{36/(3.14 \times 0.144^2)}{0.548 \times 5.71 \times 45}} - 1\right] \times 0.144$$

$= 0.17$ m

Therefore, the required base thickness for the geogrid-reinforced case is 0.17 m.

Example 10.7 Design of Paved Roads

A pavement for a rural highway (two lanes each direction) needs to be designed for a 20-year life. Assume there is no subbase layer. The average subgrade CBR is 3%. Expected traffic at the end of the design life is 500000 ESALs. The design terminal present serviceability index is 2.5. The reliability level is 95% and standard deviation is 0.45. The elastic modulus of the asphalt concrete at 20°C is 3100 MPa. The granular base course CBR is 50%. Assume the drainage condition of this pavement is "Good" and the pavement is exposed to saturation moisture more than 25% of the time. Geogrid is used to reinforce the base course and the MIF is 1.5. Design this pavement section without and with geogrid.

Solution

Since the elastic modulus of the asphalt concrete at 20°C is 3100 MPa, the layer coefficient is

$$a_1 = 0.398\log_{10}E_{ac} - 0.953$$
$$= 0.398\log_{10}3100 - 0.953 = 0.44$$

The granular base course CBR of 50% corresponds to the elastic modulus and the layer coefficient as follows:

$$E_{bc}(\text{MPa}) = 29.4\text{CBR}_{bc}^{0.4358} = 29.4(50)^{0.4358} = 162 \text{ MPa}$$

$$a_2 = 0.249\log_{10}\left(\frac{E_{bc}}{0.0069}\right) - 0.977$$
$$= 0.249\log_{10}\left(\frac{162}{0.0069}\right) - 0.977$$
$$= 0.11$$

Since the drainage condition of this pavement is "Good" and the pavement is exposed to saturation moisture more than 25% of the time, the drainage factor is $m_2 = 1.0$.

Based on the 1993 AASHTO *Design Guide*, the structural number can be calculated as follows:

$$\log_{10}(W_{18}) = Z_R s_0 + 9.36\log_{10}(SN+1) - 0.20 + \frac{\log_{10}\left[\frac{\Delta PSI}{4.2-1.5}\right]}{0.40+\frac{1094}{(SN+1)^{5.19}}} + 2.32\log_{10}\left(\frac{M_R}{0.0069}\right) - 8.07$$

For SN_1, 95% reliability (i.e., $Z_R = -1.645$), $s_0 = 0.45$, $\Delta PSI = 4.2 - 2.5 = 1.7$, $M_R = E_{bc} = 162$ MPa, W_{18} = ESALs = 500,000. From the above equation, $SN_1 = 2.10$.

The required thickness of the asphalt layer is

$$h_1 = \frac{SN_1}{0.04a_1} = \frac{2.1}{0.04 \times 0.44} = 120 \text{ mm}$$

Select $h_1^* = 125$ mm and $SN_1^* = 0.04 \times 0.44 \times 125 = 2.20$. This thickness is also greater than the minimum thickness of 63 mm required by the 1993 AASHTO design guide based on the condition (2,000,001 < ESALs < 7,000,000).

For SN_2, 95% reliability (i.e., $Z_R = -1.645$), $s_0 = 0.45$, ΔPSI $= 4.2 - 2.5 = 1.7$, $M_R = 10.35$ CBR $= 10.35 \times 3 = 31.05$ MPa, W_{18} = ESALs = 500,000. From the same equation, $SN_2 = 4.00$. The required unreinforced base course thickness is

$$h_2 = \frac{SN_2 - SN_1^*}{0.04a_2m_2} = \frac{4.00 - 2.20}{0.04 \times 0.11 \times 1.0} = 409 \text{ mm}$$

Select $h_2^* = 420$ mm. This thickness is also greater than the minimum base thickness of 150 mm required by the 1993 AASHTO design guide based on the condition (2,000,001 < ESALs < 7,000,000).

Therefore, the design pavement section without geosynthetics is as follows: 125-mm-thick asphalt layer and 420-mm base course.

Assume the same asphalt thickness for the reinforced section (i.e., $h_1^* = 125$ mm). Considering MIF = 1.5, the LCR can be calculated as follows:

$$\mathrm{LCR} = \frac{0.249\log_{10}(\mathrm{MIF} \cdot E_{bc}/0.0069) - 0.977}{0.249\log_{10}(E_{bc}/0.0069) - 0.977}$$

$$= \frac{0.249 \times \log_{10}(1.5 \times 162/0.0069) - 0.977}{0.249\log_{10}(162/0.0069) - 0.977} = 1.4$$

Therefore, the required thickness of the base course is

$$h_{2r} = \frac{\mathrm{SN} - 0.04a_1h_1^*}{\mathrm{LCR}(0.04a_2m_2)}$$

$$= \frac{4.00 - 0.04 \times 0.44 \times 125}{1.5 \times (0.04 \times 0.11 \times 1.0)} = 273 \text{ mm}$$

Select $h_{2r} = 280$ mm. Therefore, the design pavement section with geogrid is as follows: 125-mm-thick asphalt layer and 280-mm-thick base course. The use of geogrid results in a base thickness reduction of 140 mm.

10.7.8 Construction

Construction of geosynthetic-reinforced unpaved and paved roads consists of site preparation, geosynthetic installation, and fill placement and compaction. Below is the typical procedure:

- Remove large and sharp objects (e.g., trees) and unsuitable materials (such as organic matters, vegetation, etc.) and level the ground to a desired elevation.
- Anchor the geosynthetic at the center and corners before unrolling it. Anchoring can be achieved by small piles of fill, pins, and stapes into the underlying soil.
- Unroll the geosynthetic and manually apply tension to the geosynthetic to minimize wrinkles.
- Geosynthetic can be overlapped or sewn together side by side or end to end. The overlap of geosynthetics depends on underlying soil strength as shown in Table 10.14. The adjacent geosynthetic to be overlapped will be placed underneath the current geosynthetic sheet in the direction of anticipated fill spreading to avoid separation of geosynthetics at overlaps. A minimum overlap of 0.3 m is required even if geotextiles are sewn.
- On firm subgrade (CBR > 3%), most standard road construction procedures can be applied. However, no tracked vehicle is allowed to directly drive on the geosynthetic unless at least 100-mm-thick fill is placed above the geosynthetic for protection. On weak subgrade (CBR < 3%), low ground pressure equipment is recommended to spread fill over weak subgrade. No turns and sudden stops of vehicles are allowed during fill spreading.
- Fill material should be placed in lift thickness and compacted according to design requirements. Any rutting developed during spreading and compaction should be filled with additional fill material. Do not grade out ruts.

Table 10.14 Recommended Geogrid Overlap

Subgrade CBR (%)	Overlap (m)
>3	0.3
1 − 3	0.6
<1	0.9

10.7.9 Quality Control and Assurance

The quality control and assurance include the following items:

- Verify the quality and gradation of the base material and the grade and properties of geosynthetic reinforcement.
- Inspect subgrade strength or modulus by vane shear tests, dynamic cone penetrometer tests or lightweight deflectomter tests.
- Verify the dry density of the base course and its corresponding moisture content.
- Evaluate the modulus of the base course by dynamic cone penetrometer tests or lightweight deflectomter tests.
- Inspect geosynthetic overlap or seam.
- After the completion of road construction, plate load tests or trafficking tests may be conducted to evaluate the performance of the constructed road section.

PROBLEMS

10.1. What are the mechanisms of fill reinforcement?

10.2. In terms of the lateral restraint function, what can geocell reinforcement do differently as compared with geogrid and woven geotextile?

10.3. Can nonwoven geotextile function as geosynthetic reinforcement?

10.4. When a slope angle is 65°, what kind of facing should be considered for design?

10.5. In the current design practice, what theories are used for geosynthetic-reinforced slopes and walls? Are they different?

10.6. Why can geosynthetic reinforcement at the base of an embankment increase the ultimate bearing capacity of a soft foundation?

10.7. What is percent coverage of geosynthetic reinforcement?

10.8. When can the front pullout of geosynthetic reinforcement become a controlling factor in the analysis of geosynthetic-reinforced slopes?

10.9. What is the effective way to increase the factor of safety against global failure in geosynthetic-reinforced slopes?

10.10. When can the geosynthetic-reinforced column-supported embankment become an attractive solution?

10.11. What are the basic assumptions for soil arching theories above columns?

10.12. Explain why it is important to control the critical height above columns.

10.13. Explain why metallic-reinforced earth walls have different internal slip surfaces from geosynthetic-reinforced earth walls.

10.14. Explain why the drainage layer behind the wall facing is important for the stability of MSE walls.

10.15. If the punching of a footing through a geosynthetic-reinforced foundation is the most critical failure mode, what options can be taken to prevent this failure mode?

10.16. Explain why geogrid-reinforced unpaved roads have higher subgrade bearing capacity than geotextile-reinforced unpaved roads.

10.17. What is TBR? How can it be determined?

10.18. When 1.5-m-wide geosynthetic sheets are placed parallel to each other at 1.0 m apart, what is the percent coverage of the geosynthetics?

10.19. A 3-m-long geosynthetic reinforcement is located at a depth of 2 m from the flat top of the slope. The slope is 5 m high and has a slope angle of 26.7°. The unit weight of the slope fill is 19 kN/m^3. The interaction coefficient between the reinforcement and the fill is 0.8. The critical slip surface intersects with the reinforcement at a distance of 1.4 m from the slope facing. Calculate the pullout capacities of the reinforcement from the front and the rear.

10.20. A 5-m-high steep slope with a slope angle of 60° is to be built on a firm foundation. The fill has a unit weight of 19.5 kN/m^3, effective cohesion of 0 kPa, and effective friction angle of 34°. The surcharge on top of the slope is 12 kPa. Ten layers of geosynthetic reinforcement are uniformly distributed with the slope height. The required factor of safety is 1.5. Use the design charts by Schmertmann et al. (1987) to develop the cross section of the geosynthetic-reinforced slope.

10.21. A geosynthetic-reinforced slope is shown below. Seven geosynthetic layers are spaced equally. The center of failure circle is 5 m directly above the toe of the slope. Calculate the factors of safety along this slip surface without and with geosynthetic reinforcements using the ordinary method.

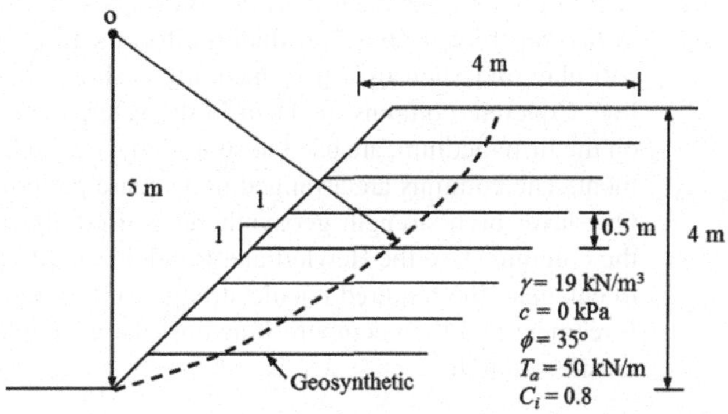

10.22. In Problem 10.21, if the horizontal seismic coefficient, $k_{hs} = 0.15$, what are the factors of safety along the same slip surface without and with geosynthetic reinforcements?

10.23. In Problem 10.21, if the face of the slope within a depth of 1 m is saturated (assume the saturated soil unit weight is 20 kN/m^3), what are the factors of safety against the surficial failure without and with geosynthetic reinforcements?

10.24. A 2-m-high embankment is to be constructed on a 2-m-thick uniform saturated soft clay underlain by a firm medium as shown below. A high-strength woven geotextile is placed at the base of the embankment as a basal reinforcement. The required factor of safety against all modes of failure is 1.5. (1) Calculate the factors of safety against local failure without and with geotextile, (2) calculate the factors of safety against general bearing failure without and with geotextile, (3) calculate the factor of safety against lateral sliding on or below the geotextile, and (4) use the simplified method (Milligan and Busbridge, 1983) to calculate the required tensile strength of geosynthetic reinforcement and then compare it with the available tensile strength.

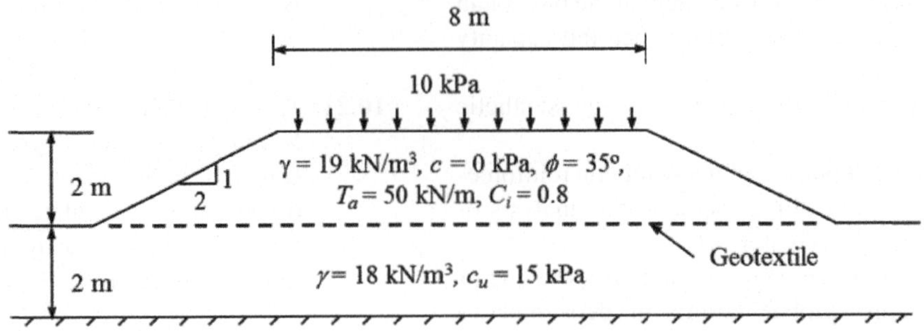

10.25. A 2-m-high embankment with a cohesionless fill is to be constructed on a 5-m-thick uniform saturated soft clay underlain by a firm medium as shown below. Concrete columns of 0.6 m in diameter, seated on the firm medium, are used to support the embankment. The columns are arranged in a square pattern. One layer high-strength geosynthetic is used above the columns. Use the Hewlett and Randolph method to calculate the required tensile strength of the reinforcement and then compare it against the available tensile strength.

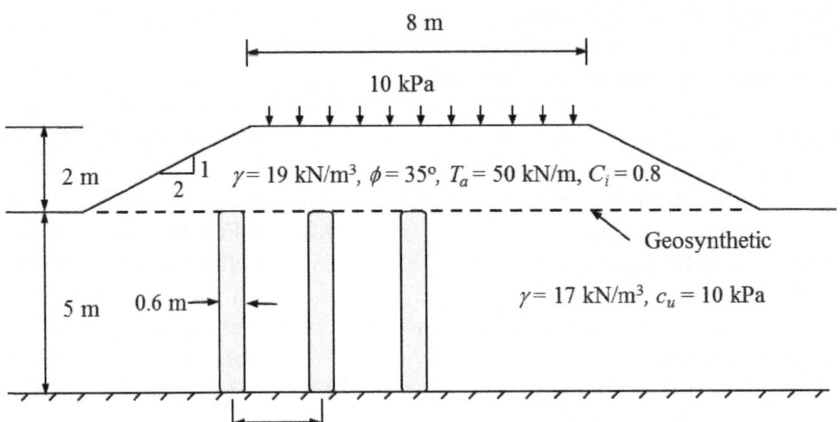

10.26. A vertical geosynthetic-reinforced earth wall is shown below. Geosynthetic reinforcement are equally spaced. The foundation soil, reinforced fill, and retained fill have the same properties. Determine the factors of safety against internal and external failures.

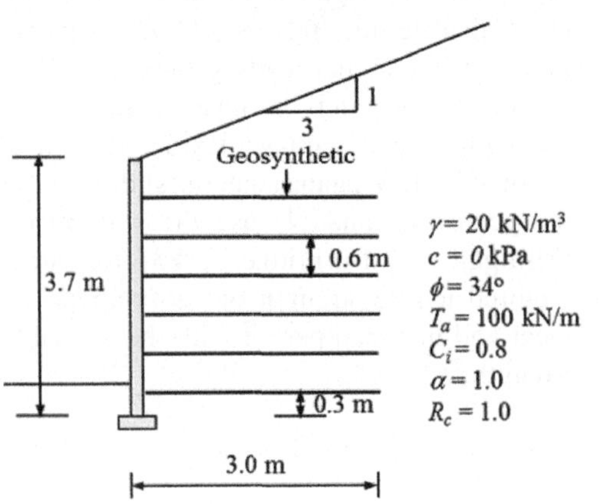

10.27. A 1-m concrete continuous footing is designed to support a vertical wall load of 100 kN/m and embedded at a depth of 1 m below the ground surface. The in situ soil consists of a thick deposit of soft clay with the properties of $c_u = 20$ kPa and $\gamma = 17 \text{kN/m}^3$. The groundwater is at the base of the footing. Overexcavation/replacement with two biaxial geogrids is proposed to improve the soil. Medium sand is used as the backfill material for the replaced zone with the properties: $c' = 0$, $\phi' = 35°$, and $\gamma = 18.5 \text{ kN/m}^3$ (above groundwater table) and 20 kN/m^3 (below groundwater table). The replaced zone is 2.5 m wide and 0.6 m thick. One geogrid is placed at the base of the reinforced zone while another geogrid is placed in the middle of the reinforced zone. The tensile strength of the geogrids at 3% is 12 kN/m and the interaction coefficient is 1.0. Calculate the factors of safety against shallow bearing failure, general bearing failure in the reinforced zone, and punching failure through the reinforced zone.

10.28. An unpaved haul road needs to be designed for 2000 passes of trucks with an axle load of 80 kN and tire pressure of 550 kPa. The subgrade CBR is 2% and the base CBR is 30%. The allowable rut depth is 70 mm. Design this unpaved road section without geosynthetic, with geotextile, and with geogrid using the Giroud and Han method (assume $J_a = 0.32$ m-N/deg).

10.29. The design ESALs of a highway pavement is 0.8 million. Assume there is no subbase layer. The average subgrade CBR is 5%. The design terminal present serviceability index is 3.0. The reliability level is 96% and standard deviation is 0.35. The elastic modulus of the asphalt concrete at 20°C is 3500 MPa. The granular base course CBR is 80%. Assume the drainage condition of this pavement is "Good" and the pavement is exposed to saturation moisture more than 25% of the time. The thickness of the asphalt layer is determined to be 125 mm. Determine the base course thickness for the road section without and with one layer of geogrid (assume LCR = 1.5).

REFERENCES

AASHTO. (1993). *AASHTO Guide for Design of Pavement Structures*. American Association of State Highway and Transportation Officials, Washington, DC.

AASHTO. (2008). *Mechanistic-Empirical Pavement Design Guide, Interim Edition: A Manual of Practice*. America Association of State Highway and Transportation Officials, Washington, DC.

AASHTO. (2012). *AASHTO LRFD Bridge Design Specifications*. Section 11: Walls, Abutments, and Piers. America Association of State Highway and Transportation Officials, Washington, DC.

Acharya, B., Han, J., Thakur, J.K., and Parsons, R.L. (2012). *Onsite Use of Recycled Asphalt Pavement Materials and Geocells to Reconstruct Damaged Pavements by Heavy Trucks*. Final Report MATC-KU: 462, the Mid-America Transportation Research Center, Lincoln, NE.

Adams, M.T. and Collin, J.G. (1997). "Large model spread footing load tests on geosynthetic reinforced soil foundations." *J. Geotechn. Geoenviron. Eng.*, 123(1): 66–72.

Alzamora, D.E., Wayne, M.H., and Han, J. (2000). "Performance of SRW supported by geogrids and jet grout columns." *Performance Confirmation of Constructed Geotechnical Facilities*, ASCE Geotechnical Special Publication No. 94, A.J. Lutenegger and D.J. DeGroot (eds.). ASCE Press, Reston, VA 456–466.

Berg, R.R., Christopher, B.R., and Perkins, S.W. (2000). *Geosynthetic Reinforcement of the Aggregate Base/Subbase Courses of Flexible Pavement Structures* – GMA White Paper II, Geosynthetic Materials Association, Roseville, MN.

Berg, R.R., Christopher, B.R., and Samtani, N.C. (2009). *Design of Mechanically Stabilized Earth Walls and Reinforced Soil Slopes*. Volume I, FHWA-NHI-10-024, FHWA GEC 011-Vol I.

Bhandari, A. (2010). *Micromechanical Analysis of Geosynthetic-Soil Interaction under Cyclic Loading*. Ph.D. dissertation, the University of Kansas.

Binquet, J. and Lee, K.L. (1975). "Bearing capacity analysis on reinforced earth slabs." *J. Geotech. Eng. Div.*, 101(12): 1257–1276.

Bishop, A.W. (1955). "The use of the slip circle in the stability analysis of slopes." *Geotechnique*, 5(1): 7–17.

Bonaparte, R., Holtz, R. D., and Giroud, J. P. (1987). "Soil reinforcement design using geotextiles and geogrids," *Geotextile Testing and the Design Engineer*, ASTM, STP No. 952: 69–116.

British Standard BS8006-1. (2010). *Code of Practice for Strengthened/Reinforced Soils and Other Fills*. British Standards Institution, London.

Brown, S.F., Kwan, J., and Thom, N.H. (2007). "Identifying the key parameters that influence geogrid reinforcement of railway ballast." *Geotext. Geomembr.*, 25: 326–335.

Burmister, D.M. (1958). "Evaluation of pavement systems of the WASHO road test by layered systems method." *Bulletin 177*, Highway Research Board, 26–54.

California Department of Transportation (2012). *Aggregate Base Enhancement with Biaxial Geogrids for Flexible Pavements Guidelines for Project Selection and Design*. California Department of Transportation, Sacramento, CA.

Carlsson, B. (1987). *Reinforced Soil, Principles for Calculation*. Terratema, Linköping (in Swedish).

Chen, Q. (2007). An Experimental Study on Characteristics and Behavior of Reinforced Soil Foundation. Ph.D. dissertation, Louisiana State University.

Chen, Y.M., Cao, W.P., and Chen, R.P. (2008a). "An experimental investigation of soil arching within basal reinforced and unreinforced piled embankments." *Geotextiles and Geomembranes* 26: 164–174.

Chen, R.P., Chen, Y.M., Han, J., and Xu, Z.Z. (2008b). "A theoretical solution for pile-supported embankments on soft soil." *Canad. Geotechn. J.*, 45(5): 611–623.

Chen, R.P., Xu, Z.Z, Chen, Y.M., Ling, D.S., and Zhu, B. (2010). Field tests of pile-supported embankments on soft ground. *J. of Geotechnical and Geoenvironmental Eng.* 136(6): 777–785.

Coleman, D.M. (1990). *Use of Geogrids in Railroad Track: A Literature Review and Synopsis*. Final Report, U.S. Army Engineer District, Omaha.

Collin, J.G. (1996). "Controlling surficial stability problems on reinforced steepened slopes." In *Geotechnical Fabrics Report*, IFAI, Roseville, MN.

Collin, J.G., Kinney, T.C., and Fu, X. (1996). "Full scale highway load test of flexible pavement systems with geogrid reinforced base courses." *Geosynthetics International*, 3(4): 537–549.

Collin, J.G. (2003). *NHI Ground Improvement Manual – Technical Summary #10: Column Supported Embankments*. Federal Highway Administration, Washington, DC.

Collin, J.G. (2004). "Column supported embankment design considerations." *Proc. University of Minnesota 52nd Annual geotechnical Engineering Conference*, Minneapolis, MN.

Das, B.M. (1998). *Principles of Foundation Engineering*, 4th ed. PWS Publishing, Boston, MA.

Demir, A., Laman, M., Yildiz, A., and Ornek, M. (2013). "Large scale field tests on geogrid-reinforced granular fill underlain by clay soil." *Geotext. Geomembr.*, 38: 1–15.

Dong, Y.L., Han, J., Qian, Y., and Bai, X.H. (2010). "Behavior of triaxial geogrid-reinforced bases under static loading." In *Geosynthetics for a Challenging World,* E.M. Palmeira, D.M. Vidal, A.S.J.F. Sayao, and M. Ehrlich (eds.). *Proceedings of the 9th International Conference on Geosynthetics*, Guarujá, Brazil, May 23–27, 2010, 1547–1550.

Duncan, J.M. and Wright, S.G. (2005). *Soil Strength and Slope Stability*. Wiley, Hoboken, NJ.

Elias, V., Christopher, B.R., and Berg, R.R. (2001). *Mechanically Stabilized Earth Walls and Reinforced Soil Slopes, Design and Construction Guidelines*, FHWA-NHI-00-043. Federal Highway Administration, Washington, DC.

Elias, V., Welsh, J., Warren, J., Lukas, R., Collin, J.G., and Berg, R.B. (2006). *Ground Improvement Methods: Reference Manual*—Volume II. FHWA NHI-06-020, National Highway Institute, Washington, DC.

Evans, D.A. (1972). *Slope Stability Report*. Slope Stability Committee, Dept. of Build. and Safety, Los Angeles, CA.

Filz, G.M. and Smith, M.E. (2006). *Design of Bridging Layers in Geosynthetic-Reinforced, Column-Supported Embankments*. Virginia Transportation Research Council, Charlottesville, VA.

Filz, G., Sloan, J., McGuire, M.P., Collin, J., and Smith, M. (2012). "Column-supported embankments: settlement and load transfer." *Geotechnical Engineering State of the Art and Practice*, ASCE GSP No. 226: 54–77.

Fowler, J. and Koerner, R.M. (1987). "Stabilization of very soft soil using geosynthetics." *Proceedings of Geosynthetics'87 Conference*, New Orleans, 289–299.

Gabr, M. (2001). *Cyclic Plate Loading Tests on Geogrid Reinforced Roads*. Research Report to Tensar Earth Technologies, Inc., North Carolina State University, Raleigh, NC.

Gabr, M.A. and Hart, J.H. (1996). *Stress Distribution Characteristics of Geogrid-Reinforced Soil Using Plate Load Tests*. Research Report Prepared for Tensar Earth Technologies, Department of Civil Engineering, West Virginia University, Morgantown, WV.

Giroud, J.P. (1995). "Determination of geosynthetic strain due to deflection." *Geosynth. Int.*, 2(3): 635–641.

Giroud, J.P., Bonaparte, R., Beech, J.F., and Gross, B.A. (1990). "Design of soil layer-geosynthetic systems overlying voids." *Geotext. Geomembr.*, 9(1): 11–50.

Giroud, J.P. and Han, J. (2004a). "Design method for geogrid-reinforced unpaved roads — Part I: theoretical development." *J. Geotechn. Geoenviron. Eng.*, 130(8): 776–786.

Giroud, J.P. and Han, J. (2004b). "Design method for geogrid-reinforced unpaved roads—Part II: calibration and verification." *J. Geotechn. Geoenviron. Eng.*, 130(8): 787–797.

Giroud, J.P. and Han, J. (2011). "The Giroud-Han design method for geosynthetic-reinforced unpaved roads. Part I. The method development and its calibration." *Geosynthetics*, 30(1): 40–49.

Giroud, J.P. and Han, J. (2013). Design of Geosynthetic-Reinforced Unpaved and Paved Roads. Short course, Geosynthetics 2013, Long Beach, CA, April 4.

Giroud, J.P. and Noiray, L. (1981). "Geotextiles-reinforced unpaved road design." *J. Geotech. Eng.*, 107(9): 1233–1253.

Guido, V.A., Chang, D.K., and Sweeney, M.A. (1986). "Comparison of geogrid and geotextile reinforced earth slabs," *Canad. Geotechn. J.*, 3(4): 435–440.

Haliburton, T.A., Douglas, P.A., and Fowler, J. (1977). *Feasibility of Pinto Island as a Long-Term Dredged Material Disposal Site*. Miscellaneous paper, D-77-3, U.S. Army Waterways Experiment Station, Vicksburg, MS.

Hammitt, G.M. (1970). "Thickness requirement for unsurfaced roads and airfields, bare base support, Project 3782-65." Technical report S-70-5, U.S. Army Engineer Waterways Experiment Station, CE, Vicksburg, MS.

Han, J. (1999). "Design and construction of embankments on geosynthetic reinforced platforms supported by piles." Proceedings of 1999 ASCE/PaDOT Geotechnical Seminar. Central Pennsylvania Section, ASCE and Pennsylvania Department of Transportation, Hershey, PA, 66–84.

Han, J., Acharya, B., Thakur, J.K., and Parsons, R.L. (2012a). *Onsite Use of Recycled Asphalt Pavement Materials and Geocells to Reconstruct Damaged Pavements by Heavy Trucks*. Final Report MATC-KU: 462, the Mid-America Transportation Research Center, Lincoln, NE.

Han, J., Bhandari, A., and Wang, F. (2012b). "DEM analysis of stresses and deformations of geogrid-reinforced embankments over piles." *Int. J. Geomech.*, 12(4): 340–350.

Han, J. and Gabr, M.A. (2002). "A numerical study of load transfer mechanisms in geosynthetic reinforced and pile supported embankments over soft soil." *J. Geotech. Geoenviron. Eng.*, 128(1): 44–53.

Han, J. and Giroud, J.P. (2012). "The Giroud-Han design method for geosynthetic-reinforced unpaved roads. Part II Recommendations for the proper use of the method." *Geosynthetics*, 30(2): 12–19.

Han, J. and Jiang, Y. (2013). "Use of geosynthetics for performance enhancement of earth structures in cold regions." *Sciences in Cold and Arid Regions*, 5(5): 517–529.

Han, J. and Leshchinsky, D. (2006). "General analytical framework for design of flexible reinforced earth structures." *J. Geotech. Geoenviron. Eng.*,132(11): 1427–1435.

Han, J. and Pokharel, S.K. (2015). "Design method for geocell-reinforced unpaved roads over weak subgrade." Paper to be submitted for publication.

Han, J., Pokharel, S.K., Yang, X., Manandhar, C., Leshchinsky, D., Halahmi, I., and Parsons, R.L. (2011a). "Performance of geocell-reinforced RAP bases over weak subgrade under full-scale moving wheel loads." *J. Mat. Civil Eng.*, 23(11): 1525–1534.

Han, J. and Thakur, J.K. (2014). "Sustainable roadway construction using recycled aggregates with geosynthetics." *Sustainable Cities and Society* 14, 342–350; available online 18 January 2014 at doi:10.1016/j.scs.2013.11.011.

Han, J., Thakur, J.K., Parsons, R.L., Pokharel, S.K., Leshchinsky, D., and Yang, X. (2013). "A summary of research on geocell-reinforced base courses." Ling, H.I., Gottardi, G., Cazzuffi, D., Han, J., and Tatsuoka, F. (eds.) (2013). *Design and Practice of Geosynthetic-Reinforced Soil Structures*. Honoring Research Achievement of Professor Dov Leshchinsky, October 14–16, Bologna, Italy.

Han, J. and Wayne, M.H. (2000). "Pile-soil-geosynthetic interactions in geosynthetic reinforced/piled embankments over soft soil." Presentation and CD-Rom Paper at 79th Annual TRB Meeting, Washington, DC.

Han, J. and Ye, S.L. (1991). "Analysis of characteristics to composite grounds." *Proc. of First Young Asian Geotechnical Engineers Conference*, Thailand, 197–206.

Han, J., Zhang, Y.Z., and Parsons, R.L. (2011b). "Experimentally quantifying the influence of geosynthetics on performance of reinforced granular bases." *J. Geotech. Eng.*, Southeast Asian Geotechnical Society, 42(1): 74–83.

Hewlett, W.J. and Randolph, M.F. (1988). "Analysis of piled embankments." *Ground Eng.*, 21(3): 12–18.

Holtz, R.D., Christopher, B.R., and Berg, R.R. (2008). *Geosynthetic Design and Construction Guidelines—Reference Manual*. Publication No. FHWA NHI-07-092, National Highway Institute, Washington, DC.

Houlsby, G.T. and Jewell, R.A. (1988). "Analysis of unreinforced and reinforced embankments on soft clay by plasticity theory." In *Numerical Methods in Geomechanics*, Swoboda (ed.). Balkema, Rotterdam, 1443–1448.

Huang, J. and Han, J. (2010). "Two-dimensional coupled hydraulic and mechanical modeling of geosynthetic-reinforced column-supported embankments." *Comput. Geotech.*, 37: 638–648.

Huang, J., Han, J., and Collin, J.G. (2005). "Geogrid-reinforced pile-supported railway embankments—three dimensional numerical analysis." *J. Transport. Res. Board*, 1936: 221–229.

Huang, J., Han, J., Parsons, R.L., and Pierson, M. (2013). "Refined numerical modeling of a laterally-loaded drilled shaft in an MSE wall." *Geotextiles and Geomembranes*, 37: 61–73.

Huang, C.C. and Menq, F.Y. (1997). "Deep-footing and wide-slab effects in reinforced sandy ground." *J. Geotech. Geoenviron. Eng.*, 123(1): 30–36.

Huang, J., Parsons, R.L., Han, J., and Pierson, M.C. (2011). "Numerical analysis of a laterally-loaded shaft constructed within an MSE wall." *Geotext. Geomembr.,* 29: 233–241.

Huang, C.C. and Tatsuoka, F. (1990). "Bearing capacity of reinforced horizontal sandy ground." *Geotext. Geomenbr.*, 9: 51–82.

Jewell, R.A. (1988). "The mechanics of reinforced embankments on soft soils." *Geotext. Geomembr.*, 7: 237–273.

John, N.W.M. (1987). *Geotextiles*. Blackie, Glasgow and London.

Jones, C.J.F.P., Lawson, C.R., and Ayres, D.J. (1990). "Geotextile reinforced piled embankments." In *Geotextiles, Geomembranes, and Related Products*, Den Hoedt (ed.). Balkema, Rotterdam, 155–160.

Kempfert, H.-G., Gobel, C., Alexiew, D., and Heitz, C. (2004). "German recommendations for reinforced embankments on pile-similar elements." EuroGeo3—Third European Geosynthetics Conference, Geotechnical Engineering with Geosynthetics Munich, Germany 279–284.

Kinney, T.C. (2000). *Standard Test Method for Determining the "Aperture Stability Modulus" of a Geogrid*. Shannon & Wilson, Seattle, WA.

Kwon, J., Tutumluer, E., and Kim, M. (2007). "Development of a mechanistic model for geosynthetic-reinforced flexible pavements." *Geosynth. Int.*, 12(6): 310–320.

Leshchinsky, D. (1999). "Stability of geosynthetic reinforced steep slopes." In *Slope Stability Engineering*, Yagi, Yamagami, and Jiang (eds.). Proceedings of International Symposium on Slope Stability Engineering—IS-Shikoku'99, Matsuyama, Shikoku, Japan, 8–11 November, Balkema, Rotterdam, 49–66.

Leshchinsky, D. and Boedeker, R.H. (1989). "Geosynthetic reinforced soil structures." *J. Geotechn. Eng.*, 115(10): 1459–1478.

Leshchinsky, D. and Han, J. (2004). "Geosynthetic reinforced multitiered mechanically walls." *J. Geotechn. Geoenviron. Eng.*, 130(12): 1225–1235.

Leshchinsky, D., Hu, Y.H., and Han, J. (2004). "Limited reinforced space in segmental retaining walls." *Geotext. Geomembr.*, 22(6): 543–553.

Leshchinsky, B. and Ling, H.I. (2013). "Numerical modeling of behavior of railway ballasted structure with geocell confinement." *Geotext. Geomembr.*, 36: 33–43.

Magnan, J. (1994). "Methods to reduce the settlement of embankments on soil clay: A review." In *Foundations and Embankments Deformations*, Geotechnical Special Publication No. 40, ASCE, Washington, DC, 77–90.

Mandel, J. and Salencon, J. (1972). "Force portante d'un sol une assise rigide." *Geotechnique*, 22: 79–93.

McGuire, M.P. and Filz, G.M. (2008). "Quantitative comparison of theories for geosynthetic reinforcement on column-supported embankments." *Proceedings of the First American Geosynthetics Conference & Exhibition*, Cancun, Mexico, 2–5, March, 1303–1312.

McNulty, J.W. (1965). *An Experimental Study of Arching in Sand.* Rep. No. I-674, U.S. Army Engineer Waterways Experiment Station, Corps of Engineers, Vicksburg, MS.

Mengelt, M.J., Edil, T.B., and Benson, C.H. (2000). *Reinforcement of Flexible Pavements Using Geocells*. Geo Engineering Report No. 00-04, University of Wisconsin-Madison, Madison, WI.

Meyerhof, G.G. and Hanna, A.M. (1978). "Ultimate bearing capacity of foundation on layered soils under inclined load." *Canad. Geotechn. J.*, 15: 565–572.

Miki, H. (1997). "*Design of deep mixing method of stabilization with low improvement ratio*." The First Seminar on Ground Improvement in Highways, Bangkok, Thailand, August, 197–204.

Milligan, V. and Busbridge, J.R. (1983). *Guidelines for the Use of Tensar in Reinforcement of Fills over Weak Foundations*. Golder Associates Report to the Tensar Corp., Mississauga, Ontario, Canada.

Miura, N., Sakai, A., Fujikawa, K., and Shivashankar, R. (1995). "Reinforced gravel foundation for box culvert construction on soft and subsiding ground." *Geosynth. Int.*, 2(2): 473–493.

Moghaddas-Nejad, F. and Small, J.C. (2003). "Resilient and permanent characteristics of reinforced granular materials by repeated load triaxial tests." *Geotechn. Testing J.*, 26(2): 152–166.

Montanelli, F., Zhao, A., and Rimoldi, P. (1997). "Geosynthetics-reinforced pavement system: Testing and design." Proceeding of Geosynthetics '97, IFAI, Long Beach, 619–632.

Nazzal, M. (2007). Laboratory Characterization and Numerical Modeling of Geogrid Renforced Bases in Flexible Pavements. Ph.D. Dissertation, Louisiana State University, Baton Rouge, LA.

National Concrete Masonry Association (2010). *Design Manual for Segmental Retaining Walls* (3rd ed.), Herndon, VA, National Concrete Masonry Association.

NCHRP (National Cooperative Highway Research Program) (2004). *Guide for Mechanistic-Empirical Pavement Design of New and Rehabilitated Pavement Structure*, NCHRP 1-37A. National Cooperative Highway Research Program, Washington, DC.

Perkins, S.W., Christopher, B.R., Cuelho, E.L., Eiksund, G.R., Hoff, I., Schwatz, C.W., Svano, G., and Watn, A. (2004). *Development of Design Methods for Geosynthetic Reinforced Flexible Pavements*. Report No. DTFH61-01-X-00068. Montana State University, Bozeman, MT.

Perkins, S.W. and Ismeik, M. (1997). "A synthesis and evaluation of geosynthetic-reinforced base layers in flexible pavements: Part I." *Geosynth. Int.*, 4(6): 549–604.

Pierson, M.C., Parsons, R.L, Han, J., and Brennan, J.J. (2009). "Capacities and deflections of laterally loaded shafts behind an MSE wall." *Transport. Res. Rec., J. Transport. Res. Board, 2116, Soil Mechanics* 2009: 62–69.

Pierson, M.C., Parsons, R.L., Han, J., and Brennan, J.J. (2011). "Laterally loaded shaft group capacities and deflections behind an MSE wall." *J. Geotechn. Geoenviron. Eng.*, 137(10): 882–889.

Pokharel, S., Han, J., Leshchinsky, D., Parsons, R.L., and Halahmi, I. (2010). "Investigation of factors influencing behavior of single geocell-reinforced bases under static loading." *Geotext. Geomembr.*, 28(6): 570–578.

Qian, Y., Han, J., Pokharel, S.K., and Parsons, R.L. (2013). "Performance of triangular aperture geogrid-reinforced base courses over weak subgrade under cyclic loading." *J. Mat. Civil Eng.*, 25(8): 1013–1021.

Rathmayer, H. (1975). "Piled embankment supported by single pile caps." In *Proceedings,* Istanbul Conference on Soil Mechanics and Foundation Engineering (March 31–April 4), 283–290.

Rogbeck, Y., Gustavsson, S., Sodergren, I., and Lindquist, D. (1998). "Reinforced piled embankments in Sweden: Design aspects." In *Proceedings of the Sixth International Conference on Geosynthetics*, R.K. Rowe (ed.), 755–762.

Russell, D. and Pierpoint N. (1997). "An assessment of design methods for piled embankments." *Ground Engineering* (November): 39–44.

Schmertmann, J.H., Chouery-Curtis, V.E., Johnson, R.D., and Bonaparte, R. (1987). "Design charts for geogrid-reinforced soil slopes." *Proceedings of Geosynthetics'87 Conference, 1, New Orleans*, 108–120.

Silvestri, V. (1983). "The bearing capacity of dykes and fills founded on soft soils of limited thickness." *Canad. Geotech. J.*, 20(3): 428–436.

Sloan, J.A., Filz, G.M., and Collin, J.G. (2011). "A generalized formulation of the adapted Terzaghi method of arching in column-supported embankments." *Proceedings of Geo-Frontiers*, J. Han and D. Alzamora (eds.). ASCE Press, Reston, VA 798–805.

Spencer, E. (1981). "*Slip circles and critical shear planes.*" *J. Geotechn. Eng. Div.*, 107(GT7): 929–942.

Steward, J., Williamson, R., and Mohney, J. (1977). *Guidelines for Use of Fabrics in Construction and Maintenance of Low-Volume Roads*. Report FHWA-TS-78-205, United States Department of Transportation, Federal Highway Administration, Washington DC.

Terzaghi, K. (1943). *Theoretical Soil Mechanics*, Wiley, New York, 66–75.

Terzaghi, K. (1936). "Stress distribution in dry and in saturated sand above a yielding trap door." *Proc., 1st Int. Conf. on Soil Mechanics and Foundation Engineering*, Harvard University, Cambridge, MA, 307–311.

Tingle, J.S. and Webster, S.L. (2003). *Review of Corps of Engineers Design of Geosynthetic Reinforced Unpaved Roads*. Presentation at the TRB 82nd Annual Meeting (January 12–16), Washington, DC.

Tseng, K. and Lytton, R. (1989). "Prediction of permanent deformation in flexible pavement materials." *Implication of Aggregates in the Design, Construction, and Performance of Flexible Pavements*, ASTM STP 1016, 154–172. American Society for Testing and Materials, West Conshohocken, PA.

van Eekelen, S.J.M., Bezuijen, A., and van Tol, A.F. (2011). "Analysis and modification of the British Standard BS8006 for the design of piled embankments." *Geotext. Geomembr.*, 29: 345–359.

van Eekelen, S.J.M. and Bezuijen, A. (2012). "Dutch research on basal reinforced piled embankments." *Proceedings of Geo-Congress* 2013, 1838–1847.

van Eekelen, S.J.M. and Bezuijen, A. (2013). "Dutch research on basal reinforced piled embankments." In *Geo-Congress 2013: Performance of Slopes and Embankments III*, C.L. Meehan et al., eds., pp. 1831-1840. Geotechnical Special Publications (GSP) 31. American Society of Civil Engineers, Reston, VA.

Wayne, M.H., Boudreau, R.L., and Kown, J. (2011). "Characterization of mechanically stabilized layer by resilient modulus and permanent deformation testing." *Transport. Res. Rec.: J. Transport. Res. Board*, 2204: 76–82.

Wayne, M.H., Han, J., and Akins, K. (1998). "The design of geosynthetic reinforced foundations." *Design and Construction of Retaining Systems*, ASCE Geo-Institute Geotechnical Special Publication, No. 76, John J. Bowders et al. (eds.). ASCE Press, Reston, VA, 1–18.

Webster. S.L. (1992). *Geogrid Reinforced Base Courses for Flexible Pavements for Light Aircraft: Test Section Construction, Behavior under Traffic, Laboratory Tests, and Design Criteria*. Final report, DOT/FAA/RD-92/25, U.S. Department of Transportation and Federal Aviation Administration, Washington, DC.

Yang, X. and Han, J. (2013). "An analytical model for resilient modulus and permanent deformation of geosynthetic-reinforced unbound granular materials." *J. Geotech. Geoenviron. Eng.*, 139(9).

Yang, X., Han, J., Leshchinsky, D., and Parsons, R.L. (2013). "A three-dimensional mechanistic-empirical model for geocell-reinforced unpaved roads." *Acta Geotechn.*, 8(2): 201–213.

Yang, X., Han, J., Pokharel, S.K., Manandhar, C., Parsons, R.L., Leshchinsky, D., and Halahmi, I. (2012). "Accelerated pavement testing of unpaved roads with geocell-reinforced sand bases." *Geotextil. Geomembr.*, 32: 95–103.

Yetimoglu, T., Wu, J.T.H., and Saglamer, A. (1994). "Bearing capacity of rectangular footings on geogrid-reinforced sand." *J. Geotech. Eng.*, 120(12): 2083–2099.

INDEX

Note: Italic page numbers indicate figures.

A

B

C

D

E

F

G

H

I

J

L

M

N

O

P

Q

R

S

T

U

V

W

X

Y

Wiley The manufacturer's authorized representative according to the EU General Product Safety Regulation is Wiley-VCH GmbH, Boschstr. 12, 69469 Weinheim, Germany, e-mail: Product_Safety@wiley.com.

Printed and bound by CPI Group (UK) Ltd, Croydon, CR0 4YY
15/09/2026
02217281-0001